# Notable Mathematicians

# Notable Mathematicians

From
Ancient
Times
to the
Present

Robyn V. Young,
Editor

Zoran Minderovic,
Associate Editor

GALE

DETROIT · NEW YORK · TORONTO · LONDON

## STAFF

Robyn V. Young, *Editor*

Zoran Minderovic, *Associate Editor*

Nicole Beatty, Christine B. Jeryan, Kyung Lim Kalasky,
Donna Olendorf, Pamela Profitt, Bridget Travers
*Assisting Editors*

Mary Beth Trimper, *Production Director*
Evi Seoud, *Assistant Production Manager*
Shanna Heilveil, *Production Assistant*

Cynthia Baldwin, *Product Design Manager*
Pamela A. E. Galbreath, *Senior Art Director*
Barbara Yarrow, *Graphic Services Manager*
Randy Bassett, *Image Database Supervisor*
Robert Duncan, Mike Logusz, *Imaging Specialists*
Pamela Reed, *Photography Coordinator*

Susan Trosky, *Permissions Manager*
Margaret A. Chamberlain, *Permissions Specialist*
Michelle Lonoconus, *Permissions Associate*
Mary Grimes, *Image Cataloger*

Jeffrey Muhr, *Editorial Technical Consultant*

ISBN 0-7876-3071-3
Printed in the United States of America
10 9 8 7 6 5 4 3 2

**Library of Congress Cataloging-in-Publication Data**

Notable mathematicians: from ancient times to the present / Robyn V. Young, editor.
 p.     cm.
 Includes bibliographical references (p. -      ) and index.
 ISBN 0-7876-3071-3 (alk. paper)
 1. Mathematicians--Biography.     I. Young, Robyn V., 1958-
QA28.N66     1997
510'.92'2--dc21
[B]
                                                    97-33662
                                                      CIP

# Contents

# Introduction

*Notable Mathematicians* joins the line of other Gale Notable titles such as *Notable Twentieth-Century Scientists* and *Notable Black American Women*. *Notable Mathematicians* provides students, educators, librarians, researchers, and general readers with an affordable, comprehensive source of biographical information on 303 mathematicians throughout recorded history. Coverage ranges from the well-known mathematical giants of ancient Greece and Arabia to contemporary mathematicians whose work and research in such fields as topology, logic and computing, and combinatorics place them at the cutting edge of science and technology in the late twentieth century.

*Notable Mathematicians* also highlights the achievements of more than 50 women, 15 Asian American, African American and Hispanic American mathematicians, and almost 30 mathematicians from countries outside North America and Western Europe. While the editors have made every attempt to include representative figures throughout mathematics history in this edition, we make no claim to having included every "notable mathematician," certainly an impossible goal. We hope that we are providing a basis for future research on the lives and contributions of those important but historically marginalized segments of the mathematical community.

## Inclusion Criteria

A preliminary subject list was compiled by the editors from a wide variety of sources, including such traditional print references as the *Biographical Dictionary of Scientists: Mathematicians* and the *Dictionary of Scientific Biography*. On the World Wide Web, we found many web sites devoted to mathematics, including the *MacTutor History of Mathematics Archive* administered by the University of St. Andrews, Scotland and the *Biographies of Women Mathematicians* web site sponsored by Agnes Scott College in Decatur, Georgia, as well as the home pages of the American Mathematical Society and the Mathematical Association of America. The advisory board, consisting of educators currently teaching in high schools and universities, evaluated the candidates and offered valuable suggestions. The final selection was made by the editors on the basis of the following criteria:

- •Discoveries, inventions, overall contributions, influence, and/or impact on mathematical progress.

- •Receipt of major awards in mathematics, including the Fields Medal and the Wolf Prize in Mathematics.

- •Involvement or influence in education or organizational leadership.

- •Familiarity to the general public.

- •Notable "first" achievements, including degrees earned, positions held, or organizations founded.

## Quick Access to Information

Entries in *Notable Mathematicians* are arranged alphabetically by surname. A typical entry provides the following information:

- •**Entry head**—offers "at-a-glance" information: name, birth and death dates, nationality and ethnicity when applicable, and primary field(s) of specialization.

•**Biographical essay**—ranges from 450 to 1500 words and provides basic biographical information and mathematical endeavors and achievements explained in prose accessible to high school students and readers without a mathematical background. Intra-textual headings within the essays highlight the significant events in the listee's life and career, allowing readers to find information they seek quickly and easily. In addition, names appearing in bold type direct readers to entries on the listee's colleagues, predecessors, or contemporaries also found in *Notable Mathematicians*.

•**Selected Writings** by the listee—includes representative publications, including important individual papers, research works, lectures, textbooks, and autobiographies.

•**Further Reading**—provides bibliographic citations of print and electronic sources (including biographies, interviews, periodicals, obituaries, Internet home pages), and other sources about the listee for readers seeking additional information.

## Appendix Offers Supplementary Information

The Appendix, located in the back of this edition, is divided into three sections:

•A **timeline** listing important events in mathematics history.

•A **selected list of major mathematical awards and prizes** and their recipients through 1997.

•A **selected bibliography** of books, periodicals, and Internet web sites of general interest in mathematics.

## Indexes Provide Additional Points of Access

Readers seeking the names of individuals of a given nation, ethnic groups, gender, or mathematical specialty can consult the following indexes for additional listings:

•**Fields of Specialization Index**—groups listees according to the mathematical fields to which they have contributed.

•**Gender Index**—provides lists of the men and women covered in this edition.

•**Nationality/Ethnicity Index**—arranges listees by country of birth and/or citizenship and/or ethnic heritage.

•**Subject Index**—provides page references for proper names and terms used in the text. Includes cross references.

## Photographs and Other Images

Over 150 entries in *Notable Mathematicians* are enhanced by the presence of photographs and/or line drawings of the listee.

## Acknowledgments

The editors would like to thank, in addition to the advisory board, the following individuals for their assistance with various aspects of the production of *Notable Mathematicians*: Harry Tunis, director of publications at the National Council of Teachers of Mathematics, for his early, informal critique of the entry list, Loretta Hall of The Write Equation, Zoran Minderovic, and Margaret Patton for their fine job of revising and polishing a number of manuscripts, and Jeffrey Muhr for his technical assistance. Special thanks are due to Florence Fasanelli for her Foreword, and to Dr. Larry Riddle and Eric Wisniewski for spending a part of their summer vacations reviewing selected manuscripts.

# Advisory Board

# Contributors

Karl Leif Bates, Nicole Beatty, Barbara Boughton, Tammy J. Bronson, Leonard C. Bruno, Gerard J. Buskes, Jill Carpenter, Chris Cavette, Miriana Ceric, Tom Chen, Jane Stewart Cook, Rowan L. Dorfick, Thomas Drucker, William T. Fletcher, C. J. Giroux, Bridget K. Hall, Loretta Hall, Ralph Hampton, Fran Nicolson Hodgkins, Kelley Reynolds Jacquez, Roger Jaffe, Jeanne Spriter James, Corrine G. Johnson, J. Sydney Jones, Jennifer Swift Kramer, C. D. Lord, Avril McDonald, Laura Mangan-Grenier, Robert Messer, Fei Fei Wang Metzler, Zoran Minderovic, Sally M. Moite, Patrick Moore, Angie Mullig, David E. Newton, Kristin Palm, Nicholas Pease, David A. Petechuk, Annette Petruso, Lewis Pyerson, Leslie Reinherz, Karen Sands, Neeraja Sankaran, Michael Sims, Joel Simon, Linda Wasmer Smith, Dorothy Spencer, Monica Stevens, Maureen L. Tan, Karen Wilhelm, Katherine Williams, Rodolfo A. Windhausen, Cathleen Zucco

# Photo and Illustration Credits

Photographs and Illustrations Appearing in *Notable Mathematicians* Were Received from the Following Sources:

**Abel, Niels Henrik**, screen print. Corbis-Bettmann. Reproduced by permission.
**Agnesi, Maria Gaëtana**, engraving. Corbis-Bettmann. Reproduced by permission.
**Ahlfors, Lars V.**, photograph. Mathematisches Forschungsinstitut, Oberwolfach. Reproduced by permission.
**Aiken, Howard**, photograph. Corbis-Bettmann. Reproduced by permission.
**Aleksandrov, Pavel S.**, photograph. Mathematisches Forschungsinstitut Oberwolfach. Reproduced by permission.
**Anaxagoras of Clazomenae**, engraving. Corbis-Bettmann. Reproduced by permission.
**Archimedes of Syracuse**, sculpture. Corbis-Bettmann. Reproduced by permission.
**Archytas of Tarentum**, engraving. Corbis-Bettmann. Reproduced by permission.
**Askey, Richard**, photograph. Mathematisches Forschungsinstitut Oberwolfach. Reproduced by permission.
**Atiyah, Sir Michael**, photograph. Mathematisches Forschungsinstitut Oberwolfach. Reproduced by permission.
**Babbage, Charles**, engraving. The Library of Congress.
**Banneker, Benjamin** (on commemorative stamp), photograph. UPI/Corbis-Bettmann. Reproduced by permission.
**Barrow, Isaac**, engraving. The Library of Congress.
**Bellow, Alexandra**, photograph. Mathematisches Forschungsinstitut Oberwolfach. Reproduced by permission.
**Bernays, Paul**, 1976, photograph. Mathematisches Forschungsinstitut Oberwolfach. Reproduced by permission.
**Bernoulli, Daniel**, woodcut. Corbis-Bettmann. Reproduced by permission.
**Bernoulli, Jakob**, engraving by P. Dupin. Corbis-Bettmann. Reproduced by permission.
**Bernoulli, Johann**, engraving. Corbis-Bettmann. Reproduced by permission.
**Bessel, Friedrich**, engraving after a painting. The Library of Congress.
**Birkhoff, Garrett**, photograph. Mathematisches Forschungsinstitut Oberwolfach. Reproduced by permission.
**Birkhoff, George David**, photograph. The Library of Congress.
**Blackwell, David Harold**, photograph by Jean Libby. Reproduced by permission of David Blackwell.
**Blum, Lenore**, photograph. Mathematisches Forschungsinstitut Oberwolfach. Reproduced by permission.
**Boole, George**, illustration. The Library of Congress.
**Borel, Émile**, photograph. Mathematisches Forschungsinstitut Oberwolfach. Reproduced by permission.
**Bott, Raoul**, photograph. Reproduced by permission of Raoul Bott.
**Bremermann, Hans-Joachim**, photograph. Mathematisches Forschungsinstitut Oberwolfach. Reproduced by permission.
**Brounker, William, Viscount**, drawing. Archive Photos, Inc. Reproduced by permission.
**Browne, Marjorie Lee**, photograph. Reproduced by permission of Patricia Kenschaft.
**Bürgi, Joost**, photograph. Mathematisches Forschungsinstitut Oberwolfach. Reproduced by permission.
**Byron, Ada, Countess of Lovelace**, daguerrotype. Doris Langley Moore Collection.
**Calderón, Alberto P.**, photograph. Wolf Foundation. Reproduced by permission.
**Cantor, Georg**, photograph. The Library of Congress.
**Cardano, Girolamo**, painting. The Library of Congress.
**Carnot, Lazare Nicolas Marguérite**, engraving. The Library of Congress
**Cartan, Élie Joseph**, photograph. AP/Wide World Photos, Inc. Reproduced by permission.
**Cauchy, Augustin-Louis**, photograph. The Library of Congress.
**Cayley, Arthur**, photograph. The Library of Congress.
**Chandrasekhar, Subrahmanyan**, photograph. AP/Wide World Photos. Reproduced by permission.
**Chang, Sun-Yung Alice**, photograph. Mathematisches Forschungsinstitut, Oberwolfach. Reproduced by permission.

**Chung, Fan R. K.** (standing), photograph. Mathematisches Forschungsinstitut Oberwolfach. Reproduced by permission.

**Courant, Richard,** photograph. AP/Wide World Photos, Inc. Reproduced by permission.

**Cramer, Gabriel,** painting. The Granger Collection, New York. Reproduced by permission.

**De Morgan, Augustus,** photograph by Ernest Edwards. Corbis-Bettmann. Reproduced by permission.

**Deligné, Pierre René,** photograph by Neal Koblitz. Mathematisches Forschungsinstitut Oberwolfach. Reproduced by permission.

**Descartes, René,** photograph. The Library of Congress.

**Dirichlet, Peter Gustav Lejeune,** engraving. Archive Photos, Inc. Reproduced by permission.

**Einstein, Albert,** photograph. The Library of Congress.

**Erdös, Paul,** photograph. Mathematisches Forschungsinstitut Oberwolfach. Reproduced by permission.

**"Euclid of Alexandria,"** painting by Justus van Ghent. Corbis-Bettmann. Reproduced by permission.

**Euler, Leonhard,** engraving. Mathematisches Forschungsinstitut Oberwolfach. Reproduced by permission.

**Falconer, Etta Zuber,** photograph. Reproduced by permission of Etta Falconer.

**Fasenmyer, Sister Mary Celine,** photograph. A.K. Peters Ltd. Reproduced by permission.

**Fermat, Pierre de,** illustration. Corbis-Bettmann. Reproduced by permission.

**Fibonacci, Leonardo Pisano,** engraving. The Granger Collection, New York. Reproduced by permission.

**Fields, John Charles,** photograph. Mathematisches Forschungsinstitut Oberwolfach. Reproduced by permission.

**Fisher, Ronald Aylmer,** photograph. The Library of Congress.

**Fredholm, Erik Ivar,** photograph. Mathematisches Forschungsinstitut Oberwolfach. Reproduced by permission.

**Freedman, Dr. Michael H.,** photograph. AP/Wide World Photos, Inc. Reproduced by permission.

**Galileo** with telescope, illustration. The Library of Congress.

**Galois, Évariste,** drawing. Corbis-Bettmann. Reproduced by permission.

**Gauss, Carl Friedrich,** drawing. Corbis-Bettmann. Reproduced by permission.

**Gelfond, Aleksandr O.,** drawing. Mathematisches Forschungsinstitut Oberwolfach. Reproduced by permission.

**Germain, Sophie,** engraving after a life mask. The Granger Collection, New York. Reproduced by permission.

**Gibbs, Josiah Willard,** engraving. The Library of Congress.

**Godel, Kurt,** photograph. AP/Wide World Photos, Inc. Reproduced by permission.

**Granville, Evelyn,** photograph. The University of Texas at Tyler. Reproduced by permission.

**Gregory, James,** engraving. The Library of Congress.

**Grothendieck, Alexander,** photograph. Mathematisches Forschungsinstitut Oberwolfach. Reproduced by permission.

**Hadamard, Jacques,** photograph. Reuters/Corbis-Bettmann. Reproduced by permission.

**Halley, Edmond,** engraving. The Library of Congress.

**Hamilton, Sir William R.,** photograph. The Library of Congress.

**Hardy, Godfrey Harold,** photograph. Mathematisches Forschungsinstitut Oberwolfach. Reproduced by permission.

**Hausdorf, Felix,** photograph. Mathematisches Forschungsinstitut Oberwolfach. Reproduced by permission.

**Hawking, Dr. Stephen W.,** photograph by Miriam Berkley. Reproduced by permission.

**Heine, Heinrich Edward,** photograph. Mathematisches Forschungsinstitut Oberwolfach. Reproduced by permission.

**Hermite, Charles,** photograph. Corbis-Bettmann. Reproduced by permission.

**Herschel, Caroline,** engraving. The Library of Congress.

**Hilbert, David,** photograph Corbis-Bettmann. Reproduced by permission.

**Hill, George,** photograph. The Library of Congress.

**Hipparchus of Chios,** engraving. Archive Photos, Inc. Reproduced by permission.

**Hopper, Grace ,** photograph. AP/Wide World Photos, Inc. Reproduced by permission.

**Huygens, Christiaan,** painting. The Library of Congress.

**Hypatia of Alexandria,** conte crayon drawing. Corbis-Bettmann. Reproduced by permission.

**Jacobi, Karl Gustav,** painting. The Granger Collection, New York. Reproduced by permission.

**Keen, Linda**, photograph. Reproduced by permission of Linda Keen.

**Kepler, Johannes**, engraving. The Library of Congress.

**Khayyam, Omar**, drawing. Corbis-Bettmann. Reproduced by permission.

**Klein, Christian Felix**, illustration. Mathematisches Forschungsinstitut Oberwolfach. Reproduced by permission.

**Knuth, Donald**, photograph. UPI/ Corbis-Bettmann. Reproduced by permission.

**Kolmogorov, Andrei**, photograph. UPI/ Corbis-Bettmann. Reproduced by permission.

**Kovalevskaya, Sonya**, engraving. Corbis-Bettmann. Reproduced by permission.

**Kunihiko Kodaira**, photograph. Mathematisches Forschungsinstitut Oberwolfach. Reproduced by permission.

**Kurtz, Thomas E.**, photograph. Reproduced by permission of Thomas E. Kurtz.

**Ladd-Franklin, Christine**, photograph. Archives of the History of American Psychology. Reproduced by permission.

**Lagrange, Joseph-Louis**, painting. The Library of Congress.

**Langlands, Robert**, photograph. AP/Wide World Photos, Inc. Reproduced by permission.

**Laplace, Pierre-Simon**, photograph. The Library of Congress.

**Leibniz, Gottfried Wilhelm**, engraving. Archive Photos, Inc. Reproduced by permission.

**Maclaurin, Colin**, illustration. Corbis-Bettmann. Reproduced by permission.

**MacPherson, Robert**, photograph © by Randall Hagadorn. Reproduced by permission of Robert MacPherson.

**Margulis, Gregori**, photograph. Mathematisches Forschungsinstitut Oberwolfach. Reproduced by permission.

**Markov, Andrei**, photograph. Courtesy of The Library of Congress.

**Marquise du Châtelet, (Gabrielle-Émilie Le Tonnelier de Breteuil)** wood engraving. The Granger Collection, New York. Reproduced by permission.

**McAfee, Walter S.**, photograph. UPI/Corbis-Bettmann. Reproduced by permission.

**McDuff, Margaret Dusa**, photograph. Reproduced by permission of Dusa McDuff.

**Mercator, Gerardus**, engraving by Larmessan. Corbis-Bettmann. Reproduced by permission.

**Mersenne, Marin**, engraving by P. Dupin. The Granger Collection, New York. Reproduced by permission.

**Milnor, John**, photograph. The Library of Congress.

**Minkowski, Hermann**, photograph. Corbis-Bettmann. Reproduced by permission.

**Monge, Gaspard**, photograph. The Library of Congress.

**Mordell, Louis J.**, photograph. Mathematisches Forschungsinstitut Oberwolfach. Reproduced by permission.

**Moufang, Ruth**, photograph. Mathematisches Forschungsinstitut Oberwolfach. Reproduced by permission.

**Napier, John**, illustration. The Library of Congress.

**Nash, John Forbes**, photograph by Robert P. Matthews. Reproduced by permission of John Nash.

**Neumann, Hanna**, photograph. Mathematisches Forschungsinstitut Oberwolfach. Reproduced by permission.

**Newton, Sir Isaac**, engraving. Archive Photos, Inc. Reproduced by permission.

**Noether, Emmy**, photograph. Mathematisches Forschungsinstitut Oberwolfach. Reproduced by permission.

**Noether, Max**, photograph. Mathematisches Forschungsinstitut Oberwolfach. Reproduced by permission.

**Novikov, Sergei**, photograph. Mathematisches Forschungsinstitut Oberwolfach. Reproduced by permission.

**Oleinik, Olga**, photograph. Mathematisches Forschungsinstitut Oberwolfach. Reproduced by permission.

**Pascal, Blaise**, painting. The Library of Congress.

**Peano, Giuseppe**, photograph. Mathematisches Forschungsinstitut Oberwolfach. Reproduced by permission.

**Pearson, Karl**, drawing. Mathematisches Forschungsinstitut Oberwolfach. Reproduced by permission.

**Peirce, Charles Sanders**, photograph. Corbis-Bettmann. Reproduced by permission.

**Plücker, Julius**, photograph. Mathematisches Forschungsinstitut Oberwolfach. Reproduced by permission.

**Poincaré, M. Henri**, photograph. Hulton-Deutsch Collection/Corbis-Bettmann. Reproduced by permission.

**Pólya, Dr. George**, photograph. AP/Wide World Photos, Inc. Reproduced by permission.

**Poncelet, Jean Victor**, marble sculpture. Corbis-Bettmann. Reproduced by permission.

**Ptolemy of Alexandria**, engraving. The Library of Congress.

**Quetelet, Adolphe**, steel engraving. The Library of Congress.

**Ramanujan, Srinivasa** (frontal, head and shoulders), photograph. The Granger Collection, New York. Reproduced by permission.

**Rees, Mina S.**, photograph. The Library of Congress.

**Regiomontanus, Johannes**, illustration. Corbis-Bettmann. Reproduced by permission.

**Riemann, G.F.B.**, engraving after a photograph. Corbis-Bettmann. Reproduced by permission.

**Robinson, Julia Bowman**, photograph. Mathematisches Forschungs- institut Oberwolfach. Reproduced by permission.

**Russell, Bertrand**, photograph. The Library of Congress.

**Scott, Charlotte**, photograph. Reproduced by permission of Patricia Kenschaft.

**Serre, Jean-Pierre**, photograph. Mathematisches Forschungsinstitut Oberwolfach. Reproduced by permission.

**Smale, Stephen**, photograph. AP/Wide World Photos, Inc. Reproduced by permission.

**Snell, Willebrord**, painting. Photo Researchers, Inc. Reproduced by permission.

**Somerville, Mary**, engraving. Corbis-Bettmann. Reproduced by permission.

**Stanley, Richard P.**, photograph. Reproduced by permission of Richard P. Stanley.

**Staudt, Karl Georg Christian von**, painting. Mathematisches Forschungsinstitut Oberwolfach. Reproduced by permission.

**Steiner, Jakob**, drawing. Mathematisches Forschungsinstitut Oberwolfach. Reproduced by permission.

**Stevin, Simon**, painting. Corbis-Bettmann. Reproduced by permission.

**Stokes, Sir George Gabriel**, illustration. Archive Photos, Inc. Reproduced by permission.

**Sylvester, James Joseph**, photograph. The Library of Congress.

**Szegö, Gabor**, photograph. Mathematisches Forschungsinstitut Oberwolfach. Reproduced by permission.

**Tarski, Alfred**, photograph. Mathematisches Forschungsinstitut Oberwolfach. Reproduced by permission.

**Tartaglia, Niccolò Fontana**, photograph. The Library of Congress.

**Taussky-Todd, Olga**, photograph. The Estate of Olga Taussky-Todd. Reproduced by permission.

**Taylor, Brook**, engraving. Archive Photos, Inc. Reproduced by permission.

**Thales of Miletus**, engraving by Ambroise Tardieu. Corbis-Bettmann. Reproduced by permission.

**Thom, René**, photograph. Mathematisches Forschungsinstitut Oberwolfach. Reproduced by permission.

**Tosio, Kato**, photograph by George Bergman. Mathematisches Forschungsinstitut, Oberwolfach. Reproduced by permission.

**Turing, Alan**, photograph. The Granger Collection, New York. Reproduced by permission.

**Uhlenbeck, Karen**, photograph. Mathematisches Forschungsinstitut Oberwolfach. Reproduced by permission.

**Velez-Rodriguez, Argelia**, photograph. Reproduced by permission.

**Viète, François**, drawing. The Library of Congress.

**Von Mises, Prof. Richard**, photograph. UPI/Corbis- Bettmann. Reproduced by permission.

**Von Neumann, John**, photograph. AP/Wide World Photos. Reproduced by permission.

**Wallis, John**, painting. The Library of Congress.

**Weierstrass, Karl**, etching by Hans Thoma. Corbis-Bettmann. Reproduced by permission.

**Weyl, Hermann**, photograph. AP/Wide World Photos, Inc. Reproduced by permission.

**Whitehead, Alfred North**, photograph. AP/Wide World Photos, Inc. Reproduced by permission.

**Wiener, Norbert**, photograph. The Library of Congress.

**Wiles, Andrew J.**, photograph. Wolf Foundation. Reproduced by permission.

**Yau Shing-Tung**, photograph. Mathematisches Forschungsinstitut Oberwolfach. Reproduced by permission.

**Young, Grace Chisholm**, photograph. Reproduced by permission.

# Foreword

It is the rare student who reads biographies of mathematicians and learns of their inventiveness and consequent achievement. As students, we are taught mathematics as a formal subject without our having any awareness of the human endeavor which generated the discipline. This is a paradox which can be eliminated by learning how problems in textbooks and mathematical theories arise. To discern that mathematics is created by men and women from diverse backgrounds and historical periods and to delve into their personal as well as academic lives, enriches our imagination, for we may be able to identify ourselves with their lives and struggles. This is the gift of biographies.

In 1989, the National Council of Teachers of Mathematics called for a change in the way mathematics is currently taught and learned in the United States. In order to learn to value mathematics, the Council recommends that students are taught the history of the subject and the diverse cultures in which it developed so they can fully appreciate the role of mathematics in the development of human society. Mathematics, then, can be treated as a broad subject—not just the memorization of numerical formulas. Once a student begins to understand the depth of mathematics, then problem solving itself will bring to his or her own efforts a broader perspective of the purpose of mathematics.

*Notable Mathematicians: From Ancient Times to the Present* was conceived to fulfill this need. The reader will find in this work biographies of 303 mathematicians, some of whom are famous for their groundbreaking work and research, others because their names have appeared numerous times in history texts or have lent their names to theorems. The reader will also find biographical sketches of prominent female mathematicians as well as members of several ethnic groups often considered underrepresented in the mathematical community.

The biographies of Majorie Lee Browne and Evelyn Boyd Granville, the first two African American women to earn a Ph.D. in mathematics, inspire the reader. The essays on Ingrid Daubechies, Richard Tapia, Georg Cantor, and Srinivasa Ramanujan can entice one to learn a little bit more about wavelet theory, parallel computing, set theory, and number theory. Essays about such ancient mathematicians as al-Battānī, Euclid of Alexandria, and Zeno of Elea describe a different world than we live in now for the problems they solved in astronomy and geometry seem ordinary to us. The reader will also find biographical sketches about those mathematicians who first developed many familiar school subjects, including imaginary numbers (Rafaello Bombelli); determinants (Gabriel Cramer); analytic geometry (René Descartes); fibonacci numbers (Leonardo of Pisa); and fractals (Benoit Mandelbrot).

The most prestigious prize given to mathematicians under the age of forty is the Fields Medal. Among the mathematicians featured in this edition are 16 Fields Medalists, including biographical sketches of Paul Cohen, who proved the independence of the continuum hypothesis; Kunihiko Kodaira, who pioneered research in algebraic varieties; and René Thom, who created catastrophe theory. The individuals included in *Notable Mathematicians* exist not only for the purpose of maintaining historical records but to allow each reader the opportunity to identify with the lives of the mathematicians and, ultimately, to narrow the gap between those who solve problems and those who created the theories to solve them.

Florence Fasanelli
October 1997

*Florence Fasanelli's career has been in mathematics education as a teacher, a Program Officer at the National Science Foundation, and, in her current position, the Director of Intervention Programs at the Mathematical Association of America. She was founding Chair of the International Study Group on the History of Pedagogy of Mathematics. Ms. Fasanelli speaks regularly on the relation between the history of art and the history of mathematics.*

# Entry List

## A

Abel, Niels Henrik
Agnesi, Maria Gaetana
Ahlfors, Lars V.
Aiken, Howard
Aleksandrov, Pavel S.
Anaxogoras of Clazomenae
Antonclli, Kay McNulty
  Mauchy
Apian, Peter
Apollonius of Perga
Arbogast, Louis François
  Antoine
Archimedes of Syracuse
Archytas of Tarentum
Argand, Jean Robert
Artisaeus the Elder
Aristarchus of Samos
Aryabhata the Elder
Askey, Richard
Atiyah, Michael Francis

## B

Babbage, Charles
Backus, John
Banach, Stefan
Banneker, Benjamin
Bari, Nina
Bari, Ruth Aaronson
Barrow, Isaac
al-Battānī
Baxter, Agnes
Bayes, Thomas
Bellow, Alexandra
Beltrami, Eugenio
Berkeley, George
Bernays, Paul
Bernoulli, Daniel
Bernoulli, Jakob
Bernoulli, Johann
Bernstein, Dorothy Lewis
Bessel, Friedrich Wilhelm
Bézout, Étienne
Birkhoff, Garrett
Birkhoff, George David
Blackwell, David Harold

Blum, Lenore
Bolyai, János
Bombelli, Rafaello
Boole, George
Boole, Mary Everett
Borel, Émile
Bott, Raoul
Bouelles, Charles de
Bremermann, Hans- Joachim
Breteuil, Gabrielle-Émilie
  Le Tonnelier de
Brounker, William
Brouwer, Luitzen Egbertus
  Jan
Browne, Majorie Lee
Bürgi, Joost
Byron, Ada, Countess of
  Lovelace

## C

Calderón, Alberto P.
Cantor, Georg
Cardano, Girolamo
Carnot, Lazare Nicolas
  Marguérite
Cartan, Élie Joseph
Cartan, Henri Paul
Cartwright, Mary Lucy
Cataldi, Pietro Antonio
Cauchy, Augustin-Louis
Cavalieri, Bonaventura
Cayley, Arthur
Ceva, Giovanni
Chandrasekhar,
  Subrahmanyan
Chang, Sun-Yung Alice
Chasles, Michel
Chebyshev, Pafnuty
  Lvovich
Chung, Fan R. K.
Chuquet, Nicolas
Church, Alonzo
Clairaut, Alexis Claude
Cohen, Paul
Conon of Samos
Cotes, Roger
Courant, Richard

Cox, Elbert Frank
Cox, Gertrude Mary
Cramer, Gabriel

## D

D'Alembert, Jean le Rond
Dal Ferro, Scipione
Dantzig, George Bernard
Daubechies, Ingrid
De Bessy, Bernard Frenicle
Dedekind, (Julius Wilhelm)
  Richard
Deligné, Pierre René
De Morgan, Augustus
Desargues, Girard
Descartes, René
Dinostratus
Diocles
Dirichlet, Johann Peter
  Gustav Lejeune
Donaldson, Simon K.

## E

Einstein, Albert
Eratosthenes of Cyrene
Erdös, Paul
Erlang, Agner Krarup
Euclid of Alexandria
Euler, Leonhard

## F

Falconer, Etta Zuber
Fasenmyer, Mary Celine
Fenchel, Käte
Fermat, Pierre de
Feuerbach, Karl Wilhelm
Fibonacci, Leonardo Pisano
Fields, John Charles
Fisher, Ronald Aylmer
Flügge-Lotz, Irmgard
Fréchet, Maurice
Fredholm, Erik Ivar
Freedman, Michael H.
Frege, Gottlob

## G

Galileo
Galois, Évariste
Gauss, Johann Karl Friedrich
Geiringer, Hilda
Gelfond, Aleksandr
  Osipovich
Geminus
Gentry, Ruth
Germain, Sophie
Gibbs, Josiah Willard
Girard, Albert
Gödel, Kurt Friedrich
Granville, Evelyn Boyd
Grassmann, Hermann
  Günter
Green, George
Gregory, James
Grothendieck, Alexander

## H

Hadamard, Jacques
Halley, Edmond
Hamilton, William Rowan
Hardy, Godfrey Harold
Harriot, Thomas
Hausdorff, Felix
Hawking, Stephen W.
Hay, Louise Schmir
Hayes, Ellen Amanda
Hazlett, Olive Clio
Heine, Heinrich Edward
Hermite, Charles
Hero of Alexandria
Herschel, Caroline Lucretia
Hilbert, David
Hill, George William
Hipparchus of Rhodes
Hippocrates of Chios
Hopper, Grace
Huygens, Christiaan
Hypatia of Alexandria

## J

Jacobi, Karl Gustav Jacob
Janovskaja, Sof'ja
  Aleksandrovna
Jones, William

## K

Kato, Tosio
Keen, Linda
Kepler, Johannes
Khayyam, Omar
al-Khwarizmi
Klein, Christian Felix
Knuth, Donald E.
Kochina, Pelageya
  Yakovlevna
  Polubarinova
Kodaira, Kunihiko
Kolmogorov, Andrei
  Nikolaevich
Kovalevskaya, Sonya
  Vasilievna
Krieger, Cecilia
Kuperberg, Krystyna
Kurtz, Thomas Eugene

## L

Ladd-Franklin, Christine
Lagrange, Joseph-Louis
Lambert, Johann Heinrich
Landau, Edmund Georg
  Hermann
Langlands, Robert P.
Laplace, Pierre Simon
Lassar, Edna Ernestine
  Kramer
Lebesgue, Henri
Legendre, Adrien-Marie
Leibniz, Gottfried Wilhelm
  von
Levi-Civita, Tullio
Lie, Marius Sophus
Lin, Chia-Chiao
Lindemann, Carl Louis
  Ferdinand von
Liouville, Joseph
Litvinova, Elizaveta
  Fedorovna
Lobachevsky, Nikolai
  Ivanovich

## M

Macintyre, Sheila Scott

Maclaurin, Colin
MacPherson, Robert
Maddison, Ada Isabel
Malone-Mayes, Vivienne
Mandelbrot, Benoit B.
Margulis, Gregori
  Aleksandrovitch
Markov, Andrei
  Andreevich
McAfee, Walter S.
McDuff, Margaret Dusa
Menaechmus
Méray, Hugues Charles
  Robert
Mercator, Nicolaus
Merrill, Helen Abbot
Merrill, Winifred Edgerton
Mersenne, Marin
Milnor, John
Minkowski, Hermann
Moivre, Abraham de
Monge, Gaspard
Morawetz, Cathleen Synge
Mordell, Louis Joel
Mori, Shiegfumi
Moufang, Ruth

## N

Napier, John
Nash, John Forbes
Nelson, Evelyn
Neumann, Hanna
Newton, Issac
Nicholas of Cusa
Nicomachus of Gerasa
Nicomedes
Noether, Emmy
Noether, Max
Novikov, Sergei

## O

Oleinik, Olga Arsen'enva
Oresme, Nicole d'
Oughtred, William

## P

Pascal, Blaise

# Notable Mathematicians

*Niels Henrik Abel*

# Niels Henrik Abel
## 1802–1829
### Norwegian algebraist

Niels Henrik Abel was a Norwegian mathematician who is most famous for having proved that fifth and higher order equations have no algebraic solution. Had he not died prematurely at the age of 26, it is speculated that he might have become one of the most prominent mathematicians of the 19th century. He provided the first general proof of the binomial theorem and made significant discoveries concerning elliptic functions.

Abel was born in Finnöy, on the southwestern coast of Norway, on August 5, 1802. He was the second son of Sören Georg Abel, a Lutheran minister, and Anne Marie nee Sorensen, the daughter of a wealthy merchant. In 1804, Abel's father was appointed to a new parish and the family moved to Gjerstad, in southern Norway. Abel received his early education from his father, and in 1815 he was sent to the Cathedral School in Oslo. There, he rapidly developed a passion for mathematics. In 1818, a new instructor,

Berndt Holmboe, arrived at the school and fueled Abel's interest further, introducing him to the works of such European masters as **Isaac Newton, Joseph–Louis Lagrange,** and **Leonhard Euler**. Holmboe was to become a lifelong friend and advocate, eventually helping to raise money that allowed Abel to travel abroad and meet the leading mathematicians of Germany and France.

Abel graduated from the Cathedral School in 1821. His father had died a year earlier and his older brother had developed mental illness. The responsibility of providing for his mother and four younger siblings fell largely on Abel and to make ends meet he began tutoring. Meanwhile, he took the entrance examination for the university. His performance in geometry and arithmetic was distinguished and he was offered a free dormitory room. In an exceptional move, the mathematics faculty, who were already aware of Abel's promise, contributed personal funds to cover his other expenses. Abel enrolled at the University of Kristiania (Oslo) at the age of 19. Within a year he had completed his basic courses and was a degree candidate.

### Proves the Impossibility of Solutions for the Quintic Problem

During his final year at the Cathedral School, Abel had become intrigued by a challenge that had occupied some of the best mathematical minds since the 16th century, that of finding a solution to the "quintic" problem. A quintic equation is one in which the unknown appears to the fifth power. Abel believed he had discovered a general solution and presented his results to his teacher Holmboe, who was wise enough to realize Abel's mathematical reasoning was already beyond his full comprehension. Holmboe sent the solution to the Danish mathematician Ferdinand Degen, who expressed skepticism but was unable to determine whether Abel's argument was flawed. Degen asked Abel to provide examples of his general solution, who, in the end, discovered the error in his approach. Abel would remain obsessed with the quintic problem for the next few years. Finally, at Christmas time in 1823, he hit upon the realization and derived a proof that an algebraic solution was impossible. Abel sent a paper describing his proof to **Johann Karl Friedrich Gauss**, who reportedly ignored the treatise. Meanwhile, Abel began working on what would become the first proof of an integral equation, and went on to provide the first general proof of the binomial theorem, which until then had only been proved for special cases. He also investigated elliptic integrals and developed a novel way of examining them through the use of inverse functions.

In 1825, Abel left home and traveled to Berlin, where he met August Leopold Crelle, a civil engineer and the builder of the first German railroad. Crelle had a strong reverence for mathematics, and was about to publish the first edition of *Journal for Pure and Applied Mathematics*, the first periodical devoted entirely to mathematical research. Recognizing in Abel a man of genius, Crelle asked if the young man would contribute to the premiere edition. Abel obliged, providing Crelle with a manuscript that described his proof that an algebraic solution to the general equation of the fifth and higher degrees was impossible. The paper would insure both Abel's fame and the success of Crelle's fledgling journal. From Germany, Abel toured southern Europe, then traveled to France, where he made the acquaintance of **Adrien Marie Legendre, Augustin Louis Cauchy,** and others. In their company, he wrote the *Memoir on a General Property of a Very Extensive Class of Transcendental Functions*, which was submitted to the Paris Académie Royale des Sciences. The memoir expounded on Abel's earlier work on elliptical functions, and proposed what has come to be known as Abel's theorem. Unfortunately, it was received poorly, rejected by Legendre because it was "illegible," then temporarily lost by Cauchy. Two years after Abel's death, the manuscript finally resurfaced, but it was not published until 1841.

By 1827, Abel had run out of money and was forced to return to Norway. He had hoped to take up a university post, but could only find work as a tutor. Meanwhile, he had discovered he had contracted tuberculosis. Later in 1827 he wrote a lengthy paper on elliptic functions for Crelle's journal and began working for Crelle as an editor.

Abel died on April 6, 1829, while visiting his Danish fiancée, Christine Kemp, who was living then in Froland. A few days later, as yet unaware of Abel's death, Crelle wrote to say he had secured a position for him at the University of Berlin. Abel was honored posthumously, in 1830, when the French Académie awarded him the Grand Prix, a prize he shared with **Karl Jacobi.**

## SELECTED WRITINGS BY ABEL

*Oeuvres complètes de N.H. Abel.* Edited and annotated by B. Holmboe, 1839

## FURTHER READING

### Books

Bell, E.T. "Genius and Poverty," in *Men of Mathematics.* New York: Simon and Schuster, 1986, pp. 17, 307–26.

Ore, Oystein. *Niels Henrik Abel: Mathematician Extraordinary.* Minneapolis: University of Minnesota Press, 1957.

### Other

"Niels Henrik Abel." *MacTutor History of Mathematics Archive.* http://www–groups.dcs.st–and.ac.uk/~history/Mathematics/abel.html (March 1997).

—*Sketch by Leslie Reinherz*

# Maria Gaëtana Agnesi
## 1718–1799
### Italian algebraist, geometer, logician, physicist, and philosopher

One of the great figures of Italian science, Maria Gaëtana Agnesi was born and died in Milan, an Italian city under Habsburg rule. In early childhood, she demonstrated extraordinarily intellectual abilities, learning several languages, including Greek, Latin, and Hebrew.

Agnesi's father, who taught mathematics at the University of Bologna, hired a university professor to tutor her in mathematics.

### Fascinates Visitors by Displays of Erudition

While still a child, Agnesi took part in learned discussions with noted intellectuals who visited her parents' home. Her knowledge encompassed various fields of science, and to any foreign visitor who was not a Latinist (the discussions were held in Latin), she spoke fluently in his language. Her brilliance as a multilingual and erudite conversationalist was matched by her fluency as a writer. When she was 17 years old, Agnesi wrote a memoir about the marquis de l'Hospital's 1687 article on conic sections. Her *Propositiones philosophicae*, a book of essays published in 1738, examines a variety of scientific topics, including philosophy, logic, and physics. Among the subjects discussed is **Isaac Newton**'s theory of universal gravitation.

Following her mother's death, Agnesi wished to enter a convent, but her father decided that as the oldest child, she should supervise the education of her numerous younger siblings. As an educator, Agnesi recognized the educational needs of young people, and eloquently advocated the education of women.

*Maria Gaëtana Agnesi*

### Witch of Agnesi

Agnesi's principal work, *Instituzione analitiche ad uso della gioventu' italiana* (1748), known in English as her *Analytical Institutions*, is a veritable compendium of mathematics, written, as the Italian title indicates, for the edification of Italian youth. The work introduces the reader to algebra and analysis, providing elucidations of both and of integral and differential calculus. Praised for its lucid style, Agnesi's book was translated into English by John Colson, Lucasian Professor of Mathematics at Cambridge University. Colson, who learned Italian for the express purpose of translating Agnesi's book, had already translated Newton's *Principia mathematica* into English. Among the prominent features of Agnesi's work is her discussion of a curve, subsequently named the "Witch of Agnesi," due in part by an infelictious confusion of terms. (The Italian word *versiera*, derived from the Latin *vertere*, meaning *to turn*, became associated with *avversiera*, which in Italian means *devil's wife*, or *witch*.) Studied previously by **Pierre de Fermat** and by Guido Grandi, the "Witch of Agnesi" is a cubic curve represented by the Cartesian equation $y(x^2 + a^2) = a^3$, where $a$ represents a parameter, or constant. For $a = 2$, as an example, the maximum value of $y$ will be 2. As $y$ tends toward 0, $x$ will tend, asymptotically, toward $\pm\infty$.

### Receives Papal Recognition and Accepts Religious Vocation

In 1750, Pope Benedict XIV named Agnesi professor of mathematics and natural philosophy at the University of Bologna. As David M. Burton explains, it is not quite clear whether she accepted the appointment. Considering the fact that her father was gravely ill by 1750, there is speculation that she would have found the appointment difficult to accept. At any rate, after her father's death in 1752, Agnesi apparently lost all interest in scientific work, devoting herself to a religious life. She directed charitable projects, taking charge of a home for the poor and infirm in 1771, a task to which she devoted the rest of her life.

### SELECTED WRITINGS BY AGNESI

*Propositiones philosophicae, quas crebris disputationibus domi habitis coram clarissimis viris explicabat extempore*, 1738 (Translated in 1801 as *Analytical Institutions* by John Colson)

### FURTHER READING

**Books**

Alic, Margaret. *Hypatia's Heritage: A History of Women in Science from Antiquity through the Nineteenth Century*. Boston: Beacon Press, 1986.

Burton, David M. *Burton's History of Mathematics: An Introduction*. Dubuque, IA: Wm. C. Brown, 1995.

Kramer, Edna E. "Maria Gaetana Agnesi," in *Dictionary of Scientific Biography*. Volume I. Edited by Charles Coulston Gillispie. New York: Charles Scribner's Sons, 1970, pp. 75–77.

Olsen, Lynn M. *Women in Mathematics*. Cambridge, MA: MIT Press, 1974.

—*Sketch by Zoran Minderovic*

# Lars V. Ahlfors
## 1907-1996
### Finnish-born American analyst

Lars V. Ahlfors made major contributions to the field of complex analysis. In 1936 he was one of the first two people to receive a Fields Medal, a mathe-

*Lars V. Ahlfors*

matics award that has since acquired a status approximating that of the Nobel Prize. Ahlfors received this recognition for his work on Riemann surfaces, which are schematic devices for mapping the relation between complex numbers according to an analytic function. Ahlfors' results led to new developments in the field of meromorphic functions (functions that are analytic everywhere in a region except for a finite number of poles); the methods he developed to obtain these results created an entirely new field of analysis.

Lars Valerian Ahlfors was born on April 18, 1907, in Helsingfors, Finland. His mother, Sievä Helander Ahlfors, died giving birth to him. His father, Axel Ahlfors, was a mechanical engineering professor at the Polytechnical Institute. Even as a child, Ahlfors was interested in mathematics; his high school did not offer calculus courses, but Ahlfors taught himself by reading his father's engineering books.

With little background in mathematical theory, Ahlors began his studies in 1924 at the University of Helsingfors, where he was taught by two highly capable mathematicians Ernst Lindelöf and Rolf Nevanlinna. Lindelöf specialized in complex analysis and was known as the "father" of mathematics in Finland because, in the 1920s, all Finnish mathematicians were his students. In the commentary of his *Collected Works*, Ahlfors recalled that Lindelöf was a master teacher, and recalled some 45 years later, "many Saturday mornings when I had to visit him in his home at 8 a.m. to be praised or scolded—as the

case may have been." Ahlfors received his baccalaureate degree in the spring of 1928 and promptly began his graduate work. Although there were no official graduate courses in mathematics at the university, Lindelöf supervised students' advanced readings.

### Proposes Geometric Interpretation of Nevanlinna Theory

Ahlfors took his first formal graduate course in mathematics in the fall of 1928, when he accompanied Nevanlinna to Zürich, where the teacher replaced **Hermann Weyl** during a leave of absence. The class Nevanlinna taught was on contemporary function theory. Topics included the major parts of Nevanlinna's theory of meromorphic functions (complex functions whose only singularities are isolated, nonremovable, and nonessential) and Denjoy's conjecture on the number of asymptotic values of an entire function, as well as Carleman's partial proof of it. During his study of this subject, Ahlfors proved the full Denjoy conjecture after he discovered a new approach based on conformal mapping. A conformal map is a function in which, if two curves intersect at an angle, then the images of the curves in the map will also intersect at the same angle.

This visit to Zürich marked Ahlfors' first trip outside Finland, and he later wrote that it was his first exposure to "live mathematics." His interactions with Nevanlinna (who was only 33 years old at the time) and **George Pólya** both stimulated his creative thinking and bolstered his self-confidence. When the course ended, Ahlfors traveled to Paris, where he continued his work for three months before returning to Finland. His research there led to a geometric interpretation of the Nevanlinna theory, which he would publish in 1935. Although this interpretation was also discovered independently in Japan, it was the beginning of Ahlfors' concentration on meromorphic functions.

When he returned to Finland, Ahlfors was given the position of lecturer at Åbo Akademi, a Swedish-language university. He also began work on his thesis, the subject of which was conformal mapping and entire functions. Ahlors had finished his thesis by the spring of 1930, and received his Ph.D. in 1932. Ahlfors was named a fellow of the Rockefeller Institute in 1932, which allowed him to live and do research in Paris for a year. In 1933, he returned to the University of Helsingfors as an adjunct professor. That same year he married Erna Lehnert, who had emigrated to Finland from Vienna with her parents; the young couple would ultimately have three daughters. In the fall of 1935, Ahlfors began a three-year assignment as an assistant professor at Harvard University. During these international trips, he later wrote, "my name was becoming known, and I met many of the leading mathematicians." He found the

Harvard experience to be particularly helpful in broadening his horizons beyond the classical analysis he had thus far concentrated on. After acquiring a greater appreciation for the overall unity of mathematics, he mentioned in the initial **John von Neumann** Lecture in 1960 that he did not "believe in separating real and complex analysis."

### Receives Fields Medal

In 1936, Ahlfors attended the conference of the International Congress of Mathematicians (ICM) in Oslo. That year, the ICM had decided to institute a new award, the Fields Medal, that would recognize a young mathematician under the age of 40 who had produced important work. A few hours before the ceremony, Ahlfors was surprised to learn that he would be receiving the inaugural award, along with Jesse Douglas of New York. The award cited Ahlfors' "research on covering surfaces related to Riemann surfaces of inverse functions of entire and meromorphic functions." Speaking to the congress about Ahlfors' work, German mathematician Constantin Carathéodory specifically noted the contribution of Ahlfors' paper "On the Theory of Covering Surfaces," which explained the methods Ahlfors had developed in his work on Riemann surfaces. Carathéodory pointed out that these methods were also the start of a new branch of analysis, which he termed "metrical topology." Later in his career, Ahlfors wrote that the importance of the awards he had received lay not in the consequent fame, but in the confidence it gave him in his work.

In the spring of 1938, Ahlfors left the United States and returned to Finland to take a position as a professor at the University of Helsinki. World War II soon spread to Finland, however, and the university closed because there were not enough students. Although his wife and two small children were evacuated and went to live with relatives in Sweden, Ahlfors stayed in Helsinki. He was not called for military duty because of a physical condition, but he participated in a communications organization. His duties included lengthy periods in air-raid shelters with no substantial assignments, so he decided to pass the time by studying meromorphic curves.

In the summer of 1944, the University of Zürich offered Ahlfors a professorship, and he accepted the position in order to be reunited with his family. After an arduous journey through war zones via train and unpressurized airplane, the Ahlfors family finally reached Switzerland in the summer of 1945. He was not happy there, feeling displaced in a scarred Europe; when Harvard University asked him to return, he gladly accepted. He began teaching there in the fall of 1946 and became a naturalized U. S. citizen in 1952. In 1953, Ahlfors' book *Complex Analysis* was published. Now in its third edition, it is still widely used as a basic text in graduate courses. The three textbooks he authored earned him a 1982 Steele Prize from the American Mathematical Society.

Ahlfors remained at Harvard, holding the title of professor emeritus after his 1977 retirement. He died of pneumonia on October 11, 1996, near his longtime Boston home.

## SELECTED WRITINGS BY AHLFORS

*Conformal Invariants,* 1973
*Complex Analysis: An Introduction to the Theory of Analytic Functions of One Complex Variable.* Third edition, 1979
*Lectures on Quasiconformal Mappings,* 1987
*Lars Valerian Ahlfors: Collected Papers,* Volumes I and II. 1982

## FURTHER READING

"Lars V. Ahlfors, 89, Pioneer in the Outer Reaches of Higher Math." Obituary. *The New York Times* (October 20, 1996): 40.
Long, Tom. "Lars V. Ahlfors, Mathematician Who Won First Fields Medal; 89." Obituary. *Boston Globe* (October 17, 1996).

## Other

http://www.math.hut.fi/teaching/y3/ahlfors.txt (June 1, 1997).

*—Sketch by Laura Mangan–Grenier and Loretta Hall*

# Howard Aiken
## 1900-1973
### American computer scientist and inventor

A noted physicist and Harvard professor, Howard Aiken designed and built the Mark I calculator in the late 1930s and early 1940s. The first large-scale digital calculator, the Mark I provided the impetus for larger and more advanced computing machines. Aiken's later conceptions, the Mark II, Mark III, and Mark IV, each surpassed its previous model in terms of speed and calculating capacity.

Howard Hathaway Aiken was born on March 8, 1900, in Hoboken, New Jersey, and was raised in Indianapolis, Indiana. Because of his family's limited resources, he had to go to work after completing the eighth grade. He worked twelve-hour shifts at night,

*Howard Aiken*

seven days a week, as a switchboard operator for the Indianapolis Light and Heat Company. During the day he attended Arsenal Technical High School. When the school superintendent learned of his round-the-clock work and study schedule, he arranged a series of special tests that enabled Aiken to graduate early. In 1919 Aiken entered the University of Wisconsin at Madison and worked part-time for Madison's gas company while he attended classes. He received his bachelor of science degree in 1923 and upon graduation was immediately promoted to chief engineer of the gas company. Over the next twelve years he became a professor at the University of Miami and later went into business for himself. By 1935, however, he had decided that he wanted to return to school to work on his Ph.D. He began his graduate studies at the University of Chicago before going on to Harvard. He received a master's degree in physics in 1937 and was made an instructor. He wrote his dissertation while he was teaching and received his doctorate in 1939.

## Proposes Design for First Modern Computer

As a graduate student in physics, Aiken completed a great deal of work requiring many hours of long and tedious calculations; it was at that time that he began to think seriously about improving calculating machines to reduce the time needed for figuring large numerical sequences. In 1937, while at Harvard, Aiken wrote a 22-page memorandum proposing the initial design for his computer. His idea was to build a computer from existing hardware with electromagnetic components controlled by coded sequences of instructions, and one that would operate automatically after a particular process had been developed. Aiken proposed that the punched-card calculators then in use (which could carry out only one arithmetic operation at a time) could be modified to become fully automated and to carry out a wide range of arithmetic and mathematical functions. His original design was inspired by a description of a more powerful calculator in the work of English mathematician Charles Babbage, who devoted nearly forty years to developing a calculating machine.

Although Aiken was by then an instructor at Harvard (and was to become an associate professor of applied mathematics in 1941 and a full professor in 1946), the university offered little support for his initial idea. He therefore turned to private industry for assistance. Although his first attempt to muster corporate support was turned down by the Monroe Calculating Machine Company, its chief engineer, G. C. Chase, approved of Aiken's proposal and suggested he contact Theodore Brown, a professor at the Harvard Business School; Brown, in turn, put Aiken in touch with IBM. Aiken's idea impressed IBM enough that the company agreed to back the construction of his Mark I. In 1939 IBM President Thomas Watson, Sr., signed a contract that stated that IBM would build the computer under Aiken's supervision and with additional financial backing from the U.S. Navy. At the time IBM only manufactured office machines, but its management wanted to encourage research in new and promising areas and was eager to establish a connection with Harvard. During that same year, Aiken became a school officer of the Naval Warfare School at Yorktown, and when the Mark I contract was worked out he was made officer in charge of the U.S. Navy Computing Project. The Navy agreed to support Aiken's computer because the Mark I offered a great deal of potential for expediting the complex mathematical calculations involved in aiming long-range guns onboard ship. The Mark I provided a solution to the problem by calculating gun trajectories in a matter of minutes.

## Builds Mark I-IV

With a grant from IBM and a Navy contract, Aiken and a team headed by IBM's Clair D. Lake began work at IBM's laboratories in Endicott, New York. Aiken's machine was electromechanical—mechanical parts, electrically controlled—and used ordinary telephone relays that enabled electrical currents to be switched on or off. The computer consisted of thousands of relays and other components, all assembled in a 51-foot-long and 8-foot-high (1554 cm x 243 cm) stainless steel and glass frame that was completed in 1943 and installed at Harvard a year later. The heart of this huge machine was formed by 72 rotating registers, each of which could store a positive or negative 23-digit number. The telephone relays established communication between the registers. Instructions and data input were entered into the computer by means of continuous strips of IBM punch-card paper. Output was printed by two electrical typewriters hooked up to the machine. The Mark I did not resemble modern computers, either in appearance or in principles of operation. The machine had no keyboard, for instance, but was operated with approximately 1,400 rotary switches that had to be adjusted to set up a run. Seemingly clumsy by today's computer standards, the Mark I nevertheless was a powerful improvement over its predecessors in terms of the speed at which it performed a host of complex mathematical calculations. Many scientists and engineers were eager for time on the machine, underscoring the project's success and giving added impetus for continued work on improved models. However, a dispute developed with IBM over credit for the computer, and subsequently the company withdrew support for all further efforts. A more powerful model was soon undertaken under pressure from competition from ENIAC, the much faster computer then being built at Columbia University.

Mark I was to have three successors, Mark II through IV. It was with the Mark III that Aiken began building electronic machines. Aiken had a conservative outlook with respect to electronic engineering and sacrificed the speed associated with electronic technology for the dependability of mechanics; only after World War II did he begin to feel comfortable using electronic hardware. In 1949 Aiken finished the Mark III with the incorporation of electronic components. Data and instructions were stored on magnetic drums with a capacity of 4,350 sixteen-bit words and roughly 4,000 instructions. With Aiken's continued concern for reliability over speed, he called his Mark III "the slowest all-electronic machine in the world," as quoted by David Ritchie in *The Computer Pioneers: The Making of the Modern Computer*. The Mark III's final version, however, was not completely electronic; it still contained about 2,000 mechanical relays in addition to its electronic components. The Mark IV followed on the heels of the Mark III and was considerably faster.

Aiken contributed to the early computing years by demonstrating that a large, calculating computer could not only be built but could also provide the scientific world with high-powered, speedy mathematical solutions to a plethora of problems. Aiken remained at Harvard until 1961, when he moved to Fort Lauderdale, Florida. He went on to help the University of Miami set up a computer science program and a computing center and became Distinguished Professor of Information there. At the same time he founded a New York-based consulting firm,

Howard Aiken Industries Incorporated. Aiken disliked the idea of patents and was known for sharing his work with others. He died on March 14, 1973.

## FURTHER READING

### Books

Augarten, Stan. *Bit by Bit*. New York: Ticknor & Fields, 1984.

Fang, Irving E. *The Computer Story*. St. Paul, MN: Rada Press, 1988.

Moreau, R. *The Computer Comes of Age*. Cambridge, MA: MIT Press, 1984.

Ritchie, David. *The Computer Pioneers: The Making of the Modern Computer*. New York: Simon and Schuster, 1986.

Slater, Robert. *Portraits in Silicon*. Cambridge, MA: MIT Press, 1987.

Stine, Harry G. *The Untold Story of the Computer Revolution: Bits, Bytes, Bauds, and Brains*. Arbor House, 1985.

Wulforst, Harry. *Breakthrough to the Computer Age*. New York: Charles Scribner's Sons, 1982.

—*Sketch by Dorothy Spencer*

# Pavel S. Aleksandrov
## 1896-1982
### Russian topologist

Pavel S. Aleksandrov laid the foundation for the field of mathematics known as topology. In addition to writing the first comprehensive textbook on the subject, Aleksandrov introduced several basic concepts of topology and its offshoots, homology and cohomology, which blend topology and algebra. His important work in defining and exploring bicompact (compact or locally compact) spaces laid the groundwork for research done by other mathematicians in these fields.

The youngest of the six children of Sergei Aleksandrovich Aleksandrov and Tsezariia Akimovna Zdanovskaia, Pavel Sergeevich Aleksandrov was born in Bogorodsk, Russia, on May 7, 1896. A year later the family moved to Smolensk, where Aleksandrov's father became head doctor in the state hospital. Although educated mainly in public schools, Aleksandrov learned German and French from his mother, who was skilled in languages.

In grammar school Aleksandrov developed an interest in mathematics under the guidance of Aleksandr Eiges, his arithmetic teacher. Aleksandrov

*Pavel S. Aleksandrov*

entered the University of Moscow in 1913 as a mathematics student, and achieved early success when he proved the importance of Borel sets after hearing a lecture by Nikolai Nikolaevich Luzin in 1914. Aleksandrov graduated in 1917 and planned to continue his studies. However, after failing to reach similar results on his next project—**Georg Cantor**'s continuum hypothesis (since acknowledged unsolvable; that is, it can be neither proved nor disproved)—Aleksandrov dropped out of the mathematical community and formed a theater group in Chernigov, a city situated seventy-seven miles north of Kiev, in the Ukraine. Besides participating in the theater group, he lectured publicly on various topics in literature and mathematics. He also was involved in political support of the new Soviet government, for which he was jailed briefly in 1919 by counterrevolutionaries.

### Key Friendship Leads to Lifelong Career

Later that same year, Aleksandrov suffered a lengthy illness, during which he decided to return to Moscow and mathematics. To help himself catch up, he enlisted the help of another young graduate student, Pavel Samuilovich Uryson. The two immediately became close friends and colleagues. After a brief, unsuccessful marriage in 1921 to his former teacher's sister, Ekaterina Romanovna Eiges, Aleksandrov joined some fellow graduate students in renting a summer cottage. There, he and Uryson

began their study of the new field of topology, the branch of mathematics that deals with properties of figures related directly to their shape and invariant under continuous transformation (that is, without cutting or tearing). In topology, often called rubber-sheet geometry, a cylinder and a sphere are equivalent, because one can be shaped (or transformed) into the other. A doughnut, however, is not equivalent to a sphere, because it cannot be shaped or stretched into a sphere. No textbooks were available on the subject, only articles by **Maurice Fréchet**, **Felix Hausdorff**, and a few others. Nonetheless, from these articles, Uryson and Aleksandrov came up with their first major topological discovery: the theorem of metrization. Metrization is the process of deriving a specific measurement for the abstract concept of a topological space. In order to do this, Aleksandrov and Uryson first had to develop definitions of topological spaces. They initially defined a *bicompact* space (now known as compact and locally compact spaces), whose property is that for any collection of open sets (or groups of elements) that contains it (the interior of a sphere is an example of an open set). There is a subset of the collection with a finite number of elements that also contains it. Prior to their work, the concept of space was too abstract to be applicable to other mathematical fields; Aleksandrov and Uryson's research led to the acceptance of topology as a valid field of mathematical study.

With this result, the pair rose to fame within the mathematical community, gaining the approval of such notable scholars as **Emmy Noether**, **Richard Courant**, and **David Hilbert**. In 1924 Uryson and Aleksandrov went to Holland and visited with **Luitzen Egbertus Jan Brouwer**, who suggested that they publish their studies on topology. Aleksandrov and Uryson went on to the seaside in France for a spell of work and relaxation that ended tragically when Uryson drowned while swimming. In the aftermath of his friend's death, Aleksandrov lost himself in his work, conducting a seminar on topology that he and Uryson had begun organizing in 1924, and spending 1925 to 1926 working with Brouwer in an attempt to get his research into a form suitable for publication. During this time he further developed his theories of topology and compact space, with an eye to applying topology to the investigation of complex problems.

### Defines New Mathematical Field of Cohomology

In 1927 Aleksandrov left Europe for a year to continue his work with a new friend and colleague, Heinz Hopf, at Princeton. Aleksandrov had met Hopf during the summer of 1926 in Göttingen, which along with Paris was considered to be the mathematical hub of Europe. It was in Göttingen in 1923 that Aleksandrov and Uryson first presented their results outside the U.S.S.R., and it was Aleksandrov's preferred summer residence until 1932. There he worked with others, including Noether, who gave the topological work of Aleksandrov and Hopf its algebraic bent. This may have led to Aleksandrov's growing interest in homology, the offshoot of topology incorporating algebra. Homology had first been developed by the French mathematician **Jules Henri Poincaré**, but only for certain types of topological spaces. In 1928 Aleksandrov made a major step in expanding the field when he was able to generalize homology to other topological spaces.

In 1934 Aleksandrov at last received his doctorate from the University of Moscow. The next year, he would issue his most famous work. After much difficult research, the first volume of Aleksandrov and Hopf's still-classic work *Topologie* was published (the remaining two volumes would not be published until after World War II, though they were completed sooner). In the tome they outlined, often for the first time, many basic concepts of this branch of mathematics. They also introduced the definition of cohomology, which is the "dual" theory, or mirror image, of homology. Cohomologists consider the same topics as homologists, but from a different vantage point, providing different results. The publication achieved, Aleksandrov settled in a small town outside of Moscow with his friend and colleague **Andrei Nikolayevich Kolmogorov**. They stayed together here, teaching at the University of Moscow, for the rest of their lives.

Always concerned with the younger generation of mathematicians, Aleksandrov in later years crafted ground-breaking textbooks in the fields of topology, homology, and group theory, which studies the properties of certain kinds of sets. He guided his students—noted mathematicians such as A. Kuros, L. Pontriagin, and A. Tikhonov—to great heights. He also led the mathematical community in Moscow, presiding over that city's mathematical society for more than thirty years. In 1979 Aleksandrov wrote his autobiography. He died three years later in Moscow on November 16, 1982.

## SELECTED WRITINGS BY ALEKSANDROV

### Books

(With Heinz Hopf) *Topologie I,* 1935
*Combinatorial Topology,* three volumes, 1956–60
*An Introduction to the Theory of Groups,* 1959

### Periodicals

"Pages from an Autobiography." *Russian Mathematical Surveys,* Volume 34, number 6 (1979): 267–302, and Volume 35, number 3 (1980): 315–358.

## FURTHER READING

### Books

Boyer, Carl, and Uta C. Merzbach *A History of Mathematics.* Second edition, New York: Wiley, 1989.
Brown, Ronald. *Elements of Modern Topology.* New York: McGraw-Hill, 1968.
Fang, J. *Mathematicians from Antiquity to Today.* Volume 1, New York: Paidcia Press, 1972, p. 156.
Temple, George. *100 Years of Mathematics.* New York: Springer-Verlag, 1981.

### Periodicals

Arkhangelskii, A. V., and others. "Pavel Sergeevich Aleksandrov (On His 80th Birthday)." *Russian Mathematical Surveys* 31, number 5 (1976): 1–13.

*—Sketch by Karen Sands*

# Anaxagoras of Clazomenae
## 500 B.C.–428 B.C.
### Greek geometer and philosopher

Anaxagoras of Clazomenae was a Greek philosopher who made contributions in astronomy and physics. He was the first philosopher to live in Athens and the first to propose several important theories about the cosmos. Among these are theories about the Earth and moon, including the reason for an eclipse and accurate descriptions of the lunar surface. In mathematics, Anaxagoras was the first to attempt squaring the circle.

Not much is known about Anaxagoras life, though records of his theories are preserved. The Greek philosopher was born on the Ionian coast in the town of Clazomenae in Asia Minor (in what is now Turkey). His parents were wealthy but Anaxagoras chose to forsake his life of leisure to study philosophy. In 462 B.C., he moved to Athens, which was rapidly becoming an intellectual center. There he attracted the attention of the politician Pericles as well as the playwright Euripedes. Pericles welcomed Anaxagoras into his circle of friends.

Anaxagoras was the first to propose a molecular theory of matter. He believed that matter was infinitely divisible. The universe, he said, began as a great whirling jumble of matter, which was controlled by the Mind. The Mind, however, was not a god, or a spiritual or mental essence. "It was the most delicate

*Anaxagoras of Clazomenae*

and purest of all things," Anaxagoras wrote. In Anaxagoras' theory of "nous" the Mind caused the dark and the light to form, creating air, water, and earth. In another stage, animal and plant seeds (which were part of the original mixture) came together to form flesh and vegetation. The growth of all living things occurred because they had portions of the Mind within them, which could attract nourishment.

Anaxagoras believed that objects of the natural world were elemental. That is, they could not have been derived from elements simpler than themselves, from things that were not made of the same material. Every single piece of matter came from something like itself. "How could hair come to be from what is not hair and flesh from what is not flesh," he asks in his writings. He believed that every element in the world—hair, skin, bone, and an infinite number of other things—preexisted in our food

The Greek philosopher was also a great astronomer. He was the first to propose the reason for an eclipse and the first to theorize that the moon shone by reflected light. He also described the moon's surface accurately as a series of flat areas and depressions. The moon was not a sphere, Anaxagoras said. This was a contention that was confirmed 2,000 years later when **Galileo** trained his telescope on the moon. Direct proof of this theory also came through the American astronauts' first trip to the moon.

Anaxagoras, however, created real problems for himself and his friends when he proposed that the sun

was a red hot stone. All the planets and stars were, in fact, made of stone, he said. His belief may have been suggested by the fall of a huge meteorite near his home when he was young.

Anaxagoras' belief about the sun, however, made him a prime target for his and Pericles' enemies. They resented the attempts of Pericles, philosophers like Anaxagoras and artists like Pheidias to bring a higher culture to Athens. As Pericles grew aged, his enemies began to attack his friends. They accused Pheidias of stealing some of the gold used on his artistic statues. They campaigned for a law that permitted prosecution of those who did not believe in religion and taught theories about celestial bodies. Under this law, they brought Anaxagoras to trial.

It's not certain what was the result of the trial (records are not preserved), but we do know that while he was in jail, Anaxagoras made the first attempt to square the circle. In other words, he used a compass and a ruler to try to construct a square with the same area as a certain circle. This was the first time that such an effort had been made and preserved on record.

Pericles was able to get Anaxagoras released from prison. But Anaxagoras was forced to return to Ionia. There he started a school and was celebrated as a hero. He died in 428 B.C. and the anniversary of his death was celebrated for a century afterward in Ionia.

## FURTHER READING

Cleve, F.M. *The Philosophy of Anaxagoras.* New York, 1949.
Russell, Bertrand. *History of Western Philosophy.* New York: Simon and Schuster, Inc., 1945, pp. 79–81.

—*Sketch by Barbara Boughton*

# Kay McNulty Mauchly Antonelli
## 19??–
### American applied mathematician

Kay McNulty Mauchly Antonelli was one of about 75 young women mathematicians hired to serve as "computers" for United States government military projects during World War II. As such, these women calculated weapons firing and bombing trajectories for the U. S. Army at the University of Pennsylvania. With the development of the first electronic digital computer (ENIAC), those calculations became automated, and the women switched from tedious hand computing (each trajectory calculation took five days) to configuring the monster machine for every trajectory calculation.

Kay McNulty graduated from Chestnut Hill College in Philadelphia. The young math major heard that the Moore School of Engineering at the University of Pennsylvania in Aberdeen, a pioneer in computer development, was hiring mathematicians. It was there that she began her "computing" work. When the brilliant physicist team of John Mauchly and Presper Eckert created the ENIAC, McNulty became one of its programmers.

Thomas Petzinger, Jr., in a *Wall Street Journal* article, talks about the enormous complexity of the women's work: "The first task was breaking down complex differential equations into the smallest possible steps. Each of these had to be routed to the proper bank of electronics and performed in sequence — not simply a linear progression but a parallel one, for the ENIAC, amazingly, could conduct many operations simultaneously. Every datum and instruction had to reach the correct location in time for the operation that depended on it, to within 1/5,000th of a second."

That the Army considered this complex task "women's work" is evident in the job rating — SP, meaning subprofessional. In the beginning, the women were not allowed into the room which housed the ENIAC. Regarded as security risks, they had to work from wiring diagrams. Possessing no operating system or computer language, they were guided only by human logic.

It is ironic that history has largely ignored the contributions of the "ENIAC girls." The huge machine (100 feet long, 10 feet high, built of 17,480 vacuum tubes)—the hardware—became the dramatic focus of all who witnessed its flawless premier performance in February 1946.

Today, it is the programmers and software that are recognized as the "brains" behind computers. Quoted in the *Wall Street Journal* article, Kathryn Kleiman, who has produced a documentary film about the ENIAC programmers, says, "I absolutely think that computing and programming would be different today [without their contributions]."

McNulty married John Mauchly in 1948. They lived on a farm in Ambler, Pennsylvania. Several years after Mauchly's death in 1980, she married Severo Antonelli.

## FURTHER READING

### Books

Ritchie, David. *The Computer Pioneers: The Making of the Modern Computer.* New York: Simon and Schuster, 1986.

Shurkin, Joel. *Engines of the Mind: A History of the Computer.* New York: W. W. Norton & Company, 1984.

### Periodicals

Petzinger, Thomas Jr. "History of Software Begins with the Work of Some Brainy Women."
*Wall Street Journal* (November 1996).

### Other

"Kay McNulty Mauchly Antonelli." *Past Notable Women of Computing.* http://www.cs.yale.edu/homes/tap/past–women–cs.html (July 22, 1997).

—*Sketch by Jane Stewart Cook*

---

# Peter Apian (also known as Petrus Apianus and Peter Biennewitz)
## 1495–1552
### German mathematician, astronomer, and educator

Peter Apian was a Renaissance university professor who applied mathematics to the works of **Claudius Ptolemy** to develop new systems for surveying and mapmaking, to predict the position and movement of celestial bodies, and to calculate detailed sine tables.

Apian was born in Leisnig, Germany, April 16, 1495, the son of Martin and Gertrud Biennewitz, during an era in which Europe had shaken off the lethargy of the Middle Ages and plunged into the flurry of achievements that was the Renaissance. Mysticism and superstition were yielding to new understanding in the natural sciences; Nicholas Copernicus was working on his heliocentric system which placed the Earth and other planets as satellites of the sun; literature, art, and architecture flowered; and Christopher Columbus changed the social and economic shape of Europe with his discovery of the New World.

Apian studied mathematics and astronomy at Leipzig and Vienna, where he earned a reputation as a scholar. His first major work, *Cosmograpia seu descriptio totius orbis*, published in 1524, defined terrestrial grids, described the use of maps and basic surveying techniques, defined weather and climate, and included crude sketches of the continents. This work was later revised and reissued as *Cosmographia*. The work was translated into all major European languages, and its importance to navigation and the trade it made possible made it one of the most widely distributed books of its time. Apian also wrote and published an arithmetic book for use in business in 1527. This practical guidebook included on the title page what would later be known as the Pascal triangle, the first time the triangle appeared in print in Europe.

Apian's second major work, *Astronomicon Caesareum*, published in 1540, suggested that solar eclipses could be used to measure longitude. In *Astronomicon*, Apian was the first to publish the observation that comets always point their tails away from the sun. He included information on new uses of simple mechanical devices to plot the position and movement of stars and planets—helpful advice for who sailors relied on celestial navigation to guide vessels on existing trade routes, to find faster, safer, and more profitable routes, to explore previously untapped sources of goods, and to establish new markets.

Apian's most important work, published in 1534, was *Instrumentum sinuum sive primi mobilis*. Using Ptolemy's concept of minutes—one-sixtieth of a degree of an arc of a circle—Apian calculated and tabulated sines for every minute. These tables made it possible to quickly and accurately identify the ratio between the side opposite a given acute angle in a right triangle and the hypotenuse. Like his math book for business, Apian's *Instrumentum* had a host of practical applications and simplified the calculating tasks required in fields ranging from architecture to astronomy to navigation.

As a mapmaker, Apian compiled and published a cordiform world map and maps of Hungary and France which still survive. In 1520, he distributed what is believed to be the earliest map of the Old and New worlds in which the name "America" was used. His large-scale map of Europe, the first to be published, was printed in 1534; the original no longer exists.

Apian's work in mathematics focused more on application than theory. The practical benefits derived from his observations, calculations and mapmaking earned him recognition, and he was knighted by Charles V, emperor of the Holy Roman Empire.

From 1527 until his death in 1552, Apian served as a professor at the University of Ingolstadt, teaching astronomy and mathematics. Although most university professors taught in Latin, Apian preferred to instruct his students in German.

Apian's son, Phillip, followed in his father's footsteps. Born in 1531, Phillip was appointed to a post of professor of mathematics at the University of Ingolstadt in 1552 (the year of his father's death), and

from 1569 until his death in 1584, taught at the University of Tubingen.

## SELECTED WRITINGS BY APIAN

*Cosmographia seu descriptio totius orbis*, 1524
*Rechnung*, 1527
*Instrumentum sinuum sive primi mobilis*, 1534
*Astronomicon Caesareum*, 1540

## FURTHER READING

### Books

Fang, J. *Mathematicians From Antiquity to Today*. Volume 1. Memphis State University: Paideia, 1972, pp. 164–65.
Kish, George. "Peter Apian," in *Dictionary of Scientific Biography*. Volume II. Edited by Charles Coulston Gillispie. New York: Charles Scribner's Sons, 1970, pp. 178–79.
Gunther, S. "Peter und Phillipp Apian: Zwei Deutsche Mathematiker und Kartopraphen," in *Abhandlungen der Koniglich bohmischen Gesellschaft der Wissenschaften*, 6th ser., 11 (1882)
Smith, David E. *History of Mathematics*. Volume 1. New York: Dover Publications, 1951, pp. 333–334.
Whitrow, G.J. "Why Did Mathematics Begin to Take Off in the Sixteenth Century?" in *Mathematics from Manuscript to Print: 1300 –1600*. Edited by Cynthia Hay. Oxford: Clarendon Press, 1988, pp. 264–29.

### Periodicals

Ortroy, F. Van. "Bibliographie de l'oeuvre de Pierre Apian." *Bibliographie modern* (March–October 1901).

*—Sketch by A. Mullig*

# Apollonius of Perga
## c. 262–240 – c. 190–170 B.C.
### Greek geometer

Apollonius was one of the founding fathers of mathematical astronomy in ancient Greece. He also originated the geometric shapes and terms that would become central to Newtonian astronomical physics nearly 20 centuries later. He may have even prefigured **Christiaan Huygens'** 1673 use of the "evolute," the locus of the centers of curvature in a given curve. Certainly the projective geometry of **Gérard Desargues** and **Blaise Pascal** owe their genesis to Apollonius. He also invented his own counting system for very large numbers. Still considered the greatest achievement of Greek geometry, the *Conics* earned Apollonius the moniker "The Great Geometer," according to a later mathematician, Eutocius. These books quickly supplanted the works of **Euclid** as authoritative texts.

Estimations of the time frame in which Apollonius lived have varied over the years, so much so that one reference will only place him in the second half of third century B.C. Others place him according to the reigns of Ptolemy Euergetes, who was king of Egypt beginning around 247 B.C., or of Ptolemy Philopator ending in 210–205 B.C. Apollonius was born in what was then the Greek town of Perga, south Asia Minor, now part of Turkey. He was apparently a second generation Euclidean scholar in Alexandria. Legend has it he was nicknamed "Epsilon" there, because that Greek letter looks like the half moon he studied. Apollonius is also said to have visited Pergamum, where there was a new library and museum like the ones in Alexandria, and traveled to Ephesus. At least one source speculates that he may have been employed as the treasurer–general to Ptolemy Philadelphus. While he lived around the same time as **Archimedes**, there is no direct proof that the two either influenced each other or had contact. Apollonius did, however, improved upon Archimedes' calculation of the value of $\pi$.

## A New Mind–Set

Apollonius set forth in his eight books on conic sections, along with roughly 400 theorems, a new idea on how to subdivide the cone to produce circles. He also catalogued new kinds of closed curves that he named ellipses, parabolas, and hyperbolas. The Pythagorean distaste for infinities, infinitesimals, and infinite sets was put aside in this new frame of mind, which paved the way for the eventual discovery of the infinitesimal calculus. Additionally, his epicircles, epicycles, and eccentrics replaced the concentric spheres of Eudoxus and influenced Ptolemaic cosmology. That framework would stand until **Johannes Kepler** finally reformed the geometry of astronomical modeling for the current day.

The first half of the *Conics* surveys and completes all inherited Greek geometry, including early efforts of Euclid. Apollonius boasted that a Euclidean problem such as finding the locus relative to three or four lines was completely solvable for the first time, thanks to his new propositions. It is perhaps this style of presentation that led Pappus to accuse him of envy, and for Archimedes' biographer Heracleides to accuse him of plagiarism. The material on conic sections took up the last four volumes, laying the foundation for modern–day astronomy, ballistics, rocketry, and space science. Conic sections, it was only discovered many hundreds of years afterwards, are the shapes formed by the paths or "loci" of projectiles and other objects in orbit.

The *Conics* cover both pure and applied geometry. In it, Apollonius considered the problem of finding "normals" on points along curves, which involves trigonometry, though he could not apparently figure the focus of a parabolic curve the way he could for an ellipse or hyperbola. He also presented a method of figuring at what points a planetary orbit takes on apparent retrograde motion. Finally, his still famous "problem of Apollonius" calls for the construction of a circle tangent to three given circles. His most important contribution in pure terms was how he generalized the means of production. From one cone he could derive all conic sections, whether perpendicular to it or not. Apollonius used this standard cone in a way that prefigured analytic geometry by splitting it along two fixed lines called the *latus transversum* and the *latus erectum*. These "conjugate diameters" became a coordinate system and frame of reference, making geometry do the work now done by algebra.

## The Lost Works

A number of writers have attempted to restore or recreate the lost eighth book of the *Conics* or Apollonius' other writings, including Alhazen, **Edmond Halley** and **Pierre de Fermat**. The *Conics* were all that survived, perhaps because most of Apollonius' writings were considered too obscure or outrageous to be worth preserving by his contemporaries. That one masterwork influenced the next generation of mathematicians such as **Hipparchus**, and later commentators, including **Hypatia of Alexandria** and Eutocius, reinforced its reputation. Some of Apollonius' ideas and writings are mentioned in other ancient writings, which document some of his other conclusions. In his work on "burning mirrors," for instance, Apollonius disproved the notion that parallel rays of light could be focused by a spherical mirror, and he also noted properties of the parabolic mirror. Titles of his lost works include *Quick Delivery*, *Vergings*, and *Plane Loci*, as well as *Cutting–off of a Ratio* and *Cutting–off of an Area*. The subjects of these and some of their formulae and comments were summed up by Pappus.

Although his works were undervalued by many commentators over the years, beginning with his Greek contemporaries, Apollonius has recently undergone a revisionist examination. One academic, Wolfgang Vogel of Massey University in New Zealand, believes that after two millennia the ideas of Apollonius can be applied to current, significant problems related to intersecting conics.

## SELECTED WRITINGS BY APOLLONIUS

*Treatise on Conic Sections.* Edited by T.L. Heath, 1961

## FURTHER READING

### Books

Abbott, David, editor. *The Biographical Dictionary of Scientists.* New York: Peter Bedrick Books, 1986, pp. 9–10.

Asimov, Isaac. *Asimov's Biographical Encyclopedia of Science and Technology.* Second revised edition. Garden City, NY: Doubleday, 1982, p. 33.

Boyer, Carl B. "Apollonius of Perga," in *A History of Mathematics.* New York: John Wiley & Sons, Inc., 1968, pp. 157–75.

Bunt, Lucas N.H., et al. *The Historical Roots of Elementary Mathematics.* Englewood Cliffs, NJ: Prentice Hall, Inc, 1976, pp. 197–98.

Cajori, Florian. *A History of Mathematics.* Second edition, revised and enlarged. New York: Macmillan, 1919, pp. 38–41.

Calinger, Ronald, editor. *Classics of Mathematics.* Oak Park, IL: Moore Pubs. Co., Inc., 1982, pp. 148–54.

Eaves, Howard. *Great Moments in Mathematics (Before 1650).* Washington, D.C.: Mathematical Association of America, Inc., 1980, p. 219.

————. *An Introduction to the History of Mathematics.* Fourth edition. Holt, Rinehart & Winston, 1976, pp. 139–43.

Knorr, Wilbur R. "Apollonius of Perga." *The Great Scientists.* Edited by Frank N. Magill. Danbury, CT: Grolier, 1989, pp. 84–90.

Morgan, Bryan. *Men and Discoveries in Mathematics.* London: John Murray Pubs., 1972, pp. 46–51.

Ronan, Colin. *Astronomers Royal.* Garden City, NY: Doubleday & Co., Inc., 1969, p. 67.

Van Der Waerden, B.L. *Science Awakenings.* Translated by Arnold Dresden. New York: Oxford University Press, 1961, pp. 237–39.

Williams, Trevor I., editor. *A Biographical Dictionary of Scientists.* Third edition. New York: John Wiley & Sons, 1982, pp. 13–14.

### Periodicals

Coxeter, H.S.M. "The Problem of Apollonius." *American Mathematical Monthly* 75 (1968): 5–15.

Neugebauer, O. "Eccentric and Epicyclic Motion According to Apollonius." *Scripta Mathematica* 24 (1959): 5–21.

### Other

"Apollonius of Perga." *MacTutor History of Mathematics Archive.* http://www–groups.dcs.st–and.ac.uk/~history/Mathematicians/index.html (July 1997).

Vogel, Wolfgang. "Appreciating Apollonius 2,000 Years Later." (Inaugural Lecture). Department of Mathematics, Massey University, New Zealand. http://fmis–www.massey.ac.nz.maths/NZMS64/local_news.html

*—Sketch by Jennifer Kramer*

# Louis François Antoine Arbogast
## 1759–1803
### French number theorist and mathematical historian

Louis François Antoine Arbogast made a three-fold contribution to mathematics. He is primarily known as a mathematics historian who organized **Marin Mersenne**'s papers, as well as letters and miscellany of other scientists. Arbogast was also a noteworthy mathematician in his own right; he did the earliest work in discontinuous fractions and predated some developments in calculus. He also participated in local politics (including the Commune of Strasbourg) and was influential in the development of certain schools and their curriculums in the 1790s.

Much of Arbogast's early life and education is unclear other than his date of birth, October 4, 1759 in Mutzig, Alsace. It is known that in 1780, he was listed as a lawyer in Alsace, and he also taught mathematics at the Collège de Colmar in 1787. Arbogast's life becomes more documented after moving to Strasbourg, France, in 1789. In that same year, he became a mathematics instructor at l'École d'Artillerie and a physics professor at Collège Royal. When the Collège became nationalized, he was director for seven months in 1791. In 1794, Arbogast was appointed to a professorship in Paris, at l'École Central de Paris, though he only taught at l'École Préparatoire, a temporary institution of higher learning. He began planning its replacement, l'École Centrale du Bas–Rhin, in 1795. (Arbogast was given this responsibility in part because he had experience in this area. He designed a program for public schools to a legislative assembly in Alsace circa 1792.) After the school's establishment in 1796, he served as mathematics chair until 1802.

As a mathematician, Arbogast's lasting contribution was in the collection and arrangement of manuscripts by other scientists. He hand copied the writings of such people as Marin Mersenne, **René Descartes, Jean Bernoulli**, Guillaume de L'Hosptial, Pierre Varignon, and **Pierre de Fermat.** Arbogast's collection are now in the Bibliothèque Nationale in Paris and the Laurenziana Library in Florence.

Arbogast made his own mark on mathematics in 1787 with his work on discontinuous fractions and arbitrary functions. In 1789, he wrote a report that was never published outlining new principles of differential calculus. In his mathematical writings, he demonstrated what came to be known as operational calculus. In this area, he was years ahead of his time, implicitly demonstrating that operation and function are inherent differences in calculus. This is one of many mathematical areas Arbogast anticipated.

Arbogast's contributions to mathematics did not go unnoticed by his colleagues. In 1792, the Académie des Sciences made him a corresponding member. In 1796, the Institut National elected him an associate nonresident member. Arbogast died in Strasbourg seven years later on April 18, 1803.

## SELECTED WRITINGS BY ARBOGAST

*Du calcul des derivations*, 1800

## FURTHER READING

Itard, Jean. "Louis François Antoine Arbogast," in *Dictionary of Scientific Biography*. Volume I. Edited by Charles Coulston Gillispie. New York: Charles Scribner's Sons, 1970, pp. 206–08.

*—Sketch by Annette Petruso*

# Archimedes of Syracuse
## 287–212 B.C.
### Greek geometer

Archimedes of Syracuse is considered one of the greatest thinks of the ancient world. He established the principles of plane and solid geometry, discovered the concept of specific gravity, conducted experiments on buoyancy, demonstrated the power of mechanical advantage, and invented the Archimedes Screw, an auger–like device for raising water.

Archimedes was born in the Greek city of Syracuse, on the island of Sicily, in 287 B.C. He was

*Archimedes of Syracuse*

the son of Phidias, the astronomer and mathematician. What we know of Archimedes' life comes from his extant writings, and from the histories authored by Plutarch, Cicero, and other historians several centuries after his death. Due to the length of time between Archimedes' death and his biographers' accounts, as well as inconsistencies among their writings, details of his life must remain subject to question.

Plutarch records that Archimedes was a relative of King Hieron, but Cicero claims he was of low birth. It is believed he obtained his early schooling in Syracuse, then traveled to Alexandria to study with **Conon**, the Egyptian mathematician and astronomer. Archimedes was close friends with both Conon and the custodian of the Alexandrian library, **Eratosthenes.** He corresponded with them about his mathematical and scientific discoveries long after he had completed his formal studies and returned to Syracuse. Although much of Archimedes' work was applied to practical ends, he himself was more interested in pure thought, and supposedly believed things connected with daily needs were ignoble and vulgar. Reports of his personal habits reflect his lack of concern with the mundane. He was known for becoming so engrossed in his thoughts that he would forget to eat and bathe. Many of the equations he developed were scratched out first in ashes or traced with after–bath oil on his skin.

### Sets Force Principles of Plane and Solid Geometry

Archimedes' contributions to mathematical knowledge were diverse. On the subject of plane geometry three of the treatises he wrote have survived, *Measurement of a Circle*, *Quadrature of the Parabola*, and *On Spirals*. In *Measurement of a Circle*, he described his method for calculating π, the ratio between the circumference of a circle and its diameter. By a method that involved measuring the perimeter of inscribed and circumscribed polygons, Archimedes correctly determined that the value of π was somewhere between 3.1408 and 3.1428. In the same treatise he set forth the formula π $r^2$ for determining a circle's area. In *Quadrature of the Parabola* and *On Spirals*, Archimedes advanced his technique for determining the area under curves, a sophisticated version of the method of exhaustion, originally developed by the Egyptians. His use of this technique, elaborated upon in another volume, The *Method*, anticipated the development of integral calculus by two thousand years.

Archimedes dealt with the topic of solid geometry in his writings *On the Sphere and Cylinder* and *On Conoids and Spheroids*. *On the Sphere and Cylinder* contains several famous proofs, including his demonstration that the volume of a sphere is equal to $4/3 \pi r^3$. Archimedes also showed in this work that the volume of a sphere is two–thirds the volume of a cylinder surrounding it, as long as the cylinder's height and width are equal to the sphere's diameter. So proud was he of this latter discovery that he requested its illustration be engraved on his tombstone, a wish that was eventually fulfilled.

On the subject of arithmetic Archimedes wrote several essays, of which only *The Sand Reckoner* remains. Addressed to the son of King Hieron, it proposed the problem of determining the number of grains of sand in the universe, and contained a special notation for estimating and expressing very large numbers. Archimedes was famous for another complicated arithmetic puzzle called the Cattle Problem, in which one had to determine the number of bulls and cows of various colors, given that each cattle color was represented in a particular ratio to the others. There are an infinite number of possible solutions to the problem, but deriving them was especially challenging to the ancient Greeks, who had no knowledge of algebra.

### Establishes Fundamental Rules of Mechanics

According to the ancient historians, Archimedes was frequently called upon by King Hieron to solve practical problems. Perhaps the most famous of these was the task of determining whether King Hieron's crown was made of pure gold. Hieron believed his jeweler had stolen a portion of the gold intended for the crown, substituting an equal weight of another metal. Legend has it that Archimedes stepped into a bath and, noticing the displacement of water, conceived of a vertical buoyancy force. Realizing that this

concept would allow him to measure the density of the king's crown in comparison to pure gold, he supposedly jumped from the tub and ran naked through the streets yelling "eureka, eureka!" Archimedes not only proved the crown was tainted, but went on to describe the idea of specific gravity and develop a generalized concept of buoyancy known today as Archimedes' Principle.

The Archimedes Screw was reportedly invented in order to empty water from the hold of one of King Hieron's ships. This device consisted of a screw, encased in a cylinder, that was turned by a hand–crank. As the screw spiraled upward it carried water. A similar device is still used today to lift water in the Nile Delta of Egypt. Archimedes recognized the mechanical advantage that could be gained by using levers, and it is said he boasted that with a long enough lever he could move the Earth. He determined the inverse mathematical relationship between the effort required to raise a load with a lever and the distance of the load from the lever's pivot point or fulcrum. The story is told that King Hieron, skeptical of the power of mechanical advantage, challenged Archimedes to move a three–mast ship, laden with passengers and freight, that lay aground near Syracuse Harbor. To meet the challenge Archimedes is said to have designed a system of compound pulleys. With a relatively effortless pull of a rope he was able to guide the vessel into the water. Archimedes was also interested in astronomy and made several accomplishments in this field. He built a device to estimate the size of the sun and constructed a model planetarium to demonstrate the motion of the planets.

### Invents War Machines to Aid Syracuse Against the Roman Siege

During Archimedes' lifetime the first two of the three Punic Wars between the Romans and the Carthaginians were fought. Syracuse allied itself with Carthage, and when the Roman general Marcellus began a siege on the city in 214 B.C., Archimedes was called upon by King Hieron to aid in its defense. The historical accounts of Archimedes' war–faring inventions are vivid and possibly exaggerated. It is claimed that he devised catapult launchers that threw heavy beams at the Roman ships, grappling cranes that hoisted ships out of water, and burning–glasses that reflected the sun's rays and set ships on fire. Marcellus had given orders that when Syracuse was finally conquered, Archimedes, whose reputation was widely known, should be taken alive. When the Romans finally sacked the city in 212 B.C., a soldier found Archimedes quietly etching equations in the sand, absorbed in a mathematical problem. Reportedly, Archimedes ordered the soldier not to disturb the figures in the sand. Enraged, the soldier drew his sword and impaled him.

*Archytas of Tarentum*

## FURTHER READING

### Books

Cajori, Florian. *A History of Mathematics*, New York: Chelsea Publishing Co., 1991.

Eves, Howard. *An Introduction to the History of Mathematics*. New York: Holt, Rinehart and Winston, 1960.

Ibsen, D.C. *Archimedes: Greatest Scientist of the Ancient World.* Enslow Pubs., 1989.

Lafferty, Peter. *Pioneers of Science: Archimedes.* New York: The Bookwright Press, 1991.

Moffatt, Michael. *The Ages of Mathematics: The Origins.* Volume 1. New York: Doubleday, 1977.

*—Sketch by Leslie Reinherz*

# Archytas of Tarentum
## c. 428–c. 350 B.C.
### Greek geometer, philosopher, and statesman

Archytas of Tarentum was a Greek mathematician of the Pythagorean school who formulated the harmonic mean and was the first to integrate mathematics and mechanics. He also developed an ingenious geometric solution for the ancient Greek problem of doubling the cube. A contemporary of the famous Greek philosopher Plato, Archytas was also famous in his own time as a philosopher, statesman, and military leader.

Only a few fragments identified as the work of Archytas have survived. As a result, most of what we know about his life and work comes from ancient Greek writers, such as Aristotle and Proclus. Archytas was probably born in Tarentum (now Taranto, Italy) around 428 B.C., possibly into an aristocratic family. Nothing else is known about his early life.

In the beginning of the fourth century B.C., Dionysius the Elder, a tyrant of Syracuse, had driven the Pythagoreans out of most southern Italy's cities. Tarentum was the last city in the region where the Pythagorean school of philosophy and mathematics maintained a strong presence in education and politics. A close friend of Plato, Archytas may have been his chief teacher of Pythagorean science and philosophy. More importantly, Archytas saved this great philosopher's life by obtaining a pardon for him through a letter he wrote to Dionysius the Younger, who wanted to execute Plato for subversive activities.

Archytas was a powerful statesman and an influential leader in Tarentum and throughout the Greek city–states. Immensely popular, he served as general of his city's citizen army for seven years, despite a law forbidding anyone to hold the position for more than one year. The confederation of Hellenic cities of Magna Grecia also appointed him commander, with full autocratic authority, over the confederation's armies. Troops under his command reportedly were never defeated in battle. According to some accounts, Archytas eventually gave up his command because of envious detractors, and his troops were immediately captured. Archytas was also greatly admired for his virtues and noble character, which included a love of children and the just and kind treatment of his slaves.

### Makes Major Contributions in Mathematics and Geometry

Archytas' creativity and ingeniousness were grounded in his recognition of how the sciences interconnect, especially the disciplines of mathematics, geometry, music, and astronomy. For example, he was the first to apply mathematics to the realm of mechanics and wrote a systematic treatise on the subject. He invented the simple pulley and screw and wrote on the mathematical basis of astronomy. Through his work in the theory of means and proportions he differentiated three basic means: the arithmetic mean, the geometric mean, and the harmonic mean. Seven other means were eventually added by Archytas and others. In his theory of music, Archytas developed numerical ratios representing intervals of the tetrachord on which he based his three musical scales: the enharmonic, the chromatic, and the diatonic. **Claudius Ptolemy** the astronomer credited Archytas as the most important Pythagorean to delve into the theory of music.

According to Proclus, Archytas also increased the number of theorems in geometry, developing them into a systematic body of knowledge, and influenced many other Greek mathematicians. Archytas is credited with most of the geometry contained in Book VIII of **Euclid**'s *Elements*, which served as the primary textbook of elementary geometry and logic for more than a thousand years.

Archytas' most famous mathematical achievement was to provide an elegant geometric solution to the Delian problem, known as duplicating or doubling the cube, or enlarging a cube according to a given ratio. Although Plato had complained that the Greeks knew little about three–dimensional geometry, Archytas exhibited a comprehensive knowledge of this area in his remarkable solution to the problem. By inventing a new type of three–dimensional curve through the intersection of a cylinder, a cone, and a torus (or doughnut shape), he was able to find the two mean proportionals (or geometric means) between two lines, a method first proposed by **Hippocrates** of Chios for doubling the cube.

Although he was a pioneer in mathematics and geometry, Archytas has been criticized for not following his contemporaries' lead in applying clear and logical explanations for his theories. His inability to cope well with the logical aspects of his work also affected other interests. For example, in the largest fragment of his extant works, Archytas proposed a complicated theory of sound. He correctly theorized that faster motion produces higher sounds or notes. However, based solely on empirical observations without the application of mathematical theories, Archytas wrongly concluded that higher sounds reach the listener faster than lower sounds.

Archytas' philosophy was probably based on his training as a Pythagorean, leading him to hold a strong belief in numbers as a mystical and basic part of nature. In a small fragment of one of his works, Archytas also reveals that he believed the universe was infinite in extent. Although Aristotle wrote three books on the philosophy of Archytas, none of them have survived the passing of time.

While Archytas is rightly remembered today for his seminal contributions in mathematics and geometry, he should also be admired as a man of action who applied his keen intellect to affairs of state. A successful army general, Archytas was an influential leader who helped forge alliances among Greek city–states to provide greater protection against for-eign powers. Archytas also had his playful side; his interest in mechanics led him to create two mechanical devices: a mechanical wooden pigeon that could fly and, according to Aristotle, a type of rattle to amuse and occupy infants. In an ode, the famous Greek poet Horace recounts Archytas' death by shipwreck in the Adriatic Sea.

## FURTHER READING

Allman, George Johnston. *Greek Geometry from Thales to Euclid*. London: Longmans, Green, & Co, 1889, pp. 102–127.

Fritz, Kurt von. "Archytas of Tarentum," in *Dictionary of Scientific Biography*. Volume I. Edited by Charles Coulston Gillispie. New York: Charles Scribner's Sons, 1975, pp. 231–33.

Terry, Leon. *The Mathmen*. New York: McGraw–Hill Book Company, 1964, pp.74–75; 128–192.

Van Der Waerden, L. Van. *Science Awakening*. New York, John Wiley & Sons, Inc., 1963, pp. 149–159.

—*Sketch by David A. Petechuk*

# Jean Robert Argand
## 1768–1822
### Swiss number theorist and geometer

Jean Robert Argand invented a method of geometrically representing complex numbers and their operations. The Argand diagram is a graphic representation of complex numbers as points on a plane and their additions. He is also credited with giving proof, although with a few gaps, of the fundamental theorem of algebra. In 1814, Argand published a proof of the fundamental theorem of algebra which may be the simplest of all proofs of this theorem.

Argand was born in Geneva, Switzerland, on July 18, 1768, to Jacques Argand and Eves Canac. Historians have limited knowledge of Argand's background and nothing is known of his education. Apparently, he was a self–taught mathematician, belonging to no mathematical societies or organizations. Mathematics appeared to be just a hobby to Argand. He has often been confused with Aime Argand, the physicist and chemist who had invented the Argand lamp, however, they are not related.

In 1806, Argand, his wife and children moved to Paris. He was working as an accountant when he published his method in a book entitled *Essai sur une maiere de representer les quantities imaginaires.* The book had been published in a small, privately printed edition but it did not include Argand's name. No one knew he wrote the book until sometime later.

How it came to light that Argand was the author is an unusual story. He may never have been credited with writing the book, had a set of curious circumstances not occurred. Around the time the book was published, two other mathematicians, **Casper Wessel**, a Norwegian and **Karl Gauss**, from Germany, were working on the same idea. However, neither had put their ideas into print, and Gauss is sometimes credited with writing the *Essai*.

Argand had spoken of his new method to **Adrien M. Legendre** before the book was published. Legendre spoke of the method in a letter to the brother of J.R. Français. Français was a lecturer at the Imperial College of Artillery at Paris. Français found the letter after his brother's death. In an essay published in the journal *Annales de Mathematiques*, Français discussed the idea of the new method and even developed it further. At the end of the piece, Français called for the author of the book to come forward and be recognized. Argand acknowledged his works by writing an article that was published in a later edition of the same journal.

Argand died in Paris on August 13, 1822, having contributed nothing more to the science of mathematics.

## SELECTED WRITINGS BY ARGAND

### Books

*Essai sur unde maniaere de representor les quantities imaginaires dans les constructions geomaetriques,* 1971

*Imaginary quantities; their geometrical interpretation, tr. From the French of M. Argand. By Prof. A.S. Hardy . . .,* 1881

## FURTHER READING

### Books

Jones, Phillip. "Jean Robert Argand," in *Dictionary of Scientific Biography*. Volume I. Edited by Charles Coulston Gillispie. New York: Charles Scribner's Sons, 1970, pp. 237–40.

Porter, Roy, editor. *The Biographical Dictionary of Scientists*. New York: Oxford University Press, 1994, pp. 19–20.

Wyckoff, Jerome, editor. *The Harper Encyclopedia of Science*. Washington, D.C.: Harper & Row, Publishers, 1967.

### Other

"Abstract Linear Spaces." http://www–groups.dcs.st–and.ac.uk/~history/HistTopics/Abstract_linear_spaces.html#15 (May 6, 1997).

"The fundamental theorem of algebra." http://www–groups.dcs.st–and.ac.uk/~history/HistTopics/Fund_theorem_of_algebra.html (May 6, 1997).

"Jean Robert Argand." *MacTutor History of Mathematics.* http://www–groups.dcs.st–and.ac.uk/~history/Mathematicians/Argand.html (May 6, 1997).

"Johann Carl Friedrich Gauss." *MacTutor History of Mathematics.* http//www–groups.dcs.st–and.ac.uk:80/~history/Mathematicians/Gauss.html (June 2, 1997).

Library of Congress. http://lcweb.loc.gov/cgi–bin/zgate (May 7, 1997).

*—Sketch by Monica L. Stevens*

---

# Aristarchus of Samos
## c. 320–310–c. 250–230 B.C.
### Greek geometer and astronomer

As with many of his contemporaries, the only extant facts about the life of Aristarchus involve remarks about him and his work written by others. Only one of Aristarchus' writings survived, *On the Magnitudes and Distances of the Sun and Moon,* but in it he articulated the reasoning behind what later became modern trigonometry, and how it might be employed in astronomy and navigation. Typical of Greek mathematics was Aristarchus' primarily geometric method of approximating this strategy of triangulation. From **Archimedes**, it is known that Aristarchus had proposed the sun be considered a fixed star, with the Earth circulating around it. This view was ridiculed at the time and remained dormant until Nicolaus Copernicus devised his heliocentric theory.

Birth and death dates for Aristarchus vary, but it is agreed that he was born on the island of Samos in Greece and studied under Strato (or Straton) of Lampsacus in Alexandria, Egypt. Strato went on to succeed Theophrastus as head of the lyceum founded by Aristotle at Athens, so it is probable that Aristarchus circulated among highly intellectual and influential men. Certain of his activities can be roughly dated, as he made observations of the summer solstice around 281–280 B.C., according to Ptolemy.

## Myriad–myriads

Aristarchus favored a mathematical approach to astronomy over the descriptive one, which tended to rely on intuition and rhetoric rather than observation. An example of his commitment to observation is his reported correction of Callippus' estimate of the length of the year, adding 1/1623 of a day to it. His own observations led to six astronomical hypotheses, from which Aristarchus drew eighteen propositions. These regarded measuring the sizes and distances of various celestial objects relative to the known diameter of the Earth. He correctly concluded that the orbit of the Earth was dwarfed by the overall size of the universe and distance of its furthermost visible stars. Archimedes proved this immensity was calculable with his famous "sand reckoning," counted at the time in myriad–myriads.

By studying the relative positions of the sun, moon, and Earth, Aristarchus concluded that during the half–moon each of them occupy respective points on a right triangle. He then reasoned that the Pythagorean theorem could be applied to determine the ratio of the sun–Earth distance and the moon–Earth distance. In fact, his proof of this is best expressed today as a trigonometric formula.

Because Aristarchus did not have the tools to measure angular distances of heavenly bodies, he consequently underestimated these distances. Likewise, his estimate of the size of the moon relative to the Earth, and the size of the sun relative to the moon were inaccurate as well. Those figures were improved during the next century by **Hipparchus**, though it is only later that we learned Aristarchus underestimated the sun's size by nearly 400 times his original estimation.

Unfortunately, the idealistic Greek model called for circular orbits, which did not account for the unevenly distributed changing of seasons. These are now attributable to elliptical orbits. While Aristarchus could not completely free himself of the Greek intellectual loyalty to mathematical harmonies, he went a long way towards letting experiment rule theory rather than idealism.

## Beyond Ideals

It is from the writings of Archimedes and Plutarch that Aristarchus' heliocentric hypothesis of 260 B.C. became known. As articulated by Aristarchus, the hypothesis accounted for the apparent motion of the heavenly bodies and diurnal motion of the stars. He not only proposed that the sun is fixed and that the Earth revolves around it, but also that the Earth rotates on its own axis. Aristarchus was roundly criticized—his contemporaries marshaled Aristotelian logic to refute his premise as untenable—although he was apparently never persecuted.

Aristarchus died at the earliest estimate around 250 B.C. in Alexandria. Debunked in his own time, his contributions as a scientist and mathematician have since been reevaluated. He may have also been an inventor, making an improved design of sundial called a "skaphe" or "scaphion," which seems to have placed the shadow–throwing pointer within a hollow hemispherical base.

Aristarchus was not acknowledged by Copernicus himself, having struck out a passage referring to his distant precursor during the editing of his manuscript *De revoluntionibus orbium coelestium*. Aristarchus' most ambitious ideas could not be confirmed or denied until the time of **Isaac Newton**, when it became possible to test the effects of the rotation of the Earth and the phenomenon of stellar aberration.

## SELECTED WRITINGS BY ARISTARCHUS

*Aristarchus of Samos on the Sizes and Distances of the Sun and Moon.* Translated and with notes by Sir Thomas Heath, 1938

## FURTHER READING

### Books

Asimov, Isaac. *Asimov's Biographical Encyclopedia of Science and Technology.* Second Revised Edition. Garden City, NY: Doubleday, 1982, pp. 26–27.

*Biographical Encyclopedia of Scientists.* Edited by Urdang and Associates. New York: Facts on File, 1981, p. 26.

Boyer, Carl B. *A History of Mathematics.* NY: John Wiley & Sons, Inc., 1968, pp. 138–39.

*The Columbia Encyclopedia.* Fifth Edition. New York: Columbia University Press, 1993, p. 2091.

Eaves, Howard. *Great Moments in Mathematics (Before 1650).* Mathematical Association of America, Inc., 1980, p. 96.

Greider, Ken. *Invitation to Physics.* New York: Harcourt, Brace Jovanovich, Inc., 1973, pp. 16, 28.

Hogben, Lancelot. *Mathematics in the Making.* Garden City, NY: Doubleday, 1962, pp. 126–9.

Ronan, Colin. *Astronomers Royal.* Garden City, NY: Doubleday & Co., Inc., 1969, pp. 5–6.

Van Der Waerden, B.L. *Science Awakening.* Translated by Arnold Dresden. New York: Oxford University Press, 1961, pp. 202–4.

Williams, Trevor I., editor. *A Biographical Dictionary of Scientists.* Third Edition. New York: John Wiley & Sons, 1982, pp. 16–17.

*World Who's Who in Science.* First edition. Edited by Allen G. Debus. Chicago: Marquis, 1968, p. 61.

**Other**

"Aristarchus of Samos." *MacTutor History of Mathematics Archive.* http://www-groups.dcs.st-and.ac.uk/~history/Mathematicians/Aristarchus.html (July 1997).

*—Sketch by Jennifer Kramer*

# Aristaeus the Elder
## c. 360–c. 300 B.C.
### Greek geometer

Aristaeus the Elder was an important Greek mathematician most widely known as one of the first developers of conics and the conic section theory. **Euclid**, who is known as the founder of geometry, reportedly recognized Aristaeus as a "worthy mathematician" because of his work in conics.

Nothing is known about the life of Aristaeus except that he was an older contemporary of Euclid and lived circa 360 to 300 B.C. The primary source of information about him comes from Pappus, who lived nearly six hundred years after Aristaeus and wrote a commentary, *Treasury of Analysis*, on all of the Greek mathematics known up to his time. Since Pappus refers to him as Aristaeus the "Elder," there may have been another Greek mathematician of the same name. A reference by Iamblichus that Aristaeus was the son-in-law of the 6th century B.C. Greek mathematician and philosopher **Pythagoras** is highly unlikely, since the two lived more than two centuries apart.

### Writes Influential Book on Conics

Although no writings of Aristaeus survive today, Pappus did possess a copy of a work by Aristaeus called the *Five Books Concerning Solid Loci.* The book's title indicates that Aristaeus focused primarily on conics (and the curves, lines, and points that they can contain) regarded as loci. According to Pappus, Euclid credits Aristaeus for much of his own work in conics. Most historians believe that the thirteenth book of Euclid's *Elements*, which served as the definitive textbook of elementary geometry and logic for over 1,000 years, is partially a recapitulation of Aristaeus' work. As a result, they surmised that Aristaeus wrote the more original work on conics,

which Euclid then compiled, rearranged, and improved upon. Pappus also notes that the terms "section of the acute-angled cone," "section of right-angled cone," and "section of the obtuse-angled cone" were introduced by Aristaeus in his book.

Another ancient Greek author and editor of Book XIV of the *Elements*, Hypsicles, also credits Aristaeus with a book called the *Comparison of the Five Regular Solids.* Hypsicles refers to a theorem developed by Aristaeus in this book and later given by **Apollonius**, who followed Aristaeus a century or so later and introduced a comprehensive theory of conic sections.

Historians have also debated whether or not Aristaeus authored another book called *Five Books of the Elements of Conic Sections.* If he did write the book, it may have come after *Five Books Concerning Solid Loci* and have been written to further facilitate the study of the theorems and knowledge presented in that book.

Based on his *Five Books Concerning Solid Loci* and its tremendous influence on Euclid, Aristaeus is rightly considered the most important geometer of ancient Greece before Euclid. He is also credited, along with Euclid and Apollonius, as one of primary cultivators of the method, or doctrine, of analysis for conic sections. The fact that Euclid based his work in conics on the work of Aristaeus and that Pappus identifies Aristaeus as an important contributor to early mathematics demonstrates that Aristaeus was held in high regard by ancient Greek mathematicians and historians.

### FURTHER READING

Allman, George Johnston. *Greek Geometry from Thales to Euclid.* London: Longmans, Green, & Co., 1889, pp. 194–205.

Coolidge, Julian Coolidge. *A History of the Conic Sections and Quadric Surfaces.* Oxford: Clarendon Press, 1945, pp. 5–6.

Vogel, Kurt. "Aristaeus the Elder," in *Dictionary of Scientific Biography.* Volume I. Edited by Charles Coulston Gillispie. New York: Charles Scribner's Sons, 1975, pp. 245–46.

*—Sketch by David A. Petechuk*

# Aryabhata the Elder
## 476(?)–550
### Hindu mathematical astronomer

In a time and place where people believed certain distant stars, called "asuras," possessed malevolent powers capable of inflicting harm on Earth, Aryabhata the Elder took the first steps towards separating scientific explication from folklore and superstition. His *Aryabhatiya* was the first major book on Hindu mathematics, which summarized knowledge of his predecessors. While covering many aspects of arithmetic, algebra, and numerical notation, the majority of *The Aryabhatiya* dealt with trigonometric tables and formulae for use in astronomy. Any astronomical observations Aryabhata made were most likely completely unaided. Although *The Aryabhatiya* contained errors, it was translated and reproduced as *Zij al-Arjabhar* by the Arabs. One of Aryabhata's methods, a solution to the indeterminate quadratic $xy = ax + by + c$, was rediscovered by **Leonhard Euler** in the 18th century.

### The City of Flowers

Aryabhata the Elder was born near what is now the city of Patna in India. His year of birth is sometimes cited as 475 but also possibly 476. Aryabhata's hometown was called Kusumapura, or The City of Flowers. The two major centers in the area represented the two intellectual threads he contended with in his lifetime. At that time, Patna was a royal seat, and according to legend it was founded by a knight with magic powers in honor of his princess, and blessed by the Buddha. Further away in Ujjian, the study of science and astronomy began to flourish, and this knowledge was being disseminated in the form of rhymed, romantic stories. Various mathematical problems, similar to those used in textbooks today, were solved in verse form used for social amusements. The public challenge and the romantic forms are combined in the most famous quotation from *The Aryabhatiya*, as Aryabhata the Elder commands a "beautiful maiden" to answer a problem that requires inversion.

*The Aryabhatiya* was produced in the year 499. It described the Indian numerical system with nine symbols, and listed various rules for arithmetic and trigonometric calculations. It also made use of continued fractions, square and cube roots, and the sine function when needed. Solutions were given for linear and quadratic equations and diophantine equations of the first degree. These involved one of the first recorded uses of algebra and decimal place-value. Unlike the Greeks, the Hindus solved diophantines for all possible integral solutions, as they were more tolerant of negative, irrational, and other such numbers. For instance, one value for $\pi$ given by Aryabhata is the square root of 10, generally called "the Hindu value."

### Pebbles and Crystals

The combination of correct and incorrect answers to the major questions of its time led one Arabic commentator, al-Biruni, to describe Hindu mathematics as a mixture of "common pebbles and costly crystals." Aryabhata gave an accurate approximation of $\pi$, although he overestimated the length of the year by 12 minutes and 30 seconds. He was singular in describing the orbits of the planets as ellipses. In the *Ganita*, a poem composed in 33 couplets, he correctly states the formulae for areas of a triangle and a circle. However, when Aryabhata attempts to extrapolate those to figure the volumes of three-dimensional shapes in the same couplets, he is not successful. Nonetheless, Aryabhata the Elder's commitment to general methods caused him to apply what is now nicknamed "the pulverizer." This rule finds the greatest common divisor of $a$ and $b$ by division, equivalent to Euler's later version of reducing $a$ over $b$ to a continued fraction. While "the pulverizer" has also been known as the Diophantine method, the fact that Diophantus himself never used it renders the term a misnomer.

Aryabhata died around 550, though it is not known where. As an astronomer he argued—against Vedic tradition—that the Earth was round and rotated daily. He correctly explained why equinoxes, solstices, and eclipses occur. These ideas were not accepted in Aryabhata's lifetime, but his mathematics had set the foundation for developments in the Eastern and Western worlds for centuries to come. In India particularly, Aryabhata the Elder marked the end of the sacred or "S'ulvasutra" period, during which mathematics was used primarily by priests for temple architecture. He ushered in the "astronomical period" that lasted until the year 1200.

*The Aryabhatiya* held the same stature in India that **Euclid**'s *Elements* did in ancient Greece. Bhaskara I wrote a commentary on this work in 629. Aryabhata also influenced the work of Hindu astronomer Brahmagupta.

## SELECTED WRITINGS BY ARYABHATA

### Books

*The Aryabhatiya of Aryabhata.* Edited by W.E. Clark, 1930.

## FURTHER READING

### Books

Boyer, Carl B. *A History of Mathematics.* New York: John Wiley & Sons, 1968, pp. 229–241.
Cajori, Florian. "The Hindus." *A History of Mathematics.* Second Edition, revised and enlarged. New York: Macmillan, 1919, pp. 83–98.

Eaves, Howard. *Great Moments in Mathematics (Before 1650).* Mathematical Association of America, Inc., 1980, pp. 21–2, 105.

Eaves, Howard. *An Introduction to the History of Mathematics.* Fourth Edition. Holt, Rinehart & Winston, 1976, pp. 180–87.

Wu, Frank. "Aryabhata." *The Great Scientists.* Edited by Frank N. Magill. Danbury, CT: Grolier, 1989, pp. 137–141.

**Periodicals**

Datta, B. "Two Aryabhatas of al–Biruni." *Bulletin of Calcutta Mathematical Society* 17 (1926): 59–74.

**Other**

O'Connor, John J. and Robertson, Edmund F. "Aryabhata the Elder." *MacTutor History of Mathematics Archive.* http://www–groups.dcs.st–and.ac.uk/~history/Mathematicians/Aryabhata.html (July 1997).

*—Sketch by Jennifer Kramer*

*Richard Askey*

# Richard Askey
## 1933–

### American analyst

Richard Askey has kept up the momentum of classical mathematics in an age where much attention has been placed to abstract mathematics. He is best known for providing a crucial element in the proof of a longstanding conjecture in the field of complex analysis.

Richard Allen Askey was born on June 4, 1933 in St. Louis, Missouri, to Philip Edwin and Bessie May Yates Askey. He received his undergraduate education there at Washington University, where the mathematics department held a strong tradition in analysis. Askey carried that strength and interest with him to Harvard University, where he earned a master's degree in 1956 and a doctorate from Princeton University in 1961.

Askey taught at the University of Chicago for two years before joining the faculty at the University of Wisconsin in 1963, where he has remained. His work has been largely in the area of special functions (generalizations of the ordinary trigonometric and exponential functions to solve differential equations) and orthogonal polynomials (classes of polynomials that are used to represent more complicated types of functions). This has provided for plenty of exposure to results from classical mathematics, the importance of which Askey never tires of stressing to his students. As a faculty member he has been known to roam the halls of the mathematics department at Wisconsin, looking for students unsure of their area of specialization.

### Provides Link to Bieberbach Conjecture

Askey attracted attention in 1984, when a paper of his provided a necessary link in the proof by Louis de Branges of the Bieberbach conjecture, a problem having to do with functions of a complex variable. The German mathematician Ludwig Bieberbach had claimed that if a function of a complex variable were well–behaved in the circle of radius one in the complex plane, then there was a limit as to how large the coefficients of the terms in the polynomial representation of the function could be. Askey's results proved to be what de Branges needed to complete his proof.

Askey was a Guggenheim Fellow during 1969–70 and has served as vice president of the American Mathematical Society in 1986–87.

### SELECTED WRITINGS BY ASKEY

*Orthogonal Polynomials and Special Functions,* 1975

*Michael Francis Atiyah*

*The Bieberbach Conjecture: Proceedings of the Symposium on the Occasion of the Proof.* Edited by Albert Baernstein II, et al, editors. New York: American Mathematical Society, 1986, pp. 7–32, 213–215.

## FURTHER READING

Aspray, William, and Philip Kitcher, editors. *History and Philosophy of Modern Mathematics.* Minneapolis: University of Minnesota Press, 1988, pp. 201–217.

—*Sketch by Thomas Drucker*

# Michael Francis Atiyah
## 1929-
### English topologist

Michael Francis Atiyah has had a remarkably long and productive career that is based in topology but encompasses such diverse fields as algebraic geometry, differential equations, and theoretical physics. In recognition of three major theorems he developed during the first decade of his career, he was

awarded the Fields Medal in 1966. Atiyah is also one of the pioneers of string theory, a new way of looking at the structure of matter. Claude LeBrun wrote in *American Scientist* that Atiyah has "played the role of the great unifier for a swath of mathematical subfields that had been developing into autonomous petty fiefdoms, oblivious to the outside world." He has also served in various capacities to help guide the scientific policy of the British government and has been an articulate spokesperson for the importance of theoretical research.

Atiyah was born in London on April 22, 1929. His father was Edward Selim Atiyah, originally from Lebanon, and his mother was the former Jean Levens, an English citizen. Atiyah was educated in Egypt, where his father was a broadcaster and commentator for the British Broadcasting Company. After completing his secondary education at Victoria College in Egypt, Atiyah enrolled at Manchester Grammar School, a preparatory school in England. After a year, he was accepted as a mathematics student at Trinity College, Cambridge.

During his second undergraduate year, Atiyah published his first paper, which dealt with higher-dimensional projective geometry. He earned his doctorate from Trinity in 1955, the same year he married Lily Brown, who would be the mother of his three sons. His first postdoctoral position was a fellowship at the Institute for Advanced Studies (IAS) in Princeton, New Jersey. He later told Glyn Jones in an interview for *New Scientist* that he went to the IAS (which had been founded by **Albert Einstein** and **John von Neumann**) to "get new ideas, meet new people, and open up new avenues." Over the next two decades, Atiyah held a variety of teaching and research positions on both sides of the Atlantic, at Cambridge University, Harvard University, Oxford University and the IAS. In 1972, he returned to Oxford, where he stayed as a Royal Society Research Professor for 18 years. In 1990, he was elected president of what is arguably the most prestigious scientific association in the world, the Royal Society of London. That same year, he was chosen for another post of significant influence: mastership of Trinity College, a 400-year-old institution that developed its mathematical reputation through such scholars as **Isaac Newton, Bertrand Russell, Alfred North Whitehead** and **Godfrey Hardy**.

Atiyah's major field of interest in mathematics has been topology, which is the study of properties do not change under continuous deformation (stretching and bending). It is sometimes referred as "rubber-sheet geometry." Because the principles of topology apply to such a wide variety of conditions, it has evolved as a fundamental field that unifies many other seemingly unrelated fields of mathematics. Topology has evolved from the geometrical side of mathematics. Atiyah titled his presidential address to the Mathematical Association "What Is Geometry?" In it, he said that "geometry is that part of mathematics in which visual thought is dominant whereas algebra is that part in which sequential thought is dominant. . . . Geometry is not so much a branch of mathematics as a way of thinking that permeates all branches."

## A Trio of Early Theorems

The discovery of three theorems during his first decade of research was cited as the basis for Atiyah's 1966 Fields Medal. The first of these accomplishments was the development of a purely topological version of K-theory, which he derived with Friedrich Hizebruch. The topic concerns systems of linear equations that depend continuously on auxiliary parameters. Such systems arise naturally in many different contexts, and K-theory has become a powerful and useful tool. Topological spaces are often characterized by the number of holes they exhibit; that number is invariant (it does not change when the object is deformed in a continuous way). K-theory revealed was a new topologically invariant quantity. Upon receiving the Feltrinelli Prize in 1984, Atiyah told the Accademia Nazionale dei Lincei that "[K-theory] led easily to the solution of many difficult problems, most notably that of the vector-field problem on spheres. This geometric problem, simple to state but difficult to prove, had long been regarded as a test case for new techniques, and it was finally solved by Frank Adams in 1962 using K-theory."

With the collaboration of Isadore Singer, Atiyah used K-theory to develop what is now known as the Atiyah-Singer Index Theorem. Elliptic differential equations describe various physical situations, but they are difficult to solve. In fact, it is even hard to tell how many independent solutions such an equation will have. However, if a pair of elliptic equations have $n$ and $m$ independent solutions, respectively, the index theorem provides a formula for calculating $n-m$ (called the index of the pair). It turns out that the index can be used by physicists to determine the difference between the number of right-handed and left-handed particles in a system.

The index theorem led to another major result, which Atiyah developed with **Raoul Bott**. Fixed-point theorems deal with the number of points of a topological object that remain unchanged under a certain type of transformation, and they have many practical applications. Atiyah and Bott found a way to calculate the number of fixed points that exist under a transformation that preserves an elliptic system.

## Helps Develop String Theory

Beginning in the 1970s, Atiyah became interested in a new field of research known as string theory.

Physicists have traditionally constructed theories about the nature of matter based on the assumption that the fundamental particles of matter can be thought of as discrete, dimensionless points. String theory adopts a radically new assumption that the fundamental units of matter do have a dimension (length) and can be thought of as stringlike objects. This approach to the study of matter has evolved out of mathematical theories than out of experimental observations. As a result, string theory tends to be both more complex and less easily interpreted in physical terms than traditional theories of matter. For example, one consequence of string theory is that matter has to be thought of in terms of many (often, more than a dozen) dimensions. Many scientists find it unnatural and difficult to speak about objects in, say, 14 dimensions. It is hardly surprising, then, that string theory has been received with something less than enthusiasm by many physicists, although it has become an influential and fruitful theory.

For his work in topology and string theory Atiyah has garnered a number of awards and honors. In addition to the 1966 Fields Medal, the highest honor in mathematics, he has received the Royal Medal and the Copley Medal of the Royal Society, the De Morgan Medal of the London Mathematical Society, the Feltrinelli Prize of the Accademia Nazionale dei Lincei, and the King Faisal International Prize for Science. He has receieved honorary doctorates from 20 universities, including those of Bonn, Dublin, Chicago, Helsinki, Rutgers, and Montreal. Atiyah was knighted by Queen Elizabeth II in 1983.

## Educated in Person and in Print

In an effort to make concise resources available to young mathematicians (particularly in China), in 1985 Shiing-shen Chern convinced Atiyah to compile and publish his collected papers. Atiyah commented in the preface on the appropriateness of publishing such a compilation prior to the author's death. With characteristic humor, he wrote, "there are several clear advantages to all parties: posterity is saved the trouble of undertaking the collection, while the author can add some personal touches in the way of a commentary. There are also disadvantages: the commentary will be biased, and the author may feel that he is being pensioned off." On the contrary, Atiyah has continued to generate important material in the decade since the collection was published.

Besides being a prolific researcher, Atiyah is a highly respected teacher. Even at times of his career when he does not conduct classes, he educates through the clear exposition of his books and articles. In his *Collected Works*, he wrote that he has often been asked to speak to a broad range of audiences: "In some cases I was talking to professional mathematics but in others I was almost the only mathematician in the room. Giving such general talks. . . requires much greater thought on the material and presentation than for a normal seminar, but it is a worthwhile and important activity." It is one that Atiyah performs remarkably well, in print as well as in person. Two different reviewers described Atiyah's 1990 book *The Geometry and Physics of Knots* as "poetic," and his article "Geometry and Physics" is a very readable survey of highly theoretical topics.

For many years, Atiyah has also contributed to professional organizations in mathematics and science. He served as president of the London Mathematical Society and the Mathematical Association. From 1984 to 1989 he was a member of the British government's Science and Engineering Research Council. In 1991, Atiyah established the Isaac Newton Institute for Mathematical Studies in Cambridge, where researchers can work for six-months terms on their chosen topics.

## SELECTED WRITINGS BY ATIYAH

### Books

*K-Theory,* 1966
*Collected Works,* 5 volumes 1988.
*Geometry and Dynamics of Magnetic Monopoles,* 1988
*The Geometry and Physics of Knots,* 1990

### Periodicals

"What Is Geometry?" *The Mathematical Gazette* 66 (October 1982): 179–84.
"Geometry and Physics" *The Mathematical Gazette* 80 (March 1996): 78–82.

## FURTHER READING

### Periodicals

Bott, Raoul. Review of *The Geometry and Physics of Knots,* in *Bulletin of the American Mathematical Society* 26 (January 1992): 182–87.
Bown, William. "Parcelling Maths and Physics Up with String." *New Scientist* 129 (January 19, 1991): 44.
Dickson, David. "Scientists Must Assert Their Independence.'" *Nature* 378 (December 7, 1995): 525.
Jones, Glyn. "Topologist at the Top." *New Scientist* (January 19, 1991): 42–45.
LeBrun, Claude. Review of *Michael Atiyah: Collected Works,* " in *American Scientist* (May-June 1991): 283.
Stasheff, Jim. Review of *The Geometry and Physics of Knots,*" in *American Scientist* (November-December, 1991):568–569.

*—Sketch by David E. Newton and Loretta Hall*

*Charles Babbage*

# Charles Babbage
## 1792–1871

**English mathematician, inventor, and philosopher**

Charles Babbage is considered the creator of modern computers. A mathematician and 19th century British intellectual, he conceived of a steam driven Difference Engine that could automatically calculate and print error-free mathematical tables. He later developed the idea of the Analytical Engine which could be programmed to make calculations and could store results in a memory unit. Babbage also invented the first automated typesetter for printing the results of computations. Although construction of his calculating machines was never completed in his lifetime, his concepts were used as the basis for the Harvard Mark I Calculator, the prototype of the modern digital computer, built by **Howard Aiken** in 1944.

Charles Babbage was born on December 26, 1792, in Teignmouth, Devon, England. He was the son of a London banker, Benjamin Babbage, from whom he inherited a sizable fortune, enabling him to devote his life to intellectual pursuits. As a child Babbage suffered several bouts of violent fever that interfered with his early education. He was placed in the care of a clergyman who ran a school in Devonshire, with instructions not to tax his health with too much knowledge. In his early teens Babbage attended a boarding school in London, where he developed a keen interest in algebra. He spent much of his leisure time studying mathematical works, and was especially influenced by Ward's *Young Mathematician's Guide*, a text he had found in his school library. In 1810 Babbage entered Trinity College at Cambridge University, and soon discovered the knowledge of mathematics he had obtained through self-instruction, which exceeded that of his tutors. Together with a circle of friends Babbage founded the Analytical Society, whose purpose was to promote mathematics. A co-founder was to became a lifelong companion, John Herschel, a noted astronomer. Together, in 1813, they published a translation of LaCroix's *Differential and Integral Calculus*, accompanied by several volumes of mathematical examples. Babbage graduated from Cambridge in 1814, and the following year wrote several papers on the calculus of functions for the Royal Society of London. In 1816 he was elected a Fellow of the Society.

## Conceives of a Steam Driven Calculator

While at Cambridge, Babbage had begun thinking about the possibility of building a machine that could compute arithmetic tables. In the early 19th century, actuaries, bankers, navigators, engineers, and others relied heavily on published numerical values of mathematical formulas and functions. Errors in calculation and transcription were so common that the tables were often accompanied not only by a list of corrections, but a list of corrections to the corrections. Mechanical calculators had existed since the time of **Blaise Pascal**, but they worked slowly and were only capable of performing single arithmetic calculations. Babbage designed a steam-driven machine he called the Difference Engine, which could rapidly calculate and automatically print the results of large numbers of mathematical operations. The first version of his Difference Engine was intended to compute values of squares and quadratic functions. It worked on the principle known as the method of finite differences, a technique which employs only addition to determine successive values for polynomial functions. In 1822 Babbage built a prototype that made accurate calculations up to five–place numbers. That same year, with backing from the Royal Society, he convinced the British government to provide funds for building a full–scale Difference Engine, with a capacity to work with numbers up to one million with 20 decimal places. The project, expected to take three years, was abandoned after a decade. Babbage's design called for a series of gear wheels on shafts that would be turned by cranks. The machine was to contain 25,000 die–cast pewter and precision gauged brass and steel parts. If finished, it would have weighed over two tons. Historians speculate its realization may have been beyond the engineering capacity of the era. The project was also hapered by Babbage's constant revisions of design. Construction of the Difference Engine was far from completion when financial arguments between Babbage and his chief engineer brought the project to a halt. By then, Babbage had come upon a better idea.

### Invents the Precursor to Modern Computers

Babbage's work on the Difference Engine led to the evolution of his new invention, the Analytical Engine. While the former machine was designed to work straight through a computational problem, the latter was designed to make calculations, store the results, analyze what to do next, then return to complete the problem. Babbage's design for the Analytical Engine had four key components: the mill, the card reader, the store, and the typesetter. The mill was the heart of the machine, where the four basic arithmetic operations could be performed with an accuracy of up to 50 decimal places. It received instructions and numerical data from punched cards that were deciphered by the card reader. Babbage borrowed the idea for this input system from the weaving industry, which in the mid–1700s had begun using punched cards as hand–held guides for creating different patterns in cloth. An automatic card reader that controlled a power loom had been invented in 1801 by the French carpet–maker Joseph Marie Jacquard. Babbage's Jacquard cards, as he referred to them, could provide instructions and data not only for the machine's mill, but for the store, a place where numbers were retained in memory for future use. The store consisted of a bank of one thousand registers, each of which could hold a 50–digit number. Finally, after mathematical operations had been performed, the Jacquard cards could instruct the machine to typeset the results for printing.

In 1834 Babbage began an eight–year campaign to convince the government to fund construction of the Analytical Engine, but was unsuccessful. Britain had already spent seventeen thousand pounds on the Difference Engine to no avail. Babbage had contributed a comparable amount of money to the Difference Engine project, depleting most of his personal fortune. In 1848, Babbage drew up plans for a scaled–down version of the Analytical Engine, called the Difference Engine No. 2, but once again was unable to obtain funds for construction. The scaled–down version was finally built, nearly a century and a half later, in honor of Babbage's bicentenary, by the Science Museum of London. It weighed three tons and worked flawlessly.

### Applies Intellect to Diverse Scientific and Practical Problems

Although the primary focus of Babbage's intellectual pursuits was his calculating engines, he devoted time to many other areas of scientific and practical interest. He published a paper with Herschel in 1825, on magnetization arising from rotation, and made contributions to the fields of geology, anthropology, and astronomy. In 1820 Babbage helped found the Royal Astronomical Society, in 1831 the British Association for the Advancement of Science, and in 1834 the Statistical Society of London. Babbage once descended into the crater of Mt. Vesuvius to research its volcanic activity. He studied glaciers, and suggested a way of learning about past climatic conditions by measuring tree ring growth in fossilized wood. Babbage even designed a colored lighting system for theaters, an idea he reportedly dreamed up while bored during an opera performance. Concerned with economic efficiencies, in 1832 he wrote a pamphlet called "Economy of Manufactures and Machinery." Babbage advised the British postal service, consulted for the British rail system, and was the inventor of the "cowcatcher," the track clearing safety devise that protrudes from the front of a train engine. He ran twice, unsuccessfully, for a seat in Parliament. From 1828 until 1839, Babbage held the Lucasian Chair of Mathematics at Cambridge, although during his tenure he never taught or lived at the university.

Babbage was widely known in London's social circles, hosting regular Saturday night parties at his home at 1 Dorset Street. He was friends with naturalist Charles Darwin, German naturalist and statesman Alexander Humboldt, and **Ada Byron, Countess of Lovelace**, who published articles explaining Babbage's engines and authored the first computer program. Babbage is described as a man who in his final years was embittered and disappointed over the lack of support for his calculating engines. He was critical of the scientific establishment and of governmental funding policies, and published papers on what he described as the decline of science in England. Eventually, Babbage developed a reputation as an eccentric, launching a campaign to ban organ grinders as street nuisances. When he died in London on October 18, 1871, his *London Times* obituary commented that he lived to be almost eighty "in spite of organ–grinding persecutions."

Babbage married Georgiana Whitmore shortly after he graduated from Cambridge. Their eldest son was Herschel Babbage. Another son, Henry Babbage, attempted to carry on in his father's tradition, presenting a paper on calculating engines to the British Association in 1888, and a year later editing a volume about his father's works.

## FURTHER READING

### Books

Gonick, Larry. *The Cartoon Guide to the Computer*. New York: Harper Perennial, 1991.

Hyman, Anthony. *Charles Babbage: Pioneer of the Computer*. Princeton, NJ: Princeton University Press, 1982.

Morrison, Philip and Emily Morrison, editor. *Charles Babbage: On the Principles and Development of the Calculator*. New York: Dover Publications, 1961.

Wade, Mary Dodson. *The Lady and the Computer*. New York: Dillon Press, 1994.

### Periodicals

Swade, D. "Redeeming Charles Babbage's Mechanical Computer." *Scientific American* 268, no. 2 (February 1993): 86–91.

*—Sketch by Leslie Reinherz*

# John Backus
## 1924-
### American computer scientist

Winner of the 1993 Charles Stark Draper Prize, a prestigious engineering award, the 1977 Association for Computing Machinery's (ACM) Turing Award, and the 1975 National Medal of Science, John Backus headed a pioneering group of IBM engineers, who in the 1950s developed FORTRAN, the first widely used programming language. FORTRAN, which gave programmers the freedom from the tedious task of writing out instructions as strings of 1s and 0s, is the precursor of nearly all contemporary computer languages.

Backus was born in Philadelphia, Pennsylvania, in 1924, and grew up in Wilmington, Delaware. Planning to major in chemical engineering, he enrolled in the University of Virginia in 1942, but was thrown out after one semester for cutting classes. He was drafted into the army in early 1943, where he first served in an antiaircraft program. From September 1943 until March of 1944 he studied engineering at the University of Pittsburgh as part of the army's specialized training. This was followed by six months of premedical training in a hospital in Atlantic City, New Jersey, and an additional six months at Flower and Fifth Avenue Medical School in New York City. By May of 1946 Backus had left behind both the army and his interest in a medical career.

Remaining in New York City he entered the Radio Television Institute, a training school for radio and television repairmen. It was here that he developed an interest in mathematics and began taking courses in math at Columbia University. He earned a bachelor of arts degree in 1949, and the following year he received a master's degree, also from Columbia, and also in mathematics. Upon graduation, he went directly to work for IBM even though he knew very little about computers. One of the very few programmers in the computer industry in the early 1950s, Backus soon earned a reputation as a trailblazer in the field.

### Develops and Promotes the Use of FORTRAN

In 1952 he led the group of IBM researchers who produced the Speedcoding system for the IBM 701 computer. The following year Backus, while a project manager, wrote a memo to his boss, Cuthbert Hurd, outlining the need for a general-purpose, high-level computer programming language. The programming language was called FORTRAN (an acronym for formula translator) and was designed to perform mathematical, scientific, and engineering computations on the IBM 704 computer. More importantly, FORTRAN was developed to serve as a translator between the human user and the computer brain, which at that time could only think in zeroes and ones. In the 1950s computers were somewhat rare and prohibitively expensive, also, three-quarters of the cost of running a computer was given to debugging and programming; FORTRAN was created to address these problems.

In an interview in 1978 Backus noted: "In the early 1950s, because of the lack of high-level languages, the cost of programming was at least equal to that of the equipment, and this held back the development of computers." To overcome this difficulty, Backus and his team of fellow researchers at IBM, pursued the idea of developing languages that were easy to use and efficient translators. Although the odds against success in developing such a program were great, Backus was able to convince the IBM directors that it was possible. The language proved to be a useful tool for IBM, helpful in promoting its computers. In addition, FORTRAN, despite being written for one IBM computer, was quickly adopted to be used with other systems, and continues to be used today.

The FORTRAN compiler (a compiler is a computer program designed to translate high-level language statements into a form that can directly activate the computer hardware), considered to be the forerunner of all modern compilers, was the first to have the power and scope to perform the complicated computing tasks that had previously been done by handwritten machine-code programs. With its innovative capabilities, FORTRAN quickly became the most important innovation in the history of programming languages. People using FORTRAN were able to deal with computers without knowing the internal workings of the machine and its assembly languages.

In 1954 IBM published the first version of the language, FORTRAN I. Although there were bugs in this original version, by 1955 Backus, in collaboration with R.A. Nelson and I. Ziller, began work on correcting them. Several changes were made in the original language enabling FORTRAN to evolve along lines that were suggested by the experience gained with its usage. During the late 1950s there were two opposing views on programming languages. A mainly American group contended that only specialized languages could meet the needs of users; at the same time, a European group of scientists expressed the concern that this view led to too many programming languages. Dr. F. Bauer of the University of Munich, Germany, initiated the movement to define a multipurpose language that would be completely independent of specific computers, and in which any algorithm could be clearly stated. Bauer approached the American Association of Computing Machinery (ACM) who, in turn, formed a committee to cooperate with the Europeans. In the spring of 1958 a meeting was held in Zurich, Switzerland, and the committee later published a report defining an International Algebraic Language, later called AL-GOL.

### Joins International Computer Programming and Design Team

Backus, who was part of the American group that met in Zurich, moved his research to IBM's Watson Research Center in Yorktown Heights, New York, to become part of the international programming and design team that created ALGOL. Although ALGOL never gained widespread use commercially, it had an important influence on three other widely used programming languages: Pascal, C, and Ada. In 1978 Backus, still working on ways to improve computer languages, wrote a paper suggesting that they should be restructured. "Programming languages appear to be in trouble," he wrote, "conventional languages create unnecessary confusion in the way we think about programs." By the early 1980s he had become

an IBM Fellow, which enabled him to devote his time to his own research projects, including advancing mathematical theories of programming. "The complacent acceptance most of us give to these enormous, weak languages has puzzled and disturbed me for a long time," he said, "I have tried to analyze some of the basic defects of conventional languages and show that those defects cannot be resolved unless we discover a new kind of language framework."

Backus has spent the last ten years pursuing his search for a more efficient programming language. Designed with great care and attention to logic, his "functional" language is constructed from ones already defined, thereby eliminating the programmer's need to spell out every instruction in minute detail. "Our goal is to produce a functional language . . . so that you can run functional programs on personal computers," he has explained, describing his language. He retired from IBM in 1991 after 41 years with the company. He lives in San Francisco where he continues to work on programming research and to keep an eye on his functional programming language, which is still gaining in popularity.

### FURTHER READING

#### Books

Moreau, R. *The Computer Comes of Age.* Cambridge, MA: MIT Press, 1984.
Slater, Robert. *Portraits in Silicon.* Cambridge, MA: MIT Press, 1987.

#### Periodicals

"In the Beginning." *Datamation* (September 1982): 51–52.
Pauly, David, with Gerald C. Lubenow. "IBM's Mavericks in the Lab." *Newsweek* (January 10, 1983): 58.
Peterson, Ivars. "Computer Languages: In Search of a Better Bug Finder." *Science News* 124 (September 24, 1983): 202–203.

—*Sketch by Dorothy Spencer*

# Stefan Banach
## 1892-1945
### Polish analyst

In spite of his somewhat fragmented education (he never completed a formal doctoral program), Stefan Banach made important contributions to a number of fields of mathematics, including the theory of orthogonal series, topology, the theory of measure and integration, set theory, and the theory of linear spaces of an infinite number of dimensions. He is probably best remembered, however, for his work on functional analysis.

Stefan Banach was born on March 30, 1892, in Kraków, Poland. His father was named Greczek, a railway official from peasant background. He and Stefan's mother (whose name has been lost) abandoned their young child to a laundress almost immediately after his birth. The child took on his foster mother's surname of Banach, but almost nothing else is known about his early childhood. Banach apparently developed an interest in mathematics at an early age and taught himself the fundamentals of the subject. By the age of 15 he was supporting himself as a private teacher of mathematics. He also taught himself enough French to master Tannery's text on the theory of functions, *Introduction à la théorie des fonctions*. Banach attended lectures on mathematics at Jagellon University on an irregular basis before entering the Lwów Institute of Technology in the Ukraine in 1910. He did not, however, graduate from the institute.

In 1914, with the outbreak of World War I, Banach returned to Kraków. Two years later, a chance event was to change his life. While sitting on a park bench in Kraków talking with a friend about mathematics, he was overhead by the mathematician H. Steinhaus. Steinhaus later wrote that he was "so struck by the words 'the Lebesque integral'" that he heard from the two that he came closer and introduced himself to the young men. As the group talked, the conversation turned to a problem on the congruence of a Fourier series on which Steinhaus had been working. "I was greatly surprised," Steinhaus went on to say, "when, after a few days, Banach brought me a negative answer with a reservation which resulted from his ignorance of [a technical point about which he did not know]." Banach and Steinhaus were later to collaborate on a number of mathematical studies.

Banach's natural gift for mathematics soon became more widely known, and at the conclusion of the war he was offered a position as mathematical assistant at the Lwów Institute of Technology by Antoni Lomnicki. For the first time in his life Banach had some degree of financial security and he married. Beginning in 1919, Banach was assigned to lecture on mathematics and mechanics. In the same year, he was awarded his doctoral degree although he had not completed the full program of courses expected for that degree.

### Publishes Historic Paper on Integral Equations

The primary basis for Banach's degree was the paper he had written on integral equations, which had been published in *Fundamenta mathematicae* in 1922 as "Sur les opérations dans les ensembles abstraits et leur application aux équations intégrales." At about the same time he was made an instructor at the institute; in 1927 he was promoted to full professor. From 1939 to 1941 Banach also served as dean of the faculty at the institute.

The mathematical work for which Banach is best known is his book *Théorie des opérations linéaires*, which appeared in 1932 as the first volume in the Mathematical Monographs series, published in Warsaw. In this book, Banach developed a general theory for working with linear operations that proved to be a landmark in the field. Prior to this work, a number of individual, discrete methods had been developed for solving specific problems. But there was no comprehensive theory that could be applied to a great variety of problems. In his book, Banach introduced the concept of normed linear spaces, now known as Banach spaces, which, Steinhaus later wrote, can be used "to solve in a general way many problems which formerly called for special treatment and considerable ingenuity."

Banach's significance in the history of mathematics goes beyond his own research. He was also an effective teacher whose influence was spread throughout Europe and the United States by a number of brilliant students. In addition, he wrote an important popular textbook, *Differential and Integral Calculus* (1929–30) and was founder with Steinhaus of the journal *Studia mathematica*.

World War II was a personal disaster for Banach. After the German army occupied the city of Lwów, he was forced to work in a German laboratory studying infectious diseases. His job there was to feed the lice used in experiments. As degrading as this work was, Banach was able to continue teaching in underground schools and carry on his own research. By the time the war ended, however, his health had so badly deteriorated that he lived only a few more months. Banach died in Lwów on August 31, 1945. Among the honors accorded him during his lifetime were election as corresponding member of the Polish Academy of Sciences and of the Kiev Academy of Sciences. He also received the Prize of the City of Lwów in 1930 and the Prize of the Polish Academy in 1939. Upon his death, the city of Warsaw renamed one of its streets in his honor.

## SELECTED WRITINGS BY BANACH

### Books

*Differential and Integral Calculus*, 1929 and 1930
*Théorie des opérations linéaires*, 1932
*Mechanika w zakresie szkol akademickich*, 1938

**Periodicals**

"Sur le problème de la mesure." *Fundamenta mathematicae* 4 (1923): 7–33.

## FURTHER READING

### Books

Katêtov, Miroslav. "Stefan Banach," in *Dictionary of Scientific Biography*. Volume I. Edited by Charles Coulston Gillispie. New York: Charles Scribner's Sons, 1975, pp. 427–428.

### Periodicals

Steinhaus, H. "Stefan Banach." *Studia mathematica*, special series, 1 (1963): 7–15.
Ulam, S. "Stefan Banach, 1892–1945." *Bulletin of the American Mathematical Society* 52 (1946): 600–603.

— *Sketch by David E. Newton*

*Benjamin Banneker*

# Benjamin Banneker
## 1731–1806
### African–American astronomer and applied mathematician

America's first recognized black scientist, Benjamin Banneker was a mostly self–taught mathematician and astronomer who spent his life in an avid quest for knowledge. He gained renown by publishing astronomical almanacs and ephemerides. Very popular in the 18th century, these almanacs included astronomical data for each day of a given year; ephemerides—plural of *ephemeris*—were tables predicting the daily positions of celestial bodies. In addition to his scientific work, Banneker raised tobacco, played the violin and flute, worked as a surveyor, and built mechanical artifacts. His world view, it seems, successfully integrated a traditional Christian spirituality and a modern scientist's openness to the world. Finally, Banneker was acutely aware of the profound injustice of American slavery, and worked hard to discredit the belief, supported by intellectuals such as Thomas Jefferson, that the people of African descent was intellectually deficient.

Banneker was born on a farm near Baltimore, Maryland, on October 9, 1731, to Robert, a farmer and a former slave, and Mary Banneky. Banneker's maternal grandmother, Molly Welsh, was English. As a young girl living in England, Welsh was accused of stealing milk and condemned to death. Because she

was literate, however, her sentence was commuted to banishment and seven years of indentured servitude in the British Colonies. Welsh she ended up in Baltimore county, Maryland, where after regaining her freedom she acquired a farm. Welsh eventually married one of her slaves, whom she had freed. His name was Bannka, or Bannaka, and he claimed to be the son of an African chieftain. Although a free Englishwoman, and relatively prosperous, Welsh was aware of her precarious position as a white woman married to a black man. Their daughter, Mary Banneky, married Robert, a former slave, who took her surname.

Banneker attended the county Quaker school, ending his formal education when he was old enough to help his father on the farm. The most significant force in his early education and spiritual formation, however, was his grandmother. She taught him how to read and write, and insisted that he read to her from the family Bible. Showing an early aptitude for mathematics and engineering, Banneker built a wooden striking watch in 1753, using a borrowed watch as a model; the watch, which brought him local fame, ran until his death. He acquired his first book, a quarto edition of the Bible, when he was 32 years old.

Banneker eventually inherited the farm from his father and remained there his entire life. His existence was profoundly affected by the arrival of the prominent Ellicott family in the area when he was 41. The young George Ellicott and Banneker found they shared many scientific interests, and the two men developed a close friendship. A talented mathematician and astronomer, George Ellicott provided his friend with valuable books and astronomical instruments. Fascinated by the heavens, Banneker started studying astronomy in 1788. Toward the end of that year, George Ellicott lent Banneker some basic astronomy textbooks, which he mastered on his own. Banneker also absorbed Charles Leadbetter's *A Compleat System of Astronomy*, an advanced work which was considered one of the principal compendia of 18th–century astronomy. Astronomers of that era applied trigonometrical calculations and the calculus of variation in their compilations of astronomical tables, and Banneker's Almanacs, certain inconsistencies notwithstanding, conform to the scientific standards of his time.

In 1791, Banneker became involved in the survey for a new national capital. Among the leaders of that project was an Ellicott, noted surveyor Andrew Ellicott, who, being familiar with Banneker's scientific work—particularly his calculation of an astronomical almanac, employed Banneker as an assistant. The plans for the city were developed by the French–born American architect and soldier Pierre–Charles L'Enfant, who was dismissed in 1792 despite the fact that his work provoked admiration. Banneker was in charge of an astronomical clock, and he also assisted in surveying and worked in the observatory tent. However, as Banneker's biographer Silvio Bedini remarks, there is no support for the story that Banneker reconstructed from memory L'Enfant's plans, which the architect refused to leave behind. As Bedini explains, Banneker's name cannot be found in any of the records documenting the survey of the new city site, and it is hard to imagine that a feat like that would not be recorded.

Despite his great interest in surveying, Banneker decided to dedicate his energies to work on his astronomical almanac for 1792. In August 1791, Banneker sent the manuscript of his *Almanac* to Thomas Jefferson, then secretary of state, and also enclosed a respectful but strong statement against slavery. Banneker urged Jefferson to take action against racism and slavery. Jefferson responded, assuring Banneker of his wish to see the lot of black people improved, and informing the scientist that he was forwarding the manuscript of the almanac to the Marquis de Condorcet, secretary of the Académie Royale des Sciences in Paris.

Not wishing to doubt the veracity of Jefferson's statement, historians have nevertheless been baffled by the fact that there is no evidence that Condorcet ever received the manuscript. There is no mention of Banneker in Condorcet's papers; the Académie had no record of Banneker's work; and neither did the Société des amis des Noirs, an antislavery organization of which Condorcet was a founding member.

However, while a formal presentation of Banneker's work at the Académie may have been impossible to organize during the turbulent summer and autumn of 1791, it is difficult to explain the fact that not even Condorcet's colleague, the prominent cleric Henri Grégoire, bishop of Blois and also a member of the Société des amis des Noirs, knew of Banneker's *Almanac*. In fact, Grégoire does not even mention the *Almanac* in his book *De la littérature des Nègres, ou, Recherches sur leurs facultés intellectuelles* (1808), which includes a biography, probably the first one, of Banneker.

An ardent proponent of racial justice, Grégoire wrote admiringly about Banneker. Grégoire's book attracted considerable attention, and, significantly, among the intellectuals who praised it was also Thomas Jefferson, who in a letter to the author spoke highly of his work. However, in a letter, dated October 8, 1809 to the well–known American writer and diplomat Joel Barlow, Jefferson criticizes Grégoire for supposedly exaggerating the accomplishments of blacks. Suggesting that Banneker may have received help in compiling his *Almanac*, Jefferson declared: "I have a long letter from Banneker, which shows him to have had a mind of very common stature indeed."

While Banneker actively fought for racial justice his entire life, in his capacity as a scientist and a man of letters, science was his primary vocation. A solitary person (he never married), Banneker seemed indifferent to public life. This fact, however, did not prevent abolitionists for using his scientific work for political purposes, and several antislavery groups, particularly the Maryland Society for the Abolition of Slavery, welcomed Banneker's work as proof that black people are not intellectually inferior. In a sense, Banneker's *Almanac* became a political document, which, it seems, did not perturb him. For example, his 1793 *Almanac* includes a utopian document entitled "A Plan of a Peace Office for the United States." The profoundly irenic nature of the unsigned "Plan," which urged the United States government to establish a Department of Peace, suggested Banneker's authorship, and many commentators accepted the document as his creation. Subsequent research, however, showed that the author was Dr. Benjamin Rush, an eminent American physician and educator. Nevertheless, the 1793 *Almanac*, which also included Banneker's correspondence with Jefferson, was, as Bedini has written, "one of the most important publications of its time." According to Bedini, Banneker's almanacs were among the first to successfully publicize the cause of abolitionism by presenting "tangible proof of the mental equality of the races."

Although Banneker did not publish any new almanacs after 1797, they continued appearing until 1802. Commentators have ascribed the discontinuation of Banneker's *Almanac* to diminished intellectual acuity. Bedini, however, suggest a different explanation. In this historian's view, there was, in the final decade of the 18th century, a strong reaction against the abolitionist cause, as the revolutionary government, both in France and the United States, replaced their initial idealistic, agendas with political concerns. As Bedini observes, the cessation of Banneker's almanacs coincided with the dissolution of the Maryland Society for the Abolition of Slavery.

Banneker, who had sold his farm to the Ellicot family while retaining the right of residence, spent his final years in frugal obscurity, studying and performing scientific experiments. He died October 9, 1806. On the day of his funeral, a fire consumed his house which destroyed his laboratory.

## SELECTED WRITINGS BY BANNEKER

### Books

*Benjamin Banneker's Pennsylvania, Delaware, Maryland, and Virginia Almanac for the Year of the Lord. . . .*, 1792–97

## FURTHER READING

### Books

Bedini, A. Silvio. *The Life of Benjamin Banneker.* New York: Scribner's, 1972.
Grégoire, Henri–Baptiste. *An Inquiry Concerning the Intellectual and Moral Faculties and Literature of Negroes. Followed with an Account of the Life and Works of Fifteen Negroes and Mulattoes Distinguished in Science, Literature, and the Arts.* Translated by D. B. Warden. College Park, MD: McGrath, 1967.

### Periodicals

Saulny, Susan. "Banneker Kin Decry Auctioning of His Artifacts." *Washington Post* (August 16, 1996): A–01.

*—Sketch by Zoran Minderovic*

# Nina Bari
## 1901–1961
### Russian trigonometer and educator

Nina Bari's work focused on trigonometric series. She refined the constructive method of proof to prove results in function theory, and her work is regarded as the foundation of function and trigonometric series theory.

Nina Karlovna Bari was born in Moscow on November 19, 1901, the daughter of Olga and Karl Adolfovich Bari, a physician. In the Russia of her youth, education was segregated by gender and the best academic opportunities reserved for males only. Bari attended a private high school for girls, but in 1918 she defied convention and sat for—and passed—the examination for a boy's high school graduation certificate.

In 1917, Russia's political and social structure was shattered by the Russian Revolution. The power vacuum left the country at the mercy of the czarists, socialist revolutionaries, and Bolsheviks. While many of Russia's universities closed at the beginning of the Revolution, the Faculty of Physics and Mathematics of Moscow State University reopened in 1918, and began accepting applications from women. Records show that Bari was the first woman to attend the university and was probably the first woman to graduate from it. Russia's educational institutions were in the same turmoil as the society around them. Graduation exams were scheduled on a catch–as–can basis, and Bari took advantage of the disorder to sit for her examinations early. She graduated from Moscow State in 1921—just three years after entering the university.

### Finds a Mentor and a Movement

After graduation, Bari began her teaching career. She lectured at the Moscow Forestry Institute, the Moscow Polytechnic Institute, and the Sverdlov Communist Institute. Bari applied for and received the only paid research fellowship awarded by the newly created Research Institute of Mathematics and Mechanics. (Ten postgraduate students were accepted at the Research Institute; Bari won the stipend because her name appeared first on the alphabetically arranged list. According to a colleague, she shared the stipend with her fellow students.)

As a student, Bari was drawn to an elite group nicknamed the Luzitania—an informal academic and social organization. These scholars clustered around Nikolai Nikolaevich Luzin, a noted mathematician who rejected any area of mathematical study but function theory. With Luzin as her inspiration, Bari plunged into the study of trigonometric series and functions. She developed her thesis around the topic and presented the main results of her research to the Moscow Mathematical Society in 1922—the first woman to address the society. In 1926, she defended her thesis, and her work earned her the Glavnauk Prize.

In 1927, Bari took advantage of an opportunity to study in Paris at the Sorbonne and the College de

France. She then attended the Polish Mathematical Congress in Lvov, Poland; a Rockefeller grant enabled her to return to Paris to continue her studies. Bari's decision to travel may have been influenced by the disintegration of the Luzitanians. Luzin's irascible, demanding personality had alienated many of the mathematicians who had gathered around him. By 1930, all traces of the Luzitania movement had vanished, and Luzin left Moscow State for the Academy of Science's Steklov Institute.

### Gains Recognition at Home and Abroad

Bari returned to Moscow State in 1929 and in 1932 was made a full professor. In 1935, she was awarded the degree of Doctor of the Physical–Mathematical Sciences, a more prestigious research degree than the traditional Ph.D.

In 1936, during the dictatorship of Josef Stalin, Bari's mentor, Luzin, was charged with ideological sabotage. For some reason—possibly Stalin's preoccupation with more important enemies of the state—Luzin's trial was canceled. Luzin was officially reprimanded and withdrew from academia.

Bari managed to avoid the taint of association. She and D.E. Men'shov took charge of function theory work at Moscow State during the 1940s. In 1952, she published an important piece on primitive functions, and trigonometric series and their almost everywhere convergence. Bari also presented works at the 1956 Third All–Union Congress in Moscow and the 1958 International Congress of Mathematicians in Edinburg.

Mathematics was the center of Bari's intellectual life, but she enjoyed literature and the arts. She was also a mountain hiking enthusiast and tackled the Caucasus, Altai, Lamir and Tyan'shan' mountain ranges in Russia. Bari's interest in mountain hiking was inspired by her husband, Viktor Vladmirovich Nemytski, a Soviet mathematician, Moscow State professor and an avid mountain explorer. There is no documentation of their marriage available, but contemporaries believe the two married later in life.

Bari's last work—her 55th publication—was a 900–page monograph on the state of the art of trigonometric series theory, which is recognized as a standard reference work for those specializing in function and trigonometric series theory.

Bari died July 15, 1961, when she fell in front of a train at the Moscow Metro. Colleagues, however, suspect her death was suicide; they speculate she was despondent over the death of Luzin in 1950, who some believe had been not only her mentor but her lover.

## SELECTED WRITINGS BY BARI

### Books

*A Treatise on Trigonometric Series.* Translated by Margaret F. Mullins, 1964

### Periodicals

"Sur l'uncite du developpement trigonometrique." *Comptes rendus hebdomadaires des seances de l'Academie des Sciences* 177 (1923): 1195–1197.

"Sur la nature diophantique du probleme d'uncite du developpement trigonometrique." *Comptes rendus hebdomadaires des seances de l'Academie des Sciences* 202 (1936): 1901–1903.

"The Uniqueness Problem of the Representation of Functions by Trigonometric Series." *Uspekhi Matematicheskikh Nauk* 3, no. 31, (1949): 3–68. Supplement, 7 (1952): 193–196. (Translation: American Mathematical Society Translation no. 52, (1951.)

"On primitive functions and trigonometric series converging almost everywhere" (in Russian). *Matematicheskii Sbornik* 31, no. 73, (1952): 687–702.

"Trigonometric Series" (in Russian). *Trudy III Vzesoyuznogo Matematicheskoga S'ezda* 2 (1956): 25–26.

"Trigonometric Series" (in Russian). *Trudy III Vzesoyuznogo Matematicheskoga S'ezda* 3 (1957): 164–177.

(With D. E. Men'shov) "On the International Mathematical Congress in Edinburgh" (in Russian). *Upsekhi Matematicheskikh Nauk*, 14, no. 2, (1959): 235–238.

"Subsequences converging to Zero Everywhere of Partial Sums of Trigonometric Series" (in Russian). *Izvestiya Akademii Nauk SSSR. Seriya Matematika* 24 (1960): 531–548.

## FURTHER READING

### Books

Fang, J. *Mathematicians From Antiquity to Today.* Memphis State University: Paideia, 1972, p. 185.

Spetich, Joan and Douglas E. Cameron. "Nina Karlovna Bari," in *Women of Mathematics: A Bibliographic Sourcebook.* Edited by Louise S. Grinstein and Paul J. Campbell. Westport, CT.: Greenwood Press, 1987, pp. 6–12.

Zdravkovska, Smilka and Peter L. Duren, editors. *Golden Years of Moscow Mathematics, History of Mathematics*, Volume 6, Providence, RI: American Mathematical Society, 1993, pp. 35–53.

**Other**

Soublis, Giota. "Nina Karlvona Bari." *Biographies of Women Mathematicians.* June 1997. http://www.scottlan.edu/lriddle/women/chronol.htm (July 22, 1997).

*—Sketch by A. Mullig*

# Ruth Aaronson Bari
## 1917–
### Polish–American algebraist

Ruth Bari is considered the world expert on chromatic polynomials. She was born November 17, 1917 in Brooklyn, New York, the daughter of polish immigrants Israel Aaronson and Becky Gursky and has one younger sister, Ethel. Bari attended the all–girls Bay Ridge High School in Brooklyn, where it was thought that mathematics was not important for the girls' education. When Bari requested a more in depth class on algebra, her teacher offered her only a book with higher algebra problems that she had to study on her own. Bari persisted and went on to earn a medal for her mathematics work at graduation.

Bari enrolled at Brooklyn College and graduated in 1939 with a BA in mathematics. That summer she met Arthur Bari, a diamond setter, while working on Staten Island and they were married on November 22, 1940. Bari switched gears at this point in her life, taking and passing a civil service exam. She went to Washington, D.C. and worked as a statistical clerk for the Bureau of Census. At night, Bari continued to take classes and obtained a master's degree from Johns Hopkins University in 1943. During World War II, Bari worked for her department chairman grading papers for $25 a month, but money was tight with her husband away in the Marines, so she left the university and went to work for Bell Telephone Laboratories in New York City. Bari was employed as a technical assistant for a short time and was then hired as an instructor at the University of Maryland. Bari resigned when she became pregnant with her first daughter, Gina, in 1948. The Baris had two more daughters, in 1949 and 1951, respectively, before she returned to the University of Maryland to pursue her doctorate, which she received in 1966.

Bari's work with Ph.D. candidates at George Washington University has earned her the title of "doctoral mother" due to her patience and encouragement. She has also given numerous lectures on graphs and polynomials. In addition, she has been invited to lecture abroad at prestigious universities such as the Mathematical Institute at Oxford.

**SELECTED WRITINGS BY BARI**

(Editor, with Frank Harary) *Graphs and Combinatorics: Proceedings,* 1974

**FURTHER READING**

Fasanelli, Florence D. "Ruth Aaronson Bari," in *Women of Mathematics: A Biobibliographic Sourcebook.* Edited by Louise S. Grinstein and Paul J. Campbell. Westport, CT: Greenwood Press, 1987, pp. 13–16.
Harary, Frank. "Academic Roots." *The Mathematical Intelligencer* 7, no. 1, (1985): 7.

**Other**

"Ruth Aaronson Bari." *Biographies of Women Mathematicians.* June 1997. http://www.scottlan.edu/lriddle/women/chrono.htm (July 22, 1997).

*—Sketch by Nicole Beatty*

# Isaac Barrow
## 1630–1677
### English geometer and theologian

Isaac Barrow is noted for his contributions to the field of optics. He is also remembered as the professor who served as inspiration and mentor to **Isaac Newton.**

Barrow was born in London. His father, Thomas, was a merchant who served as linen draper to King Charles I. Barrow's mother, Anne, died shortly after his birth. Barrow attended Charterhouse, a school noted for its emphasis on a classical education. He was a rowdy youngster, more interested in scrapping with other students than studying. Barrow was transferred to Felsted School in Essex, where it was hoped that schoolmaster Martin Holbeach's strict discipline would correct his bad habits and promote scholastic achievement.

Barrow stayed at Felsted for four years, thriving on the social and academic discipline it offered. He studied Latin, Greek, Hebrew, French, logic, and the classics. When Barrow's father suffered financial losses when a rebellion destroyed textile trade with Ireland, Barrow began tutoring Thomas Fairfax,

*Isaac Barrow*

fourth viscount Fairfax of Emely, Ireland. In 1646, Barrow finally secured a scholarship to Trinity College, Cambridge.

Barrow earned his baccalaureate in 1648 and was elected as a college fellow in 1649. He completed an M.A. in 1652 and was named a college lecturer and university examiner.

Barrow completed what would be his first published work in 1654, *Euclidis Elementorum libri XV*, a highly regard translation of **Euclid**. Designed as an undergraduate text, the work was reissued in 1657 and eventually reached a wide public in a pocket–sized edition. In 1655, Barrow was nominated for a prestigious professorship in Greek. But this was the decade of Cromwell, and Barrow had never bothered to hide his loyalty to the monarchy. The Regius Professorship went to Ralph Widdrington, a candidate who had the backing of the university's chancellor—Oliver Cromwell.

Frustrated and angry, Barrow sold his books, applied for and won a Trinity College traveling fellowship, and left England. He traveled to France, Italy, and Turkey, lingering abroad for nearly five years. It was during his travels that his interest in mathematics intensified, as he came into contact not only with classical scholars like himself, but scientists and mathematicians.

Barrow returned to England in 1660, the year King Charles II was restored to England's throne. He immediately was ordained in the Anglican Church and was promptly appointed to the Regius Professorship of Greek.

Barrow supplemented the modest pay of a professor of classics by accepting a professorship of geometry at Gresham College and filling in as a professor of astronomy. When he was named Lucasian professor of mathematics at Cambridge in 1663, however, the stipend attached to the professorship made it possible for him to give up extra teaching appointments. In that same year Barrow was the first to be named a fellow of the newly established Royal Society of London; it was in that capacity that in 1664 he served as a scholarship examiner for Isaac Newton. Apparently Barrow's wit and knowledge made an impression, and Newton began attending Barrow's lectures on optics.

Between 1663 and 1669, Barrow developed and presented the *Lectiones geometricae*, a series of lectures on geometry in which he defined time as the measure of motion, the properties of curves generated when moving points and lines are combined, the construction of tangents and the nature of quadrature. Although the combined lectures include elements key to the fundamentals of calculus theory, the work is not original. Barrow relied on contemporary mathematicians—including **René Descartes, John Wallis**, and **James Gregory**—for the information; his genius lay in combining these works into a comprehensible whole and relaying the results to a new generation of scholars.

Barrow is credited with developing the method of finding the point of refraction at a plane interface and the point construction of the diacaustic of a spherical interface. His work seems to have served as the starting point for Newton's ideas, although the most Newton admitted was that Barrow's lectures "might put me upon considering the generation of figures by motion, tho I no now remember it."

Barrow's work in optics was quickly eclipsed by Newton's. Influenced both by Newton's genius and the tug of other interests, Barrow stepped down as Lucasian professor in favor of Newton in 1669.

Barrow devoted his energies to theology and served as royal chaplain in London. He returned to Trinity College in 1673 at the king's request and was appointed vice chancellor in 1675.

Barrow never married; he died in 1677, and contemporary accounts indicate his death at the age of 47 was the result of a drug overdose.

## SELECTED WRITINGS BY BARROW

*Lectiones XVIII Cantabrigiae in scholis publicis habitae; In qubus opticorum phaenomenon genuinae rationes investiganture, ac exponunture*, 1669

*Lectiones gemoetricae: In quibus (praesertim) generalia curvarum linearum symptomata declaranture*, 1670

*Scriptores optici*, 1823 (the collected texts of optical lectures)

*The Mathematical Works of Isaac Barrow*. Compiled by W. Wherell, 1860

## FURTHER READING

Abbott, David, editor. *The Biographical Dictionary of Scientists: Mathematicians*. New York: Peter Bedrick Books, 1986, pp. 14–15.

Daintith, John, Sarah Mitchell, and Elizabeth Tootill, editors. *A Biographical Encyclopedia of Scientists*. Volume 1. New York: Facts on File Inc., p. 51.

Eves, Howard. *An Introduction to the History of Mathematics*. New York: CBS College Publishing, 1983, pp. 300–301.

Feingold, Mordechai, editor. *Before Newton: The Life and Times of Isaac Barrow*. New York: Cambridge University Press, 1990.

Hooper, Alfred. *Makers of Mathematics*. New York: Random House, 1948, pp. 283–85.

Jenkins–Jones, Sara, editor. *The Hutchinson Dictionary of Scientists*. Oxford, England: Helicon Publishing Ltd., 1996, pp. 34–35.

Murray, Jane. *The Kings and Queens of England*. New York: Charles Scribner's Sons, 1974, pp. 73–78.

Smith, David E. *History of Mathematics*. Volume II. New York: Dover Publications Inc., 1958, pp. 690–91.

Whiteside, D.T. "Isaac Barrow," in *Dictionary of Scientific Biography*. Volume II. Edited by Charles Coulston Gillispie. New York: Charles Scribner's Sons, 1973, pp. 473–76.

—*Sketch by A. Mullig*

# al–Battānī (also known as Albatenius, Albategni, or Albategnius)
## 858(?)–929
### Arab trigonometer and astronomer

al–Battānī was one of the greatest figures in a long line of Arab astronomers who dominated that branch of science during the 9th and 10th centuries. In addition to his extensive observations of the sun, moon, planets, and stars, al–Battānī was one of the first to use a wide variety of trigonometric functions to calculate various astronomical coefficients with great accuracy. His theorems regarding the relationship between the sides and angles of a three–dimensional spherical triangle, useful in his astronomical calculations, provided the basis for later developments in spherical trigonometry in the 15th century.

Al–Battānī's full name was Abū'Abd Allāh Muhammad ibn Jābir ibn Sinān al–Battānī, al–Raqqī, al–Harrānī, al–Sābi'. He is also known by his Latinized names of Albatenius, Albategnius, and Albategni

Al–Battānī was born in or near the city of Harran in what is now Turkey. His exact date of birth is not known, but it is believed to have been in the period of 850–858, based on the age at which he first began making his astronomical observations. The identity of his father is not certain, although it was probably the great instrument maker Jābir ibn Sinān al–Harrānī, who constructed many astronomical instruments. Although al–Battānī was a Muslim, his ancestors followed the Harranian Sabian's religion, which incorporated much of the ancient Mesopotamian knowledge of the stars.

Surrounded by scientific instruments and living in a culture rich with astral knowledge and star lore, al–Battānī quickly developed an interest in astronomy. His early education came from his father, which contributed to his later skill in constructing new instruments to study the skies. When al–Battānī was older, he moved to the city of Raqqa on the left bank of the Euphrates River in present–day Syria, where he continued his education and began his long and illustrious career.

### Refines the Work of Greek Astronomers

Al–Battānī began his observations of the skies in 877. Like many Arab astronomers of the period, he was familiar with the writings of the famous Greek astronomer and mathematician **Claudius Ptolemy**, who had done most of his work in Alexandria, Egypt, starting in about 127. Through careful observations and measurements, al–Battānī sought to further refine and correct Ptolemy's data.

Over the course of the next 42 years, al–Battānī made many significant contributions to astronomy. He determined the length of a solar year to be 365 days, 5 hours, 48 minutes, and 24 seconds, which is very close to the value we use today. He observed that the distance of the Earth's orbit from the sun (or as he would have said, the sun's orbit from the Earth) was not constant but varies, as measured by changes in the apparent angular diameter of the sun. From these observations he calculated the variation of the orbit, called the eccentricity, to a much more accurate value than his predecessors. al–Battānī inferred that this eccentricity made it possible for the moon to produce both annular and total solar eclipses, contrary to what Ptolemy believed. He also made many other accurate measurements regarding the motions of the sun, moon and the closest, visible planets.

### Incorporates Trigonometry Into His Astronomical Calculations

Part of al–Battānī's success as an astronomer was due to his careful measurements and the more precise instruments available to him. However, a significant part was due to his use of trigonometric methods to make his calculations, instead of the geometric methods used by the earlier Greeks. Trigonometry, and the concepts of sines and cosines, had been introduced to the Arabs by Indian astronomers. Arab astronomers added other trigonometric functions and developed the relationships between the various functions. Al–Battānī regularly used sines and cosines in his formulas, and described tangents, which he called *zill muntasib*, or *umbra erecta* in Latin, and cotangents, which he called *zill mabsut*, or *umbra recta* in Latin. He also introduced solutions to the relationship between the sides and angles of a three–dimensional spherical triangle by using the principles of orthographic projection involving right angles and perpendicular lines.

Al–Battānī applied the same trigonometric techniques and calculations to arrive at solutions in astrology, as well as astronomy. Although most scientists today consider astrology to be nothing more than a mild form of entertainment, it was an integral part of astronomical studies for the Arabs, who believed that the positions of the stars and planets had a direct effect on people's lives.

### Life's Work Influences European Astronomers and Mathematicians

Al–Battānī wrote the details of his many observations and calculations in a book entitled *Kitab al–Zij*, commonly referred to as *Zij*. The word *zij* originally was a Persian term for the lengthwise threads in a rug, but in Arabic had come to be a technical term for a book on astronomy. Al–Battānī's *Zij* contained 57 chapters and included discussions of trigonometric functions, a table of cotangent values, and the development of solutions to problems in spherical trigonometry. The book also included tables of his observations of the sun, moon, and the five nearest planets, as well as a catalog of the positions of almost 500 stars.

There were two editions of the *Zij*. One was completed before 900, and the other sometime after 901. The second version contains descriptions of two eclipses—one solar and one lunar—which al–Battānī observed while in the city of Antioch in 901. In his preface to the *Zij*, al–Battānī wrote a request that future generations should try to improve upon his theories and conclusions based on observations and calculations of their own, just as he had done with Ptolemy's work.

Al–Battānī continued to work and make astronomical observations until 918. In 929, he accompanied some people from Raqqa to the city of Baghdad in present–day Iraq to protest certain unfair taxes which had been placed on them. On the return trip, al–Battānī died at Qasr al–Jiss near Samarra in Iraq.

The *Zij* was translated into Latin by two different scholars in the 12th century. Only one of those translations, done by Plato of Tivoli, survives. A translation into Spanish was ordered by King Alfonso el Sabio of Spain in the 13th century and also survives. The Latin translation by Plato of Tivoli was printed twice—once in 1537 and again in 1645 with a slightly different content. Because of these translations, al–Battānī's work in astronomy and trigonometry became well known in Europe where it had significant influence up through the period of the Renaissance in the 15th and 16th centuries. His work in spherical trigonometry was the foundation for further development by Regiomontanus in the 15th century and became an important part of that branch of mathematics. The famous astronomers of the Renaissance—Copernicus, **Johannes Kepler**, and Tycho Brahe—all acknowledge the influence of al–Battānī's observations and methods in discussions of their own work.

### FURTHER READING

American Council of Learned Societies. *Biographical Dictionary of Mathematicians*. New York: Charles Scribner's Sons, 1991, pp. 185–194.

Sarton, George. *Introduction to the History of Science*. Volume 1. Huntington, NY: Robert E. Krieger Publishing Company, 1975, pp. 585–589, 602–603.

Taton, René, editor. *History of Science*. Volume 1. New York: Basic Books, Inc., 1962, pp. 409–10, 429.

—*Sketch by Chris Cavette*

# Agnes Baxter
## 1870–1917
### Canadian algebraist

Agnes Sime Baxter was the second Canadian woman to earn a doctoral degree in mathematics and the fourth to receive such a honor in North America. Her award–winning undergraduate career was capped with historic higher degrees. Even though Baxter abandoned mathematics to support her husband's academic career, her name now adorns a reading room in the Department of Mathematics, Statistics and Computing Science at her alma mater, Dalhousie University.

Baxter was born on March 18, 1870, in Halifax, Nova Scotia to Mr. and Mrs. Robert Baxter. Little is known of her early life. She enrolled at Dalhousie at the age of 17, majoring in mathematics. In 1891, Baxter was granted a B.A. degree with a first class distinction, making her the first female honors student at the university. She also won the Sir **William Young** prize, a gold medal, that year. In 1882, Baxter earned an M.A. in the same field, specializing in mathematical physics, from the same institution. She then studied at Cornell University, funded by a fellowship, and received her Ph.D. in 1895. Baxter's doctoral thesis was on Abelian integrals.

### An Early "Retirement"

As Mrs. A. Ross Hill, whom she became just a year later, Baxter assisted her husband's presidency at the University of Missouri. She apparently raised a family, hosted university functions, and quietly nursed a wasting illness never specified in the press. After a long medical struggle she seemed to be cured, but was then succumbed to an infection on March 9, 1917.

Her widower, also a Dalhousie graduate, thought Baxter's memory would best be served by his endowment of the sum of $1000, intended to fund a collection of books. Obituaries at the time noted her aptitude for a subject not widely considered as within a female's natural intellectual realm. Hill was quoted as intending his gift to bring recognition to Baxter for supporting his career, "instead of making an independent record for herself."

### Medal in Safe Keeping

The Agnes Baxter Reading Room was dedicated with an afternoon ceremony on March 15, 1988. Dr. John Roberts of the University of California, Berkeley, a grandson of the Hills, was present for the event. The gold medal Baxter had won as an undergraduate was given to the Department of Mathematics, Statistics and Computing Science soon after.

### SELECTED WRITINGS BY BAXTER

"On Abelian Integrals, a Resume of Neumann's Abelsche Integral with Comments and Applications." Unpublished thesis, circa 1895.

### FURTHER READING

#### Periodicals

A.M.M. "Agnes Sime Baxter." *The Dalhousie Gazette* (June 15, 1917): 1, 3.

Anonymous. "The Agnes Baxter Library." *The Dalhousie Gazette* (June 15, 1917): 1.

#### Other

"Agnes Baxter." *Biographies of Women Mathematicians.* June 1997. http://www.scottlan.edu/lriddle/women/chronol.htm (July 22, 1997).

*The Dalhousie President's Report.* Halifax, N. S.: Dalhousie University, 1917, p. 12.

Wilkins, Gina. "Dalhousie University Honours Alumna with Room Dedication." *Dalhousie University Public Relations Office Press Release*, March 11, 1988.

—*Sketch by Jennifer Kramer*

# Thomas Bayes
## 1702(?)–1761
### English probabilist and minister

Thomas Bayes, a Presbyterian minister, expressed a method of inductive inference in a precise and quantitative form, which lead to the development of Bayesian statistics, or Bayesian inference. His stature as a mathematician is based on only two short mathematical papers, both of which were published posthumously by the Royal Society of London. The first paper demonstrates what may be the first recognition of asymptotic behavior by series expansions. The second and far more important paper addresses a problem with continuing application in most areas of human endeavor. In this paper, Bayes discusses the estimation of future occurrences of an event, given knowledge of the history of the event— that it has occurred a number of times and failed a number of times. This work continues to spawn mathematical research, and provides the foundations for Bayesian statistical estimation, used today on such diverse problems as electoral polling or estimating time to failure of mechanical devices.

Little is known of Bayes' childhood. Some sources note that he was privately educated, while others state that he received a liberal education in preparation for the ministry. Bayes was the eldest of six children of Joshua and Ann Carpenter Bayes. His father was a Nonconformist minister, one of the first seven publicly ordained in England. Thomas' paternal grandfather, Joshua Bayes, had been a cutler and town collector in Sheffield. Thomas' place of birth is usually listed as London, but one biographer suggests that he was born in Hertfordshire, where his peripatetic father supposedly preached at the time of his birth. Unfortunately, the appropriate parish records of 1700–1706 have been lost. Thomas' epitaph in the family vault at Moorgate states his age at death as 59 in April 1761, placing his birth in 1701 or 1702.

### Educated for the Ministry

Andrew I. Dale argues persuasively that Bayes was educated at Edinburgh University. Thomas Bayes' name appears in a 1719 catalogue of manuscripts in the Edinburgh University Library, and in a number of other records at the University over the period 1720–1722, including class lists and a list of theologues. The Bayes signature at Edinburgh matches closely that of the Royal Society records. Bayes received only licensure for the ministry at Edinburgh, but he was ordained during or before 1727, and is included in Dr. John Evans' 1727 list of "Approved Ministers of the Presbyterian Denomination." Bayes assisted his father at his ministry in Leather Lane for some years from 1728 before succeeding the Rev. John Archer as minister at Tunbridge Wells, Kent. He spent the remainder of his life in Tunbridge Wells as Presbyterian minister of the Mount Sion meeting house.

### Defense of Newton Is Catalyst for Election to Royal Society

Two years after the publication of *The Analyst; or, a Discourse addressed to an Infidel Mathematician* (1734), **George Berkeley**'s famous attack of **Isaac Newton**'s work on fluxions (differentials), an anonymous tract was published that answered Berkeley and vigorously defended Newton's work. The tract, titled *An Introduction to the Doctrine of Fluxions and Defence of the Mathematicians against the Objections of the Author of the Analyst*, was widely attributed to Bayes and was probably the reason behind his election as a Fellow of the Royal Society in 1742. In it, he addresses the "business of the mathematician," and stated that "[he] is not inquiring how things are in matter of fact, but supposing things to be in a certain way, what are the consequences to be deduced from them; and all that is to be demanded of him is, that his suppositions be intelligible, and his inferences just from the suppositions he makes." The proposal for Bayes' election to the Royal Society read in part that he was "well skilled in geometry and all parts of mathematical and philosophical learning." It was signed by Eames, James Burrow, Cromwell Mortimer, Martin Folkes, and Earl Stanhope.

### Work Published Posthumously

Bayes retired from the ministry around 1750. He died at Tunbridge Wells on April 17, 1761, leaving a fairly substantial estate, and was buried in the family vault at Bunhill Fields Burial Ground at Moorgate. Upon his death, Bayes' family asked his friend, the Unitarian Reverend Richard Price, to examine his papers. Among them Price found Bayes' work on probability.

In 1764, Bayes' paper on the Stirling–De Moivre Theorem, dealing with series expansions, was published in the *Philosophical Transactions of the Royal Society*. Price declared the problem in inductive reasoning stated by Bayes to be central "to the argument taken from final causes for the existence of the Deity." The same issue of *the Philosophical Transactions* contains a second piece by Bayes, "An Essay towards Solving a Problem in the Doctrine of Chances," presented also by Price with his preface, footnotes and appendix. The problem posed in the essay, wrote Price, is "to find out a method by which we might judge concerning the probability that an event has to happen, in given circumstances, upon supposition that we know nothing concerning it but that, under the same circumstances, it has happened a certain number of times, and failed a certain other number of times." The essay was followed in the next volume by "A Demonstration of the Second Rule in the Essay. . .," a continuation of Bayes' results which were further developed by Price.

A notebook belonging to Bayes has been preserved in the London records room of the Equitable Life Assurance Society, due to the action of Price and his nephew William Morgan, an actuary. Among other curiosities, the notebook includes the key to a system of shorthand, details of an electrifying machine, lists of English weights and measures, notes on topics in mathematics, natural philosophy and celestial mechanics, and a proof of one of the rules in the "Essay" that was published after Bayes' death.

## FURTHER READING

### Books

Dale, Andrew I. *A History of Inverse Probability: From Thomas Bayes to Karl Pearson.* Studies in the History of Mathematics and Physical Sciences 16. New York: Springer Verlag, 1990.

Press, S. James. *Bayesian Statistics: Principles, Models, and Applications.* New York: John Wiley & Sons, 1989. (Includes Bayes' essays.)

Stigler, Stephen M. *The History of Statistics: The Measurement of Uncertainty Before 1900.* Cambridge, MA: Harvard University Press, 1986.

*Alexandra Bellow*

## Periodicals

Barnard, G. A. "Thomas Bayes: A Biographical Note." *Biometrika* 45 (1958): 293–315.

—*Sketch by Jill Carpenter*

# Alexandra Bellow
## 1935–

### Romanian–American ergodic theorist

Ergodic theory, Alexandra Bellow's field of specialization, deals with the long term averages of the successive values of a function on a set when the set is mapped into itself, and whether these averages equal (converge to) a reasonable function on the set. The theory applies to probability and time series, and to the concept of entropy in physics and information theory. Bellow has proved significant results in this field.

Alexandra Bellow was born in Bucharest, Romania, on August 30, 1935. Her father, Dumitru Bagdasar, had studied medicine in the United States. He was a famous neurosurgeon who founded a school of neurosurgery in Romania in that year. Her mother, Florica Bagdasar, was a psychiatrist specializing in the treatment of mentally retarded children. The Roma-

nian philosopher Nicolae Bagdasar was Bellow's uncle. After World War II, the politics of Romania were extremely unstable as the communists took control of the country. Bellow's parents supported the communists, and in 1946, her father was appointed minister of health in the Groza government. Bellow's father was soon accused of "defection," removed from the ministry, and imprisoned. He died in 1946, reportedly of cancer. Bellow's mother succeeded him as minister of health. According to the *New York Times*, she was the first woman to hold a ministerial position in Romania. However, by 1948 she was removed from that post. Bellow told Ruth Miller (a biographer of her second husband, Saul Bellow) that her mother was accused of "cosmopolitanism" and was prohibited from doing any work or practicing medicine.

In spite of these problems, Bellow studied mathematics at the University of Bucharest. In 1956, she married Cassius Ionescu Tulcea, a professor of mathematics at the university. She received a M.S. in mathematics in 1957. In that year, Bellow and her husband came to the United States to study at Yale University. He was a research associate at Yale while they were students. They both received Ph.D.s in mathematics from Yale in 1959. Bellow's thesis was titled "Ergodic Theory of a Random Sequence." After graduation, Bellow became a research associate at Yale from 1959 to 1961 and at the University of Pennsylvania from 1961 to 1962. She taught at the University of Illinois at Urbana as an assistant professor from 1962 to 1964, then as associate professor from 1964 to 1967. In 1967, Bellow went to Northwestern University as a professor. Her husband held positions at the same schools, although not exactly in the same years. Both remained at the mathematics department at Northwestern for the rest of their careers. Their book, *Topics in the Theory of Liftings*, appeared in 1969. They were divorced in that year.

Starting in 1971, Bellow published papers on ergodic theory. In 1974 she married the writer Saul Bellow, and in the following year they traveled to Israel, where she taught and worked with colleagues in mathematics at the University of Jerusalem. Saul Bellow won the Nobel Prize for literature in 1976. In the late 1970s Bellow published several papers on asymptotic martingales. Bellow edited, with D. Kolzow, the proceedings of a conference on Measure Theory held in Oberwolfach in 1975. She was an editor for the *Transactions of the American Mathematical Society* from 1974 to 1977, associate editor of the *Annals of Probability* from 1979 to 1981, and associate editor for *Advances in Mathematics* since 1979. Bellow's 1979 paper with Harry Furstenberg on applying number theory to ergodic theory contains the Bellow–Furstenberg Theorem.

Bellow was a Fairchild scholar at CalTech in 1980, and also visited the University of California at Los Angeles. She was divorced from Saul Bellow in 1986. Bellow received an award from the Alexander von Humboldt Foundation, which sponsors visits to Germany by scholars, in 1987. In 1989, she and Roger L. Jones of De Paul University organized a conference on "Almost Everywhere Convergence in Probability and Ergodic Theory" at Northwestern University and edited the conference proceedings. In that same year, Bellow married **Alberto P. Calderón**, a distinguished mathematician and civil engineer retired from the University of Chicago, whose fields of research are partial differential equations, functional analysis and harmonic analysis.

In 1991, Bellow gave the **Emmy Noether** Lecture for the Association for Women in Mathematics. Bellow, Jones, and others have collaborated on eight recent papers that deal primarily with partial sequences of observations. This work consists of their attempts to identify when averages based on partial observations are probably valid for the whole population, when they are not valid, and by how much. A 1996 paper in this series had six authors, unusual for a paper in mathematics, which was featured in the *Mathematical Reviews*. It describes averages that behave very badly. Jones describes Bellow as an excellent collaborator who is extremely knowledgeable, "very clever and very careful," and, according to another colleague, "never seems to make a mistake." Jones also reports that Bellow enjoyed her teaching and was a very good instructor who supervised at least four Ph.D. students. A conference is scheduled at Northwestern in October 1997 on the occasion of Bellow's retirement. Bellow plans to continue her research in mathematics.

## SELECTED WRITINGS BY BELLOW

### Books

(With Cassius Ionescu Tulcea) *Topics in the Theory of Lifting*, 1969

### Periodicals

(With H. Furstenberg) "An Application of Number Theory to Ergodic Theory and the Construction of Uniquely Ergodic Models." *Israel Journal of Mathematics* 33 (1979): 231–40.

(With Roger L. Jones) "A Banach Principle for $L^\infty$." *Advances in Mathematics* 120 (1996): 155–72.

(With Mustafa Akcoglu et al.) "The Strong Sweeping Out Property for Lacunary Sequences, Riemann Sums, Convolution Powers and Related Matters." *Ergodic Theory and Dynamical Systems* 16 (1996): 207–53.

## FURTHER READING

### Books

Green, Judy and Jeanne Laduke. "Women in American Mathematics: A Century of Contributions." *A Century of Mathematics in America, Part II.* Edited by Peter Duncan. Providence, RI: American Mathematical Society, 1989: pp. 379–398.

James, Robert C. *Mathematics Dictionary.* Fifth Edition. New York: Van Nostrand Reinhold, 1992.

Mathematical Society of Japan. *Encyclopedic Dictionary of Mathematics.* Second Edition. Edited by Kiyosi Ito. Cambridge, MA: MIT Press, 1987.

Petersen, Karl. *Ergodic Theory.* Cambridge, England: Cambridge University Press, 1983.

### Periodicals

Blum, Lenore. "A Brief History of the Association for Women in Mathematics: The Presidents' Perspectives." *Notices of the American Mathematical Society* (September 1991): 738–54.

### Other

"Alexandra Bellow." *Biographies of Women Mathematicians.* June 1997. http://www.scottlan.edu/lriddle/women/bellow.htm. (July 22, 1997).

American Mathematical Society. *Mathematical Reviews: MathSci Disc.* (CD–ROM). Boston: SilverPlatter Information Systems.

Jones, Roger L., interview with Sally Moite, conducted May 1–May 2, 1997.

*—Sketch by Sally M. Moite*

# Eugenio Beltrami
## 1835–1899
### Italian geometer and mathematical physicist

Eugenio Beltrami is best known for his study of differential geometry, particularly his study of curves and surfaces. He was born in Cremona, Italy, on November 16, 1835, to a family with an artistic tradition. His grandfather was an engraver who worked with precious stones, and his father painted miniatures. Beltrami studied mathematics at the University of Pavia, but was forced to abandon his studies in 1856 because of his family's financial difficulties. He obtained a post as a secretary to a railroad engineer, a position which took him to Verona and Milan.

In Milan, he resumed his mathematical studies and in 1862 published his first paper, which examined the differential geometry of curves. That same year Beltrami was named chairman of complementary algebra and analytic geometry at the University of Bologna. In 1864, he moved to Pisa and chaired geodesy; in 1866, he moved back to Bologna as the chair of rational mechanics. When Rome became the capital of Italy in 1870, Beltrami accepted an appointment as a professor of rational mechanics at the newly established University of Rome, a post he maintained until 1876. He then accepted the chair of mathematical physics at Pavia and in 1891, returned to Rome and taught at the university until his death.

Beltrami's most significant work, which focused on differential geometry, was completed before 1872. In 1865, he published a paper on how line elements on the surfaces of constant curvature could be represented by linear expressions. His approach offered a new representation of the geometry of constant curvature that was consistent with **Euclid**ean theory. Other papers established as a valid concept as Beltrami resolved apparent contradictions between Euclidean and non–Euclidean approaches.

Although Beltrami's later work focused on applied mathematics, he relied heavily on his understanding of geometry. Beltrami studied elasticity, wave theory, optics, thermodynamics, and potential theory, and was among the first to explore the concepts of hyperspace and time as a fourth dimension. His investigations in the conduction of heat led to linear partial differential equations, but his later work had far less impact on the development of mathematics than his earlier efforts.

Beltrami's artistic heritage was evident in his lifelong love affair with music and his sustained interest in connecting mathematical principles to music composition. Late in life, Beltrami accepted his role as an academic and civic leader. In 1898, he served as president of the Accademia dei Lincei, and in 1899—the year of his death—Beltrami was elected to the Italian senate.

## SELECTED WRITINGS BY BELTRAMI

### Books

*Opera mathematiche,* 4 volumes, 1902

## FURTHER READING

Cajori, Florian. *A History of Mathematics.* Second edition. New York: The Macmillan Co., 1938, p. 307.

Fang, J. *Mathematicians from Antiquity to Today.* Hauppauge, N.Y.: Paideia Press, 1972, p. 194.

Struik, D.J. "Eugenio Beltrami," in *Dictionary of Scientific Biography.* Volume I. Edited by Charles Coulston Gillispie. New York: Charles Scribner's Sons, 1970, pp. 599–600.

*—Sketch by A. Mullig*

# George Berkeley
## 1685–1753
**Irish philosopher, Anglican cleric, and mathematician**

Born in the same year as the great composers Johann Sebastian Bach, Georg Frideric Handel, and Domenico Scarlatti, Berkeley was one of the seminal figures in Western philosophy, his doctrines exerting a particularly significant influence on analytic philosophy. As a mathematician, George Berkeley is known for his thought–provoking critique of the mathematical theories of his time, particularly infinitesimal calculus.

### Devotes His Life to Religion and Philosophy

Of English descent, Berkeley was born near Kilkenny, Ireland, and always considered himself an Irishman. Educated at Trinity College, Dublin, he studied mathematical logic, and philosophy. Berkeley graduated in 1704, publishing a short Latin work on mathematics in 1707. In 1710, the year he published his famous work *A Treatise Concerning the Principles of Human Knowledge*, he was ordained priest of the Church of England. After holding several academic appointments, he was named dean of Derry in 1724. Owing to his keen interest in education, Berkeley soon left for London, hoping to receive government funding for a college in Bermuda, where he intended to provide education for English and local youths. In 1728 he married Anne Forster, the well–educated daughter of a chief justice. Soon after the wedding, Berkeley and his new wife set sail, getting as far as Newport, Rhode Island, which he then decided was a better location for his school. However, the project was abandoned when promised funding failed to materialize, and the Berkeleys returned to London in 1731.

Berkeley was appointed bishop of Cloyne in 1734, and his home there became a social and cultural center, as well as a dispensary in times of epidemics. The Berkeleys eventually had six children, four sons (one became canon of Canterbury) and two daughters.

*George Berkeley*

general words. We can utter the term 'material substance', but it does not denote any abstract general idea; and if we suppose that because we can frame the term it must signify an entity apart from the objects of perception, we are misled by words. . . . 'Matter' is not a name in the way in which William is a name, though some philosophers seem to have thought mistakenly that it is."

Berkeley accepted mathematics as a practical science and pursuit, but adamantly rejected, in accordance with his criticism of meaningless concepts, the idea of number. As J. O. Urmson explains, to Berkeley, the term *ten* may denote the fact that there are ten *individual* entities in a group, but nothing more than that, and certainly not an abstract idea of *ten*, independent of any practical context. It is also important to note that Berkeley supplemented his purely philosophical critique of infinitesimal calculus with solid mathematical arguments. What Berkeley strenuously objected to was the practice, accepted by his contemporaries, of assuming that the quantity dx, being infinitesimally small, could simply be eliminated in mathematical derivations. Thus, though Berkeley's criticism may be irrelevant to modern applications of infinitesimals, he was nevertheless right in questioning the practice of treating an infinitesimal quantity as zero.

They retired to Oxford in 1752, and after Berkeley's death in 1753, Anne Berkeley continued to defend her husband's philosophy. An indefatigable writer, polemicist, and researcher, Berkeley was also a clergyman, who took his pastoral duties very seriously, ministering to the needs of people far removed from the world of 18th–century philosophy and science.

## Rejects Contemporary Mathematics as Unclear and Logically Deficient

Berkeley denied the existence of matter. The essence of his philosophy is expressed by the statement *esse est percipi* (to exist is to be perceived), which means that an object can be said to exist only insofar as it is perceived by a spirit—finite (a human being) or infinite (God). Certainly, Berkeley, being a very practical person, did not claim that the world of physical objects should be treated as an illusion, as some of his detractors naively assumed. In essence, Berkeley asserted that the postulate suggesting that matter existed independently of a percipient observer was based on illogical and unclear thinking. However, Berkeley does not deny the validity of general terms. As Copleston has written in his discussion of Berkeley's philosophy, "A proper name such as William, signifies a particular thing, while a general word signifies indifferently a plurality of things of a certain kind. Its universality is a matter of use or function. If we once understand this, we shall be saved from hunting for mysterious entities corresponding to

## Raises Logical Objections to Newton's Calculus

In the history of mathematics, Berkeley is best known for attacking the logical foundations of **Isaac Newton's** calculus. "Newton's theory," according to Tobias Dantzig, "dealt with continuous magnitudes and yet postulated the infinite divisibility of space and time; it spoke of a flow and yet dealt with this flow as if it were a succession of minute jumps." This theory of *fluxions* (Newton's term denoting the rate of change of a variable, such as length, speed, area, etc.) was open to criticism because it attempted to reconcile a smooth flow with a series of leaps. In *The Analyst* Berkeley asked: "And what are these fluxions? The velocities of evanescent increments. And what are these same evanescent increments? They are neither finite quantities, or quantities infinitely small, nor yet nothing. May we not call them the ghosts of departed quantities?" Although Berkeley felt that Newton's calculus yielded true results (even developing a clever explanation for these correct results), he nonetheless felt compelled to point out the logical fallacy on which he believed the calculus was based. In this way, he inspired other mathematicians to focus their attention on a logical clarification of calculus.

## SELECTED WRITINGS BY BERKELEY

### Books

*The Analyst; or, a Discourse Addressed to an Infidel Mathematician wherein Is Examined*

*Whether the Object, Principles and Inferences
of the Modern Analysis are More Distinctly
Conceived, or More Evidently Deduced than
Religious Mysteries,* 1734
*A Defence of Free-Thinking in Mathematics,* 1735
*De motu,* 1721
*A Treatise Concerning the Principles of Human
Knowledge,* 1710

## FURTHER READING

### Books

Boyer, Carl B. *A History of Mathematics.* Second
edition. Revised by Uta C. Merzbach. New
York: Wiley, 1991.

Copleston, Frederick. *Modern Philosophy: The
British Philosophers from Hobbes to Hume.*
Volume 5: *A History of Philosophy.* New
York: Image Books, 1963.

Dantzig, Tobias. *Number: The Language of Science.* Fourth edition. Garden City, NY: Doubleday, 1954.

Urmson, J. O. *Berkeley.* Oxford: Oxford University Press, 1982.

Windelband, Wilhelm. *A History of Philosophy:
With Especial Reference to the Development of
Its Problems and Conceptions.* Translated by
James H. Tufts. New York: Macmillan, 1921.

—*Sketch by Zoran Minderovic*

# Paul Bernays
## 1888-1977
### English-born Swiss logician

Paul Bernays secured his reputation with a
classic treatise on mathematical logic, the *Foundations of Mathematics,* and through his refinement and
consolidation of set theory into the von Neumann-
Bernays system. Bernays was a platonic mathematician—one who thought of the world of mathematics
as separate from the world of material reality. Although Bernays's concept of mathematics as a mental
product meant that no system could be designated as
right or wrong, he believed that there were truths
within the mathematical realm that allowed for a
system to remain consistent and logical within itself.

Paul Isaac Bernays was born in London on
October 17, 1888, to Julius Bernays, a Swiss business-
man from a prominent Jewish family, and Sara
Bernays. Shortly after his birth, the family moved to
Paris and then to Berlin, where Bernays studied from

*Paul Bernays*

1895 to 1907. While studying engineering at the
Technische Hochschule (Technical High School) in
Charlottenburg, Germany, he developed an interest in
pure mathematics. This led him to transfer to the
University of Berlin where he studied for four semes-
ters under a distinguished faculty that included
philosopher Ernst Cassirer and physicist Max Planck.
Bernays then attended the University of Göttingen
where physicist Max Born and mathematician **David
Hilbert** were among his professors. In 1912 Bernays
received his doctorate degree from Göttingen under
Hilbert.

In 1912 Bernays completed his postdoctoral
thesis on modular elliptic functions at the University
of Zurich in Switzerland under the German mathematician Ernst Zermelo. Bernays remained at Zurich
until 1917 when he was invited by Hilbert to return to
Göttingen to assist with a program on the foundations
of mathematics. Bernays completed a second postdoctoral thesis at Göttingen in 1918 on propositional
logic. In addition to serving as assistant to Hilbert, he
gave lectures at Göttingen until the Nazi party's rise
to power in 1933 when Bernays's right to lecture was
withdrawn because of his Jewish background.

### Publishes *Foundations of Mathematics* and *Axiomatic Set Theory*

Bernays escaped to Zurich in 1934, eventually
teaching at the Eidgenossische Technische Hochschule. In 1935 and 1936 he participated in the

Institute for Advanced Study at Princeton University in New Jersey. Bernays published the first volume of his work on mathematical logic, *Foundations of Mathematics,* in 1934; the second volume was published in 1939. Research for this work was a collaborative effort between Bernays and Hilbert, but Bernays wrote both volumes singlehandedly. In this book Bernays and Hilbert created the mathematical discipline of proof theory, in which the correctness of a mathematical statement or theorem is demonstrated in terms of accepted axioms. E. Specker, a colleague of Bernays at the Eidgenossische Technische Hochschule, remarked in *Logic Colloquium '78* that the *Foundations of Mathematics* is unique because "it does not reduce mathematics to logic, or logic to mathematics—both are developed at the same time."

Over the years Bernays published a series of articles in the *Journal of Symbolic Logic* that was published collectively as *Axiomatic Set Theory* in 1958. Axiomatic set theory applies proof theory to set theory, the study of the properties and relationships of sets. Thus, axiomatic set theory involves the presentation of set theory in terms of fundamental axioms and logical rules of inference, rather than as a formalization of tabulated or intuitive knowledge. Classical set theory was largely established by Zermelo at the turn of the century and improved by the German mathematician Abraham Fraenkel in the 1920s. The Zermelo-Fraenkel (ZF) system was defined exclusively in terms of sets, but it could not address transfinite sets (for example, the set consisting of all possible sets). In the late 1920s the axioms of Hungarian mathematician **John von Neumann** accomplished many tasks previously left unsolved by the Zermelo-Fraenkel system. However, von Neumann's system was expressed in symbolic logic and was defined in terms of function rather than set, and it was less practical in both pure and applied mathematics.

Bernays's contribution to set theory both improved and simplified von Neumann's system. Bernays introduced a distinction between "sets" and "classes" to set theory. He did not view "classes" as mathematical objects in the normal sense. As G. H. Muller characterizes Bernays's distinction in *Mathematical Intelligencer,* a set is a collection of elements or members, a "multitude forming a proper thing." A class is a collection of objects that can be manipulated or extended, a "predicate regarded only with respect to its extension." For each set there was a corresponding class, but for each class there need not be a corresponding set. This idea created two axiomatic systems, one for sets and one for classes. The sets in the von Neumann-Bernays system operate similarly to those in the Zermelo-Fraenkel system, and thus a new system was created to allow for the construction of classes.

After World War II Bernays became Extraordinary Professor at the Eidgenossische Technische Hochschule. He also served as visiting professor at the University of Pennsylvania and at Princeton, where he was again a member of the Institute for Advanced Study in 1959–60. Bernays served as the president of the International Academy of the Philosophy of Science, as honorary chair of the German Society for Mathematical Logic and Foundation Research in the Exact Sciences, and as a corresponding member of the Academy of Science of Brussels and of Norway. He also served on the editorial boards of several journals, including *Dialectica, Journal of Symbolic Logic,* and *Archiv fur mathematische Logik und Grundlangenforschung.* Bernays received an honorary doctorate from the University of Munich in 1976 for his contributions to proof and set theory. Although he remained based in Zurich until his death from heart disease in 1977, Göttingen was always more of a home for Bernays. Bernays, who never married, lived most of his life with his mother and two sisters.

## SELECTED WRITINGS BY BERNAYS

(With D. Hilbert) *Foundations of Mathematics,* 2
   volumes, 1934–39
*Axiomatic Set Theory,* 1958
*Abhandlungen zur Philosophie der Mathematik,*
   1976
*Sets and Classes,* 1976

## FURTHER READING

### Books

Specker, E. "Paul Bernays." *Logic Colloquium '78.*
   North-Holland, 1979, pp. 381–389.

### Periodicals

Muller, G. H. "Paul J. Bernays." *Mathematical
   Intelligencer* no. 1 (1978/1979): 27–28.

*—Sketch by Fei Fei Wang Metzler*

---

# Daniel Bernoulli
## 1700–1782
### Swiss mathematical physicist and scientist

Known as the discoverer of the Bernoulli principle, which applies the law of conservation of energy to fluids, Daniel Bernoulli was a true polymath, excelling in many fields of science, and contributing

*Daniel Bernoulli*

brilliant ideas and insights which not only shaped 18th century science, but also anticipated future discoveries. In essence, Bernoulli's scientific spirit easily transcended the boundaries set by scientific disciplines, surveying the landscape of science from a height where the idea of separate fields of study seems meaningless. Nevertheless, many of his insights stemmed from his profound understanding of mathematics, and from his extraordinary ability to illuminate physical phenomena by mathematical reasoning.

Born in Gröningen, Netherlands, Bernoulli was the second son of **Johann Bernoulli**, professor of mathematics at the University of Gröningen. In 1705, when Johann took over his brother **Jakob Bernoulli's** chair of mathematics at the University of Basel, the family returned to Switzerland. A precocious student, young Daniel studied logic, philosophy, and mathematics, earning a master's degree at the age of 16. His father, who tried to keep him away from a scientific career, grudgingly allowed him to study medicine.

### Mathematician and Physician

Bernoulli finished his medical studies in 1721, writing a dissertation on breathing entitled "De respiratione." Unable to obtain a teaching position, Bernoulli continued his medical studies in Italy, also keeping up with his mathematical studies. In fact, his most important accomplishment during his stay in Italy was a mathematical work entitled *Exercitationes quaedam mathematicae*, published in Venice in 1724.

In this treatise, Bernoulli discussed a variety of subjects, including probability and fluid motion. Most notably, Bernoulli succeeded in integrating Jacopo Franceso Riccati's well-known differential equation by separating its variables. Bernoulli's *Exercitationes* so impressed the scientific community that he was immediately offered a teaching position at the St. Petersburg Academy (his older brother Nikolaus was also invited to teach mathematics). In 1725, having won his first prize from the French Académie Royale des Sciences (he would win another nine), Bernoulli moved to St. Petersburg.

### A Polymath in St. Petersburg

Although officially a professor of mathematics, Bernoulli worked in several fields. For example, his medical publications include important papers on subjects such as muscular contraction and the optic nerve, and his writings on physics include a paper on oscillation. Nevertheless, Bernoulli's mathematical genius seemed to inform most of his endeavors, a case in point being research in the areas of probability and statistics. In St. Petersburg, Bernoulli further developed his ideas on probability, demonstrating, for example, the importance of probability for economic theory. Furthermore, he extended the relevance of probability to a seemingly unrelated field such as ethics, proposing a hypothesis which asserts that if a person's material fortune increases geometrically, his or her moral fortune will increase arithmetically. Bernoulli used the moral hypothesis in an attempt to solve the famous "Petersburg paradox," which he and his brother Nikolaus concocted. The paradox emerges from an imagined game of tossing a coin. When the first toss is head, B pays A one dollar. If the first toss is tail, A gets nothing, but if the second toss is head, he gets two dollars, and so on, the idea being that if head first appears on the $n$th toss, A should get $2^{(n-1)}$ dollars. The mathematical expectation of A's winnings is infinite, but when the French naturalist Georges-Louis-Leclérc, Comte de Buffon, sought an empirical verification of Bernoulli's paradox, he found that after 2,084 games, A would have received only $10,057, an average of less than $5 per game.

In 1726, Bernoulli discussed the parallelogram of forces in paper on mechanics; the following year, he started corresponding with his compatriot **Leonhard Euler**, one of the greatest mathematicians of that era, who was also interested in mechanics. This research in mechanics, begun in St. Petersburg, laid the foundations for Bernoulli's brilliant, and even prophetic, discoveries in the field of physics.

### Returns to Basel

Although his work in St. Petersburg was highly successful, Bernoulli was so eager to obtain a post in Basel that he accepted a professorship in botany, a subject he did not particularly care for. Nevertheless, Bernoulli continued his research in mechanics, along with his multifarious scientific pursuits. Thus, for example, he delivered, in 1737, a historic lecture detailing the calculations needed to measure the work done by the heart.

### Hydrodynamica

In 1738, Bernoulli published his seminal book, *Hydrodynamica*, begun in St. Petersburg and completed in 1734. This may have inspired Bernoulli's father, Johann, who strongly resented his son's success, to publish his *Hydraulica,* predating the publication to 1732 in an effort to claim some of Bernoulli's discoveries for himself. However, *Hydrodynamica* established Bernoulli as one of the great scientists of his time. Among the important discoveries included in the book are Bernoulli's principle, one of his many insights into the behavior of fluids, which affirms that a fluid's pressure diminishes as its velocity increases. The application of Bernoulli's principle still underlies numerous industrial designs, including boats, automobiles, aircraft, and fluid conduits. Even more important, perhaps, is Bernoulli's explanation, presented in the tenth chapter, of the mechanics of gases. Gases, he posited, consist of fast and randomly moving particles. Placed in a cylinder, the particles exert constant pressure on the cylinder's walls. By applying outside pressure on a quantity of gas trapped in a cylinder by depressing a piston, Bernoulli confirmed Boyle's law, discovered in 1660 by Robert Boyle, which states that, in a closed container, gas pressure is inversely proportional to its volume, i.e., that the product $PV$ (where $P$ is pressure, and $V$ is volume) remains constant if the temperature stays unchanged. As Sheldon Glashow pointed out in his *From Alchemy to Quarks*, "Bernoulli presented a mathematical deduction of Boyle's law from the principles of mechanics." Furthermore, Bernoulli also realized that gas pressure is in direct proportion to its temperature. Connecting the idea of pressure and particle motion, Bernoulli understood the correlation of temperature and particle speed: as the temperature rises, the particles accelerate. Bernoulli's discovery was one of the great moments in the history of science, for his extraordinary insights laid the foundations for the kinetic theory of gases, which in the 19th century fully validated his profound understanding of the nature of gases. "Once the temperature of a gas," Glashow wrote, "was recognized to be proportional to the mean kinetic energy of its constituent molecules, the transformation of mechanic energy into heat was demystified. It is the conversion of ordered motion into the disordered motion of molecules."

### Studies Wave Motion and Explores the Nature of Energy

In 1743, Bernoulli became professor of physiology, a field which he much preferred to botany, bu it

was not until 1750 that he obtained the chair of Natural Philosophy, or physics, an appointment which perfectly matched his scientific pursuits. An immensely popular lecturer, Bernoulli retained this post until his retirement in 1776.

As many scholars have asserted, it was in the field of physics that Bernoulli most brilliantly applied his mathematical genius. "Daniel Bernoulli," A. Wolf wrote in his *A History of Science, Technology, and Philosophy in the Eighteenth Century*, "applied himself especially to solving, with the aid of the new analysis, difficult mechanical problems of which the geometrical methods adhered to by Huygens, or by Newton in his *Principia*, offered no prospect of a successful solution. He must be therefore regarded as one of the principal founders of that branch of science known as *Mathematical Physics*."

When Bernoulli tackled the phenomenon of kinetic energy, which was then called *vis viva*, or *living force*, he (going a step further than **Christiaan Huygens** and **Gottfried Wilhelm von Leibniz**, who studied kinetic energy in an uniform gravitational field) posited that the principle of the conservation of *vis viva* is valid for the entire universe. While he lacked the tools to verify his insight empirically, Bernoulli correctly anticipated the law of energy conservation, which was finally empirically confirmed by J. P. Joule in the 1840s. According to Donald E. Tilly and Walter Thumm in their textbook *College Physics*, since Joule's confirmation of Bernoulli's ideas, "the concept of energy, as a conserved quantity that can be transformed into many guises but never created or destroyed, has emerged perhaps as the most useful idea in all science."

Bernoulli proceeded to apply his mathematical insights to natural phenomena. While studying the nature of sound, Bernoulli discovered distinct mathematical regularities in the shape of sound waves. For example, he found that every particular body vibrates at certain natural, or *proper*, frequencies, the lowest natural frequency being the *fundamental frequency*, and the higher frequencies called the *overtones*. He also discovered that the number of *nodes*, or quiet spots where no vibration occurs, increases with higher frequency. Translating these insights into the language of mathematics, Bernoulli was able to calculate the natural frequency of a variety of musical instruments. In addition, he demonstrated that the intervals between overtones are not always harmonious.

### Shows Links between Mathematics and Music

Unlike his many predecessors, who often described music and mathematics as kindred arts forms from a purely philosophical point of view, Bernoulli demonstrated that these two arts indeed share certain fundamental traits. For example, his ideas about the nature of sound, particularly his remarkable insight that any small vibration consists of a series of natural frequencies, or modes, the higher frequency is superimposed on the lower, were expressed in the language of mathematics. It is significant that Bernoulli's seminal work in the field of acoustics directly anticipated such great discoveries of 19th–century mathematics as Jean–Baptiste–Joseph Fourier's harmonic analysis. What Bernoulli knew intuitively, namely that a sound can be represented as a trigonometric function, Fourier expressed in precise terms, as a series of sine waves. Fourier thus confirmed Bernoulli's understanding of the mathematical and physical nature of sound, particularly the extraordinary idea that each complex musical sound, such as the sound made by a musical instrument, consists of a series of simple sounds, such as those produced by a tuning fork.

Bernoulli died in Basel on March 17, 1782, five years after his retirement from the university.

## SELECTED WRITINGS BY BERNOULLI

### Books

*Hydrodynamica, sive de viribus et motibus fluidorum commentarii,* 1738

### Periodicals

"De sonis multifariis quos laminae elasticae diversimode edunt disquisitiones mechanico–geometricae experimentis acusticis illustratae et confirmatae." *Comentarii Academiae Scientarum Imperialis Petropolitanae* 12 (1740 [1750]).

"Mémoires sur les vibrations des cordes d'une épaisseur inégale." *Mémoires des l'Académie royale des science et belle lettres, Berlin* (1765[1767]).

"Oratio physiologica de vita." *Verhandlungen der Naturforschenden Gesellschaft Basel* 52 (1940–1941): 189–266.

"Remarques sur le principe de la conservation des forces vives pris dans un sens général." *Mémoires de l'Académie royale des sciences et belle lettres, Berlin* (1748[1750]).

## FURTHER READING

Boyer, Carl B. *A History of Mathematics.* Second edition. Revised by Uta C. Merzbach. New York: Wiley, 1991.

Glashow, Sheldon. *From Alchemy to Quarks.* Pacific Grove, CA: Brooks/Cole, 1994.

Kline, Morris. *Mathematics in Western Culture.* London: Oxford University Press, 1972.

*Jakob Bernoulli*

Straub, Hans. "Daniel Bernoulli," in *Dictionary of Scientific Biography.* Volume II. Edited by Charles Coulston Gillipsie. New York: Charles Scribner's Sons, 1970, pp. 36–46.

Tilley, Donald E. and Walter Thumm. *College Physics: A Text with Applications to the Life Sciences.* Menlo Park, CA: Cummings, 1971

Wolf, A. *A History of Science, Technology, and Philosophy in the Eighteenth Century.* Second revised edition. London: George Allen & Unwin, 1952.

*—Sketch by Zoran Minderovic*

# Jakob Bernoulli
## 1645–1705
### Swiss probabilist

Jakob Bernoulli, also known as Jacques, James, or Jakob I—to avoid any confusion with Jakob Bernoulli II—is one of the great names of 17th–century mathematics as well as the first member of the prodigiously mathematical Bernoulli family to attain international fame. Originally from the Spanish Netherlands, the Bernoullis moved to Basel, Switzerland, in 1583, to escape Spanish oppression; the first prominent Bernoulli was Nikolaus I, Jakob's father. A contemporary of the great German philosopher and mathematician **Gottfried Wilhelm von Leibniz**, with whom he maintained a correspondence, Bernoulli is known for his extraordinary contributions to calculus and the theory of probability.

Bernoulli earned a degree in theology in 1676, having studied mathematics and astronomy against his father's wishes. Employed as a tutor in Geneva in 1676, he later spent two years in France, where he studied the works of **René Descartes**. In 1681, Bernoulli traveled to the Netherlands and England, where he met Robert Boyle. During this period, Bernoulli wrote on a variety on scientific topics, including comets and gravity. Following his return to Basel 1682, Bernoulli founded a school for science and mathematics, where he lectured and conducted experiments in the field of mechanics. He also wrote articles for the two preeminent European scientific journals, *Journal des sçavans* and *Acta eruditorum*. In 1687, Bernoulli was named professor of mathematics at the University of Basel, a position he held until his death. Owing to his passion for learning, Bernoulli carefully studied the works of predecessors, such as Descartes, and contemporaries, such as Leibniz.

## Masters and Develops Calculus

Leibniz's first discussion of the differential calculus, a six–page article, appeared in *Acta eruditorum* in 1684 (the presentation of the integral calculus was published two years later). While the term *calculus* refers to the process of calculation in general, Leibniz's calculus dealt with infinitesimals, quantities smaller than any definable finite quantity but larger than zero, and was therefore called infinitesimal calculus. It should be pointed out, however, that the term *infinitesimal* is no longer used in mathematical terminology, infinitesimal quantities being instead named *limit values*. Therefore, when mathematicians refer to calculus, the predicate *infinitesimal* is implied.

Calculus encompasses four distinct types. *Differential calculus* calculates derivatives. *Integral calculus* is the reverse of differential calculus—in other words, we use this type of calculus to determine a function when its derivative is known. *Calculus of variation* is used to find a function for which a given integral assumes a maximal or a minimal value. *Differential equations* are equations containing derivatives. Misunderstood by most of his colleagues, Leibniz's discovery nevertheless attracted a small following of mathematicians who realized the tremendous analytical power of calculus. Among Leibniz's followers, Bernoulli was among the first who completely grasped the essence of calculus, and he proceeded, in numerous contributions to *Acta eruditorum*, to develop the foundations of calculus.

In 1689, Bernoulli formulated the famous "Bernoulli inequality"—$(1 + x)^n > 1 + nx$, where $x$ is real, $x > -1$, and $x \neq 0$, and $n > 1$—which had in fact already been presented in 1670 by **Isaac Barrow** in his *Lectiones geometriae*. Bernoulli also provided solutions for several famous mathematical problems. For example, he solved the catenary equation. When a flexible, non–elastic cable is suspended from two fixed points, the shape it assumes as a result of gravity is a catenary. Bernoulli formulated an equation to refute the traditional hypothesis among mathematicians that the catenary is a parabola. In his investigations of the isochrone, a plane curve enabling an object to fall with uniform velocity, Bernoulli found an equation which demonstrated that the necessary curve is a semicubical parabola. In fact, while working on the isochrone problem, Bernoulli used the term *integral,* in an 1690 article, to denote the inverse of the differential calculus. Leibniz later agreed that *calculus integralis* was a more concise term than the original *calculus summatorius.*

## Studies Curves

Bernoulli was among several mathematicians, including his younger brother **Johann Bernoulli** and Leibniz, who worked on the brachistochrone (the term was derived from the Greek words *brachystos*, meaning *the shortest* and *chronos*, meaning *time*) problem. Essentially, the problem challenged mathematicians to find a curve of quickest descent between two given points $A$ and $B$, assuming that $B$ does not lie right beneath $A$. While his brother correctly assumed that the required curve is a cycloid but offered an incorrect proof, Bernoulli provided the correct proof. The competition between the two brothers became so intense that Johann appropriated the brachistochrone solution as his own, which was just one episode in their bitter struggle for preeminence in mathematics.

Much of Bernoulli's work was devoted to the study of curves. The *lemniscate*, or figure eight curve (from the Latin term *lemniscus*, meaning *ribbon*), was named after him. However, he was totally fascinated by the logarithmic spiral, which in nature can be seen in a cross section of the shell of a chambered nautilus. Bernoulli noticed that the logarithmic spiral has several unique properties, including self–similarity, which means that any portion, if scaled up or down, is congruent to other parts of the curve. In fact, Bernoulli was so taken by this spiral, also called *spira mirabilis*, or *wonderful spiral*, that he requested that it be engraved on his tombstone, along with the Latin inscription *Eadem mutata resurgo* ("Though changed, I arise again the same").

## Contributes to Probability Theory

Bernoulli's research on probability is documented in his treatise *Ars conjectandi* ("The Art of

Conjecture"), published posthumously in 1713. Building on earlier writings on the subject, including **Girolamo Cardano**'s *Liber de ludo aleae* ("On Casting the Die"), published in 1663, the correspondence between **Pierre de Fermat** and **Blaise Pascal**, and **Christiaan Huygens**' 1656 book *De ratiociniis in ludo aleae* ("On Reasoning in Games of Chance"), Bernoulli created a work which scholars consider the substantial book on probability. In his book, to which he attached a treatise on infinite series (*Tractatus de seriebus infinitis*), Bernoulli presented his famous theorem which Siméon–Denis Poisson later named the "Law of Large Numbers." This law states that if a very large number of independent trials are made, then the observed proportion of successes for an event will, with probability close to 1, be very close to the theoretical probability of success for that event on each individual trial. Unfortunately, Bernoulli's treatise ends with his theorem, which he had hoped to use as the foundation of his project to apply the calculus of probability to a variety of fields, including demographics, politics, and economics.

## SELECTED WRITINGS BY BERNOULLI

*Ars conjectandi*, 1713
*Opera*, 1744

## FURTHER READING

Boyer, Carl B. *A History of Mathematics.* Second edition. Revised by Uta C. Merzbach. New York: Wiley, 1991.
Burton, David M. *Burton's History of Mathematics: An Introduction.* Third edition. Dubuque, IA: Wm. C. Brown, 1995.
Gullberg, Jan. *Mathematics: From the Birth of Numbers.* New York: W. W. Norton, 1997.
Hofmann, J.E. "Jakob (Jacques) Bernoulli," in *Dictionary of Scientific Biography.* Volume II. Edited by Charles Coulston Gillispie. New York: Charles Scribner's Sons, 1970, pp. 46–51.

—*Sketch by Zoran Minderovic*

# Johann Bernoulli
## 1667–1748
### Swiss infinitesimal mathematician

Johann Bernoulli (also known as Jean Bernoulli) was born in Basel, Switzerland. He studied physics

*Johann Bernoulli*

and medicine and wrote his doctoral dissertation on medicine. However, his primary interest was mathematics, and Bernoulli studied mathematics privately with his brother, **Jakob Bernoulli**, who was a professor of mathematics at the University of Basel. While still formally a medical student, Bernoulli became passionately involved in the emerging field of infinitesimal calculus, joining, along with Jakob, the select group of European mathematicians who fully grasped calculus. His first notable accomplishment in the field of mathematics is the solution to the catenary problem, which was posed by his brother Jakob. In fact, Bernoulli's doctoral dissertation, "De motu musculorum," although dealing with muscle contractions, is a mathematical treatise, written under the influence of the Italian mathematician and physiologist Giovanni Alfonso Borelli.

### Studies, Masters, and Teaches Infinitesimal Calculus

Already regarded as a member of the European mathematical elite, Bernoulli traveled to Paris in 1691, quickly gaining access to the circle of France's elite scientists and philosophers, who were particularly impressed by his development of a formula for the radius of curvature of a curve, actually first given by Jakob. In Paris, Bernoulli became known as a partisan and representative of **Gottfried Wilhelm von Leibniz**'s infinitesimal calculus. In 1692, he began instructing a young mathematician, the Marquis Guillaume–François–Antoine de L'Hospital, in calculus. In addition to hiring Bernoulli as a teacher, L'Hospital also bought his mathematical discoveries. The method known as "L'Hospital's Rule," for example, was among the discoveries that L'Hospital bought from his teacher. L'Hospital's Rule enabled mathematicians to determine the limiting value of a fraction in which both numerator and denominator tend toward zero. This idea, as well as other mathematical insights provided by Bernoulli, was eventually included in L'Hospital's book, *Analyse des infiniment petits* (1696), which was the first, and for almost a century the principal textbook of differential calculus. While the principal ideas used by L'Hospital came from Bernoulli, evidence suggests that the Marquis, a competent mathematician in his own right, contributed at least some ideas, such as the rectification of the logarithmic curve. That same year, in an anonymous contribution to the *Journal des sçavans*, Bernoulli used calculus to solve "Debeaune's problem." Posed by the French mathematician Florimond Debeaune, the problem consisted in determining a curve from a property of its tangent. In 1693, Bernoulli began a fruitful correspondence with Leibniz, vehemently defending the German philosopher as the true discoverer of calculus against any claims put forth by **Isaac Newton**'s partisans. During this period, Bernoulli also wrote extensively, publishing articles in *Acta eruditorum* and *Journal des sçavans*, Europe's leading scientific journals.

### Accepts Professorship in Groningen

In 1695, thanks to a recomendation by **Christiaan Huygens**, Bernoulli obtained the chair of mathematics at the University of Groningen in The Netherlands. He eagerly accepted the position, knowing that a professorship at Basel was out of the question as long as his brother Jakob was a faculty member. Bernoulli felt intense animosity toward his brother, who insisted on portraying himself as the more competent mathematician. It is true that Bernoulli learned much from his older brother; however, the student had become his master's equal.

Wishing to challenge his colleagues, Bernoulli posed the famous brachistochrone problem in the June 1696 issue of *Acta Eruditorum*, also sending it, with a six–month deadline, to a select group of mathematicians. In essence, the challenge was to find the curve ensuring the quickest descent of an object between two points at different altitudes (but not along a vertical line). Leibniz, who solved the problem on the day he received it, correctly predicted that only five people would solve the problem. Indeed, only Leibniz, Newton, the Bernoulli brothers, and L'Hospital found the solution—the cycloid curve.

### Formulates Calculus of Variations and Exponential Calculus

Bernoulli was initially unable to provide a complete solution to the isoperimetric problem (involving the comparison of different polygons with equal perimeters). It was Jakob, realizing that the isoperimetric problem is essentially variational, who produced the solution, much to his brother's chagrin. However, Bernoulli used his brother's insight to further differential geometry, which uses differential and integral calculus, by discovering the variational properties of geodesic lines—curves describing the shortest distance between two points on a convex surface. As Morris Kline explains in his book *Mathematical Thought from Ancient to Modern Times*: "[Bernoulli] wrote to Leibnitz in 1698 to point out that the osculating plane at any given point of a geodesic is perpendicular to the surface at that point." The osculating plane is determined by the osculating circle, and a circle is determined by three points. Bernoulli published a complete solution of the isoperimetric problem in 1718, laying the foundations of the calculus of variation. In addition, Bernoulli studied exponential curves (curves whose formula is an exponential function, or an equation in which the unknown is an exponent), eventually establishing the basis for differentiating and integrating exponential functions.

### Accepts Professorship at Basel

When Jakob died in 1705, Bernoulli took over his position as professor of mathematics at the University of Basel. Having attained a more prominent position in the academic world, Bernoulli continued his efforts as a champion of Leibniz, who had no academic appointment. He criticized Newton's approach to calculus, focusing on the method of fluxions. While Newton viewed variable quantities as generated by the continuous motion of points, lines, and planes, actually calling them *fluents*, and defined their derivatives as *fluxions*, or rates of changes, Leibniz's system, which eventually prevailed, used the concept of differentials.

### Makes Important Contributions to Mechanics

Bernoulli, whose authority as a mathematician (especially after Newton's death in 1727) seemed unrivaled, also furthered the field of mechanics. At this point in Bernoulli's career, there were two contending views in mechanics regarding the conservation of energy. While the Cartesians believed in the conservation of momentum, according to the Leibnizians, what was conserved was the *vis viva*, or living force. Although the Cartesian explanation was not incorrect, the Leibnizian concept prevailed, eventually having its original term replaced by the modern *kinetic energy*, defined as the energy a body possesses by virtue of its motion. As a Leibnizian, Bernoulli ardently supported the *vis viva*, fully accepting the idea put forth by Leibniz's student Christian Wolff, who postulated the universal validity of the conservation of the living force. Agreeing with Wolff, Bernoulli identified the living force as one of the fundamental principles in mechanics in his essay "De vera notione virium vivarum" in 1735. Scholars believe that Bernoulli may have been the first to realize the importance of the principle of conservation.

## SELECTED WRITINGS BY BERNOULLI

*Commercium philosophicum G. Leibnitii et Joh. Bernoullii,* 1645
*De effervescentia et fermentatione,* 1690
*De motu musculorum,* 1694
*Opera Johannis Bernoulli,* 1742
*Théorie de la maneouvre des vaisseaux,* 1714

## FURTHER READING

Boyer, Carl. B. *A History of Mathematics.* Second edition. Revised by Uta C. Merzbach. New York: Wiley, 1991.
Fellmann, E.A., and J.O. Fleckenstein. "Johann (Jean) Bernoulli," in *Dictionary of Scientific Biography.* Volume II. Edited by Charles Coulston Gillispie. New York: Charles Scribner's Sons, 1970, pp. 51–55.
Gullberg, Jan. *Mathematics: From the Birth of Numbers.* New York: W. W. Norton, 1997.
Kline, Morris. *Mathematical Thought from Ancient to Modern Times.* New York: Oxford University Press, 1972.

—*Sketch by Zoran Minderovic*

# Dorothy Lewis Bernstein
## 1914-
### American applied mathematician

Dorothy Lewis Bernstein is a distinguished mathematician and educator in the fields of applied mathematics, statistics, and computer programming. Her research focused on the Laplace transform, a mathematical function named after the French mathematician **Pierre Simon Laplace**. The Laplace transform is used in the solution of partial differential equations (equations that contain the partial derivatives of functions of two or more variables) and has been widely applied in the 20th century in conjunction with operational calculus. Bernstein was a pioneer in incorporating applied mathematics and computer science into the undergraduate mathematics curriculum. In 1979, she became the first woman

president of the Mathematical Association of America, a national association concerned with college mathematics. Bernstein was born in Chicago on April 11, 1914, to Jacob and Tillie Bernstein, who were Russian immigrants. The family lived in Milwaukee during Bernstein's youth. In 1930, Bernstein began her studies at the University of Wisconsin at Madison. During her junior and senior years, she studied mathematics under an independent curriculum. In 1934, based on an oral examination and her thesis on the complex roots of polynomials (mathematical expressions containing certain algebraic terms), she received both a bachelor's (summa cum laude) and a master's degree in mathematics.

After another year at Madison as a teaching fellow, Bernstein received a scholarship to the doctoral program in mathematics at Brown University in Rhode Island. As Ann Moskol indicates in *Women of Mathematics,* Bernstein's experiences at Brown reflect various forms of discrimination. Bernstein's graduate teaching was restricted to only three female students. When she sought advice on finding a teaching position, the graduate school dean advised her not to apply in the South because she was Jewish or in the West because of her gender. Bernstein underwent an unusually arduous doctoral examination, which her advisor later acknowledged was due to her gender and to her midwestern university credentials.

Nonetheless, Bernstein independently secured a teaching position at Mount Holyoke College in Massachusetts. Bernstein taught at Mount Holyoke from 1937 to 1940, completing her doctorate from Brown in 1939 with a thesis related to the Laplace transform. In 1941, Bernstein returned to Madison as an instructor. In the summer of 1942, she was a research associate at the University of California at Berkeley under the Polish mathematician and statistician Jerzy Neyman. In 1943, Bernstein took an instructorship at the University of Rochester in New York, where she became an assistant professor in 1946.

### Studies Computer Applications of Mathematics

At Rochester, Bernstein's research was directed toward exploiting the computational potential of digital computers (their ability to perform complex mathematical operations on large amounts of data at high speeds) in solving partial differential equations. This research, intended for military application and conducted in affiliation with the Office of Naval Research, led to the publication of Bernstein's *Existence Theorems in Partial Differential Equations* in 1950. In 1951, Bernstein was a member of the Institute for Advanced Study in Princeton, New Jersey. Bernstein became an associate professor at Rochester in 1951, and a full professor in 1957. From 1957 to 1958, she was a visiting professor at the University of California in Los Angeles.

In 1959, Bernstein assumed a professorship at Goucher College in Baltimore, Maryland, where she chaired the mathematics department from 1960 to 1970 and directed the computer center from 1961 to 1967. She served on the board of governors of the Mathematical Association of America, the professional association with which she was most closely involved, from 1965 to 1968. As a department administrator at Goucher, Bernstein brought applied mathematics and the emerging field of computer science into the undergraduate mathematics curriculum, and integrated computer programming into her own courses in statistics. Moskol notes that Bernstein "believed that applied mathematics not only made material more relevant to students, but it also motivated them to understand the axioms and theorems of pure mathematics, which could then be used in applied problems." Bernstein's practical vein was further indicated by the internship program she established for Goucher's math majors.

During her tenure at Goucher, Bernstein was also involved through the National Science Foundation in promoting computer programming instruction and the use of computers in advanced mathematics courses at area high schools. She helped establish the Maryland Association for Educational Use of Computers in 1972 and served on its governing board from 1972 to 1975. Bernstein was vice-president of the Mathematical Association of America from 1972 to 1974 and president from 1979 to 1981. She also served on the Joint Projects Committee and the Joint Committee on Women of the Mathematical Association of America, the American Mathematical Society, and the Society of Industrial and Applied Mathematics, and on the editorial board of the *Two Year College Mathematics Journal.* Bernstein retired from Goucher College in 1979.

## SELECTED WRITINGS BY BERNSTEIN

### Books

*Existence Theorems in Partial Differential Equations,* 1950

### Periodicals

"The Double Laplace Integral." *Duke Mathematical Journal* 8 (1941): 460–496.
(With Geraldine A. Coon) "Some Properties of the Double Laplace Transformation." *Transactions of the American Mathematical Society* 74 (1953): 135–70.
(With Coon) "Some General Formulas for Double Laplace Transformations." *Proceedings of the American Mathematical Society* 14 (1963): 52–59.

(With Coon) "On the Zeros of a Class of Exponential Polynomials." *Journal of Mathematical Analysis and Applications* 11 (1965): 205–212.
"The Role of Applications in Pure Mathematics." *American Mathematical Monthly* 86 (1979): 245–253.
"Women Mathematicians before 1950." *Association for Women in Mathematics Newsletter* (July-August 1979): 9–11.

## FURTHER READING

### Books

Moskol, Ann. "Dorothy Lewis Bernstein." In *Women of Mathematics,* edited by Louise S. Grinstein and Paul J. Campbell. Westport, CT: Greenwood Press, 1987, pp. 17–20.

### Periodicals

Coon, Geraldine A. "Coon on Bernstein." *Goucher Quarterly* (Fall 1979): 16–17.

### Other

"Dorothy Lewis Bernstein."*Biographies of Women Mathematicians.* http://www.scottlan.edu/lriddle/women/chronol.htm (July 22, 1997).

—*Sketch by Nicholas Pease*

# Friedrich Wilhelm Bessel
## 1784–1846
### German astronomer and mathematician

Unlike many men of achievement, Friedrich Wilhelm Bessel did little to give an indication of his forthcoming accomplishments during his early and limited education. He stumbled upon his genius by way of accepting an apprenticeship as a bookkeeper with a merchant house when he was 15 years old. Then, because Bessel wished to further his career and enter the world of foreign trade, he studied navigation as well as geography and foreign languages. This study of navigation soon blossomed into the study of astronomy and became an avocation that eventually led to a life's work affecting not only mathematics, but profoundly changing the world of astronomy. In his own lifetime, Bessel's discoveries were applauded as "inaugurating a new era of practical astronomy" and hailed as the "beginning of modern astronomy."

Bessel was born in Minden, Germany, on July 22, 1784. Not much is recorded about his family,

*Friedrich Wilhelm Bessel*

except that his father was a government employee whose salary barely kept the nine children fed, and his mother came from Rheme, where her father was a minister. In 1812, Bessel married Johanna Hagen. The couple had five children, two boys and three girls, and the marriage was said to be successful except for the loss of both sons when they were quite young. Bessel was fond of taking walks and, despite his small stature and less than robust health, enjoyed hunting. His only other form of relaxation was his correspondence with fellow astronomers and mathematicians.

### Astronomical Discoveries

Bessel's discoveries and achievements were not based on any sort of formal education; instead, he was almost self–taught. When Bessel was only 20 years old, he studied the 1607 observations of **Thomas Harriot** to calculate the orbit of Halley's Comet. After writing a paper on his calculations, Bessel sent it to astronomer Wilhelm Olbers, who was so impressed that he arranged for its publication in *Monatliche Corresondenz*. Olbers immediately recognized genius when he saw it and further prevailed on officials at the Lilienthal Observatory to make Bessel an assistant. During his four years at Lilienthal, Bessel attracted the attention of Friedrich Wilhelm III of Prussia, who was appointed him director of the Royal Observatory at Königsburg in 1810, where he spent the next 30 years.

While at the observatory, Bessel improved upon the work of past astronomers and paved the way for further discoveries for future cosmologists. Since the time of **Isaac Newton**, men of science had been attempting to calculate the distance of stars, and until Bessel they had failed. This failure can be attributed to two things—the lack of proper instruments and the lack of a proper method for going about the measurement. In 1829, telescope maker Joseph Fraunhofer of Austria designed the equipment and Bessel designed the method that would finally provide the answer.

Bessel reasoned that the easiest and most reliable way to measure the distance of a star was to measure its annual parallax. This parallax, or shift in the apparent position of an object resulting from the Earth's orbit around the sun, had to be painstakingly measured every night for a year in order for Bessel's conjectures to be taken seriously by his peers. Using the speculations of astronomer James Bradley and his own intuition, Bessel accomplished what Bradley and all who had come before him could not and became the first person to accurately determine the distance of a star from Earth. Bessel's calculations vastly expanded astronomers' abstractions of space, and turned the conception that the Earth was part of a solar system into the realization that it is actually part of a universe. Bessel astounded his fellow astronomers when he proved that one of the nearest stars to Earth, 61 Cygni, was more than 60 trillion miles away. Bessel was also the first to use the term "light years" as a way of vividly explaining this distance. Traveling at the speed of light, 186,000 miles per second, it would take 10.3 years to reach 61 Cygni. Using his newly found method of computation, Bessel further compiled a catalog of the position of 75,000 stars.

In a paper written in 1840, Bessel recounted that Uranus displayed certain small but noticeable "irregularities" in its orbit. The planet, it seems, alternately slowed down and then ran ahead of its expected positions. Bessel surmised that this could only be due to the influence of an as yet unknown planet laying somewhere beyond it. Although Bessel did not live to see his hypothesis confirmed, that planet later proved to be Neptune. In 1842, Bessel calculated the mass of Jupiter by studying the orbital period of each of its major moons and established that while it had 388 times the mass of Earth, its overall density was only 1.35 times the density of water, bringing to light that this mammoth planet was essentially very light for its size.

The same sort of "irregularities" bothered Bessel when he studied the luminous stars Sirius and Procyon. Bessel proposed that the slight but significant erratic behavior of the two stars must be caused by "invisible companions." After Bessel's death, more powerful telescopes provided the evidence to prove him correct, and we now know those "invisible companions" as Sirius B and Procyon B.

## Mathematical Contributions

Most of Bessel's gifts to the realm of mathematics materialized as a result of his astronomical breakthroughs, but that does not lessen the importance of his legacy to mathematical functions still used extensively today in such fields as geology, physics, engineering, and of course applied mathematics. These Bessel functions, also known as cylinder functions, were first devised to unlock the mystery of "planetary perturbation." This deviation of a planet from its regular orbit, usually caused by the presence of one or more bodies acting upon it, could be accurately calculated and anticipated due to Bessel's functions and later were used as solutions for a differential equation—intrinsic in the investigation of numerous problems in mathematical physics.

In addition to his duties at the Royal Observatory, Bessel was required to spend a considerable amount of time surveying. Like everything else he did, this task brought him acclaim and notoriety. Using his own idea for an improved measuring device, he commissioned the construction of an instrument that would accurately gauge base lines. Then, by using the methods of **Karl Friedrich Gauss**, he developed a new system of triangulation. These two accomplishments enabled Bessel to paint a mathematically accurate and, at that time, an astounding picture of the form and magnitude of the Earth. Bessel's work on triangulations also resulted in the eventual establishment of the International Bureau of Weights and Measures.

## FURTHER READING

### Books

*Abhandlungen von Friedrich Wilhelm Bessel.* 3 volumes. Edited by Rudolf Englemann, 1875

Asimov, Isaac. *Asimov's Biographical Encyclopedia of Science and Technology.* Garden City, NY: Doubleday & Company, Inc., 1972, pp. 265–266.

Fricke, Walter. "Friedrich Wilhelm Bessel," in *Dictionary of Scientific Biography.* Volume II. Edited by Charles Coulston Gillispie. New York: Charles Scribners' Sons, 1970, pp. 97–102.

Ley, Willy. *Watchers of the Skies: An Informal History of Astronomy from Babylon to the Space Age.* 1963.

MacPherson, Hector. *Makers of Astronomy.* 1933.

Williams, Henry Smith. *The Great Astronomers.* 1930.

—*Sketch by Kelley Reynolds Jacquez*

# Étienne Bézout
## 1739–1783
### French algebraist and author

Étienne Bézout is best known as the author of one of the most widely used mathematics textbooks of his era, the six–volume *Cours de mathematique*. He is renowned for his work dealing with the usage of determinants in algebraic elimination. Born in Nemours, France, on March 31, 1739, Bézout was the second son of Pierre and Helene Jeanne Filz Bézout. His family was well established and respected in the region; his grandfather and father both served as district magistrates. Bézout, however, preferred numbers to politics and convinced his father to permit him to study mathematics instead of law. He was greatly influenced by the work of **Leonhard Euler**, and his abilities were quickly recognized by the Académie Royale des Sciences, which elected him an adjunct member at the age of 19 and a full member at age 29.

Little of Bézout's early life is documented, but it is recorded that he married at a young age. From the few reports available the marriage was a happy one, but it seems to have had a significant impact on his career. In 1763, at the age of 24, Bézout accepted a position as a teacher and examiner for aspiring naval officers who wished to serve the Duc de Choiseul in the Gardes du Pavillon et de la Marine. As a young father, his decision to accept the appointment was undoubtedly influenced by a need for a steady income.

His work as a teacher helped shape his approach to mathematics in general. Teaching mathematics to aspiring military officers (in 1768, he took on additional teaching duties for artillery officers) demanded a practical approach. In his courses, Bézout avoided detailed theory and shied away from intimidating terms. He introduced practical geometry before algebra to make his students more comfortable with the notion of mathematical reasoning and labored to make his lectures easy to understand.

Between 1764 and 1769 Bézout compiled the material he used in his lectures and published his six–volume *Cours de mathematiques: A l'usage des Gardes du Pavillon est de la Marine.* The work was criticized by some scholars as lacking academic rigor, but was received by educators with enthusiasm. *Cours de mathematiques* sold briskly and went into numerous reprints. It was translated into English in the early 1800s for use in schools in the United States and influenced American teaching methods well into the 19th century. John Farrar, one of the translators of *Cours de mathematiques*, based his Harvard University calculus course on Bézout's textbooks, and mathematics courses offered at West Point Military Academy were also based on Bézout's work. If academicians felt disdain for Bézout's straightforward approach, others owed a debt to him for his compilation of the works of such mathematicians as Euler and **Jean le Rond d'Alembert** in an easy–to–study format. Without such texts as Bézout's, general knowledge of important mathematical advances would have disseminated much more slowly.

Bézout's teaching duties left him little time for research. To make the most of the time he had, Bézout limited himself to the study of the theory equations; his early work probed integration, but he narrowed his investigations even further, and by 1762 he was absorbed by the questions posed by algebraic equations.

### Striving for Simplicity

Bézout focused on the use of determinants in algebraic elimination. He developed artificial rules for solving simultaneous linear equations in unknowns. He is noted for developing an extension of this concept into a system of equations in one or more unknowns in which it is necessary to find the condition on coefficients for the equations to have a common solution.

In his paper *Sur plusieurs classes d'equations*, published in 1762, Bézout shifted his emphasis from solving the $n$th degree equation to elimination theory. Given $n$ equations in n unknowns, Bézout searched for a resultant equation with the fewest possible number extraneous roots. In 1764, Bézout published *Sur le degre de equations resultantes de l'evanouissement des inconnues*. In this paper, he discussed Euler's methods for finding the equation that resulted from two equations in two unknowns, but rejected Euler's method as clumsy when applied to equations of high degree. Bézout identified determinants by listing permutation of coefficients in a table and used the table to determine if simultaneous linear equations could be solved.

In 1779, Bézout published *Theorie des ezuations algebriques*, his most important work on elimination theory, which includes the statement and proof of the following: The degree of the final equation resulting from any number of complete equations in the same number of unknowns and of any degrees, is equal to the product of the degrees of the equations. He was also the first to prove that two algebraic curves of degrees $m$ and $n$ intersect in general in $mn$ points. Although this paper focused on algebraic theory, Bézout was compelled to point out his theory's geometric interpretation—a legacy of the practical concerns that dominated much of his professional life.

Some mathematical historians regard Bézout as a minor writer and mathematician. He was a product of his time and country; in 18th century France, most mathematicians were not affiliated with universities but with the Church or the military and the latter, at least, wanted practical applications. Bézout had practical concerns in his private life, as well. His family was well–regarded, but not wealthy, and his need to support his wife and children made his military teaching post his most professional priority. While Bézout's scholarly output was not large, his influence was significant. He instructed both **Gaspard Monge** and **Lazare Carnot**, and his textbooks had a profound influence on mathematics education in France and the United States.

Bézout died in 1783 in Basses–Loges, France, six years before the French Revolution rewove the country's social, academic, and intellectual fabric.

## SELECTED WRITINGS BY BÉZOUT

### Books

*Cours de mathematiques a l'usage des Gardes du Pavillon et de la Marine,* 6 volumes, 1764–69
*Theorie generale des equations algebriques,* 1779

### Papers

"Sur plusieurs classes d'equations de tous les degres qui admettent unesolution algebrique," in *Memoires de l'Academie royale des sciences* (1762): 17–52.
"Sur le degre des equations resultantes de l'evanouissement des inconnues," in *Memoires de l'Academie royale des sciences* (1764): 288–338.
"Sur la resolution des equations de tous les degres," in *Memoires de l'Academie royale des sciences* (1765): 533–552.

## FURTHER READING

Boyer, Carl. B. *A History of Mathematics.* Second edition. Revised by Uta C. Merzbach. New York: John Wiley & Sons, 1989. pp. 516–17.
Fang, J. *Mathematicians From Antiquity to Today.* Hauppauge, N.Y.: Paideia Press, 1972, p. 207.
Grabiner, Judith V. "Étienne Bézout," in *Dictionary of Scientific Biography.* Volume III. Edited by Charles Coulston Gillispie. New York: Charles Scribner's Sons, 1970, pp.111–14.
Kline, Morris. *Mathematical Thought from Ancient to Modern Times.* New York: Oxford University Press, 1972, pp. 552–53.
Smith, David E. *History of Mathematics.* New York: Dover Publications, 1951, pp. 481–82.

*—Sketch by A. Mullig*

*Garrett Birkhoff*

# Garrett Birkhoff
## 1911–1996
### American algebraist and applied mathematician

Garrett Birkhoff's most important contribution to modern mathematics is his theory of lattices, which helped simplify abstract algebraic concepts. His work in this field influenced the development of quantum theory.

Born January 19, 1911, in Princeton, New Jersey, Garrett Birkhoff was the son of noted mathematician **George David Birkhoff**, considered by many the leading mathematician of his day for his work in mathematical analysis and metric transitivity. Birkhoff was home–schooled until age eight. He then attended a public grammar school and Brown and Nichols, a private high school. His mother, a librarian, seemed to be less of an influence than his famous father; in a 1982 interview with G. L. Alexanderson and Carroll Wilde, Birkhoff said, "I think (my mother) was more social than intellectual, really . . . my father was a stimulating person at all ages. He took my sister and me on excursions, particularly when we were very young, and made life interesting for us somewhat later by telling us exciting stories, having opinions on controversial questions. . . . I sometimes think I was trained to be precocious."

Birkhoff's youth was comfortable, if not privileged. In reminiscing about his teens, he mentions that his **Euclid**ean geometry was "a little sloppy" because he joined his parents in Europe in February of his third year of high school. In order to meet language requirements for graduation, Birkhoff remained in high school for a fifth year. "My parents suggested I take my college board exams early. After passing them, I had a wonderful fifth year at Lake Placid where I did much skiing . . . while my family went around the world."

At age 17, Birkhoff enrolled at Harvard University. His father's influence was pervasive; the summer before entering college, Birkhoff's father tutored him in calculus. During his college career, several of his instructors completed their doctoral work under his father's supervision. "My Harvard undergraduate training was almost an inside job," Birkhoff said. It was not until after he graduated from Harvard in 1932 and won a Henry Fellowship to Cambridge University that Birkhoff switched his focus from mathematical physics and quantum mechanics—an interest encouraged by his father—to abstract algebra.

### Foreign Studies Spark New Perspectives

At Cambridge, Birkhoff discovered the work of B. L. van der Waerden and Constantin Caratheordory. His research supervisor, Phillip Hall, had already achieved some recognition for his work in group theory. Birkhoff gradually developed his theory of lattices, which explored the relationship of subdomains within algebraic domains and how operations of union and intersection are defined for such subgroups.

Birkhoff published his first major work on the theory of abstract integrals in 1935. He completed much of his work before discovering **Julius Dedekind**, who in 1897 developed a system to define a special type of lattice. Dedekind's work was largely ignored until Birkhoff and mathematicians of his generation pushed the boundaries of algebraic theory. "I was lucky to have gone beyond Dedekind before I discovered his work," Birkhoff said. "It would have been quite discouraging if I had discovered all my results anticipated by Dedekind."

Birkhoff completed his Cambridge fellowship in 1933; he carried the enthusiasm and new ideas absorbed during his time abroad with him to Munich in July 1933, where he continued to pursue his research in group theory and met Caratheoedory (who invited him to family tea), who encouraged him. Birkhoff continued his work in the United States. In 1936, he became a mathematics instructor at Harvard University, and father and son teaching appointments overlapped until 1944, when George Birkhoff died. Before his father's death, however, Birkhoff firmly established his international reputation with the publication of *Lattice Theory* in 1940.

## Collaborations Prove to be Productive

Birkhoff attracted more attention when he teamed up with Saunders MacLane in 1941. A Benjamin Pierce instructor at Harvard from 1934 to 1936, MacLane returned to Harvard in 1938, which coincided with Birkhoff's first experience teaching "modern" algebra at Harvard. MacLane added a different perspective on teaching to Birkhoff's understanding of theory and research; their textbook, *Survey of Modern Algebra*, first published in 1941, quickly became the standard college text. It was reprinted in 1953 and 1965. Birkhoff and MacLane developed a new text, *Algebra*, which they published in 1967.

In the late 1960s, Birkhoff entered another collaborative agreement, this time with Thomas Bartee; the result was *Modern Applied Algebra*, published in 1970. One of the earliest works to connect algebraic concepts to practical problems, the book included finite groups, Boolean algebra, lattices, combinatorial analysis, and correlated algebraic coding theory with basic abstract principles.

## Pursuit of Practical Applications

Throughout his career, Birkhoff sought to link theory with applications. "I feel strongly that it is very dangerous for mathematics to detach itself from the rest of the world; to be part of the world around one is much healthier," Birkhoff stated. As a Harvard freshman, he wrote an essay on the bounce of a spinning tennis ball; during World War II, he contributed his talents to weapons systems such as the "proximity faze" (a device to determine target distance by timing the reflection of radio waves), and antitank charges. Of his work during that era, Birkhoff said simply, "I am proud of having contributed to the defeat of Hitler."

In the 1950s, Birkhoff served as a consultant to the Westinghouse Corporation, working with scientific computing and problems of vector lattices related to nuclear reactors. Birkhoff became an ardent proponent of the development and use of computers as tools to solve mathematical puzzles. "If you have a thousand equations in a thousand unknowns, you know there exists a solution, but how do you compute it? With it [computer programming] you can do for $5 what would have cost $10,000."

Birkhoff authored numerous works on applied mathematics, including Hydrodynamics (1950) and Jets, Wakes and Cavities (1957). Throughout his career, Birkhoff continued as a consultant to military and industrial institutions, including the Los Alamos Science Laboratory, the General Motors Corporation, and the Rand Corporation.

In the 1980s, Birkhoff remained active, working with the Naval Postgraduate School in Monterey, California, conducting research, studying fluid dynamics and teaching. A member of the National Academy of Sciences, the American Academy of Arts and Sciences, and numerous other academic organizations, Birkhoff was named George Putnam Professor of pure and applied mathematics at Harvard in 1969, a post he held until he retired from teaching in 1981. Birkhoff died November 22, 1996, at his home in Water Mill, New York

## SELECTED WRITINGS BY BIRKHOFF

### Books

*Lattice Theory*, 1940
(With Saunders MacLane) *Survey of Modern Algebra*, 1941
(With E. Zarantello) *Jets, Wakes & Cavities*, 1957
(With G. C. Rota) *Ordinary Differential Equations*, 1962
(With S. MacLane) *Algebra*, 1967
(With Thomas Bartee) *Modern Applied Algebra*, 1970
*Source Book in Classical Analysis*, 1973

## FURTHER READING

### Books

Albers, Donald J and G. L. Alexanderson, editors. *Mathematical People: Profiles and Interviews.* Cambridge, MA: Birkhauser Boston, Inc., 1985, pp. 3–15.
Bell, E. T. *The Development of Mathematics.* New York: McGraw–Hill Book Co., Inc., 1945, pp. 258–265.
Fang, J. *Mathematicians from Antiquity to Today.* Hauppauge, NY: Paideia Press, 1972, pp. 211–212.

### Periodicals

*The New York Times* (obituary). Nov. 28, 1996.
*Harvard Magazine* (obituary). New England Regional Edition, January/February 1997, p.76P.

—*Sketch by A. Mullig*

# George David Birkhoff
## 1884-1944
### American algebraist

George David Birkhoff's contributions as a theoretical mathematician, a teacher, and a member of the

*George David Birkhoff*

international scientific community rank him as one of the foremost mathematicians of the 20th century. He made extensive contributions to the area of differential equations and continued the work of the great French mathematician **Jules Henri Poincaré** on celestial mechanics. He is considered the founder of the modern theory of dynamical systems.

Born in Overisel, Michigan, on March 21, 1884, Birkhoff was the eldest of six children born to David Birkhoff, a physician, and Jane Gertrude Droppers. When Birkhoff was two years old, his family moved to Chicago, where he spent most of his childhood. From 1896 to 1902 Birkhoff studied at the Lewis Institute (now the Illinois Institute of Technology). Following a year at the University of Chicago as an undergraduate, Birkhoff transferred to Harvard University in 1903. In 1904, while still an undergraduate, he wrote his first mathematics paper, on number theory. He earned a bachelor's degree at Harvard in 1905 and a master's degree in 1906. Returning to the University of Chicago for his doctorate, Birkhoff wrote a dissertation on differential equations under the guidance of Eliakim Hastings Moore. He was awarded a doctorate *summa cum laude* in 1907.

### Leaves His Mark through His Students

Birkhoff taught mathematics at the University of Wisconsin from 1907 to 1909, when he took a position as assistant professor at Princeton University. He joined the faculty of Harvard University in 1912, teaching there until his death in 1944. Birkhoff, though not a great lecturer, was an inspiring teacher. Many of the influential American mathematicians of the mid-twentieth century, including Marston Morse and Marshall Stone, studied with Birkhoff at the doctoral or post-doctoral level. Six of his former students went on to become members of the National Academy of Sciences. From 1935 to 1939 Birkhoff also served as Dean of the Faculty of Arts and Science at Harvard.

The single most important influence on Birkhoff's mathematical research was that of Poincaré. The two never met, but Birkhoff studied Poincaré's work and adopted some of the problems in differential equations and celestial mechanics Poincaré left behind at his death in 1912. In 1913 Birkhoff first attracted international attention by proving a geometrical theorem that Poincaré had proposed but not proved in his last published paper. Birkhoff's accomplishment marked a special advance in solving the problem of three bodies. The three-body problem of celestial mechanics concerns trajectories and orbits of bodies moving in systems in such a way that each body affects the motion of the others.

### Advances Understanding of Dynamical Systems

From Poincaré's theorem, Birkhoff went on to consider the entire field of dynamical systems and made contributions that have become fundamental to this branch of mathematics. In his book *Dynamical Systems* (1927), Birkhoff wrote that "the final aim of the theory of motions of a dynamical system must be directed towards the qualitative determination of all possible types of motions and the interrelation of these motions." Using ideas developed by Poincaré, Birkhoff laid the foundations for the topological theory of dynamical systems by defining and classifying possible types of dynamic motions. Another signal achievement came in 1931, when Birkhoff offered further proof of the so-called ergodic theorem, which demonstrates the conditions needed for the behavior of a large dynamical system, such as a container of a gas, to reach equilibrium. This problem had baffled scientists for more than 50 years.

Birkhoff's journal articles and books reflect the breadth of his talent and the diversity of his interests. Among his published works is a basic geometry text that for many years formed the basis of high-school geometry curricula. Birkhoff also wrote extensively about the theory of relativity and quantum mechanics. Although his ideas in this field are not widely accepted, the mathematical tools that he developed for his approach play an important role in modern relativity theory. Birkhoff's other work encompasses number theory and point-set theory and the famous four-color problem, which is concerned with the possibility of coloring any map using only four colors.

In 1933, his life-long passion for art, music, and poetry led him to write *Aesthetic Measure,* in which he attempted to create a general mathematical theory of the fine arts, starting out from the Pythagorean notion that beauty is mathematical in nature. He later extended the theory to ethics. Birkhoff was also the editor of several mathematical journals, including the *Annals of Mathematics, Transactions of the American Mathematical Society,* and the *American Journal of Mathematics.*

### Recognized Worldwide for Mathematical Achievements

During his lifetime, Birkhoff received many honors and honorary degrees from universities worldwide. Among others, he was awarded the Querini-Stampalia prize of the Royal Institute of Science, Letters and Arts, Venice (1918); the Bôcher prize of the American Mathematical Society (1923) for his research in dynamics; the annual prize of the American Association for the Advancement of Science (1926); and the biennial prize of the Pontifical Academy of Sciences (1933) for his research on systems of differential equations. Birkhoff was elected to membership in the National Academy of Sciences (1918), the American Philosophical Society, and the American Academy of Arts and Sciences. He was made an officer of the French Legion of Honor in 1936, and was an honorary member of the Edinburgh Mathematical Society, the London Mathematical Society, the Peruvian Philosophic Society, and the Scientific Society of Argentina.

Birkhoff was fluent in French, the language of his famous mathematical predecessor, Poincaré, and presented several of his fundamental papers in that language. He traveled widely, promoting his belief in international fellowship among scientists. Because of his preeminence in research, he was able to represent mathematics in international scientific circles, and he played an important role in the creation of the mathematical institutes at Göttingen and Paris after World War I.

Birkhoff married Margaret Elizabeth Grafius of Chicago on September 2, 1908, and they had three children: Barbara, Garrett, and Rodney. Garrett went on to become a professor of mathematics at Harvard. Birkhoff died of a heart attack on November 12, 1944, in Cambridge, Massachusetts.

## SELECTED WRITINGS BY BIRKHOFF

*Relativity and Modern Physics,* 1923
*Dynamical Systems,* 1927
(With Ralph Beatley) *Basic Geometry,* 1941
*Collected Mathematical Papers,* 3 volumes, 1950

"Mathematics of Aesthetics," in *The World of Mathematics,* Volume 4. New York: Simon and Schuster, 1956, pp. 2185–2195.
"A Mathematical Approach to Ethics." In *The World of Mathematics.* Volume 4. New York: Simon and Schuster, 1956, pp. 2198–2208.

## FURTHER READING

### Books

Diner, S., D. Fargue, and G. Lochak, editors. *Dynamical Systems: A Renewal of Mechanism.* World Scientific, 1986.
Fang, J. *Mathematics From Antiquity to Today.* New York: Paideia Press, 1972, pp. 212–213.
Newman, James R. *The World of Mathematics.* New York: Simon and Schuster, 1956, pp. 2182–2208.
Turner, R., editor. *Thinkers of the Twentieth Century.* New York: St. James Press, 1987, pp. 79–81.

*—Sketch by Maureen L. Tan*

# David Harold Blackwell
## 1919-
### African-American statistician and educator

David Blackwell is a theoretical statistician noted for the rigor and clarity of his work. Most of his career has been dedicated to teaching and to exploring topics in Bayesian statistics, probability theory, game theory, set theory, dynamic programming, and information theory. Blackwell is a member of the National Academy of Sciences and the National Academy of Arts and Sciences and holds honorary doctorates from 12 universities, including Carnegie-Mellon, Yale, Harvard, Howard, and the National University of Lesotho. In 1986, Blackwell received the R. A. Fisher Award from the Committee of Presidents of Statistical Societies, in recognition of his career accomplishments.

Blackwell grew up in Centralia, Illinois, where he was born on April 24, 1919, to Grover and Mabel Johnson Blackwell. His brothers, J.W. and Joseph, and his sister, Elizabeth, were younger than he. His mother was a full-time homemaker, and his father was a hostler for the Illinois Central Railroad. Although two of the city's elementary schools were racially segregated, the one Blackwell attended was integrated.

*David Harold Blackwell*

Blackwell was intrigued with games like checkers, and wondered about such questions as whether the first player could always win. His interest in mathematical topics increased in high school. The mathematics club advisor would challenge members with problems from the *School Science and Mathematics* journal and submit their solutions; Blackwell was identified three times in the magazine as having solved a problem, and one of his solutions was published.

After graduating from high school at the age of 16, Blackwell entered the University of Illinois in 1935. After his freshman year, Blackwell became concerned that his father was borrowing money to send him to college, and he decided to support himself. His jobs included washing dishes, waiting tables, and cleaning the equipment in the entomology lab. By taking summer courses and proficiency exams, Blackwell graduated in three years, and continued on at the university to earn a master's degree and a doctorate. His dissertation on Markov chains led to his first publications in 1942 and 1945. After receiving his Ph.D. in mathematics in 1941, Blackwell spent a year as a Rosenwald Fellow at the Institute for Advanced Study (IAS) in Princeton, where he became acquainted with **John von Neumann.**

Blackwell then launched a job search by writing to each of the 105 Black colleges in the country. Although he was not aware of any overt racial discrimination directed at him, he simply assumed that his role would be teaching at a Black school. In later years, he would learn of some behind-the-scenes difficulties related to his race, including opposition to his appointment as an honorary faculty member at Princeton (which was customary for members of the Institute for Advanced Study). In 1942, Jerzy Neyman interviewed Blackwell for a possible position at the University of California at Berkeley; Neyman's support apparently did not prevail over others' prejudices, and no offer was made.

One of the first of three schools to offer Blackwell a position was Southern University in Baton Rouge, and he taught there for the 1942-1943 academic year. In the following year, Blackwell was an instructor at Clark College in Atlanta. In 1944, he joined the faculty of Howard University, which was the most prestigious employer of Black scholars in the country. He was promoted to full professor in 1947 and served as head of the Mathematics Department until 1954. Two days after Christmas in 1944, Blackwell married Ann Madison. Their family life proved to be a happy one, enriched by the presence of their three sons and five daughters.

The focus of Blackwell's mathematical interests shifted to statistics in 1945, when he heard Abe Girshick lecture on sequential analysis. He was intrigued by the presentation, and later contacted Girshick with what he thought was a counterexample to a theorem presented in the lecture. That contact resulted in an enduring friendship and fruitful collaboration. The two co-authored *Theory of Games and Statistical Decisions,* which was first published in 1954 and revised in 1980. Blackwell's first statistical paper, published in 1946, contained what he saw as his first original contribution to mathematics. He produced an elegant proof that extended an important equation to a weaker set of constraints.

During the summers of 1948-1950, Blackwell worked at the Rand Corporation. He and a few colleagues, including Girshick, became interested in the theory of duels. The initial condition concerned two people advancing toward each other, each holding a gun with one bullet; if one fires and misses, he is required to continue walking toward his opponent. The problem was how a dueler should decide the optimal time to shoot. After developing the theory of that situation, Blackwell proposed and investigated the more challenging case where each gun was silent, so a dueler does not know whether his opponent has fired unless he has been hit. Pursuing such topics earned Blackwell a reputation as a pioneer in the theory of duels.

In 1954, Blackwell accepted a professorship at the University of California at Berkeley. From 1956 to 1961, he served as chairman of the Department of Statistics. During the 1973-1975 academic years, Blackwell was director of the University of California

Study Center for the United Kingdom and Ireland. While in England, he was invited to give the Rouse Ball Lecture at the University of Cambridge. In 1989, Blackwell retired from the University of California, Berkeley, where he remains a professor emeritus.

Blackwell told Donald Albers in an interview for *Mathematical People* that he had never been interested in doing research. "I'm interested in *understanding,* which is quite a different thing." He explored topics that intrigued him in many mathematical areas. He told Morris DeGroot in an interview for *Statistical Science,* "Don't worry about the overall importance of the problem; work on it if it looks interesting. I think there's probably a sufficient correlation between interest and importance." Indeed, results of his work have found application in a variety of fields, including economics and accounting. In 1979, he was awarded the John von Neumann Theory Prize for his work in dynamic programming.

One of Blackwell's most satisfying accomplishments was finding a game theory proof for the Kuratowski Reduction Theorem in topology. Some 15 years after it was published, he told Albers, "That gave me real joy, connecting these two fields that had not been previously connected."

## SELECTED WRITINGS BY BLACKWELL

*Basic Statistics,* 1970
(With M. A. Girshick) *Theory of Games and Statistical Decisions,* 1980

## FURTHER READING

### Books

Albers, Donald J., and G. L. Alexanderson, editors. *Mathematical People.* Boston: Birkhauser, 1985, pp. 18–32.
Duren, Peter, editor. *A Century of Mathematics in America,* Part III. Providence, RI: American Mathematical Society, 1989, pp. 589–615 (reprinted from *Statistical Science,* February 1986, pp. 40–53).

### Other

Blackwell, David, interview with Loretta Hall conducted January 14, 1994.

*—Sketch by Loretta Hall*

*Lenore Blum*

# Lenore Blum
## 1943–
### American algebraist, logician, and model theorist

Lenore Blum has played an integral role in increasing the participation of girls and women in mathematics. She was one of the founders of the Association for Women in Mathematics (AWM), acting as its president from 1975 to 1978. The AWM has membership totaling over 1,500 women and men. In addition to local, national and international meetings, the AWM sponsors the **Emmy Noether** Lecture series and has organized symposiums. It provides a list women who are available to speak at high schools and colleges and also contributes to the *Dictionary of Women in the Mathematical Sciences.*

### Educational Pursuits

Blum was born in 1943 and as a child, she enjoyed math, art and music. Finishing high school at 16, Blum applied to Massachusetts Institute of Technology (MIT), but was turned down, several times, in fact. After being turned down by MIT, Blum attended Carnegie Tech in Pittsburgh, Pennsylvania. She began studying architecture, then changed her major to mathematics. For her third year, Blum enrolled at Simmons, a Boston area college for women. However,

Blum found that she did not have to put forth much effort in the math classes. She then cross-registered at MIT, graduated from Simmons, and received her Ph.D. in mathematics from MIT in 1968. Blum continued her education as a postdoctorate student and lecturer at the University of California at Berkeley.

According to a biography written by Lisa Hayes, a student at Agnes Scott College, "Blum's research, from her early work in model theory, led to the formulation of her own theorems dealing with the patterns she found in trying to use new methods of logic to solve old problems in algebra." The work she did on this project became her doctoral thesis, which earned her a fellowship. Blum has also had the honor of reporting on work she did with **Stephen Smale** and Mike Shub in developing a theory of computation and complexity over real numbers.

Blum has written mathematical books with her husband, Manuel, a mathematician as well. They collaborated on a paper that proposed designing computers that had the ability to learn from example, much in the way young children learn. Blum has studied this project to discover why some computers learn the methods they do. Blum has been involved in other fields of research, in addition to working with her husband, which includes work in developing a new (homotopy) algorithm for linear programming.

When Blum was hired to teach algebra at Mills College, she was not happy with the program and sought a way to make the classes more interesting to the students and to the instructors. In 1973, she founded the Mills College Mathematics and Computer Science Department. Blum served as Head or Co-Head of the department for 13 years. While at Mills, Blum received the Letts-Villard Research professorship. Since 1988, she has been a research scientist in the Theory Group of the International Computer Science Institute (ICSI). In 1989, Blum was employed as an adjunct professor of Computer Science at Berkeley. During the 1980s, Blum became a research mathematician full-time, giving numerous talks at international conferences.

To further girls and women's participation in mathematics, Blum founded the Math/Science Network and its Expanding Your Horizons conferences. The Network began as an after-school problem-solving program. The aim of the program is to get high school girls interested in math and logic. The conference now travels nationwide. Blum served as its Co-Director from 1975-1981. Blum has written books and produced films, including "Count Me In", *The Math/Science Connection*," and "*Four Women in Science*," for the Network.

In addition to her work with the Math/Science Network, Blum is involved in the Mills College Summer Mathematics Institute for Undergraduate Women (SMI). The SMI is a six-week intensive mathematics program. Twenty-four undergraduate women are selected from across the nation to participate. According to the Mills College SMI page on the World Wide Web, the program aims "to increase the number of bright undergraduate women mathematics majors that continue on into graduate programs in the mathematical sciences and obtain advanced degrees."

## Energetic Society Member

Blum is an active member of several mathematical societies. She is a fellow of the American Association for the Advancement of Science and the American Mathematical Society (AMS), where she served as Vice President from 1990 to 1992. Blum represented the AMS at the Pan African Congress of Mathematics held in Nairobi, Kenya, in the summer of 1991. At that time she became dedicated to creating an electronic communication link between American and African mathematics communities. Blum also served as a member of the Mathematics Panel of Project 2061. The project was to determine how much a typical adult must know about science and technology to be prepared for the return of Halley's Comet. Blum also served as the first woman editor of the *International Journal of Algebra and Computation* from 1989 to 1991.

Blum served as the deputy director at the Mathematical Sciences Research Institute (MSRI) at U.C. Berkeley. She has participated in MSRI's Fermat Fest and has been an organizer of MSRI's "Conversations" between mathematics researchers and mathematics teachers.

## SELECTED WRITINGS BY BLUM

(With Manuel Blum) "Toward a Mathematical Theory of Inductive Inference." *Information and Control* 28, no. 2, (1975): 125–155.

(With Mike Shub) "Evaluating Rational Functions: Infinite Precision is Finite Cost and Tractable on Average" (Extended Abstracts). *FOCS* (1984): 261–267.

(With Manual Blum and Mike Shub) "A Simple Unpredictable Pseudo-Random Number Generator." *SIAM J. Comput.* 15 no. 2, (1986): 364–383.

(With Mike Shub) "Evaluating Rational Functions: Infinite Precision is Finite Cost and Tractable on Average." *SIAM J. Comput.* 15 no. 2, (1986): 384–398.

(With Mike Shub and Stephen Smale) "On a Theory of Computation over the Real Numbers; NP Completeness, Recursive Functions and Universal Machines" (Extended Abstracts). *FOCS* (1988): 387–397.

## FURTHER READING

### Books

Perl, Teri. *Women and Numbers: Lives of Women Mathematicians*. Wide World Publishing, 1993.

### Periodicals

Blum, Lenore. "Women in Mathematics: An International Perspective, Eight Years Later: Association for Women Mathematics Panel." *The Mathematical Intelligencer* 9, no.2, (1987): 28–32.

### Other

"DB&LP: Lenore Blum." *Database Systems & Logic Programming.* http://sunsite.ust.hk/dblp/db/indices/a–tree/b/ Blum:Lenore.html. (April 29, 1997).

Hayes, Lisa. "Lenore Blum." *Biographies of Women Mathematicians.* http://www.scottlan.edu/ lriddle/women/blum.html. (April 29, 1997).

*Mills College Summer Mathematics Institute for Undergraduate Women.* http:// aug3.augsburg.edu/pkal/resources/ptw/mills.html. (April 29, 1997).

—*Sketch by Monica L. Stevens*

# János Bolyai
## 1802–1860
### Hungarian geometer

János Bolyai is remembered as the Hungarian mathematician whose work on non–**Euclid**ean geometry was eclipsed by the work of **Nikolai Lobachevsky** in Russia and **Karl Gauss** in Germany. The son of Farkas (Wolfgang) and Susanna Arkos Bolyai, Bolyai was born in Kolosvar, Transylvania, Hungary, on December 15, 1802. Bolyai was education in Marosvasarhely at the Evangelical–Reformed College, where his father was a professor of mathematics, physics, and chemistry.

From childhood, Bolyai showed an aptitude for both mathematics and music; he was an accomplished violinist at an early age. His father—a notable mathematician in his own right who devoted considerable effort to trying to prove the Euclidean theory of parallels—wanted Bolyai to study mathematics in Germany with Gauss; instead, the young man enrolled in the imperial engineering academy in Vienna in 1818 and pursued a military education. However, he still shared his father's passion for proving the Euclidean postulate that there can be only one parallel to a line through a point outside it. The elder Bolyai, in despair about his own lack of success, wrote to his son: "For God's sake, I beseech you, give it up. Fear it no less than sensual passions because it, too, may take all your time, and deprive you of your health, peace of mind, and happiness in life."

While still a student in Vienna, Bolyai began to consider concepts that would alter the focus of studies. Inspired partly by his inability to prove Euclidean parallelism and partly by his exposure to other ideas, Bolyai began to explore the notion of a geometry constructed without the Euclidean axiom. He graduated from military college in 1822 and was commissioned as a sublieutenant. A flamboyant man, Bolyai developed a reputation as a competent swordsman and violinist. He readily accepted challenges to duel; in one account, he crossed swords with 13 consecutive opponents, stipulating only that he be permitted to pause between matches to play a violin selection. According to the story, he defeated all 13 swordsmen and was applauded for his musicianship. Bolyai still found time to pursue his mathematical inquiries, however. In 1823, he wrote to his father that he had made significant progress in his non–Euclidean geometry constructs, stating "from nothing I have created another entirely new world."

### Work Eclipsed by Elder Mathematicians

Bolyai's military duties took him to Temesvar, where he was stationed from 1823 to 1826. He managed a visit to his father in 1825 and took with him a manuscript that detailed his theory of absolute space. Although the father rejected the son's concept, he forwarded the manuscript to Gauss in 1831. In 1832, Gauss wrote to the elder Bolyai: "Now something about the work of your son. . . . The whole content of the paper, the paths that your son has taken, and the results to which he has been led, agree almost everywhere with my own meditations, which have occupied me in part already for 30 –35 years . . . now I have been saved the trouble [of writing a paper]."

The younger Bolyai was distressed and humiliated; in either 1831 or 1832, Bolyai's work was published as an addendum to a longer work of his father's, entitled *Tentamen*. Bolyai was dismayed anew on publication—neither father or son had been aware that Lobachevsky had published a paper outlining the concepts of non–Euclidean geometry three years before their work was issued. Stung by the lack of recognition and his failure to establish his own priority, Bolyai virtually abandoned his scholarly efforts in mathematics. Plagued by poor health—he suffered from chronic fevers—he accepted an invalid's pension and left the military in 1833.

Bolyai returned to Marosvasarhely to live with his father, but the two strong–willed, emotional, and disappointed men were unable to peacefully share the same house. The younger Bolyai retired to a small family estate at Domald; in 1834 he contracted an "irregular marriage" (probably a live–in arrangement with no legal standing) with Rosalie von Orban, with whom he had three children.

In 1837 father and son tried again in vain to build a place for themselves in the world of mathematics. They entered the Jablonow Society prize contest, the subject of which was the geometric construction of imaginary quantities. Several mathematicians, Gauss included, were exploring the subject at the time. Unfortunately, the Bolyais' solutions to the problem were too complicated to win, and in fact, János' solution was similar to **William Rowan Hamilton**'s, who had already published a solution that was easier to understand.

Notes and letters indicate that Bolyai continued to dabble in mathematics. He was interested in absolute geometry, the relationship between absolute trigonometry and spherical trigonometry, and the volume of tetrahedrons in absolute space. Bolyai also dabbled in philosophy, outlining what he termed salvation theory, in which he examined the concepts of individual and universal happiness and the relationship of virtue to knowledge.

Bolyai's father died in 1856, at about the same time Bolyai's arrangement with Rosalie von Oraban ended. Bolyai continued to live at the family estate as a semi–recluse, and his occasional writings from 1856 to 1860 include a memorial to his mother and a lively appreciation of the ballet company at the Vienna Opera House. He died after a lengthy illness and was buried in Marosvasarhely.

## SELECTED WRITINGS BY BOLYAI

"Appendix Explaining the Absolutely True Science of Space," in *Tentamen juventutem studiosam in elementa matheseos purae, elementaris ac sublimioris, methodo intuitiva, evidentaque huic propria introducendi, cum appendice triplici (An Attempt to Introduce Studious Youth into the Elements of Pure Mathematics, by an Intuitive Method and Appropriate Evidence, with a Threefold Appendix)*, by Farkas Bolyai, 2 volumes, 1832–33

## FURTHER READING

Asimov, Isaac. *Asimov's Biographical Encyclopedia of Science and Technology*. New York: Doubleday & Co., 1982, p. 350.

Boyer, Carl B. *A History of Mathematics*. Second edition. Revised by Uta C. Merzbach. New York: John Wiley & Sons, 1989, pp. 281–83.

Eves, Howard. *An Introduction to the History of Mathematics*. Revised edition. New York: Holt, Rinehart and Winston, 1964, pp. 127–28.

Fang, J. *Mathematicians from Antiquity to Today*. Hauppauge, NY: Paideia Press, 1972, pp. 233–34.

Struik, D. J. "János Bolyai," in *Dictionary of Scientific Biography*. Volume II. Edited by Charles Coulston Gillispie. New York: Charles Scribner's Sons, 1970, pp. 269–71.

—*Sketch by A. Mullig*

# Rafaello Bombelli
## 1526–1573
### Italian algebraist

Rafello Bombelli was the last of a long line of Italian algebraists who contributed to the theory of equations during the Renaissance. He was the first to develop a consistent theory of imaginary numbers which included the rules for the four operations on complex numbers. **Gottfried Wilheim von Leibniz** complimented Bombelli years later referring to him as an "outstanding master of the analytical art."

Bombelli was born in 1526 in Bologna, Italy. His father, Antonio Bombelli, was a wool merchant. Bombelli choose not his father's profession but instead became an engineer. He never attended any university but instead was trained by Pier Francesco Clementi of Corinaldo, who as an engineer–architect was responsible for draining swamps. For the major part of his working life, Bombelli was employed as an engineer under the patronage of Monsignor Alessandro Rufini, who later became the Bishop of Melfi. Bombelli's engineering career included two major projects; the reclaiming of the Val di Chiana marshes from 1551 to 1560, and the failed attempt to repair the Ponte Santa Maria Bridge in Rome in 1561.

### Writes Algebra Text

Bombelli wrote his *Algebra* text in 1560, just a few years after **Girolamo Cardano**'s *Ars magna* was published. Bombelli's *Algebra* was not published the first time until 1572 (and only a partial edition), and it became a very important work for several reasons. It featured 143 problems found originally in Diophantus' *Arithmetica*. Bombelli had found a copy of Diophantus' book in the Vatican Library and until that time the ancient Greek's Diophantus' work was mainly ignored. In *Algebra,* Bombelli made notable contributions to improvements in algebraic notation. He introduced an index notation for denoting powers, which he referred to by the term "dignita." Bombelli also introduced in this work a new symbol to indicate the root of an entire expression, was the predecessor of a modern bracket.

What Bombelli is best known for is his justification of conjugate imaginary roots for the "irreducible case" of cubic equations. Starting with the Cardano–**Tartaglia** formula, he developed a very detailed theory of imaginary numbers by arguing by analogy with the known rules for operating on real numbers. The irony of Bombelli's discovery was that the irreducible case actually results in three real roots, but the Cardano–Tartaglia formula produces two conjugate imaginary roots involving the square root of a negative number. Bombelli knew that this type of cubic equation has three real roots and used the results of the Cardano–Tartaglia formula to demonstrate that real numbers can be the result of operations on complex numbers.

Bombelli never realized his great contribution, since he still referred to complex numbers as useless. Nonetheless, Bombelli's *Algebra* was widely read and respected. Leibniz used it to study cubic equations and **Leonhard Euler** quoted from it in his text, *Algebra.*

## SELECTED WRITINGS BY BOMBELLI

*L'Algebra Opera,* 1572

## FURTHER READING

### Books

Burton, David M. *The History of Mathematics.* Third Edition. New York: McGraw–Hill, 1997.

Katz, Victor J. *A History of Mathematics.* Reading MA: Addison–Wesley, 1993, pp. 334–337.

Stillwell, John. *Mathematics and Its History.* New York: Springer–Verlag, 1989.

—*Sketch by Cathleen M. Zucco*

*George Boole*

binary algebra that today has broad applications in the design of computer circuits and telephone switching. He also made significant contributions to probability theory, the field of differential equations, and the calculus of finite differences.

George Boole was born in Lincoln, England, on November 2, 1815, into what was regarded at the time as a lower class family. His father, a tradesman, encouraged Boole to obtain an education in the classics, believing this might elevate him to a higher social rank. Because Latin and Greek were not offered at the National Schools, the only formal educational institutions then available for boys of Boole's social status, he studied the ancient languages on his own. At the age of 12 made a translation of a Latin ode by Horace, which his father submitted for publication in a local newspaper. When the translation appeared in print, a local Latin scholar argued that no 12–year–old could have been capable of such an accomplishment, and Boole was falsely accused of plagiarism. In addition to the classics, he was encouraged to learn mathematics. Boole's father had a personal interest in the subject and passed on what rudimentary knowledge he could from his own private study. Boole took a job as an assistant schoolteacher when he was 16, in order to help support his parents. He also began, in his spare time, to study for the clergy. In 1835, his parents' poverty worsened and Boole chose to abandon his religious pursuits. Boole opened his own elementary school and, in the course of preparing his students in mathematics, found that

# George Boole
## 1815–1864

### English logician and algebraist

George Boole was the founder of the modern science of mathematical logic. He devised a system of the available textbooks were inadequate. As he searched for better teaching materials, Boole began reading the classical mathematical works of the 17th and 18th century masters, developing an interest in algebra and calculus, and was influenced by the works of **Isaac Newton** and **Pierre Laplace**. He made a special study of *Mecanique analytique*, in which the author and French mathematician **Joseph–Louis Lagrange** set forth a purely analytical calculus of variations.

### Earns Recognition with Work on Algebraic Invariants and Calculus of Operators

By the late 1830s, Boole was ready to make original contributions in the field of analysis. He established contact with Duncan Gregory, a Scottish mathematician who edited the newly founded *Cambridge Mathematical Journal*. This relationship was significant because, uncredentialed and lacking membership in a learned society, Boole had few options for presenting his ideas to a mathematical audience. Gregory, impressed with Boole's style and originality, began publishing his work. In 1841 Boole wrote a paper on the theory of algebraic invariants which greatly extended the work of Lagrange. On the basis of this paper, Boole is often credited with the discovery of invariants, a construct that has application in theoretical physics. In 1844 Boole published a paper on the calculus of operators in the *Philosophical Transactions of the Royal Society*. For his original contribution to this field, the Royal Society of London awarded him a gold medal. This honor led to Boole's correspondence with many of the prominent British mathematicians of his day. He was encouraged to take courses at Cambridge University, but finances never allowed. Boole continued teaching at his elementary school until 1849, when, owing to the reputation he had established, he was offered a university post. Despite his lack of formal training, Boole was welcomed as the Chair of Mathematics at the newly established Queens College, Cork, in Ireland. His professorship marked the end of his economic struggles, and allowed him to devote more time to his interests in both calculus and symbolic logic.

### Develops Theory of Symbolic Logic

Boole's reputation as a prominent mathematician was secured by his work on the subject of mathematical logic. In 1847 he published a groundbreaking pamphlet, *The Mathematical Analysis of Logic*, in which he argued that logic was more closely allied with mathematics than philosophy. The pamphlet was written in reaction to an ongoing dispute between **William Hamilton**, professor of logic and metaphysics at the University of Edinburgh, and **Augustus De Morgan**, a logician and professor of mathematics at the University of London. Hamilton believed that the study of mathematics was a useless exercise, and that de Morgan would be a better philosopher if he were less of a mathematician. De Morgan, already recognized in Britain for his contributions to the study of logic, was also known for his satirical wit. He apparently made short–order of Hamilton's arguments and was in no need of additional aid from Boole. Nonetheless, Boole's pamphlet, which established that mathematical rules could be applied to logic, won de Morgan's admiration.

In 1854 Boole wrote *An Investigation of the Laws of Thought,* in which he greatly developed his ideas about logic. He reduced logical relationships to simple statements of equality, inequality, inclusion and exclusion, and expressed these statements symbolically, using a two digit or binary code. Boole then devised algebraic rules that governed the logical relationships. This bridge between mathematics and logic came to be known as Boolean algebra.

In the years following the publication of his treatise on logic, Boole turned his attention to calculus and differential equations. He wrote a standard textbook, *Treatise on Differential Equations*, in 1859, in which he investigated partial differential equations. Boole also advanced criteria for distinguishing between singular solutions and particular solutions of differential equations of the first order. In 1860 he published another textbook, *Treatise on the Calculus of Finite Differences*. Boole's two textbooks were in wide use at universities for many decades after his death.

In 1855 Boole married **Mary Everest (Boole)**, niece of Sir George Everest, a Professor of Greek at Queen's College and the man for whom the world's highest mountain is named. They had five daughters, including the mathematician **Alicia Boole Stott**. Boole died prematurely at the age of 49. Attempting to get to a class on time, he walked two miles through a drenching rain. Soon after lecturing in his wet clothes he contracted pneumonia. The story is told that his wife, believing the remedy for an illness ought to bear resemblance to its cause, put him to bed and doused him with buckets of cold water. Boole died in Ballintemple on December 8, 1864, and is buried at St. Michael's Church in Blackrock, County Cork, Ireland.

Boole was honored in his lifetime with degrees from the universities of Dublin and Oxford. He was elected a fellow of the Royal Society of London in 1857. He was also made a member of the Royal Irish Academy. Although for many years the ideas of Boole's symbolic logic were regarded mostly as philosophical curiosities, they eventually found important practical application. Much of modern computer processing is based on the binary system of Boolean algebra, as is the design of computer circuitry and telephone switching equipment.

## SELECTED WRITINGS BY BOOLE

Boole, George. *An Investigation of the Laws of Thought, on Which Are Founded the Mathematical Theories of Logic and Probabilities* 1854, reprinted 1951.

## FURTHER READING

Barry, Patrick, editor. *George Boole, A Miscellany.* Cork, Ireland: Cork University Press, 1969.

Bell, E.T. *Men of Mathematics.* New York: Simon and Schuster, 1986.

Cajori, Florian. *A History of Mathematics.* New York: Chelsea, 1991.

MacHale, Desmond. *George Boole: His Life and Work.* Profiles of Genius Series, Volume 2. Dublin, 1985.

*—Sketch by Leslie Reinherz*

# Mary Everest Boole
## 1832–1916
### English mathematical learning theorist

Mary Everest Boole lectured and wrote on the philosophy and psychology of learning, presenting a farsighted vision of mathematical education. She once wrote: "[It] is not necessary that the child should see the evidence for every hypothesis on which he works; what is necessary to mental health is a clear understanding of what constitutes evidence, and the power to distinguish between what is, and what is not, proved." She advocated developing and exercising the mathematical imagination through the practice of "mental hygiene." Boole created and patented "curve–sewing cards," learning tools to help teach children such geometrical entities as angles, lines, and spaces. While Boole held many unorthodox beliefs and opinions which may be discounted today, her ideas on educational psychology are of lasting relevance.

Boole was born in 1832 in Warwickshire, Gloucestershire, England, the eldest child of Reverend Thomas Roupell Everest and his wife, Mary Ryall. Boole and her younger brother, George, grew up in an intellectually stimulating environment. Her father's elder brother, Sir George Everest, was surveyor general of India and the man after whom the world's highest mountain is named. Her mother's brother, John Ryall, was a professor of Greek and vice president of Queen's College in Cork. Although he was considered an eccentric, Ryall's own experience

and scholarship were wide–ranging, and his friends included the Cambridge mathematicians Sir John Herschel and **Charles Babbage**. Before Boole turned seven years old, her father introduced her to Euclidean geometry.

When Boole was five years old, her ailing father left his position at Wickwar and moved the family to Poissy, France, to be near the ministrations of Samuel Hahnemann, the founder of homeopathic medicine. In Poissy, Boole was tutored by M. Deplace, whose teaching methods had a profound effect on her. Deplace's method was to state a problem and then ask question after question, requiring her to write the answers, leading her step by step to her own discovery of principles. Boole was also bilingual, and at age 11 she was reading Bonnycastle's *Algebra*.

In 1843, the family returned to Wickwar and Boole left school early to assist her father at his parish. She met the logician **George Boole** in 1850 when she visited her uncle in Cork. She was then 18 and Boole was 35; their relationship, at first avuncular, blossomed as they corresponded for several years on mathematical and scientific topics. In 1855, her father died, leaving her destitute. Shortly thereafter she and Boole were married in Wickwar and she accompanied him to Cork. Boole acted as her husband's secretary, collaborator, and editor; their most noteworthy project was George's *Law's of Thought*, in which he opined that algebraic formulas could be employed to solve qualitative as well as quantitative problems. The Booles had five daughters—Mary Ellen, Margaret, Alicia, Lucy Everest, and Ethel Lilian—who later demonstrated various talents: Alicia in mathematics, Lucy in chemistry, and Ethel Lilian as novelist. Several of Boole's grandchildren were to make brilliant contributions to various disciplines of science as well.

When George Boole died of pneumonia in 1864, Mary Boole's means were again meager. She returned to London, where she took a job as librarian at Queen's College, England's first college for women. She also ran a boarding house for students, and relatives helped in the rearing of her daughters. She organized "Sunday night conversations" where she discussed many topics of interest, including mathematics, philosophy, psychology, Darwinism, and Judaism, with the students. In 1873, her lease was terminated at the boarding house because she was considered unstable. Her pioneering book on mental hygiene, *The Message of Psychic Science to Mothers and Nurses*, was written during this time, but was not published until 1883.

Also, in 1873, Boole took a job as secretary to the ear surgeon James Hilton, who had been her father's friend. Hinton had authored a number of books on psychology; when he died in 1875, Boole continued to promote his ideas, as well as those of Thomas

Wedgwood, who was interested in the training of geniuses. She also promoted her husband's work, couched in terms of her own psychology, and the work of the French logician August Gratry. She corresponded with a wide circle of friends on many topics, including the reconciliation of religion and science. Her house on Ladbroke Road was a regular meeting place for groups of varying ideologies.

From age 50 until the time of her death, Boole produced a series of books and articles. As intellectuals ignored her ideas on psychology, particularly the unconscious, she turned her attention to the mental processes of children. Her *Lectures on the Logic of Arithmetic* and *The Preparation of the Child for Science* were published when she was in her seventies. These books contain many exercises and examples and had influenced education in England's more progressive schools during the early part of the 20th century.

Boole died in 1916 at the age of 84. In 1931, her *Collected Works*, totaling more than 1,500 pages, were published in four volumes.

## SELECTED WRITINGS BY BOOLE

*Lectures on the Logic of Arithmetic*, 1903
*The Preparation of the Child for Science*, 1904

## FURTHER READING

### Books

Cobham, E. M. *Mary Everest Boole: A Memoir with Some Letters*. Ashington, England: The C. W. Daniel Co., 1951.

MacHale, Desmond. *George Boole: His Life and Work*. Dublin: Boole Press, 1985, pp. 105–111; 156–170; 252–276.

Michalowicz, Karen Dee Ann. "Mary Everest Boole (1832–1916): An Erstwhile Pedagogist for Contemporary Times," in *Vita Mathematica*. Edited by Ronald Calinger. The Mathematical Association of America, 1996, pp. 291–299.

Tahta, D.G., editor. *A Boolean Anthology: Selected Writings of Mary Boole on Mathematical Education*. Association of Teachers of Mathematics, 1972.

### Other

Frost, Michelle. "Mary Everest Boole." *Biographies of Women Mathematicians*. http://www.scottlan.edu/lriddle/women/chronol.htm (July 23, 1997).

—*Sketch by Jill Carpenter*

*Émile Borel*

# Émile Borel
## 1871-1956
### French number theorist

Émile Borel was one of the most powerful mathematicians of the twentieth century. He displayed great virtuosity in working in a number of different areas of mathematics, particularly complex numbers and functions, but in addition served visibly as a representative of the mathematical community to the general public. He brought the influence of the mathematical spirit to bear, thanks to his popular writings and to his participation in academic and national politics. His views on mathematics and how it should be supported affected French national policy on mathematics for many years after his death.

Émile Félix-Édouard-Justin Borel was born in Saint-Affrique in the Aveyron district of France, the son of Protestant pastor Honoré Borel and Émilie Teissié-Solier, the daughter of a successful merchant. He had two older sisters, but rapidly impressed the neighborhood with his intelligence. Born on January 7, 1871, he entered a France still in disarray after the setbacks of the Franco-Prussian war. His original education was at the hands of his father, from which he passed at the age of 11 to the lycée at Montauban, the nearest cathedral town. His merits were recognized quickly enough that he was sent to Paris to study at some of the leading preparatory schools for the university. His efforts were crowned by admission to the École Normale Supérieure, an institution that was at the summit of French scientific education and with which he was to be associated throughout the rest of his career.

Academically Borel made rapid progress through the next few years. He was welcome in the family circle of the eminent geometry specialist Gaston Darboux and married mathematician Paul Appell's daughter Marguerite in 1901. Although they had no children, they adopted one of Borel's nephews, Fernand, who was later killed during the First World War. After Borel graduated first in his class at the École Normale, he was offered a teaching position at the university at Lille, exceptionally early since he had not finished his doctoral thesis. He took his doctorate in 1894 and returned a couple of years later to the École Normale. Despite his busy schedule of research and publication, Borel always took his responsibilities at the École Normale seriously, paying attention both to his own teaching and to the curriculum. The École Normale rewarded his work by instituting a special chair in the theory of functions, of which Borel was the first occupant.

Borel started his career of research by publishing three brief articles at the age of 18. From there he went to a life in which writing always played a large part, as witnessed by his over 300 papers and books. In addition to his research articles, the dedicated teacher also devoted effort to the re-creation of French textbooks in mathematics. His activities as a writer were supplemented by his editing one of the best known series of expositions, known as the Borel collection. The first of these, written when he was still only 27, *Leçons sur la théorie des fonctions* ( *Lessons on the Theory of Functions* ), went through several editions and rapidly acquired the status of a classic, for it laid the foundation for the field of measure theory. The French publishing house of Gauthier-Villars owed its preeminence in mathematics to the work that Borel did on behalf of this series.

Borel's earliest mathematical work was already distinguished by its variety, but two of the areas in which his contributions attracted attention were the study of complex functions and of the topology, or properties of geometric configuration, of real numbers. Functions serve as a type of formula in which different numbers, known as variables, are plugged into a mathematical expression for a resulting value. Just as there arc functions of real numbers that give out real numbers as values, so there are functions of complex numbers—numbers that include the imaginary unit $i$, the square root of $-1$—that give out complex numbers as values. There was a clear geometric significance attached to the existence of certain objects from calculus for real functions, but the geometry of curves representing functions of complex numbers was more complicated. Borel was able to use

the information that complex functions had derivatives—the limits of rates of change—to prove what is known as Picard's theorem, concerning the number of possible values that a complex function can take. The theorem and the methods used to prove it are fundamental to the study of complex functions.

Borel's important results were not limited to complex functions, however. It might seem easy to talk about the area of a region in two dimensions, but as the region gets more complicated, it becomes harder to define exactly what is meant by area. Borel extended older ideas and results about areas to more general kinds of regions, including those defined by an infinite sequence of operations. Such regions or sets are said to be "Borel-measurable," and the more general notion of area introduced by Borel (called "measure" from the French "mesure") became the basis of the field called "measure theory." Although the most general notion of area that was taken up by the next generation of mathematicians came from Borel's younger colleague, Frenchman **Henri Lebesgue**, Borel influenced Lebesgue's work in many ways, including personal encouragement.

In his thesis, Borel articulated another idea that is now identified with his name. The idea of compactness has been a central one in twentieth-century mathematics and comes out of topology. The term is not easy to summarize, but perhaps the simplest way to consider compactness is that it describes the extent to which even an infinite set can have some of the simpler characteristics of a finite set. Borel showed that the set of real numbers includes the property that every closed and bounded interval (such as the real numbers between 1 and 2, inclusive) is compact. This theorem is known as the Heine-Borel theorem—a somewhat misleading name, as German mathematician **Heinrich Heine** observed but never articulated the significance of the idea.

One way of getting a perspective on Borel's central role in French mathematics from early in his career is to look at a debate about the role of the infinite that took place in the year 1905. As a result of the publication of a proof involving a controversial axiom in set theory called the axiom of choice (which allowed for an infinite number of choices from a set), Borel published an article expressing distrust of the general principle involved. There was a rapid exchange of letters between leading French mathematicians including Lebesgue, not all of whom agreed with Borel's restrictions. Nevertheless, the entire discussion was centered about Borel, who kept it going by forwarding letters he had received to those who might have more to say on the issue. There are many ways of describing the philosophy of mathematics that crystallized out of this discussion, and the views of Borel had been influenced by those of **Jules Henri Poincaré** of the previous generation, but Borel's role in raising the issue and trying to reach a modicum of agreement testifies to his central position.

## Moves into Applications and Politics

Borel's interests had never been restricted entirely to mathematics. In 1906 he took the money that he had received for the Petit Prix d'Ormoy and used it to start a journal entitled *La revue du mois* ( *Monthly Review* ) that dealt with issues of general interest. Both he and his wife worked on the editing with success in the quality of the articles and the range of subscribers. It only ceased publication in 1920 as a result of the darker economic climate created by World War I.

The war also had the effect of moving Borel's interests in the direction of applied mathematics. Among the long-term consequences was a stream of publications in probability, which had become a standard subject in the mathematical curriculum. The original notions of finite probability had been developed in the seventeenth century, and the progress of calculus shortly thereafter allowed for the generalization of probabilistic notions to the real numbers as well. As in his work on measure, Borel was able to extend the notions of probability to more complicated sets of events. Some of his work in probability was fundamental to the new field called game theory, including his observation that there were situations where random behavior had its advantages. In addition to his strictly mathematical work on the subject, Borel wrote for a popular audience to make the subject of probability more widely accessible.

As an indication of the width of Borel's scientific interests, he wrote a volume in 1922 on **Albert Einstein**'s theory of relativity that was subsequently translated into English as *Space and Time*. The discussion proceeds almost entirely without equations, since Borel's concern once again was to make the subject comprehensible to the widest audience. The discussion includes both the special and general theories of relativity and is characterized by many references to familiar objects to remove the sense of strangeness. Although Borel did not contribute to the mathematical details of Einstein's theories, he took it as a public responsibility of the mathematical community to help make science comprehensible.

His sense of public responsibility extended to the political sphere as well. Although he remained a resident of Paris, he was elected mayor of Saint-Affrique, perhaps partly as a tribute to his scientific standing. He represented the area as a member of the Chamber of Deputies from 1924 to 1936 under the banner of the Radical-Socialist party, although it was neither radical nor socialist as those terms are customarily used. He also served as Minister of the Navy in 1925, but his most lasting accomplishments in the political arena concerned the funding of mathematical and scientific research. The Centre National de la Recherche Scientifique (CNRS) received funding under his aegis and has continued to be essential for the support of research in France. He helped plan the Institut Henri Poincaré and served as its director for many years, from its founding in 1928 until his death, thereby paying tribute to one of the great influences on his own approach to mathematics.

Borel was elected to the Académie des Sciences in 1921, rather later than one might have expected and perhaps out of a distrust of his political involvement. He served as president of the Académie in 1934, another expression of his stature in the wider scientific community, and received the gold medal of the CNRS on its first being awarded. He was decorated for his work during World War I, and again for having stood up to the German Gestapo during World War II. He had left his chair at the École Normale and had taken up a position at the Sorbonne (the University of Paris) instead, as his memories of the generation of French mathematicians dead on the battlefields of World War I were too painful in the familiar surroundings. He retired from the Sorbonne but remained mathematically active for the rest of his life, including regular attendance at international congresses. Part of the reason for his success in accomplishing so much was that he welcomed those who had something constructive to say and chose not to waste time on formalities and empty words. As a lecturer he was impressive, thanks to his tall, dignified manner and his air of distinction. His death on February 3, 1956, hastened by a fall on board ship returning from a conference in Brazil, was met with a sense of great loss by colleagues and pupils, but the heritage he left in his mathematics and in his writing testifies to a strong belief in the mathematician's obligation to serve the community at large.

## SELECTED WRITINGS BY BOREL

*Leçons sur la théorie des fonctions,* 1898
*Éléments de la théorie des probabilités,* 1909
*Le hasard,* 1914
*L'espace et le temps,* 1922
*La politique républicaine,* 1924
*Les probabiliités et la vie,* 1943
*Les nombres inaccessibles,* 1951
*L'imaginaire et le réel en mathématiques et en physique,* 1952
*Émile Borel: Philosophe et homme d'action,* 1967
*Oeuvres de Émile Borel,* 4 volumes, 1972

*Raoul Bott*

**FURTHER READING**

May, Kenneth O. "Émile (Félix-Édouard-Justin) Borel," in *Dictionary of Scientific Biography.* Volume II. Edited by Charles C. Gillispie. New York: Charles Scribner's Sons, 1970, pp. 302–305.

Moore, Gregory H. *Zermelo's Axiom of Choice.* New York: Springer-Verlag, 1982.

—*Sketch by Thomas Drucker*

# Raoul Bott
## 1923–
### Hungarian–born American topologist

Raoul Bott has made numerous contributions to 20th century mathematics. One aspect of his research has focused on topology of Lie groups and differential geometry, as well as special research in network theory and global analysis. In fact, one of his most celebrated accomplishments was taking the topology of Lie groups and applying Morse theory to it, thereby developing an important periodicity theorem. As Bott's career has progressed, he has done important research on differential operators, then foliations, and most recently mathematics as it relates to physics.

Bott's students might argue that his mentoring skills and accomplishments as a charismatic teacher have proved nearly as exceptional as his research.

Bott was born in Budapest, Hungary, on September 24, 1923. He was the son of Rudolph and Margit (nee Kovacs) Bott, who divorced soon after his birth. Bott's father was of Austrian Catholic descent, while his mother was Jewish and the daughter of a Hungarian patriot; Bott was raised as a Catholic. Both of his parents died of cancer at an early age: his mother died in 1935, while his father died in 1937 of liver cancer. Despite these early losses, Bott was raised by his stepfather after his mother's death in an upper–middle class environment.

Bott and his family escaped to England in 1939 because of the tensions in Europe caused by the Nazis and their anti–Jewish policies. A year later, Bott and his family emigrated to Canada. Bott entered McGill University in Montreal and studied engineering. After receiving his bachelor's degree in 1945, Bott served as a volunteer in the Canadian Army for the summer. After the bombing of Hiroshima in August 1945, which effectively ended World War II, Bott left the military. That fall he entered a Master's in Engineering program at McGill, and began teaching calculus there. After earning his masters degree in 1946 with a thesis topic on impedance matching, Bott emigrated to the United States in 1947. In that same year, he married Phyllis Hazell Aikman. They are the parents of four children (Anthony, Jocelyn, Renee, and Candace). In 1949 he earned his doctorate of science in applied mathematics from Carnegie Institute of Technology (now called Carnegie–Mellon University), in Pittsburgh.

Immediately after graduation, Bott become a member of the Institute for Advanced Study at Princeton for two years. Bott did pure mathematics there while increasing his knowledge of topology. After serving as an instructor at the University of Michigan for one school year (1951–1952), and as an assistant professor from 1952 to 1955, Bott returned to the Institute at Princeton for two more years, from 1955 to 1957. Bott served as a full professor at the University of Michigan from 1957 to 1959. He became a naturalized American citizen in 1959.

During this decade in New Jersey and Michigan, Bott's research proved quite fruitful. Much of his work in the 1950s concerned the topology of Lie groups and Morse theory. He combined these two concepts to develop his fundamental periodicity theorem. Bott's periodicity theorem proved useful and applicable in many areas of mathematics, especially as a powerful tool for describing phenomena in homotopy theory.

One of Bott's early important papers, "Homogeneous Vector Bundles," was published in 1957 in the journal *Annals of Mathematics.* This influential paper touched a wide range of mathematics, including representation theory, differential geometry, algebraic geometry, and algebra. Bott became an associate editor at the *Annals of Mathematics* in 1958, a post he retained through the 1980s and early 1990s. Bott's work at the *Annals* was not his only editorial experience; he also served as an editor of *Topology*.

In 1959, Bott left Michigan to became a professor of mathematics at Harvard University, where he is still a professor as of 1997. He took a sabbatical from Harvard to return to the Institute at Princeton for one year, 1970–71. While at Harvard, in the 1960s, Bott's work focused on differential operators. He proved (with **Michael Atiyah**), the Atiyah–Bott fixed point theorem, which combines fixed point formulas for elliptic operators with K–theory, using his periodicity theorem. K–theory was used in much of his work and the work of other mathematicians in the 1960s. Bott also explored other topics, including the relationships between analysis and topology, and studied partial differential equations in hyperbolic rather than elliptic terms.

Bott's research in the 1970s focused on foliations. He first wrote on foliation theory in a paper published in 1966. Among the many areas of foliations he studied were the integrability of foliations and their obstructions as affected by characteristic classes; the singularities of holomorphic foliations and a formula for their residue; and he studied foliations from a quantitative angle. Bott's work on foliations brought together the Gelfand–Fuks Cohomology of vector fields and the foliations' algebraic topology aspects.

Bott's personal warmth led to he and his wife serving as the master in an undergraduate dormitory, Dunsterhouse, at Harvard University, from 1977 to 1983. His mathematics students also benefited from his personal enthusiasm for learning and life. One former student, Bob MacPherson, says in the introduction to volume two of *Bott's Collected Works*: "In fact, it seems to me that his influence as a teacher and role model may be nearly comparable in importance to the influence of his mathematical research on the history of mathematics."

During the 1980s Bott moved onto a new area of research, mathematics as it relates to physics. He employed the methodology he developed during his research on topology in this area. The depth and breadth of Bott's mathematical career was recognized a number of times. Among the many prizes and honors Bott has received are the Veblen prize in geometry from the American Mathematical Society, 1964; the National Medal of Science, 1987; and the Steele Career Prize from the American Mathematical Society in 1990.

## SELECTED WRITINGS BY BOTT

*Lectures on K(X)*, 1969
*Collected Papers*, Volumes 1–4. Edited by Robert D. MacPherson, 1993–1995. (This collection also contains articles about Bott).

## FURTHER READING

Baum, Paul. "Working with Bott (Rocking and Rolling with Raoul)," in *Collected Papers of Raoul Bott*. Volume 3: Foliations. Edited by Robert D. MacPherson. Boston: Birkhäuser, 1995,
pp. xxii–xxiii.
Conlon, Lawrence. "Raoul Bott, Foliations, and Characteristic Classes: An Appreciation," in *Collected Papers* of *Raoul Bott*. Volume 3: Foliations. Edited by Robert D. MacPherson. Boston: Birkhäuser, 1995, pp. xxiv–xxvi.
MacPherson, Robert D. "Introduction to *Collected Papers of Raoul Bott* . Volume 2: Differential Operators. Edited by Robert D. MacPherson. Boston: Birkhäuser, 1994, pp. vii–xi."

—*Sketch by Annette Petruso*

# Charles de Bouelles (also known as Charles Bouvelles or Carolus Bovillus)
## c. 1470–c. 1553
### French geometer and number theorist

Charles de Bouelles made several significant contributions to mathematics, although his primary occupation was the priesthood. He wrote a text on geometry which was published in several languages in the 16th century. In it, de Bouelles attempted to solve an age–old dilemma concerning the quadrature of a circle. De Bouelles also did an early study on the cycloid (which is generated by a point on a circle that rolls along a straight line) and published an important treatise on perfect numbers.

De Bouelles was born in Saucourt, Picardy, France, in approximately 1470. There is only limited information on de Bouelles' early life save that his father was an aristocrat. De Bouelles was educated in Paris at least until 1495, when he left to avoid the Plague. While in Paris, he was a student of Le Fevre d'Etaples. De Bouelles traveled off and on for several years at the beginning of the 16th century. In approximately 1503–1507, he journeyed across Europe, visiting Switzerland, Germany, Italy, and Spain. De Bouelles returned to France and became a priest. As a member of the clergy, he was a canon first in the cathedral at Saint Quentin, then at Noyon. He also served as a theology professor at Noyon. De Bouelles benefited from patronage from Charles de Hangest, an ecclesiastic official in Noyon, who gave de Bouelles both time and space to work on his mathematical publications.

In 1503, de Bouelles first published his *Goemetricae introductionis* (also known as *Metaphysicum introducorium*). Though the first edition was in Latin, the book proved popular enough to be published in translation. It was published in French in 1542 (as *Livre singulier et utile*) and in Dutch in 1547 (as *Boeck aenghaende de Conste en de Practycke van* Geometrie). In *Goemetricae,* de Bouelles attempted to solve a problem that had been circulating for many years, that of the quadrature of a circle. The controversy did not end with de Bouelles' answer—in fact, he made a number of attempts to answer it—but the problem continued to circulate throughout the 16th century.

In 1510, de Bouelles published a treatise dealing with perfect numbers entitled *Liver de XII numbers.* Perfect numbers are equal to the sum of all their possible factors. De Bouelles proposed a few solid theories on perfect numbers although he did not always include a proof. A year later, de Bouelles wrote what was probably the earliest work on geometry written in French. It was called *Le Livre de l'art et science de géométrie* (also known as *Geometrie en francoy*).

Before de Bouelles died in Noyon, he published two more significant books: *Prover biorum vulgarium libri tres* in 1531 and *Liber de differentia vulgarium linguarum et gallici sermonis varietate* in 1533.

## SELECTED WRITINGS BY DE BOUELLES

*Goemetricae introductionis*, 1503
*Liver de XII numbers*, 1510
*Le livre de l'art et science de géométrie*, 1511
*Prover biorum vulgarium libri tres*, 1531
*Liber de differentia vulgarium linguarum et gallici sermonis varietate*, 1533

## FURTHER READING

Busard, H.L.L. "Charles Bouvelles," in *Dictionary of Scientific Biography*. Volume II. Edited by Charles Coulston Gillispie. New York: Charles Scribner's Sons, 1970, pp. 360–61.

—*Sketch by Annette Petruso*

*Hans–Joachim Bremermann*

# Hans–Joachim Bremermann
## 1926–1996
### German–born American mathematical physicist and biologist

Hans–Joachim Bremermann made his home in many American universities throughout his career. To them all he brought the benefits of his training in the famed Müenster school of thought in complex analysis, as well as his enthusiasm for new fields ripe for mathematical modeling. Although he began his work in pure mathematics, Bremermann published papers on such varied and controversial subjects as artificial intelligence, human evolution, and the AIDS crisis. He was a leader among those inventing new buzzwords for the dawn of the 21st century: complexity theory; fuzzy logic; neural nets. He even speculated on an eponymous "Bremermann limit" to the ultimate computational capacity of all matter in the universe—a sort of intellectual thermodynamics.

Bremermann was born to Bernard and Berta (nee Wicke) Bremermann in Bremen, Germany, on September 14, 1926. His education was apparently unremarkable, perhaps even interrupted during World War II, but that changed in the late 1940s. Once Bremermann turned 20 and began his doctoral work at the University of Müenster, he joined the circle of analysts led by Heinrich Behnke. The year 1949 was an especially active one, in which many

German mathematicians and physicists returned to Germany or emigrated to other European countries. They brought with them a mix of new ideas and inventions.

By the early 1950s, Bremermann had devised a general solution to what was known as the Levi problem, previously only solvable in two dimensional forms. He had also emigrated to the United States and married Maria Isabel Lopez Perez–Ojeda. Their marriage would last 42 years, until his death.

A new set of functions introduced by Pierre Lelong and Kiyoshi Oka in the early 1940s set the foundation for Bremermann's continued investigations. He disproved a generalization of the Bochner and Martin conjecture of 1948, offering a simpler proof instead. It was later incorporated into a 1966 textbook on complex analysis that remained in use for many years. By 1959, Bremermann had attacked an even older problem of **Peter Dirichlet**'s, involving continuous functions, proving it could be solved in two classes of domains.

## MANIAC

Bremermann and a few other ambitious colleagues were already experimenting with aspects of quantum field theory. By the 1960s, he was involved in biology and computer science as well. Graduate classes in programming and Turing machines had sparked Bremermann's interest in the new technology, but a more practical motivation came in the form of **John von Neumann**'s computer, nicknamed MANIAC. While attempting to program the machine, Bremermann quickly saw the need for the introduction of more subtle algorithms into MANIAC's brute computational processes. He was inspired to set out an agenda for the development of artificial intelligence, in a publication funded by the Office of Naval Research and distributed throughout Europe and the United States. One tactic Bremermann used was to devise and employ evolutionary or "genetic" search procedures. He saw their future in training what he called "perceptrons," and finally witnessed their implementation in automated reasoning or "neural nets" in 1989.

### The Red Queen Hypothesis

During the 1980s Bremermann had moved from genetic algorithms in computing to computer models of evolution. Arguing against group–selection models, he supported the Red Queen hypothesis. This still–controversial vision of the human body posits a complex system as host to a churning population of rapidly mutating parasites. Biology and computer analysis continued to overlap as his "Bremermann optimizer" method came to be used in the fields of genetics, bioscience, and cybernetics.

Bremermann took memberships in a variety of scientific clubs and organizations devoted to mathematics both pure and applied. He published in just as wide a variety of publications including one he co–founded, the *Journal of Mathematical Biology.*

Aside from teaching and conducting research at Stanford and Harvard universities, the University of Washington, and the Institute for Advanced Study at Princeton, New Jersey. Bremermann returned to the University of Müenster to teach, before settling permanently in California. He held joint professorships in mathematics and biophysics at the Berkeley campus of the University of California until his retirement in 1991. Bremermann's wife served at San Francisco State as professor emeritus of romance language literature. She survived him upon his death from cancer on February 21, 1996 in Berkeley. The year before Bremermann's death he was feted with such honors as the Evolutionary Programming Society's lifetime achievement award, and an invitation to speak at the Dalai Lama's 60th birthday celebration.

## SELECTED WRITINGS BY BREMERMANN

### Books

*Distributions, complex variables, and Fourier transforms,* 1965

### Periodicals

"On the Conjecture of the Equivalence of the Plurisubharmonic Functions and the Hartogs Functions." *Mathematics Annual* 131 (1956): 76–86.
"Parasites at the Origin of Life." *Journal of Mathematical Biology* 16, no. 2 (1983): 165–180.
"The Adaptive Significance of Sexuality." *Experientia* (1985): 1245–1253.

### Other

"Cybernetic Functionals and Fuzzy Sets." *Proceedings of the IEEE Conference on Man, Systems and Cybernetics.* Anaheim, CA: IEEE, 1971.

## FURTHER READING

Anderson, R.W. and Conrad, M. "Hans J. Bremermann: A Pioneer in Mathematical Biology." *Biosystems* 34 (1995):1–10.
Conrad, Michael. "Interview with Bremermann." *Society for Mathematical Biology Newsletter* (April 1992).

**Other**

Anderson, Russell W. and Range, R. Michael. "Hans–Joachim Bremermann, 1926–1996." *Notices of the AMS* (September 1996). http://e–math.ams.org/notices/199609/comm–bremer.html

BHN. "99.31 Deaths." *IMU Canberra Circular no. 99.* http://wwwmaths.anu.edu/au/imu/99/personal.html

Jackson, Allyn. "Biographical Sketch." *Notices of the AMS* (September 1996). http://e–math.ams.org/notices/199609/comm–bremer.html

*—Sketch by Jennifer Kramer*

*Gabrielle–Émilie Le Tonnelier de Breteuil, the Marquise du Châtelet*

# Gabrielle–Émilie Le Tonnelier de Breteuil, the marquise du Châtelet 1706–1749

**French physicist, chemist, and translator**

Born Gabrielle–Émilie Le Tonnelier de Breteuil in Paris on December 17, 1706 into an aristocratic family, the marquise du Châtelet received an exceptional education at home, which included scientific, musical, and literary studies. In 1725, she married the marquis du Châtelet, who was also the count of Lomont. It was a marriage of convenience, but she nevertheless had three children with him. After spending some years with her husband, whose political and military career kept him away from Paris, Châtelet returned to the capital in 1730.

Initially leading a busy social life, Châtelet became the lover of the philosopher François–Marie Arouet de Voltaire in 1733. One of the greatest intellectual figures of 18th–century France, Voltaire recognized her exceptional talent for science, and encouraged her to develop her intellect. Châtelet consequently embarked on a study of mathematics, taking private lessons from the prominent French philosopher and scientist Pierre–Louis Moreau de Maupertuis. Both Voltaire and Maupertuis were enthusiastic supporters of **Isaac Newton**'s scientific theories and world view, and it seems that the marquise was, as a result, imbued by the spirit of Newtonian philosophy.

## Creates Intellectual Center at the Chateau de Cirey

In 1734, Voltaire, who faced arrest because of his criticism of the monarchy, was offered sanctuary at Châtelet's chateau at Cirey, in Lorraine, where they spent many productive years. The two welcomed Europe's intellectual elite, thus creating a remarkable cultural center away from Paris. Châtelet was involved in a variety of literary and philosophical projects, eventually concentrating on the study of Newton's philosophy. She assisted Voltaire in the preparation of his 1738 book, *Elements of Newton's Philosophy*.

In 1737, Châtelet, like many other 18th–century scientists, attempted to explain the nature of combustion, submitting an essay entitled *Dissertation sur la nature et la propagation du feu*, as an entry for a contest organized by the Académie Royale des Sciences. Voltaire also participated in the contest, but was unaware of her work. When **Leonhard Euler** and two other scientists were declared the winners, Voltaire arranged that Châtelet's essay be published with the winning entries. In her study, she correctly argued that heat was not a substance, a view defended by the proponents of the phlogiston theory, which the great French chemist Antoine–Laurent Lavoisier empirically disproved in 1788. Furthermore, Châtelet put forth the original idea that light and heat were essentially the same substance.

## Incorporates Leibniz's Ideas into Work on Newtonian Physics

While writing her *Institutions de physique*, a work on Newtonian physics and mechanics, Châtelet became acquainted with the ideas of **Gottfried Leibniz**, particularly his conception of *forces vives*, which she accepted as true. While **René Descartes** described the physical world geometrically as extended matter, to which force can be applied as an external agent, Leibniz defined force as a distinctive quality of matter. In view of Châtelet's general Newtonian orientation as a scientist, her passionate interest in Leibnizian metaphysics, which essentially contradicts the Newtonian world view, may seem odd. However, as Margaret Alic argues, the marquise sought a synthesis of the two world views. "*Institutions,*" Alic has written, "remained faithful to Newtonian physics, but Newton's purely scientific, materialistic philosophy did not completely satisfy the marquise. She believed that scientific theory demanded a foundation in metaphysics and this she found in Leibniz. . . . She never doubted that Leibnizian metaphysics was reconcilable with Newtonian physics, as long as the implications of the Newtonian system were limited to empirical physical phenomena." Châtelet's acceptance of the metaphysical foundations of science was an implicit rejection of any mechanistic world view, Cartesian or Newtonian. Naturally, French scientists, most of whom tacitly accepted the Cartesian scientific paradigm, found the marquise's ideas offensive. For example, the eminent Cartesian physicist and mathematician Jean–Baptist Dortous de Mairan, whom she had singled out for criticism, responded sharply in 1741, representing a majority view which Châtelet was unable to refute alone.

### Translates Newton's Masterpiece into French

Retreating from the philosophical war between the Cartesians and the Leibnizians, Châtelet focused on her Newtonian studies, particularly the huge task of translating Newton's *Principia mathematica* into French, an undertaking which she devoted the rest of her life. An excellent Latinist with a deep understanding of Newtonian physics, she was ideally suited for the project. Despite many obstacles, which included a busy social life and an unwanted pregnancy at the age of 42, Châtelet finished her translation. On September 4, 1749, she gave birth to a daughter, and died of puerperal fever shortly thereafter. Her translation of Newton's work remains one of the monuments of French scholarship.

## SELECTED WRITINGS BY THE MARQUISE DU CHÂTELET

### Books

*Dissertation sur la nature et la propagation du feu,* 1739
*Institutions de physique,* 1740

*Principes mathématiques de la philosophie naturelle,* 1759 (Translation of *Principia mathematica,* by Isaac Newton)

### Periodicals

"Lettre sur la philosophie de Newton." *Journal des sçavans.* (September 1738): 534–41.

## FURTHER READING

### Books

Alic, Margaret. *Hypatia's Heritage: A History of Women in Science from Antiquity through the Nineteenth Century.* Boston: Beacon Press, 1986.

Copleston, Frederick. *Modern Philosophy: From Descartes to Leibniz.* Volume 4: *A History of Philosophy.* New York: Image Books, 1960.

Klens, Ulrike. *Mathematikerinnen im 18. Jahrhundert: Maria Gaëtana Agnesi, Gabrielle–Émilie du Châtelet, Sophie Germain.* Pfaffenweiler: Centaurus–Verlagsgesellschaft, 1994.

Mitford, Nancy. *Voltaire in Love.* New York: Greenwood Press, 1957.

Olsen, Lynn M. *Women in Mathematics.* Cambridge: The MIT Press, 1974.

Smelding, Anda von. *Die göttliche Emilie.* Berlin: Schlieffen Verlag, 1933.

Vaillot, René. *Madame du Châtelet.* Paris: Albin Michel, 1978.

Wolf, A. *A History of Science, Technology, and Philosophy in the Eighteenth Century.* Second revised edition. London: George Allen & Unwin, 1952.

*—Sketch by Zoran Minderovic*

# William Brouncker
# 1620–1684
### English geometer

Viscount William Brouncker is noted as the first English mathematician to use continued fractions to express π and the quadrature of a rectangular hyperbola. A son of the aristocracy, Brouncker was born in Castle Lyons, Ireland, in 1620, to Sir William and Lady Winefrid Brouncker. Little is recorded about his childhood and early education, but Brouncker's parents probably followed the custom of the times and engaged private tutors for William and his brother, Henry. Brouncker was 16 years old when he began his studies at Oxford University in London. He showed

*Viscount William Brouncker*

an aptitude for mathematics, languages, and medicine, as well as an ability and fondness for music; at the age of 27, he was granted the degree Doctor of Physick. In 1645, King Charles I named Brouncker's father a viscount, but the newly elevated lord died in the same year. The title passed to young William, who became a high-ranking peer of the realm before he earned a doctorate.

In 1647, the year Brouncker completed his Oxford studies, Charles I was deposed by Oliver Cromwell. Two years later the former monarch was executed, and Cromwell firmly established himself as England's ruler. Cromwell's rejection of the divine right of kings and his strong anti aristocracy sentiments made his rule a tense and sometimes dangerous decade for titled nobles. But young Viscount Brouncker, immersed in his private studies of mathematics, attracted little notice and remained out of politics and the public eye.

There is little recorded of Brouncker's life or activities between 1647 and 1660, except for the publication of the translation of **René Descartes'** *Musicae compendium*, in which he attempted to perfect a mathematical formula to divide the diapason (the full range of an instrument or voice) into 17 equal semitones. This is Brouncker's only published work; what is known of his work in mathematics has been published as parts of other mathematician's histories, and various collections of manuscripts and correspondence.

That Brouncker's sudden emergence into public life in 1660 coincided with the restoration of Charles II to the throne, hints that his 11-year seclusion during the Cromwell era was deliberate. In the same year Charles II was crowned, Brouncker entered Parliament. In 1664, he became president of Gresham College and served as commissioner for the navy from 1664 to 1668. Brouncker was appointed comptroller of the treasurer's accounts in 1668 and served as master of St. Catherine's Hospital in London from 1681 to 1684.

Charles II appreciated and rewarded loyalty. When Brouncker proposed a special institute to promote scientific discussion, the Royal Society of London was chartered in 1662. When the king nominated Brouncker as the Society's first president, there was no opposition—and Brouncker was reconfirmed annually as its president until he chose to resign the office in 1677.

Brouncker corresponded with noted mathematicians and scientists of the day, and much of his work in mathematics was an attempt to solve problems devised by others. His work with continued fractions and the expression of $\pi$, for example, was a response to a request from **John Wallis** to develop an expression of $\pi$ other than an infinite product. Brouncker and the French mathematician **Pierre de Fermat** contributed to the development of the Pell Equation, and **James Gregory**'s papers indicate Brouncker's work advanced the understanding of binomial series.

Brouncker never married. When he died in 1684, his brother Henry inherited the title of viscount. Henry, too, was a bachelor; when he died in 1687, the title died with him.

## FURTHER READING

Eves, Howard. *An Introduction to the History of Mathematics.* New York: CBS College Publishing, 1983, pp. 87, 277.

Daintith, John, Sarah Mitchell, and Elizabeth Tootill, editors. *A Biographical Encyclopedia of Scientists.* New York: Facts on File Inc., 1981, p. 113.

Fang, J. *Mathematicians from Antiquity to Today.* Hauppauge, NY: Paideia Press, 1972, p. 247.

Dubbey, J. "William Brouncker," in *Dictionary of Scientific Biography.* Volume II. Edited by Charles Coulston Gillispie. New York: Charles Scribner's Sons, 1973, pp. 506–07.

Murray, Jane. *The Kings and Queens of England.* New York: Charles Scribner's Sons, 1974, pp. 71–77.

Smith, David E. *History of Mathematics.* New York: Dover Publications Inc., 1958, pp. 410–11.

*—Sketch by A. Mullig*

# Luitzen Egbertus Jan Brouwer
## 1881-1966
### Dutch logician

The Dutch mathematician Luitzen Egbertus Jan Brouwer made contributions in the fields of topology and logic. He founded the school of thought known as intuitionism, which is based on the notion that the only dependable basis of mathematics consists of proofs that can actually be constructed in the real world. In addition, he developed a fixed-point theorem and demonstrated the connection between two previously distinct fields of topology—point-set topology and combinatorial topology.

Brouwer was born on February 27, 1881, in Overschie, the Netherlands. His parents were Egbert Brouwer and Henderika Poutsma. Brouwer completed high school in the town of Hoorn at the age of fourteen and attended the Haarlem Gymnasium where he satisfied the Greek and Latin requirements needed to enter a Dutch university in 1897.

At the University of Amsterdam Brouwer easily moved through the traditional mathematics curriculum and began some original studies on four-dimensional space that were published by the Royal Academy of Science in 1904. A year later he published his first book, *Leven, Kunst, en Mystiek*, a philosophical treatise in which he considers the role of humans in society. In 1907 Brouwer presented his doctoral thesis and was granted his doctor of science degree by the University of Amsterdam. He began teaching at Amsterdam in 1909 and spent his entire academic career at the university.

### Develops the Fundamental Concepts of Intuitionism

His doctoral thesis, "On the Foundations of Mathematics," outlined a field of research that occupied Brouwer on and off for the rest of his life. In the early twentieth century the two primary schools of mathematics were logicism and formalism. Logicism is based on the premise that fundamental concepts in mathematics, such as lines and points, have an existence independent of the human mind. The job of mathematicians is to derive theorems from these concepts. Formalism is less concerned with the nature of fundamental concepts, but insists that those concepts be manipulated according to very strict rules.

Brouwer proposed a third concept of mathematics, later given the name intuitionism (also known as constructivism or finitism). The basic argument of intuitionism, according to Richard von Mises in *World of Mathematics,* is that "the simplest mathematical ideas are implied in the customary lines of thought of everyday life and all sciences make use of

them; the mathematician is distinguished by the fact that he is conscious of these ideas, points them out clearly, and completes them. The only source of mathematical knowledge" in intuitionism, von Mises continues, is "the intuition that makes us recognize certain concepts and conclusions as absolutely evident, clear and indubitable."

Brouwer's intuitionist school was not particularly influential when it was first proposed in the 1910s. According to Victor M. Cassidy, in *Thinkers of the Twentieth Century,* "Brouwer made few converts during his lifetime, and Intuitionism has only a tiny number of adherents today."

### Makes Contributions in the Field of Topology

The 1910s were a period of intense activity in the field of topology, the mathematical discipline concerned with geometric point sets. In 1912 Brouwer announced perhaps his most famous theorem, the fixed-point theorem, also known as Brouwer's theorem. This theorem stated that during any transformation of all points in a circle or on a sphere, at least one point must remain unchanged. Brouwer was later able to extend this theorem to figures of more than three dimensions.

Brouwer's first appointment at the University of Amsterdam in 1909 was as a tutor. Three years later, he was promoted to professor of mathematics, a position he held for thirty-nine years. In 1951 he retired and was given the title of Professor Emeritus. He died in Blaricum, the Netherlands, on December 2, 1966. Brouwer had been married in 1904 to Reinharda Bernadina Frederica Elisabeth de Holl. She predeceased Brouwer in 1959. The couple had no children.

Brouwer received honorary doctorates from the universities of Oslo (1929) and Cambridge (1955) and was awarded a knighthood in the Order of the Dutch Lion in 1932. He had also been elected to membership in the Royal Dutch Academy of Sciences (1912), the German Academy of Science (1919), the American Philosophical Society (1943), and the Royal Society of London (1948).

## SELECTED WRITINGS BY BROUWER

### Books

*Leven, Kunst, en Mystiek,* 1905
*Over de Grondslagen der Wiskunde,* 1907
*Wiskunde, Waarheid, Werkelijkheid,* 1919
*Collected Works,* edited by A. Heyting, 1975

### Periodicals

"Intuitionism and Formalism." *Bulletin of the American Mathematical Society* 20, (1913): 81–96.

"Consciousness, Philosophy and Mathematics."
*Proceedings of the Tenth International
Congress of Philosophy* I, (1949): 1235–1249.

## FURTHER READING

Daintith, John, et al. *A Biographical Encyclopedia
of Scientists*. Volume 1 New York: Facts on
File, 1981, pp. 113–114.

Turner, Roland, editor. *Thinkers of the Twentieth
Century*. New York: St. James Press, 1987,
pp. 116–118.

Van Rootselaar, B. "Luitzen Brouwer," in *Dictionary of Scientific Biography*. Volume II. Edited
by Charles Coulston Gillispie. New York:
Charles Scribner's Sons, 1975, pp. 512–514.

Von Mises, Richard. "Mathematical Postulates
and Human Understanding," in *World of
Mathematics*. Edited by James Newman. New
York: Simon & Schuster, 1956.

*—Sketch by David E. Newton*

*Marjorie Lee Browne*

# Marjorie Lee Browne
## 1914-1979
### African-American topologist

In 1949 Marjorie Lee Browne, along with **Evelyn
Boyd Granville**, was one of the first African American
women to receive a Ph.D. degree. By training a
topologist (specializing in a branch of mathematics
that deals with certain geometric aspects of spaces and
shapes), Browne made her greatest contributions in
the areas of teaching and university administration.
She also provided a leadership role in seeking funding
for better educational opportunities; her goals included the strengthening of mathematical preparation for
science and mathematics teachers in secondary
schools and the increased presence of females and
minorities in the mathematical sciences.

Browne was born in Memphis, Tennessee, the
second child of Lawrence Johnson Lee, a transportation mail clerk; her stepmother, Lottie Taylor Lee,
was a school teacher. As a young woman growing up
in Memphis and New Orleans, she was an expert
tennis player, a singer, an avid reader—a trait she
inherited from her father—and a gifted mathematics
student. Browne graduated from LeMoyne High
School in Memphis in 1931. In 1935 she received a
B.S. degree cum laude in mathematics from Howard
University, then earned a M.S. in mathematics in
1939 and a Ph.D. in 1949 from the University of

Michigan. She wrote her doctoral dissertation on one-
parameter subgroups in certain topological and matrix groups, and her dissertation served as the basis
for one of her major publications on the classical
groups in 1955.

Browne began her teaching career in 1935 at
Gilbert Academy in New Orleans, Louisiana, where
she taught physics and mathematics for a year. From
1942 to 1945, she served as an instructor at Wiley
College in Marshall, Texas. In 1949, Browne was
appointed to the faculty in the department of mathematics at North Carolina Central University (NCCU),
where she rose to the rank of professor and became
the first chair of the department from 1951 to 1970.
She served as a principal investigator, coordinator of
the mathematics section, and lecturer for the Summer
Institute for Secondary School Science and Mathematics Teachers, a program funded by the first
National Science Foundation grant awarded to
NCCU in 1957. Browne continued this role until
1970.

Browne was acutely aware of the obstacles which
women and minorities faced in pursuing scientific
careers. Shortly after receiving her doctorate in 1949
she sought, unsuccessfully, to obtain an instructorship
at several major research institutions. After receiving
many polite letters of rejection, she decided to remain
in the South and resolved that her greatest contributions would be directing programs designed to
strengthen the mathematical preparation of secondary

school mathematics teachers and to increase the presence of minorities and females in mathematical science careers.

Thus, Browne spent her summers teaching secondary school teachers, and her objective was to insure that the teachers whom she taught would be able to understand and teach their students the so-called "modern math" or "new math." Browne's teaching standards were exacting and her methods were thorough; her demands for excellence and concise, clear ideas contributed greatly to the academic growth and development of her students, and many of her students have made significant contributions in a number of professions. Nine of them have earned doctorates in the mathematical sciences or related disciplines. Browne was also a steady, outspoken critic of racism and the discriminatory practices prevalent among funding agencies relative to minorities and predominantly minority universities and colleges. She was an ardent advocate for the integration of the previously segregated meetings of the national mathematics organizations of which she was a member.

For her work in mathematics education, Browne was awarded the first W. W. Rankin Memorial Award from the North Carolina Council of Teachers of Mathematics in 1974. During her acceptance speech, she described herself as "a pre-sputnik mathematician." She was referring to the purist nature of her advanced mathematical preparation and the practice of many American industries and businesses, prior to the launching of the first Russian satellite in 1957, to allow scientists and mathematicians to pursue research projects that had no immediate real world or job-related applications. The launching of Sputnik had a tremendous impact on mathematics education in the United States. America was viewed as having fallen behind the Russians in space explorations, and as a result there was a shift in emphasis from pure abstract mathematical and scientific research to investigations that were of an applied nature. Browne, however, remained a mathematical purist, and like many great mathematical philosophers of the nineteenth century she viewed mathematics as an intellectual quest, free from the limitations of the physical universe.

Browne was a member of the Woman's Research Society, American Mathematical Society, Mathematical Association of America, and the International Congress of Mathematicians as well as the author of several articles in professional and scholarly journals. In 1960, she received a sixty thousand dollar grant from IBM to establish one of the first electronic digital computer centers at a predominantly minority university, and in 1969, she received the first of seven Shell Foundation Scholarship Grants, awarded to mathematics students for outstanding academic achievements. The director of the first Undergraduate Research Participation Program at NCCU during 1964 and 1965—which was sponsored by the National Science Foundation—Browne was also one of the first African American females to serve on the advisory panel to the National Science Foundation Undergraduate Scientific Equipment Program in 1966, 1967, and 1973. In addition, she served as a Faculty Consultant in Mathematics for the Ford Foundation from 1968 to 1969 at their New York office. Browne was awarded numerous fellowships, including one from the Ford Foundation at Cambridge University in England from 1952 to 1953.

Browne died of an apparent heart attack at her home in Durham, North Carolina, in 1979. At the time of her death, she was preparing a monograph on the development of the real number system from a postulational approach. Browne was a generous humanitarian who believed that no good student should go without an education simply because he or she lacked the financial resources to pay for it. Thus, it was not uncommon for her to assume the financial responsibilities for many able students whose families were unable to provide tuition, books, board, or transportation for them. To continue the philanthropic legacy which she began, four of her former students established the Marjorie Lee Browne Trust Fund at North Carolina Central University in 1979. This fund supports two major activities in the mathematics and computer science department: the Marjorie Lee Browne Memorial Scholarship—which is awarded annually to the student who best exemplifies those traits which Browne sought to instill in young people—and the annual Marjorie Lee Browne Distinguished Alumni Lecture Series.

## SELECTED WRITINGS BY BROWNE

"A Note on the Classical Groups." *American Mathematical Monthly.* (August, 1955).

## FURTHER READING

### Periodicals

"$57,500 Granted for High School Teachers Summer Institute at NCC." *Durham Sun,* December 21, 1956.

### Other

Personal information supplied to William T. Fletcher by the department of mathematics and computer science of North Carolina Central University, 1994.

Fogg, Erica, Cecilia Davis, and Jennifer Sutton. "Marjorie Lee Browne." Biographies of Women

*Joost Bürgi*

Mathematicians. http://www.scottlan.edu/lriddle/women/chronol.htm (July 22, 1997).

—*Sketch by William T. Fletcher*

# Joost Bürgi
## 1552–1632
### German algebraist, instrument maker, and astronomer

Joost Bürgi is best remembered for developing algebraic logarithm tables, but his interest in mathematics was a by–product of his work in astronomy and instrument making.

There are few records of Bürgi's early life. He was born in Liechtenstein in 1552. Bürgi was unable to read Latin, which was at that time was the universal language of education and science, so he probably received little or no formal education as a young man. In 1579, Bürgi was appointed court watchmaker to Duke Wilhelm IV in the west German city of Kassel. The appointment was a prestigious one; in the 16th century, growth of trade, the first hints of industrialization, and increasingly sophisticated methods of scientific observation required accurate and reliable ways to measure time. Bürgi did much more than

craft clocks. The duke had an observatory, and Bürgi built several instruments to aid astronomical observations. He invented or improved devices for practical geometry observations and calculations and developed a proportional compass that rivaled that produced by **Galileo.**

Bürgi's reputation for instrument making attracted the attention of Rudolf II (1552–1612), emperor of the Holy Roman Empire. Rudolf was preoccupied—or obsessed—with science. He founded a science center in Prague and recruited noted scientists like **Johannes Kepler** to his service. (Rudolf also placed alchemists under house arrest and ordered them to make gold. An unstable man, Rudolf was forced to delegate imperial power to his brother, Matthias, in 1606.)

When Duke Wilhelm died, Bürgi accepted Rudolf's invitation (which amounted to an imperial summons) to join the scientists at the Prague center. Bürgi arrived in Prague in 1603, and he was again given the official designation of court watchmaker. But just as he had done in Duke Wilhelm's court, Bürgi devoted much of his time on astronomy rather than clock making. He worked with Kepler as an assistant, refining instruments and performing mathematical calculations. It was this interest in astronomical calculations that inspired Bürgi to develop faster, more accurate ways to compute. In an early manuscript found in Kepler's unpublished papers, Bürgi used the decimal point and refined a method of approximating the calculation of the roots of algebraic equations.

Bürgi was not a philosopher or theoretician. He focused on making the application of math to astronomy faster and more reliable. He saw the need for an easier method of multiplying large numbers to facilitate processing astronomical data. In 1584, he began efforts to improve "prosthaphairesis," a system of applying trigonometric formulas to problems so that multiplication could be converted to addition.

During his service with Duke Wilhelm, Bürgi had access to scholarly materials and educated people. He was exposed to a wealth of ideas, and he chanced on the idea of logarithms. Bürgi computed tables that correspond to natural logarithms, and it seems he completed his work in logarithms before he moved to Prague, because he brought the manuscript with him in 1603.

Bürgi apparently had little interest in establishing himself as a mathematician, because he did not publish his logarithm tables—titled *Arithmetische und geometrische Progress–Tabulen, sambt grundlichem Unterrich, wie solche nutzlich in allerly Rechnungen zu gebrauchen, und erstanden werden sol*—until 1620. **John Napier**, the Scottish mathematician, published his first work on logarithms, *Mirifici Logarithmorum Canonis Descriptio*, in 1614; Napier's second work on logarithms, *Mirifici Logarithmorum Canonis Construction*, was published in 1619, two years after his death.

Although Rudolf II ceded power to Matthias in 1606, Bürgi retained his imperial appointment. Ferdinand II succeeded Matthias as Holy Roman Emperor in 1619, and Bürgi remained at the emperor's court until 1631. Bürgi left the court and returned to Kassel in 1631 shortly before his death in 1632.

Because Napier's work was published first, Bürgi's work had little impact; the only complete copy of Bürgi's publication known to exist is in a library in Danzig, Poland.

## FURTHER READING

Bell, E.T. *The Development of Mathematics*. York, PA: Maple Press Co., 1945, p.162

Boyer, C.B. *A History of Mathematics*. Second edition. Revised by Uta C. Merzbach. Princeton, NJ: Princeton University Press, 1985, pp. 309–ll, 327–28, 345–50.

Eves, Howard. *An Introduction to the History of Mathematics*. New York: CBS College Publishing, 1983, pp. 228–29.

Hooper, Alfred. *Makers of Mathematics*. New York: Random House, 1948, p. 246.

Kline, Morris. *Mathematical Thought from Ancient to Modern Times*. Volume 1. New York: Oxford University Press, 1972, pp. 258–59.

Luboš, Nový. "Joost Bürgi," in *Dictionary of Scientific Biography*. Volume II. Edited by Charles Coulston Gillispie. New York: Charles Scribner's Sons, 1970, pp. 602–03.

Smith, David E.. *History of Mathematics*. Volume 2. New York: Dover Publications, 1951, p. 523.

Struik, D.J., editor. *A Source Book in Mathematics, 1200–1800*, Cambridge, MA: Harvard University Press, 1969, pp. 12–13.

—*Sketch by A. Mullig*

# Ada Byron, Countess of Lovelace
## 1815–1852
### English applied mathematician

Ada Byron, Countess of Lovelace is best known for her early contributions to the field of computing. A friend and devotee of **Charles Babbage**, she published detailed descriptions of his calculating

*Ada Byron, Countess of Lovelace*

machines. Byron is credited with having written the world's first computer program, a set of instructions for Babbage's Analytical Engine. She was a visionary who speculated on how calculating machines might someday be put to practical use. Byron's full potential as a mathematician was probably never realized, for her personal life was plagued by drug addiction and compulsive gambling.

Byron was born in London, December 10, 1815, to an aristocratic family. Named Augusta Ada Byron, she was the only legitimate daughter of George Gordon, Lord Byron, the English poet. Her mother, Lady Byron, was born Anne Isabella Milbanke, a wealthy intellectual and eccentric, who had been tutored in mathematics. According to historians, Lord Byron was unprepared for the commitments of marriage and fatherhood. In the hours before his daughter's birth he threw furniture around the room. When Ada Byron was only a month old, her mother took her and fled from his household. The young Byron never again saw her father, who died of fever when she was eight years old. Byron was raised by her mother, maternal grandparents, and nannies. Her mother was strongly opinionated about child–rearing practices. While she encouraged Byron's formal education, uncommon among the wealthy during the era, she also believed in such practices as teaching young children to lie perfectly still for long periods of time. Byron spent many hours of her childhood lying flat on a wooden plank, forbidden to move even a finger. She was a precocious child, able to add six rows of numbers and spell two–syllable words at the age of five. Her mother decided science and mathematics should be emphasized in her education, believing these disciplines would counter the romanticism and lack of self–control Byron had supposedly inherited from her father. Byron was tutored in science and mathematics by William Frend, a controversial peace advocate, William King, the family's physician, and **Mary Somerville**, an astronomer and the first female to be elected to Britain's prestigious scientific group, the Royal Society. Byron also studied and excelled in drawing, music, and foreign languages. Her fluency in French would play a role in her contribution to computer history.

In the spring of 1833, Byron attended a party at the home of Charles Babbage, inventor of several steam–driven calculating machines that were to become the precursors of modern computers. Byron quickly took interest in a small working model of Babbage's first Difference Engine, a contraption that could mechanically compete values of quadratic functions. She studied Babbage's plan for the construction of his Analytical Engine, a machine that received instructions and numerical data from punched cards and could make and analyze mathematical calculations. Babbage became a close friend and intellectual mentor to Byron, directing her in 1840 to **Augustus de Morgan**, a professor of mathematics at the University of London. Under De Morgan's tutelage Byron began advanced studies in mathematics, equivalent to what men were receiving at the time from Cambridge University. De Morgan was reportedly impressed by Byron's intellect, but feared her studies might strain her delicate female nervous system.

### Writes "First" Computer Program

The same year that Byron began her advanced studies, Babbage traveled to Turin, Italy, to give a presentation about his Analytical Engine. In attendance was a young military engineer, Luigi Federico Manabrea. Manabrea, who would eventually become prime minister of Italy, was impressed with Babbage's invention, and described its operation for a Swiss journal. His article, written in French, was published in October of 1842. Babbage, who had been so preoccupied with the development and fund–raising for construction of his engines, had never bothered to publish his own descriptions. Charles Wheatstone, pioneer of the telegraph, read Manabrea's article and convinced Byron to translate the article for the British journal *Taylor's Scientific Memoirs*. Wheatstone was a close family friend and aware of Byron's facility in both French and mathematics. When Byron discovered the original article described only the mathematical concepts by which the engine would work, she decided to append a series of notes to the translation. The notes were remarkable in that she not only produced the first clear mechanical explanation of

Babbage's Analytical Engine, but provided illustrations of how it might be instructed to perform particular tasks. In so doing, Byron created the world's first computer program. She invented the idea of a subroutine, a set of instructions that are used repeatedly in a variety of contexts. Byron also anticipated the process she called "backing," which is equivalent to the modern day concept of looping, and she described the notion of a conditional jump, in which the machine responds to "if–then" statements. In her final note, Byron produced a diagram which showed how Bernoulli numbers could be derived through mechanical computation. Her program provided instructions on where to set and how to display calculations in the engine. Byron's eight lengthy notes were written in just under a year, during which time she corresponded heavily with Babbage. In July of 1843 the translation and notes, titled "Sketch of the Analytical Engine Invented by Charles Babbage, Esq.," appeared in print. Although the publication was to constitute Byron's most important contribution to mathematics, she chose to sign her work with only the initials A.A.L., for Augusta Ada Lovelace. It has been speculated she did so to preserve the image of a modest and proper Victorian lady. Many of her friends knew nothing of her intense interest in mathematics.

### Creates Scandal for Her Family

While some aspects of Byron's lifestyle did conform to the expectations placed on her by Victorian society, in the end her behavior brought scandal to her family. In 1835 Byron had met and married William King, unrelated to her childhood tutor of the same name. King took a seat in the House of Lords in 1838 and adopted the title first earl of Lovelace. Byron became countess of Lovelace. The couple shared a love of horses and spent considerable time riding on their estates. Their social circle, in addition to Charles Babbage, included physicist Michael Faraday, astronomer Sir John Herschel, and inventor of the kaleidoscope, Sir David Brewster. They were also close friends with Charles Dickens. The couple had three children: Byron, Anne Isabella, and Ralph. In 1837, shortly after giving birth to Anne Isabella, Byron developed cholera. Although she survived, her health never fully recovered. She suffered from asthma and digestive problems, and was prescribed laudanum, opium, and morphine, which she took with wine. Historians believe that Byron was unaware of an addiction, but her personal letters reveal that she experienced the symptoms of withdrawal. She became known for her bizarre mood swings and reportedly had hallucinations. In the decade before her death Byron also took up gambling. An avid fan of horse–racing, she squandered much of the family fortune on a flawed betting scheme. This, and an alleged affair with a gambling accomplice, John

Crosse, were the subject of gossip among her social peers. In 1851 Byron was diagnosed with uterine cancer. She died on November 27, 1852, and was laid to rest in a church near Newstead, next to her father, Lord Byron.

## Honored Posthumously

In 1953, after digital computing had become a reality, Byron's notes were rediscovered. They appeared in a volume by B.Y. Bowden, titled *Faster Than Thought: A Symposium on Digital Computing Machines*. The computer revolution underway, her contribution to the field would finally receive public recognition. In 1974, the U.S. Defense Department decided to standardize its computer operations by choosing a single computer language for all its tasks. In 1980, on what would have been Byron's 165th birthday, the Ada Joint Program Office was created for the purpose of introducing the Ada language. Three years later, the American National Standards Institute approved Ada as a national all–purpose standard. It was given the code name MIL–STD–181 5, the last four digits honoring her birth year.

## FURTHER READING

### Books

Baum, Joan. *The Calculating Passion of Ada Byron*. Hamden CT: Archon Books, 1986.

Moore, Doris Langley–Levy. *Ada, Countess of Lovelace: Byron's Legitimate Daughter*. New York: Harper and Row, 1977.

Stein, Dorothy. *Ada: A Life and a Legacy*. Cambridge, MA: MIT Press, 1985.

Toole, Betty Alexandra, editor. *Ada, the Enchantress of Numbers: A Selection from the Letters of Lord Byron's Daughter and Her Description of the First Computer*. Mill Valley, CA: Strawberry Press, 1992.

Wade, Mary Dodson, *Ada Byron Lovelace: The Lady and the Computer*. New York: Dillon Press, & Toronto: Maxwell Macmillan Canada, 1994.

### Other

Toole, Betty. "Ada Byron." *Biographies of Women Mathematicians*. June 1997. http://www.scottlan.edu/lriddle/women/chronol.htm (July 22, 1997).

*—Sketch by Leslie Reinherz*

*Alberto P. Calderón*

# Alberto P. Calderón
## 1920-
### Argentine-born American analyst

Calderón is known for his revolutionary work in the field of analysis, particularly in the area of partial differential equations. His influence helped turn the 1950s trend toward abstract mathematics back into a more practical direction. His award-winning research in the area of integral operators is an example of his impact on contemporary mathematical analysis. During his 45-year career, Calderón has produced a number of seminal results and techniques. His accomplishments have been recognized by such prestigious awards as the National Medal of Science and the Wolf Prize in mathematics, and he is a member of the national academies of science of five countries as well as the Third World Academy of Sciences based in Italy. When Calderón was awarded the Steele Prize for a Fundamental Paper in mathematics, his work was described in the *Notices of the American Mathematical Society* as "a major progenitor of the modern theory of microlocal analysis."

Alberto Pedro Calderón was born on September 14, 1920, in Mendoza, Argentina, a small town at the foot of the Andes. His father, a descendant of notable 19th-century politicians and military officers, was a renowned medical doctor who helped organize the General Central Hospital of Mendoza. Calderón completed his secondary education in his hometown and in Zug, Switzerland. After graduating from the School of Engineering of the National University of Buenos Aires in 1947, he studied mathematics under Alberto Gonzalez Dominguez and Antoni Zygmund a renowned mathematician who was a visiting professor in Buenos Aires in 1948).

A Rockefeller Foundation fellowship brought Calderón to the United States, where he received his Ph.D. in mathematics at the University of Chicago three years after earning his bachelor's degree in civil engineering. However, his teaching career had begun in 1948 when he was made an assistant to the chair of electric circuit theory in Buenos Aires. He went on to work as a visiting associate professor at Ohio State University from 1950 to 1953, moving to the Institute for Advanced Study (IAS) in Princeton in 1954-55. Calderón then spent four years at the Massachusetts Institute of Technology (MIT) as an associate professor. Returning to the University of Chicago, he served as professor of mathematics from 1959 to 1968, Louis Block professor of mathematics from 1968 to 1972, and chairperson of the mathematics department from 1970 to 1972.

### Reverses Trends in Mathematical Research

By the early 1970s, Calderón's reputation was well established in scientific circles, and his research in collaboration with his longtime mentor Zygmund had already been dubbed "The Chicago School of Mathematics," also known today as "the Calderón-Zygmund School of Analysis." This movement significantly impacted contemporary mathematics by countering a predominant trend toward abstract mathematics and returning to basic questions of real and complex analysis. This work, completed in tandem with Zygmund, came to be known as the Calderón-Zygmund Theory of Mathematics.

A landmark in Calderón's scientific career was his 1958 paper titled "Uniqueness of the Cauchy Problem for Partial Differential Equations." Two years later he used the same method to build a complete theory of hyperbolic partial differential equations. His theory of singular operators, which is used to estimate solutions to geometrical equations, contributed to linking together several different branches of mathematics. It also had practical applications in many areas, including physics and aerodynamic engineering. This theory has dominated contemporary mathematics and has made important inroads in other scientific fields, including quantum physics. Calderón had created what is now commonly known as pseudodifferential calculus.

Calderón's article on the uniqueness of the Cauchy Problem had such a profound effect on the research community that in 1989 its legacy earned him both the Steele Prize from the American Mathematical Society and the Mathematics Prize of the Karl Wolf Foundation of Israel. In 1991, President George Bush awarded him the National Medal of Science for his career accomplishments.

### Creates Techniques with Broad Applications

The techniques devised by Calderón have transformed contemporary mathematical analysis. In addition to his pseudodifferentials work, he also conducted fundamental studies in interpolation theory—the idea that any algebraic problem that has a rational solution is solvable. He was responsible (with R. Arens) for what is considered one of the best theorems in Banach algebras. Calderón also put forth an approach to energy estimates that has been of fundamental importance in dozens of subsequent investigations, and has provided a model for general research in his field.

A summary of Calderón's work was written by three Yale University mathematicians for publication in the *Notices of the American Mathematical Society* in 1992. The authors assert that Calderón, working in collaboration with Zygmund, invented techniques that "have been absorbed as standard tools of harmonic analysis and are now propagating into nonlinear analysis, partial differential equations, complex analysis, and even signal processing and numerical analysis." They comment that after developing powerful methods in real analysis, "Calderón was not reluctant to return to the methods of complex analysis. . . [which] had fallen out of favor, mostly due to the power of Calderón-Zygmund theory!" The resulting blend of complex and real analysis has been quite fruitful. In 1979 it earned him the Bôcher Memorial Prize, which is awarded once every five years for a notable research paper in analysis, in this case "Cauchy Integrals on Lipschitz Curves and Related Operators."

In the early 1970s, Calderón briefly returned to his home country to serve as a visiting lecturer and conduct mathematical Ph.D. dissertation studies at his alma mater, the National University of Buenos Aires. He has continued to encourage mathematics students from Latin America and the United States to pursue their doctoral degrees, in many instances directly sponsoring them. Some of his pupils, in turn, have become reputed mathematicians. For example, Robert T. Seeley applied Calderón-Zygmund techniques to singular integral operators on manifolds, laying the foundation for the Atiyah-Singer index theorem.

After his stay in Argentina, Calderón returned to MIT as a professor of mathematics and, in 1975, he became university professor of mathematics at the University of Chicago, a special position he held until his retirement in 1985. He subsequently served as a professor emeritus with a post-retirement appointment at that same institution. Beginning in 1975, he was also an honorary professor at the University of Buenos Aires. He lectured extensively around the world, and held occasional visiting professorships at several universities, including Cornell, Stanford, Bogotá, Madrid, Rome, Göttingen, and the Sorbonne.

An active member of the American Mathematical Society for more than 40 years, Calderón served as a member–at–large of its council in the mid–1960s and on several of its committees. He has also been an associate editor of various scientific publications, including the *Duke Mathematical Journal, Journal of Functional Analysis,* and *Advances in Mathematics.*

Calderón married **Alexandra Bellow**, a well-known mathematician in her own right and a professor at Northwestern University. Their daughter, Maria Josefina, holds a doctorate in French literature from the University of Chicago and their son, Pablo Alberto, is a mathematician who has studied in Buenos Aires and New York.

### SELECTED WRITINGS BY CALDERÓN

"Uniqueness in the Cauchy Problem for Partial Differential Equations." *American Journal of Mathematics* 80 (1958): 16–36.
"Cauchy Integrals on Lipschitz Curves and Related Operators." *Proceedings of the National Academy of Sciences* 74 (1977): 1324–27.

### FURTHER READING

Beals, Richard W., et al. "Alberto Calderón Receives National Medal of Science." *Notices of the American Mathematical Society* 39 (April 1992): 283–85.
Ewing, J. H., et al. "American Mathematics from 1940 to the Day Before Yesterday." *American Mathematical Monthly* 83 (1976): 503–516.
"1989 Steele Prizes: Fundamental Paper." *Notices of the American Mathematical Society* 36 (September 1989): 833–35.

*—Sketch by Rodolfo A. Windhausen and Loretta Hall*

*Georg Cantor*

# Georg Cantor
## 1845-1918
### Russian-born German algebraist and analyst

Georg Cantor, a German mathematician, developed a number of ideas that profoundly influenced 20th-century mathematics. Among other accomplishments, he introduced the idea of a completed infinity, an innovation that earned him recognition as the founder and creator of set theory. His revolutionary insights, however, were accepted only gradually and not without opposition during his lifetime. The praise for his work was best epitomized by the famous mathematician **David Hilbert**, who said that "Cantor has created a paradise from which no one shall expel us." Besides being the founder of set theory, Cantor also made significant contributions to classical analysis. In addition, he did innovative work on real numbers and was the first to define irrational numbers by sequences of rational numbers.

Georg Ferdinand Ludwig Philipp Cantor was born on March 3, 1845, in St. Petersburg, Russia, the first child of Georg Woldemar Cantor and Maria Böhm. The family moved to Frankfurt, Germany, in 1856 when the father became ill. His father, born in Copenhagen, had moved to St. Petersburg at a young age and had become a successful stockbroker there. His mother came from an artistic family. Cantor's brother Constantin was an accomplished piano player

and his sister Sophie had drawing talents. Cantor himself sometimes expressed regret that he had not become a violinist. Of Jewish descent on both sides, Cantor was nevertheless raised in an intensely Christian atmosphere. The breadth and depth of his knowledge of the old masters, theologians, and philosophers was brought about by his religious upbringing and became evident in his more philosophical writings.

At a young age, while still in St. Petersburg, Cantor showed clear signs of mathematical talent. Though he wanted to become a mathematician, his father had charted out an engineering career for him. He attended several schools along the lines of his father's wishes, including the Gymnasium in Wiesbaden and, from 1860, a Technical College in Darmstadt. Cantor finally received parental approval to study mathematics in 1862. He started his studies in the fall of that year in Zurich, but moved to Berlin after one semester. Cantor was a solid student. He spent a summer semester in Göttingen in 1866 and successfully defended a Ph.D. thesis in number theory on December 14, 1867, in Berlin. Cantor then moved to the University of Halle as a *Privatdozent,* becoming an associate there in 1872 and a professor in 1879. He remained at Halle for his entire career.

A friend of his sister's, Vally Guttmann, became his wife in 1874. During their honeymoon in Switzerland, the couple met **Richard Dedekind**, from then on a friend and mathematical confidant of Cantor's. Georg Cantor and Vally Guttmann had six children.

### Discovers Set Theory

When Cantor arrived at Halle, the leading mathematician there was Heinrich Heine, under whose influence Cantor began to study Fourier series. His analysis of the convergence of these trigonometric series eventually led to far-reaching innovations. What started as a slight improvement of a theorem on the uniform convergence of Fourier series contained the first seeds of set theory. Cantor's first paper on set theory proper was published in 1874 under the title *Über eine Eigenschaft des Inbegriffes aller reellen algebraischen Zahlen* and dealt with algebraic numbers. An algebraic number is any real number that is a solution to an equation with integer coefficients. Cantor's paper contained the proof that the set of all algebraic numbers can be put in a one-to-one correspondence with the set of all positive integers. Moreover, Cantor proved that the set of all real numbers cannot be put into a one-to-one correspondence with the positive integers. As he later explained it, the set of positive integers has the same power (called *Mächtigkeit* in German) as the set of algebraic numbers, while the power of the set of real numbers is different from either. The 1874 paper was accepted for publication only after Dedekind's intercession.

The set of all algebraic numbers, containing, for example, the square root of 2, is properly larger than the set of all rational numbers (that is, quotients of integers). In turn, the set of rationals contains infinitely many more elements than the set of positive integers. In spite of that, Cantor showed that the three sets—the rationals, the algebraic numbers, and the positive integers—have the same power. Sets like the algebraic numbers or the rationals are said to be countable, and Cantor furnished proof that this is so. However, he discovered that the set of all real numbers is not countable. Encouraged by these successes, Cantor introduced the notion of equipollency of sets in his next paper, written in 1878. Two sets are equipollent if a one-to-one correspondence exists between them. Where he had previously shown that the set of algebraic numbers is equipollent with the set of positive integers, Cantor then proved that the set of points on any surface such as a plane is equipollent to the set of all real numbers. He finished the 1878 paper with the conjecture that every infinite subset of the set of real numbers is either countable or equipollent to the set of all real numbers. That conjecture became known as the continuum hypothesis. The possibility of a one-to-one correspondence between an infinite set and one of its proper subsets had been observed earlier by scientists such as **Galileo** and **Gottfried Leibniz**. The novelty and courage of Cantor's contributions are in his refusal to consider this a contradiction and in using it to define infinite sets of equal power.

### Consolidates Set Theory

Cantor's next paper was published in six installments between 1879 and 1884. Where he had previously come to grips with the countable and had realized the gap between the countable and the continuum, his ideas about infinite sets in general had ripened. The paper broaches the idea of a proof of the continuum hypothesis. In 1882, Cantor defined another main concept of set theory, that of well-ordering. In 1883 he wrote that we may assume as a law of thought that every set can be well-ordered. Earlier, in 1878, Cantor had stated, without proof, "If two point-sets M and N are not equipollent then either M will be equipollent to a proper subset of N or N will be equipollent to a subset of N." This principle later became known as the trichotomy of cardinals. However, Cantor was not able to give a solution to the continuum hypothesis or a proof of the trichotomy of cardinals.

In 1884, Cantor had a nervous breakdown; several of such mental crises would follow. He had applied for a professorship in Berlin but was turned down, strongly opposed by Leopold Kronecker, a former teacher of his. In spite of his illness, Cantor remained active. He worked to institute the German Mathematical Society, founded in 1889, and was instrumental in establishing the first International Congress of Mathematicians in 1897 in Zurich. Between 1895 and 1897 Cantor published his last paper, *Beiträge zur Begründung der transfiniten Mengenlehre* ("Contributions to the Foundation of Transfinite Set Theory") in two parts. In these he defined the transfinite numbers that measure the magnitude of infinite sets.

While there was little enthusiasm for his discoveries within his own country of Germany at this time, Cantor's ideas were gaining support in the world mathematical community. The eventual recognition of sets as a notion underlying all of mathematics led to new fields like topology, measure theory, and set theory itself. Developments at the turn of the century reflected the importance of Cantor's work. At the Second International Congress of Mathematicians in Paris in 1900, the continuum hypothesis was first among 23 problems that Hilbert proposed as central to the development of twentieth-century mathematics. Not much later, in 1904, the mathematician Ernst Zermelo established that every set can be well-ordered, using the so-called axiom of choice. From it followed the trichotomy of cardinals. The earlier controversy between Kronecker and Cantor intensified into a new rage about what was and what was not permitted in mathematics. Much of the debate was later settled by the work of **Kurt Friedrich Gödel** and **Paul Cohen**. The first book on set theory was published in 1906 by **William H. Young** and **Grace C. Young**. These years also showed the beginnings of the study of topology. In 1911, **L. E. J. Brouwer** proved the topological invariance of dimension, at which Cantor himself had tried his hand earlier. In 1914, Hausdorff published the first book on topology, entitled *Grundzüge der Mengenlehre* ("Principles of Set Theory").

Toward the end of his career Cantor's achievements were recognized with various honors. He became honorary member of the London Mathematical Society (1901) and the Mathematical Society of Kharkov and obtained honorary degrees at several universities abroad. A bust of Cantor was placed at the University of Halle in 1928. Perhaps more fittingly, one special subset of the real numbers that he introduced is now known under his name, the Cantor set. Cantor died on January 6, 1918, at the psychiatric hospital in Halle.

## SELECTED WRITINGS BY CANTOR

*Gesammelte Abhandlungen Mathematischen Und Philosophischen Inhalts,* edited by Ernst Zermelo (with a biography by Adolf Fraenkel), 1966

## FURTHER READING

### Books

Bell, E. T. *Men of Mathematics.* New York: Simon & Schuster, 1986.

Meschkowski, H. "Georg Cantor," in *Dictionary of Scientific Biography.* Volume III. Edited by Charles C. Gillispie. New York: Charles Scribner's Sons, 1971, pp. 52–58.

Itô, Kiyosi, editor. *Encyclopedic Dictionary of Mathematics.* Volume I. Cambridge, MA: MIT Press, 1987, p. 169.

Moore, Gregory, H. *Zermelo's Axiom of Choice, Its Origins, Development, and Influence.* New York: Springer Verlag, 1982.

Noether, Emmy and Jean Cavaillès. *Briefwechsel Cantor-Dedekind.* Hermann, 1937.

Young, Laurence. *Mathematicians and Their Times.* North Holland Publishing Company, 1981.

### Periodicals

Stern, Manfred. "Memorial Places of Georg Cantor in Halle." *Mathematical Intelligencer* 10 (1988): 48–49.

—*Sketch by Gerard J. Buskes*

*Girolamo Cardano*

# Girolamo Cardano
## 1501–1576
### Italian algebraist and physician

It would be incomplete to simply list Girolamo Cardano as an Italian physician and mathematician. To reflect the true character of his life, one would have to add that he was the illegitimate son of a noted lawyer, a compulsive gambler, a popular astrologer, a one–time prisoner of the Inquisition, and the father of a convicted murderer. Despite the sordid nature of many aspects of his life, Cardano was also a brilliant physician and mathematician, as well as the author of more than 200 works on human medicine, mathematics, physics, philosophy, religion, natural science, and music. His contributions to mathematics were mainly in the area of algebra and included the first generalized solution to cubic (third degree polynomial) equations as well as the solution to certain quartic (fourth degree polynomial) equations, even though some of the solutions had been borrowed from others. Cardano was also one of the first to recognize the existence of the square root of negative numbers, now called imaginary numbers, although he did not know how to deal with them.

Girolamo Cardano was born in Pavia, Italy, on September 24, 1501. His father was Fazio Cardano, a well–known lawyer and a friend of Leonardo da Vinci. His mother was Chiara Micheri. Cardano's parents were not married at the time he was born, and the stigma of being an illegitimate child followed him through life. As a youth, Cardano was often sick and mistreated. Although his parents did eventually marry, his father did not live with the family until Cardano was seven years old.

When Cardano became of age, his father encouraged him to study the classics, mathematics, and astrology. He entered the University of Pavia, where he completed his undergraduate studies. In 1524, his father died and left him a house and a small inheritance. Cardano returned to study in Pavia, where he earned a doctorate in medicine in 1526. Shortly thereafter he took up a medical practice in Saccolongo near Padua, where he married Lucia Bandareni in 1531. In time they had two sons and a daughter.

### Begins a Second Career in Mathematics

In 1534, friends of his father helped Cardano obtain a teaching position in mathematics at a school in Milan. He continued to practice medicine in addition to his teaching duties, and his success in treating several influential patients soon made him the most sought–after physician in Milan. By 1536, Cardano's thriving medical practice allowed him to leave his job as a teacher, although his interest in mathematics continued.

Cardano published his first book on mathematics, *Practica arithmetice et mensurandi singularis*, in 1539. In this book, he first demonstrated his superior mathematical skills in approaching problems in algebra. In one example, he was able to solve a specific cubic equation by manipulating the terms to reduce the problem to a second–degree quadratic equation that could be solved.

An algebraic solution to cubic and quartic equations had long eluded mathematicians, and Cardano's work in his *Practica arithmetice,* although limited to only certain cubic equations, was impressive. Cardano felt little satisfaction in his accomplishments, however, because he knew that fellow mathematician Nicolò Fontana, also known as **Tartaglia**, had achieved solutions to even more difficult cubic equations four years earlier, but had refused to divulge his methods. Cardano had repeatedly beseeched Tartaglia for his secret to no avail. Finally, just as Cardano's book was being published in 1539, Tartaglia agreed to share his methods, but only on the condition that Cardano take an oath that he would not reveal them until Tartaglia had written his own book.

## Makes a Breakthrough in Advanced Algebra

At this point fate smiled on Cardano. A young man named Ludovico Ferrari came asking for a job as a servant. Cardano took him in and quickly discovered that his new servant possessed a brilliant mind. Eager to share his enthusiasm for mathematics, Cardano taught Ferrari algebra and eventually revealed Tartaglia's secret to him. One of the cubic equations Tartaglia had learned to solve was the so–called "depressed cubic," which lacked a second–power term. Working together, Cardano and Ferrari discovered a method to reduce any generalized cubic equation to a depressed cubic. Using Tartaglia's methods, they could then solve the equation. Their elation in making this monumental advancement in algebra was dampened by the knowledge that Cardano had given his oath not to reveal Tartaglia's methods. What was worse, Tartaglia seemed in no hurry to publish his long–promised book, which would have freed Cardano from his oath.

In 1543, Cardano and Ferrari traveled to Bologna where they went over the papers of another mathematician, **Scipione dal Ferro**, who had died in 1526. They discovered that dal Ferro had solved the depressed cubic equation in 1515, but had kept it secret until just before his death, when he revealed it to his student, Antonio Fior. Fior had foolishly used his new–found knowledge to challenge Tartaglia to a contest, only to have Tartaglia discover the method for himself during the competition.

Armed with the knowledge that it was dal Ferro who had originally solved the depressed cubic equation, Cardano felt his obligation to Tartaglia was removed. Thus, in 1545, Cardano published his second book on mathematics entitled *Artis magnae sive de regulis algebraicis liber unus*, commonly called *Ars magna*, ("Great Art"). In it, he revealed not only the solution to the generalized cubic equation, but also the solution to the biquadratic quartic equation, which had been developed by Ferrari.

Although Cardano gave full credit to dal Ferro, Fior, and Tartaglia for their work on cubic equations, Tartaglia was outraged. He claimed Cardano had broken his oath, and he began a long and vicious letter–writing campaign denouncing Cardano as a scoundrel. Despite this dispute, Cardano's book was widely acclaimed. Besides the solutions to cubic and quartic equations, Cardano presented many other new ideas in algebra that became the basis for the theory of algebraic equations.

## Suffers a Series of Personal Tragedies

In 1546, Cardano's wife died at the age of 31, leaving him with three children. His eldest son, Giambattista, was a promising scholar and it appeared he would become a successful physician like his father. Then, in 1557, Giambattista, poisoned his wife. Despite his father's appeals and influence, Giambattista was convicted of murder and was beheaded in 1560. In 1562, Cardano left Milan and took a position teaching medicine at the University of Bologna. Tragedy struck again in 1565, when his devoted student and collaborator Ferrari was poisoned.

In 1570, Cardano was accused of heresy for having cast the horoscope of Jesus Christ and for attributing the events of Christ's life to the influence of the stars. He was imprisoned by the Inquisition for several months before he was released on the condition that he abandon teaching. At the advice of friends, Cardano moved to Rome in 1571 and asked for protection from the Pope. Pope Pius V refused to help him because of his conviction for heresy, but Pope Gregory XIII was more lenient and granted Cardano a lifetime pension in 1573.

## Influences Scholars in Many Disciplines

Cardano revised his *Ars magna* in 1570 to include a section dealing with solutions to the cubic equation that involve the square roots of negative numbers. Today, these numbers are known as imaginary numbers. Cardano had no such knowledge of imaginary numbers, but the fact that he recognized their existence led to further work by others. His work on solving cubic and quartic equations stimulated work by **Thomas Harriot, Leonhard Euler, René**

**Descartes**, and others over the next several hundred years.

Cardano also wrote several books that presented new ideas in many other disciplines. He studied the physics of projectiles in motion and correctly observed that their trajectories resembled a parabolic curve. Cardano was the first to deduce that the ratio of the distances of a projectile shot through air and water is inversely proportional to the ratio of the densities of the two mediums. In hydrodynamics, he observed and measured the flow of streams and stated that the velocity of the water was greater at the surface than near the bottom, contrary to what most people believed. Cardano's work in geology was also influential.

Cardano retired on his pension and lived quietly in Rome. He died on September 21, 1576, just three days short of his 75th birthday. A century after Cardano's death, **Gottfried Leibniz** summed up Cardano's turbulent life when he wrote "Cardano was a great man with all his faults; without them, he would have been incomparable."

## SELECTED WRITINGS BY CARDANO

*Practica arithmetice et mensurandi singularis,* 1539
*Artis magnae, sive de regulis algebraicis liber unus,* 1545
*De subtilitate liber XXI* 1550; 6th edition, 1560

## FURTHER READING

American Council of Learned Societies. *Biographical Dictionary of Mathematicians.* New York: Charles Scribners' Sons, 1991, pp. 410–413.

Bergamini, David. *Mathematics.* New York: Time Incorporated, 1963, pp. 69–70.

Dunham, W. *Journey through Genius: The Great Theorems of Mathematics.* New York: Wiley, 1990, pp. 133–154.

Struik, D.J., Editor. *A Source Book in Mathematics, 1200–1800.* Princeton, NJ: Princeton University Press, 1986, pp. 62–69.

—*Sketch by Chris Cavette*

# Lazare Nicolas Marguérite Carnot
## 1753–1823
### French geometer

Lazare Nicolas Marguérite Carnot is known for his work in science as well as for his political works.

*Lazare Nicolas Marguerite Carnot*

In the field of science, Carnot worked on engineering mechanics and has published several books on mechanics, geometry, and calculus. Carnot also worked with Napoleon Bonaparte during the French Revolution and is known as the "Organizer of Victory."

Carnot was born in Norlay, Cote–d'Or, France. Norlay was a small town near Beaune in the region of Burgundy, in central France. The ancestral home is still in the possession of Carnot's descendants. Upon finishing his early education in the Oratorian *college* at Autun, Carnot's father enrolled him in a tutoring school in Paris. The school specialized in preparing students for entrance exams to service schools, which trained cadets for the navy, artillery and the Royal Corps of Engineers.

After graduating from the service school, Carnot obtained a commission in the Engineers Corps of Conde. While in the corps, he continued his interest in mathematical studies. In 1783, Carnot wrote an essay, entitled, *"Essai sur les machines en general,"* which W.W. Rouse Ball explains in *A Short Account of the History of Mathematics* ". . . [the work] contains a statement which foreshadows the principle of energy as applied to a falling weight, and the earliest proof of the fact that kinetic energy is lost in the collision of imperfectly elastic bodies."

In 1784, Carnot entered an essay in an Academy of Berlin's competition for a "clear and precise" justification of the infinitesimal calculus. The essay earned Carnot an honorable mention award. The essay served another purpose as well, forming the crux of a book Carnot published later, *Reflexions sur la metaphysique du calcul infinitesimal.*

Carnot's early essays were not widely known. It was not until after he became famous for his military and political pursuits that his mathematical and scientific adroitness were recognized. Carnot began his political career in 1791 by being elected a deputy to the Legislative Assembly from the Pas–de–Calais. During the next year, the monarchy came under scrutiny and Carnot's republican views became apparent. Carnot had strong convictions, taking on a more humane view of war. He believed that warfare should be in defense of civilization, not the destruction of the enemy. His views were due, at least in part, to his social class and family background.

With the outbreak of the French Revolution in April 1792, things began to change for Carnot. As an engineer, he had the exceptional ability to improvise and organize. After the overthrow of the monarchy in August 1792, Carnot took charge and forced the authority of the republic on officers and agents who were loyal to the Crown. In the spring of 1793, having earned a reputation as a tough and reliable patriot, Carnot was elected to an emergency Commission of Public Safety. During this time, Carnot went against his republican beliefs and fought in the war. He is known as the "Organizer of Victory" because he led the French armies to victory against the monarchy.

Carnot served on the Committee until 1797, when he was extricated from the government by a leftist coup d'etat. He then fled to Geneva, Switzerland, where he published *Reflexions sur la metaphysique du calcul infinitesimal,* which took its basis from the essay he wrote for the competition at the Academy of Berlin.

From Geneva, Carnot journeyed to Germany, then back to France, not long after Napoleon Bonaparte appropriated power. Although Napoleon named Carnot his minister of war, he resigned after only a few months. In 1796, Carnot qualified for membership in the Institut de France, so after resigning as minister of war, he returned to his other interests, technology, and science. Because of his prominence as a military and political figure, Carnot's early writings went into print about this time. During Napoleon's reign, Carnot was appointed to numerous commissions by the Institut to analyze the qualities of many mechanical inventions. However, as Napoleon's reign began to collapse, Carnot again offered his services to him. Napoleon designated Carnot as the governor of Antwerp. Later he served as Napoleon's last minister of the interior.

Carnot, along with **Gaspard Monge,** were key figures in the establishment of the École Polytechnique. The school was an engineering school, nevertheless, it had many prominent mathematicians, including **Pierre Laplace,** who made important contributions in astronomy, mechanics, and probability.

Carnot educated his son, Nicolas Leonard Sadi Carnot, at the school. Sadi, as he was called, graduated in 1814, and later conceived the Carnot Cycle, a cycle of operations that established the ultimate limits on heat engines. He also contributed the theory of thermodynamics.

Carnot published various works during his lifetime, ranging from mechanics to geometry to the foundations of calculus. He revised *Essai sur les machines en general* and published it under the title *Principes fondamentaux de l'equilbre et du mouvement.* Not until the 1820s when Sadi, along with his contemporaries, put the theoretical practice to work, did the book have any effect on the actual treatment of problems. The purpose of this work was to explain, in a general way, the optimal conditions for the operation of all kinds of machines.

Carnot and his son were not the only notable Carnots. Carnot's grandson served as president of the French Republic from 1887 to 1894.

Carnot was forced into exile after the monarchy was restored. He fled to Magdeburg, Germany, taking his younger son, Hippolyte, with him yet leaving Sadi behind in France. Carnot died in Magdeburg, Germany on August 2, 1823.

## SELECTED WRITINGS BY CARNOT

*Reflexions sur la metaphysique du calcul infinitesimal,* 1797
*De la defense des places fortes,* 1812
*Essai sur les machines en general,* 1783
*Geometrie de position,* 1803

### Collected Papers

*Proces–verbaux des seances de l'Académie tenues depuis la foundation de l'Institut jusqu'au mois d'aout 1835,* Volumes I–V, 1910–1914

## FURTHER READING

### Books

Gillispie, Charles Coulston. *Lazare Carnot Savant: A Monograph Treating Carnot's Scientific Work, with Facsimile Reproduction of His Unpublished Writings on Mechanics and on the Calculus, and an Essay Concerning the Latter by A.P. Youschkevitch.* Princeton: NJ: Princeton University Press, 1971.

_____. "Lazare Nicolas Carnot," in *Dictionary of Scientific Biography*. Volume III. Edited by Charles Coulston Gillispie. New York: Charles Scribner's Sons, 1971, pp. 70–79.

Newman, James R., editor. *The Harper Encyclopedia of Science*. Washington, D.C., 1967, pp. 724–727.

**Periodicals**

Thiele, R. "A French Officer in Prussian Magdeburg." *The Mathematical Intelligencer* 15, no. 1 (1993): 53–57.

**Other**

"Lazare Nicolas Marguerite Carnot." *MacTutor History of Mathematicians.* http://www-groups.dcs.st-and.ac.uk:80/~history/Mathematicians/Carnot.html. (May 5, 1997).

Ball, W.W. Rouse. "French Contemporaries of Lagrange and Laplace." *A Short Account of the History of Mathematics.* http://www.maths.tcd.ie/pub/HistMath/Peop . . . leonic/RouseBall/RB_FrAlgGeom.html#Carnot.

(May 9, 1997).

—*Sketch by Monica L. Stevens*

*Élie Joseph Cartan*

# Élie Joseph Cartan
## 1869–1951

**French number theorist, topologist, and geometer**

Élie Joseph Cartan is one of the most important mathematical figures in the first half of the 20th century. Although recognition for his many accomplishments did not come until late in his career, his intellectual influence is still felt as modern mathematics develops. From the earliest point in his career, Cartan further developed Norwegian mathematician **Marius Sophus Lie**'s group theory which concerned continuous groups and symmetries within them. He interworked Lie's theory and the original means in which he studied its global properties with differential geometry, classical geometry, and topology. These combined areas are still a vital part of contemporary mathematics. Cartan also formulated many original theories based on his studies. Despite making such vital contributions to mathematics, Cartan was quite modest, good humored, and easy going. He was also a gifted and well-liked teacher who could break down his intricate theories for the consumption of an average student.

Cartan was born April 9, 1869, in Dolomieu Isère, France, a village in the Alps. Of peasant descent, he was the son of the village blacksmith, Joseph, and his wife, Anne (Cottaz) Cartan. Cartan was the second oldest of four children. An inspector of primary schools, Antonin Dubost, noticed Cartan's impressive scholastic aptitude while on a visit to Cartan's school. Though most children from poor families were not put on the track to attend university, Dubost helped Cartan to win a scholarship to attend lycée (secondary school). Cartan's success as a student inspired his youngest sister, Anna, to follow in his intellectual footsteps, and become a mathematics teacher. She also published several texts on geometry.

After attending three lycées, Cartan went on to the l'Ecole Normale Supérieure in Paris in 1888, obtaining his doctorate in 1894. In his graduate thesis, Cartan began the first phase of his life's research, that of Lie's theory of continuous groups, a topic neglected by most of his contemporaries. In his thesis Cartan completed the classification of semisimple algebras begun by Wilhelm Killing during the last two decades of the 19th century, and gave a rigorous foundation for the types of Lie algebras that Killing had shown to be possible.

Before pursuing a career as a mathematician, Cartan was drafted into the French Army for one year after his graduation, rising to the rank of sergeant before his discharge. Cartan then served as a lecturer at two successive institutions, the University of Montpellier (1894 to 1896), and the University of Lyons (1896 to 1903). While a lecturer, he continued his mathematical studies based on Lie's theory, and began bringing together the four disparate disciplines that became the hallmark of Cartan's work: differential geometry, classical geometry, topology, and Lie theory. Cartan began by exploring the structure for associative algebras, then moved onto semisimple Lie groups and their representations. In this period (roughly from 1894 to 1904), Cartan also helped create and develop the calculus of exterior differential forms. This calculus became an important tool Cartan used in his research and he applied it to various differential geometric problems.

In 1903, Cartan's life changed in two fundamental ways. He married Marie–Louise Bianconi in that year. With her he had four children (three sons and one daughter), and enjoyed a happy home life. His eldest son, **Henri Paul Cartan**, born in 1904, became a prominent mathematician in his own right. His daughter Hélène also became a mathematician who taught at lycées and published some mathematical papers. Cartan's two other sons died tragic deaths. Jean, a composer, died at 25 of tuberculosis; Louis, a physicist, was arrested and imprisoned by the Nazis for his activities in the French Resistance. Louis Cartan was executed by the Nazis in 1943, but his family did not learn of his death until 1945.

In 1903, Cartan also became a professor at the University of Nancy, where he worked until 1909. Cartan moved on to the University of Paris (the Sorbonne) in 1909, where he was a lecturer from 1909 to 1912 before becoming a full professor in 1912. He remained a professor there until he retired in 1940. In his first years at the Sorbonne, in 1913, Cartan made one of his most significant contributions to math when he discovered spinors. The spinors are complex vectors used to make two–dimensional representations out of three–dimensional rotations. The spinors were important in the development of quantum mechanics.

Though Cartan continued to intertwine the four subjects mentioned earlier, at the height of his career, he continued to look at them from different perspectives. For example, after 1916, Cartan's publications focused mainly on differential geometry. Cartan developed a moving frames theory and a generalization of Daboux's kinematical theory. In 1925, Cartan refocused his attention to the study of topology. In his paper "La géométrie des espaces de Riemann" ("The Geometry of Riemann Spaces"), Cartan came up with innovative ways to study Lie groups' global properties. He demonstrated, in topological terms, that a Euclidean space and a compact group can produce a connected Lie group. And, from 1926 to 1932, Cartan published treatises concerning his geometric theory of symmetric spaces.

Cartan's continual output of important work led to his appointment as a member of the Académie Royale des Sciences in 1931, one of many honors he received late in his career. About the same time, Cartan began collaborating with his son Henri on mathematical problems, building on his son's theorems. On Cartan's 70th birthday, in 1939, the Sorbonne honored him with a celebratory symposium which praised his many mathematical accomplishments. Although Cartan retired from the Sorbonne in 1940, he remained an honorary professor there until his death. He also continued to publish mathematical treatises and his love of classroom instruction led him to teach math at the l'Ecole Normale Supérieure for Girls. Cartan died in Paris on May 6, 1951, after a long illness.

## SELECTED WRITINGS BY CARTAN

"La géométrie des espaces de Riemann," in *Mémorial des Sciences Mathématiques.* Volume 9, 1925

## FURTHER READING

### Books

Akivis, M.A., and B.A. Rosenfeld. *Translations of Mathematical Monographs: Élie Cartan.* Volume 123. Providence, RI: American Mathematical Society, 1991.
Daintith, John, Sarah Mitchell, and Elizabeth Tootill, editors. *A Biographical Encyclopedia of Scientists.* Volume 1. New York: Facts on File, Inc., 1981, pp. 144–45.
Dieudonné, Jean. "Élie Cartan." *Dictionary of Scientific Biography.* Volume 3. Edited by Charles Coulston Gillispie. New York: Charles Scribner's Sons, 1970, pp. 95–96.

### Periodicals

Shiing–Shen and Claude Chevalley. "Élie Cartan and his Mathematical Work." *Bulletin of the American Mathematical Society* (March 1952): 217–250.

*—Sketch by Annette Petruso*

# Henri Paul Cartan
## 1904–
### French algebraist

Henri Cartan has made monumental contributions in essentially every field of algebraic topology, including analytical functions, the theory of sheaves, homological theory, and potential theory. His most important works include *Homological Algebra* (1956) and *Elementary Theory of Analytic Functions of One or Several Complex Variables* (1963). He also worked on a definitive convergence theorem on decreasing potentials of positive masses, which became a fundamental instrument for improving potential theory. Cartan is the recipient of the 1980 Wolf Prize in Mathematics.

Cartan was born in Nancy, France, in 1904. His father was **Élie Joseph Cartan**, a French mathematician who made significant contributions to the theory of subalgebras. Cartan was educated at the École Normal Superieure in Paris, one of the finest schools for mathematics. Cartan attended the École from 1923-1926. He was a protégée of **Jules Henri Poincaré**, who is considered the founder of algebraic topology and the theory of analytic functions of several complex variables. Along with **Albert Einstein** and Hendrik Lorentz, Poincare is considered a co–discoverer of the special theory of relativity. Cartan's studies with Poincare may explain his own interest in algebraic topology.

### The "Birth" of Nicholas Bourbaki

In 1935, Cartan was one of the founders of a group of young French mathematicians who had all graduated from the École Normal Superieure. Other members included Claude Chevalley (or Chevallier), Jean Dieudonné, Jean Deslarte, and **Andre Weil.** All members were brilliant in their own right, however, Weil was the only "universalist" among them, accomplished in every area of mathematics. At the beginning, Cartan was not a universalist, although he did become more accomplished later in his career.

The group wrote under the pseudonym of Nicolas Bourbaki. In "Nicholas Bourbaki, Collective Mathematician," Claude Chevalley, as told to Denis Guedj, explains how Bourbaki got his name: "Weil had spent two years in India and for the thesis of one of his pupils he needed a result he couldn't find anywhere in the literature. He was convinced of its validity, but he was too lazy to write out the proof. His pupil, however, was content to put a note at the bottom of the page which referred to 'Nicolas Bourbaki, of the Royal Academy of Poldavia'." The "real" Bourbaki liked to pretend he was in a secret society when he was young. From this, Weil tried to get the members of the group to stay anonymous. Members refused to answer questions about other members and of projects they were working, and even how the name of Bourbaki originated. The group did not remain anonymous for very long.

The Bourbaki group produced *Elements de mathematique*, a 30–volume textbook on analysis geared toward French university students, in 1939. It was intended to replace the class analysis textbook used in France written by Edouard Goursat. Bourbaki's book aimed to achieve the same high standards that Goursat set forth in his text, yet it also included mathematical advances. By 1968, there were 33 volumes in print. Early editions of *Elements* did not include credit to other contributors, so in 1960, *Elements d'historique des mathematiques* was published to rectify the situation.

The group sought the best way to produce the work together. They did not just want to assign topics to the person most qualified to write it, believing if they did, the book would turn out like an encyclopedia, which they did not want. Each member studied the same areas of mathematics, from the beginning. During the Bourbaki–congres, as their thrice yearly meetings were called, the group would discuss sections of the book, making numerous revisions. Bourbaki members had to retire from the group by the age of 50. This assurance was made to keep Bourbaki from "growing old," keeping him young in spirit. Women were not allowed to join the Bourbaki group. Through the Bourbaki members, a new kind of algebraic topology was born. Bourbaki played a key role in the rethinking of structural mathematics.

### IMU and Academia

Cartan became president of the International Mathematical Union (IMU) in 1967, serving in that capacity until 1970. He then became the past–president, holding a seat on the executive committee for the following four years. The IMU is a scientific organization with the purpose of promoting international cooperation in mathematics and is part of the International Council of Scientific Unions. Founded in 1919, the IMU was disbanded in 1936, then reconstituted in 1951. Cartan was an invited speaker for a May 1995 IMU conference taking place at the Institut Henri Poincaré.

Cartan's academic career began at the Lyceé Caen in 1928, where he was a professor of mathematics. He left the Lycee in 1929, and was appointed a deputy professor at the University of Lille. From there, he went to he University of Strasbourg, which named him a professor of mathematics in 1931. From 1940 until 1969, Cartan was on the faculty at the University of Paris. After the University of Paris, Cartan taught at Orsay for about five years, until 1975.

### A Warning to Humanity

In 1992, Cartan was part of a group of international scientists issuing a warning to humanity. The warning urged changes to protect the living world. According to the Worlds' Scientists Warning to Humanity, found on the World Wide Web, "Human beings and the natural world are on a collision course. . . . If not checked, many of our current practices put at serious risk the future that we wish for human society and the plant and animal kingdoms, and may so alter the living world that it will be unable to sustain life in the manner that we know. . . . Fundamental changes are urgent if we are to avoid the collision our present course will bring about." The warning lists several environments in need of help. Cartan's name is among the 1,500 signatories, all of which are members of regional, national, and international science academies.

## SELECTED WRITINGS BY CARTAN

### Books

*Homological Algebra,* 1956
*Elementary Theory of Analytic Functions of One or Several Complex Variables,* 1963
*Differential Forms,* 1971

### Periodicals

"Colloque 'Analyse et Topologie' en l'Honneur de Henri Cartan." *Asterisque 32–33, Soc. Math. France* (1976): 3–4.
"Sixteen French Intellectuals On: The Joys of Being Jewish in the USSR." *The New Leader* (November 28, 1983): 5–6.

### As Nicolas Bourbaki

*Elements de Mathematique,* 1939
*âElaements d'histoire des mathaematiques,* 1960, 1969
*General Topology,* 1966
*Cummutative Algebra,* 1972

## FURTHER READING

### Books

Fang. J. *Mathematicians From Antiquity to Today.* Hauppauge, NY: Paideia Press, 1972, pp. 269–270.

### Periodicals

"Breve analyse des travaux de Henri Cartan, Colloque 'Analyse et Topologie' en l'Honneur de Henri Cartan." *Asterisque 32–33, Soc. Math. France* (1976): 5–21.
Dimiev, S. and R. Lazov. "Henri Cartan: On the Occasion of His 80th Birthday." *Fiz.–Mat. Spis. Bulgar. Akdad. Nauk.* 26 (59) (2) (1984): 222–225.

Guedj, Denis. "Nicholas Bourbaki, Collective Mathematician, An Interview with Claude Chevalley." Translated by Jeremy Gray. *The Mathematical Intelligencer* 7, no. 2, (1985): 18–22.
"Liste des travaux de Henri Cartan." *Asterisque 32–33, Soc. Math. France* (1976): 22–27.

### Other

Conlay, Mike. "Nicolas Bourbaki." http://www.mcs.csuhayward.edu/~malek/Mathlinks/Bourbaki.html (May 9, 1997).
"The Bourbaki View." http://www.rbjones.com/rbjpub/logic/jrh0105.html (May 9, 1997).
"Henri Paul Cartan." *MacTutor History of Mathematician.* http://www–groups.dcs.st–and.ac.uk/~history/Mathematicians/Cartan_Henri.html. (April 25, 1997).
"The International Mathematical Union (IMU)." Online. Available at http://130.225.112.194:8000/0x82e170c2_0x00000b59;sk=311225BD (May 16, 1997).
"IMU Executive Committees 1952–1998." http://130.225.112.194:8000/0x82e170c2_0x00000b59a;sk=311225BD (May 16, 1997).
"World Scientists' Warning to Humanity." http://www.livelinks.com/sumeria/warning.html (April 29, 1997).

*—Sketch by Monica L. Stevens*

# Mary Lucy Cartwright
## 1900–
### English analyst

Dame Mary Lucy Cartwright's research and contributions to the field of mathematics have spanned more than seven decades. England claims her as one of its most brilliant citizens and has honored her for the past 50 years. As late as 1996, she was featured, along with two other women scientists, on British television, modestly answering questions about her impressive achievements.

Cartwright was born in Aynho, Northamptonshire, England, on December 17, 1900. Her father, William Digby Cartwright, served as a curate to his uncle, who was a rector at a church. Her mother was Lucy Harriette Maud Bury. The youngest of three children, Cartwright's two other brothers, John and Nigel, were both killed in World War I. When Cartwright was 11 years old, she was sent to live with her maternal uncle, Fred Bury, and his wife Annie, so she could attend Leamington High School, now known as the Kingsley School. There, she learned mathematics from a teacher she remembers as Miss Hancock, to whom she still pays tribute as "an excellent teacher of mathematics." In 1919, Cartwright entered St. Hughes College at Oxford and remained there until 1923. At the age of 23 she graduated from the University of Oxford, then taught first at the Alice Ottley School in Worcester, then at the Wycombie Abbey School for four years before returning to Oxford to complete her doctorate in mathematics. In 1925 Cartwright received the Hurry Prize for mathematics.

One of only five students, Cartwright began study and research with **Godfrey Hardy** in 1928 and it was under his supervision (as well as of E.C. Titchmarsh) that she pursued her degree. Initially, Hardy was Cartwright's advisor on her thesis, then Titchmarsh took over when Hardy left to spend some time at Princeton University. Hardy is best known for the Hardy–Weinberg law, which resolved the controversy over what proportions of dominant and recessive genetic traits would be reproduced in a large mixed population. Cartwright had great admiration for Hardy, both for his work and for the pains he took in instructing his students. Hardy also collaborated with John E. Littlewood on a series of papers that contributed fundamentally to the theory of Diophantine analysis, divergent series summation, and Fourier series.

### Collaboration and Accomplishment

Littlewood was Cartwright's examiner for her Ph.D. in 1930. She was the first woman to read for finals in mathematics at Oxford. The meeting was the first in a long series of conferences that would eventually lead to more than ten years of collaboration on a number of projects. Together, Cartwright and Littlewood published four papers on large parameters, differential equations combining topological and analytical methods, solutions for the Van der Pol equation, and chaos theory. They published several other papers, but the content was based on their collaborative work. Based on her own research, Cartwright also published a number of papers concerning classical analysis. Cartwright was a prolific writer; besides mathematical papers, she produced biographical essays on other women mathematicians such as **Shelia Scott Macintyre** and **Grace Chisholm Young** for the London Mathematical Society's *Journal*.

Also in 1930 Cartwright became a Yarrow Research Fellow of Girton. During her fellowship she attended lectures given by Edward Collingwood concerning integral and metamorphic functions. She worked closely with Collingwood on cluster sets in the theory of functions of one complex variable and these papers were also published. Other publications by Cartwright include *Religion and the Scientist*, Specialization in Education, The Mathematical Mind, and *Integral Functions*.

In 1935, Cartwright was given a Faculty Assistant Lectureship at Cambridge University and her Yarrow Research Fellowship was extended. Cartwright contributes the appointments to a paper she published entitled "Mathematische Zeitschrift" in early 1935. An earlier version of the paper had been shown to Hardy and Littlewood, and Cartwright believes it was upon their joint recommendations that she was given her position at Cambridge.

During World War II, Cartwright volunteered and served with the British Red Cross Detachment from 1940 to 1944. Cartwright was elected to the Council of the London Mathematical Society for the first time in 1933 and served as a member until 1938. She was again elected in 1961 and served as president until 1963.

In 1949, she was appointed Mistress of Girton College and held that position until her retirement in 1968. During that time she received many honors, including an honorary Doctor of Laws degree from Edinburgh in 1953 and Doctor of Science degrees from Leeds in 1958 and from the University of Wales in 1967. Cartwright was elected to the Royal Society of London in 1955 and received the Sylvester Medal of the Royal Society in 1964. Succeeding her Yarrow Fellowships and Lectureship appointments, she became a Reader in Theory of Functions in 1959. At the same time, she additionally served as director of studies in mathematics. In 1968, she was awarded the De Morgan Medal of the London Mathematical Society.

### Earns Worldwide Recognition

Cartwright has visited numerous countries around the world, sharing and acquiring knowledge. She was a consultant on the United States Navy Mathematical Research Project at Stanford and Princeton Universities in 1949. Cartwright returned to the United States in 1968 after her retirement and spent two years at Brown University in Providence, Rhode Island, as a Resident Fellow. While at Brown, she also lectured at Clairmont Graduate School and Case Western Reserve University. Cartwright's years in America made a lasting impression on her. She wrote extensively in her memoirs of the killing of three students at Kent State University and of the many college protests she witnessed, including those at the University of California at Berkeley and the Madison Army Research Centre.

In 1973, she received recognition for her lifetime achievements from the University of Jyvaskyla, Finland. The culmination of Cartwright's recognition came in 1969, when she was ordained Dame Commander of the British Empire by Queen Elizabeth II. Cartwright has henceforth been known as Dame Cartwright. She currently leads a sequestered life in Cambridge.

## SELECTED WRITINGS BY CARTWRIGHT

"Moments in a Girl's Life," in *Bulletin of the Institute of Mathematics and Its Applications* 25 (March–April 1989): 63–67.

## FURTHER READING

Kay, Ernest. *The World's Who's Who of Women.* Volume 1. Cambridge and London: Melrose Press Limited, 1973, p. 140.
McMurran, Shawnee L. and James Tattersall. "Mr. Littlewood and I": The Mathematical Collaboration of M.L. Cartwright and J.E. Littlewood. *The American Mathematical Monthly* 103, no. 10 (December 1996): 833–43.

### Other

"Dame Mary Lucy Cartwright." *Biographies of Women Mathematicians.* http:// www.scottlan.edu/lriddle/women/chronol.htm

—*Sketch by Kelley Reynolds Jacquez*

# Pietro Antonio Cataldi
## 1548–1626
### Italian number theorist, algebraist, and astronomer

Pietro Antonio Cataldi published 30 mathematical works in the course of his life. The most important of these treatises, published in 1613, delineated some of the earliest work on continued fractions, including its definition, common form, and symbolism using standard notation. Cataldi also used continued fractions to discover numerical square roots. Cataldi's mathematical accomplishments were not limited to continued fractions: he published treatises concerning algebra, perfect numbers and arithmetic, and served as an editor of other mathematical books. An influential teacher as well as a mathematician, Cataldi taught at higher learning institutions in Perugia and Bologna for most of his career.

Cataldi was born in Bologna on April 15, 1552. After studying at the Academy of Design in Florence, Cataldi began teaching math at the age of 17. He taught there from 1569 to 1570, although some sources state that he taught mathematics and astronomy in Perugia that year. He preferred to lecture in Italian rather than the traditional Latin, and continued to teach both subjects in Italian throughout his teaching career. By 1570 (some sources cite 1572), Cataldi was in Perugia, and he held positions at the University of Perugia and the Perugia Academy of Design until 1584.

Cataldi returned to Bologna in 1584, and he received his doctorate that same year. He remained in Bologna teaching mathematics and astronomy at the Studio di Bologna (the University of Bologna in some sources) until his death. Cataldi exerted a looming influence on mathematics and how it was taught while employed at the university, but it was not his only source of support. Most of Cataldi's influential publications were written upon his return to Bologna. If the books' dedications are any indication, Cataldi may have had many different patrons while at this post. Among his apparent supporters were important officials, both ecclesiastic and court–based, a city magistrate and various aristocrats, some of whom were members of the Bolognian Senate.

Cataldi's first noteworthy publication, *Practica aritmetica*, was published in four parts, beginning in 1606. One source indicates that Cataldi paid for its production, although the book was dedicated to the Senate of Bologna. The rest of the series was published in 1606, 1616, and 1617. Cataldi arranged for copies of *Practica aritmetica* to be handed out at no cost by Franciscan monks to underprivileged children and some religious organizations.

In 1613, Cataldi published *Trattato del modo brevissimo di trovar la radice quadra delli numeri*, arguably his most important mathematical contribution. The text of *Trattato* was apparently complete by 1597, but was not published for 16 years. In the text, Cataldi explores continued fractions (also known as infinite algorithms). His work specifically employs infinite series and unlimited continued fractions in finding a number's square root. Cataldi's work in continued fraction theory in *Trattato* predates other mathematicians' work, and aided in the exploration of the nature of numbers at least through the 19th century.

Cataldi also addressed problems related to the military, specifically concerning the possibilities of certain artillery's range in *Trattato*. Five years later, in 1618, Cataldi published another book that brought together military and mathematics. *Operetta di ordinanze quadre* addresses algebra as it relates to the military.

Cataldi's contributions to mathematics were not limited to published treatises. He served as editor of a 1620 edition of the first six books of **Euclid**'s *Elements*. In another of Cataldi's own publications, *Operetta delle linee rette equidistanti et non equidistanti*, he expounded on Euclid's fifth postulate, using remainders to demonstrate it. This book was influential in the development of non–Euclidean geometry. Cataldi also attempted to organize a Bologna–based mathematics academy. Sources disagree on his success: some believe political opposition prevented its formation, while others say he succeeded.

Cataldi died in Bologna on February 11, 1626. In his will, he requested that his home was to be used as the base for a new school for mathematics and sciences. This wish was never fulfilled.

## SELECTED WRITINGS BY CATALDI

*Trattato del modo brevissimo di trouare la radice delli numeri*, 1613

## FURTHER READING

### Books

Carruccio, Ettore. "Pietro Antonio Cataldi," in *Dictionary of Scientific Biography*. Volume III. Edited by Charles Coulston Gillispie. New York: Charles Scribner's Sons, 1971, pp. 125–29.

—*Sketch by Annette Petruso*

*Augustin–Louis Cauchy*

# Augustin–Louis Cauchy
## 1789–1857
### French number theorist

Augustin–Louis Cauchy brought formalism to mathematics in the 19th century and defined the concepts of limits, continuity, and derivatives, familiar to modern–day students of calculus. He made contributions to number theory, developed ideas on the convergence of infinite series, and contributed to the field of astronomy. Cauchy's most significant contribution to mathematics, however, is as the co-founder, along with **Karl Gauss**, of complex analysis.

Less than two months after the fall of the Bastille, on August 21, 1789, Augustin–Louis Cauchy, the son of Louis–France Cauchy and Marie–Madeleine Desestre, was born in Paris. It was a time of great turmoil in France, and Cauchy's father, having served the recently overthrown monarchy as a parliamentary lawyer and a lieutenant of police, thought his new family would be safer if they moved to the countryside. Shortly after Cauchy's birth, they took up residence in the village of Arceuil. Cauchy spent the first 11 years of his life there, enduring food scarcity and suffering from malnutrition. His frail condition kept him from being physically active and he spent much of his childhood studiously poring over books. Because schools had been shut down after the revolution, Cauchy was educated primarily by his father, whose lessons emphasized the study of religion and French and Latin verse. Cauchy's mathematical talents were recognized and encouraged by a family friend, one of the leading mathematicians of his day, **Pierre Laplace.** Laplace, too, had retreated to the countryside after the storming of the Bastille, and owned a large estate not far from the Cauchy's cottage.

In 1800, Cauchy's father accepted a position in the new French government as Secretary of the Senate. The family returned to Paris, where Cauchy soon became exposed to another leading mathematician, **Joseph–Louis Lagrange.** Impressed with the young boy's mathematical ability, Lagrange predicted that Cauchy would someday supplant both he and Laplace in mathematical prominence. Cauchy entered the Central School of the Pantheon at the age of 13, taking top honors. In 1805, he enrolled at the École Polytechnique, and two years later entered a college of civil engineering. By 1810 his superior academic performance had earned him a high ranking commission in Cherbourg as a military engineer.

## Co–founds Complex Analysis and Brings Rigor to Calculus

Although Cauchy's first job was demanding, he still found time to begin an exhaustive study of mathematics and astronomy. By 1813, when he moved back to Paris, he had already written one important treatise on symmetrical functions and another on polyhedra. In 1814, he published a manuscript on definite integrals with complex–number limits, independently deriving and more fully developing the theory of functions of a complex variable that had been established by Gauss only a few years earlier. This 1814 publication was to lay the foundation for a more lengthy treatment of complex analysis 12 years later.

In 1815, Cauchy proved a theorem of **Pierre de Fermat**'s that had defeated many of his distinguished predecessors, including **Leonhard Euler**, Lagrange, and **Adrien–Marie Legendre**. That same year he took a post at the École Polytechnique, lecturing on analysis. Then he turned himself to applied mathematics, winning the Grand Prize of the French Académie Royale des Sciences in 1816 for his work on the propagation of waves. Shortly afterward Cauchy was appointed a member of the Académie and was promoted to the position of professor at the Ecole Polytechnique. He also received an appointment at the College de France and the Sorbonne. Cauchy soon became known for the clarity of his mathematical presentations, especially on the subject of calculus. He was not yet 30 years old, and mathematicians from across Europe were traveling to Paris to attend his classes. In 1821, at the urging of Lagrange and others, Cauchy began publishing his lectures. His efforts resulted in three major textbooks, *Cours d'Analyse de l'Ecole Polytechnique* in 1821, *Resume des Lecons sur le Calcul Infinitesimal* in 1823, and *Lecons sur les Applications de Calcul Infinitesimal a la Geometrie*, from 1826 to 1828. In these works, Cauchy established the definitions of limits and continuity and discussed conditions for convergence of infinite series. In his third book he also developed the field of complex analysis, which he had first written about in 1814. The material Cauchy set forth in his texts has stood the test of time and would be familiar to any student of calculus today. Through his definitions and proofs of theorems, Cauchy brought rigor to the field of calculus. Many believe these three volumes are the most important of Cauchy's writings.

During his mathematical career, Cauchy established a reputation not only for his lectures but for his prodigious output. He sometimes generated two full–length manuscripts a week. In 1826, he founded a personal journal, *Exercises de Mathematique*, superseded in 1830 by *Exercise d'Analyse Mathematique et de Physique*. In these publications Cauchy presented his expositions on various topics in mathematics and physics. His articles often ran several hundred pages in length. In 1835, when the Académie began publishing its weekly bulletin, the *Comptes Rendu*, Cauchy began flooding this publication with his memoirs as well. Eventually printing costs got out of hand and the Académie had to limit manuscripts to four pages. Over the course of his career, Cauchy wrote more mathematical works than anyone in the history of mathematics, other than Leonhard Euler.

## Imposes Self–Exile, Then Fights for Freedom of Conscience

In 1830, King Charles X was unseated from the throne in France and went into exile. Cauchy, who was loyal to the king, refused to take an oath of allegiance to the new government. He too went into exile, first to Switzerland, where he lived in a Jesuit community, then to Italy, where he was appointed professor of mathematical physics at the University of Turin. In 1833, Charles invited Cauchy, supposedly as a reward for his loyalty, to become the tutor to his 13–year–old grandson, the Duke of Bordeaux. Charles also bestowed upon Cauchy the title of Baron. Leaving Turin to tutor the Duke in Prague, Cauchy now entered a relatively unproductive phase of his mathematical career. His most important work during this time was a memoir in physics, pertaining to the dispersion of light. Cauchy returned to France in 1838 and resumed his post at the Académie in Paris, where a special dispensation from oaths of allegiance had, by now, been granted for its members. Cauchy's productivity soared. Over the remaining 19 years of his life, he was to publish approximately 500 manuscripts in areas ranging from pure mathematics to astronomy to mechanics.

Soon after Cauchy had returned to the Académie, a position at the College de France became available, which he was asked to fill. There was no special dispensation at the College regarding oaths of allegiance, so Cauchy turned down the post and went to work instead for the Bureau des Longitudes. There, the fact that he had not taken an oath of allegiance was temporarily overlooked. Over the next four years Cauchy made a variety of contributions to the field of astronomy, including a description of the motion of the asteroid Pallus. Eventually, the government attempted to oust him from the Bureau and Cauchy became embroiled in a political stand–off. The situation culminated in an open letter to the French citizenry regarding freedom of conscience and the government eventually retreated. Cauchy joined the Sorbonne faculty in 1848, where he remained until 1852.

On May 23, 1857, in the countryside near Paris, Cauchy died. He had left the city in order to recuperate from a bronchial condition, but developed a fever that eventually consumed him. Cauchy married Aloise de Bure, the daughter of a cultured family, in 1818. Together they had two children.

## FURTHER READING

### Books

Belhoste, B. *Augustin–Louis Cauchy: A Biography.* New York: Springer–Verlag, 1991.
Bell, E.T. *Men of Mathematics.* New York: Simon and Schuster, 1986, pp. 270–293.
Cajori, Florian. *A History of Mathematics.* New York: Chelsea Publishing Co., 1991.

*—Sketch by Leslie Reinherz*

# Bonaventura Cavalieri
## 1598–1647
### Italian geometer, physicist, and theologian

Bonaventura Cavalieri refined early Greek work on the concept of indivisibles. His work served as a stepping stone to the concept of infinitesimals and was the foundation of **Isaac Newton**'s development of the calculus. Cavalieri was born in Milan, Italy, in 1598. His actual birth date is uncertain; even his first name is unknown, because he entered a monastic order at an early age and adopted the first name Bonaventura as his religious name.

Cavalieri entered the Jesuatis (a Roman Catholic order founded on the rule of St. Augustine and not to be confused with the Jesuits or Society of Jesus) and took minor orders in 1615 at a monastery in Milan. In 1616, he transferred to a monastery in Pisa, where he met Benedetto Castelli, a former pupil of **Galileo**. Castelli introduced Cavalieri to the study of geometry and later introduced him to Galileo himself. From that point on, Cavalieri considered Galileo his mentor and teacher. During their long association, Cavalieri wrote more than a hundred letters to Galileo, which have survived in the national edition of the *Le opere di Galileo Galilei.*

Cavalieri's entry into religious life was typical of the route many aspiring scholars took in that era. Monasteries were centers of learning and operated the most comprehensive libraries; the Roman Catholic Church wielded tremendous power over education and the dissemination of information. (For instance, Galileo was forced by the Inquisition to repudiate his

1632 treatise that clarified the Copernican theory on the movement of the Earth around the sun. The Church did not officially lift the ban on treatises that treated as fact such celestial movement until 1828.) As a cleric, Cavalieri had access to the classic texts of **Euclid, Archimedes,** and **Apollonius,** and his status as a monk served as an introduction to the finest minds of his time.

### One Theory, Indivisible

In 1621, Cavalieri was ordained as a deacon to Cardinal Federigo Borromeo, whose regard for scholarship and appreciation of mathematics encouraged Cavalieri. The cardinal's esteem for Cavalieri's abilities made it possible for him to teach theology at the monastery of San Girolamo in Milan while still in his early twenties. It was during this time (1620 to 1623), that Cavalieri first began his work on the method of indivisibles. From 1623 to 1629 he served as prior of St. Peter's at Lodi and at the Jesuati monastery in Parma. On a trip to Milan, he fell ill with an attack of gout and was confined to Milan for several months. Perhaps the forced relaxation from his duties as prior gave him the opportunity to concentrate on his works in mathematics, because he told Galileo and Cardinal Borromeo that he had completed his *Geometria* in December of 1627.

In 1628, Cavalieri sought Galileo's help in securing a teaching post at the University of Bologna. Galileo wrote to a patron of the institution that "few, if any, since Archimedes, have delved as far and as deep into the science of geometry." Cavalieri won the academic appointment and was named the first chair in mathematics in 1629, a post he held until his death.

At the same time, his order appointed him prior of the Church of Santa Maria della Mascarella in the city, combining his responsibilities to the Church with convenient access to the university. Cavalieri was able to pursue his academic and theological ambitions, and this period was a fruitful one. While in Bologna, he published 11 books, including *Geometria* in 1635.

Cavalieri's inspiration was Archimedes, who first proposed the notion—unexamined for centuries— that indivisibles could be used to determine areas, volumes, and centers of gravity. Cavalieri developed a rational system that employed indivisibles to determine area and volume, a method that made calculations easier and quicker than the ancient Greek method of exhaustion. Cavalieri's Theorem states that if two solids have equal altitudes, and if sections made by planes parallel to the bases and at equal distances from the bases always has a given ratio, then volumes of the solids have the same ratio to each other. Cavalieri also refined a general proof of Guldin's theorem related to the area of a surface and the volume of rotating solids.

Cavalieri's work had tremendous impact on the works of his contemporaries. Torricelli, who expanded on Cavalieri's concepts, wrote "the geometry of indivisible was, indeed, the mathematical briar bush, the so–called royal road, and one that Cavalieri first opened and laid out for the public as a device of marvelous invention." **Pierre de Fermat, Blaise Pascal, Issac Barrow** and Newton were influenced by Cavalieri's work, which, according to Isaac Asimov, was "a stepping–stone toward . . . the development of the calculus by Newton, which is the dividing line between classical and modern mathematics."

## SELECTED WRITINGS BY CAVALIERI

*Directorium generale uranometricum*, 1632
*Geometria indivisibilbus continuorum nova quadam ratione promota*, 1635
*Ckompendio delle regole dei triangoli con le loro dimostrationi*, 1639
*Centuria di varii problemi*, 1639
*Nuova pratica astrologica*, 1639
*Tavola prima logaritimica. Tavola seconda logaritimica. Annotationi nell'opera, e correttioni de gli errori piu notabili*, (date unknown)
*Trigonometria plana, et sphaerica, linearis et logarithmica*, 1643
*Tratato della ruoaato planetaria perpetua*, 1646
*Exercitationes geometricae sex*, 1647

## FURTHER READING

Asimov, Isaac. *Asimov's Biographical Encyclopedia of Science and Technology*. New York: Doubleday & Co., 1982, p. 119.
Boyer, Carl B. *A History of Mathematics*. Second edition. Revised by Uta C. Merzbach. New York: John Wiley & Sons, 1989, pp. 366–70.
Cajori, Florian. *A History of Mathematics*. New York: Macmillan Co., 1938, p. 161.
Callinger, Ronald. *Classics of Mathematics*. Englewood Cliffs, NJ: Prentice Hall, 1995, pp. 314–15.
Carruccio, Ettore. *Mathematics and Logic in History and in Contemporary Thought*. Translated by I. Quigly. Chicago: Aldine Publishing Co., 1964, pp. 204–07.
Carruccio, Ettore. "Bonaventura Cavalieri," in *Dictionary of Scientific Biography*. Volume III. Edited by Charles Coulston Gillispie. New York: Charles Scribner's Sons, 1970, pp. 149–53.
Eves, Howard. *An Introduction to the History of Mathematics*. Revised edition. New York: Holt, Rinehart and Winston, 1964, pp. 327–29.

*Arthur Cayley*

Fang, J. *Mathematicians from Antiquity to Today*. Hauppauge, NY: Paideia Press, 1972, pp. 277– 278.
Kline, Morris. *Mathematical Thought from Ancient to Modern Times*. New York: Oxford University Press, 1972, pp. 349–350.

—*Sketch by A. Mullig*

# Arthur Cayley
## 1821–1895
### English algebraist and geometer

Arthur Cayley was one of the most prolific mathematicians of the 19th century. He was one of the individuals responsible for elevating English mathematics of the era to a position of visibility and authority. Although his best known for his work in algebra, Cayley was influential in generalizing geometry, particularly n–dimensional geometry and non–Euclidean geometry, in British mathematics and mathematical education.

Cayley was born on August 16, 1821, in Richmond, Surrey, England and spent his early childhood in Russia. His father, Henry Cayley, was a merchant who was based in St. Petersburg, Russia, and it has been claimed that Cayley's mother, Maria Antonia Doughty, was of Russian ancestry. On his father's side of the family, Cayley could trace his ancestry back to a Norman who had come to England with William the Conqueror in the 11th century.

Cayley's academic career was uniformly distinguished. At the age of 14 he entered King's College School and proceeded to Trinity College, Cambridge, when he was 17. In 1842, he was Senior Wrangler, placing first on the final university examinations in mathematics. So convincing was Cayley's performance that the examiners did not subject him to the standard oral examination. At age 20 he was elected a fellow of Trinity, the youngest of any candidate in that century.

Cayley was reluctant to take holy orders, however, and in light of the regulations of the time he was required to give up his fellowship after a certain period. Cayley entered Lincoln's Inn, one of the Inns of Court for the training of prospective lawyers in 1846 and was admitted to the bar in 1849. For the duration of his legal career, Cayley limited his practice to conveyancing, and wrote about 300 mathematical papers during the 14 years of his legal practice. In 1852, Cayley was elected to the Royal Society.

The branch of mathematics to which Cayley devoted the most time was invariant theory. This involved looking at algebraic expressions, transforming them by classes of functions, and seeing what remained the same after the transformation. The subject lent itself to enormous amounts of calculation, an area in which Cayley excelled. In a series of writings dated from 1854 to 1878, Cayley not only performed a vast number of calculations but interpreted them for usage in other areas.

### Offered Sadlerian Chair

In 1863, Cayley was offered the Sadlerian chair of pure mathematics at Cambridge University, which he held until his death. It is a tribute to his sense of duty to the university that Cayley put his knowledge of the law to help draft new statutes as necessary. That same year he married Susan Moline, by whom he had a son and a daughter.

Cayley's name is associated with some of the changes that were sweeping the mathematical community of the era. He was the first to use the term "n–dimensional geometry," recognizing

the extent to which going beyond three dimensions was simply a matter of adding another variable to an algebraic expression. In geometry, perhaps Cayley's greatest contribution was developing a method for calculating distances in projective geometry. This quantity, called a cross–ratio, made it easier to consider projective geometry as a generalization of ordinary Euclidean geometry rather than as an entirely different field.

Cayley essentially created the theory of matrices, which has proved to be fundamental to mathematics and physics ever since. A matrix is a way of representing a linear transformation by an array of numbers, and Cayley developed an algebra of matrices to represent combinations of such transformations. Every matrix is associated by a polynomial whose roots reflected the behavior of the linear transformation that the matrix expressed. One of Cayley's striking results was that if one treated the matrix itself as an algebraic object, it was a solution of the equation derived by setting this characteristic polynomial to zero. From this theorem all sorts of matrix calculations can be simplified.

Cayley launched the new subject of group theory in a more abstract way than had been recognized previously. Aspects of group theory can be traced back to the French mathematician **Èvariste Galois**, but the groups Galois and his French successors considered were of particular kinds of objects. Cayley recognized that all types of objects could make up a group. A basic result which bears his name is that every group with a finite number of members has the same structure as a group of permutations. There was no need for such a theorem in the days when all groups were considered to be made up of permutations.

Cayley wrote on a variety of subjects that filled the 13 volumes of his collected works. He was quite a pure mathematician, but still pursued problems in dynamics as a source of problems in pure mathematics. Cayley received many awards and honors, including the De Morgan medal given by the London Mathematical Society and the Copley medal of the Royal Society in 1881. He was a guest lecturer at the Johns Hopkins University in the United States in 1881 and served as president of the British Association in 1883. Cayley was also an avid mountain climber and enjoyed painting in watercolors.

Cayley died on January 26, 1895. In light of his prolific composition, it is not surprising that there is some repetition in his papers. Nevertheless, the number of contributions that Cayley made to mathematics is staggering. His ability to reduce complicated expressions to manageable size made invariant theory the talk of the mathematical community and carried over to many other disciplines as well.

## SELECTED WRITINGS BY CAYLEY

*Collected Mathematical Papers,* 13 volumes, 1889–98

## FURTHER READING

Bell, Eric T. *Men of Mathematics*. New York: Simon and Schuster, 1937, pp. 378–405.

Forsyth, A.R. "Arthur Cayley, in *Dictionary of National Biography*. Volume 22. Edited by Leslie Stephen and Sidney Lee. Oxford, UK: Oxford University Press, 1968, pp. 401–402."

North, J.R. "Arthur Cayley," in *Dictionary of Scientific Biography*. Volume III. Edited by Charles Coulston Gillispie. New York: Charles Scribner's Sons, 1973, pp. 162–170.

Richards, Joan L. *Mathematical Visions*. San Diego: Academic Press, 1988.

*—Sketch by Thomas Drucker*

# Giovanni Ceva
## 1647(?)–1734
### Italian geometer and engineer

Giovanni Ceva was an authority on his era's geometric problems with an vast interest in pure geometry. He proved theorems on transversals and developed what is now known as Ceva's theorem, which concerns when lines from the vertices of a triangle to the opposite sides intersect at a common point. In Ceva's most important work, *De lineis rectis*, he combined his background in mechanics and geometry to attack the problems of geometric systems. Ceva also demonstrated applications of geometry statics and wrote an early treatise on mathematical economics.

Ceva was born into a rich, prominent Italian family. His younger brother, Tommaso Ceva, was also a mathematician who taught at Brera College, Milan, as well as a poet and a philosopher. Ceva studied in Pisa. He might have studied under a professor of logic there, one Donato Rossetti, as Ceva praises him in the preface to one of his books. In other works, he mentions two mathematics professors at Pisa, Alessandro Marchetti and his son, Angiolo.

In approximately 1686, Ceva became employed by the Duke of Mantua and worked in Mantua for most of his career. As an employee of the duke, he worked in various capacities, including court official, city magistrate, and ecclesiastic official. Ceva was also a professor at the University of Mantua.

As a mathematician, the two books of *De lineis rectis* ("Concerning Straight Lines," 1678) comprise Ceva's best known work. *De lineis* is important in part because Ceva brings together mechanics and geometry to study the properties and. In the text Ceva exploited properties of the center of gravity of a system of points and other applications of mechanics to make his calculations about geometric figures. The theorem now named after him was an important result on the synthetic geometry of the triangle. Ceva's next important work was published in 1682, the four–part *Opuscula mathematica* ("A Short Mathematical Work"). This study furthered the work he began in *De lineis rectis*; Ceva applied his ideas about the center of gravity to other shapes. *Opuscula* also contained a famous mistake of Ceva's concerning pendulum laws. He wrongly deduced that the periods of oscillation pendulums are in the same ratio as their lengths, and subsequently corrected this error in *Geometrica motus*.

Ceva was married and fathered a daughter, born three years after the publication of *Opuscula*. In 1692, Ceva published *Geometrica motus* ("The Geometry of Motion"). Using geometry as his basis, Ceva studied the nature of motion and also anticipated some of the elements of infinitesimal calculus in this work. Another published study of note, *De Re Numeraria* ("Concerning Money Matters," 1711), was one of the first texts devoted to mathematical economics.

Throughout his career, Ceva weighed in on controversial physical problems. In Mantua, Ceva also worked as a hydraulic engineer. As a representative of Mantua in the early 18th century, he opposed efforts to divert the Reno river into the Po, and as a result, and the project was abandoned.

Though Ceva was regarded as a master of contemporary geometrical problems throughout his career, he claimed his later work was distracted by family needs. At the time of his death on February 3, 1737, Ceva was on salary at the duke's royal court. He was buried at a local church, St. Teresa de' Carmelitani Scalzi.

## SELECTED WRITINGS BY CEVA

*De lineis rectris*, 1678
*Opuscula mathematica*, 1682
*Geometrica motus*, 1692
*De Re Numeraria*, 1711

## FURTHER READING

### Books

Boyer, Carl B. *A History of Mathematics*. Second edition. Revised by Uta C. Merzbach. New York: John Wiley & Sons, 1989, p. 476.

Oettel, Herbert. "Giovanni Ceva," in *Dictionary of Scientific Biography*. Volume III. Edited by Charles Coulston Gillispie. New York: Charles Scribner's Sons, 1970, pp. 181–83.

*Subrahmanyan Chandrasekhar*

## Periodicals

Byrkit, Donald R. "A Corollary to Ceva's Theorum and Some of its Consequences." *School Science and Mathematics* (December 1990): 683–84.

Grunbaum, Branko and G.C. Shepard. "Ceva, Menelaus, and the Area Principle." *Mathematics Magazine* (October 1995): 254–68.

Klamkin, Murray and Andy Liu. "Simultaneous Generalizations of the Theorems of Ceva and Menelaus." *Mathematics Magazine* (February 1992): 48–52.

*—Sketch by Annette Petruso*

# Subrahmanyan Chandrasekhar
## 1910-1995

### Indian-born American astrophysicist and applied mathematician

Subrahmanyan Chandrasekhar was an Indian-born American astrophysicist and applied mathematician whose work on the origins, structure, and dynamics of stars has secured him a prominent place in the annals of science. His most celebrated work concerns the radiation of energy from stars, particularly white dwarf stars, which are the dying fragments of stars. Chandrasekhar demonstrated that the radius of a white dwarf star is related to its mass: the greater its mass, the smaller its radius. Chandrasekhar has made numerous other contributions to astrophysics. His expansive research and published papers and books include topics such as the system of energy transfer within stars, stellar evolution, stellar structure, and theories of planetary and stellar atmospheres. For nearly 20 years, he served as the editor-in-chief of the *Astrophysical Journal,* the leading publication of its kind in the world. For his immense contribution to science, Chandrasekhar has received numerous awards and distinctions, most notably, the 1983 Nobel Prize for Physics for his research into the depths of aged stars.

Chandrasekhar, better known as Chandra, was born on October 19, 1910, in Lahore, India (now part of Pakistan), the first son of C. Subrahmanyan Ayyar and Sitalakshmi nee (Divan Bahadur) Balakrishnan. Chandra came from a large family: he had two older sisters, four younger sisters, and three younger brothers. As the firstborn son, Chandra inherited his paternal grandfather's name, Chandrasekhar. His uncle was the Nobel Prize-winning Indian physicist, Sir C. V. Raman.

Chandra received his early education at home, beginning when he was five. From his mother he learned Tamil, from his father, English and arithmetic. He set his sights upon becoming a scientist at an early age, and to this end, undertook at his own initiative some independent study of calculus and physics. The family moved north to Lucknow in Uttar Pradesh when Chandra was six. In 1918, the family moved again, this time south to Madras. Chandrasekhar was taught by private tutors until 1921, when he enrolled in the Hindu High School in Triplicane. With typical drive and motivation, he studied on his own and steamed ahead of the class, completing school by the age of fifteen.

After high school, Chandra attended Presidency College in Madras. For the first two years, he studied physics, chemistry, English, and Sanskrit. For his B.A. honors degree he wished to take pure mathematics but his father insisted that he take physics. Chandra resolved this conflict by registering as an honors physics student but attending mathematics lectures. Recognizing his brilliance, his lecturers went out of their way to accommodate Chandra. Chandra also took part in sporting activities and joined the debating team. A highlight of his college years was the publication of his paper, "The Compton Scattering and the New Statistics." These and other early successes while he was still an 18-year-old undergraduate only strengthened Chandra's resolve to pursue a career in scientific research, despite his father's wish that he join the Indian civil service. A meeting the following year with the German physicist Werner Heisenberg, whom Chandra, as the secretary of the student science association, had the honor of showing around Madras, and Chandra's attendance at the Indian Science Congress Association Meeting in early 1930, where his work was hailed, doubled his determination.

### Leaves India for Cambridge, England

Upon graduating with a M.A. in 1930, Chandra set off for Trinity College, Cambridge, as a research student, courtesy of an Indian government scholarship created especially for him (with the stipulation that upon his return to India, he would serve for five years in the Madras government service). At Cambridge, Chandra turned to astrophysics, inspired by a theory of stellar evolution that had occurred to him as he made the long boat journey from India to Cambridge. It would preoccupy him for the next ten years. He also worked on other aspects of astrophysics and published many papers.

In the summer of 1931, he worked with physicist Max Born at the Institut für Theoretische Physik at Göttingen in Germany. There, he studied group theory and quantum mechanics (the mathematical theory that relates matter and radiation) and produced work on the theory of stellar atmospheres. During this period, Chandra was often tempted to leave astrophysics for pure mathematics, his first love, or at least for physics. He was worried, though, that with less than a year to go before his thesis exam, a change might cost him his degree. Other factors influenced his decision to stay with astrophysics, most importantly, the encouragement shown him by astrophysicist Edward Arthur Milne. In August 1932, Chandra left Cambridge to continue his studies in Denmark under physicist Niels Bohr. In Copenhagen, he was able to devote more of his energies to pure physics. A series of Chandra's lectures on astrophysics given at the University of Liège, in Belgium, in February 1933 received a warm reception. Before returning to Cambridge in May 1933 to sit his doctorate exams, he went back to Copenhagen to work on his thesis.

Chandrasekhar's uncertainty about his future was assuaged when he was awarded a fellowship at Trinity College, Cambridge. During a four-week trip to Russia in 1934, where he met physicists Lev Davidovich Landau, B. P. Geraismovic, and Viktor Ambartsumian, he returned to the work that had led him into astrophysics to begin with, white dwarfs. Upon returning to Cambridge, he took up research of white dwarfs again in earnest.

As a member of the Royal Astronomical Society since 1932, Chandra was entitled to present papers at its twice monthly meetings. It was at one of these that Chandra, in 1935, announced the results of the work that would later make his name. As stars evolve, he told the assembled audience, they emit energy generated by their conversion of hydrogen into helium and even heavier elements. As they reach the end of their life, stars have progressively less hydrogen left to convert and emit less energy in the form of radiation. They eventually reach a stage when they are no longer able to generate the pressure needed to sustain their size against their own gravitational pull and they begin to contract. As their density increases during the contraction process, stars build up sufficient internal energy to collapse their atomic structure into a degenerate state. They begin to collapse into themselves. Their electrons become so tightly packed that their normal activity is suppressed and they become white dwarfs, tiny objects of enormous density. The greater the mass of a white dwarf, the smaller its radius, according to Chandrasekhar. However, not all stars end their lives as stable white dwarfs. If the mass of evolving stars increases beyond a certain limit, eventually named the *Chandrasekhar limit* and calculated as 1.4 times the mass of the sun, evolving stars cannot become stable white dwarfs. A star with a mass above the limit has to either lose mass to become a white dwarf or take an alternative evolutionary path and become a supernova, which releases its excess energy in the form of an explosion. What mass remains after this spectacular event may become a white dwarf but more likely will form a neutron star. The neutron star has even greater density than a white dwarf and an average radius of about 9 mi (15 km). It has since been independently proven that all white dwarf stars fall within Chandrasekhar's predicted limit, which has been revised to equal 1.2 solar masses.

### Theory of Stellar Evolution Unexpectedly Ridiculed

Unfortunately, although his theory would later be vindicated, Chandra's ideas were unexpectedly undermined and ridiculed by no less a scientific figure than astronomer and physicist Sir Arthur Stanley Eddington, who dismissed as absurd Chandra's notion that stars can evolve into anything other than white dwarfs. Eddington's status and authority in the community of astronomers carried the day, and Chandra, as the junior, was not given the benefit of the doubt. Twenty years passed before his theory gained general acceptance among astrophysicists, although it was quickly recognized as valid by physicists as noteworthy as Wolfgang Pauli, Niels Bohr, Ralph H. Fowler, and Paul Dirac. Rather than continue sparring with Eddington at scientific meeting after meeting, Chandra collected his thoughts on the matter into his first book, *An Introduction to the Study of Stellar Structure*, and departed the fray to take up new research around stellar dynamics. An unfortunate result of the scientific quarrel, however, was to postpone the discovery of black holes and neutron stars by at least 20 years and Chandra's receipt of a Nobel Prize for his white dwarf work by 50 years. Surprisingly, despite their scientific differences, he retained a close personal relationship with Eddington.

Chandra spent from December 1935 until March 1936 at Harvard University as a visiting lecturer in cosmic physics. While in the United States, he was offered a research associate position at Yerkes Observatory at Williams Bay, Wisconsin, staring in January 1937. Before taking up this post, Chandra returned home to India to marry the woman who had waited for him patiently for six years. He had known Lalitha Doraiswamy, daughter of Captain and Mrs. Savitri Doraiswamy, since they had been students together at Madras University. After graduation, she had undertaken a masters degree. At the time of their marriage, she was a headmistress. Although their marriage of love was unusual, as both came from fairly progressive families and were both of the Brahman caste, neither of their families had any real objections. After a whirlwind courtship and wedding, the young bride and groom set out for the United States. They intended to stay no more than a few years, but, as luck would have it, it became their permanent home.

## Joins Staff of Yerkes Observatory in the United States

At the Yerkes Observatory, Chandra was charged with developing a graduate program in astronomy and astrophysics and with teaching some of the courses. His reputation as a teacher soon attracted top students to the observatory's graduate school. He also continued researching stellar evolution, stellar structure, and the transfer of energy within stars. In 1938, he was promoted to assistant professor of astrophysics. During this time Chandra revealed his conclusions regarding the life paths of stars.

During the Second World War, Chandra was employed at the Aberdeen Proving Grounds in Maryland, working on ballistic tests, the theory of shock waves, the Mach effect, and transport problems related to neutron diffusion. In 1942, he was promoted to associate professor of astrophysics at the University of Chicago and in 1943, to professor. Around 1944, he switched his research from stellar dynamics to radiative transfer. Of all his research, the latter gave him, he recalled later, more fulfillment. That year, he also achieved a lifelong ambition when he was elected to the Royal Society of London. In 1946, he was elevated to Distinguished Service Professor. In 1952, he became Morton D. Hull Distinguished Service Professor of Astrophysics in the departments of astronomy and physics, as well as at

the Institute for Nuclear Physics at the University of Chicago's Yerkes Observatory. Later the same year, he was appointed managing editor of the *Astrophysical Journal,* a position he held until 1971. He transformed the journal from a private publication of the University of Chicago to the national journal of the American Astronomical Society. The price he paid for his editorial impartiality, however, was isolation from the astrophysical community.

Chandra became a United States citizen in 1953. Despite receiving numerous offers from other universities, in the United States and overseas, Chandra never left the University of Chicago, although, owing to a disagreement with Bengt Strömgren, the head of Yerkes, he stopped teaching astrophysics and astronomy and began lecturing in mathematical physics at the University of Chicago campus. Chandra voluntarily retired from the University of Chicago in 1980, although he remained on as a post-retirement researcher. In 1983, he published a classic work on the mathematical theory of black holes. Since then, he has studied colliding waves and the Newtonian two-center problem in the framework of the general theory of relativity. His semi-retirement has also left him with more time to pursue his hobbies and interests: literature and music, particularly orchestral, chamber, and South Indian. Chandrasekhar died in 1995.

## Receives Numerous Honors and Awards

During his long career, Chandrasekhar has received many awards. In 1947, Cambridge University awarded him its Adams Prize. In 1952, he received the Bruce Medal of the Astronomical Society of the Pacific, and the following year, the Gold Medal of the Royal Astronomical Society. In 1955, Chandrasekhar became a Member of the National Academy of Sciences. The Royal Society of London bestowed upon him its Royal Medal seven years later. In 1962, he was also presented with the Srinivasa Ramanujan Medal of the Indian National Science Academy. The National Medal of Science of the United States was conferred upon Chandra in 1966; and the Padma Vibhushan Medal of India in 1968. Chandra received the Henry Draper Medal of the National Academy of Sciences in 1971 and the Smoluchowski Medal of the Polish Physical Society in 1973. The American Physical Society gave him its Dannie Heineman Prize in 1974. The crowning glory of his career came nine years later when the Royal Swedish Academy awarded Chandrasekhar the Nobel Prize for Physics. ETH of Zurich gave the Indian astrophysicist its Dr. Tomalla Prize in 1984, while the Royal Society of London presented him with its Copley Prize later that year. Chandra also received the R. D. Birla Memorial Award of the Indian Physics Association in 1984. In 1985, the Vainu Bappu Memorial Award of the Indian National Science Academy was conferred upon Chandrasekhar. In May 1993, Chandra received the state of Illinois's highest honor, Lincoln Academy Award, for his outstanding contributions to science.

While his contribution to astrophysics has been immense, Chandra has always preferred to remain outside the mainstream of research. He described himself to his biographer, Kameshar C. Wali, as "a lonely wanderer in the byways of science." Throughout his life, Chandra has striven to acquire knowledge and understanding, according to an autobiographical essay published with his Nobel lecture, motivated "principally by a quest after perspectives."

## SELECTED WRITINGS BY CHANDRASEKHAR

### Books

*An Introduction to the Study of Stellar Evolution,* 1939, reprinted, 1967
*Principles of Stellar Dynamics,* 1943, reprinted, 1960
*Radiative Transfer,* 1950, reprinted, 1960
*Plasma Physics,* 1960
*Hydrodynamic and Hydromagnetic Stability,* 1961, reprinted, 1987
*Ellipsoidal Figures of Equilibrium,* 1968, reprinted, 1987
*The Mathematical Theory of Black Holes,* 1983
*Eddington: The Most Distinguished Astrophysicist of His Time,* 1983
*Truth and Beauty: Aesthetics and Motivations in Science,* 1987
Selected Papers (6 volumes), 1989–90

### Periodicals

"The Compton Scattering and the New Statistics." *Proceedings of the Royal Society* 125 (1929).
"Stochastic Problems in Physics and Astronomy" *Review of Modern Physics* 15 (1943): 1–89. Reprinted in *Selected Papers on Noise and Stochastic Processes,* edited by Nelson Wax. New York: Dover Publications, 1954.

## FURTHER READING

*The Biographical Dictionary of Scientists, Astronomers.* London: Blond Educational Company, 1984, pp. 36.
*Chambers Biographical Encyclopedia of Scientists.* New York: Facts on File, 1981.
Goldsmith, Donald. *The Astronomers.* New York: St. Martin's Press, 1991.
*Great American Scientists.* Englewood Cliffs, NJ: Prentice-Hall, 1960.
Land, Kenneth R., and Owen Gingerich, editors. *A Sourcebook in Astronomy and Astrophysics.* Cambridge, MA: Harvard University Press, 1979.

*Sun–Yung Alice Chang*

*Modern Men of Science.* New York: McGraw-Hill, 1966, p. 97.
Wali, Kameshwar C. *Chandra: A Biography of S. Chandrasekhar.* Chicago: Chicago University Press, 1991.

*—Sketch by Avril McDonald*

# Sun–Yung Alice Chang
## 1948–
### Chinese–born American analyst

Sun–Yung Alice Chang, working with Paul Yang, Tom Branson, and Matt Gursky, has produced what the American Mathematical Society has termed "deep contributions" to the study of partial differential equations in relation to geometry and topology.

Sun–Yung Alice Chang was born in Ci–an, China, on March 24, 1948, and studied for her bachelor's degree at the National University of Taiwan, which she received in 1970. Chang emigrated to the United States for graduate work and a series of teaching jobs. In 1974, Chang earned her Ph.D. from the University of California at Berkeley. She has served as assistant professor at SUNY–Buffalo, the University of California at Los Angeles, and the

University of Maryland. Chang returned to UCLA in 1980 as an associate professor and was later promoted to full professor at the same institution. Her most visible performance was as a speaker at the International Congress of Mathematicians, held in Berkeley in 1986.

### Wins Satter Prize

In 1995, Chang was awarded the third the Ruth Lyttle Satter Prize at the American Mathematical Society's 101st annual meeting in San Francisco (the previous two recipients were **Lai–Sung Young** and **Margaret McDuff**). Young was on the selection committee that recommended Chang for that year's prize. In her acceptance speech, Chang acknowledged her debt to her collaborators and promised to "derive further geometric consequences" in various problems currently under study.

Reflecting momentarily on her own school years, Chang admitted that it had been important for her to have role models and female companionship. However, she stated, the deciding factor in the future will be to have more women proving theorems and contributing to mathematics as a whole. To that end, she has joined with a number of steering committees and advisory panels at the national level. After being a Sloan Fellow for the National Academy of Sciences in 1980, she returned ten years later as a member of their Board of Mathematical Sciences. Chang also advised the National Science Foundation and the Association for Women in Mathematics throughout the early 1990s.

Chang always finds time to involve her students in the sometimes arcane world of her specialty. At the University of Texas, she took part in their Distinguished Lecturer Series of 1996–97, a program that successfully targets an audience of young graduate students. She has been most active with the American Mathematical Society, working on a range of committees and speaking at a number of their meetings. Chang's most current professional positions include a three–year term on the Editorial Boards Committee of the AMS, expiring in 1998.

Chang was married in 1973 and has two children.

### FURTHER READING

"1995 Ruth Lyttle Satter Prize in Mathematics." *Notices of the AMS* (April 1995) http://e-math.ams.org/notices/199504/prize-satter.html

"Sun–Yung Alice Chang." *Biographies of Women Mathematicians.* http://www.scottlan.edu/lriddle/women/chronol.htm (July 1997).

*—Sketch by Jennifer Kramer*

# Michel Chasles
## 1793–1880
### French geometer and mathematics historian

With his work defined by its practical fusion of means and intention, Michel Chasles was a mathematician primarily focused on mathematical history and geometry. His major geometric contribution concerned the development of synthetic projective geometry. Synthetic geometry is the study of the properties of a geometric line or plane figure that are unchanged by the projection of that figure onto a plane from a point independent of the plane and the figure. While **Jakob Steiner** also developed many of the same properties independently, Chasles' historical work led to his appointment as a professor.

Chasles was born in Épernon, France, on November 15, 1793, into a somewhat wealthy Catholic family. He was son of Charles–Henri, a contractor and lumber dealer. Chasles was born with the first name Floréal, but it was legally changed in 1809 to Michel. He entered the École Polytechnique in Paris in 1812, after studying at the Lycée Impérial. Chasles earned a Master's Degree in Engineering and studied math at Chartres, both in 1815.

To appease his father, Chasles worked in a stock brokerage firm, but soon gave it up to devote his undivided attention to the study of mathematics and mathematics history. Chasles published his mathematical research, though he was not attached to a university until 1841. The most well known of these early publications is entitled *Aperçu historique sur l'origine et le développment des méthodes en géométrie* ("Historical Survey of the Origin and Development of Geometric Methods"), which was published in 1837. This book was important for Chasles' career as it gave him recognition in the field of geometry and as a mathematical historian, and is still used as a standard historical reference book.

Chasles' publishing activity led to his appointment as a professor of applied mathematics at the École Polytechnique in 1841. He taught astronomy, applied mechanics, and geodesy before resigning this post in 1851. The Sorbonne in Paris created a chair of higher geometry specially for Chasles in 1846. He kept this post until his death. Chasles wrote two textbooks specifically for the classes he taught at the Sorbonne. *Traité de géométrie* ("Treatise on Higher Geometry") was published in 1852. In this book, Chasles delineated the importance of synthetic geometry. He ushered many original mathematical ideas such as involutions, cross ratio, and pensils. The second text, *Traité des sections coniques* ("Treatise on Conic Sections"), published in 1865, used the theories promulgated in *Traité de géométrie* on conic sections.

Chasles' mathematical accomplishments were noted by his peers. He became a member of the French Académie Royale des Sciences in 1851, serving as president in 1860. Chasles also served as the chair in advanced geometry division for the Royal Society of London in 1854. Outside of mathematics, Chasles was an eager collector of autographs and correspondence by famous personages. This hobby led him to pay approximately 200,000 francs from 1861 to 1869 for various forged letters from famous men of science and other notable figures in history. The fraud, perpetuated by the famous forger Denis Vrain–Lucas, caused quite a stir at the time.

Collecting these letters was one of the few interests Chasles had outside of mathematics and related activities. A lifelong bachelor, Chasles died in Paris on December 18, 1880.

## SELECTED WRITINGS BY CHASLES

*Rapport sur le progres de la geometrie,* 1870

## FURTHER READING

Koppelman, Elaine. "Michel Chasles," in *Dictionary of Scientific Biography*. Volume III. Edited by Charles Coulston Gillispie. New York: Charles Scribner's Sons, 1970, pp. 212–15.

*—Sketch by Annette Petruso*

# Pafunty Lvovich Chebyshev
## 1821–1894
### Russian probabilist and analyst

Chebyshev has given his name to results in probability and analysis, one of the first Russian mathematicians by birth to be so recognized. His work reflected a great deal of mathematical sophistication, making connections between different areas and generalizing techniques. He played a primary role in establishing a viable mathematical curriculum at St. Petersburg University, which laid the foundations for subsequent achievements in Russian mathematics.

Pafnuty Lvovich Chebyshev was born on May 16, 1821 in Okatovo in the Kaluga region of Russia. His father, Lev Pavlovich Chebyshev, was a former army officer. In 1832, the family moved to Moscow, where Chebyshev was educated at home under the instruction of an author of popular arithmetics. As a result, he was well prepared when he enrolled in

Moscow University in 1837. There, Chebyshev studied physics and mathematics.

In 1841, Chebyshev graduated with a bachelor's degree in mathematics and within two years he had passed his master's examination. Chebyshev's thesis, entitled "An Essay on an Elementary Analysis of the Theory of Probability," dealt with the derivation of a law of large numbers (one of a whole group of results that indicated the increasing reliability of experimental results the larger the number of trials) using the methods of analysis of which he had already shown himself a master. In general, Chebyshev was looking for derivations of the leading results of probability by methods that could not be faulted for rigor, but which were not dependent on mathematical ideas that seemed out of proportion to the depth of the subject.

Chebyshev could not find employment in Moscow. As in the German university system, it was necessary to produce a thesis to earn the privilege of teaching and Chebyshev's thesis examined integration by means of logarithms, a topic straight out of analysis. After its acceptance, Chebyshev joined the faculty at St. Petersburg University in 1847, lecturing on algebra, number theory, and probability. In addition to his teaching, he helped prepare a new edition of **Leonhard Euler**'s papers on number theory. Not far removed from this task was the subject of the theory of congruences, which Chebyshev dealt with in his doctoral thesis. He defended the thesis in 1849 and received a prize from the Russian Academy of Sciences.

Between the years 1850 and 1860 Chebyshev spent much of his time working on questions of mechanical engineering. During this period he moved up the academic ladder and by 1859 he was a senior academician in the St. Petersburg Academy. The subject of hinges led Chebyshev to consider problems of best approximation to a function, one of the results of which was later known as the Chebyshev polynomials. In addition to his own polynomials, he studied other systems of what are called orthogonal polynomials as well (the orthogonality refers to an independence they have in being needed to represent a given function).

### Establishes Russian Tradition in Probability

There had been work done in Russia on probability before Chebyshev, but the number and quality of the students picked up considerably as a result of his efforts. One contributing factor was his generosity with his time, which explains the impressive list of students who chose to study with him. He kept an open house for students and continued to do so even after his retirement. In his own work, Chebyshev preferred to use elementary methods, and that may have given his work an appearance of comprehensibility.

Among the subjects to which Chebyshev contributed was the distribution of prime numbers. There had been much work devoted to the question of whether the apparent irregularities in the distribution of prime numbers (any number only divisible by one and itself) disappeared as one looked at larger initial segments of the positive whole numbers. Chebyshev was able to get a decent approximation for the number of prime numbers less than a fixed number compared to known functions of that fixed number, but he did not prove that there was a limiting value. He did, however, demonstrate a conjecture that if n is greater than 3, there is always at least one prime between n and $2n$–2.

In addition to his own teaching, Chebyshev was also active in improving the quality of mathematics and physics teaching on a national basis. He built a calculating machine in the late1870s, although this was more as a demonstration of the potential usefulness of mechanical devices than as a genuine aid to calculation. He was perhaps the creator of the tradition in Russia of making probability a part of the general mathematical curriculum. In light of the profound contributions made to the subject by **Andrei Kolmogorov**, this was no slight step in the development of probability.

Chebyshev's virtuosity as an analyst enabled him to derive results in probability that had previously been used without being well understood. He was able to take advantage of material developed by the French probabilist I.J. Bienaymé to produce a convincing demonstration of the law of large numbers. Chebyshev was recognized at home and abroad, being the first Russian to be elected a foreign member of the Paris Académie des Sciences. At the time of his death on December 8, 1894, the Russian mathematical tradition was stronger than ever before, thanks to his work and that of his students.

## SELECTED WRITINGS BY CHEBYSHEV

Oeuvres. Edited by A.A. Markov and N.Y. Sonin, 1899–1907

## FURTHER READING

Kline, Morris. *Mathematical Thought from Ancient to Modern Times.* New York: Oxford University Press, 1972, pp. 830–831.
Young, Laurence. Mathematicians and Their Times. Amsterdam: North–Holland, 1981, pp. 194, 318.
Youschkevitch, A.P. "Pafnuty Lvovich Chebyshev," in *Dictionary of Scientific Biography.* Volume III. Edited by Charles Coulston Gillispie.

*Fan R. K. Chung*

New York: Charles Scribner's Sons, 1973, pp. 222–32.

*—Sketch by Thomas Drucker*

# Fan R. K. Chung
## 1949(?)–
### Taiwan–born American number theorist

For Fan Chung, mathematics is more than a career. It is sometimes an obsession, and it is always fun. During her 20 years in a corporate research environment at Bell Laboratories or in her academic position as an endowed professor at the University of Pennsylvania, Chung seeks interesting problems in mathematics that make new connections between fields. She has made another kind of connections as well, doing work that helped lay the foundation for modern computer networks and wireless communication.

### A Math–Friendly Upbringing

Chung's father was an engineer who encouraged her to pursue mathematics while she was growing up in Kaoshiung, Taiwan. As an undergraduate at the

National Taiwan University, many of her classmates were women and their successes at mathematics further encouraged her. She was drawn to combinatorics, a field that focuses on manipulating sets of numbers, which is useful in figuring permutations and some statistical analyses. Chung earned her bachelors degree in Taiwan in 1970 and then came to the United States for her graduate education, earning a masters in 1972 and a Ph.D. in 1974 from the University of Pennsylvania.

Chung began working at Bell Laboratories in Murray Hill, New Jersey, which at the time had a staff that included several Nobel prize winners. In an 1995 interview for the journal *Math Horizons*, she told Don Albers that she was initially intimidated by the caliber of her co-workers. "But I got over it. And very soon I discovered that if you just put your hands out in the hallway, you'd catch a problem." Listening to what others were working on and making those connections between fields that she always seeks furthered her research as a member of the technical staff at Bell Labs.

Following the breakup of the Bell Telephone Company in 1984, Chung was promoted to research manager at Bell Communications Research in Morristown, New Jersey, and aided in the formation of a new Discrete Mathematics Research Group there directed by her mentor from Bell Labs. Within two years, she had been promoted again, to division manager for Mathematics, Information Sciences and Operations Research. In addition to her own research, Chung has to supervise the work of other mathematicians, and help recruit new talent to the lab. "Usually with positions in management you obtain more influence and you certainly have more power to make decisions," she told Albers. "But I do not want people to respect me because of that power. I'd rather win their admiration because of the mathematics I'm doing."

Chung's work at Bell Labs focused on practical problems and applications, though the fun of mathematics was never lost for her. Chung devised a method of encoding and decoding signals that is crucial to digital cellular phones which use "code division multiple access," or CDMA. This scheme, which was patented in 1988, allows several conversations to share the same radio frequency with complete security, because each call is encoded for sending to the cellular antenna. The encoding and decoding has to be done very rapidly and accurately to ensure that the caller's voice sounds natural. Chung also holds a patent on a method for routing network traffic, issued in 1993.

### A Second Career in Academics

In 1989, Chung began teaching as a visiting professor in the computer science department at nearby Princeton University. In 1990, she became a Bellcore Fellow, studying at Harvard University. She also served as a visiting professor at Harvard from 1991 to 1993.

In 1995, Chung accepted an endowed chair in the mathematics department at the University of Pennsylvania. She began teaching as a full professor in both the math and computer science departments. Chung encourages her students to explore and to make connections between fields, as she has done. Too much exposure to academic mathematics, she contends, makes one feel isolated. "Professors train students to work in their areas and thus, there is a danger of narrowing down instead of broadening and making connections," she told Albers. The curriculum has not always kept pace with the fast-changing world, either, Chung noted. "It is essential for the students to be able to connect the mathematics you learn in the classroom to problems we face in this information age."

Chung has authored or coauthored more than 180 papers, including several with her second husband, mathematician Ron Graham, whom she met at Bell Labs. Her sense of fun is reflected in a 1993 paper in *American Scientist*, in which she and Shlomo Sternberg included a paper cutout for readers to make their own model of the molecule Carbon-60, or the Buckyball, after they read about its mathematical properties.

Chung has two children by her first marriage, which ended in divorce in 1982. They were born in 1974 and 1977.

Chung is an editor for nine academic journals and has served as editor-in-chief of the *Journal of Graph Theory* and chair of the editorial board committee for the American Mathematical Society. In 1990, she was awarded the Allendoerfer Award from the Mathematical Association of America.

## SELECTED WRITINGS BY CHUNG

"Should You Prepare Differently for a Non-Academic Career?" *The Notices of the American Mathematics Society* 38, no. 6, (August 1991): 560–61.

(With Shlomo Sternberg) "Mathematics and the Buckyball." *American Scientist* 81, no. 1, (1993): 56–71.

## FURTHER READING

Albers, Don. "Making Connections: A Profile of Fan Chung." *Math Horizons* (September 1995): 14–18.

"Fan Chung." *Biographies of Women Mathematicians*. June 1997. http://www.scottlan.edu/lriddle/women/chronol.htm (July 22, 1977).

—*Sketch by Karl Leif Bates*

# Nicolas Chuquet
## c. 1455–c. 1500
### French arithmetician, algebraist, and physician

Born in Paris, Chuquet worked in Lyon as a physician. He is primarily known for his *Triparty en la science des nombres*, which, according to Carl Boyer, was the most important mathematical work since **Leonardo Fibonacci**'s *Liber abaci*. It is important to point out that Chuquet, working with the intellectual and notational apparatus of medieval mathematics, lacked what modern mathematicians would define as the fundamental methodological tools of their science. In other words, Chuquet's thinking was far ahead of his methodology. For example, having no mathematical symbols for the basic operations—addition, subtraction, multiplication, and division—Chuquet used symbols based on the appropriate French terms and expressions: *plus, moins, multiplier par*, and *partyr par*. The first two terms were usually abbreviated as $\bar{p}$ and $\bar{m}$. The symbol he used for a square root was R)$^2$, so, for example, in Chuquet's notation, the square root of nine would be written R)$^2$9.

Fascinated, like many of his predecessors, by average numbers, Chuquet formulated a rule, which he called *regle des nombres moyens*. Thus, if $a$, $b$, $c$, and $d$, are positive numbers, $(a + b)/(b + d)$ will be larger than $a / b$, but smaller than $c / d$.

### Makes Important Discovery in Algebra

In the third part of his book, Chuquet turns to algebraic problems, discussing different types of equations. He introduces his own term for the unknown, *premier*, which may seem strange to his contemporaries, who used the traditional French (*chose*) or Latin (*res*) terms. Similarly, he abandoned the accepted Latin term for the second power, *census*, in favor of his own *champs*. The third power he called *cubiez*, and the fourth, *champs des champs*.

Chuquet's great discovery was notation for exponents. Although lacking convenient symbols for variables, Chuquet used expressions which would in modern notation translate as $5x$ and $6x^2$ and $10x^3$. In his book, these expressions are written .5.$^1$ and .6.$^2$ and .10.$^3$. Significantly, Chuquet accepted zero and negative integers as exponents, thus greatly expanding the field of algebra.

### Anticipates Invention of Logarithms

In his analysis of the powers of 2, Chuquet discovered the correspondence between the sum of two indices and the product of their powers. For example, if we look at two powers of 2, 2$^2$ (which equals 4) and 2$^3$ (which equals 8), we will notice that $2^2 \times 2^3 = 2^{2+3}$. Indeed, $4 \times 8 = 32$. Specifically, Chuquet created a table from 0 to 20, working out all the powers of 2. For example, $2^0 = 1$; $2^1 = 2$; $2^2 = 4$; $2^3 = 8$; $2^4 = 16$; $2^5 = 32$, etc. It is clear, given that a logarithm is the power to which a base number needs to be raised in order to produce a particular number, that, using Chuquet's table, one can see that, if $x = 32$, $5 = \log_2 x$.

Among Chuquet's solutions to equations is, in modern notation, $4x = -2$. Here, as Boyer has observed, "Chuquet was for the first time expressing an isolated negative number in an algebraic equation." Some of the equations analyzed in Chuquet's book imply imaginary numbers, but he did not discuss them as viable.

## SELECTED WRITINGS BY CHUQUET

*Triparty en la science des nombres*, 1488
*La geometrie*, 1979

## FURTHER READING

Boyer, Carl B. *A History of Mathematics*. Second edition. Revised by Uta C. Merzbach. New York: Wiley, 1991.
Flegg, Graham. *Nicolas Chuquet Renaissance Mathematician: A Study with Extensive Translation of Chuquet's Mathematical Manuscript Completed in 1488*. Boston: Reidel, 1985.

—*Sketch by Zoran Minderovic*

# Alonzo Church
## 1903-1995
### American logician

Alonzo Church was an American mathematician and logician who provided significant innovations in number theory and in the decision theory that is the foundation of computer programming. His most important contributions focus on the degrees of decidability and solvability in logic and mathematics.

Church was born in Washington, D.C., on June 14, 1903, to Samuel Robbins Church and Mildred Hannah Letterman Church. He took his undergraduate degree from Princeton University in 1924. On August 25, 1925, he married Mary Julia Kuczinski. They had three children: Alonzo, Mary Ann, and Mildred Warner. Church completed his Ph.D. in mathematics at Princeton in 1927. After receiving his doctorate, he was a fellow at Harvard from 1927 to 1928. He studied in Europe from 1928 to 1929 at the University of Göttingen, a prestigious center for the study of mathematics and physics. He taught mathematics and philosophy at Princeton from 1929 to 1967. Among his Ph.D. students at Princeton was the British mathematician **Alan Turing**, who was to crack the German's World War II secret code, called Enigma, which played a key role in the defeat Nazi Germany. Church was a professor of mathematics and philosophy at the University of California at Los Angeles from 1967 until his retirement in 1990. He also edited the *Journal of Symbolic Logic* from 1936 to 1979. His wife died in February, 1976.

Church's private life is very quiet and unremarkable. As Andrew Hodges said in his biography of Alan Turing (Church's famed student who killed himself in 1954 after being arrested on homosexual charges), Church "[is] a retiring man himself, not given to a great deal of discussion."

### Cracking the Decidability Problem

One of the key problems in the foundations of mathematics was stated by the German mathematician **David Hilbert**: Is mathematics decidable? That is, as Andrew Hodges explains in his biography of Alan Turing, "did there exist a definite method which could, in principle, be applied to any assertion, and which was guaranteed to produce a correct decision as to whether that assertion was true"? Although Hilbert thought the answer would be yes, Church's answer was no. Church's theorem says in effect that there is no method to guarantee in advance that a mathematical assertion will be correct or incorrect. Specific mathematical assertions may be found to be correct or incorrect, but there is no general method that will work in advance for all mathematical assertions.

What Church's proof—and the proofs of other mathematicians such as **Kurt Friedrich Gödel**—showed was that mathematics in general was not as tidy, logical, and airtight as people had always thought it was. And, to make matters worse, mathematics could *never* be perfectly tidy, logical, or airtight. There would always be some statements that were undecidable, inconsistent, and incomplete. Church and other mathematicians of his time showed that like everything else, mathematics was fallible.

### Laying the Foundation of Computer Programming

For computer programs to run, programmers have to be able to reduce all problems to the kinds of simple binary logical (or on/off) statements that can be processed by the electronic circuits inside the computer. For a problem to be solvable by a computer, it must be possible to break it down into an operational set of rules and terms. Next it must be possible to apply these rules recursively—that is repeatedly—to the problem until it is solved in terms of the existing set of rules. In short, a computer's binary circuits can only solve a problem under three conditions: (1) if the problem can be expressed as a meaningful set of rules (i.e., meaningful to the computer); (2) if the result of each step is also meaningful in terms of the computer's predefined set of rules; (3) if the computer's set of rules can be applied repeatedly to the problem. For example, in a simple addition or subtraction computer program, it must be possible for a small number (e.g., 1) to be repeatedly added to or subtracted from a larger number (e.g., 100) to get some result, say 10 or 10,000. If any of these three conditions mentioned above is absent, then a computer program cannot solve the problem.

Church's contribution to the foundation of computer programming is that he discovered—as did Alan Turing and Emil Post simultaneously and independently—the importance of recursiveness in solving logical problems. That is, for calculations to take place, some actions (e.g., adding or subtracting) have to be repeated a certain number of times. Church's deceptively simple thesis (which is often called the Church-Turing thesis) is that a function is computable or calculable if it is recursive. That is, the idea of recursiveness (repeatability) is tightly bound up with computability. Church's thesis is important because the repetition of a simple action can result in significant changes. It also means that one simple action can be useful over a broad range of problems, and at different levels of a problem.

Church's contributions to decidability theory have led to many honors, including induction into the National Academy of Science and the American Academy of Arts and Sciences. He has received honorary doctorates from Case Western Reserve University in 1969, Princeton University in 1985, and the State University of New York at Buffalo in 1990. Church died in 1995.

## SELECTED WRITINGS BY CHURCH

### Books

*Introduction to Mathematical Logic,* Volume 1, 1956

**Periodicals**

"An Unsolvable Problem of Elementary Number Theory." *American Journal of Mathematics* 58 (1936): 345–363.

## FURTHER READING

### Books

Hodges, Andrew. *Alan Turing: The Enigma.* New York: Simon & Schuster, 1983.

Hofstadter, Douglas R. *Gödel, Escher, and Bach: An Eternal Golden Braid.* New York: Basic Books, 1979.

*—Sketch by Patrick Moore*

# Alexis–Claude Clairaut
## 1713–1765
**French geometer**

A child prodigy, Alexis Claude Clairaut studied calculus at age 10, wrote mathematical papers at 13, and published a mathematical work on the gauche curve at age 18. Clairaut surpassed even **Isaac Newton** in his analysis of the effects of gravity and centrifugal force on a rotating body such as Earth, now known as Clairaut's theorem.

Clairaut was born in Paris, France, on May 7, 1713. His father, Jean–Baptiste, was a mathematics teacher who recognized his child's precociousness and guided his studies. Clairaut received his entire education at home; his father tutored him in algebra and geometry.

His mother, Catherine Petit, gave birth to 20 children, but most of them did not survive. Clairaut never married, but led an active social life. He and several other young mathematicians formed a society that served as a mathematical training ground for its members, and he often assisted his friends with their studies. He also visited and corresponded with most of the leading mathematicians of the age, including **Leonhard Euler** and **Johann Bernoulli**, and members of the Académie Royale des Sciences.

Clairaut's book, *Théorie de la figure de la terre*, which he published in 1743, was said to be responsible to a great degree for the acceptance of **Isaac Newton**'s gravitational theories. The book was the result of Clairaut's journey to Lapland in 1736, where he assisted Pierre Louis Moreau de Maupertuis, director of the exploration, in measuring the curvature of the Earth inside the arctic circle. Their successful attempt at a meridian arc measurement proved Newton's theory that the Earth's shape was an oblate ellipsoid. Clairaut demonstrated, through an experiment which timed the swings of a pendulum, how the Earth's shape could be determined.

Clairaut was also interested in celestial mechanics. This field of study resulted in the first accurate determination of the size of the planet Venus (two–thirds the size of Earth). His studies of that planet also calculated its gravitational effects on the Earth as compared to the moon. Clairaut's work enabled him to determine a new figure for the size of the moon in relation to the Earth. He also was able to predict how close to the sun Halley's comet would come in its orbit, and correctly predicted the comet's return in 1759.

Clairaut was elected to the Académie Royale des Sciences in 1731, when he was 18 years old. The Académie awarded him a prize for his work on tides in 1740, and he became associate director of the Académie in 1743. He was also made a Fellow of the Royal Society of London in 1737, the Académie's counterpart in England.

A prolific writer, Clairaut published the first complete book on solid analytical geometry, *Recherches sur les courbes à double courbure*, in 1731 at age 18. In addition to *Théorie de la figure de la terre*, he also published books on motions of the moon (*Théorie de la lune*, 1752; *Tables de la lune*, 1754) and the comets (*Théorie du mouvement des comètes*, 1760). Clairaut died at age fifty–two on May 17, 1765, following a brief illness.

## SELECTED WRITINGS BY CLAIRAUT

*Recherches sur les courbes à double courbure*, 1731
*Élémens de géométrie*, 1741
*Théorie de la figure de la terre*, 1743
*Élémens d' algébre*, 1746
*Théorie de la lune*, 1752
*Tables de la lune*, 1754
*Théorie du mouvement des cométes*, 1760

## FURTHER READING

### Books

Asimov, Isaac. *Asimov's Biographical Encyclopedia of Science and Technology.* Garden City, New York: Doubleday & Company, Inc., 1972.

Derbus, Allen G., editor. *World Who's Who in Science.* Chicago: Marquis Who's Who, Inc., 1968, p. 338.

Itard, Jean. "Alexis–Claude Clairaut," in *Dictionary of Scientific Biography.* Volume III. Edited by Charles Coulston Gillespie. New York: Charles Scribner's Sons, 1971, pp. 281–286.

*—Sketch by Jane Stewart Cook*

# Paul Cohen
## 1934-
### American logician and analyst

Paul Cohen's reputation as a mathematician has been earned at least partly because of his ability to work successfully in a number of very different fields of mathematics. He received the highly regarded Bôcher Prize of the American Mathematical Society, for example, in 1964 for his research on the Littlewood problem. Two years later he was awarded perhaps the most prestigious prize in mathematics, the Fields Medal, for his research on one of **David Hilbert**'s "23 most important problems" in mathematics, proving the independence of the continuum hypothesis.

Paul Joseph Cohen was born in Long Branch, New Jersey, on April 2, 1934, but his childhood and adolescence were spent in Brooklyn, New York. His parents were Abraham Cohen and the former Minnie Kaplan. Both parents had immigrated to the United States from western Russia (now part of Poland) while they were still teenagers. Cohen's father became a successful grocery jobber in Brooklyn.

Cohen appears to have had a natural and precocious interest in mathematics from an early age. To a large extent, he was self-educated, depending on books that he could find in the public library or that his elder sister Sylvia was able to borrow for him from Brooklyn College. He told interviewers Donald J. Albers and Constance Reid for their book *More Mathematical People* that "by the time I was in the sixth grade I understood algebra and geometry fairly well. I knew the rudiments of calculus and a smattering of number theory."

For his secondary education, Cohen attended the Stuyvesant High School in lower Manhattan, widely regarded as one of the two (along with the Bronx High School of Science) best mathematics and science high schools in the United States. In 1950, having skipped "a few grades," as he told Albers and Reid, he graduated from Stuyvesant at the age of 16. He ranked sixth in his class and received one of the 40 national Westinghouse Science Talent Search awards given that year. He then enrolled at Brooklyn College, where he remained for two years. In 1952 he was offered a scholarship at the University of Chicago, from which he received his M.S. in mathematics in 1954 and his Ph.D. in 1958.

### Solves the Littlewood Problem of Harmonic Analysis

Until he reached Chicago, Cohen had a relatively unstructured and diverse background in mathematics. He was fairly knowledgeable in some areas that interested him especially and that he had been able to teach himself. But he was still naive about some important areas of mathematics, such as logic, in which he had never had a formal course or even any informal training. Partly through the influence of one of his professors at Chicago, Antoni Zygmund, Cohen became interested in a classical problem in harmonic analysis commonly known as the Littlewood problem, named for the English mathematician John Edensor Littlewood. Cohen's solution to this problem won him the American Mathematical Society's Bôcher Prize in 1964.

On receiving his degree from Chicago, Cohen accepted a position as instructor of mathematics at the Massachusetts Institute of Technology (MIT). A year later he moved to the Institute for Advanced Studies at Princeton, New Jersey, where he was a fellow from 1959 to 1961. At MIT and Princeton Cohen continued to work on problems of analysis and seemed to have found a field to which he could devote his career. That illusion soon evaporated, however. As Cohen later told Albers and Reid, he has a restless mind and is constantly looking for new fields to conquer. "I [have been] told by many people that I should stick to one thing," he said, "but I have always been too restless."

### Solution of the Consistency Proof Problem Brings the Fields Medal

An occasion for shifting gears presented itself to Cohen soon after he was appointed assistant professor at Stanford in 1961. At a departmental lunch, Cohen's colleagues were discussing the problems of developing a "consistency proof" in logic, first suggested by **Georg Cantor** in the late 19th century. The term *consistency* in mathematics refers to the condition that any mathematical theorem be free from contradiction. Developing a consistency proof had been listed as number one on David Hilbert's 1900 list of the 23 most important problems in mathematics for the twentieth century. Although he had no specific background in the field of logic, in which the consistency proof is particularly relevant, Cohen was intrigued by the challenge. He saw it as a way of providing convincing evidence "that set theory is based on some kind of truth," as he told Albers and Reid.

Cohen's work on the consistency proof went forward in fits and starts over the next two years. During one period he became so discouraged that he set the work aside and concentrated on other problems. He seems to have had a glimpse of the general approach for solving his problem during a vacation with his future wife to the Grand Canyon in late 1962. Still, it was another four months before the details of that approach were worked out and a solution produced. Two years later, Cohen received his second major award in mathematics, the International Mathematics Union's Field Prize, for his work on the consistency proof.

In 1964 Cohen was promoted to the post of professor of mathematics at Stanford, a position he has held since. He continues to work on a variety of problems, including those in the fields of analysis and logic. Cohen was married to Christina Karls, a native of Sweden, in 1963. They have three sons, Steven, Charles, and Eric. In addition to the Bôcher Prize and the Fields Medal, Cohen was awarded the Research Corporation of America Award in 1964 and the National Medal of Science in 1967.

## SELECTED WRITINGS BY COHEN

*Set Theory and the Continuum Hypothesis,* 1966

## FURTHER READING

Albers, Donald J., Gerald L. Alexanderson, and Constance Reid. *More Mathematical People.* New York: Harcourt, 1991, pp. 43–58.

—*Sketch by David E. Newton and
Thomas Drucker*

# Conon of Samos
## fl. 245 B.C.
### Greek astronomer and meteorologist

Known in antiquity for his contributions to mathematics and astronomy, Conon is remembered primarily as the discoverer of a famous constellation, *Coma Berenices.*

Conon was born on the Greek island of Samos (the birthplace of **Pythagoras**), worked in Sicily, and finally settled in Alexandria, Egypt, where he became the court astronomer to Ptolemy III Euergetēs. Information about Conon's life and work comes from later sources, some of which may be unreliable. None of

Conon's writings have survived. In his *Phases of the Fixed Stars,* **Claudius Ptolemy** discusses Conon's astronomical and meteorological work. According to Seneca's work, *Naturales quaestiones,* Conon studied Egyptian records of solar eclipses. In his *Conics* (*kōnika*), **Apollonius of Perga** claims that Conon wrote a treatise, *Pròs thrasudaîon,* in which he discussed, apparently inaccurately, the intersection points of conic sections—the intersections being between conic sections, and between a conic section and a circle. The Roman critic and grammarian Probus credits Conon with a work entitled, in Latin translation, *De astrologia* ("On Astrology").

When Ptolemy III returned from a campaign in Syria in 246, his wife, Berenice, cut off a lock of her hair and offered it to the temple of Arsinoë Zephyritis. The lock mysteriously disappeared, but Conon "found" it in the heavens, between the constellations Virgo, Leo, and Boötes. In fact, what Conon identified was a constellation of seven faint stars near Leo's tail. Known thereafter as "the lock Berenice," the constellation was celebrated by poets, eventually becoming a literary topos. The Greek poet Callimachus immortalized Conon's discovery by composing a poem, *Berenīkēs plokamos* ("The Lock of Berenice"), in which the lock speaks to the reader.

To Roman poets of the Golden Age, Conon became something of a mythical figure, the quintessential astronomer. Catullus translated Callimachus's poem into Latin as *Coma Berenices,* and this version was the model for Alexander Pope's poem *Rape of the Lock,* in which Belinda's lock appears in the sky as a new star.

## FURTHER READING

### Books

Boyer, Carl B. *A History of Mathematics.* Second edition. Revised by Uta C. Merzbach. New York: John Wiley, 1991.
Gow, James. *A Short History of Greek Mathematics.* New York: Stechert, 1923.

—*Sketch by Zoran Minderovic*

# Roger Cotes
## 1682–1726
### English trigonometer and astronomer

Roger Cotes advanced the understanding of trigonometric functions through his original work on integration, functions, and the *nth* roots of unity. He is also remembered for his contributions to **Isaac Newton**'s work on universal gravitation.

Cotes was born in Burbage, England, the son of the Reverend Robert Cotes and his wife, Grace. He attended the Leicester School; his uncle, the Reverend John Smith, was so impressed by the boy's ability in mathematics that he decided to oversee Cotes' early education. When Cotes left Leicester School for St. Paul's School in London, he and his uncle regularly corresponded about science and mathematics. Cotes was admitted as a scholarship student to Trinity College in Cambridge in 1699. He completed his undergraduate work in 1702 and earned a master's degree in 1706.

### Looking to the Stars

Like many mathematicians of his era, Cotes was also an astronomer. The universe offered fascinating mathematical puzzles, and astronomical studies had an important practical benefit in an era when successful trade—especially for island nations like England—depended on sailing vessels that charted course by the stars.

In letters to Newton, Cotes described a heliostat telescope that used a clock mechanism to make an interior mirror revolve. He refined existing solar and planetary tables and made notes on the total solar eclipse in 1715.

Cotes was named as a fellow of Trinity College in 1705. In 1706—at the age of 24—he was appointed as the first Plumian professor of astronomy and natural philosophy at Cambridge. Cotes immediately began to solicit donations for an observatory at Trinity. Complete with living quarters, the observatory was built over King's Gate, and Cotes lived there with his cousin Robert Smith. The observatory itself was not completed in Cotes' lifetime and was demolished in 1797.

Cotes established a school of physical sciences at Trinity, and he and colleague William Whiston began a series of experiments in 1707. The details of these experiments were published after Cotes' death as *Hydrostatical and Pneumatical Lectures by Roger Cotes* in 1738.

In 1709, Cotes threw himself into the preparation of the second edition of Newton's *Philosophiae naturalis principia mathematica*, in which Newton refined his theory of universal gravitation. He and Newton worked closely for more than three years on the second edition. Cotes wrote a preface for the work that defended Newton's hypothesis against competing ideas. He argued that Newton's hypothesis, based on observation, was accurate and superior to theories based on description unaccompanied by rational explanation and, certainly, to theories based on superstition and the occult. Cotes' work on this edition was a labor of love—he received 12 copies of the book as payment for more than three years of work.

### The Work of a Lifetime

Cotes published only one paper on mathematics during his lifetime. His *Logometria* was released in 1714, and this paper is evidence not only of Cotes' brilliance, but his persistent and organized approach to the study of mathematics. Cotes calculated the natural base for a system of logarithms and introduced two inventive methods for calculating Briggsian logarithms for a number and interpolating intermediate values. He applied integration to problems related to quadratures, arc lengths areas of surfaces of revolution, and atmospheric density. In attempting to evaluate the surface area of an ellipsoid of revolution, Cotes identified not one, but two approaches to the problem, using both logarithms and arc sines to develop formulas to calculate area.

*Logometria* was included in a book of Cotes' papers compiled after his death by Robert Smith and published in 1722 as *Harmonia mensurarum*. The lengthiest section of the book includes Cotes' work in systematic integration. His work was based on a geometrical result involving $n$ equally-spaced points on a circle—now known as Cotes' theorem—that is equivalent to finding all the factors of $x^n - a^n$ when $n$ is a positive integer. Another section of *Harmonia mensurarum* includes miscellaneous papers on estimating errors, Newton's differential method, the descent of heavy bodies, and cycloidal motion. Cotes' solution to determining the area under a curve, in its modern version, is known as the Newton–Cotes formula.

Cotes died from a fever at the age of 33. "Had Cotes lived," Newton mourned, "we might have known something."

### SELECTED WRITINGS BY COTES

Preface to Isaac Newton's *Philosophiae naturalis principia mathematica*, 1713; reprinted, 1934
"Logometria." *Philosophical Transactions of the Royal Society*, 29 (1714): 5–47.
*Harmonium mensurarum, sive Analysis et synthesis per rationum et angulorum mensuras promotae: Addedunt alia opuscula mathematica per Rogerum Cotesium*, 1722
*Hydrostatical and Pneumatical Lectures by Roger Cotes A.M.* Edited by R. Smith, 1738.

*Richard Courant*

## FURTHER READING

Dubbey, J. M. "Roger Cotes," in *Dictionary of Scientific Biography.* Volume III. Edited by Charels Coulston Gillispie. New York: 1973, Charles Scribner's Sons, pp. 430–34.

Fang, J. *Mathematicians.* Hauppauge, NY: Paideia Press, 1972, p. 303.

Murray, Jane. *The Kings and Queens of England.* New York: Charles Scribner's Sons, 1974, pp. 59–78.

Smith, David E. *History of Mathematics.* New York: Dover Publications Inc., 1958, pp. 447–48.

—*Sketch by A. Mullig*

# Richard Courant
## 1888-1972
### German American algebraic geometer

Richard Courant received worldwide recognition as one of the foremost organizers of mathematical research and teaching in the twentieth century. Most of Courant's work was in variational calculus and its applications to physics, computer science, and other fields. He contributed significantly to the resurgence of applied mathematics in the twentieth century. While the Mathematics Institute in Göttingen, Germany, and the Courant Institute of Mathematical Sciences at New York University stand as monuments to his organizing and fund-raising abilities, his numerous honorary degrees and awards, as well as the achievements of his students, testify to his noteworthy contributions to mathematics and other sciences.

Courant, the first of three sons, was born on January 8, 1888, in Lublinitz, a small town in Upper Silesia that was then German but later Polish. The family moved to Glatz when he was three; when he was nine they moved to the Silesian capital, Breslau (now Wroclaw). He was enrolled in Breslau's König Wilhelm Gymnasium, preparing to attend a university. At the age of 14, Courant felt a need to become self-supporting and started tutoring students for the high-school math finals, which he himself had not yet taken. He was asked to leave the school for this reason in 1905, and he began attending lectures in mathematics and physics at the local university. He passed the high-school finals later that year and became a full-time student at the University of Breslau. Unhappy with the lecture methods of his physics instructors, he began to concentrate on mathematics.

In 1907 Courant enrolled at the University of Göttingen to take courses with mathematician **David Hilbert** , a professor there. Soon Courant became an assistant to Hilbert, working principally on subjects in analysis, an area of mathematics with a close relationship to physics. Under Hilbert, Courant obtained his Ph.D. in 1910 for a dissertation in variational calculus.

### Military Service

In the fall of 1910, Courant was called up for a year of compulsory military service, during which he became a noncommissioned officer. After Courant completed his tour of duty, Hilbert encouraged him to come back to Göttingen for the *Habilitation,* an examination that qualified him for a license as a *privatdozent,* an unsalaried university lecturer or teacher remunerated directly by students' fees. In 1912 Courant received his license to teach at Göttingen.

Two years of teaching and other mathematical work in Göttingen came to an abrupt halt when Courant received his orders to serve in the Army on July 30, 1914. The Kaiser declared war the next day. Courant, like many others, thought the war would be over quickly and was eager to serve. He believed that Germany's cause was right and that his country would be victorious. After about a year of fighting in the trenches, Courant was wounded and was subsequently deployed in the wireless communications department. Courant proposed that the use of mirrors to obtain visibility of what was going on above ground would help save lives. He also proposed the use of earth telegraphy, a means of communication that would use the earth as a conduit. Both ideas were utilized by the army.

## Between the Wars

About two weeks after the Armistice, which was signed November 11, 1918, in the midst of tremendous political turmoil, Courant managed to sign a contract with Ferdinand Springer to serve as editor for a series of mathematics books. Courant had made the original proposal for this series to Springer a year earlier. He envisioned timely mathematical treatises that would be especially pertinent to physics. These yellow-jacketed books became known worldwide as the *Yellow Series,* and their publication continued after Courant resigned as editor.

Courant returned to Göttingen in December of 1918 and resumed teaching as *Privatdozent* in the spring of 1919. During the summer of 1919, he completed a lengthy paper on the theory of eigenvalues of partial differential equations (of importance in quantum mechanics). After teaching at the University of Münster for a year, he returned as a professor to Göttingen in 1920; in addition to teaching, he was expected to take care of the informal administrative duties of the mathematics department. During the period from 1920 to 1925, Courant succeeded in making Göttingen an international center of theoretical and applied mathematics. Courant's emphasis on applied mathematics attracted physicists from all over the world, making the university a hub of research in quantum mechanics. His tireless efforts as a researcher, teacher, and organizer finally resulted in the creation of the Mathematics Institute of the University of Göttingen. Defining his vision of the future of mathematics, Courant said, "The ultimate justification of our institute rests in our belief in the indestructible vitality of mathematical scholarship. Everywhere there are signs to indicate that mathematics is on the threshold of a new breakthrough which may deepen its relationship with the other sciences and demand their mathematical penetration in a manner quite beyond our present understanding."

In the 1920s and 1930s, Courant worked with Hilbert on his most important publication, *Methoden der mathematischen Physik,* later translated into English as *Methods of Mathematical Physics.* The text was tremendously successful because it laid out the basic mathematical techniques that would play a role in the new quantum theory and nuclear physics. The Great Depression of the early 1930s created a need for university faculties to cut expenses, and there was an order to discharge most of the younger assistants. Courant successfully helped lead the fight of members of the mathematics and natural science faculty to pass

a proposal that professors themselves pay the salaries of the assistants who were to be dismissed. He also helped students get scholarships and arranged for some to become part of his household to assist them financially. Those from wealthy families were encouraged to work without pay so that stipends could be available for needy students.

Courant took a leave of absence from Göttingen during the spring and summer of 1932 to lecture in the United States. His positions as professor and director of the Göttingen Mathematics Institute ended on May 5, 1933, when he and five other Göttingen professors received official word that they were on leave until further notice. The move reflected the National Socialist government's escalating campaign against German Jews as well as its displeasure with the university, which had become a locus of independent liberal thought. As Americans and other foreigners were leaving Göttingen at this time, Courant observed that the spirit of the institute had already been destroyed. During the 1933–1934 academic year he became a visiting lecturer at Cambridge in England. In January of 1934, he accepted an offer of a two-year contract with New York University.

## New York Years

In 1936, when his temporary position ended, he was appointed professor and head of New York University's mathematics department. In that position, he did for New York University what he had done for Göttingen by creating a center of mathematics and science of international importance. In recognition of his work, Courant was made director of the new mathematics institute at New York University, later named the Courant Institute of Mathematical Sciences. Courant's success as an organizer was largely due to his ability to attract promising young mathematicians. He was always available to them as a teacher, helped them to publish their work, and organized financial support for them if they needed it. His students often remained loyal to him for the rest of their lives and tended to stay in his orbit.

During World War II, Courant was a member of the Applied Mathematics Panel, which assisted scientists involved with military projects and contracted for specific research with universities throughout the country. While the group at New York University under contract with the panel made important contributions to the war effort, the contract in turn played a vital role in setting up a scientific center at New York University. Courant's mathematical work in numerical analysis and partial difference equations played a vital role in the development of computer applications to scientific work. Courant was also instrumental in getting the Atomic Energy Commission to place its experimental computer UNIVAC at New York University in 1953. Courant retired in 1958, the same year he was honored with the Navy Distinguished Public Service Award and the Knight-Commander's Cross and Star of the Order of Merit of the Federal Republic of Germany. In 1965, he received an award for distinguished service to mathematics from the Mathematical Association of America.

Married in 1912 to a woman he had tutored as an adolescent, Courant was divorced in 1916. In 1919 he married Nerina (Nina) Runge, and they had two sons and two daughters. He enjoyed skiing and hiking, and played the piano. In November 1971, Courant suffered a stroke. He died on January 27, 1972, a few weeks after his 84th birthday. According to a *New York Times* obituary by Harry Schwartz, Nobel laureate in physics Niels Bohr once remarked that "every physicist is in Dr. Courant's debt for the vast insight he has given us into mathematical methods for comprehending nature and the physical world." At a memorial in Courant's honor, the mathematician Kurt O. Friedrichs said of him: "One cannot appreciate Courant's scientific achievements simply by enumerating his published work. To be sure, this work was original, significant, beautiful; but it had a very particular flavor: it never stood alone; it was always connected with problems and methods of other fields of science, drawing inspiration from them, and in turn inspiring them."

## SELECTED WRITINGS BY COURANT

### Books

(With Herbert Robbins) *What Is Mathematics?* 1941

(With Kurt O. Friedrichs) *Supersonic Flow and Shock Waves,* 1948

*Dirichlet's Principle, Conformal Mapping, and Minimal Surfaces,* 1950

(With David Hilbert) *Methods of Mathematical Physics* (translation of *Methoden der mathematischen Physik* ), 2 volumes, 1953, 1962

*Differential and Integral Calculus,* 2 volumes, 1965, 1974

(With Fritz John) *Introduction to Calculus and Analysis,* 2 volumes, 1965, 1974

### Periodicals

"Objectives of Applied Mathematics Education." *Society for Industrial and Applied Mathematics (SIAM) Review* 9 (1967): 303–05.

(With Friedrichs and Hans Lewy) "On the Partial Difference Equations of Mathematical Physics." *IBM Journal of Research and Development* (March 1967): 215–34.

## FURTHER READING

### Books

Albers, Donald J., and G. L. Alexanderson, editors. *Mathematical People: Profiles and Interviews.* New York: Birkhäuser, 1985.

Beyerchen, Alan D. *Scientists under Hitler: Politics and the Physics Community in the Third Reich.* New Haven, CT: Yale University Press, 1977.

*Courant Anniversary Volume: Studies and Essays Presented to Richard Courant on his 60th Birthday.* Interscience Publishers, 1948.

Reid, Constance. *Courant in Göttingen and New York: The Story of an Improbable Mathematician.* New York: Springer-Verlag, 1976.

Struik, Dirk, J., *A Concise History of Mathematics.* 4th rev. ed. New York: Dover Publications, 1987.

### Periodicals

*Journal of Mathematical and Physical Sciences: A Journal of the Indian Institute of Madras* (March 1973): i–iv.

Schwartz, Harry. *New York Times* January 29, 1972, p. 32.

### Other

Friedrichs, Kurt O., Peter D. Lax, K. Müller, Karl-Friedrich Still, Richard Emery, Jerome Berkowitz, and James M. Hester. "Richard Courant, 1888–1972: Remarks Delivered at the Memorial, February 18, 1972, and at the Meeting of the Graduate Faculty, March 15, 1972." New York University.

### Other

*Richard Courant in Göttingen and New York* (video). MAA Video Classics No. 4. Washington, DC: Mathematical Association of America, 1966.

*—Sketch by Jeanne Spriter James*

# Elbert Frank Cox
## 1895-1969
### African-American pure mathematician and educator

Elbert Frank Cox was the first African American to earn a Ph.D. in pure mathematics. Cox entered the teaching profession as a high school instructor and eventually rose to become the head of Howard University's mathematics department. In addition to his contributions to abstract mathematics, he made his mark as an educator by helping craft Howard's grading system in 1947 and by advising numerous successful masters degree candidates in mathematics.

Cox was born in Evansville, Indiana, on December 5, 1895. He was the oldest of three boys born to Johnson D. Cox, an elementary school principal, and his wife, Eugenia D. Cox. Close knit and highly religious, the Cox family had a respect for learning that reflected the father's educational career. When young Elbert demonstrated unusual ability in high school mathematics and physics, he was directed toward Indiana University. While at Indiana, he was elected to undergraduate offices and joined the Kappa Alpha Psi fraternity. After graduation in 1917, Cox entered the U.S. Army as a private during World War I and was promoted to staff sergeant in six months. Upon discharge, he became an instructor of mathematics at a high school in Henderson, Kentucky.

In 1920 or 1921 (sources vary) Cox joined the faculty of Shaw University in Raleigh, North Carolina, and left there two years later to attend Cornell University with a full scholarship. In the summer of 1925, when Cox received his doctorate from Cornell, he became the first black to earn such a degree in pure mathematics, a field concerned with mathematical theory rather than with practice or application. The topic of Cox's dissertation concerned polynomial solutions of difference equations.

In the fall of 1925, Cox became the head of the mathematics and physics department at West Virginia State College. Four years later, he moved to Washington, D.C., to join the faculty of Howard University. In 1947, Cox became chair of Howard's department of mathematics, a position he held until 1961 (a university rule mandated that all department heads resign at the age of 65). He remained a full professor in the department until his retirement in 1966.

During his career, Cox specialized in difference equations, interpolation theory, and differential equations. Among his professional accolades were memberships in such educational societies as Beta Kappa Chi, Pi Mu Epsilon, and Sigma Pi Sigma. He was also active in the American Mathematical Society, the American Physical Society, and the American Physics Institute. He married Beulah P. Kaufman, an elementary school teacher, on September 14, 1927. They had three sons, James, Eugene, and Elbert. Cox died at Cafritz Memorial Hospital on November 28, 1969, after a brief illness.

## FURTHER READING

"Dr. Elbert F. Cox, 73, Howard U. Professor." Obituary. *The Washington Post* (December 2, 1969): C6.

—*Sketch by Leonard C. Bruno and Loretta Hall*

# Gertrude Mary Cox
## 1900-1978
### American statistician

Gertrude Cox organized and directed several agencies dedicated to research and teaching in statistics. "By her missionary zeal, her organizational ability and her appreciation of the need for a practical approach to the statistical needs of agricultural, biological and medical research workers she did much to counter the confused mass of theory emanating from mathematical statisticians, particularly in the United States, who had little contact with scientific research," eulogized Frank Yates in the *Journal of the Royal Statistical Society.*

Cox was born on a farm near Dayton, Iowa, on January 13, 1900, to John William Allen and Emmaline (Maddy) Cox. After graduating from Perry (Iowa) High School in 1918, she devoted several years to social service and training for the role of deaconess in the Methodist Episcopal Church. She spent part of that time caring for children in a Montana orphanage.

In 1925, in preparation for advancement to superintendent of the orphanage, Cox entered Iowa State College in Ames. Although she took courses in psychology, sociology, and other topics that would advance her intended career, she majored in mathematics because of her talent for it. She graduated in 1929 and registered for graduate work under the direction of Professor George Snedecor, a proponent of **Ronald A. Fisher**'s statistical methods. Cox and Fisher became friends when he worked at Iowa State during the summers of 1931 and 1936. In 1931, she earned Iowa State's first M.S. degree in statistics, and for the next two years Cox worked as a graduate assistant at the University of California, Berkeley, studying psychological statistics.

In 1933, Snedecor asked Cox to return to work at Iowa State's new Statistical Laboratory, where she built a reputation of expertise in experimental design. By 1939, Cox had become an assistant professor at Iowa State, although her teaching and consulting activities never allowed her time to write a doctoral dissertation. Eventually, in 1958, Cox was awarded an honorary Doctor of Science degree by Iowa State.

When Snedecor was asked to recommend nominees to head the new Department of Experimental Statistics being formed at North Carolina State (NCS) College's School of Agriculture, he showed his list to Cox, who asked why her name was not included. He then added a footnote to his letter: "Of course if you would consider a woman for this position I would recommend Gertrude Cox of my staff." She was hired in 1940, becoming the first woman to head a department at North Carolina State.

In 1944, Cox assumed additional duties as director of the NCS Institute of Statistics, which she had organized. By 1946, the University of North Carolina (UNC) joined the Institute, taking responsibility for teaching statistical theory while NCS provided courses in methodology. Cox saw the Institute's mission as developing strong statistical programs throughout the South, a vision described by colleagues as "spreading the gospel according to St. Gertrude." Always insistent that good statistical analysis depended on adequate computations, Cox unhesitatingly embraced the computer era. NCS became one of the first colleges in the country to install an IBM 650 computer, and statisticians in Cox's organizations developed powerful statistical software programs. Cox helped create the Biometric Society in 1947 and edited its journal *Biometrics* from 1947 to 1955. In 1949, she became the first female member of the International Statistical Institute, and was elected president of the American Statistical Association seven years later.

In 1950, Cox and her colleague William Cochran published *Experimental Designs* which was intended to be a reference book for research workers with little technical knowledge. In fact, it became a widely-used textbook that Frank Yates described nearly 30 years later as "still the best practical book on the design and analysis of replicated experiments." In her own experimental design classes, Cox taught by focusing on specific examples gleaned from her years of consulting experience.

Although Cox made substantial contributions to the theory of statistics and experimental design, Richard Anderson wrote in *Biographical Memoirs* that her most valuable contribution to science was organizing and administering programs. She was exceptionally successful in generating financial support for research. For example, one large grant Cox obtained from the General Education Board established a revolving fund that supported fundamental statistical research for many years. Cox also played an integral role in planning what would become the Research Triangle Institute (RTI) for consulting and research, uniting the resources of NCS, UNC, and Duke University. In 1960, she retired from NCS and became the first director of RTI's Statistics Section.

Cox loved world travel, and during her lifetime she made 23 trips to various international destinations. After retiring a second time in 1965, she spent a year in Egypt establishing the University of Cairo's Institute of Statistics. On five different occasions, Cox worked on statistical assistance programs in Thailand. At the age of 76, she toured Alaska and the Yukon Territory by bus, train, and boat.

Although she received numerous honors, including her 1975 election to the National Academy of Sciences, Cox was particularly pleased with the dedication of the statistics building at North Carolina State University as "Cox Hall" in 1970 and the establishment by her former students of the $200,000 Gertrude M. Cox Fellowship Fund for outstanding students in statistics at NCS in 1977.

Cox died of leukemia on October 17, 1978, at Duke University Medical Center in Durham. During the preceding year, she had kept meticulous records of her treatment and response, making herself the subject of her final experiment.

## SELECTED WRITINGS BY COX

(With William G. Cochran) *Experimental Designs*. Second Edition. 1992

## FURTHER READING

### Books

Nichols, Maryjo. "Gertrude Mary Cox," in *Women of Mathematics*. Edited by Louise S. Grinstein and Paul J. Campbell. New York: Greenwood Press, 1987, pp. 26–29.

O'Neill, Lois Decker, editor. *The Women's Book of World Records and Achievements*. Garden City, NY: Doubleday, 1979, p. 176.

### Periodicals

Anderson, R.L., R.J. Monroe, and L.A. Nelson. "Gertrude M. Cox: A Modern Pioneer in Statistics." *Biometrics* (March 1979): 3–7.

Anderson, Richard L. "Gertrude Mary Cox." *Biographical Memoirs*. National Academy of Sciences 59 (1990): 117–34.

Cochran, William G. "Gertrude Mary Cox, 1900-1978." *International Statistical Review* (April 1979): 97–8.

"Some Reflections." *Biometrics* (March 1979): 1–2.

Monroe, Robert J., and Francis E. McVay. "Gertrude Mary Cox, 1900-1978." *The American Statistician* (February 1980): 48.

*Gabriel Cramer*

Yates, Frank. "Gertrude Mary Cox, 1900-1978." *Journal of the Royal Statistical Society* 142, Part 4 (1979): 516-17.

**Other**

"Gertrude Mary Cox." *Biographies of Women Mathematicians.* June 1997. http://www.scottlan.edu/lriddle/women/chronol.htm (July 22, 1997).

—*Sketch by Loretta Hall*

# Gabriel Cramer
## 1704–1752
### Swiss geometer and probability theorist

Gabriel Cramer labored in the shadow of his more well-known mathematical contemporaries. Cramer added to mathematical knowledge in the areas of analysis, determinants, and geometry. Both Cramer's rule and Cramer's paradox, discussed below, were not completely new ideas, but Cramer contributed significantly to both. Cramer also introduced the idea of utility to mathematics. Some of Cramer's most important work concerned the history of mathematics, as Cramer often served as the editor of other mathematicians' writings.

Cramer was born July 31, 1704, in Geneva, Switzerland, to physician Jean Isaac Cramer and his wife, Anne Mallet. Educated in Geneva, Cramer had two brothers. One, Jean–Antoine, also became a doctor; the other, Jean, was a law professor. All three brothers took an interest in local government and its inner workings. Cramer was a lifelong bachelor.

At the age of 18, Cramer defended his thesis with a topic about sound. At age 20, he was appointed co–chair of mathematics at the Académie de la Rive, and led the geometry and mechanics classes. His co–chair was Giovanni Ludovico Calandrini, a friend who taught algebra and astronomy. They also split the salary designated for one chair. As professors, Cramer and Calandrini broke with tradition and allowed recitations in French instead of the traditional Latin, so that students without a Latin background could participate.

The pair were encouraged to travel to enhance their scholarship. Cramer went to Basle for five months in 1727, where he met and befriended, among others, **Johann Bernoulli**. Cramer continued to travel until 1729, visiting London, Leiden, and Paris, before returning to Geneva. In 1734, Cramer became full chair of the department when Calandrini was appointed to a philosophy professorship. In the same year, Cramer's activity in local government was marked by his service on the Conseil des Deux–Cents.

Cramer's sense of community extended to his mathematical colleagues. He edited the collected works of fellow mathematician Johann Bernoulli in four volumes, which were published in 1742. Cramer's other editorial projects included the works of Johann's brother, **Jakob Bernoulli**, and Christian Wolff. His accomplishments as a mathematics editor were vital for the support and circulation of mathematical knowledge, and he was one of the first scholars of note to contribute to mathematics in this manner.

Cramer's most important mathematical work was published in 1750, the same year he was appointed professor of philosophy at Académie de la Rive when Caladrini left to work for the Swiss government. Within the four volumes of *Introduction à l'analyse des lignes courbes algebriques*, Cramer delineated what came to be known as Cramer's rule, which provided a mechanism for linear equation solutions. Although deteminants were discovered by **Gottfried Wilhelm Leibniz** in 1693, it is Cramer who brought determinants and their uses to wide spread attention. Also in this text is Cramer's discussion of curve analysis which led to their rediscovery, which led to other mathematicians improving on his analysis.

Another important element featured in Cramer's text is what came to be known as Cramer's paradox.

The paradox clarified a theorem first proposed by **Colin Maclaurin**, who pointed out that two separate cubic curves can meet at nine different points. Cramer added that a single cubic curve is itself defined by nine different points. Because Cramer's explanation for this phenomenon was lacking, other mathematicians fleshed it out. Cramer is also responsible for the concept of utility, which today links probability theory with mathematical economics.

In 1751, Cramer fell from a carriage and was confined to his bed for two months. His doctor recommended a rest in the south of France for his health because, in addition to the fall, he was overworked. Cramer died en route in Bagnoles, France, on January 4, 1752.

## SELECTED WRITINGS BY CRAMER

*Introduction a l'analyse des lignes courbes algebriques,* 1750

## FURTHER READING

Abbott, David, editor. *The Biographical Dictionary of Scientists: Mathematicians.* New York: Peter Bedrick Books, 1986, pp. 34–35.

Boyer, Charles B. *History of Analytical Geometry.* New York: 1956, pp. 194–97.

Jones, Phillip S. "Gabriel Cramer," in *Dictionary of Scientific Biography.* Volume III. Edited by Charles Coulston Gillispie. New York: Charles Scribner's Sons, 1970, pp. 459–62.

*—Sketch by Annette Petruso*

nancy as an unwelcome accident and later abandoned her infant son on the steps of the church of Saint–Jean–le–Rond. The foundling was baptized with the name of the church, then sent to a foster home in Picardy.

After Destouches returned from military service, he brought his son back to Paris and arranged for him to be raised by one Madame Rousseau, the wife of a glazier. D'Alembert always regarded this woman as his real mother, and he continued to live in her home until he was 47 years old. Destouches provided for his son and paid for his education. When Destouches died in 1726, he left the boy a legacy that gave d'Alembert a modest lifetime income of 1,200 livres a year. D'Alembert always cherished the independence this income provided.

### The Enlightened Mathematician

D'Alembert attended the Collège des Quatre–Nations (also called Mazarin College), a school run by Jansenists, members of a religious sect. While at college, he adopted the name Jean–Baptiste Daremberg, later shortened to d'Alembert. Despite the urging of teachers, however, d'Alembert rejected the religious life. After receiving a bachelor's degree in 1735, he went on to study law, receiving a license to practice in 1738. He also studied medicine for a year. Neither the law nor medicine held much lasting appeal for d'Alembert. He finally settled upon a career in mathematics, a vocation for which he had much natural talent.

In 1739, d'Alembert submitted his first paper to the French Académie Royale des Sciences, a critique of a mathematics book written by Father Charles Reyneau. Over the next two years, d'Alembert sent the Académie additional papers on such topics as fluid mechanics and the integration of differential equations. After several failed attempts to join the Académie, he was finally admitted in May 1741.

During the 1740s, d'Alembert became a fixture in French intellectual and social salons, where he was known for his gaiety and wit. He took his place among leading *philosophes*, thinkers of the Enlightenment, a philosophical movement marked by an emphasis on human reason and a rejection of traditional religious and political ideas. Although d'Alembert never married, he shared a close relationship for many years with Julie de Lespinasse, a popular salon hostess.

*Jean le Rond d'Alembert*

# Jean le Rond d'Alembert
## 1717–1783
### French calculist and physicist

Jean le Rond d'Alembert was a mathematician and physicist who applied his considerable genius to solving problems in mechanics. This is the branch of physics that deals with the effect of forces on matter, either at rest or in motion. His most important contribution was d'Alembert's principle, which states that the forces in an object that resist acceleration must be equal and opposite to the forces that produce the acceleration. D'Alembert was a pioneer in the development of calculus. He also served as science editor of Denis Diderot's *Encyclopédie*.

D'Alembert was born in Paris on November 17, 1717. He was the illegitimate son of Claudine–Alexandrine Guérin, Marquise de Tencin, an intelligent and unprincipled woman who broke her vows as a nun and became the mistress of many powerful men. D'Alembert's father was the Chevalier Louis–Camus Destouches, a military officer. D'Alembert's mother apparently regarded her preg-

## Moving Bodies and Fluid Motion

From 1741 through 1743, d'Alembert studied various problems in dynamics, the branch of mechanics that deals with the effect of forces on the motion of bodies. His writings were hastily collected into a book, *Traité de dynamique* (1743), that became his most important scientific work. This book introduced the famous principle that bears d'Alembert's name. The principle was actually an extension of **Isaac Newton**'s third law of motion, which states that for every force exerted on a static body, there is an equal and opposite force from that body. D'Alembert maintained that the law applied not only to bodies at rest, but to bodies in motion as well. This was the dawn of a new era in the science of mechanics. The next year, d'Alembert published *Traité de l'équilibre et du mouvement des fluides* (1744), in which he applied his principle to the motion of fluids.

D'Alembert published *Réflexions sur la cause générale des vents* in 1747, a treatise on winds that won a prize from the Prussian Academy. This paper marked the first general use of partial differential equations in mathematical physics. That same year, d'Alembert published an article on the motion of vibrating strings. This paper is notable for the first use of a wave equation in physics. While d'Alembert pioneered both partial differential equations and wave equations, it was left to his Swiss contemporary **Leonhard Euler** to develop these concepts more fully.

Next, d'Alembert turned his attention to astronomy, applying calculus to celestial mechanics. In 1749 he issued *Recherches sur la précession des équinoxes et sur la nutation de la terre*. This book dealt with the precession of the equinoxes; that is, the slow, gradual westward motion of the equinoxes due to the movement of the Earth's axis. D'Alembert's research on astronomy continued in *Recherches sur différens points importants du systeme du monde*. This three–volume work, published in 1754–1756, dealt mainly with the motion of the moon.

D'Alembert issued one more scientific publication in the 1750s, returning to the subject of fluid mechanics in *Essai d'une nouvelle théorie de la résistance des fluides* (1752). In this essay, he introduced such important concepts as the components of fluid velocity and acceleration. With *Élémens de musique théorique et pratique suivant les principes de M. Rameau* (1752), however, d'Alembert departed from science to indulge an interest in music. In this work, he described the new theory of musical structure advanced by French composer Jean–Philippe Rameau.

## An Encyclopedic Knowledge

Much of the 1750s was devoted to work on Diderot's *Encyclopédie*. This monumental work was conceived by Diderot as a synthesis of all human knowledge, with an emphasis on new ideas and scientific discoveries. D'Alembert's first task was writing the *Discours préliminaire* (1751), an introduction that sought to show the links between disciplines and to trace the progress of thought, culminating in the philosophy of the Enlightenment. The discourse was widely praised. Its publication led to d'Alembert's acceptance into the French Académie in 1754. He later became very active in that organization, eventually being elected its permanent secretary.

D'Alembert wrote 1,500 articles for the *Encyclopédie*. While many of these articles discussed mathematics and science, others addressed philosophy and the arts. In fact, d'Alembert was increasingly drawn to nonscientific topics. Between 1753 and 1767, he published five volumes of *Mélanges de littérature et de philosophie*. This collection contained essays on music, law, and religion; a treatise on philosophy; translations of Tacitus; and a hodgepodge of other material.

In 1757, d'Alembert visited the French writer Voltaire, his closest friend among the *philosophes*. One result of the visit was an article on Geneva, Switzerland, which appeared in the seventh volume of the *Encyclopédie*. This article caused a furor with its depiction of Protestant ministers, managing to offend Roman Catholics and Calvinists alike. The uproar caused d'Alembert to resign as an editor of the project.

## Views on Science and Religion

During the following decades, d'Alembert resumed scientific publication. From 1761 to 1780, he issued eight volumes of *Opuscules mathématiques*, which included essays on hydrodynamics, astronomy, and lenses. At this time, d'Alembert was almost alone in regarding the differential as the limit of a function, a key concept in modern calculus. However, he could never rise above the traditional focus on geometry, which prevented him from ever putting the idea of the limit into a purely algorithmic form.

In 1764, d'Alembert spent three months in the court of Frederick the Great. Although he was offered the presidency of the Prussian Academy, d'Alembert declined. He also turned down an offer from Russian Empress Catherine the Great to tutor her son for 100,000 livres a year. Above all, d'Alembert prized his financial independence and the intellectual freedom it afforded. D'Alembert published a work on religion the following year, in which he called for the suppression of both the Jesuits and their rivals, the Jansenists. This book, not one of d'Alembert's better efforts, was written at Voltaire's behest. It was issued anonymously, but the author's identity was known.

D'Alembert's final years were not easy. He became seriously ill in 1765 and moved into the home of Julie de Lespinasse, who nursed him back to health. He continued to live with Lespinasse until her death in 1776. After she died, d'Alembert discovered evidence of love affairs with other men among Lespinasse's effects, which made her loss doubly painful. He withdrew into a lonely, bitter retirement, living in a small apartment provided by the French Académie. D'Alembert died in Paris on October 29, 1783. He had outlived many of his fellow *philosophes*, but the scientific and philosophical legacy of this remarkable group of thinkers survives.

## SELECTED WRITINGS BY D'ALEMBERT

*Preliminary Discourse to the Encyclopedia of Diderot.* Translated by Richard N. Schwab, 1995

## FURTHER READING

### Books

Ball, W.W. Rouse. *A Short Account of the History of Mathematics.* New York: Dover Publications, 1960, pp. 374–77.

Briggs, J. Morton. "Jean le Rond d'alembert." *Biographical Dictionary of Mathematicians.* Volume 1. New York: Charles Scribner's Sons, 1991, pp. 42–50.

Diderot, Denis. "Biographical Note on the Characters." *Rameau's Nephew and d'Alembert's Dream.* Translated by Leonard Tancock. New York: Penguin, 1966, pp. 141–47.

Magill, Frank N., editor. *Great Lives from History: Renaissance to 1900 Series.* Volume 1. Pasadena, CA: Salem Press, 1989, pp. 50–55.

Porter, Roy, editor. *The Biographical Dictionary of Scientists.* Second edition. New York: Oxford University Press, 1994, pp. 160–61.

*— Sketch by Linda Wasmer Smith*

# Scipione Dal Ferro
## 1465–1526
### Italian algebraist

Scipione dal Ferro may be one of the most overlooked and unappreciated mathematicians of his day. Little is known of his life, and none of his writings are extant. Dal Ferro's most significant contribution to the development of algebra, the solution to the so-called depressed cubic equation, was never published during his lifetime and was later credited to another mathematician.

Dal Ferro was born in Bologna, Italy, on February 6, 1465. His father was Floriano Ferro, who worked as a papermaker. His mother was named Filippa. Scipione's last name is written variously as Ferro, Ferreo, dal Ferro, or del Ferro. Nothing is known of his childhood or education, nor of his early adult life.

In 1496, at the age of 31, dal Ferro secured a position as a lecturer in arithmetic and geometry at the University of Bologna, where he remained until his death. At some point dal Ferro fathered a daughter, who he named Filippa, although there is no mention of his wife or of any other children. His daughter married one of her father's students, Annibale dalla Nave, who went on to take over dal Ferro's teaching duties at the university after dal Ferro's death.

### Finds a Solution to Cubic Equations

Up until the early 1500s, no one had been able to develop a method for solving cubic, or third–power, equations. In fact, in 1494, the Italian mathematician Luca Pacioli had written a book entitled *Summa de Arithmetica* in which he expressed the belief that the solution to a cubic equation was impossible, given the current state of mathematics.

Pacioli's statement was taken as a challenge by the Italian mathematical community, and the next 50 years was a period of competitions, accusations, and counter–accusations as one mathematician after another sought credit for developing a solution. Unbeknownst to all of them, dal Ferro had found a solution to the so–called depressed cubic equation, in which the second–power term was missing, but had kept his discovery a secret. Dal Ferro kept it so secret that the exact date of his discovery is not known, but it is believed to have been sometime in the first two decades of the century, possibly about 1505–1515.

Although dal Ferro's solution only applied to the depressed cubic equation and not the generalized cubic equation in which all the terms are present, it was still a monumental discovery and would have instantly established him as a celebrity among his academic peers. No one knows his reasons for keeping it a secret. There were certainly many other mathematicians who would have been eager to publish the solution and enjoy the accolades that went with it. One reason may have been the practice of public challenges, which were a constant threat to those holding positions in a university. Continued employment depended on personal and political patronage, as well as successful performance when publicly challenged by a rival scholar. These challenges could come at any time and in any form, and failure to prevail meant public humiliation and the possible loss of a job. Dal Ferro may have viewed his solution to the depressed cubic equation as his "ace–in–the–hole" and opted to withhold it in case he was seriously challenged.

In any case, dal Ferro never published his solution, but committed it to a handwritten manuscript which he kept among his personal papers. It was only on his deathbed that he revealed his secret to one of his pupils, Antonio Fior. Dal Ferro died sometime between October 29 and November 16 in 1526 (some sources cite November 5), without ever having received public recognition of his discovery.

### Others Receive Credit For His Work

Dal Ferro made a lamentable choice when he entrusted his secret to Antonio Fior. Fior was an average mathematician at best and was even more fearful of being publicly challenged than dal Ferro. He kept the secret to himself until 1535, when he rashly used it to challenge **Tartaglia** (Nicolò Fontana), a newly–appointed professor of mathematics in Venice. Tartaglia already knew the solution to cubic equations that lacked the first power term, and so he readily accepted Fior's challenge. Tartaglia gave Fior problems covering a wide range of mathematical subjects, while Fior gave Tartaglia nothing but depressed cubic equations to solve. Thus, Tartaglia was faced with the possibility of total defeat if he could not come up with the secret solution for himself. With the deadline drawing near, Tartaglia discovered the same method used by dal Ferro and solved all the problems. Fior, on the other hand, lived up to his lackluster reputation and solved only a few of Tartaglia's problems.

**Girolamo Cardano** learned of Tartaglia's victory and repeatedly beseeched him for the secret. After many years, in 1539, Tartaglia agreed to share his knowledge with Cardano, but made him promise not to reveal the secret until Tartaglia had published it first. In time, Cardano discovered the solution to a generalized cubic equation using Tartaglia's methods as an intermediate step, but was unable to make his discovery public because of his promise. Finally, in 1543, Cardano discovered dal Ferro's handwritten manuscript and learned that it was dal Ferro, not Tartaglia, who had first solved the depressed cubic equation. Believing that this removed his pledge of silence to Tartaglia, Cardano published his solution in a book entitled *Ars magna* in 1545. In his book, he gave full credit to dal Ferro and Tartaglia for their contributions. Regarding dal Ferro, Cardano wrote that his solution of the depressed cubic was "proof of the power of reason, and so illustrious that whoever attains it may believe himself capable of solving any problem."

Despite Cardano's acknowledgment of dal Ferro's achievement, history awarded the recognition to Cardano, and dal Ferro's solution to the depressed cubic became known as Cardano's formula.

### FURTHER READING

#### Books

American Council of Learned Societies. *Biographical Dictionary of Mathematicians.* New York: Charles Scribner's Sons, 1991, pp. 788–790.

Dunham, W. *Journey Through Genius: The Great Theorems of Mathematics.* New York: Wiley, 1990, pp. 134–135, 141.

Struik, D. J., editor. *A Source Book in Mathematics: 1200–1800.* Princeton, NJ: Princeton University Press, 1986, pp. 62–63.

*—Sketch by Chris Cavette*

# George Bernard Dantzig
## 1914-
### American applied mathematician

George Bernard Dantzig is a mathematician and the founder of linear programming, a mathematical technique that has had extensive scientific and technical applications in such areas as computer programming, logistics, and scheduling. Applicable to such endeavors as military research, industrial engineering, and business and managerial studies, linear programming is a method for formulating solutions to problems of how to optimally allocate resources among competitive activities. For example, linear programming could be used to develop a diet that contains all the necessary minimal quantities of dietary elements at a minimum cost by factoring in such variables as calories, protein, vitamins, and the prices of food. Dantzig also discovered the simplex method, an algorithm that was remarkably efficient for use in the linear programming of computers. It has been largely through Dantzig's vision that mathematical programming has become a field in which deep interactions between mathematics, computation, and application models are probed and developed. Dantzig is also coauthor of the book *Compact City*, which suggests improved approaches to urban development, including the use of computer programming that takes into account the "socioeconomic as well as physical aspects of complex urban systems."

Dantzig was born on November 8, 1914, in Portland, Oregon, to Tobias and Anja (Ourisson)

Dantzig. Dantzig's father was born in Russia and participated in a failed revolution in 1905. After spending nine months in a Russian prison, Tobias Dantzig went to Paris and studied mathematics at the Sorbonne before immigrating to the United States in 1909. A well-known mathematician in his own right, Tobias Dantzig wrote the influential book, *Number, the Language of Science,* which focused on the concept of the evolution of numbers as related to the growth of the human mind.

Following in his father's footsteps, Dantzig attended the University of Maryland to study mathematics and physics. Upon graduation in 1936, Dantzig was appointed a Horace Rackham Scholar at the University of Michigan, where he earned his M.A. in mathematics in 1938. For the next two years, Dantzig worked as a junior statistician for the U.S. Bureau of Labor Statistics before enrolling in the mathematics doctoral program at the University of California, Berkeley; his studies were interrupted, however, when the United States entered World War II. Dantzig left Berkeley in 1941 to become chief of the combat analysis branch of the U.S. Air Force's statistical control headquarters. In 1944, he received the War Department Exceptional Civilian Service Medal for his efforts. In the meantime, Dantzig had returned to his doctoral studies at Berkeley, studying under Jerzy Neyman, a major contributor to modern mathematical statistics. He received his Ph.D. in mathematics in 1946.

## Makes Groundbreaking Discovery in Linear Programming

Rapid advances in technology combined with the effects of World War II and urban development brought on a new era of large-scale planning tasks. With his valuable war-time military experience, Dantzig was asked to continue working for the Air Force and, in 1946, was appointed chief mathematical adviser on the staff of the Air Force Comptroller. At this time, the Air Force had begun Project SCOOP (Scientific Computation of Optimum Programs), which was designed to increase the mechanization and speed for planning and deploying military forces. Focusing primarily on the planning segment of the project, Dantzig discovered that linear program s could be used to solve a wide range of planning problems. Conceptually, this discovery was an important step toward a mathematical approach to many planning and management difficulties, but it was Dantzig's simultaneous discovery of the simplex method—an algorithm that could be efficiently used to solve programming problems—that revealed the enormous power of linear programming. Dantzig's discovery was facilitated by the fact that the modern era of computer research was also getting underway. The development of technology that could rapidly solve complicated equations—equations that other-wise could take years to complete—made linear computing programming a practical resource for use in such areas as industry and economics, which could now quickly compare the many factors involved in interdependent courses of action.

The key to linear programming and the simplex method is the use of a "best value" or set of best values for many variables involved in a certain problem. Linear programming works most efficiently when a quantity can be optimized, or made as perfect and functional as possible. This quantity, called the objective function, for example, could be the most economical way to produce and distribute a product taking into account various "system" factors, such as product composition, production scheduling, and distribution. A key to the programming's success is to develop proportional values for these factors, such as their linear interdependency, in which at least one linear combination of an element equals zero when the coefficients are taken from another given set and at least one of its coefficients is not equal to zero.

In 1952, Dantzig went to work with the RAND Corporation, one of the first private industries to use computer technology. As a research mathematician at RAND, Dantzig played a major role in developing the new discipline of operations research using linear programming, and became a pioneer in identifying its exhaustive uses. In Michael Olinick's book *An Introduction to Mathematical Models in the Social and Life Sciences,* Dantzig notes: "Industrial production, the flow of resources in the economy, the exertion of military effort in a war theater—all are complexes of numerous interrelated activities. Differences may exist in the goals to be achieved, the particular processes involved and the magnitude of effort. Nevertheless, it is possible to abstract the underlying similarities in the management of these seemingly disparate systems."

## Helps Establish the Field of Linear Programming

Over the years, Dantzig helped refine linear programming and contributed to establishing the field in both industry and academia. By the 1970s, decision-making software based on the principles of linear programming was being marketed for both technical and nontechnical users. As the growing importance of this field became apparent, universities began developing academic studies of operations research, also referred to as mathematical decision making, in such areas as business science, industrial engineering, and mathematical computing. In 1960, Dantzig left private industry and joined the University of California, Berkeley, as chairman of the Operations Research Center. Located in the heart of the "silicon valley," home of the computer programming and software industry, Dantzig was ideally situated to continue his studies. In 1963, he published the highly influential book *Linear Programming and Extensions,* which includes discussions of the origins of linear programming; according to Dantzig, the theories of linear programming dates back to Jean Baptiste Joseph Fourier, a French mathematician known for his research into numerical equations and the conduction of heat. In the book, Dantzig also delves into how the field was developed both on a theoretical basis and by real-life problems presented in the military and economics. His book has become a classic in the field.

In 1966, Dantzig became a professor of operations research and computer science at Stanford and served as acting chairman of the Operations Research Department from 1969 to 1970. He contributed to the development of such major areas of mathematical programming and operations research as quadratic programming, complementary pivot theory, nonlinear equations, convex programming, integer programming, stochastic programming, dynamic programming, game theory, and optimal control theory. With the mathematician Philip Wolfe, he also originated the decomposition principle, a method for solving large systems by exploiting the special characteristics of their block-diagonal structure. This procedure was successfully used in 1971 to solve an equation containing 282,468 variables and 50,215 equations in just 2.5 hours; such an equation would have otherwise required 37 years to complete. Throughout his career, Dantzig consulted on the development of large-scale management planning models and created mathematical models of chemical and biological processes. He also utilized computers as a fundamental aspect of mathematical programming—for example, he participated in the development of a computer language and compiler which was designed to facilitate experimentation on mathematical programming algorithms.

### Co-develops Plans for High-Tech Urban Development

One of Dantzig's primary interests was in the development of analytical models of transportation systems; in 1974, he was the recipient of an endowed chair at Stanford, the C. A. Criley Chair of Transportation. Dantzig's interest in transportation and the efficient and most economical use of resources through mathematical programming led him to write *Compact City, A Plan for a Livable Urban Environment* with Thomas L. Saaty. Dantzig and Saaty set out to learn more about city planning by consulting with a range of experts, including engineers, economists, social workers, sociologists, seismologists, waste-removal engineers, and environmentalists. Concerned with such urban crises as the shortage of energy, growth of slums, congestion, and pollution, the book focuses on finding more advanced ways of developing urban areas while increasing the standard of living and minimizing the consumption of nonrenewable resources.

*Compact City* describes a new concept of living in which as many as two million people could live in ideal weather in spacious homes and gardens and walk to work within a few minutes. An integral part of the planning process was to transform urban development from "flat, predominantly two-dimensional cities to four-dimensional cities in which vertical space and time are exploited." In addition to simplifying transportation systems and alleviating the burden on energy consumption, Dantzig and Saaty were also concerned with "bringing the community together" and offering new opportunities for the underprivileged. Computers and linear programming played an integral part in the planning by taking into account not only the physical aspects but also the socioeconomic aspects of urban development.

As the conceptual developer of linear programming, Dantzig was invited to lecture around the world. He went on a one-year sabbatical in 1974 as head of the methodology group at the International Institute for Applied Systems Analysis, in Laxenburg, Austria, and received an honorary degree of doctor of science from the Israel Institute of Technology in 1973. During his career, Dantzig also received honorary degrees from the University of Linköping in Sweden, the University of Maryland, and Yale University. In 1975, U.S. President Gerald Ford awarded him the National Medal of Science in recognition of his inventing linear programming, developing methods that allowed it to be applied widely in industry and science, and using computers to incorporate mathematical theory. On November 1, 1976, California passed State Resolution No. 1748, honoring Dantzig's contributions to applied science.

In addition to his many honors, including being elected to the National Academy of Sciences in 1971, Dantzig served in many scientific societies and was the founder of the Mathematical Programming Society. Dantzig married Anne S. Shmumer on August 23, 1936. They have three children, David Franklin, Jessica Rose, and Paul Michael.

## SELECTED WRITINGS BY DANTZIG

*Linear Programming and Extensions,* 1963
(With Thomas L. Saaty) *Compact City, A Plan for a Livable Urban Environment,* 1973

## FURTHER READING

Abbott, David. *Biographical Dictionary of Scientists: Mathematicians,* New York: Peter Bedrick Books, 1986, pp. 36–37.
Cortada, James W. *Historical Dictionary of Data Processing.* Westport, CT: Greenwood Press, 1987, pp. 68–70.

*McGraw-Hill Modern Engineers and Scientists.* New York: McGraw, 1980, pp. 262–263.

Olinick, Michael. *An Introduction to Mathematical Models in the Social and Life Sciences.* New York: Addison-Wesley, 1978, pp. 164–167.

*—Sketch by David Petechuk*

# Ingrid Daubechies
## 1954–

### Belgium–born American applied mathematician and educator

Ingrid Daubechies was born August 17, 1954, in Houthalen, Belgium. Her father, Marcel Daubechies, is a retired civil engineer and her mother, Simone, is a retired criminologist. Daubechies credits her parents with giving her a love of learning and her mother with teaching her by example to be her own person. Her father always encouraged her to pursue her interest in science. She has one brother.

As a small child, Daubechies displayed an insatiable interest in how things worked and in making things with her hands. She took up the hobbies of weaving and pottery at a young age and continues to produce objets d'art in both crafts. At the age of eight or nine Daubechies' favorite hobby was to sew clothes for her dolls because it fascinated her that flat pieces of material could be worked into curved surfaces that fit the angles of the doll's body. But she also fascinated with machinery and mathematical axioms. Daubechies used to lie in bed and compute the powers of two, or test the mathematical law that any number divisible by nine produces another number divisible by nine when the digits are added together. Reading has been a lifelong hobby.

Daubechies spent her entire childhood and school years in Belgium. She was educated at the Free University Brussels, earning a B.S. degree in 1975 and a Ph.D. in 1980, both in physics. Her thesis was entitled "Representation of Quantum Mechanical Operators by Kernels on Hilbert Spaces of Analytic Functions." Between 1978 and 1980 Daubechies wrote ten articles based on her own original research. While pursuing her own studies, she taught at the Free University Brussels a total of 12 years. Daubechies first visited the United States in 1981, staying for two years, then returned to Belgium believing she would not come back to America.

In 1984, Daubechies was the recipient of the Louis Empain Prize for physics. The prize is given every five years to a Belgian scientist for scientific contributions done before the age of 29. She returned to the United States in 1987 and joined AT&T Bell Laboratories, where she was a technical staff member for the Mathematics Research Center. During her employment with AT&T, she concurrently took leaves of absences to teach at the University of Michigan and later at Rutgers University. In 1993, Daubechies became a full professor at Princeton University in the Mathematics Department and Program in Applied and Computational Mathematics, where shc has remained to date. Daubechies is the first woman to obtain this position at Princeton. Her responsibilities include teaching both undergraduate and graduate courses, directing Ph.D. students in thesis work, and collaborating with postdoctoral fellows in research. She has also devoted much time to creating mathematics curriculums for grades kindergarten through 12th grade that reflect present–day applications of mathematics.

### The Physicist Who Became a Mathematician

Daubechies' original intent was to become a physicist (particularly in the field of engineering). But she involved in mathematical work which was very theoretical in nature. She soon found herself caught up in mathematical applications. Her designation as a mathematician was sealed through her brilliant and innovative work in wavelet theory.

In 1987, Daubechies made one of the biggest breakthroughs in wave analysis in the past two hundred years. Prior to the development of Daubechies' theorem, signal processing was accomplished by using French mathematician Jean–Baptiste Fourier's series of trigonometric functions, breaking down the signal into combinations of sine waves. Sine waves can measure the amplitude and frequency of a signal, but they can't measure both at the same time. Daubechies changed all that when she discovered a way to break signals down into wavelets instead of breaking them down into their components; a task thought by most mathematicians to be impossible.

This discovery has changed the image–processing techniques used by the Federal Bureau of Investigation for transmitting and retrieving the information contained in their massive database of fingerprints. With more than 200 million fingerprints on file, the technique also allows for data compression without loss of information, and eliminates extraneous data that slows or clutters the procedure. Of more significance to Daubechies is the application of her discovery to the field of biomedicine. She likens a wavelets transform to "a musical score which tells the musician which note to play at what time," and this is of particular importance to medical science. Through the analysis of signals used in electrocardiograms, electroencephalograms, and other processes used in medical imaging, the medical world hopes to employ Daubechies' development to detect disease and abnormalities in patients much sooner than is presently possible. The development and implementation of wavelet imagery in medicine would improve the ability of an ECG from a simple recording of a heartbeat to a digitized record of complete heart function.

Other applications for wavelets still in the research stage include video and speech compression, sound enhancement, statistical analysis, and partial differential equations involving shock waves and turbulence, to name only a few.

### Leaving a Legacy in Her Own Time

Daubechies' work have not gone unnoticed by her peers. She has been a fellow of the John D. and Catherine T. MacArthur Foundation from 1992 to 1997 and an elected member of the American Academy of Arts and Sciences since 1993. She was the recipient of the American Mathematical Steele Prize for Exposition for her "Ten Lectures on Wavelets" in 1994, and received the Ruth Lyttle Satter Prize in 1997. Daubechies is also a member of the American Mathematical Society, the Mathematical Association of America, the Society for Industrial and Applied Mathematics, and the Institute of Electrical and Electronics Engineers.

Daubechies has written more than 70 articles and papers during her career, more than 20 of them dealing with the nature, application, and interdisciplinary use of wavelets. She has held memberships in more than 17 professional organizations and committees, including her current memberships with the United States National Committee on Mathematics and the European Mathematical Society's Commission on the Applications of Mathematics. Daubechies has been a guest editor or member of the editorial board for ten professional journals and has served as editor–in–chief for the publication *Applied and Computation Harmonic Analysis.*

Daubechies married A. Robert Calderbank, a mathematician, in 1987 and has two children.

### SELECTED WRITINGS BY DAUBECHIES

"Ten Lectures on Wavelets." CBMS–NSF Lecture Notes nr. 61, *SIAM*, 1992.
(With S. Maes) "A Nonlinear Squeezing of the Continuous Wavelet Transform Based on Auditory Nerve Models," in *Wavelets in Medicine and Biology*, edited by A. Aldroubi and M. Unser, 1996.
"Where Do Wavelets Come From? A Personal Point of View," in *Proceedings of the IEEE Special Issue on Wavelets* 84, no. 4 (April 1996): 510–13.

### FURTHER READING

#### Periodicals

Von Baeyer, Hans Christian. "Wave of the Future." *Discover* (May 1995): 69–74.
*What's Happening in the Mathematical Sciences* 2 (1994): 23.

#### Other

Daubechies, Ingrid with Kelley Reynolds Jacquez conducted May 16, 1997.

*—Sketch by Kelley Reynolds Jacquez*

# Bernhard Frenicle De Bessy
## 1605(?)–1675
### French number theorist

The contributions of Bernhard Frenicle De Bessy to mathematics belie his status as an "amateur." His legacy is twofold. De Bessy's correspondence with the leading mathematical and scientific minds of his day (among them **René Descartes** and **Pierre de Fermat**) illuminate the course of mathematical thought, especially concerning number theory. De Bessy made a number of mathematics discoveries in his own right, many of them dealing with magic squares, and also solved some mathematical problems publicly proposed by Fermat, and, in return, proposed new ones.

Little is known about this gifted amateur mathematician's life other than the approximate year of his birth, 1605, and that he served as a court official, a counselor at the Cour of Monnais, in Paris, for most of his life. De Bessy began corresponding with fellow mathematicians as early as 1634. By 1640, De Bessy was corresponding with Fermat on number theory and perfect numbers. This correspondence led to Fermat's articulation of what became known as Fermat's theorem. In 1657, Fermat came up with a series of mathematical propositions. De Bessy solved them all immediately and published his results, along with four more mathematical questions, in what is arguably his most important publication, *Solutio duorum problematum circa numeros cubos et quadratos.*

Among De Bessy's other publications is an article, "Des quassez ou tables magiques," which concerns magic squares. Magic squares are a square array of numbers such that the sums along each row, column, and diagonal are equal. De Bessy's contributions to magic numbers include a rule for recording even order magic squares (the script for the odds had been proven in 1612 by Bachet de Méziriac) and the idea that as the order of magic squares increases, the number of magic squares increases.

De Bessy also contributed to Pythagorean numbers theory (right angle triangles and the numbers that form their sides) in several publications. The most important of these treatises is *Traité des triangles rectangles en nombres*. In that text, De Bessy proved that each side of a right triangle cannot be a square; therefore, its area cannot be two times that of a square. Additionally, there is evidence that he might have had a hand in the authorship of public commentary on **Galileo**'s *Dialogue*.

Although De Bessy was not a professor or professional mathematician, he became a member of the French Académie Royale des Sciences in 1666 by appointment of King Louis XIV. De Bessy died nine years later on January 17, 1675, in Paris. The Académie published some of his work posthumously in two publications, *Divers ouvarages de mathématqiue et de physique* in 1693 and *Mémoires de l'Académie royale des sciences* in 1729.

## SELECTED WRITINGS BY DE BESSY

*Solutio duorum problematum circa numeros cubos et quadratos*, 1657

## FURTHER READING

### Books

Busard, H.L.L. "Bernard Frenicle De Bessy," in *Dictionary of Scientific Biography*. Volume V. Edited by Charles Coulston Gillispie. New York: Charles Scribner's Sons, 1971, pp. 158–60.

### Periodicals

Fletcher, Colin R. "A Reconstruction of Frenicle–Fermat Correspondence of 1640." *Historica Mathematica* (1991): 344–51.

—*Sketch by Annette Petruso*

# (Julius Wilhelm) Richard Dedekind
## 1831–1916
### German number theorist

Richard Dedekind is best known for his work in number theory. He redefined irrational numbers,

proposing that rational and irrational numbers form a continuum in which real numbers are located by "cuts" in the realm of rational numbers. He also introduced the notion of an ideal (for example, the collection of all integer multiples of a given integer), which allowed wider application of factorization and is fundamental to modern ring theory. In addition to Dedekind cuts, about a dozen mathematical concepts carry his name. Although he was among the most capable and original mathematicians of his day, Dedekind was a modest man, spending most of his professional life as a teacher at the technical high school in his hometown of Brunswick. He was a gifted teacher, and his teaching was an integral part of his mathematical thinking.

Dedeking was born Julius Wilhelm Richard Dedekind on October 6, 1831, in Brunswick (Braunschweig), now Germany, the last of four children. As an adult he dropped his first two names. His father, Julius Levin Ulrich Dedekind, was a lawyer and professor at Caroline College in Brunswick and the son of a physician and chemist. His mother, Caroline Marie Hanriette Emperius Dedekind, was a daughter of a professor at the College and a granddaughter of an imperial postmaster. Dedekind's brother, Adolf, became a district court president; his sister, Julie, became a novelist.

From age seven to age sixteen, Dedekind studied at the Gymnasium in Brunswick. At first, he concentrated on physics and chemistry, considering mathematics merely a scientific tool; however, he eventually became enthralled with the logic of mathematics. From 1848 to 1859, Dedekind attended Caroline College where he studied analytic geometry, advanced algebra, the calculus, and higher mechanics, and gave private lessons. In 1850, at age nineteen, he entered the University of Göttingen, where he became the last doctoral student trained by **Karl Gauss**. At Göttingen, Dedekind studied calculus, elements of higher arithmetic, least squares, higher geodesy, and experimental physics. Dedekind's doctoral thesis on Eulerian integrals, completed after only four semesters at Göttingen, was a solid but uninspired piece of work. However, Gauss praised his knowledge and independence, and predicted future success for him.

### Posts at Göttingen, Zürich, and Brunswick

Dedekind continued to study and attend lectures, and in 1854 he was appointed lecturer (*privatdozent*) at Göttingen, just a few weeks after his friend **Georg Riemann**, who also studied under Gauss, received a similar appointment. Some time later Dedekind and Riemann traveled together to Berlin to meet with the mathematical community there. When Gauss died in 1855, Dedekind served as a pallbearer at the funeral service. When **Peter Gustav Lejeune Dirichlet** came to Göttingen from the University of Berlin to take

Gauss' place, he and Dedekind became close friends and colleagues. Dedekind attended Dirichlet's lectures, and their discussions inspired Dedekind's investigations in new directions, making a "new man" of him. Dedekind was among the first to recognize the application of Galois groups in algebra and arithmetic, and in 1857–58 gave a course to two students on **Évariste Galois'** theory of equations.

In 1858, Dedekind was invited to succeed Joseph Ludwig Raabe at the Polytechnic School in Zürich. In recommending him for the position, Dirichlet described Dedekind as "an exceptional pedagogue." A position in Zürich was traditionally a first step toward a professorship in Germany. However, after five years in Zürich, in 1862 Dedekind succeeded Wilhelm Julius Uhde as professor of higher mathematics at the technical high school in Brunswick. He stayed there the remainder of his life; Dedekind directed the school from 1872 to 1875, and was named professor emeritus in 1894.

In assuming the position at the technical school, which had been created under the auspices of Carolina College, Dedekind was following in his father's administrative footsteps. In Brunswick, he lived in close association with his family and did not aspire to a greater position. He was an accomplished cellist and pianist, and composed a chamber opera for his brother's libretto. Dedekind lived with his sister Julie until her death in 1914. He died in Brunswick on February 12, 1916.

### Works Focus on Number Theory

Dedekind's work focused almost totally on the area of numbers. As a result of attempts to answer questions about real numbers, several ideas had been put forth, all involving infinite sets or sequences. The simplest idea was Dedekind's. He defined a real number to be a partition, or "cut," of the rational numbers into two sets, so that each member of one set is less than all numbers of the other. These Dedekind cuts—which he said occurred to him on November 24, 1858—gave a precise model for the continuous number line, since they filled all the gaps in the rationals. Other formulations followed from his definition. In 1872, Dedekind defined an infinite set in his paper *Stetigkeit und irrationale Zahlen*. In 1888, he expanded these ideas in a book, *Was sind und was sollen die Zahlen?* With his 1872 paper he had joined **Karl Weierstrauss** and **Georg Cantor** in defining a new mathematical area.

In 1879, Dedekind published *Über die Theorie der ganzen algebraischen Zahlen*, in which he introduced the notion of an ideal, which is fundamental to ring theory. Dedekind formulated his theory in the ring of integers of an algebraic number field. His idea was later extended by **David Hilbert** and **Emmy Noether**. Dedekind also collected, explained, extended, and published the works of those mathematicians who had influenced him: Gauss (1863), Riemann (1876), and Dirichlet (1863, 1871). His work on Dirichlet's lectures led him to a theory of generalized complex numbers and forms that can be resolved into linear factors. His work was characterized by exceptional clarity, and he has been credited with creating a "style" of mathematics.

Dedekind was a corresponding member of the Göttingen Academy (1862), the Berlin Academy (1880), and the Paris Académie des Sciences (1910). He was a member of the Leopoldino–Carolina Naturae Curiosorum Academia and the Academy of Rome. Dedekind also received many honors, including honorary doctorates from Brunswick and the University of Oslo. On one occasion, his death was listed on a *Calendar for Mathematicians* as September 4, 1899; an amused Dedekind wrote the publisher that he had spent that day talking with his friend Georg Cantor.

### SELECTED WRITINGS BY DEDEKIND

*Essays on the Theory of Numbers.* Translated by Wooster Woodruff Beman, 1924

### FURTHER READING

Bell, E. T. *Men of Mathematics.* New York: Simon & Schuster, 1937, pp. 516–525.

Biermann, Kurt-R. "Julius Wilhelm Richard Dedekind," in *Dictionary of Scientific Biography.* Volume IV. Edited by Charles Coulston Gillispie. New York: Charles Scribner's Sons, 1971, pp. 1–5.

Boyer, Carl B. *A History of Mathematics.* New York: John Wiley & Sons, 1968, pp.604–617.

Edwards, Harold M. "Dedekind's Invention of Ideals," in *Studies in the History of Mathematics.* Volume 26. Edited by Esther R. Phillips. The Mathematical Association of America, 1987, pp. 8–20.

*—Sketch by Jill Carpenter*

# Pierre René Deligné
## 1944-
### Belgian algebraic geometer and number theorist

Pierre Deligné is a research mathematician who has excelled at making connections between various

*Pierre René Deligné*

fields of mathematics. His research has led to several important discoveries, the most critical of which is the proof of three famous conjectures made by the mathematician **André Weil.** For this work, Deligné received both the Fields Medal, the highest honor in mathematics, and the Crafoord Prize. In recognition of his reception of the Fields Medal, David Mumford and John Tate, both of the Harvard University Department of Mathematics, wrote in *Science* magazine that "There are few [mathematical] subjects that [Deligné's] questions and comments do not clarify, for he combines powerful technique, broad knowledge, daring imagination, and unfailing instinct for the key idea."

Pierre René Deligné was born on October 3, 1944, in Brussels, Belgium, where he and his parents, Albert and Renee Bodart Deligné, lived throughout his childhood. The young Deligné showed an early affinity for mathematics, and his interest was encouraged by M. J. Nijs, his high school teacher. Nijs loaned Deligné several books by Nicolas Bourbaki that introduced concepts of modern mathematics, such as topology, long before discussing the topics traditionally studied first. Despite the unfamiliar and complicated terminology, Deligné's understanding of mathematics flourished, and after completing high school he enrolled at the University of Brussels. Deligné obtained his degree in mathematics there in 1966 and remained for graduate study.

Deligné's adviser at the University of Brussels, group theorist Jaques Tits, suggested in 1965 that

Deligné travel to Paris. Since Deligné was interested in algebraic geometry, Tits felt that he should study where some of the most important researchers in that field were teaching and researching—mathematicians such as **Jean-Pierre Serre** and **Alexander Grothendieck.** Deligné went and met both Serre and Grothendieck; his association with them would strongly influence his career. After returning to Brussels to complete work on his dissertation, he received his Ph.D. in 1968.

Following completion of his doctorate, Deligné took up residence in Bures-sur-Yvette, a small community south of Paris, where the Institut des Hautes Etudes Scientifiques (Institute for Advanced Scientific Study—IHES) is located. He had been appointed a visiting member of this organization so that he could continue his research with Grothendieck; he became a permanent member of the IHES in 1970. For several years, Grothendieck had been working to generalize and update the field of algebraic geometry by making it more compatible with recent abstract mathematical theories. Deligné admired and learned from Grothendieck's work, although he followed a different approach. Whereas Grothendieck tried to connect algebraic geometry with all other fields by creating new theories or rules, Deligné instead worked to uncover connections already implied by previous work in these fields. Contrasting the two men's styles, Mumford and Tate observed, "One could say that Grothendieck liked to cross a valley by filling it in, Deligné by building a suspension bridge."

### Conquers the Weil Conjectures

A prime example of Deligné's methods is his work on the Weil conjectures. Proposed in 1949 by the mathematician André Weil, these three conjectures state that it should be possible to determine the number of solutions for certain systems of equations by predicting the shapes of the graphs of the solutions. In other words, by using certain topological concepts, algebraic results can be obtained. As explained by **Michael Atiyah** in his 1975 Bakerian Lecture, this amounted to finding an algebraic technique for identifying holes in the manifold of complex solutions of an equation. Although Weil felt certain that he was correct, he was never able to prove his conjectures. Over a period of several years, Deligné whittled away at the conjectures. Combining the new theory of étale cohomology (a branch of topology), which had been developed by Grothendieck, and a related conjecture by the Indian number theorist **S. I. Ramanujan,** he completed the final proof in 1973.

Deligné's work has been valued not only because he solved an important problem in mathematics, but also because he proved that seemingly disparate subjects can be connected. Referring to Deligné's use of a 1939 paper on Ramanujan's conjecture along with the new étale cohomology, Mumford and Tate wrote, "It is hard to imagine two mathematical schools more different in spirit and outlook than were those of the British analytic number theorists in the 1930s and of the French algebraic geometers in the 1960s. That Deligné's proof is a blend of ideas from both is an indication of the universality of his mathematical taste and understanding." For this reason, as much as for actually proving the Weil conjectures, the International Mathematics Union in 1978 awarded Deligné its highest honor, the Fields Medal, noting that his work "did much to unify algebraic geometry and algebraic number theory."

### Continues to Develop Algebraic Geometry

Deligné continued to study the Weil conjectures even after his initial success, attempting to use automorphic forms (equations involving multiple functions) and prime numbers to determine more and more exact solutions. He worked on several problems proposed by the American mathematician **Robert Langlands,** who was leading a major research program in the area of automorphic forms at Princeton's Institute for Advanced Studies (IAS). At Langlands' invitation, Deligné traveled to the United States in 1977 to help organize a conference at Oregon State University.

Also, in the late 1970s, Deligné gave a series of lectures in étale cohomology with the help of Grothendieck and others in the field of algebraic geometry. These lectures were considered definitive in describing this relatively new field, but Deligné's contributions were not limited to lecturing. He added significantly to the content by his work with Shimura varieties and by his proofs of some conjectures proposed by William V. D. Hodge. In 1988, the Royal Swedish Academy of Sciences awarded Deligné and Grothendieck the Crafoord Prize for Mathematics for their work in defining étale cohomology and applying it to algebraic geometry (Grothendieck declined his prize).

In 1980, Deligné married Elena Vladimirovna Alexeeva, who would become the mother of his two children. He enjoys simple pleasures such as vegetable gardening, bicycle riding, and hiking. He brought his young family to the United States in 1984 to continue his mathematical research at the IAS, where he has remained. In 1993, Deligné and G. Daniel Mostow coauthored a book titled *Commensurabilities Among Lattices in PU(1,n).* Reviewing this book for the *Bulletin of the American Mathematical Society,* P. Beazely Cohen and F. Hirzebruch wrote that the authors "extract the best aspects of the previous techniques of algebraic and differential geometry . . . together with function theory, giving an overall coherent presentation yielding new results." A quarter of a century after cracking the Weil conjectures,

Deligné has not lost his powerful touch for creating mathematics.

## SELECTED WRITINGS BY DELIGNÉ

*Cohomologie Etale,* 1977
(With G. Daniel Mostow) *Commensurabilities Among Lattices in PU(1,n),* 1993

## FURTHER READING

Atiyah, M. F. "Bakerian Lecture, 1975: Global Geometry." *Proceedings of the Royal Society of London* A347 (1976): 291–99.
Cohen, P. Beazely, and F. Hirzebruch. Review of *Commensurabilities Among Lattices in PU(1,n),* in *Bulletin of the American Mathematical Society* n.s., 32 (January 1995): 88–104.
Ewing, J. H., et al. "American Mathematics From 1940 to the Day Before Yesterday." *American Mathematical Monthly* 83 (1976): 503–16.
Mumford, David, and John Tate. "Fields Medals (IV): An Instinct for the Key Idea." *Science* 202 (November 17, 1978): 737–39.

*—Sketch by Karen Sands and Loretta Hall*

# Augustus De Morgan
## 1806–1871
### English algebraist and logician

Augustus De Morgan entered the English mathematical scene during a period of inactivity and by the time of his death it had regained the stature it had since the time of **Isaac Newton.** Although De Morgan did not devote himself wholeheartedly to the pursuit of mathematics, he is credited for promoting its study by his publications and his teaching. He worked outside the established universities and was able to appeal to a wider audience than merely the mathematics graduates of Oxford and Cambridge.

De Morgan was born in June 1806 in Madurai, a picturesque town in southern India. Shortly after his birth he lost the sight of his right eye. The De Morgans moved back to England when he was seven months old. His father, a colonel in the Indian army, continued to spend time in India and died on St. Helena in 1816. De Morgan was educated at a series of private schools before enrolling at Trinity College, Cambridge, in 1823 and graduated as fourth wrangler.

*Augustus De Morgan*

After graduation, De Morgan entered Lincoln's Inn, one of the Inns of Court intended to prepare students for a legal career. He may have been discouraged by some aspects of the mathematics curriculum at Cambridge at the time, which was still recovering from the inertia of the 18th century. Partly in response to the quarrel between Isaac Newton and **Gottfried Leibniz** over the invention of calculus, the English mathematical community began to isolate itself from the continent in the 18th century, idolizing Newton and reluctant to change. The result was a petrification of both the foundations of the calculus and of its notation, an area in which Leibniz's version was clearly an improvement over Newton's. Cambridge was drifting along serenely, unaware of the progress of mathematics outside its walls until the arrival of a group of students who were known as the Analytical Society. They devoted themselves to the reform of mathematical education.

De Morgan soon found the legal profession unappealling and applied for the chair of mathematics at University College, London. He was offered the chair in 1828 on the strength of recommendations from his former tutors, who included some members of the Analytical Society. As a teacher, De Morgan was devoted to the presentation of ideas and principles rather than techniques, and his pupils included some of the most distinguished British mathematicians of the next generation. In addition, he also produced a series of textbooks on arithmetic, algebra, trigonometry, calculus, complex numbers, probabili-

ty, and logic. These were written clearly and with attention to giving an intuitive understanding as well as one based on calculation.

## Extends the Boundaries of Logic

In his own mathematical work, De Morgan made major contributions in the area of logic. The Aristotelian tradition of logic had become fossilized during the Middle Ages, and instruction in logic frequently resorted to memorizing a few lines of low Latin and a little caution about the misuse of rhetoric. If reasoning about mathematics was being carried out in ordinary language, there was not much advantage to studying mathematics in approaching questions of logic.

One of the main interests of the English mathematical community at that time was the status of the laws of algebra. It was clear that some of the laws applied to all the systems of numbers then known, but other laws did not apply beyond a restricted domain. The quaternions discovered by Sir **William Rowan Hamilton**, for example, (which are related to vectors in three dimensions) had an operation of multiplication which was not commutative (that is, $a \times b$ was not equal to $b \times a$). An obvious question was that of which laws are automatically satisfied by any objects whatever, which could be considered laws of logic.

The crucial respect in which De Morgan sought to improve on the traditional logic of Aristotle was in the treatment of the logic of relations. In Aristotelian logic, all statements had to be analyzed into the form "A is (or is not) B," with the possible inclusion of "all" and "some." It was not clear how this could be used to handle statements like "A is taller than B" or "A is closer to B than C." De Morgan's work on the logic of relations did not become part of the mainstream, due to the shortcomings of his notation. More successful in the reform of logic was **George Boole**, author of *The Mathematical Analysis of Logic* and creator of a superior notation. Boole acknowledged his debt to De Morgan, whose name remains attached to two laws of Boolean algebra involving the negations of compound expressions.

De Morgan also wrote on probability. As far as the interpretation of probabilities was concerned, De Morgan fell in the "subjectivist" rather than the "frequentist" school. In other words, probability statements were reflections of the degree of belief attached to propositions rather than features of the natural world itself.

In 1831 De Morgan resigned his chair at University College. By 1836, following the death of his successor, however, De Morgan returned to his position. The next year he married Sophia Elizabeth Frend, who wrote De Morgan's biography after his death. Although De Morgan found his family life a comfort after some of the controversies in the academic world, his later years were saddened by the deaths of a couple of his children.

De Morgan wrote many articles for the popular press. This type of writing did not command much respect, and it is worth noting that De Morgan never became a Fellow of the Royal Society, of whom he criticized for being too open to social influence, as indicated by the proportion of nobility among its members. Much more to De Morgan's taste was the London Mathematical Society, which he cofounded and served as first president. This type of mathematical organization was more fitting for someone who had contributed 850 articles for one reference work alone.

De Morgan resigned a second time from University College in 1866 and on this occasion could not be tempted to return. He died in London on March 18, 1871. Not so much by his research as by his pedagogical efforts had De Morgan transformed the mathematical community in England into a setting for the discussion of the current topics of interest in mathematics both English and European.

## SELECTED WRITINGS BY DE MORGAN

*Formal Logic*, 1847
*A Budget of Paradoxes*, 1872

## FURTHER READING

Dubbey, John. "Augustus De Morgan," in *Dictionary of Scientific Biography*. Volume IV. Edited by Charles Coulston Gillispie. New York: Charles Scribner's Sons, 1973, pp. 35–37.

Merrill, Daniel D. *Augustus De Morgan and the Logic of Relations*. Dordrecht, Holland: Kluwer Academic Publishers, 1990.

Smith, C.C., editor. *The Boole–De Morgan Correspondence*. Oxford: Oxford University Press, 1982.

Stephen, Leslie. "Augustus De Morgan," in *Dictionary of National Biography*. Volume 5. Edited by Leslie Stephen and Sidney Lee. Oxford: Oxford University Press, 1968, pp. 781–784.

—*Sketch by Thomas Drucker*

# Girard Desargues
## 1591–1661
### French geometer and engineer

Little regarded in his own time, Girard Desargues developed a treatise and theorem that formed the basis for the development of projective geometry, breaking with the straight Euclidian traditions that had informed geometry since the Hellenistic age. However, it was not until almost two centuries after his death that Desargues was rediscovered and his work on conic sections gained its rightful place in mathematics. Although there is scant knowledge about Desargues' personal life, it is recorded that he was, as an amateur mathematician, a member of a renowned group of Parisian mathematicians, and that his work influenced the young **Blaise Pascal**. An engineer and inventor, Desargues also developed a pump device and wrote on such practical subjects as stonecutting, the production of sundials, and music composition.

Desargues was born on February 21, 1591 in Lyons, France, one of nine children of Girard Desargues and Jeanne Croppet. The elder Desargues was a tithe collector, and the family owned several large houses in Lyons as well as a chateau and vineyards near the city. Not much is recorded of his early education, but it would appear that the family was sufficiently well off that Desargues could indulge in various pursuits, including the sciences. An early report places him in Paris in 1626, proposing to the municipality that it raise the level of the Seine by the use of machines so as to be able to pump water through the city. As an engineer, Desargues is also reported as having participated in the 1628 siege of La Rochelle, where he met the philosopher and mathematician **René Descartes**. He is also supposed to have been an engineer and technical advisor for the French government under Cardinal Richelieu.

### Desargues's Theorem

By about 1630, Desargues was spending more time in Paris and became part of the intellectual circle of **Marin Mersenne,** Étienne Pascal and his son Blaise Pascal, and René Descartes. Desargues published two papers in 1636, one a musical treatise on harmony, and the other entitled *Traite de la section perspective* ("Treatise on the Perspective Section"), which laid out his universal method of perspective and contained the initial ideas that led to the theorem named after him. This theorem was later published by Desargues's friend and confidante, Abraham Bosse, and has since been attributed directly to Desargues. The theorem, which holds for either two or three dimensions, states that two triangles may be positioned so that the three lines joining corresponding vertices meet in a point if and only if the three lines containing pairs of corresponding sides intersect in three collinear points. It was this theorem that the French mathematician **Jean–Victor Poncelet** rediscovered in the 19th century and used to help modify Euclidian geometry into projective geometry. With this universal method, Desargues hoped to develop a completely geometric study of perspective, and therefore stands between the artist, Albrecht Dürer, and the French geometer, **Gaspard Monge**, in the development of a graphical representation of perspective.

As with all the work in his lifetime, this early theorem of Desargues's received little attention. Such neglect, however, did not keep Desargues from his studies, and in 1639 he privately printed 50 copies of what is his most famous work, *Brouillon project d'une atteinte aux evenemens des rencontres d'une cone avec un plan* ("Proposed Draft of an Attempt to Deal with the Events of the Meeting of a Cone with a Plane"), a treatise which developed new practices in projective geometry, especially as applied to conic sections. As René Taton wrote in *Dictionary of Scientific Biography,* the treatise was "a daring projective presentation of the theory of conic sections. . . . But the use of an original vocabulary and the refusal to resort to Cartesian symbolism make the reading of his essay rather difficult and partially explains its meager success." Julian Lowell Coolidge, in his *A History of Geometrical Methods,* is more to the point: "The style and nomenclature are weird beyond imagining." Desargues employed botanical terms for his mathematical concepts, including among others, "tree" to denote a straight line with three pairs of points of an involution, and a "stump" to refer to the mate of an infinite point. Such nomenclature makes the essay incredibly difficult to understand, but beneath this arcane language Desargues had done no less than present a unified theory of conics. It took another two centuries for this legacy to be accepted, however.

### Desargues the Craftsman and Engineer

Possessing an inquisitive mind, Desargues did not simply focus on geometry for his life's work. He was also an accomplished architect and applied his spatial capacities to creating useful designs. Desargues is particularly noted for his spiraling staircases, the construction of which was aided by his theories of projection and perspective. About 1645 he began designing houses in Paris, and then returning to his birthplace of Lyons, he supervised the construction of several more houses and mansions, both private and public, between 1650 and 1657, after which time he returned to Paris.

As an engineer, Desargues is especially noteworthy for his early use of a cycloidal or epicycloidal teeth for gear wheels. These wheels were used in a system for raising water at the chateau of Beaulieu near Paris, one that Desargues installed in the 1630s. Desargues also extended the pragmatic uses of his method of perspective to the craft of stonecutting and developed principles to simplify the construction of sundials.

### Endures Attacks

Desargues was not, however, free from criticism in his day. Regarding his optimism that graphical representation alone was sufficient to breathe new life into geometry, Descartes took the mathematician to task. A strong proponent of algebraic geometry, Descartes doubted such an assertion. The mathematician Jean de Beaugrand was a more ardent critic, asserting that many of Desargues's supposedly original ideas on conic sections were taken directly from the ancient geometer, **Apollonius**. The two exchanged broadsides, in the form of critical essays, for several years.

It was not only his theoretical work that drew criticism, however. Desargues's methods applied to stonecutting brought him into direct conflict with the trade guilds, which had responsibility for such matters. Also, his methods outraged practitioners of older methods of perspective. Such attacks eventually took their toll on Desargues, and increasingly he kept out of the public eye, entrusting the dissemination of his works and ideas to his disciple and friend, the engraver Bosse. It was several of Bosse's publications which kept Desargues's name alive; that, and a manuscript copy of Desargues's *Brouillon project* which another colleague, Phillipe de la Hire, made. It was not until 1951 that an original copy of that treatise was uncovered. When Desargues died in October of 1661, there was little to indicate that history would grant him immortality. As Taton noted in *Dictionary of Scientific Biography,* "Desargues's work was rediscovered and fully appreciated by the geometers of the nineteenth century. Thus, like that of all precursors, his work revealed its fruitfulness much more by its remote extensions than by its immediate repercussions."

## SELECTED WRITINGS BY DESARGUES

### Books

*Oeuvres de Desargues reunies et analysees,* 1864
*L'oeuvre mathematique de Desargues,* 1951

## FURTHER READING

### Books

Coolidge, Julian, Lowell. *A History of Geometrical Methods.* New York: Dover Publications, 1963, p. 89.
Field, J. V., and J. J. Gray. *The Geometrical Works of Girard Desargues.* New York: Springer-Verlag, 1987.
Taton, René. "Girard Desargues," in *Dictionary of Scientific Biography.* Volume IV. Edited by Charles Coulston Gillispie. New York: Charles Scribner's Sons, 1971, pp. 46–51.

*René Descartes*

**Periodicals**

Ivins, M. W. "The First Two Editions of Desargues." *Bulletin of the Metropolitan Museum of Art* 1(1942): 33–45.
"A Note on Desargues's Theorem." *Scripta mathematica* 13 (1947): 202–10.

—*Sketch by J. Sydney Jones*

# René Descartes
## 1596–1650
### French geometer, algebraist, and philosopher

René Descartes was an analytical genius. He conceived and articulated ideas about the nature of knowledge that were essential to the Enlightenment and created the philosophical underpinnings for the development of modern science, which included the idea that laws of nature are constant and are sufficient to explain natural phenomena. Descartes felt that truth was clear and accessible to the ordinary human intellect, if the search for truth was directed properly. Two of his writings, *Rules for the Direction of the Mind* and *Discourse on the Method of Rightly Conducting the Reason* defined ways of obtaining knowledge. The latter work contained *Geometry*, that introduced the Cartesian coordinate system and marked the birth of analytic geometry, in which geometric relationships are investigated by means of algebra. Descartes also contributed to areas of music theory, mechanics, physics, optics, anatomy, and physiology.

René du Perron Descartes was born on March 31, 1596 in La Haye (now Descartes), in the province of Touraine, France. He was born into the gentry, a well–to–do class of landowners between the nobility and the bourgeoisie. His father, Joachim, was a councilor to the high court at Rennes in Brittany. From his mother, Jeanne Brochard, Descartes received the property that gave him his financial independence. Descartes was her third and last surviving child. She died in childbirth in 1597 and he and his older brother and sister were brought up by their maternal grandmother, Jeanne Sain. In 1600, Descartes' father remarried and moved to Chételleraut. Descartes seems not to have had enduring relationships with his father or siblings; however, the elder Descartes early on recognized his youngest child's curiosity, referring to him as "my little philosopher."

### Educated at La Fléche

In 1606, Descartes was sent to La Fléche, the Jesuit school at Anjou. Descartes' health was considered delicate and the rector, Father Charlet, allowed him to spend mornings in bed in contemplation, a habit he continued throughout most of his life. Descartes spent nine years at La Fléche, where he perfected his Latin, studied humanities, philosophy, and mathematics, and was introduced to new developments in optics in astronomy. Although Descartes expressed high regard for his education, it was at La Fléche he realized that, with the exception of mathematics and geometry, he had learned nothing that was absolute truth. He first saw mathematics only as the servant of mechanics, but was struck "by the certainty of its proofs and the evidence of its reasonings" and was surprised that nothing loftier had been erected upon its foundations.

Descartes moved to a house outside Paris in 1614, where he shut himself off from others. Although he was self–assured and expected admiration from others, scholars have suggested that he suffered from depression. He spent the year 1615–1616 at the University of Poitiers, where he earned a law degree.

### Becomes Gentleman Soldier

The law did not interest Descartes; he chose instead to become a gentleman soldier. He had resolved "to seek no knowledge other than that which could be found in myself or else in the great book of the world," and in the summer of 1618 he traveled to

Holland, where he joined the army of Prince Maurice of Nassau as an unpaid volunteer. In Breda he met Isaac Beeckman, the Dutch philosopher, doctor and physicist. Descartes' discussions with Beeckman rekindled his interest in applying mathematical reasoning to problems in physics. Descartes' first work, *Compendium Musicae*, an arithmetical account of sound, was dedicated to Beeckman and given to him as a New Year's gift in 1619. At this time, Descartes also worked on problems in falling bodies, hydrostatics, a proportional compass, and a theory of proportional magnitudes.

Descartes resigned from Maurice's army, and traveled to Bavaria to join the Bavarian Army. Stationed at Ulm in Neuburg, he met the Rosicrucian and mathematician Johannes Faulhaber. On November 10, 1619, in a "stove–heated room," Descartes had the mystical experience that set his life's course. He had been searching for a method of obtaining knowledge, and in a state of delirium had three vivid dreams in succession. Much has been made of the dreams (even the suggestion that they were symptomatic of migraine headaches), but their result was to convince Descartes of his divine mission to found a new philosophical system, in which he would reduce physics to geometry and connect all sciences through a chain of mathematical logic.

Descartes subsequently gave up the military life and traveled widely for several years, visiting Italy, Germany, and Holland, where along the way he studied glaciers, made meteorological observations, and computed the heights of mountains. From 1625 to 1628, he lived in Paris and became friends with Marin Mersenne, a Franciscan friar who had also attended La Fléche. In Paris, Descartes produced *Regulae* ("Rules for the Direction of the Mind"), which was published in 1701, after his death.

### The *Discourse* and the Birth of Analytic Geometry

In 1629 Descartes retired to Holland, where he devoted the next 20 years to studies of science and philosophy. During this time he made three trips back to Paris, where Mersenne acted as his editor and agent. The tolerant Protestant climate of Holland protected Descartes from academic and theological disputes, at least in the beginning, and he moved frequently to avoid visitors. He was not a recluse, however; he visited universities and talked with mathematicians, philosophers, and physicians. Descartes studied anatomy and frequently visited butcher shops to obtain animal carcasses for dissection. In 1633, he completed *Le monde* ("Of the World"), which included his theories in physiology, perception, and a heliocentric cosmology. When Descartes learned that **Galileo** had been condemned by the Inquisition for embracing Copernicus' ideas, he withheld *Le monde* from publication. He modified information from *Le monde* for use in his 1637 masterpiece, *A Discourse on the Method of rightly conducting the Reason and seeking Truth in the Sciences. Further, the Dioptric, Meteors, and Geometry, essays in this Method.* The *Meteors* was the first attempt to give a scientific theory of the weather. The *Dioptric* explained rainbows, and contained the law of refraction, describing the behavior of light rays transmitted from one medium to another.

Descartes' *Geometry,* essentially an appendix to the *Discourse on Method*, revolutionized mathematics and provided the foundation for what is now known as analytic geometry. It enabled the use of algebra, a relatively new branch of mathematics, for the discovery and investigation of geometrical theorems. He introduced the use of coordinates, by which is possible to begin with equations of any degree of complexity and interpret their algebraic and analytic properties geometrically. In the *Geometry,* Descartes introduced algebraic notation that is still in use today, dealt with the problem of Pappus, and provided a systematic definition of curves.

In Amsterdam, Descartes had formed a liaison with his serving girl, Héléne, who bore him a daughter, Francine, on July 19, 1635. Héléne and Francine came to live with him in Santpoort, and he made arrangements for Francine to be educated in France. Unfortunately, she died in 1640, probably of scarlet fever.

Descartes published his major metaphysical work, *Meditations on First Philosophy*, in which the Existence of God and the Distinction between Mind and Body are Demonstrated in 1641. Although he quickly published an edition containing solicited objections and his replies to them, he was particularly criticized and attacked by the president of the University of Utrecht, Gisbert Voet, and published his lengthy defense as *Episula at Voetium.*

In 1643 Descartes began a long–lasting correspondence with 24–year–old Princess Elizabeth of Bohemia, who lived in exile in Holland. In his letters, Descartes discussed his philosophy of the mind and its relation to the body, and the relationship between reason and the passions. He dedicated his 1644 *Principles of Philosophy*, which contains a naturalistic theory of the solar system, to Princess Elizabeth.

### Dies in Stockholm

Descartes accepted an invitation to tutor 20–year–old Queen Christina of Sweden in 1649. After much hesitation he left for Sweden on September 1, where the energetic Queen put him to work writing verses and a pastoral comedy, and planning a Swedish academy of science. She insisted that he meet with her at five in the morning when her mind was most active. The lessons began in mid–January 1650, but the early hours and the record cold winter quickly took their toll on Descartes. On February 1, he contracted pneumonia. He refused to see the royal physician and instead relied on his own remedy, wine flavored with tobacco. He died in Stockholm on February 11, 1650.

In 1666, Descartes' remains were exhumed and returned to France, where they were moved several times before being permanently placed in the chapel of the Sacré Coeur in the church of St. Germain–des–Prés in 1819. At the time of the original exhumation, the French ambassador was given permission to cut off Descartes' right forefinger. Descartes' skull was said to have been removed by a guard and it was sold several times, coming into the possession of Georges Cuvier in 1821. Although it has not been authenticated, the skull is on display at the Musée de l'Homme in the Palais de Chaillot.

## SELECTED WRITINGS BY DESCARTES

*The Philosophical Works of Descartes,* 2 volumes. Translated by Elizabeth S. Haldane and G. R. T. Roff, 1931

## FURTHER READING

### Books

Bos, Henk J. M. *Lectures in the History of Mathematics.* History of Mathematics, Volume 7. Providence, RI: American Mathematical Society, 1993, pp. 37–53.

Cottingham, John. *Descartes.* New York: Basil Blackwell, 1986.

Dunham, William. *The Mathematical Universe.* New York: John Wiley & Sons, Inc., 1994, pp. 273–285.

Gaukroger, Stephen. *Descartes: An Intellectual Biography.* Oxford: Clarendon Press, 1995.

Gaukroger, Stephen. *Descartes: Philosophy, Mathematics and Physics.* Sussex, England: The Harvester Press, 1980.

Crombie, A.C. "René Descartes," in *Dictionary of Scientific Biography.* Volume IV. Edited by Charles Coulston Gillispie. New York: Charles Scribner's Sons, 1971, pp. 51–65.

Newman, James R., editor. *The World of Mathematics.* Volume 1. New York: Simon & Schuster, 1956, pp. 235–253. (Includes a facsimile, with translation, of the first eight pages of *La géométrie*.)

Pearl, Leon. *Descartes.* Boston: Twayne Publishers, 1977.

Stillwell, John. *Mathematics and Its History.* New York: Springer–Verlag, 1989, pp. 66–77.

Struik, D. J., editor. *A Source Book in Mathematics, 1200–1800.* Princeton, NJ: Princeton University Press, 1986, pp. 87–93, 150–157.

—*Sketch by Jill Carpenter*

# Dinostratus
## c. 390–c. 320 B.C.
### Greek geometer

Hardly anything is known of Dinostratus' life. Proclus, in his *Commentary on Book I of* **Euclid**'s Elements, praises Dinostratus, whom he places within Plato's circle in Athens, for his contributions to geometry. According to Pappus's *Collection*, Dinostratus was among the mathematicians to employ the *quadratrix*, a special curve discovered by Hippias, to square the circle, that is, to construct a square exactly equal in area to a given circle. Despite the belief that Dinostratus was first to square the circle, Greek sources do not support this claim.

A description of the quadratrix can be found in Pappus's *Collection*. Suppose *ABCD* is a square and *BED* is a quarter arc of a circle with *A* as center. If the radius *AE* of the circle moves uniformly from *AB* to *AD,* and, simultaneously, *BC* descends from its initial position towards *AD,* then the radius and the line *BC* will intersect at a point *F*. As the radius and the line continue moving so that both will arrive at *AD* at the same time, they will keep forming points of intersection. We get the quadratrix by connecting the points *F* of intersection. The point where this curve reaches *AD* will be called *G*. According to Pappus, once point *G* is established, it can be shown that *BED* : *AB* = *AB* : *AG*, that is, the length of arc *BED* is in the same proportion to the length of *AB* as the length of *AB* is to the length of *AG*.

Since only one of the four terms, the arc *BED*, in this proportion is not a straight line, one can construct a straight line corresponding in length to the arc using a simple geometric construction. This line can then be used to square the circle whose radius is the length of segment *AB*. To prove that *BED* : *AB* = *AB* : *AG* is true, Pappus relied on an indirect proof, or *reductio ad absurdum*. In other words, he showed that the premise that *BED* : *AG* = *AB* : *AG* is false led to incorrect conclusions.

Despite a lack of clear evidence that Dinostratus applied the construction described by Pappus, including the indirect proof, the squaring of the circle has traditionally been attributed to him. It should be noted, however, that the method described here uses more than just compass and straightedge, and so does not strictly solve this classical problem of antiquity using just these Euclidean tools. Indeed, it was not until 1882 when **Ferdinand von Lindemann** showed that $\pi$ is a transcendental number that mathematicians were able to prove that it is impossible to square the circle using only compass and straightedge.

## FURTHER READING

### Books

Boyer, Carl B. *A History of Mathematics.* Second edition. Revised by Uta C. Merzbach. New York: Wiley, 1991.

Bulmer–Thomas, Ivor. "Dinostratus," in *Dictionary of Scientific Biography.* Volume IV. Edited by Charles Coulston Gillispie. New York: Charles Scribner's Sons, 1972, pp. 103–05.

Smith, D. E. *History of Mathematics.* Volume II. New York: Dover, 1953.

*—Sketch by Zoran Minderovic*

# Diocles
## c.240–c.180 B.C.
### Greek geometer

Diocles, about whose life very little is known, probably lived after **Archimedes**, who died in 212 B.C., and before Geminos of Rhodes, who flourished around 70 B.C. Two fragments of Diocles's work *On Burning Mirrors* can be found in Eutocius's commentary on Archimedes's book *On the Sphere and Cylinder.* William of Moerbeke, the Flemish cleric and classical scholar, translated the two fragments from Diocles into Latin.

One of these fragments contains the solution to the problem of doubling the cube, with which Greek mathematicians had struggled for centuries. According to a story mentioned by **Eratosthenes** in his book *Platonics,* the priestess of Apollo's sanctuary at Delos, who beseeched Apollo to stop an outbreak of the plague, learned that the god wanted his cube–shaped altar doubled. When mathematicians attempted to solve this enigma, also known as the Delian problem, by using a straightedge and compass, they were unsuccessful (it was proved to be unsolvable using only straightedge and compass in the late 19th century). The challenge was, given a cube whose side was of length $a$, find a line of length $x$, so that $x^3 = 2a^3$. Diocles introduced a special curve, later named the *cissoid* (*kissoeidēs*, from the Greek word *kissos*, mean-

ing *ivy*), which led to the solution of the problem, a solution, however, using more than just the Euclidean tools of a straightedge and compass.

Suppose $AB$ and $CD$ are perpendicular diameters of a circle and $EB$ and $BZ$ are equal arcs. Draw $ZH$ perpendicular to $CD$ and then draw $ED$. The intersection of $ZH$ and $ED$ gives a point $P$ on the cissoid. The cissoid is the locus of all such points $P$ determined by all positions of $E$ on arc $BC$ and $Z$ on arc $BD$ with arcs $BE$ and $BZ$ of equal length. One can prove that $CH : HZ = HZ : HD = HD : HP$. If the point $H$ is chosen so that $HP$ is of length $a$ and $CH$ is of length $2a$, then the work of **Hippocrates of Chios** shows that thc line $HD$ will double the cube of length $a$.

The other fragment from Diocles' work offers the solution to a problem presented in Book II of Archimedes' *On the Sphere and Cylinder.* The problem consisted in cutting a sphere with a plane so that the volumes of the two sections are in a particular ratio to each other. Diocles solved the problem by using the intersection of an ellipse and a hyperbola.

## FURTHER READING

Boyer, Carl. *A History of Mathematics.* Second edition. Revised by Uta C. Merzbach. New York: Wiley, 1991.

Dannenfeldt, Karl H. "Diocles," in *Dictionary of Scientific Biography.* Volume IV. Edited by Charles Coulston Gillispie. New York: Charles Scribner's Sons, 1972, p. 105.

Kline, Morris. *Mathematical Thought from Ancient to Modern Times.* Volume 1. New York: Oxford University Press, 1972.

Sarton, George. *A History of Science: Hellenistic Science and Culture in the Last Three Centuries B.C.* Cambridge, MA: Harvard University Press, 1959.

*—Sketch by Zoran Minderovic*

# Johann Peter Gustav Lejeune Dirichlet
## 1805–1859
### German number theorist and analyst

Johann Peter Gustav Lejeune Dirichlet was born in 1805, the son of the town postmaster of Düren (then part of the French empire). He was initially educated at public schools, then at a private school which stressed Latin. Interest in mathematics sur-

*Johann Peter Gustav Lejeune Dirichlet*

purchasing mathematics textbooks. Dirichlet enrolled at Bonn's Gymnasium in 1817, where he showed great interest in mathematics and history.

Two years later, Dirichlet's parents sent him to a Jesuit college in Cologne. He was a student of physicist George Simon Ohm, under whom he received thorough training in theoretical physics. At the young age of 16, Dirichlet completed his Abitur examination. His parents wanted him to study law, but he was already well on his way in the field of mathematics.

## In Pursuit of Mathematics

Other than **Karl Gauss,** Germany, at that time, had no notable mathematicians. Paris, on the other hand, boasted such luminaries as **Pierre Simon Laplace, Adrien–Marie Legendre**, and Jospeh Fourier. In 1822, Dirichlet visited Paris. He was not there long before he caught a mild case of smallpox, but it was not severe enough to keep him from continuing classes at the College de France, and the Faculte des Sciences. In 1823, he was appointed to a well–paid position as a tutor to General Maximilian Fay's children. Fay was a national hero of the Napoleonic wars and a liberal opposition leader in the Chamber of Deputies. Dirichlet was treated as a member of the family, thus meeting prominent French intellectuals, including mathematician Joseph Fourier. Fourier's ideas influenced Dirichlet's later works on trigonometric series and mathematical physics.

Dirichlet's main interest was number theory, which was first ignited through an early study of Gauss' *Disquistiones arthmeticae* (1801). In June 1825, Dirichlet presented his first paper on mathematics, "Memoire sur l'impossibilite de quelques equations indeterminees du cinquieme degre," to the French Académie Royale des Sciences. The paper explored Diophantine equations of the form $x^5 + y^5 = Az^5$ using algebraic number theory, Dirichlet's favorite area of study. Legendre extended these results to give a proof of Fermat's last theorem for n=5.

## Returns to Germany

With General Fay's death in 1825, Dirichlet returned to Germany. Fellow German scientist Alexander von Humboldt strongly supported Dirichlet's return, as Germany needed strengthening in the natural sciences. Although Dirichlet did not have the required doctorate, he was permitted to qualify at the University of Breslau for the habilitation required to teach at a German university.

Breslau was not an inspiring environment for scientific work, so in 1828, again with Humbolt's assistance, Dirichlet moved to Berlin and began teaching mathematics at the military academy. At age 23, he was first appointed to a temporary position at the University of Berlin and in 1831 became a member of the Berlin Academy of Sciences. That same year Dirichlet married Rebecca Mendelssohn–Bartholdy, granddaughter of philosopher Moses Mendelssohn, and sister to composer Felix Mendelssohn. In 1832, he published a proof of **Pierre de Fermat**'s Last Theorem for n=14.

During his 27 years as professor in Berlin, Dirichlet influenced the development of German mathematics through his lectures, pupils, and scientific papers. He taught with great clarity and his published scientific papers were of the highest quality.

Dirichlet was a shy, modest man, who rarely made public appearances, or spoke at meetings. His lifelong friend was mathematician **Karl Jacobi**. Both mathematicians influenced each other's work, particularly in number theory. In 1843, Jacobi moved to Rome for health reasons. This prompted Dirichlet to request a leave of absence to move his family to Rome with Jacobi. Dirichlet remained in Italy for a year and a half, visited Sicily, and spent the winter in Florence.

## Proof of Fundamental Theorem

At a meeting of the Academy of Science, held on July 27, 1837, Dirichlet presented a paper on analytic number theory. The paper offers proof of the fundamental theorem that bears his name: any arithmetic sequence of integers: an + b, n = 0, 1, 2 . . ., where a and b are relatively prime, must include an infinite number of primes. This paper was followed in 1838 and 1839 by a two–part paper on analytic number theory, "Recherches sur diverses applications de l'analyse infinitesimale a la theorie des nombres." After publication of his fundamental papers, the importance of his number theory work declined, but Dirichlet continued to publish papers in other areas.

Dirichlet is best known for his work on trigonometric series and mathematical physics. In an 1828 paper in *Crelle's Journal*, he gave the first rigorous proof of sufficient conditions for the convergence of the Fourier series for a function. His investigations of equilibrium of systems and potential theory gave rise to what is now called the Dirichlet problem about formulating and solving a class of partial differential equations that arise from the flow of heat, electricity, and fluids subject to given boundary conditions.

In 1837 Dirichlet proposed the modern definition of a function: if a variable $y$ is so related to a variable $x$ that whenever a numerical value is assigned to $x$, there is a rule according to which a unique value of $y$ is determined, then $y$ is said to be a function of the independent variable $x$. In another 1837 paper, Dirichlet proved that in an absolutely convergent series, one may rearrange the order in which terms are added in whatever way one wishes and not change the sum of the series. While this is immediate for a finite sum, the fact that an infinite sum of numbers could always be rearranged without changing the value was surprising. He also gave examples of conditionally convergent series in which the sum *was* altered by rearrangement of the terms. Almost 20 years later, **Georg Riemann** proved that the terms of a conditionally convergent series could be rearranged to yield a sum of any desired value.

At the golden jubilee celebration of his doctorate, Dirichlet's teacher Karl Gauss tried to light his pipe with a piece of his original manuscript *Disquisitiones arithmeticae*. Dirichlet was overcome by the sacrilege of such an action. He rescued the piece of paper from Gauss' fire. Dirichlet treasured the paper for his remaining years, and his editors found it among his papers after his death.

With Gauss's death in 1855, the University of Göttingen sought a successor of great distinction and chose Dirichlet. His current position at the military academy was unappealing and lacked scientific stimulation and he was required to lecture 13 times a week. Dirichlet accepted the university's offer.

### Short–Lived Contentment

Dirichlet moved to Göttingen in 1855, where he purchased a house with a garden. He enjoyed a quiet life there with excellent students and the time avail-

able for research. But his new contentment did not last long. During a speech in Montreaux, Switzerland, he suffered a heart attack and barely made it home. During his illness, his wife died of a stroke, and Dirichlet subsequently died the following spring in 1863.

After his death, Dirichlet's pupil and friend, **Julius Dedekind**, published Dirichlet's *Vorlesungen über Zahlentheorie*, adding several supplements of his own investigations on algebraic number theory. The addenda are regarded as one of the most important sources for the creation of the theory of ideals, and are the core of algebraic number theory.

## FURTHER READING

Eves, Howard. *An Introduction to the History of Mathematics.* The Saunders Series, 1983, pp. 369–370.

Ore, Oystein. "Gustav Peter Lejeune Dirichlet," in *Dictionary of Scientific Biography*. Volume IV. Edited by Charles Coulston Gillispie. New York: Charles Scribner's Sons, 1974, pp. 123–127.

Taton René, editor. *History of Science: Science in the Nineteenth Century.* New York: Basic Books, 1961, pp. 17; 49; 59–67.

—*Sketch by Corinne Johnson*

# Simon K. Donaldson
## 1957-
### English geometer and topologist

Simon Donaldson shocked the mathematical world during the 1980s with a series of papers on the structure of four-dimensional spaces. Researchers had produced a collection of results during the previous decade that outlined a general understanding of the properties of spaces of five or more dimensions, and of course, the cases of one- and two-dimensional spaces were well known. Ironically, three- and four-dimensional spaces were the hardest to interpret, even though they are the most applicable to physical space (if time is considered to be the fourth dimension). Great progress was made in three-dimensions by William Thurston, who received a Fields Medal in 1982 for his efforts. That same year, Donaldson published his most remarkable result: four-dimensional space has highly unusual properties that are found in no other dimension. Speaking on the occasion of Donaldson's presentation with the 1986 Fields Medal, **Michael Atiyah** commented, "When Donaldson produced his first few results on four-manifolds [four-dimensional topological surfaces], the ideas were so new and foreign to geometers and topologists that they merely gazed in bewildered admiration . . . Donaldson has opened up an entirely new area; unexpected and mysterious phenomena about the geometry of four dimensions have been discovered."

Simon K. Donaldson was born on August 20, 1957, in Cambridge, England. He attended Pembroke College in Cambridge University and received his B.A. degree in 1979. During his second year of graduate studies at Worcester College in Oxford University, Donaldson made the spectacular discovery of "exotic" or nonstandard differential structures of four-dimensional Euclidean space. In other words, he found that there were different ways of orienting a mathematical structure in ordinary space with the addition of a fourth dimension. Because the standard differential structure is the only one possible in all other dimensions, the mathematical community was amazed at the exceptions created by the addition of the fourth dimension. After completing his doctorate in 1984, Donaldson spent a year at Princeton University's Institute for Advanced Study (IAS) and was a visiting scholar at Harvard University during the spring of 1985. He then returned to England where he holds an appointment at the Mathematics Institute in Oxford.

## Reverses Yang-Mills Equations

In 1954, Chen Ning Yang and Robert Mills collaborated on derivations of mathematical formulas that combined the branch of mathematics known as topology (the study of the ways in which coordinate structures attach at a point) and the branch of physics called quantum electrodynamics (the study of electromagnetic phenomena under the rules of quantum mechanics). In doing so, they built upon the work of James Clerk Maxwell, who had introduced equations in the 19th century to describe the behavior of electromagnetic waves. The Yang-Mills equations generalize Maxwell's equations to more complex spaces. Since they are nonlinear partial differential equations, the Yang-Mills equations are very difficult to solve, even for specific cases. Work by Atiyah and others during the 1970s led to important connections between the equations and techniques from differential and algebraic geometry.

While others were working on methods for solving the Yang-Mills equations, Donaldson approached the topic from a fresh viewpoint. As described by John D. S. Jones in Nature, "Donaldson argues as follows: if we know something about the solutions of the Yang-Mills equations then we must be able to extract information about the underlying space . . . Donaldson starts by treating the solutions of the equations as, in some sense, the known quantity."

It is common for mathematicians to look to theoretical physics for problems to investigate. In this unusual reversal, Donaldson used the tools of physics to explore purely mathematical ideas. In four-dimensional Euclidean space, the absolute minimum solutions (under certain boundary conditions at infinity) of the Yang-Mills equations are called "instantons." Donaldson's inspiration was to look at the nonlinear space of parameters for these instantons as a lens through which he could examine the space on which the equations are defined.

One of the basic goals of topology is to classify multidimensional spaces into categories that have the same basic structure. Jones described this as being similar to the taxonomic classification system in biology. In mathematics, spaces can be classified in terms of their topology (connectedness) or their smoothness (lack of corners, as shown by continuity of derivatives). For example, the surface of a sphere belongs to a different topological class than that of a torus (doughnut–like shape) because of the existence of the hole. In three dimensions, there is no difference between the results of classifying spaces topologically or smoothly. In five or more dimensions, there are relatively minor differences that are well understood. **Michael Freedman** obtained clear results for topological classification of four-dimensional spaces about the same time that Donaldson was establishing very different results using smoothness criteria. In the words of Atiyah, this "shows that the differentiable and topological situations are totally different"—a situation that occurs only in spaces of four dimensions.

Another way of describing Donaldson's results was offered by John Baez in *This Week's Finds in Mathematical Physics.* He stated the basic question as being whether n-dimensional Euclidean space allows any smooth structure other than the usual one. He concluded, "The answer is no—EXCEPT if n=4, where there are uncountably many smooth structures!" This unexpected result generated great excitement within the mathematical community and earned Donaldson the Fields Medal, the most prestigious international mathematics award, in 1986.

## Delves Deeper Into Four–Dimensional Space

Donaldson used intersection matrices as a tool for exploring four–dimensional spaces. Any four–dimensional space can be characterized by a matrix of integers in a way that describes how two–dimensional spaces intersect within it. This symmetric, invariant matrix will be the same for all topologically equivalent spaces. This means that if one finds two spaces with nonequivalent intersection matrices, those spaces will be topologically distinct. Conversely, Freedman showed that at most two topologically distinct spaces can be represented by equivalent intersection matrices. Examining the situation from a smoothness perspective, Donaldson found that there are unlimited possibilities for distinct spaces with equivalent intersection matrices.

It was in calculating these intersection matrices that Donaldson used the Yang–Mills equations. According to Jones, the instanton solutions are "concentrated in a very small ball and they behave like particles placed at the centre of the ball. This gives a way of recovering the points of the space from the solutions of the equations." With continued work, Donaldson was able to identify other invariants capable of distinguishing between two smoothly different, topologically equivalent manifolds.

One of the dramatic byproducts of Donaldson's discoveries was the description of exotic four–spaces. In Atiyah's description, these remarkable spaces "contain compact sets which cannot be contained inside any differentiably embedded 3–sphere!" In addition to producing startling results about the mathematics of four-dimensional spaces, Donaldson's work has also generated useful information for physicists. For instance, the earliest link between topology and quantum theory was Paul Dirac's idea that the electric charge of a particle (which is an integral multiple of the charge of a single electron) has a basis in magnetism. He described this in terms of a hypothetical "magnetic monopole'—a basic particle that radiates a magnetic field just as a charged particle radiates an electrical field. Donaldson established a direct link between the parameter space of monopoles having magnetic charge $k$ and the space of rational functions of a complex variable of degree $k$.

In addition to continuing his research, Donaldson currently serves as one of nine voting members of the executive committee of the International Mathematical Union (IMU). He is a fellow of the Royal Society of London, from which he received the Royal Medal in 1992. In 1994 he shared with **Shing-Tung Yau** the Swedish Academy of Science's Crafoord Prize, which is presented once every six years in the field of mathematics to provide financial support for research in an area of "particular interest and considerable activity."

## SELECTED WRITINGS BY DONALDSON

### Books

(With P. B. Kronheimer) *The Geometry of Four-Manifolds,* 1990

### Periodicals

"Self-Dual Connections and the Topology of Smooth 4-Manifolds," in *Bulletin of the American Mathematical Society* 8 (1983): 81–3.
"The Geometry of 4-Manifolds,"in *Proceedings of the International Congress of Mathematicians,* [Berkeley, CA] (1986): 43–54.

## FURTHER READING

### Periodicals

Atiyah, Michael. "On the Work of Simon Donaldson." *Proceedings of the International Congress of Mathematicians,* [Berkeley, CA] (1986): 3–6.
————. "The Work of Simon Donaldson." *Notices of the American Mathematical Society* 33 (November 1986): 900–01.
Jones, John D. S. "Mysteries of Four Dimensions." *Nature* 332 (April 7, 1988): 488–89.

### Other

Baez, John. "Special Edition: The End of Donaldson Theory?" *This Week's Finds in Mathematical Physics (Week 44).* http://math.ucr.edu/home/baez/wee-k44.html (June 1, 1997).
Donaldson, Simon K. *International Mathematical Union Executive Committee.* http://elib.zib.de:8000/IMU/EC/DonaldsonSK.html;internal&sk=010FF3CDF (July 8, 1997).
Jackson, Allyn. "A Revolution in Mathematics." *Mathematical News* July/August 1996, http://www.ams.org/general/news1.html (June 1, 1997).

—*Sketch by Loretta Hall and Robert Messer*

E

*Albert Einstein*

# Albert Einstein
## 1879-1955
### German-born American physicist

Albert Einstein ranks as one of the most remarkable theoreticians in the history of science. During a single year, 1905, he produced three papers that are among the most important in 20th-century physics, and perhaps in all of the recorded history of science, for they revolutionized the way scientists looked at the nature of space, time, and matter. These papers dealt with the nature of particle movement known as Brownian motion, the quantum nature of electromagnetic radiation as demonstrated by the photoelectric effect, and the special theory of relativity. Although Einstein is probably best known for the last of these works, it was for his quantum explanation of the photoelectric effect that he was awarded the 1921 Nobel Prize in physics. In 1915, Einstein extended his special theory of relativity to include certain cases of accelerated motion, resulting in the more general theory of relativity.

Einstein was born in Ulm, Germany, on March 14, 1879, the only son of Hermann and Pauline Koch Einstein. Both sides of his family had long-established roots in southern Germany, and, at the time of Einstein's birth, his father and uncle Jakob owned a small electrical equipment plant. When that business failed around 1880, Hermann Einstein moved his family to Munich to make a new beginning. A year after their arrival in Munich, Einstein's only sister, Maja, was born.

Although his family was Jewish, Einstein was sent to a Catholic elementary school from 1884 to 1889. He was then enrolled at the Luitpold Gymnasium in Munich. During these years, Einstein began to develop some of his earliest interests in science and mathematics, but he gave little outward indication of any special aptitude in these fields. Indeed, he did not begin to talk until the age of three and, by the age of nine, was still not fluent in his native language. His parents were actually concerned that he might be somewhat mentally retarded.

### Leaves School Early and Moves to Italy

In 1894, Hermann Einstein's business failed again, and the family moved once more, this time to Pavia, near Milan, Italy. Einstein was left behind in Munich to allow him to finish school. Such was not to be the case, however, since he left the *gymnasium* after only six more months. Einstein's biographer, Philipp Frank, explains that Einstein so thoroughly despised formal schooling that he devised a scheme by which he received a medical excuse from school on the basis of a potential nervous breakdown. He then convinced a mathematics teacher to certify that he was adequately prepared to begin his college studies without a high school diploma. Other biographies, however, say that Einstein was expelled from the *gymnasium* on the grounds that he was a disruptive influence at the school.

In any case, Einstein then rejoined his family in Italy. One of his first acts upon reaching Pavia was to give up his German citizenship. He was so unhappy with his native land that he wanted to sever all formal connections with it; in addition, by renouncing his citizenship, he could later return to Germany without being arrested as a draft dodger. As a result, Einstein remained without an official citizenship until he became a Swiss citizen at the age of 21. For most of his first year in Italy, Einstein spent his time traveling, relaxing, and teaching himself calculus and higher mathematics. In 1895, he thought himself ready to take the entrance examination for the Eidgenössische Technische Hochschule (the ETH, Swiss Federal

Polytechnic School, or Swiss Federal Institute of Technology), where he planned to major in electrical engineering. When he failed that examination, Einstein enrolled at a Swiss cantonal high school in Aarau. He found the more democratic style of instruction at Aarau much more enjoyable than his experience in Munich and soon began to make rapid progress. He took the entrance examination for the ETH a second time in 1896, passed, and was admitted to the school. (In *Einstein,* however, Jeremy Bernstein writes that Einstein was admitted without examination on the basis of his diploma from Aarau.)

The program at ETH had nearly as little appeal for Einstein as had his schooling in Munich, however. He apparently hated studying for examinations and was not especially interested in attending classes on a regular basis. He devoted much of this time to reading on his own, specializing in the works of Gustav Kirchhoff, Heinrich Hertz, James Clerk Maxwell, Ernst Mach, and other classical physicists. When Einstein graduated with a teaching degree in 1900, he was unable to find a regular teaching job. Instead, he supported himself as a tutor in a private school in Schaffhausen. In 1901, Einstein also published his first scientific paper, "Consequences of Capillary Phenomena."

In February, 1902, Einstein moved to Bern and applied for a job with the Swiss Patent Office. He was given a probationary appointment to begin in June of that year and was promoted to the position of technical expert, third class, a few months later. The seven years Einstein spent at the Patent Office were the most productive years of his life. The demands of his work were relatively modest and he was able to devote a great deal of time to his own research.

The promise of a steady income at the Patent Office also made it possible for Einstein to marry. Mileva Marić (also given as Maritsch) was a fellow student in physics at ETH, and Einstein had fallen in love with her even though his parents strongly objected to the match. Marić had originally come from Hungary and was of Serbian and Greek Orthodox heritage. The couple married on January 6, 1903, and later had two sons, Hans Albert and Edward. A previous child, Liserl, was born in 1902 at the home of Marić's parents in Hungary, but there is no further mention or trace of her after 1903 since she was given up for adoption.

## Explains Brownian Movement and the Photoelectric Effect

In 1905, Einstein published a series of papers, any one of which would have assured his fame in history. One, "On the Movement of Small Particles Suspended in a Stationary Liquid Demanded by the Molecular-Kinetic Theory of Heat," dealt with a phenomenon first observed by the Scottish botanist

Robert Brown in 1827. Brown had reported that tiny particles, such as dust particles, move about with a rapid and random zigzag motion when suspended in a liquid.

Einstein hypothesized that the visible motion of particles was caused by the random movement of molecules that make up the liquid. He derived a mathematical formula that predicted the distance traveled by particles and their relative speed. This formula was confirmed experimentally by the French physicist Jean Baptiste Perrin in 1908. Einstein's work on the Brownian movement is generally regarded as the first direct experimental evidence of the existence of molecules.

A second paper, "On a Heuristic Viewpoint concerning the Production and Transformation of Light," dealt with another puzzle in physics, the photoelectric effect. First observed by Heinrich Hertz in 1888, the photoelectric effect involves the release of electrons from a metal that occurs when light is shined on the metal. The puzzling aspect of the photoelectric effect was that the number of electrons released is not a function of the light's intensity, but of the color (that is, the wavelength) of the light.

To solve this problem, Einstein made use of a concept developed only a few years before, in 1900, by the German physicist Max Planck, the quantum hypothesis. Einstein assumed that light travels in tiny discrete bundles, or "quanta," of energy. The energy of any given light quantum (later renamed the photon), Einstein said, is a function of its wavelength. Thus, when light falls on a metal, electrons in the metal absorb specific quanta of energy, giving them enough energy to escape from the surface of the metal. But the number of electrons released will be determined not by the number of quanta (that is, the intensity) of the light, but by its energy (that is, its wavelength). Einstein's hypothesis was confirmed by several experiments and laid the foundation for the fields of quantitative photoelectric chemistry and quantum mechanics. As recognition for this work, Einstein was awarded the 1921 Nobel Prize in physics.

## Refines the Theory of Relativity

A third 1905 paper by Einstein, almost certainly the one for which he became best known, details his special theory of relativity. In essence, "On the Electrodynamics of Moving Bodies" discusses the relationship between measurements made by observers in two separate systems moving at constant velocity with respect to each other.

Einstein's work on relativity was by no means the first in the field. The French mathematician and physicist **Jules Henri Poincaré,** the Irish physicist George Francis FitzGerald, and the Dutch physicist

Hendrik Lorentz had already analyzed in some detail the problem attacked by Einstein in his 1905 paper. Each had developed mathematical formulas that described the effect of motion on various types of measurement. Indeed, the record of pre-Einsteinian thought on relativity is so extensive that one historian of science once wrote a two-volume work on the subject that devoted only a single sentence to Einstein's work. Still, there is little question that Einstein provided the most complete analysis of this subject. He began by making two assumptions. First, he said that the laws of physics are the same in all frames of reference. Second, he declared that the velocity of light is always the same, regardless of the conditions under which it is measured.

Using only these two assumptions, Einstein proceeded to uncover an unexpectedly extensive description of the properties of bodies that are in uniform motion. For example, he showed that the length and mass of an object are dependent upon their movement relative to an observer. He derived a mathematical relationship between the length of an object and its velocity that had previously been suggested by both FitzGerald and Lorentz. Einstein's theory was revolutionary, for previously scientists had believed that basic quantities of measurement such as time, mass, and length were absolute and unchanging. Einstein's work established the opposite—that these measurements could change, depending on the relative motion of the observer.

In addition to his masterpieces on the photoelectric effect, Brownian movement, and relativity, Einstein wrote two more papers in 1905. One, "Does the Inertia of a Body Depend on Its Energy Content?," dealt with an extension of his earlier work on relativity. He came to the conclusion in this paper that the energy and mass of a body are closely interrelated. Two years later he specifically stated that relationship in a formula, $E=mc^2$ (energy equals mass times the speed of light squared), that is now familiar to both scientists and non-scientists alike. His final paper, the most modest of the five, was "A New Determination of Molecular Dimensions." It was this paper that Einstein submitted as his doctoral dissertation, for which the University of Zurich awarded him a Ph.D. in 1905.

Fame did not come to Einstein immediately as a result of his five 1905 papers. Indeed, he submitted his paper on relativity to the University of Bern in support of his application to become a *privatdozent,* or unsalaried instructor, but the paper and application were rejected. His work was too important to be long ignored, however, and a second application three years later was accepted. Einstein spent only a year at Bern, however, before taking a job as professor of physics at the University of Zurich in 1909. He then went on to the German University of Prague for a year and a half before returning to Zurich and a

position at ETH in 1912. A year later Einstein was made director of scientific research at the Kaiser Wilhelm Institute for Physics in Berlin, a post he held from 1914 to 1933.

## Debate Centers on the Role of Einstein's Wife in His Work

In recent years, the role of Mileva Einstein-Marić in her husband's early work has been the subject of some controversy. The more traditional view among Einstein's biographers is that of A. P. French in his "Condensed Biography" in *Einstein: A Centenary Volume.* French argues that although "little is recorded about his [Einstein's] domestic life, it certainly did not inhibit his scientific activity." In perhaps the most substantial of all Einstein biographies, Philipp Frank writes that "For Einstein life with her was not always a source of peace and happiness. When he wanted to discuss with her his ideas, which came to him in great abundance, her response was so slight that he was often unable to decide whether or not she was interested."

A quite different view of the relationship between Einstein and Marić is presented in a 1990 paper by Senta Troemel-Ploetz in *Women's Studies International Forum.* Based on a biography of Marić originally published in Yugoslavia, Troemel-Ploetz argues that Marić gave to her husband "her companionship, her diligence, her endurance, her mathematical genius, and her mathematical devotion." Indeed, Troemel-Ploetz builds a case that it was Marić who did a significant portion of the mathematical calculations involved in much of Einstein's early work. She begins by repeating a famous remark by Einstein himself to the effect that "My wife solves all my mathematical problems." In addition, Troemel-Ploetz cites many of Einstein's own letters of 1900 and 1901 (available in *Collected Papers* ) that allude to Marić's role in the development of "our papers," including one letter to Marić in which Einstein noted: "How happy and proud I will be when both of us together will have brought our work on relative motion to a successful end." The author also points out the somewhat unexpected fact that Einstein gave the money he received from the 1921 Nobel Prize to Marić, although the two had been divorced two years earlier. Nevertheless, Einstein never publicly acknowledged any contributions by his wife to his work.

Any mathematical efforts Mileva Einstein-Marić may have contributed to Einstein's work greatly decreased after the birth of their second son in 1910. Einstein was increasingly occupied with his career and his wife with managing their household; upon moving to Berlin in 1914, the couple grew even more distant. With the outbreak of World War I, Einstein's wife and two children returned to Zurich. The two were never reconciled; in 1919, they were formally divorced. Towards the end of the war, Einstein became very ill and was nursed back to health by his cousin Elsa. Not long after Einstein's divorce from Marić, he was married to Elsa, a widow. The two had no children of their own, although Elsa brought two daughters, Ilse and Margot, to the marriage.

### Announces the General Theory of Relativity

The war years also marked the culmination of Einstein's attempt to extend his 1905 theory of relativity to a broader context, specifically to systems with non-zero acceleration. Under the general theory of relativity, motions no longer had to be uniform and relative velocities no longer constant. Einstein was able to write mathematical expressions that describe the relationships between measurements made in *any* two systems in motion relative to each other, even if the motion is accelerated in one or both. One of the fundamental features of the general theory is the concept of a space-time continuum in which space is curved. That concept means that a body affects the shape of the space that surrounds it so that a second body moving near the first body will travel in a curved path.

Einstein's new theory was too radical to be immediately accepted, for not only were the mathematics behind it extremely complex, it replaced Newton's theory of gravitation that had been accepted for two centuries. So, Einstein offered three proofs for his theory that could be tested: first, that relativity would cause Mercury's perihelion, or point of orbit closest to the sun, to advance slightly more than was predicted by Newton's laws. Second, Einstein predicted that light from a star will be bent as it passes close to a massive body, such as the sun. Last, the physicist suggested that relativity would also affect light by changing its wavelength, a phenomenon known as the redshift effect. Observations of the planet Mercury bore out Einstein's hypothesis and calculations, but astronomers and physicists had yet to test the other two proofs.

Einstein had calculated that the amount of light bent by the sun would amount to 1.7 seconds of an arc, a small but detectable effect. In 1919, during an eclipse of the sun, English astronomer Arthur Eddington measured the deflection of starlight and found it to be 1.61 seconds of an arc, well within experimental error. The publication of this proof made Einstein an instant celebrity and made "relativity" a household word, although it was not until 1924 that Eddington proved the final hypothesis concerning redshift with a spectral analysis of the star Sirius B. This phenomenon, that light would be shifted to a longer wavelength in the presence of a strong gravitational field, became known as the "Einstein shift."

Einstein's publication of his general theory in 1916, the  essentially brought to a close the revolu-

tionary period of his scientific career. In many ways, Einstein had begun to fall out of phase with the rapid changes taking place in physics during the 1920s. Even though Einstein's own work on the photoelectric effect helped set the stage for the development of quantum theory, he was never able to accept some of its concepts, particularly the uncertainty principle. In one of the most-quoted comments in the history of science, he claimed that quantum mechanics, which could only calculate the probabilities of physical events, could not be correct because "God does not play dice." Instead, Einstein devoted his efforts for the remaining years of his life to the search for a unified field theory, a single theory that would encompass all physical fields, particularly gravitation and electromagnetism.

### Becomes Involved in Political Issues

Since the outbreak of World War I, Einstein had been opposed to war, and used his notoriety to lecture against it during the 1920s and 1930s. With the rise of National Socialism in Germany in the early 1930s, Einstein's position became difficult. Although he had renewed his German citizenship, he was suspect as both a Jew and a pacifist. In addition, his writings about relativity were in conflict with the absolutist teachings of the Nazi party. Fortunately, by 1930, Einstein had become internationally famous and had traveled widely throughout the world. A number of institutions were eager to add his name to their faculties.

In early 1933, Einstein made a decision. He was out of Germany when Hitler rose to power, and he decided not to return. Instead he accepted an appointment at the Institute for Advanced Studies  in Princeton, New Jersey, where he spent the rest of his life. In addition to his continued work on unified field theory, Einstein was in demand as a speaker and wrote extensively on many topics, especially peace. The growing fascism and anti-Semitism of Hitler's regime, however, convinced him in 1939 to sign his name to a letter written by American physicist Leo Szilard informing President Franklin D. Roosevelt of the possibility of an atomic bomb. This letter led to the formation of the Manhattan Project for the construction of the world's first nuclear weapons. Although Einstein's work on relativity, particularly his formulation of the equation $E = mc^2$, was essential to the development of the atomic bomb, Einstein himself did not participate in the project. He was considered a security risk, although he had renounced his German citizenship and become a U.S. citizen in 1940.

After World War II and the bombing of Japan, Einstein became an ardent supporter of nuclear disarmament. He also lent his support to the efforts to establish a world government and to the Zionist movement to establish a Jewish state. In 1952, after the death of Israel's first president, Chaim Weizmann, Einstein was invited to succeed him as president; he declined the offer. Among the many other honors given to Einstein were the Barnard Medal of Columbia University in 1920, the Copley Medal of the Royal Society in 1925, the Gold Medal of the Royal Astronomical Society in 1926, the Max Planck Medal of the German Physical Society in 1929, and the Franklin Medal of the Franklin Institute in 1935. Einstein died at his home in Princeton on April 18, 1955, after suffering an aortic aneurysm. At the time of his death, he was the world's most widely admired scientist and his name was synonymous with genius. Yet Einstein declined to become enamored of the admiration of others. He wrote in his book *The World as I See It:* "Let every man be respected as an individual and no man idolized. It is an irony of fate that I myself have been the recipient of excessive admiration and respect from my fellows through no fault, and no merit, of my own. The cause of this may well be the desire, unattainable for many, to understand the one or two ideas to which I have with my feeble powers attained through ceaseless struggle."

## SELECTED WRITINGS BY EINSTEIN

### Books

*On the Method of Theoretical Physics,* 1933
*Essays on Science,* 1934
*The World as I See It,* 1935
(With L. Infeld) *The Evolution of Physics: The Growth of Ideas from Early Concepts to Relativity and Quanta,* 1938
*The Meaning of Relativity,* 1950
*Out of My Later Years,* 1950
(With others) *The Principle of Relativity,* 1952
*Ideas and Opinions,* 1954
*Investigations on the Theory of the Brownian Movement,* edited by R. Fürth, 1956
*Einstein on Peace,* edited by Otto Nathan and Heinz Norden, 1960
*Relativity: The Special and General Theory,* 1961
*The Collected Papers of Albert Einstein,* Volume 1, 1987, Volume 2, 1989, Volume 3, 1993

### Periodicals

"On the Movement of Small Particles Suspended in a Stationary Liquid Demanded by the Molecular-Kinetic Theory of Heat." *Annalen der Physik* (1905).
"On a Heuristic Viewpoint concerning the Production and Transformation of Light." *Annalen der Physik* (1905).
"On the Electrodynamics of Moving Bodies." *Annalen der Physik,* (1905).

"Does the Inertia of a Body Depend on Its Energy Content?" *Annalen der Physik* (1905).

## FURTHER READING

### Books

Bernstein, Jeremy. *Einstein.* Fontana, 1973.
Clark, Ronald W. *Einstein: The Life and Times* World Publishing, 1971.
Feldman, Anthony, and Peter Ford. *Scientists & Inventors.* New York: Facts on File, 1979, pp. 264–265.
Frank, Philipp, *Einstein: His Life and Times.* New York: Knopf, 1947.
French, A. P. *Einstein: A Centenary Volume.* Cambridge, MA: Harvard University Press, 1979.
Highfield, Roger, and Paul Carter. *The Private Lives of Albert Einstein.* Wincheseter, MA: Faber, 1993.
Hoffmann, Banesh. *Albert Einstein: Creator and Rebel.* New York: Viking, 1972.
Infeld, Leopold. *Quest.* New York: Doubleday, Doran, 1941.
Pais, Abraham. *Subtle Is the Lord.* New York: Oxford University Press, 1982.
Seelig, Carl. *Albert Einstein: A Documentary Biography.* Staples Press, 1956.
Will, Clifford M. *Was Einstein Right?: Putting General Relativity to the Test.* New York: Basic Books, 1986.

### Periodicals

Shankland, Robert S. "Conversations with Albert Einstein." *American Journal of Physics* 31 (1963): 37–47.
Troemel-Ploetz, Senta. "Mileva Einstein-Marić: The Woman Who Did Einstein's Math." *Women's Studies International Forum* 13, number 5, (1990): 415–432.

*—Sketch by David E. Newton*

---

# Eratosthenes of Cyrene
## c.285 B.C.–c.205 B.C.
### Greek mathematician, astronomer, geographer, philosopher, and poet

Eratosthenes was one of the great thinkers of ancient Greece whose accomplishments included accurately measuring the Earth's circumference and discovering a simple method for finding prime numbers. In addition to his contributions in mathematics, geography, and astronomy, Eratosthenes was a poet, philosopher, and educator whose wide ranging interests served him well as director of the great ancient library at Alexandria. Eratosthenes was nicknamed "Beta" (Greek for two) by some of his contemporaries. This name may refer to his status as second only to Plato in knowledge, or it may also have been a derogatory term used by detractors to indicate that Eratosthenes had not achieved top standing in any discipline of knowledge.

Eratosthenes was born circa 285 B.C. in Cyrene, which is now Libya, in northern Africa. His father's name was Aglaus, but little else is known about his family or its station in Greek society. At that time, Cyrene was a prosperous Hellenistic Greek city–state known for its culture and as a successful commercial harbor on the Mediterranean coast. Eratosthenes is believed to have studied under Lysanias, a renowned grammarian, and Callimachus, a poet. He developed an early interest in literature and philosophy and, in his teens, traveled to Athens, where he probably studied in the famed Academy and Lyceum.

After completing his studies, Eratosthenes turned his attention to writing, especially poetry. None of his poems from this period survives, but references to them include the poem *Hermes*, which describes the astronomical "heavens" and is a homage to the Greek god, and *Ergion*, about a young woman's suicide. It is a testament to Eratosthenes intellect that he achieved wide renown as a poet before his scientific accomplishments. His literary abilities likely attracted the attention of Ptolemy III, who invited Eratosthenes to Alexandria to tutor his son, the crown prince. At the age of 40, he also garnered the prestigious appointment as director of the library at Alexandria. He then embarked on a career unparalleled at that time for its breadth of scientific, artistic, and academic contributions.

## Makes Important Contributions to Mathematics and Geometry

Eratosthenes was a consummate scholar with wide ranging interests, but his most important contributions, including those in geography and astronomy, were based on his mathematical acumen. In the field of mathematics, Eratosthenes' most famous achievement is his simple but foolproof method, or algorithm, for finding prime numbers. Known as the sieve of Eratosthenes, the approach allowed him to "sift" out prime numbers (divisible only by one and the number itself) from a composite list of natural numbers arranged in order. Eratosthenes' discovery has led to centuries of number theory research.

In the area of geometry, Eratosthenes is credited with developing a practical method for finding an infinite number of mean proportionals between two

given straight lines. He developed the mean–finder, or mesolabe, which led to solving the Delian problem, otherwise known as doubling the cube. A letter to Ptolemy III, which is repeated in the commentary of Eutocius on **Archimedes**, is the only remaining writing on geometry by Eratosthenes. In the letter, Eratosthenes describes the mesolabe, a mechanical device constructed of three equal rectangular frames, each with a diagonal that creates two triangles within each frame. These frames could be pushed along grooves back and forth, and over and under each other. With this device, Eratosthenes was able to demonstrate a mathematical method for doubling the cube. This unique approach to solving problems of proportions was vitally important to the Greeks, since they used proportions to solve geometrical problems. Eratosthenes was so proud of this accomplishment that he gave a bronze model of his invention with an inscription of how it worked as a gift to the king and the people of Alexandria.

Eratosthenes also wrote several books or treatises on mathematics, although none of them survive today. In one book, he wrote on the fundamentals of mathematics in relation to the philosophy of Plato, including definitions in geometry and arithmetic and theories of proportions. His most famous and comprehensive mathematical work is called *On Means*, which was considered important enough to be included by Pappus in his *Treasury of Analysis*. However, the actual content of *On Means* is unknown. Perhaps the greatest testament to Eratosthenes' mathematical abilities was made by Archimedes, who sent him a manuscript of his *Method*, stating that his treatise, which explains the methods he used to make his mathematical discoveries, was written especially for Eratosthenes.

## Measures the Earth and Creates the Science of Geodesy

Eratosthenes' accurate measurement of the Earth's circumference is his most renowned achievement. It marks the first important step in the science of geodesy, a branch of mathematics concerned with determining the Earth's size and shape and location of points on its surface. Prior to this, the Earth's circumference was estimated to be approximately 300,000 stades. Incorporating two geometrical propositions set forth by **Euclid**, Eratosthenes first measured the distance from Alexandria to Syene, Egypt (located south of Alexandria) at 5,000 stades. He then used calculations of the position of the sun during the summer solstice over both cities to estimate the distance from the two cities to represent 1/50th of the Earth's surface. With some minor adjustments to correct errors in his readings, Eratosthenes came up with his final estimate of 252,000 stades (24,662 miles), which is within 1 percent of modern estimates of the Earth's circumference.

Eratosthenes' most famous book on geography, *On the Measurement of the Earth* , probably included a detailed explanation of how he measured the Earth's circumference. The lost book is believed to have focused on mathematical geography and included calculations of the distance of the Earth from the moon and sun, the distance between the tropic circles, and the size of the sun. His three–volume *Geographica* included a history of geography to his time. It then discussed specifics of mathematical geography based on the belief that the Earth was round or spherical. It was largely through these books that Eratosthenes established the science of geodesy. Eratosthenes used this newly found science to make some of the most accurate maps of the world in his time, including the first maps to incorporate the system of latitude and longitude.

### Establishes Chronological Foundation for Greek History

In addition to his commitment to precise measurements in geography and astronomy, Eratosthenes also established the first scientific chronology for recording history in Greece. As a scholar, he wanted to accurately record his country's past based solely on documented records and without the inclusion of legends. In his two books on the subject, *Ilympionikai* and *Chronographiai* (Chronological Tables), he develops and details his method, which is based largely on using existing records of the Olympiads (the precursor of the modern day Olympic Games) to set an event in time.

Among Eratosthenes' other notable scientific accomplishments is his accurate measurement of the tilt of the Earth's axis and a comprehensive catalogue of 675 stars. He is also believed to have created the improved calendar adopted by Julius Caesar. Similar to the modern–day calendar, Eratosthenes' calendar incorporated the addition of one day every four years, which is now known as Leap Year.

A man of boundless intellectual curiosity, Eratosthenes also wrote a number of literary subjects, including books *Good and Evil* and *Comedy*, which discussed authors of that genre. His last known written work, *Biography of Arsinoe III*, revealed some of his intimate knowledge of the royal court.

Befitting his versatility and genius, Eratosthenes was referred to as a philologist, or lover of learning. Whether Eratosthenes himself created this title or it was bestowed upon him by others is unknown. Eratosthenes went blind in his later years. For a man committed to knowledge and learning, his loss of sight must have been a terrible blow, barring him from ever again using the greatest library in the world. Eratosthenes chose a path of voluntary starvation and died at the age of 80.

## FURTHER READING

### Books

Gow, James. *A Short History of Greek Mathematics*. New York: Chelsea Publishing Company, 1968, pp. 244–246.

Smith, David Eugene. *History of Mathematics*. Volume I. New York: Dover Publications, 1958, pp. 343–346.

Terry, Leon. *The Mathmen*. New York: McGraw–Hill Book Company, 1964, pp. 185–196.

Van Der Waerden, L. Van. *Science Awakening*. New York, John Wiley & Sons, Inc., 1963, pp. 228–236

### Periodicals

Rawlins, D. "Eratosthenes' Geodest Unraveled: Was There a High–Accuracy in Hellenistic Astronomy?" *Isis* 73 (1982): 259–265.

—*Sketch by David A. Petechuk*

# Paul Erdös
## 1913–1996
### Hungarian number theorist

For Paul Erdös, mathematics was life. Number theory, combinatorics (a branch of mathematics concerning the arrangement of finite sets), and discrete mathematics were his consuming passions. Everything else was of no interest: property, money, clothes, intimate relationships, social pleasantries—all were looked on as encumbrances to his mathematical pursuits. A genius in the true sense of the word, Erdös traveled the world, living out of a suitcase, to problem solve — and problem pose—with his mathematical peers. A small, hyperactive man, he would arrive at a university or research center confident of his welcome. While he was their guest, it was a host's task to lodge him, feed him, do his laundry, make sure he caught his plane to the next meeting, and sometimes even do his income taxes. Cosseted by his mother and by household servants, he was not brought up to fend for himself. Gina Bari Kolata, writing in *Science* magazine, reports that Erdös said he "never even buttered his own bread until he was 21 years old."

Yet this man, whom Paul Hoffman called "probably the most eccentric mathematician in the world" in the *Atlantic Monthly*, more than repaid his colleagues' care of him by giving them a wealth of new and challenging problems—and brilliant methods for

*Paul Erdös*

solving them. Erdös laid the foundation of computer science by establishing the field of discrete mathematics. A number theorist from the beginning, he was just 20 years old when he discovered a proof for Chebyshev's theorem, which says that for each integer greater than one, there is always at least one prime number between it and its double.

Erdös was born in Budapest, Hungary, on March 26, 1913. His parents, Lajos and Anna Erdös, were high school mathematics teachers. His two older sisters died of scarlet fever when he was a newborn baby, leaving him an only child with a very protective mother. Erdös was schooled at home by his parents and a governess, and his gift for mathematics was recognized at an early age. It is said that Erdös could multiply three–digit numbers in his head at age three, and discovered the fact of negative numbers when he was four. He received his higher education from the University of Budapest, entering at age 17 and graduating four years later with a Ph. D. in mathematics. He completed a postdoctoral fellowship in Manchester, England, leaving Hungary in the midst of political unrest in 1934. As a Jew, Hungary was then a dangerous place for him to be. During the ensuing Nazi reign of power, four of Erdös' relatives were murdered, and his father died of a heart attack in 1942.

In 1938, Erdös came to the United States, but because of the political situation in Hungary, he had difficulty receiving permission from the U. S. govern-

ment to come and go freely between America and Europe. He settled in Israel and did not return to the United States until the 1960s. While in the country, he attended mathematical conferences, met with top U. S. mathematicians such as Ronald Graham, Ernst Straus and Stanislaw Ulam, and lectured at prestigious universities. His appearances were irregular, owing to the fact that he had no formal arrangements with any of the schools he visited. He would come for a few months, receive payment for his work, and move on. He was known to fly to as many as fifteen places in one month—remarking that he was unaffected by jet lag. Because he never renounced his Hungarian citizenship, he was able to receive a small salary from the Hungarian Academy of Sciences.

## An Erdös Number Conveys Prestige

So esteemed was Erdös by his colleagues that they invented the term "Erdös number" to describe their close connections with him. For example, if someone had coauthored a paper with Erdös, they were said to have an Erdös number of one. If someone had worked with another who had worked with Erdös, their Erdös number was two, and so on. According to his obituary in the *New York Times*, 458 persons had an Erdös number of one; an additional 4,500 could claim an Erdös number of two. It is said that **Albert Einstein** had an Erdös number of two. Ronald Graham, director of information sciences at AT&T Laboratories, once said that research was done to determine the highest Erdös number, which was thought to be 12. As Graham recalled, "It's hard to get a large Erdös number, because you keep coming back to Erdös." This "claim to fame" exercise underscores Erdös' monumental publishing output of more than 1,500 papers, and is not only a tribute to his genius but also to his widespread mathematical network.

Throughout his career, Erdös sought out younger mathematicians, encouraging them to work on problems he had not solved. He created an awards system as an incentive, paying amounts from $10 to $3,000 for solutions. He also established prizes in Hungary and Israel to recognize outstanding young mathematicians. In 1983, Erdös was awarded the renowned Wolf Prize in Mathematics. Much of the $50,000 prize money he received endowed scholarships made in the name of his parents. He also helped to establish an endowed lectureship, called the Turán Memorial Lectureship, in Hungary.

## Perfect Proofs from God's "Great Book"

Erdös' mathematical interests were vast and varied, although his great love remained number theory. He was fascinated with solving problems that looked—but were not—deceptively simple. Difficult problems involving number relationships were Erdös' special forte. He was convinced that discovery, not invention, was the way to mathematical truth. He often spoke in jest of "God's Great Mathematics Book in the Sky," which contained the proofs to all mathematical problems. Hoffman in the *Atlantic Monthly* says "The strongest compliment Erdös can give to a colleague's work is to say, 'It's straight from the Book'."

### Mother's Death Brings on Depression

Erdös' mother was an important figure in his life. When she was 84 years old, she began traveling with him, even though she disliked traveling and did not speak English. When she died of complications from a bleeding ulcer in 1971, Erdös became extremely depressed and began taking amphetamines. This habit would continue for many years, and some of his extreme actions and his hyperactivity were attributed to his addiction. Graham and others worried about his habit and prevailed upon him to quit, apparently with little result. Even though Erdös would say, "there is plenty of time to rest in the grave," he often talked about death. In the eccentric and personal language he liked to use, God was known as S. F. (Supreme Fascist). His idea of the perfect death was to "fall over dead" during a lecture on mathematics.

Erdös' "perfect death" almost happened. He died of a heart attack in Warsaw, Poland, on September 20, 1996, while attending a mathematics meeting. As news of his death began to reach the world's mathematicians, the accolades began. Ronald Graham, who had assumed a primary role in looking after Erdös after his mother's death, said he received many electronic-mail messages from all over the world saying, "Tell me it isn't so." Erdös' colleagues considered him one of the 20th century's greatest mathematicians. Ulam remarked that it was said "You are not a real mathematician if you don't know Paul Erdös." Straus, who had worked with Einstein as well as Erdös, called him "the prince of problem solvers and the absolute monarch of problem posers," and compared him with the great 18th-century mathematician **Leonhard Euler**. And, Graham remarked, "He died with his boots on, in hand-to-hand combat with one more problem. It was the way he wanted to go."

## SELECTED WRITINGS BY ERDÖS

At his death, Paul Erdös had written more than 1,500 mathematical papers. He had collaborated with so many mathematicians still at work on problems that it was expected that an additional fifty to 100 papers would be published after his death. Erdös was once asked in the early 1980s if he had any particular mathematical works of which he was especially proud. He replied that two were the

Erdös-Kac and the Erdös-Wintner theorems. He also mentioned a theory on elementary geometry. Sources which discuss these theorems are listed below:

Kac, M. *Statistical Independence in Probability, Analysis and Number Theory.* Mathematical Association of America, 1959 (Carus Monograph #12).

Kubilius, J. *Probabilistic Methods in the Theory of Numbers.* American Mathematical Society, 1964 (Translation of Mathematical Monographs, #11).

## FURTHER READING

### Books

Albers, Donald J. and G. L. Alexanderson, editors. *Mathematical People, Profiles and Interviews.* Chicago: Contemporary Books, Inc., 1985.

Honsberger, R. "Stories in Combinatorial Geometry: A Theorem of Erdös." *Two-Year College Mathematics Journal* 10 (1979): 344–347.

Ulam, S. M. *Adventures of a Mathematician.* New York: Charles Scribner's Sons, 1976.

### Periodicals

Hoffman, Paul. "The Man Who Loves Only Numbers." *The Atlantic Monthly* (November 1987): 60–74.

Kolata, Gina Bari. "Mathematician Paul Erdös: Total Devotion to the Subject." *Science* (April 8, 1977): 144–145.

————. "Paul Erdös, 83, a Wayfarer in Math's Vanguard, Is Dead." *The New York Times.* September 24, 1996, A: 1:5 ; B: 8: 5.

### Other

"In Memoriam: Paul Erdös." February 11, 1997. http://www.cs.uchicago.edu/groups/theory/erdos.html (July 20, 1997). This web site includes hyperlinks, images, obituaries, and explications of Erdös' most important theories.

*—Sketch by Jane Stewart Cook*

# Agner Krarup Erlang
## 1878–1929
### Danish mathematician

Erlang is regarded as the founder of queuing theory and of operations research. His formulas, designed and published in 1917, enabled early telephone switching systems to become operational. These formulas give the probability that a user will encounter a busy signal instead of a dial tone, or the length of waiting time for a system that can hold calls. They are used to calculate the number of circuits needed to give a specified level of service. Erlang's formulas may be applied to any system with a limited number of servers and customers that arrive at random times.

Agner Krarup Erlang was born on January 1, 1878, at Lonborg, near Tarm in Jutland, the mainland of Denmark. His parents were Hans Nielson Erlang, the parish clerk and schoolmaster, and Magdalene Krarup Erlang. Many of his mother's family were clergymen. Erlang had an older brother, Frederik, and two younger sisters, Marie and Ingeborg. Erlang studied at his father's school, then was tutored at home by his father and the assistant school teacher, P.J. Pedersen, for his preliminary examination. Erlang passed with distinction, although at age 14 he needed special permission to take the examination. Erlang served as assistant teacher at his father's school for two years, then stayed for two years with M. Funch, in Hillerod, to prepare for his university examination at the Frederikborg Grammar–school. In 1896, Erlang passed this examination and attended the University of Copenhagen. He studied mathematics, astronomy, physics and chemistry, and completed an M.A. in 1901.

After graduation, Erlang taught at a number of schools. He joined a Christian students' association where he met his friend H.C. Nybolle, who later became a professor of statistics at Copenhagen University. Erlang also was a member of the Mathematics Association. In 1904, he won a distinction for his answer to the University mathematics prize question about **Christiaan Huygen**'s methods of solving infinitesimal problems. At the Mathematics Association, Erlang met J.L.W.V. Jensen of the Copenhagen Telephone Company. Jensen introduced Erlang to Fr. Johannsen, his managing director, who hired Erlang in 1908 as scientific collaborator and leader of the laboratory.

Johannsen had already published two essays on the barred access and waiting time problems inherent in telephone systems. He suggested that Erlang study these problems further. Erlang demonstrated in a 1909 paper that the number of calls to arrive during a period of time follows a Poisson distribution, and treated the problem of waiting time when holding times are constant, for the simplest case of one circuit. In 1917, Erlang's most important paper was published, in which he gives his B–formula for the probability of barred access, or a busy signal, for a group of circuits, and formulas for waiting time. His proof of the B–formula is based on the idea of statistical equilibrium, that transitions between pairs of states are in balance. In 1922 and 1926 Erlang published lectures on the contents of his earlier papers; the 1922 paper containing a new interconnection formula. In 1924 he wrote about a principal of K. Moe for deciding whether to add circuits to large or small groups of circuits. Erlang often presented his results as tables as well as formulas. He wrote several papers about producing accurate tables and published tables produced by his methods.

Erlang also wrote about cables, the induction coil in a telephone, and about a device for measuring transmission in cables. Also interested in more theoretical matters, Erlang used the idea of statistical equilibrium to prove Maxwell's Law in the kinetic theory of gases, and wrote short papers on geometry and on proportional representation in voting.

Erlang, who never married, lived with his sister Ingeborg in Copenhagen. He had a large collection of books on science and mathematics. Erlang's sister founded a home for mentally ill women, which he supported generously. Erlang died after a brief abdominal illness on February 3, 1929, at 51 years of age.

In addition to his widely used formulas, Erlang's name is attached to the gamma probability distribution. In 1946, the C.C.I.F. (La comite consultatif des communications telephoniques a grande distance) adopted the Erlang as the unit of telephone traffic. The average traffic in Erlangs is the sum of the lengths of calls originating during an interval of time, divided by the length of the time interval. ERLANG is also the name of a programming language which was developed at Ericsson and Ellemtel Computer Science Laboratories for programming telecommunications switching systems.

## SELECTED WRITINGS BY ERLANG

"Principal Works of A. K. Erlang." *Transactions of the Danish Academy of Technical Sciences* 2 (1948): 131–267.

## FURTHER READING

### Books

*The Biographical Dictionary of Scientists.* Second Edition. New York: Oxford University Press, 1994.
*Dansk Biografisk Leksikon.* Kobenhavn: Gyldendal, 1980.
James, Robert C. *Mathematics Dictionary.* Fifth Edition. New York: Van Nostrand Reinhold, 1992.
Mathematical Society of Japan. *Encyclopedic Dictionary of Mathematics.* Second Edition. Edited by Kiyosi Ito. Cambridge, MA: MIT Press, 1987.

*Euclid of Alexandria*

**Periodicals**

Brockmeyer, E. and H. L. Halstrom. "The Life of A. K. Erlang." *Transactions of the Danish Academy of Technical Sciences* 2 (1948): 9–24.

Brockmeyer, E. "A Survey of A. K. Erlang's Mathematical Works." *Transactions of the Danish Academy of Technical Sciences* 2 (1948): 101–126.

Halstrom, H. L. "A Survey of A. K. Erlang's Electrotechnical Works." *Transactions of the Danish Academy of Technical Sciences* 2 (1948): 127–30.

*—Sketch by Sally M. Moite*

# Euclid of Alexandria
## c. 300 B.C.
### Greek geometer

Euclid was a preeminent Greek mathematician who is often to referred to as the founder of geometry. His historical stature as one of the most influential mathematicians of all time stems from his authorship of the *Elements*, a textbook of elementary geometry and logic. The accumulated knowledge set forth in the *Elements* was so exhaustive that the book has been a standard text on geometry throughout the centuries.

Little is known about Euclid's life. Most of what can be inferred concerning Euclid stems from a brief summary by Proclus, who wrote a commentary on the *Elements*. It is generally agreed that Euclid lived about 300 B.C. Although ancient Arab authors claim that Euclid was Greek and born in Tyre, most historians are certain that Euclid was either Greek or Egyptian. He is believed to have been a student at Plato's Academy in Athens, where most of the accomplished geometers of Euclid's time studied. Little is known of Euclid's personality or beliefs other than he was completely devoted to the study of mathematics and was considered a wise, patient, and kind teacher. His acknowledgment that he derives most of the knowledge in the *Elements* from his predecessors reveals Euclid's sense of fair play and respect for the mathematicians who came before him.

Perhaps because of civic unrest or at the request of Ptolemy I, Euclid eventually traveled to Alexandria, Egypt, around 322 B.C. and established a famous school of mathematics. Built by Alexander the Great between two arms of the Nile River, Alexandria became an intellectual center of education and learning unparalleled in the Hellenistic Age. Not long after Euclid established his school, Ptolemy founded a museum that became the first national university. It included the most comprehensive library in the ancient world, housing more than 600,000 "books," actually papyrus rolls. Euclid was the museum's first teacher of mathematics.

When it came to mathematics, however, Euclid was unyielding in his insistence on the careful study of the discipline and its worth. According to one story, Ptolemy was observing one of Euclid's geometry classes and, afterwards, asked Euclid if there was a "shorter road to its mastery." Euclid replied, "There is no royal road to geometry." Another story relates the tale of a student who, upon learning a new proposition asked his master ". . . what will I get by learning these difficult things?" Euclid called for his slave and said, "Give this man an *obol* [a Greek coin], for he must make gain from what he learns."

### The *Elements*

The genius and monumental impact of Euclid's *Elements* is unquestioned. Not only was it immediately recognized as the definitive book on geometry in Euclid's own time, but it has also been a standard mathematical text for more than 2,000 years, forming the foundation of all geometry taught until the beginning of the 20th century. To write the *Elements*, Euclid drew upon three centuries of mathematical thought, including the work of **Pythagoras, Hippocrates,** and **Menaechmus**. As the book's reputation grew and handwritten copies spread throughout the ancient world, the *Elements* stimulated mathematical thought and discovery in ancient Greece, Egypt, Persia, Arabia, and India. The first printed edition appeared in 1482, and the first complete English edition in 1570.

Euclid himself is attributed with only developing a few of the theorems in the *Elements*. Composed of thirteen books, the genius of the *Elements* stems from its systematic presentation of the basic principles of geometry and the accompanying statements and theorem proofs. With astounding clarity, Euclid states his theorems and problems and then logically proceeds to present rigorous propositions.

Containing more than 450 propositions, the 13 books of the *Elements* begins with definitions, postulates, and axioms. In Book I, Euclid sets forth the definitions of points, lines, planes, angles, circles, triangles, quadrilaterals, and parallel lines. Book II focuses primarily on rectangles and squares and contains many problems now treated algebraically. Book II considers the geometry of the circle, including the relationship between circles that intersect or touch. Book IV continues with Euclid's examination of the circle and of polygons. Book V sets forth a theory of proportion and areas, and Book VI applies this theory to plane geometry. Books VII, VIII, and IX focus on arithmetic, irrational numbers, and the theory of rational numbers. Book XI, XII, and XIII deals with three dimensional, or solid, geometry.

In addition to containing the accumulated knowledge in mathematics up to Euclid's time, these 13 books include some original analyses and proofs. For example, Euclid developed a new proof for the Pythagorean theorem and, in the process, proved the existence of irrational numbers. Euclid also devised an ingenious proof showing that the number of primes is infinite. His development of the methods of exhaustion for measuring areas and volumes were later used by **Archimedes**.

### The Fifth Postulate

One of the most important contributions made by Euclid in the *Elements* is the five postulates, which embody the distinctive principles of Euclidean geometry. Postulates one through three focus on construction concerning the straight line and the circle. Postulate four states that all right angles are equal. However, it is the Fifth Postulate, which Euclid most certainly invented, that stands out as evidence of both Euclid's mathematical genius and the *Elements* greatest stumbling block.

The Fifth Postulate, also known as the "parallel postulate," has remained unchanged for 23 centuries. It states: If a straight line falling on two straight lines make the interior angles on the same side less than two right angles, the two straight lines, if extended

indefinitely, meet on that side on which the angles less than two right angles are. It is the Fifth Postulate that lays the foundation for the theory of parallel lines, including the theory that only one parallel to a line can be drawn through any point external to the line.

Ironically, Euclid and the mathematicians who followed him recognized that this postulate had no proof from the other four postulates. As a result, Euclid himself avoided using it as much as possible. In the following centuries, several mathematicians developed alternatives to the Fifth Postulate. One of the most famous is the axiom developed by Proclus stating: "Through a given point in a plane, one and only one line can be drawn parallel to a given line." This postulate became known as Playfair's Axiom in 1795, when John Playfair proposed that it replace the Fifth Postulate.

Numerous other attempts have been made to prove the Fifth Postulate, primarily from the other four postulates. While some of these proofs were accepted for periods of time, they eventually were found to contain mistakes and be untrue. However, like Playfair's Axiom, some of the substitutes have merit.

Ironically, the problems associated with the Fifth Postulate led to the creation of non–Euclidean geometry, which is based on the assumption that the Fifth Postulate is not true. First publicly proposed by **Nikolai I. Lobachevsky** in 1826 and then by **Eugenio Beltrami** in 1868, non–Euclidean geometry has contributed greatly to the study of physics and relativity.

### Euclid's Other Works

The *Elements* was so comprehensive and successful that Euclid's other works have been nearly forgotten. However, a few of these works have survived the ages. *Data*, which is most closely aligned with the *Elements*, deals with plane geometry and contains 94 propositions. An Arabic translation of Euclid's work in pure geometry, called *On Division*, also survives and was published in 1851. Euclid's other surviving works belong to applied mathematics, including the *Phenomena*, which focuses on spherical geometry. *Optics*, an elementary treatise on perspectives, may have been used to warn against paradoxical astronomical theories like the belief that the stars and other heavenly bodies were actually the size that they appear to be to the human eye.

Unfortunately, several of Euclid's works have been lost. The *Pseudaria*, or as Proclus called it, "The Book of Fallacies," presented fallacies in some of the elemental geometry of Euclid's day side by side with valid theorems. The loss of *Porisms* is believed to be the greatest loss of all Euclid's books. Focusing on higher geometry, the *Porisms* was reported to be a comprehensive and difficult work containing types of propositions intermediate between theorems and problems. Another work, the *Conics*, was reported to consist of four books, with **Apollonius** adding another four books. The *Surface-Loci* was a treatise in two books that focused on the relationships of points on surfaces.

Many works have also been falsely attributed to Euclid, including two books added to the *Elements*, the so-called Books XIV and XV. Others include the *Elements of Music*, the *Sectio Canonis*, and the *Introduction to Harmony*. Ancient Arab scholars also attributed several works on mechanics to Euclid, including the "Book of Heavy and Light." However, there is little corroborating evidence that Euclid ever wrote on mechanics.

As the author of the most successful textbook ever written, Euclid remains the most influential figure in the history of geometry and, perhaps, all mathematics. Euclid's impact on science included influencing such luminaries as **Galileo** and Sir **Isaac Newton**. **Eratosthenes** also used Euclid's theorems to make accurate measurements of the Earth's circumference, confirming its shape as spheroid, or round. From the very beginning, the *Elements* inspired Hellenistic mathematicians to write numerous commentaries, both praising it and pointing out mistakes and inconsistencies. Still, Euclid achieved the status of hero or god in his own time.

## FURTHER READING

Delacey, Estelle Allen. *Euclid and Geometry*. New York: Franklin Watts, 1963.

Heath, Sir Thomas. *A History of Greek Mathematics*. Volume I. Oxford: Clarendon Press, 1921, pp. 352–446

Magill, Frank N., editor. *The Great Scientists*. Volume 4. Danbury, CT: Grolier Educational Corporation, 1989, pp. 142–146.

Smith, David Eugene. *History of Mathematics*. Volume I. New York: Dover Publications, 1958, pp. 102–107.

Terry, Leon. *The Mathmen*. New York: McGraw-Hill Book Company, 1964, pp. 147–162.

*—Sketch by David A. Petechuk*

# Leonhard Euler
## 1707–1783
### Swiss geometer and number theorist

Leonhard Euler advanced every known field of mathematics in his day. A prolific author, among his

*Leonhard Euler*

greatest writings are treatises on analytic geometry, differential and integral calculus, and the calculus of variations. Euler developed spherical trigonometry, demonstrated the importance of convergence in algebraic series, proved important assertions in number theory, and made contributions to hydrodynamics, celestial mechanics, and optics. Euler also brought into common usage such mathematical notations "e" for the base of the natural logarithm, "i" for the square root of negative 1, and $f(x)$ for a function of x. A variety of mathematical concepts bear his name, including Euler's characteristic in topology, Euler's triangle in geometry, Euler's polynomials, Euler's integrals, and Euler's constant. His accomplishments are especially remarkable in that many were made during the last quarter of his life, when he was totally blind.

Euler was born in Basel, Switzerland, on April 17, 1707 to Paul Euler and Marguerite Brucker. In 1708, his family moved to the nearby village of Reichen, where his father, a Calvinist pastor, had taken a parish. Before joining the clergy, Euler's father had studied mathematics under the tutelage of **Jacob Bernoulli**. Following in his father's footsteps, Euler also took his formal education in religion and mathematics, studying theology and Hebrew at the University of Basel, and taking weekly mathematics lessons from **Johann Bernoulli**, Jacob Bernoulli's younger brother. The Bernoullis recognized Euler's talent and when he received his master's degree from the University of Basel at age 17, they advised him to pursue a mathematical career. The advice was met with resistance from Euler's father, who wished for his son to inherit the pastorship in Reichen. Euler was to remain a devout Calvinist throughout his life, but the Bernoullis eventually convinced his father that Euler's true destiny was not with the church.

When Euler was 19 years old, he produced his first mathematical work, entering a contest sponsored by the French Académie Royale des Sciences. The object of the contest was to solve a problem related to the optimum placement of masts on sailing ships. Euler received an honorable mention for his effort, his solution suffering primarily in the area of practicality. Having not yet traveled outside of Switzerland, he had never seen a ship. Over the course of his career, Euler would eventually receive a total of twelve prizes from the French Académie for his mathematical solutions.

## Inherits Top Mathematical Position at St. Petersburg Academy

Around the time that Euler was attempting to solve the ship mast problem, he was also trying, unsuccessfully, to obtain a post as a professor of mathematics at the University of Basel. Determined to hold an academic position, he corresponded with friends **Daniel Bernoulli** and his cousin, Nicolaus, who were members of the newly established St. Petersburg Academy of Sciences. They wrote to him about a post available in the medical section of the Academy, and, hoping to qualify, Euler immediately began studying physiology. Within three months he was considered sufficiently prepared for the medical post, and in 1727 he traveled to St. Petersburg to join the Academy. Euler's arrival in Russia coincided with the death of Catherine, the wife of Peter the Great. A period of political oppression ensued, lasting several decades, and in the initial turmoil Euler slipped quietly into the Academy's mathematical section. Academic and political freedoms eventually became so stifled that, in 1733, Daniel Bernoulli decided to leave Russia and return to Switzerland. Euler, then 26 years old, inherited Bernoulli's post, the top mathematical position in St. Petersburg. Two years later, he lost the vision in his right eye. According to some historians, Euler developed an eye infection while solving an astronomical problem that had been put forth by the French Académie. It is possible that he injured his eye by staring into the sun while working on the problem. Euler derived a solution in the course of only three days, and won the Académie's contest.

In 1736, Euler wrote a paper on the solution of the Königsburg Bridge Problem, a puzzle concerning attempts to cross seven different bridges in one journey. This work led to the development of the modern field of graph theory. Between 1736 and 1737, Euler wrote *Mechanica*, in which he demon-strated that mathematical analysis could be applied to Newtonian dynamics. This treatise and the wealth of articles he had already published secured his mathematical prominence. By the end of the 1730s, he had also established a reputation as a gifted educator, having written both elementary and advanced mathematical textbooks for the Russian schools. As a member of the Russian Academy, Euler was called upon to solve many practical problems for the benefit of the Russian government. He created a test for determining the accuracy of scales, developed a system of weights and measures, and supervised the government's department of geography. Although political oppression in Russia continued, Euler was never restricted in the pursuit of his own mathematical interests. However, he was growing increasingly weary of the injustices that surrounded him. In 1740, Euler accepted an invitation from Frederick the Great, the Prussian king, to join the Berlin Academy. He left Russia on sufficiently good terms, however, that throughout his tenure in Berlin the St. Petersburg Academy provided part of his salary. He was to remain at his Berlin post for the next 24 years.

## Develops the Calculus of Variations

Frederick the Great, while lacking in his own mathematical ability, did appreciate the utility of mathematics. He directed Euler to work on calculations related to diverse practical matters, including pension plans, navigation, water supply systems, and the national coinage. In Berlin, Euler accomplished what many consider his most important work. He wrote *Methodus inveniendi lineas curvas maximi minimive proprietate gaudentes*, on the calculus of variations in 1744, and its publication led, in 1746, to his election as a Fellow of the Royal Society of London. This masterpiece was followed by several texts on calculus that were to become instant classics. These included *Introductio in analysin infinitorum*, in 1748, and *Institutiones calculi differentialis*, in 1755.

While in Berlin, Euler corresponded with many of his mathematical contemporaries, including **Johann Lagrange** and **Jean d'Alembert**. He was introduced to the field of number theory by Christian Goldbach, who presented him with the various challenges of **Pierre de Fermat**. Euler proved many assertions in the field of number theory and was the first mathematician to make serious progress in solving Fermat's Last Theorem. In a letter to Goldbach in 1753 he described a partial proof, ultimately shown to contain a fallacy, but which laid the foundation for its eventual solution.

## Despite Blindness, Enters Most Prolific Period

In 1766, at the age of 59, Euler returned to Russia. He had fallen into gradual disfavor in King Frederick's court because of the positions he took in metaphysical arguments with contemporaries such as Voltaire. The king eventually concluded that Euler was unsophisticated, and took to calling him a "mathematical Cyclops," in reference to his partial loss of vision. When Euler went back to St. Petersburg, he was greeted with much greater esteem. Catherine the Great, now in power, provided him with a large estate and one of her personal cooks.

Not long after resettling in Russia, Euler developed a cataract in his left eye and totally lost his vision. He nonetheless entered one of the most prolific periods of his career. Nearly half of the 886 books and manuscripts Euler wrote were composed during this second tenure at the St. Petersburg Academy. From 1768 to 1770, he drafted a classical treatise on integral calculus, *Institutiones calculi integralis*. He went on to tackle the lunar theory problem, researching the phases of the moon and the tidal fluctuations on Earth. His calculations relating to the gravitational interactions among the moon, sun, and Earth won him a 300 pound prize from the British government.

Euler was known for his remarkable memory. As a boy, he had memorized the entire text of Virgil's *Aeneid*, and 50 years later could still recite it. His ability to perform complex calculations in his head was also renown. Once, when two of his students disagreed on the answer to a problem that required they sum a complicated convergent series to 17 terms, Euler settled the matter using only mental arithmetic. His memory and mental calculation skills undoubtedly allowed him to cope with the blindness during the latter part of his life.

Euler was married in 1733 to Catharina Gsell, the daughter of the Swiss painter Gsell, that Peter the Great had brought to Russia. They had 13 children, of whom only three sons and two daughters survived beyond their early years. Catharina died in 1776, and a year later Euler married her aunt and half–sister, Salome Abigail Gsell. He was known as a kind and generous man. Euler was especially fond of children, often writing mathematical treatises with a child on his lap. On September 18, 1783, while playing with his grandson, he suffered a stroke and died. Just before his death he had calculated the orbit of the newly discovered planet Uranus.

## SELECTED WRITINGS BY EULER

*Letters of Euler on Different Subjects in Natural Philosophy Addressed to A German Princess. With Notes and A Life of Euler*, by David Brewster. Volumes I & II, 1833–37

## FURTHER READING

### Books

Bell, E.T. *Men of Mathematics.* New York: Simon and Schuster, 1986, pp. 139–152.

Dunham, W. *The Mathematical Universe: An Alphabetical Journey through the Great Proofs, Problems, and Personalities.* New York: John Wiley and Sons, Inc. 1994.

### Periodicals

Calinger, R. "Leonhard Euler: The First St. Petersburg Years (1727–1741)." *Historia Mathematica* 23 (1996): 121–166.

*—Sketch by Leslie Reinherz*

*Etta Zuber Falconer*

# Etta Zuber Falconer
## 1933–

### African–American algebraist and educator

Etta Zuber Falconer has encouraged hundreds of young people, particularly African–American women, to study mathematics and the sciences through her classroom teaching and program work at Spelman College in Atlanta, where, she told Fran Hodgkins in an interview, she "was able to crystallize my desire to change the prevailing pattern of limited access and limited success for African American women in mathematics." Falconer received the **Louise Hay** Award for her contributions to mathematics education in 1995.

Falconer was born in Tupelo, Mississippi, in 1933, the younger of two daughters of Dr. Walter A. Zuber and Zadie L. (Montgomery) Zuber. She attended Tupelo public schools and graduated from George Washington High School in 1949. At Fisk University in Nashville, she found two of her three life mentors: Dr. **Evelyn Boyd Granville** and Dr. Lee Lorch. Granville taught just one year at Fisk. For Falconer,

seeing an African American woman teaching at the college level was inspiring; most instructors were men. She had intended to teach high–school mathematics after graduation, but Lorch, who served as chair of the mathematics department, encouraged her to go on to graduate school. Falconer graduated *summa cum laude* with a bachelor of science degree in mathematics in 1953, and went on to earn her master's degree in mathematics from the University of Wisconsin in 1954.

From 1954 to 1963, she taught mathematics at Okolona Junior College in Okolona, Mississippi. While teaching there, she met and married her husband, the late Dolan Falconer. They have three children, Dolan Jr., an engineer; Alice (Falconer) Wilson, a physician; and Walter, also a physician.

In 1963, Falconer left Okolona to teach at Chattanooga Public School. Two years later, she joined the faculty of Spelman College in Atlanta as an instructor. In 1969, she earned her Ph.D. in mathematics from Emory University, where she studied under her third mentor, Trevor Evans, who encouraged her growth in algebra and her study of quasigroups and loops. Her dissertation was titled "Quasigroup Identities Invariant Under Isotopy." Out of her dissertation came two published papers, "Isotopy Invariants in Quasigroups" and "Isotopes of Some Special Quasigroup Varieties." After receiving her doctorate, Falconer held an associate professorship at Norfolk State College from 1971 to 1972. She received a master's degree in computer science from Atlanta University in 1982, and also attended the National Science Foundation Teacher Training Institute at the University of Illinois from 1962 to 1965.

Falconer has spent most of her professional career at Spelman College, her mother's alma mater. She has held the positions of instructor/associate professor (1965–71), professor of mathematics and chair of the mathematics department (1972–82), chair of the division of natural sciences (1982–90), director of science programs and policy (1990), and associate provost for science programs and policy (1991–present). Falconer has been Spelman's Fuller E. Callaway Professor of Mathematics since 1990.

Falconer has devoted her career to encouraging African American students, particularly women, to study mathematics and the sciences. She is the director of Spelman College's NASA Women in Science and Engineering Scholars program (WISE), which fosters promising women students and encourages them to continue into graduate school. Among students' fields of study are applied mathematics, chemistry, and industrial engineering. Approximately 150 women have taken part in the program since its inception in 1987, and the program will soon celebrate the first of its alumnae receiving her Ph.D. Falconer also coordinates the university's NASA Undergraduate Scholar Awards program, which, like WISE, allows undergraduate students to conduct research at NASA facilities. In addition, she is one of the founders of the National Association of Mathematicians, which promotes the concerns of African American students and mathematicians, and the Atlanta Minority Women in Science Network.

Falconer has received many awards in recognition of her work on behalf of the next generation of scientists. In addition to the Louise Hay Award from the Association for Women in Mathematics (AWM), she also received the Giants in Science Award from the Quality Education for Minorities Network (1995). Her other honors include: Spelman College Presidential Faculty Award for Distinguished Service (1994); Spelman College Presidential Award for Excellence in Teaching (1988); United Negro College Fund Distinguished Faculty Award (1986–87); Achievement and Service Award, presented by the Atlanta Minority Women in Science Network and the Auxiliary to the Atlanta Medical Association; and the National Association of Mathematicians' Distinguished Service Award (1994). Falconer also received an honorary doctor of science degree from the University of Wisconsin at Madison in 1996. She is a member of Phi Beta Kappa, Pi Mu Epsilon (honorary mathematics fraternity), and Beta Kappa Chi (honorary scientific society). In addition, she has served in a variety of roles in the following organizations: the American Association for the Advancement of Science, the American Mathematical Society, the AWM, the Mathematical Association of America, the National Association of Mathematicians, and the National Science Foundation.

## SELECTED WRITINGS BY FALCONER

"Isotopy Invariants in Quasigroups." *Transactions of the American Mathematical Society* 151 (1970): 511–526.
"Isotopes of Some Special Quasigroup Varieties." *Acta Mathematica* 22 (1971): 73–79.
"Women in Science at Spelman College." *Signs* 4 (1978): 176–177.
"A Story of Success: The Sciences at Spelman College." *SAGE* 6 (1989): 36–38.
"Views of an African American Woman on Mathematics Meetings." *A Century of Mathematics Meetings.* Edited by Bettye Anne Case. American Mathematical Society, 1996.

*Sister Mary Celine Fasenmyer*

## FURTHER READING

Bailey, Lakiea. "Etta Falconer." *Biographies of Women Mathematicians.* June 1997. http:/www/scottlan.edu/lriddle/women/chronol.htm (July 21, 1997).
Falconer, Etta Zuber, interview with Fran Hodgkins conducted May 1, 1997.

*—Sketch by Fran Hodgkins*

# Mary Celine Fasenmyer
## 1906–1996
### American computer analyst and educator

Sister Fasenmyer spent most of her professional life teaching at various schools in Pennsylvania, but even her doctoral research was already setting the stage for the brave new world of computer-driven mathematical proofs.

Fasenmyer was born in central Pennsylvania on October 4, 1906. Her parents, George and Cecilia, lived in Crown, where George had an oil lease he tended as a business. Cecilia died only a year after Fasenmyer was born, but George remarried a few years later to a much younger woman named Josephine. Fasenmyer's earliest formal education took place less than 30 miles from her hometown, in Titusville, at a school called St. Joseph's Academy. She graduated at age 17 and began teaching.

### The Community

Fasenmyer was strongly impressed by her Catholic education, as she wound up a ten–year phase of her life teaching her strongest subject by gaining an A.B. degree from Mercyhurst, a Catholic college in Erie, Pennsylanvia. She was already pledged to the Sisters of Mercy there by 1933, as she soon embarked to teach at St. Justin's, a high school in Pittsburgh, at the order's behest. Her community of sisters later decided to send Fasenmyer to the University of Pittsburgh for an M.A. This she earned in 1937 with a major in mathematics and a minor in physics. A change of pace came between 1942 and 1946, during which Fasenmyer studied under Earl Rainville at the University of Michigan.

Rainville directed Fasenmyer's doctoral thesis on the subject of algorithmic deduction. Algorithms have been studied since medieval times, but here they were applied to more novel mathematical entities known as hypergeometric polynomial sequences. Thanks to a postdoctoral paper published in the *American Mathematical Monthly*, Fasenmyer's refined methodology reached a wide audience of experimenters. The burgeoning field of computer technology was already being drawn upon to aid the superhuman tasks of high mathematics computations.

### Discrete Math

Even today, the new science of proving hypergeometric identities via computer programming is influenced by Fasenmyer's work. Herbert Wilf, Marko Petkovsek, and Doron Zeilberger acknowledge her influence on their project in their 1996 book *A=B*. Wilf, of the University of Pennsylvania, was responsible for posting the only testimonial on the Internet devoted to Fasenmyer after his book's publication. Wilf only found out about Fasenmyer's sudden death on December 27, 1996, just after New Year's, 1997. Fasenmyer's colleagues at Mercyhurst College, where Fasenmyer held a full professorship for decades, had been going through her papers when they discovered a memento from a Discrete Math convention in Florida in 1994. Sister M. Eustace Taylor was moved to write the three authors, to thank them for continuing to apply Fasenmyer's methods in their work.

Fasenmyer was not highly celebrated in her day, though she clearly molded generations of students and engaged her peers through the Mathematical Association of America. With the world of computers still growing in terms of capacity and ability, however, her reputation may well grow along with it.

### SELECTED WRITINGS BY FASENMYER

Fasenmyer, Sister Mary Celine. "On Recurrence Relations," *American Mathematical Monthly* 56 (1949): 14.

### FURTHER READING

*American Men & Women of Science.* Thirteenth edition. New York: R.R. Bowker, Co., 1976.

**Other**

Wilf, Herbert. "Remarks on the life of Sister Mary Celine Fasenmyer." http://www.cis.upenn.edu/~wilf/celine (July 1997).

*—Sketch by Jennifer Kramer*

# Käte Fenchel
## 1905–1983
### German–born Danish algebraist

Without benefit of more than an undergraduate education in mathematics, Käte Fenchel published four treatises on pure mathematics in her lifetime. Suffering through poverty compounded by persecution because of her Jewish heritage, Fenchel maintained her passion for mathematics by studying and researching on her own. Her primary interests lie in finite nonabelian groups and algebra. Fenchel's accomplishments leave a profound legacy of perseverance for women mathematicians as well as Danish mathematics.

Fenchel was born Käte Sperling in Berlin, Germany, on December 21, 1905, to Otto and Rusza Sperling. Fenchel's father worked in publishing and her mother was a bookkeeper. Her parents separated when she was young, and she and her older sister grew up in semi–poverty. Though she grew up in less fortunate circumstances, Fenchel won scholarships that allowed her to attended a private girls school. In high school Fenchel realized she wanted to be a mathematician.

From 1924 to 1928, Fenchel received the rest of her formal education at the University of Berlin's Mathematical Institute. In addition to her primary area of interest, pure mathematics, she studied philosophy and physics. Although Fenchel could have written a thesis, and perhaps gone on to earn a doctorate, her economic circumstances combined with an astute perception that she probably would not be employed in research because of her gender dissuaded her from further study of this type.

Instead, Fenchel did the course work required to become a high school mathematics teacher. She taught for two years (1931–1933) before she was dismissed because of the growing Nazi oppression of Jews. After tutoring students privately for a short time, she and her fiance, fellow mathematician and Jew Werner Fenchel, escaped to Denmark. The couple married in Denmark in December 1933.

Although Fenchel's formal education had long ceased, she kept working on her algebra research on her own time. In Denmark, she worked as a secretary for a Danish mathematics professor from 1933 to 1943. In 1937, Fenchel published her first mathematical paper (on vectormodules) in a mathematics journal. Three years later, she gave birth to her only child, a son, Tom. As the Germans closed in on Denmark during World War II, Fenchel's employment and her mathematical activities temporarily ceased when the Fenchel family eluded Nazi persecution by moving to Sweden. The family moved back to Denmark in 1945 after the war ended.

Fenchel did not publish on mathematics again until 1962. In that year, she published two papers. One examined odd order groups and group decomposition; the other focused on structure matrix and group representation theory. Fenchel returned to teaching soon thereafter, and served as a lecturer at Aarhus University in Denmark, from 1965 to 1970. She published her last paper in 1978 on Frobenius's theorem and group theory, five years before her death during the night of December 18–19, 1983.

## SELECTED WRITINGS BY FENCHEL

"An Everywhere Dense Vectormodule with Discrete One–Dimensional Submodules." *Matematisk Tidsskrift* (1937): 94–96.
"Eine Bemerkung über Gruppen ungerader Ordnung." *Mathematica Scandinavica* (1962): 182–88.
"Beziehungen zwischen der Struktur einer endlichen Gruppe und einer speziellen Darstellung." *Monatshefte für Mathematik* (1962): 397–409.
"On a Theorem of Frobenius." *Mathematica Scandinavica* (1978): 243–50.

## FURTHER READING

### Books

Høyrup, Else. "Käte Fenchel," in *Women of Mathematics: A Bibliographic Sourcebook.* Edited by Louise S. Grinstein and Paul J. Campbell. New York: Greenwood Press, 1987, pp. 30–32.

*Pierre de Fermat*

### Other

"Käte Fenchel." *Biographies of Women Mathematicians.* June 1997. http://www.scottland.edu/lriddle/women/chrono.htm (July 22, 1997).

—*Sketch by Annette Petruso*

# Pierre de Fermat
## 1601–1665
### French number theorist

Pierre de Fermat, a lawyer and jurist by profession, made major contributions to every field of mathematics that existed in the seventeenth century. He developed the principles of analytic geometry independently from his contemporary **René Descartes**. Fermat is regarded by some mathematicians as the inventor of differential calculus. He created modern number theory, and, together with **Blaise Pascal**, developed the theory of probability. Fermat's wide range of accomplishments is especially remarkable considering that mathematics was for him only a hobby.

Pierre Fermat was baptized and most likely born on August 20, 1601, in Beaumont–de–Lomagne. His father, Dominique Fermat, was bourgeois second consul of Beaumont-de-Lomagne and a prosperous leather merchant. His mother, Claire de Long, came from a prominent family of parliamentary lawyers. Fermat had three siblings, a brother, Clement, and two sisters, Louise and Marie. His early education was most likely received at the Franciscan monastery in Beaumont. During the 1620s he attended the University of Toulouse, then moved for a few years to Bordeaux, where he undertook an informal study of mathematics and made contact with students of **François Viète**, the noted French algebraist. Fermat then entered the University of Orleans, where he received a Bachelor of Civil Law. In 1631 he began practicing law, purchasing positions as councilor of the Parliament of Toulouse and Commissioner of Requests. It was then that he added "de" to his name, an indication of his social standing afforded by his offices. In the same year Fermat married his fourth cousin, Louise de Long. Together they had five children, Clement-Samuel, Jean, Claire, Catherine and Louise. In Fermat's profession as a jurist, he rose gradually through the ranks, being named to the criminal court in 1638 and promoted in 1648 to a King's councilorship. He retained the latter post until his death in 1665.

### Develops Analytical Geometry and Stirs Controversy in Parisian Mathematical Circles

Within a few years of taking his first parliamentary post, Fermat began corresponding with several prominent Parisian mathematicians, including **Marin Mersenne**, Gille Personne de Roberval, and Étienne Pascal, father of probability theorist Blaise Pascal. In his first communications he described his work on geostatics. He proposed that **Galileo** had been incorrect in stating that a freely falling cannonball should follow a semicircular path. In the course of proving the path would be spiral, Fermat developed a new method of quadrature for curves. He also posed several analytical problems to the Parisian group, which he claimed to have already solved on his own. Roberval and Mersenne found those and subsequent problems difficult and eventually requested that Fermat describe the techniques by which he had derived their solution. In response, Fermat sent a paper called *Method for Determining Maxima and Minima and Tangents to Curves Lines*, along with his restoration of the Greek mathematician **Apollonius**'s *Plane Loci*. These papers, received in Paris in 1636, set out the fundamentals of differential calculus. In 1637 Fermat wrote a manuscript which the Parisian mathematicians circulated called *An Introduction to Plane and Solid Loci*. At about the same time René Descartes had sent Mersenne the galley proofs of his *Discourse on the Method* and accompanying *Essays*. The group in Paris quickly realized that Fermat and Descartes had independently developed the principles of analytic geometry, deriving the same basic technique for

treating geometric locus problems algebraically. While Fermat's treatise arrived first in Paris, Descartes had laid his foundation earlier. A bitter and protracted argument arose between the two men, ignited by issues of priority, but eventually focusing on the subject of maxima and minima. The dispute engaged most of the mathematicians in Paris and forced Fermat to prove the generality of his methods. Descartes finally admitted defeat regarding the mathematical controversy, but Fermat's reputation was damaged. While he had established himself as one of the best mathematicians in Europe, he was resented by some of his colleagues for communicating his work in piecemeal and sending problems in the form of challenges. He was even accused of posing problems that had no solutions, in order to expose his rivals.

### Fails to Arouse Interest in Number Theory

In the 1640s Fermat began his most important work, the development of modern number theory. Civic duties, however, prevented him from any meaningful mathematical correspondence between 1643 and 1654. In 1648 Fermat was occupied with the Fronde, a civil war, and in 1649, the Spanish raid on Languedoc. In 1651 the plague struck Toulouse, and Fermat became so ill that in 1653 his death was mistakenly reported. During a decade of isolation from the mathematical community he focused attention on developing a method of determining whether numbers were prime, and if not, on finding their divisors. The culmination of this work is today known as Fermat's Theorem. It states that if $n$ is any whole number and $p$ any prime, then $n^p - n$ is divisible by $p$. Fermat then went on to explore the concept of decomposition of primes of various forms into sums of their squares.

In the spring of 1654, Fermat received a letter from Blaise Pascal, asking for advice on a problem involving the consecutive throws of a die. The question to be resolved was how to divide the stakes in a dice game between two players, when their game was prematurely interrupted. The exchange that ensued between Pascal and Fermat over the course of just a few months helped lay the foundations of probability theory. By August 1654 Fermat was trying to divert Pascal's attention to his work on the theory of numbers. But Pascal, like most of his French contemporaries, saw little importance in this topic. In 1656 Fermat began to correspond about number theory with the Dutch physicist and astronomer **Christiaan Huygens**, expositor of the wave theory of light and inventor of the pendulum clock. Fermat described to him his method of infinite descent or "reduction analysis," in which larger problems are broken down into groups of problems more readily solvable. Though Huygens admired the seminal contributions of Fermat to calculus and analytical geometry, he believed Fermat's number theory had no practical application and that Fermat had become out of touch with important mathematical questions. Fermat tried to engage the two English mathematicians **John Wallis** and **William Brouncker** in discussion of his new theories, also to little avail. Finally, between 1658 and 1662 Fermat turned to the topic of optics. Using the method of maxima and minima, he investigated the laws of reflection and refraction. In the course of his work he discovered that light travels by the path of least duration, a concept now known as Fermat's Principle.

After Fermat's bout with the plague of 1651 he suffered from frequent illnesses, and in 1662 he ended all scientific and mathematical correspondence. He died on January 12, 1665, and was buried in the Church of St. Dominique in Castres.

### Leaves Legacy with Last Theorem

During his lifetime Fermat had frequently been asked by his mathematical contemporaries to publish descriptions of his work. He refused, expressing in his correspondence the desire to remain anonymous. His colleagues had resorted to including descriptions of Fermat's theories in their own writings, and circulating Fermat's letters and manuscripts by hand. Fermat's eldest son Clement-Samuel, published much of Fermat's work posthumously. In the process of collating the work, he examined his father's copy of the Latin translation of Diophantus of Alexandria's Arithmetic. In a page margin Fermat had written a theorem accompanied by a brief note which read, "I have discovered a truly remarkable proof which this margin is too small to contain." Known as Fermat's Last Theorem, it stated that for the equation $x^n + y^n = z^n$ there are no positive integer solutions for $x$, $y$, and $z$, when $n$ is greater than two. The proof of the theorem was to become one of the most famous mathematical challenges of the last three centuries. In 1994, an English-born American mathematician, **Andrew J. Wiles**, was credited with providing the first acceptable proof. The complexity of the problem was such that some historians question whether indeed Fermat had ever actually derived the correct proof himself.

### FURTHER READING

### Books

Bell, E.T. *Men of Mathematics*. New York: Simon and Schuster, 1986.

Mahoney, Michael S. *The Mathematical Career of Pierre de Fermat*. Princeton, NJ: Princeton University Press, 1994.

*—Sketch by Leslie Reinherz*

# Karl Wilhelm Feuerbach
## 1800–1834
### German geometer and educator

Karl Wilhelm Feuerbach made a significant impact on geometry in his tragically short life, and likely would have stood as one of the discipline's foremost practitioners had he lived longer. For some time, his best-known contribution to the field was the discovery of the nine-point circle of a triangle. He was later recognized for the introduction, at the same time as **August Möbius**, of homogeneous coordinates in analytic geometry.

Born on May 30, 1800, in Jena, Germany, Feuerbach was the third of 11 children of Paul Johann Anselm Feuerbach, a judge, and Eva Wilhelmina Maria Troster. A gifted student, Feuerbach earned a Ph.D. in mathematics at the age of 22 and was appointed to a professorship at the Gymnasium at Erlangen. By that time, he had also published a seminal mathematics paper, *Eigenschaften einiger merkwürdigen Punkte des geradlinigen Dreiecks und mehrerer durch sie bestimmten Linien und Figuren*, in which he revealed what came to be termed the theorem of Feuerbach. The theorem was considered a major contribution to **Euclid**ian geometry and stated that: "The circle which passes through the feet of the altitudes of a triangle touches all four of the circles which are tangent to the three sides of the triangle; it is internally tangent to the inscribed circle and externally tangent to each of the circles which touch the sides of the triangle externally." While the nine-point circle of a triangle had been described in 1821 by both Charles Jules Brianchon and **Jean-Victor Poncelet**, neither of the mathematicians had named it. Therefore, Feuerbach received credit for the theorem and the point where the incircle and the nine-point circle touch are referred to as the Feuerbach point.

In 1827, Feuerbach published a second major work which, after careful scrutiny by Moritz Cantor, was determined to have introduced the concept of homogeneous coordinates. Möbius independently introduced the same concept in his *Der barycentrische Calkul*, published that same year, so both mathematicians are credited with the development of the concept.

Tragically, Feuerbach retired from teaching the following year at the age of 28 after contracting a serious illness. He lived the next six years as a recluse in Erlangen and died there on March 12, 1834.

### Selected Writing by Feuerbach

(R.A. Johnson, translator) "Feuerbach on the Theorem Which Bears His Name," in *A Source Book in Mathematics*. Edited by D.E. Smith, 1929

*Leonardo Pisano Fibonacci*

## FURTHER READING

### Books

Johnson, R.A. *Modern Geometry: An Elementary Treatise on the Geometry of the Triangle and the Circle.* New York: Houghton–Mifflin, 1929.

### Periodicals

Guggenbuhl, L. "Karl Wilhelm Feuerbach, Mathematician." *Scientific Monthly* 81 (1955): 71.
Mackay, J.S. "History of the Nine Point Circle." *Proceedings of the Edinburgh Mathematical Society* 11 (1892): 19.

—*Sketch by Kristin Palm*

# Leonardo Pisano Fibonacci
## 1170(?)–1250(?)
### Italian number theorist

Leonardo Pisano Fibonacci is considered one of the most talented mathematicians of the Middle Ages. He is credited with introducing the Hindu–Arabic numbering system into western European culture at the beginning of the 13th century. His series of books on mathematical subjects helped revive the tradition of ancient mathematics and laid the foundation for the development of number theory. While Fibonacci's introduction of the modern day numbering system had a profound impact on the subsequent history of mathematics, his fame as a mathematician is perhaps more often associated with his development of the Fibonacci sequence, a series of numbers that he derived in order to solve a riddle about the reproduction of rabbits.

Also known as Leonardo of Pisa, Fibonacci, was born sometime during the latter half of the 12th century in Pisa, Italy. His father, William Bonacci, was a merchant and a government representative of Pisa, then an independent city–state. The name "Fibonacci" is believed to have been derived from the contraction of *Filiorum Bonacci,* or possibly, *Filius Bonacci,* meaning respectively, "of the family of Bonacci" and "Bonacci's son." What is known of Fibonacci's life has been gleaned mostly from the brief autobiographical notes he included in the introduction to his first mathematical treatise, *Liber Abaci,* in 1202, from the dedication in his *Liber Quadratorum,* written in 1225, and from his only surviving letter, written to the Emperor Frederick II's philosopher, Magister Theodoris. Based on incidents described in these writings, Fibonacci's year of birth is approximated at 1170.

### Introduces Hindu–Arabic Numbering System to Western Europe

During the 12th century, Pisa had about 10,000 inhabitants, and was an important center of commerce. Its merchants traded throughout the Mediterranean region and maintained warehouses in the coastal cities. When Fibonacci was a boy, his father was appointed as the head of a warehouse in the city of Bugia, on the North African coast. Fibonacci traveled there to join his father and to receive a business education. He studied arithmetic under the tutelage of a Moorish schoolmaster, and learned to make calculations using the Hindu–Arabic numerals 0 through 9. Through Italy and western Europe the seven Roman symbols, I, V, X, L, C, D, and M were still used in their various combinations to express all possible integers. These symbols represented respectively: one, five, ten, fifty, one hundred, five hundred, and one thousand. Because the Roman numeral system lacked the concepts of zero and place–value, multiplication and division were cumbersome and virtually impossible operations. Performing such calculations required the use of an abacus, a mechanical device which allowed for no written verification of the result.

Fibonacci was undoubtedly taught the merits of the Hindu–Arabic numbering system and the methods of using it to perform various mathematical operations. It is not clear how much more advanced his studies were, or how long Fibonacci remained in Bugia. He returned to Italy in about 1200, after having traveled extensively throughout the Mediterranean. Fibonacci's travels brought him in contact with leading scholars of the day and exposed him to the monetary systems of Egypt, Syria, Constantinople, Greece, Sicily, and France.

In 1202, Fibonacci wrote *Liber abaci* ("Book of Calculations"). Its intent was to introduce Hindu–Arabic numerals to Western culture and to explain the utility of the new numbering system in business transactions. Divided into 15 chapters, Fibonacci's book covered a broad range of topics. There was a chapter on how to read and write the numerals, as well as separate chapters on how the numerals could be used practically to make additions, subtractions, multiplications, and divisions. The concepts of fractions and of squared and cubic roots were explained. A series of chapters dealt with such business practices as pricing, bartering, and partnership. His final and perhaps most important chapter dealt with the more sophisticated topics of geometry and algebra. Throughout the text, Fibonacci posed and showed solutions for various mathematical puzzles and riddles.

Fibonacci wrote *Practica geometriae* in 1220. This manuscript dealt with practical problems in geometry and the measurement of objects. It also covered algebraic and trigonometric operations and the use of square and cubic roots. These writings revealed Fibonacci's familiarity with the works of **Euclid** and other mathematicians of antiquity.

### Wins Mathematical Tournament

Fibonacci's reputation as an influential mathematician became widespread. The Holy Roman Emperor Frederick II had read and was impressed with *Liber Abaci*, and in 1225 he traveled to Pisa to conduct a mathematical tournament as a test of Fibonacci's skills. Johannes of Palermo, a member of the emperor's staff, composed three tournament questions, sent in advance to Fibonacci and several competitors. At the emperor's court in Pisa, Fibonacci demonstrated his mathematical ability by deriving correct answers to each of the questions. His competitors withdrew, unable to provide any of the solutions.

The first question posed by Johannes of Palermo was a second–degree problem, that is, one involving squares. Specifically, the contestants were asked to determine values of $x$ and $y$, such that $x^2 + 5 = y^2$, when $x^2 - 5 = y^2$. The next question was one of the third degree, involving cubes. The third question, a first–degree problem, was posed in the form of a riddle. Three men owned, respectively a half, a third and a sixth of an unknown quantity of money. Each

man took an unspecified amount of the money, leaving none left. Then each man, respectively, returned a half, a third, and a sixth of what he had first taken. The returned money was divided into equal thirds and redistribute to the men. This resulted in each man acquiring his fair share. The contestants were asked to determine the quantity of money owned by each man.

Having displayed his skills as a mathematician before the emperor, Fibonacci went on that same year to write his *Liber quadratorum*. The manuscript was dedicated to the emperor and in it Fibonacci described both their meeting and the particulars of the contest. *Liber quadratorum* contained a collection of theorems about indeterminate analysis, specifically relating to second degree equations, and its introduction included a description of the second–degree contest problem. The first and third–degree contest problems were described in a separate manuscript of Fibonacci's, entitled *Flos*. The originality of his work and the power of his methods for solution have caused Fibonacci to be ranked among the most important seminal figures in the field of number theory.

Fibonacci revised *Liber abaci* in 1228, and it is this text that eventually became most widely distributed in Europe. It is dedicated to Michael Scot, a friend of Fibonacci's who wrote science texts and served as chief astrologer to Emperor Frederick II. As in the first edition, Fibonacci argues strongly for the adoption of the Hindu–Arabic numbering system.

### Leaves Legacy of Fibonacci Sequence

Tourists to Pisa, Italy, are today most frequently drawn to the city by its famous leaning tower, designed and partially built by a contemporary of Fibonacci's, Bonnano Pisano. Across the Arno River from the leaning tower is a lesser known monument, a statue representing Fibonacci. Since no drawings of Fibonacci has survived, the statue was created in the likeness of a "generic" Pisan of the 12th century. While the potential lack of resemblance might strike Fibonacci as an odd legacy, the enduring association of his name with a sequence of numbers he generated, in order to solve a puzzle about rabbits, might seem even more peculiar. In his *Liber abaci*, he described his solution to a well–known mathematical problem of his time. A pair of rabbits is kept in an enclosure and begins producing offspring at the rate of one pair per month, beginning in the second month. Each new pair reproduces at that same rate after its second month. Assuming there is no mortality, how many rabbits will exist at the end of a year? The numbers in the Fibonacci sequence represent the quantity of rabbit pairs at the end of each month, namely 1, 1, 2, 3, 5, 8, 13, 21, 34, 55, and so on. Fibonacci was aware that the sequence was recursive, that is, one in which the relationship between successive terms can be expressed with a formula. Various modern–day mathematical societies bearing Fibonacci's name, have devoted themselves to exploring the interesting features of this sequence.

Fibonacci is believed to have died around 1250, when Pisa was defeated by Genoa in a naval battle.

## FURTHER READING

**Books**

Gies, Joseph and Francis Gies. *Leonardo of Pisa and The New Mathematics of The Middle Ages.* New York: Thomas Y. Crowell Company, 1969.

Swetz, Frank J., editor. *From Five Fingers to Infinity.* Chicago: Open Court Press, 1994.

*—Sketch by Leslie Reinherz*

*John Charles Fields*

# John Charles Fields
## 1863–1932
### Canadian algebraist

John Charles Fields was a noted algebraist who is primarily known for the legacy he left after his death. Before he died, he established the International Medal for Outstanding Discoveries in Mathematics, now known as the Fields Medal, which is considered to be the Nobel Prize of mathematics. The Fields Medal is awarded every four years to mathematicians under the age of 40 for exceptional research in mathematics. The first Fields Medal was granted to **Lars Ahlfors** in 1936.

Fields was born in Hamilton, Ontario, Canada in 1863. His father, John Charles Fields, died when his son was 11 years old, and his mother, Harriet Bowes passed away when Fields was 18. Fields attended the University of Toronto, where he received a B.A. along with a gold medal in mathematics. At Johns Hopkins University in Baltimore, Maryland, Fields earned the Ph.D. in mathematics.

Fields landed his first research position when he was appointed to a professorship at Allegheny College in 1889. There, he began investigating his first love: algebraic functions. Later, he traveled to Europe, where he was influenced by some of the great mathematicians of his day, including Immanuel Lazarus Fuchs and Georg Ferdinand Frobenius. Fields' papers from 1901 to 1904 show his preference for algebraic functions rather than geometrical calcula-

tions. His research and equations are known for their elegance and simplicity.

In 1902, Fields was named special lecturer at the University of Toronto, becoming the first renowned mathematical researcher at that institution, and later achieving the status of research professor. Fields was internationally known during his day and was active in such prestigious scientific societies as the Royal Society of London and the Associations for the Advancement of Science, both British and American divisions, where he also held office. In 1924, Fields organized the International Congress of Mathematics in Toronto, which was a great success because of his exhaustive efforts.

Fields remained as a teacher and researcher at the University of Toronto until his death in 1932.

A few months before he died, Fields proposed that a medal of distinction be awarded to mathematicians every four years. Since there was no Nobel Prize in mathematics, Fields believed such an award would recognize great contributions to mathematics. In 1932, the International Congress of Mathematicians at Zurich accepted his proposal, and the first medal was awarded four years later. Fields recommended that the medal should be awarded to great achievers in mathematics as well as promising young research mathematicians. In a letter proposing the award, Fields suggested that funding should by taken from finances left over from the International Mathematical Congress of 1924. He emphasized that the medal should not be linked with Canada in any way, or to any other country, insisting that the medals should be as "international and impersonal as possible."

The gold–plated Fields Medals are made by the Royal Canadian Mint and are 11 inches (28 cm) in diameter.

Fields believed that establishing the medal would give much needed recognition to the mathematicians of the world and create a spirit of international harmony. "It means a new departure in the matter of international scientific cooperation and is likely to be the precursor of moves along like lines in other sciences than mathematics," he says in his proposal letter. With his legacy, Fields advanced the stature of mathematics and made it possible for even young researchers to receive needed encouragement.

## SELECTED WRITINGS BY FIELDS

*Theory of the Algebraic Functions of a Complex Variable,* 1906

## FURTHER READING

Synge, J.D. "John Charles Fields." *Journal of the London Mathematical Society* 8 (1933): 153–160.
————. "Obituary Notice of John Charles Fields." *Obituary Notices of Fellows of the Royal Society of London* 2 (1933): 129–135.

—*Sketch by Barbara Boughton*

# Ronald Aylmer Fisher
## 1890-1962
### English statistician and geneticist

Sir Ronald Aylmer Fisher formalized and extended the field of statistics and revolutionized the concept of experimental design. He worked for 14 years as a research statistician and later held professorships in genetics, another field to which he made significant contributions. Fisher wrote about 300 papers and seven books during his prodigious career.

The youngest of seven children, Ronald Fisher was born on February 17, 1890, in a northern suburb of London. His father, George Fisher, was a partner in a fine arts auction firm. Because of poor eyesight, the young Fisher was not allowed to read or write under artificial light. Consequently, he rarely took notes at lectures he attended, and he preferred to solve problems mentally rather than on paper. He developed a facility for visualizing complex geometrical relationships in his mind. This ability later proved fruitful, as his geometrical interpretation of statistics led him to previously unattainable results.

In 1909, Fisher earned a scholarship to attend Gonville and Caius College in Cambridge, where he specialized in mathematics and theoretical physics while also studying genetics. He graduated in 1912. For the next six years, Fisher searched for the right type of occupation, even working briefly as a farm laborer in Canada. Primarily, however, he worked as a statistician for the Mercantile and General Investment Company in London (1913 to 1915) and as a public school teacher (1915 to 1919). He was unhappy and, apparently, ineffective as a teacher—throughout his career, he was recognized as a brilliant thinker who had difficulty explaining his ideas to others. In 1917, he married Ruth Eileen Guinness; they had eight children and eventually separated.

Even though his jobs did not support research, Fisher published several papers during this period. One of his earliest accomplishments in statistics was determining the distribution of the correlation coefficient in normal samples, a problem he solved by formulating it in terms of n–dimensional Euclidean space with n representing the sample size. He wrote two papers on eugenics (the science of improving the human race through selective mating); his concern that the less talented lower classes produced offspring at a faster rate than the more capable upper classes influenced his personal choice to have a large family.

### Established Renowned Statistical Research Center

Because of his growing reputation as a mathematician, in 1919 Fisher was offered a job analyzing a 66–year accumulation of statistical data at the Rothamsted Experimental Station, an agricultural research laboratory about 25 miles north of London. For the next 14 years, Fisher took advantage of the huge data resources at Rothamsted to derive new analysis techniques as well as agricultural results. He formulated the analysis of variance, which is now a fundamental tool of statistical analysis; it isolates the effects of several variables in an experiment, showing what contribution each made to the results. Consequently, he advocated factorial experimentation (in which several factors can vary simultaneously) rather than attempting to vary only one factor at a time. This not only speeds results by gathering information on the effects of several variables at one time, but it also allows investigation of interactions of variables that differ from any of the variables acting alone.

In another innovation of experimental design, Fisher advocated random arrangement of samples receiving different treatments. Traditional agricultural experiments arranged samples according to elaborate placement schemes on checkerboard plots in an attempt to avoid bias from extraneous factors such as variations in soil and exposure to weather. Fisher demonstrated that by assigning these positions randomly, rather than according to a systematic pattern, facilitated statistical analysis of the results. His 1925 textbook, *Statistical Methods for Research Workers*, is considered a landmark work in the field. Unfortunately, it is so difficult to read that Fisher's friend and colleague M. G. Kendall wrote in *Studies in the History of Statistics and Probability*, "Somebody once said that no student should attempt to read it unless he had read it before."

During the course of his career, Fisher's theoretical work also included improvements to the Helmert–Pearson 2 distribution (including the addition of degrees of freedom) and the Student's t distribution, and development of what would eventually be called the F–distribution in his honor. He introduced the concept of the "null hypothesis" and developed procedures for decision-making based on the percentage difference between experimental results and the null hypothesis. He derived the distributions of numerous statistical functions including partial and multiple correlation coefficients and the regression coefficient, and he clarified the concept of the maximum likelihood estimate.

Fisher became a Fellow of the Royal Society in 1929, the same year he published a paper on sampling moments that would provide the foundation for future development of that topic. During the 1930s, he wrote several substantial papers on the logic of inductive inference.

## Employed Statistical Methods in Genetics Research

Fisher left Rothamsted in 1933 to occupy the Galton Chair of Eugenics at University College in London, only to return during World War II when his department was evacuated to the Experimental Station. In 1935, he established a blood–typing department in the Galton Laboratory, which developed important information on Rh factor inheritance. That same year, he published *Design of Experiments*, another landmark text in statistical science. The following year, he published his first paper on discriminant analysis, which is now used in such areas as weather forecasting, medical research, and educational testing. During summer lectureships at Iowa State College's agricultural research center at Ames in 1931 and 1936, Fisher established contacts that helped popularize his techniques among American educators and psychologists as well as agriculturalists.

In 1943, Fisher joined the University of Cambridge as Balfour Professor of Genetics. He was knighted in 1952, and served as president of the Royal Society from 1952 until 1954. Both the Royal Society and the Royal Statistical Society awarded him

several prestigious medals during his tenure at the University of Cambridge. In 1950, Fisher published *Contributions to Mathematical Statistics*, an annotated collection of 43 of his most significant papers, many of which had originally appeared in rather obscure journals. He formally retired in 1957, but continued working until a successor was found in 1959.

When Fisher left Cambridge in 1959, he moved to Adelaide, Australia, to join several of his former students and work as a statistical researcher for the Commonwealth Scientific and Industrial Research Organization. He died on July 29, 1962, as a result of an embolism following an intestinal disorder.

## SELECTED WRITINGS BY FISHER

### Books

*Contributions to Mathematical Statistics*, 1950
*Statistical Methods for Research Workers*, 1973
*The Design of Experiments.* Eighth edition. 1974
*Statistical Methods, Experimental Design, and Scientific Inference*, 1990

## FURTHER READING

### Books

Box, Joan Fisher. *R. A. Fisher: The Life of a Scientist.* New York: Wiley, 1985.
Fienberg, Stephen E. *R. A. Fisher, An Appreciation.* New York: Springer-Verlag, 1990.
Gridgeman, Norman T. "Ronald A. Fisher," in *Dictionary of Scientific Biography.* Volume V. Edited by Charles Coulston Gillispie. New York: Charles Scribners Sons, 1974, pp. 7–11.
*Modern Scientists and Engineers.* Volume 1. New York: McGraw-Hill, 1980, pp. 375–77.
Pearson, E. S., and M. G. Kendall, editors. *Studies in the History of Statistics and Probability.* New York: Hafner Press, 1970, pp. 439–53.
Tankard, James W. *The Statistical Pioneers.* Cambridge, MA: Schenkman Publishing Co., 1984, pp. 111–33.

*—Sketch by Loretta Hall*

# Irmgard Flügge-Lotz
## 1903-1974
### German-born American applied mathematician

Flügge-Lotz conducted pioneering studies of aircraft wing lift distribution and made significant contributions to modern aeronautic design. She served as an advisor to the National Aeronautics and Space Administration (NASA) as well as to German and French research institutes. During an era when there were few women engineers, Flügge-Lotz was named the first female professor in Stanford University's College of Engineering. Describing her 20-year career at Stanford, John R. Spreiter, and Wilhelm Flügge wrote in *Women of Mathematics,* "her work in fluid mechanics was directed toward developing numerical methods for the accurate solution of problems in compressible boundary-layer theory. She pioneered the use of finite-difference methods for such purposes and was quick to employ the emerging capability of computers to deal with the large computations inherent in the use of these methods . . . She applied these methods to solve a series of important and previously unsolved problems in compressible boundary-layer theory."

Flügge-Lotz was born on July 16, 1903, in Hameln, Germany. Her father, Oskar Lotz, was a journalist and amateur mathematician. Her mother, Dora (Grupe) Lotz, came from a family that had been in the construction business for generations. Visiting the firm's building sites, in addition to watching Count von Zeppelin conduct airship tests, fueled her interest in engineering. When she was a teenager, Flügge-Lotz's father was drafted into the German army during World War I. To help support her mother and younger sister, she began tutoring fellow students in mathematics and Latin. She continued this work after her father returned from the war in ill health. After graduating from high school in 1923, she studied applied mathematics and engineering at the Technical University of Hanover. In 1929, she earned a doctorate in engineering. Her dissertation explored the mathematical theory of circular cylinders and heat conduction.

## Wins Supervisory Position

Opportunities for women in engineering were limited in the 1930s, and Flügge-Lotz had difficulty finding a level of employment that was commensurate with her education. When she began working at the Aerodynamische Veruchsanstalt (AVA) research institute in Göttingen, she spent half of her time as a cataloguer. Perceptions of Flügge-Lotz's abilities changed dramatically after she solved a problem that had stymied Ludwig Prandtl and Albert Betz, two leading aerodynamicists at the AVA (Betz was the director of the institute). A decade earlier, Prandtl had developed a differential equation for his theory about the lift distribution of an airplane wing. However, he had made little progress in solving the equation. Flügge-Lotz tackled the problem and solved the equation for the general case. Continuing to work with the equation, she developed it so that it had widespread practical applications. Her cataloguing

days ended, and she was named supervisor of a group of engineers who researched theoretical aerodynamics within the AVA.

In 1931, Flügge-Lotz published a technique she had developed for calculating the lift distribution on aircraft wings. Later dubbed the "Lotz method," it is still used today. Continuing to delve into wing theory, she added to the knowledge base of the effects of control surfaces, propeller slipstream, and wind-tunnel wall interference.

The course of Flügge-Lotz's career changed in 1938 when she married Wilhelm Flügge, a civil engineer from Göttingen. The husband and wife team went to work at Berlin's Deutsche Versuchsansalt für Luftfahrt (DVL), a German agency similar to NASA. Beginning work as a consultant in flight and aerodynamics, Flügge-Lotz conducted groundbreaking research in automatic control theory, especially pertaining to on-off controls. Subsequently, these controls came into widespread use because they were reliable and inexpensive to build.

In 1944 Flügge-Lotz and her husband moved to the small town of Saulgau, where they continued to work for the DVL. After Germany's defeat in World War II, the town and its surrounding area became part of France. The Flügges joined the staff of the French National Office for Aeronautical Research (ONERA) in Paris. Flügge-Lotz headed a research group in theoretical aerodynamics and continued her work in automatic control theory. In 1948 the couple accepted positions at Stanford University in California.

Flügge was hired as a professor at Stanford. Nepotism regulations prohibited his wife from holding a similar position, so she was hired as a lecturer. Flügge-Lotz developed graduate and undergraduate courses in mathematical hydro- and aerodynamics and automatic control theory. She also designed a weekly seminar in fluid mechanics that was attended not only by Stanford students but also by young engineers from the National Advisory Committee for Aeronautics—NASA's predecessor. The seminar has continued to serve as an important forum for faculty and students of varying specializations to share their findings. The Flügges had no children, but they frequently invited their students to their home for dinner parties with other faculty members and visitors.

## Advances Automatic Control Theory

In addition to teaching, Flügge-Lotz also continued her research while at Stanford. She applied numerical methods and analog computer simulations to boundary layer problems in fluid dynamics and continued studying automatic control theory. Her 1953 book, *Discontinuous Automatic Control,* presented the theoretical foundation for using discontinuous controls for flight paths of missiles. Specifically, the problem concerns a missile in flight encountering some force (such as a wind gust) that causes it to begin oscillating. An automatic control mechanism could be used to counteract the vibration. Flügge-Lotz was particularly interested in discontinuous controls, which could be activated only when needed. A simple analogy is a furnace thermostat; rather than continuously adjusting the amount of heat generated by a furnace, the thermostat simply turns the furnace on when needed and off when the proper temperature is achieved. The fact that discontinuous automatic controls were simpler and cheaper to make than continuous-control mechanisms made them particularly desirable for one-time use in missiles that would crash after flight.

Flügge-Lotz's second book, published in 1968, presented techniques for optimizing discontinuous automatic controls for objects such as airplanes, rockets, and satellites. In the preface to *Discontinuous and Optimal Control*, she noted that optimal control theory literature was difficult for practicing engineers to read because it was so mathematically rigorous. "The purpose of this book," she wrote, "is to acquaint the reader with the problem of discontinuous control by presenting the essential phenomena in simple examples before guiding him to an understanding of systems of higher order."

Flügge-Lotz attended the First Congress of the International Federation of Automatic Control in Moscow as the only female delegate from the United States. Upon her return from that congress, Stanford University finally offered her a professorship in 1961. In fact, she was granted that rank in two departments—aeronautics and astronautics, and engineering mechanics. She retired from teaching seven years later. Flügge-Lotz continued her research activities, however, studying heat transfer and the control of satellites. Her retirement years were physically difficult, due to progressive arthritis.

The Society of Women Engineers gave Flügge-Lotz its Achievement Award in 1970. The American Institute of Aeronautics and Astronautics (AIAA) chose her to deliver the prestigious annual von Kármán Lecture in 1971 and elected her as its first woman fellow. The University of Maryland awarded her an honorary degree, citing "contributions [that] have spanned a lifetime during which she demonstrated, in a field dominated by men, the value and quality of a woman's intuitive approach in searching for and discovering solutions to complex engineering problems." Flügge-Lotz died a year later, on May 22, 1974.

## SELECTED WRITINGS BY FLÜGGE-LOTZ

### Books

*Discontinuous Automatic Control,* 1953

*Discontinuous and Optimal Control,* 1968
(With Gary Daniel Wolske) *Minimum Fuel Attitude Control of a Nonlinear Satellite with Bounded Control by a Method Based on Linear Programming,* 1969

### Periodicals

(With A. F. Johnson) "Laminar Compressible Boundary Layer Along a Curved, Insulated Surface." *Journal of the Aeronautical Sciences* 22 (1955): 445–454.
(With M. D. Maltz) "Attitude Stabilization Using a Contactor Control System with a Linear Switching Criterion." *Automatica* 2 (1963): 255–274.
"Trends in the Field of Automatic Control in the Last Two Decades." (1971 von Kármán lecture) *AIAA Journal* 10 (1972): 721–26.

## FURTHER READING

### Books

Spreiter, John R., and Wilhelm Flügge. "Irmgard Flügge-Lotz," in *Women of Mathematics.* Edited by Louise S. Grinstein and Paul J. Campbell. Westport, CT: Greenwood Press, 1987, pp. 33–40.

### Periodicals

"A Life Full of Work—The Flügges." *Stanford Engineering News* 68 (May 1969).
Spreiter, John R., Milton D. Van Dyke, and Walter G. Vincenti. "In Memoriam: Irmgard Flügge-Lotz, 1903-1974." *IEEE Transactions on Automatic Control* AC-20 (April 1975): 183a–83b.

### Other

Cooper, Julie, and Maria Banderas. "Irmgard Flügge-Lotz."*Biographies of Women Mathematicians.* http://www. scottlan.edu/lriddle/women/chronol.htm (June 1, 1997).

*—Sketch by Loretta Hall and Karen Wilhelm*

# Maurice Fréchet
## 1878-1973
### French topologist

Maurice Fréchet was one of the creators of the 20th-century discipline of topology, which deals with the properties of geometric configurations; his research added a new degree of abstractness to the mathematical advances of the previous generation. He profited from a rich mathematical environment during his studies and in turn passed on a wealth of ideas to his students over a long career. Some of the mathematicians who had learned their skills before Fréchet's work appeared could not help wondering whether there were advantages to the new degree of generality in his work. The answer lay in the fruitfulness of the methods of Fréchet for addressing problems whose solution included concrete problems of long standing.

René Maurice Fréchet was born on the 10th of September in 1878 in Maligny, a small town in provincial France, where his father Jacques directed an orphanage. Soon after Fréchet's birth the family moved to Paris, much to his advantage in terms of mathematical environment. His mother Zoé was responsible for a boardinghouse for foreigners, which early put Fréchet in contact with a cosmopolitan community. This may be reflected in his subsequent hospitality toward students and collaborators from all over the world. At his lycée (high school), he was singularly fortunate in learning mathematics from **Jacques Hadamard**, already a mathematician of distinction who would shortly thereafter provide a proof of a central result of number theory called the prime number theorem.

Fréchet's talents blossomed under Hadamard's encouragement and he was well prepared to enter the École Normale Supérieure, the great French scientific university, in 1900. Hadamard was not the only mathematical influence on the young Fréchet. After graduating from the École Normale, he began to work with mathematician **Émile Borel**, who was only seven years older than Fréchet but who had started his career so early that he may have seemed to belong to an earlier generation. Fréchet collaborated with Borel on the publication of a series of the latter's lectures and continued to be involved with the publishing of the so-called Borel collection for Gauthier-Villars. Even though Borel's role in the collection was primarily an editor's, he also wrote all the volumes to begin with, the first exception being one written by Fréchet. In turn, Fréchet undertook the editing of a series on general analysis published by Hermann (the other great mathematical publisher in Paris) and undertook the writing of several of the volumes as well.

Fréchet wrote his thesis under Hadamard, who had returned to Paris, and then followed Hadamard in teaching at the level of the lycée for a few years. His marriage in 1908 to Suzanne Carrive produced four children, whom he supported with professorships outside of Paris until 1928. He was officially connected with the University at Poitiers from 1910 to 1918, but World War I took him out of mathematics and into the less familiar surroundings of working as an interpreter with the British army, where his early exposure to different languages was of help. After his return from military service, he was head of the Institute of Mathematics at the University of Strasbourg, still a provincial appointment. It was not until 1928 that he was called to the University of Paris.

## Proposes Revolutionary Variations of Topology Theory

One of the reasons for the delay in the recognition of Fréchet's work by the French academic establishment was its revolutionary character. The notions of set theory as introduced by the German mathematician **Georg Cantor** in the previous century were slowly winning converts, although there were differences of opinion about which axioms ought to be accepted. What Fréchet did in his thesis and in the most influential of his subsequent work was to bring the ideas of general set theory to bear on questions of the new discipline of topology, the generalization of geometry that had been given a good deal of prominence in French mathematics by the work of **Jules Henri Poincaré**. The questions that Poincaré had raised were new, but they were in the context of classical mathematics, centered on space with standard Cartesian coordinates (those points commonly expressed as located along x, y, and z axes), although perhaps in more than three dimensions.

This much of a revolution the mathematical community had come to accept, but Fréchet's thesis pushed the level of abstractness to new heights. Rather than looking just at sets of points in Cartesian space, he was prepared to handle sets of points in arbitrary spaces—so-called abstract spaces. The important tool that he used to handle such sets was a distance function. The ordinary distance function for sets of points with Cartesian coordinates (x, y) comes from the Pythagorean theorem and involves taking the square root of the sums of the squares of the differences in each coordinate. Since in abstract spaces there weren't necessarily any coordinates to assign to points, the distance function had to be more general and governed by some of the principles that applied to the Cartesian version.

The advantage of the new approach of Fréchet was that complicated algebraic expressions could be replaced by general considerations about distance. Spaces with a distance function were called metric spaces and proved to be the setting for expressing many of the results hitherto considered limited to spaces with real numbers as coordinates. Having once introduced these ideas into topology, Fréchet proceeded to look at calculus in metric spaces, an area that became known as functional analysis. Again, the basis for progress on long-standing problems was the avoidance of the complicated calculations that had bedeviled earlier work and the application of general notions from topology instead. Fréchet extended the notions of derivatives and integrals from standard calculus so that they could be used in the setting of a metric space; in addition, he introduced new types of functions called functionals, which took real numbers as values but could operate on the points of abstract spaces. Much of his work from his thesis onward was summarized in *Les espaces abstraits,* published in 1926.

Fréchet taught at the University of Paris until 1949, and a good deal of his time there was spent on questions of probability. Just as general questions about calculus could be asked in the setting of abstract metric spaces, so the techniques of probability could be moved there as well. The application of probability to continuous quantities, as opposed to discrete quantities that took only a finite number of values, had always been dependent on calculus, and Fréchet's results showed that the extension to the abstract setting could be fruitful as well. As with functional analysis in general, the more one could move away from messy computations, the more one could hope that the idea behind a proof could be visible.

Another possible reason for Fréchet's move into probability was the hope that a more concrete area would make the techniques of abstract spaces more palatable to the part of the mathematical community uneasy about getting too far from applications. If so, the efforts proved largely unavailing, at least in France, although the level of abstractness introduced by Fréchet was one of the inspirations for the Polish mathematical school between World Wars I and II. It is perhaps indicative of the relative opinions of his work that Fréchet was elected to the Polish Academy of Sciences in 1929 but not to the French Académie Royale des Sciences until 1956. He was recognized as a member of the Legion of Honor, and some accumulation of praise could hardly be avoided as he lived into his nineties. He died in Paris on the 4th of June in 1973, having earned belated recognition of his role in bringing mathematics into the twentieth century on the wings of abstractness.

## SELECTED WRITINGS BY FRÉCHET

*Les espaces abstraits,* 1926
*L'arithmétique de l'infini,* 1934
(With Ky Fan) *Initiation to Combinatorial Topology,* translated by Howard M. Eves, 1967

## FURTHER READING

Bell, E. T. *The Development of Mathematics.* Second edition. New York: McGraw-Hill, 1945.
Temple, George. *100 Years of Mathematics.* New York: Springer-Verlag, 1981.

*Erik Ivar Fredholm*

Young, Laurence. *Mathematicians and Their Times.* North-Holland, 1981.

—*Sketch by Thomas Drucker*

# Erik Ivar Fredholm
## 1866–1927
### Swedish number theorist, mathematical physicist, and educator

Erik Ivar Fredholm developed the modern theory of integral equations. His work served as the foundation for later critical research performed by **David Hilbert**, and several concepts and theorems are attributed to him. Fredholm was born on April 7, 1866, in Stockholm, Sweden. His family was upper–middle–class; his father was a well–to–do merchant, and his mother came from a cultured background. He was privy to the highest quality education available in his country and proved gifted. In 1885 he began studies at the Polytechnic Institute in Stockholm, where he developed what turned out to be a lifelong interest in problems of practical mechanics. He remained at the Institute for only one year and enrolled in the University of Uppsala in 1886, receiving his bachelor's degree in 1888.

Fredholm received his doctorate from Uppsala in 1898, although he conducted the bulk of his studies under Mittag–Leffler at the University of Stockholm. At that time, Uppsala was the only university in Sweden that offered a doctoral degree. Fredholm conducted his doctoral thesis on partial differential equations, and his work became significant to the study of deformation of anisotropic media, such as crystals.

After receiving his doctoral degree Fredholm accepted a position as lecturer in mathematical physics at the University of Stockholm and in 1906 became a professor of rational mechanics and mathematical physics. The research he conducted during this time yielded a fundamental integral equation that now bears his name. The equation, which is highly relevant in physics, was contained in a seminal research paper for which Fredholm was honored with the Wallmark Prize of the Swedish Academy of Sciences and the Poncelet Prize of the Académie des Sciences.

Much of Fredholm's research on integral equations was based on the work of American astronomer **George William Hill**. Fredholm laid the foundation for this renowned research in a 1900 paper, *Sur une nouvelle méthode pour la résolution du problème de Dirichlet*. It was in this paper that Fredholm developed the essential component of the theory that led to what is now called Fredholm's Integral Equation. Fredholm then went on to develop what came to be known as the Fredholm Equation of the second type, which involved a definite integral. He also discovered the algebraic analog of his theory of integral equations. While Fredholm's contributions to mathematics and physics were significant, his research resume is sparse. Biographers attribute his small output to the mathematician's strict attention to detail, a characteristic that earned Fredholm an excellent reputation throughout Europe.

Fredholm's work was carried on by David Hilbert, who learned of Fredholm's work through Erik Holmgren, a colleague of Fredholm's whom Hilbert met in Göttingen. Hilbert incorporated Fredholm's ideas into his own theories, including the theory of eigen–values and the theory of spaces involving an infinite number of dimensions. These theories, in turn, laid the foundation for the study of quantum theory and the discovery of what are now termed Hilbert spaces.

Fredholm remained at the University of Stockholm until his death on August 17, 1927.

## SELECTED WRITINGS BY FREDHOLM

*Sur nouvelle méthode pour la résolution du problèm de Dirichlet*, 1900

*Michael H. Freedman*

*Oeuvres complètes de Ivar Fredholm,* 1955

## FURTHER READING

Bernkopf, M. "Ivar Fredholm," in *Dictionary of Scientific Biography*. Volume V. Edited by Charles Coulston Gillispie. New York: Charles Scribner's Sons, 1970, pp. 150–52.

*—Sketch by Kristin Palm*

# Michael H. Freedman
## 1951-
### American topologist

Michael H. Freedman has been recognized by the American Mathematical Society, the International Congress of Mathematicians, the United States Government, and the MacArthur Foundation for his research breakthroughs in topology, a branch of mathematics that deals with the invariant properties of geometric objects rather than their sizes and shapes. Freedman's work has been fundamental in making progress with some of the most difficult problems in four–dimensional geometry and topology. He is perhaps best known for his proof of the four–dimensional Poincaré conjecture, a problem dating from 1904.

Michael Hartley Freedman was born in Los Angeles on April 21, 1951, to Benedict Freedman and Nancy Mars Freedman. Freedman began his post–secondary education with a year at the Berkeley campus of the University of California in 1968. He then transferred to Princeton University, where he received his Ph.D. under William Browder four years later. While in college, he pursued his hobby of rock climbing, scaling the northeast ridge of Mount Williamson alone in 1970. A decade later, he won the Great Western boulder climbing championship.

Freedman joined the Department of Mathematics at the University of California, Berkeley as a lecturer in 1972. In 1974, he spent a year at the Institute for Advanced Study (IAS) in Princeton. Then he returned to California, this time to the University of California, San Diego (UCSD) campus, where he quickly progressed through the ranks of assistant professor, associate professor, and full professor. In 1985, Freedman was appointed by UCSD as the first professor to hold the newly endowed Charles Lee Powell chair of mathematics.

## Proves the Four-Dimensional Poincaré Conjecture

Throughout the 20th century, mathematicians have made progress in understanding geometric objects in terms of associated algebraic operations. In particular, topologists have tried to use algebra to classify manifolds (multidimensional surfaces). Visualizing surfaces in more than three-dimensional space is difficult. Four dimensions are somewhat intuitive if one considers the fourth dimension to be time. In his *Fortune* magazine description of Freedman's work, Gene Bylinsky suggested thinking of an eight-dimensional sphere as a ball with attached information about its age, color, temperature, weight, and bounciness.

In 1904, French mathematician **Jules Henri Poincaré** designed a system to classify manifolds. He imagined a loop of string wrapped around a surface and determined how far the loop could be shrunk. On a sphere, the loop could be shrunken to a single point. On a torus (doughnut-like shape), a loop encircling the hole cannot shrink smaller than the circumference of the hole. Thus, a sphere and a torus belong to different classifications.

Three-dimensional manifolds are especially difficult to classify because they can be stretched and folded in many different ways. Poincaré devised a series of tests that he believed could be used to identify any three-dimensional manifold, no matter how distorted, that was topologically equivalent to a sphere. The statement of this problem was refined over the years, but not until 1960 did **Stephen Smale**

give the first proof of the Poincaré conjecture for all dimensions greater than four. Smale and other topologists followed algebraic guidelines in cutting the manifold apart and sewing it back together as a sphere, a technique known as surgery. However, manifolds of dimension three or four do not have as much "room" for maneuvering. Thus, the necessary surgery is much more difficult, and the four-dimensional Poincaré conjecture remained unsolved for another two decades.

Finally, after seven years of work, Freedman solved the surgery problem for simply connected four-dimensional manifolds in 1982. His paper "The Topology of Four-Dimensional Manifolds" gives a complete classification of all simply connected, four-dimensional manifolds in terms of two quantities. In the course of proving this theorem, Freedman exhibited several new four-dimensional manifolds, including the first examples of such manifolds that do not support a coordinate system for calculus. These results, along with nearly 50 papers on the structure and classification of three– and four–dimensional manifolds, resolved many fundamental issues in these physically significant dimensions.

**John Milnor,** a mathematician of considerable stature, wrote in *Notices of the American Mathematical Society* that Freedman's classification theorems for important classes of four–dimensional topological manifolds "are simple to state and use, and are in marked contrast to the extreme complications that are now known to occur in the study of differentiable and piecewise linear 4-manifolds."

The four years following his proof of the famous conjecture were eventful for Freedman. In 1983, he married Leslie Blair Howland, with whom he would raise three children. In 1984, he received a five-year MacArthur Foundation Fellowship to provide financial support while he continued his research. That same year he was elected to the National Academy of Sciences, and the following year to the American Academy of Arts and Sciences. In 1986 Freedman received the Fields Medal, the highest honor in mathematics.

## Pushes the Limits

The American Mathematical Society awarded the 1986 Oswald Veblen Prize in Geometry to Freedman for his work in four–dimensional topology. In his response to this award, Freedman discussed the importance of interchange among the various branches and applications of mathematics. That statement, which was printed in the *Notices of the American Mathematical Society,* included his assertion that "Mathematics is not so much a collection of different subjects as a way of thinking. As such, it may be applied to any branch of knowledge." In particular, he praised the movement among mathematicians to voice their opinions on such topics as education, energy, economics, defense, and world peace. He noted that "Experience inside mathematics shows that it isn't necessary to be an old hand in an area to make a contribution. Outside mathematics the situation is less clear, but I can't help feeling that there, too, it is a mistake to leave important issues entirely to the experts."

Freedman pushed the limits not only in terms of his expectations of mathematicians in society, but also in his technical research. Speaking to a joint meeting of the American Mathematical Society and the Mathematical Association of America in early 1997, he described the direction of his current work on the subject of computability. Computer scientists are interested in being able to identify which problems are "hard" in the sense of not being solvable with an efficient computer algorithm. The classic question, known as the Traveling Salesman Problem, asks for the most efficient itinerary for a person to visit each of a certain number of sites. Analogous problems have practical applications in areas such as cryptography and computer data security. According to Barry Cipra's article in *Science* magazine, Freedman thinks key to deciding whether a problem is "hard" may be to look at the limiting case of such a problem as the number of choices approaches infinity. Cipra quotes Freedman as saying, "This is always an attractive situation for the pure mathematician, when there's a very clear, well-defined problem blocking understanding. It's kind of like waving a red flag at a bull!"

Although Freedman has announced the broad direction of his assault on the two-decade-old Traveling Salesman Problem, the mathematics community has yet to see a detailed description of his tactic. The considerable respect with which Freedman is regarded by fellow mathematicians generates optimism that his approach may generate at least some progress on this elusive topic.

## SELECTED WRITINGS BY FREEDMAN

### Books

(With Feng Luo) *Selected Applications of Geometry to Low Dimensional Topology,* 1990

### Periodicals

"The Topology of Four–Dimensional Manifolds." *Journal of Differential Geometry* 17 (1982): 357–454.

## Further Reading

### Periodicals

Bylinsky, Gene. "America's Hot Young Scientists." *Fortune* 122 (October 8, 1990): 56–70.

Cipra, Barry. "Taking 'Hard' Problems to the Limit." *Science* 275 (March 14, 1997): 1570.

"Michael H. Freedman Awarded 1986 Veblen Prize." *Notices of the American Mathematical Society* (March 1986): 227–228.

Milnor, John. "The Work of Michael Freedman." *Notices of the American Mathematical Society* (November 1986): 901–902.

—*Sketch by Loretta Hall and Robert Messer*

# Gottlob Frege
## 1848–1925
### German logician

Gottlob Frege made seminal contributions to the philosophy of language and of mathematics, yet his major intellectual project ran aground as it was in the last stages of being launched. He worked in a fairly narrow area of mathematics throughout his career, although his contributions to philosophy were uncommonly acute for a mathematician. Frege's understanding of language continues to be the focus for philosophical discussion and his project in mathematical logic, though unsuccessful, provided the foundation for much further work in the subject.

Friedrich Ludwig Gottlob Frege was born on November 8, 1848, in Wismar, Mecklenburg–Scherin (present–day Germany). His father, Alexander, was principal of a girls' high school. In 1869 Frege enrolled at the University of Jena, where he stayed for two years at Jena before transferring to the University of Göttingen. There, his studies were not limited to mathematics, but included physics, chemistry, and philosophy. In 1873 Frege earned his doctorate from Göttingen; his thesis examined the question of the geometric representation of imaginary structures in the plane. The choice of subject was in keeping with the geometrical aspect of German mathematical research at the time, one year after the inauguration of the Erlangen program by **Felix Klein**. A year later, Frege earned the right to teach at the University of Jena with a dissertation dealing with the subject of groups of a certain kind of function. Frege's work on group theory continued, but he never made it the focus of his courses as did some of his contemporaries.

## Reconstructs the Logical World

The goal to which Frege devoted himself was the task of reducing arithmetic to logic. There were two main currents in mathematics which led him to find the project attractive and feasible. The first was the work of **Karl Weierstrass** and his school of analysis. Weierstrass had taken the work of two centuries in the area of calculus and determined a way to rewrite the foundations arithmetically. The other current was the mathematization of logic in the work of **George Boole** and other algebraists. After centuries of little progress, logic had begun to take on some of the qualities of modern mathematics.

As a first step, Frege felt it necessary to devise a new notation in order to prevent hidden assumptions from creeping into arguments, which was published as *Begriffsschrift* ("Concept Notation"), in 1879. This notation was a tool for analyzing and representing proofs within mathematics, but it was cumbersome and typographically awkward. As a result, Frege's notation did not find many supporters, although the project was taken up by others, including **Bertrand Russell**, with more success.

The next step for Frege was to prove that the current philosophy of mathematics was flawed and in need of revamping, an argument he detailed in his 1884 book *Grundlagen der Arithmetik* ("The Foundations of Arithmetic"). Frege took on psychologism, the doctrine that numbers were merely objects in the mind, and was able to refute the claims of its current advocates. Similarly, he assailed the empiricism of John Stuart Mill, which sought to find numbers as objects in the world of sense and sound. Both psychologism and empiricism survived Frege's onslaught, but he had clearly demonstrated some of their shortcomings. The *Grundlagen* also included Frege's definition of number. It is tied to the notion of concept and deals with the question of the existence of a 1–1 correspondence between objects falling under two concepts.

Frege found it necessary to address a few philosophical issues connected with concepts, objects, and how they are represented linguistically. His essays dealing with these subjects have a continuing philosophical appeal independent of the success of his logical project. In particular, the 1892 essay, "über Sinn und Bedeutung" ("On Sense and Reference") may be the most influential paper in the history of the philosophy of language.

The first volume of the *Grundgesetze der Arithmetik* ("The Basic Laws of Arithmetic") was published in 1893 and it appeared that Frege had finally constructed a system that reduced arithmetic to logic. In 1902, however, he received a letter from Bertrand Russell, who had detected an error in Frege's system. What Russell had found was a contradiction, a derivation of both a statement and its negation. According to the rules of Frege's logic, the presence of such a contradiction rendered the system useless.

Frege's response was intellectually honest. In the second volume of the *Grundgesetze*, Frege acknowledged Russell's contribution and noted that his own system was unable to handle it. For the remainder of his career, Frege worked on issues that were connected with philosophy and logic without seeking to repair his system. He retired from the University of Jena in 1917 and died at the German resort of Bad Kleinen on July 26, 1925. The subsequent interest in Frege's work, especially among philosophical logicians, bears witness to the value of his reformulation of logic and the farsightedness of his ambition.

## SELECTED WRITINGS BY FREGE

*The Foundations of Arithmetic.* Translated by J.L. Austin, 1950
*Translations from the Philosophical Writings of Gottlob Frege.* Edited by Peter Geach and Max Black, 1960
*The Basic Laws of Arithmetic.* Translated by Montgomery Furth, 1964

## FURTHER READING

Rummett, Michael. *Frege: Philosophy of Language.* London: Ruckworth, 1973.
————. *The Interpretation of Frege's Philosophy.* London: Ruckworth, 1981.
————. *Frege and Other Philosophers.* Oxford: Oxford University Press, 1991.
————. *Frege: Philosophy of Mathematics.* London: Ruckworth, 1991.
Sluga, Hans. *Gottlob Frege.* London: Routledge and Kegan Paul, 1980.
Van Rootselaar, B. "Gottlob Frege," in *Dictionary of Scientific Biography.* Volume V. Edited by Charles Coulston Gillispie. New York: Charles Scribner's Sons, 1973, pp. 152–155.
Wright, Crispin. *Frege's Conception of Numbers as Objects.* Aberdeen: Aberdeen University Press, 1983.

—*Sketch by Thomas Drucker*

*Galileo Galilei*

# Galileo Galilei
## 1564–1642

### Italian astronomer, physicist and mathematician

Galileo Galilei, known best as simply "Galileo," was a scientist at a most difficult time in history: the time of the Inquisition, when the Roman Catholic Church was still furiously resisting evidence from new discoveries. Galileo is one of the Inquisition's most famous victims; he supported Copernicus's discovery that the Earth revolves around the Sun. Although he was not killed for his beliefs, Galileo was silenced and placed under house arrest from the time of his trial to the end of his life.

Galileo was born February 15, 1564, in Pisa, Italy. His father, Vincenzio Galilei, was a scientist, investigating acoustics and musical theory. His mother was Giulia Ammannati. Galileo, the oldest of seven children, began his studies with Jacopo Borghini, but had to leave his tutor when his father moved the family back to his native Florence about 1575.

Galileo then studied at the monastery of Santa Maria, and entered the order as a novice. His father removed the young man from the order and tried to get him a scholarship to the University of Pisa, but failed. Galileo returned to the monastery to study until entering the University of Pisa as a medical student in 1581, following his father's wishes.

### A Poor Medical Student but a Great Thinker

It was that year Galileo made the first of a lifetime of great discoveries. While sitting in church, he noticed a swinging lamp. No matter how great the swing's arc, they all seemed to take the same amount of time, he realized. After the service, Galileo began experimenting with different weights and lengths of string, and devised a simple device that would measure a patient's pulse. Pisa faculty members improved on the device and the pusilogia was used for many years afterward. Despite this invention, Galileo was not interested in medicine. He was a difficult student, often challenging his professors, who gave him the nickname "The Wrangler." Mathematics did interest him, however; quite possibly sparked by his father's work with acoustics, particularly the study of the effects of the length of musical strings on consonance. Although still officially a medical student, Galileo began his mathematical studies outside of the University with Ostillio Ricci, mathematician to the grand duke of Tuscany. Despite his father's objections and his suffering medical studies, Galileo persisted. He finally left the university in 1585 to study on his own, partly because his misbehavior made him unsuitable for a scholarship that would have allowed him to continue his medical studies.

Galileo went home to Florence and began tutoring students in mathematics. It was during this time that he became interested in **Archimedes'** experiment that disclosed that a crown was indeed not solid gold, but gold and a base metal (Archimedes had placed the crown in water and measured the amount of water it displaced, which turned out to be less than a solid–gold object of the same side would have). Galileo created a hydrostatic balance, a small scale that could perform the same measurement more accurately, and this invention led to his first published piece of scientific writing, a booklet describing it, which appeared in 1586.

Galileo apparently enjoyed debunking famous myths and regularly set about deflating the ideas of the ancient Greeks, particularly Aristotle. His work with falling bodies damaged Aristotelian physics beyond repair. Going contrary to the widely held belief of the time, Galileo demonstrated that Aristotle was incorrect in his claim that light objects fell more slowly than heavy ones. From this work arose the legend about Galileo dropping two cannon balls, one heavier than the other, off the Tower of Pisa.

In another experiment, he challenged Aristotle's idea that force had to be continually applied in order to keep a body in motion. Some had taken this idea and explained that the source of this force was hardworking angels. However, Galileo's experiment with a body rolling down an inclined plane showed that if force were continuously applied, the body would continue to accelerate.

Among Galileo's other discoveries in the field of mechanics were the fact that two forces can act on a body simultaneously, causing the body to move in a parabolic curve (the base of the mathematical science kinematics), and studies concerning the strength of materials. He discussed both of these ideas in his book *Discourses and Mathematical Demonstrations Concerning Two New Sciences*. Galileo's text forms the basis of modern physics because it shows the use of mathematics in understanding motion and the value of physical experiment and mathematical analysis in solving problems. Although the calculations he made regarding projectiles used low speeds, Galileo's tables and ideas would be further refined by others who came later, and made gunnery a science.

Throughout his investigations, Galileo's proofs used only the same geometric methods that had been available to the Greeks. Algebra and other more complicated methods of calculation would not be available until the time of **René Descartes** and **Isaac Newton**. However, Galileo believed mathematics was superior to logic. In 1592 he was named Chair of Mathematics at the University of Padua.

Galileo was not a man of all work. He had a mistress, Marina Gamba, for 10 years and had three children by her. The daughters, Virginia and Livia, entered the Franciscan convent in their mid-teens, taking the names Sister Maria Celeste and Sister Arcangela. Galileo did eventually recognize his son, Vincenzo.

### The Telescope Changes Everything

Galileo's attention was diverted from his studies of mechanics by the invention of the telescope. Over the years he built hundreds of telescopes, many of which he gave as gifts to persons of influence. Some of his telescopes that have survived to this day have nearly perfect optics.

With his telescopes, Galileo observed the face of the moon and showed that it was not smooth as had been thought, but rather rough-surfaced. He found that constellations such as Orion and formations like the Pleides comprised far more stars than had ever been guessed; in Orion alone he counted 500 stars. He also observed the planets, including Jupiter and Saturn.

Galileo's discovery of Jupiter's moons was of groundbreaking importance. On January 7, 1610, he saw three small but bright stars near the planet, two east and one west. Observations on successive nights found that the stars moved—eventually with all three west of the planet. "My confusion was transformed to amazement," he wrote. "I had now decided beyond all question that three stars were wandering around Jupiter, as do Venus and Mercury around the sun." Three nights later, he found another "star" circling Jupiter. Over the next six weeks he continued to watch, and the evidence was irrefutable: here was a perfect example of smaller heavenly bodies circling a larger one, just as Copernicus had said that the Earth circled the sun.

Quickly Galileo wrote and published his observations in a book called *Siderius nuncius* (The Starry Messenger). It became the most important book of the 17th century, and set the stage for Galileo's downfall.

### The Beginning of the End

Galileo's discoveries made him reluctant to continue teaching his students the old Ptolemaic system (Earth-centered) and he resigned his chair at Padua and turned to Florence, where he became the Grand duke of Tuscany's mathematician and philosopher and the University of Pisa's chief mathematician, a non-teaching position. By 1614, his stance on the question of an Earth-centered solar system had earned him attacks from Church leaders. In 1616, Pope Paul V summoned Galileo to Rome and ordered him to stop disseminating this theory called Coperinicism. Angry, Galileo nevertheless obeyed; there was no other choice, for if he had defied the Church he would have been imprisoned or tortured (in 1600, philosopher Giordano Bruno had been burned at the stake for refusing to recant his scientific views).

On August 6, 1623, Urban VIII was named pope. Somehow, Galileo was persuaded that Urban would be open to his ideas, and so he wrote *Dialogue on the Two Chief World Systems*. In the *Dialogue*, Galileo pitted an Aristotelian, Simplico, against a quick-minded Salviati, who got the best of the argument. Galileo wrote the book in Italian, not Latin, the language of scholars, and so it was accessible to anyone who could read, and was quickly translated into other languages.

Unfortunately for Galileo, one of his most bitter enemies, Father Christopher Scheiner, convinced Urban that the Simplico character was a buffoonish caricature of himself. Galileo was in trouble again. The Pope called him to Rome in 1633 and forced him to renounce any views that were at odds with the Church's belief.

Galileo was nearly 70 years old at this time. With the example of Bruno in his mind, he agreed. The church placed him under house arrest in his villa at Arcetri for the remainder of his life. By Christmas 1637, Galileo became completely blind. He died January 8, 1642.

On October 31, 1992, over 350 years after Galileo's trial, the Roman Catholic Church acknowledged that it had been in error and acknowledged the validity of his work.

## SELECTED WRITINGS BY GALILEO

*Le opere di Galileo Galilei.* 20 volumes. Edited by Antonio Favaro, 1965. (Contains Galileo's complete works).
*Dialogue Concerning the Two Chief World Systems.* Edited by S. Drake, 1967, revised edition.

## FURTHER READING

Abbott, David. *The Biographical Dictionary of Scientists: Astronomers.* New York: Bendrick Books, 1984.
Asimov, Isaac. *Asimov's Biographical Encyclopedia of Science and Technology.* Second Revised Edition. Garden City, NY: 1982.
*Biographical Encyclopedia of Scientists.* Second Edition. Volume 1. Bristol and Philadelphia: Institute of Physics Publishing, 1994.
Drake, Stillman. "Galileo Galilei," in *Dictionary of Scientific Biography.* Volume V. Edited by Charles Coulston Gillipsie. New York: Charles Scribner's Sons, 1972, pp. 237–49.
Reston, James Jr. *Galileo: A Life.* New York: HarperCollins, 1994

*—Sketch by Fran Hodgkins*

# Évariste Galois
## 1811–1832
### French algebraist and group theorist

Évariste Galois discovered mathematics as an adolescent and published his first original work at age 17. In his short life, Galois originated algebraic applications of finite groups, now known as Galois groups, and developed the foundations for the solva-

*Évariste Galois*

bility of algebraic equations using rational operations and extraction of roots. His mathematical works have been credited as having transformed the theory of algebraic equations.

Galois was born October 25, 1811, at Bourg–la–Reine near Paris, the second of three children of Nicholas–Gabriel and Adelaide–Marie Demante Galois. His parents were well educated, although no one in the family is known to have excelled in mathematics. His father, an ardent republican and a composer of light verse, was director of a boarding school and was elected mayor of Bourg–la–Reine in 1815. His mother, who came from a family of jurists, had been trained in religion and the classics and was the only teacher Galois knew until he was age 12.

In October 1823, Galois' parents sent him to the Lycée of Louis–le–Grand in Paris. At first, Galois did well, receiving a prize in the General Concourse and three mentions. His character was generally described as "good, but singular." In 1827, after he was demoted because of deficiencies in rhetoric, he began his first mathematics class under M. Vernier. Galois read **Adrien–Marie Legendre**'s geometry text and **Joseph Lagrange**'s original memoirs with ease, and apparently was able to work out complicated problems without pencil or paper. As he became absorbed in mathematics, he neglected his other courses. Although Vernier remarked of Galois' "zeal and progress," he described him as "closed and original" and termed his work "inconstant." Vernier urged Galois to work methodically, but he did not take his instructor's advice. Galois was eager to enter the École Polytechnique, which trained mathematicians and engineers, but without the preparation in basics, he took the entrance exam a year early and failed.

### Publishes First Paper

Later in 1827 Galois enrolled in the mathematics course of Louis-Paul-Emile Richard. Richard, a distinguished teacher, found him markedly superior to other students and believed he should be admitted to the Polytechnique without examination. In April 1829, Galois' first mathematics paper, a minor work on continued fractions, was published in the *Annales de Gergonne*. In May and June, while still only 17, he submitted two articles on the algebraic solution of equations to the Paris Académie Royale des Sciences, with **Augustin-Louis Cauchy** as referee. It has been widely written that Cauchy ignored and lost the manuscripts, but letters indicate that he had read and was impressed with them. Tony Rothman suggests that Cauchy encouraged Galois and recommended that he combine the two papers and submit them for the Académie's Grand Prize in Mathematics. In February 1830, Galois submitted such an entry to the Académie's secretary, Jean Baptiste Joseph Fourier. Unfortunately, Fourier died in May and Galois' entry was not found among his papers.

In April of 1830, Galois, now a student at the École Normale, published "An Analysis of a Memoir on the Algebraic Resolution of Equations" in the *Bulletin de Ferussac*. In June, he published "Notes on the Resolution of Numerical Equations" and "On the Theory of Numbers." These and a later memoir make up what is now called Galois theory. In July, the French revolution of 1830 came to a head, and the director of the École Normale, M. Guigniault, locked in his students so that they would not participate in the rioting. Galois tried to scale the walls, but failed, and wrote a letter critical of Guigniault to the *Gazette de Écoles,* which resulted in his expulsion. Galois then joined the Republican Artillery of the National Guard, but it was shortly abolished by royal decree because of its perceived threat to the King. Galois was probably also a member of the Society of the Friends of the People, a secret republican society. In January 1831, he organized a private class in algebra which attracted 40 students, and at the invitation of Siméon-Denis Poisson, submitted a third version of his paper to the Académie.

### Is Arrested and Jailed

On May 9, 1831, at a noisy republican banquet celebrating the acquittal of 19 guardsmen of conspiracy, Galois held a glass and an open dagger and toasted to King Louis-Phillipe. The next day he was arrested for threatening the King's life, and was imprisoned until June 15 when he was tried and acquitted. But on July 14, Bastille Day, Galois, armed with a rifle, dagger, and pistols, was again arrested, this time for wearing the illegal Artillery Guard uniform. He was sentenced to six months in prison.

In October, still incarcerated at Saint-Pélagie, he received word of Poisson's rejection of his paper. Poisson noted that Galois' argument was "neither sufficiently clear nor sufficiently developed to allow us to judge its rigor; it is not even possible for us to give an idea of this paper," and suggested a more complete account. Resolving to publish his papers privately with the aid of his friend, Auguste Chevalier, Galois gathered them and wrote a vitriolic five-page preface.

### Dies in a Duel

Because of a threatened cholera outbreak at the prison, Galois was transferred to the pension Sieur Faultrier in March 1832 and was set free on April 29. Just a month later, on May 30, he faced Pescheux d'Herbinville (a republican and one of the guards at the pension) in a duel. The circumstances leading to the duel are not known; however, it seems clear that it was not political, but rather a personal quarrel that involved Stéphanie du Motel, the daughter of a physician at the pension. Galois was shot in the abdomen and was unattended for hours until a peasant took him to the Hospital Cochin. Galois refused the services of a priest and died the following day, with his brother, Alfred, beside him. Galois was buried in a common burial ground in South Cemetery. No trace remains of his grave.

The night before the duel, Galois wrote letters and made notes and corrections on some of his papers, entrusting them to his friend Chevalier. "All I have written down here has been clear in my head for over a year," he wrote. "Make a public request of Jacobi or Gauss to give their opinions not as to the truth but as to the importance of these theorems. After that, I hope some men will find it profitable to sort out this mess."

Alfred Galois and Chevalier copied the papers and submitted them to **Karl Gauss, Karl Jacobi**, and others, and in 1846 **Joseph Liouville** edited some of Galois' manuscripts for publication in the *Journal de Mathematiques*. Liouville noted, "I saw the complete correctness of the method by which Galois proves, in particular, this beautiful theorem: In order that an irreducible equation of prime degree be solvable by radicals it is necessary and sufficient that all its roots be rational functions of any two of them."

## FURTHER READING

### Books

Bell, E. T. *Men of Mathematics.* New York: Simon and Schuster, 1937, pp. 362–377.

*Johann Karl Friedrich Gauss*

Kolmogorow, A. N. and A. P. Yushkevich. *Mathematics of the 19th Century*. Boston: Birkhauser Verlag, 1992, pp. 57–63.

Motz, Lloyd and Jefferson Hane Weaver. *The Story of Mathematics*. New York: Plenum Press, 1993, pp. 182–191.

Stewart, Ian. *Galois Theory*. London: Chapman and Hall, 1973, pp. xvii–xxii.

Stillwell, John. *Mathematics and Its History*. New York: Springer–Verlag, 1989, pp. 287–291.

**Periodicals**

Rothman, Tony. "Genius and Biographers: The Fictionalization of Évariste Galois." *The American Mathematical Monthly* (February 1982): 84–106.

—*Sketch by Jill Carpenter*

# Johann Karl Friedrich Gauss
## 1777–1855
### German geometer and astronomer

Gauss invites comparisons only to **Isaac Newton** or perhaps Johann Wolfgang von Goethe, being the sort of endlessly inventive mind that achieved results when put to any task. His contributions to the fields of pure and applied mathematics were all equally sensational in his day, and equally influential into the 20th century. Gauss' major discoveries reached back to Greek practices, either updating or employing them to novel use. His penchant for publishing only the most rigorous and polished proofs had set a standard for arguments in symbolic logic not seen before. Gauss' formulation of the complex number system advanced number theory so that all possible operations could be performed on all possible numbers without needing to create new ones. His investigations into algebra and geometry paved the way for the modern disciplines of probability theory, topology, and vector analysis. Among Gauss' inventions and collaborations include the heliotrope (a trigonometric measuring device), a prototype of the electric telegraph, and the bifilar magnetometer. Gauss' interests also ran to crystallography, optics, mechanics, and capillarity.

Johann Friedrich Carl Gauss was the only son born to Gebhard Dietrich, a laborer and merchant, and Dorothea Benze Gauss, a servant. They made their home in Brunswick, capital of the Duchy of Braunschweig, Germany. As Gauss himself later calculated he was born on April 30, 1777, eight days after Ascension of that year as his mother always told him. His mother was a functional illiterate, a fact which lent a special poignancy to Gauss' later fame, which she could only ascertain by asking others, rather than experiencing direct proof of it herself.

### Ligget se!

An arithmetic prodigy, Gauss enjoyed telling the story of catching his father's addition mistake at the age of three. The most famous vignette related to his youth involved an obnoxious teacher who instructed his class to add all the integers from one to 100. By adding them in pairs the eight–year–old boiled them down to a smaller set of 101s, fifty to be exact, and calculated the sum from there to be 5050. "Ligget se!" was all he had to say to his teacher, and showed him his slate. The formula Gauss had arrived at is given $S=n(n+1)/2$ and was actually in use during the days of **Pythagoras.** Such adroitness was initially disparaged by his father, but Gauss was eventually rewarded by a tutor's aid and admission to secondary school in 1788. He began his higher education at Caroline College in his hometown, which offered mathematical training and lessons in Latin and High German. From there, Gauss proceeded to the University of Göttingen in 1795.

Gauss never published a proof until it was airtight, but his interests ranged so far so early in his life that he preceded Bode's Law, **Janos Bolyai** and **N.I. Lobachevsky**'s non–Euclidean geometry, **Karl**

Jacobi's double–period elliptic functions, **Augustin–Louis Cauchy**'s functions of a complex variable and **William Hamilton**'s quaternions. While still a teenager, Gauss constructed a with a ruler and compass a 17–sided polygon inscribed in a circle. This was the first true innovation in Euclidean geometry since the time of the ancient Greek mathematicians.

### Achievement of Closure

Gauss also discovered the law of quadratic reciprocity and the method of least squares. In 1799, he proved the fundamental theorem of algebra: that every polynomial equation has a root in the form of a complex number a+bi. His thesis, "Disquisitiones Arithmeticae," was completed in 1798 but not published until 1801.

The University of Helmstedt awarded Gauss a doctorate in 1799. A return to the University of Göttingen allowed for his early research and later career with the help of his benefactor, the Duke of Brunswick. He held a dual post of Professor of Mathematics and Director of Göttingen Observatory by 1807, though not before enduring a period of unemployment. His most famous work in applied mathematics, *Theoria motus corporum celestium*, followed just eight years after his first major publication. In 1801, Gauss had rediscovered the "lost" orbiting asteroid or minor planet, Ceres. He successfully calculated the object's orbit according to certain observations and predicted where it would next reappear—a triumph that secured his fame. This method was refined during his subsequent tenure at Göttingen Observatory into the book–length work. He was also retained by various governments to travel, making geodetic surveys at different locations. For this, Gauss invented a new measuring device, the heliotrope.

Such applied work inspired more pure mathematics, this time differential geometry of curved space and surfaces. Gauss' third major publication, *Disquisitiones Generales Circa Superficies Curvas*, was not published until 1827 because of the subject's far–reaching implications. What he called "intrinsic" geometry would pave the way for current differential geometry. Gauss entertained the idea of the curvature of all space, an idea that would be of central importance to **Albert Einstein**'s formulation of space time as a geometric whole.

### The Newspaper Tiger

Gauss married twice, first to Johanna Osthoff on October 9, 1805. The union produced three children, the youngest of whom died soon after birth, followed by the mother. Though a recent widower, Gauss proposed to Friederica Wilhelmina Waldeck, the daughter of a fellow professor. They married August 4, 1810, and had three children before the second Mrs. Gauss died of tuberculosis. Eugene, the eldest boy from Gauss' second marriage, grew up with the same abilities as his illustrious father. However, for reasons that can only be speculated, Gauss prevented his son from following him into the mathematical field.

This reticence also held in Gauss' relationships with his students. Despite his encouragement of a protégéé named Eisenstein who died tragically young, and correspondence with the self–taught pioneer **Sophie Germain**, Gauss never really took anyone under his wing. To Gauss, lecturing would not improve a bad student nor impress a good one. He tended to consider fellow mathematicians as rivals or distractions. He was fond of newspapers and magazines, novelties of the early 19th century. Students in the university library called him the "newspaper tiger" for his habit of staring down anyone who tried to take any newspaper he wanted first.

Nearly seventy–five official honors came to Gauss throughout his life from various countries, though Gauss made light of their accompanied ceremonies, preferring instead to make curmudgeonly jokes at the speechmakers' expense. These recognitions included being installed as a Foreign Member of the Royal Society of England, but Gauss was content to be considered without question the greatest mathematician in the world and get on with his work. He was not intellectually isolated, however. At the age of 62, Gauss taught himself Russian so he could more easily read the works of Lobachevsky. Eventually his health failed, and the loss of friends and family members through death and estrangement took a toll. Gauss died February 23, 1855, of a heart attack after suffering from an enlarged heart for some time. He was buried in Göttingen next to the simple grave of his mother.

Throughout his career, Gauss repositioned pure inquiry as the ultimate test of logic, unbuttressed by geometric or theoretical assumptions and circular arguments. He considered mathematics as a science, with arithmetic as its most important subdiscipline. Gauss avoided trivialities by realizing that one cannot study a magnitude in isolation, for true mathematics lies in the study of relationships. Moreover, he envisioned new relationships to consider among infinite series, hypercycles, and pseudospheres, the sort of fanciful mathematical entities that populate the imaginations of contemporary theorists.

### SELECTED WRITINGS BY GAUSS

*Inaugural Lecture on Astronomy and Papers on the Foundation of Mathematics.* Translated by G. Waldo Dunnington. 1937

*General Investigation of Curved Surfaces.* Translated by Adam Hiltebeitel and James Morehead. 1965

*Disquisitiones Arithmeticae.* Translated by Arthur A. Clarke. 1966

## FURTHER READING

Buhler, Walter K. *Gauss: A Biographical Study.* New York: Springer–Verlag, 1981.

Dunnington, G. Waldo. *Carl Friedrich Gauss: Titan of Science.* New York: Exposition Press, 1955.

Hall, Tord. *Carl Friedrich Gauss: A Biography.* Cambridge, MA: MIT Press, 1970.

Merzbach, Uta C. *Carl Friedrich Gauss: A Bibliography.* Wilmington, DE: Scholarly Resources, 1984.

Monna, A. F. *Carl Friedrich Gauss: 1777–1855.* Utrecht, Holland: Rijkauniversiteit, 1978.

Muir, Jane. *Of Men and Numbers.* New York: Dodd, Mead and Co., 1965, pp. 157–83.

Reich, Karin. Carl Friedrich Gauss: 1777–1977. Bonn, Germany: Inter Nationes, 1977.

Schaaf, William Leonard. *Carl Friedrich Gauss: Prince of Mathematics.* New York: Watts, 1964.

*—Sketch by Jennifer Kramer*

# Hilda Geiringer
## 1893-1973
### Austrian-born American applied mathematician

Hilda Geiringer was an applied mathematician who made important contributions to the theory of plasticity of materials. She formulated the Geiringer equations for plane plastic deformations in 1930. She also pursued research in probability, statistics, genetics, and numerical methods. A refugee from Europe during World War II, Geiringer was among the European mathematicians who brought an emphasis on applied mathematics to the United States, where pure mathematics predominated. In the summer of 1942, she participated in the development of an applied mathematics program at Brown University, presenting a series of lectures on the geometric foundations of the mechanics of a rigid body. After the death of her husband, mathematician **Richard von Mises** , in 1953, Geiringer worked on the publication of new editions of his works as well as her own research.

Hilda Geiringer was born in Vienna, on September 28, 1893. She was the daughter of Ludwig, a textile manufacturer, and Martha Wertheimer Geiringer. She showed a talent and interest in mathematics at an early age. Her parents supported her studies in mathematics at the University of Vienna, where she received a Ph.D. in 1917 for her thesis on double trigonometric series. In 1919 and 1920 Geiringer assisted the editor of *Fortschritte der Mathematik* (Advances in Mathematics).

During the following year, Geiringer moved to Germany to work at the Institute of Applied Mathematics in Berlin, under Richard von Mises, a founder of mathematical aerodynamics and contributor to probability theory. This was the beginning of Geiringer's productive career in applied mathematics. She began to publish papers on probability and on the mathematical characterization of plasticity, the bending of material after deformation. In 1927 Geiringer became a lecturer at the University of Berlin.

### Flight and Refuge

Geiringer, who was Jewish, was removed from the University in 1933; she moved to Belgium and then to Turkey. From 1934 to 1939 she was Professor of Mathematics at the University of Istanbul. There, she learned Turkish for her lectures. When war broke out in 1939, Geiringer fled to the United States, where she taught at Bryn Mawr from 1939 to 1944. During this period, Geiringer published papers on probability as well as notes for her lectures at Brown.

Geiringer had married Felix Pollaczek in 1921. They had one daughter, Magda, born in 1922, but they divorced in 1925. Geiringer took Magda with her to Istanbul and then to the United States. In 1943 Geiringer married Richard von Mises, who had also come to the United States via Turkey. He became a lecturer and then Professor of Aerodynamics and Applied Mathematics at Harvard. Geiringer became a United States citizen in 1945.

From Bryn Mawr, Geiringer went to Wheaton College in Norton, Massachusetts, where she became Chairman of the Mathematics Department. In the late forties, Geiringer wrote several papers on statistics applied to Mendelian genetics and two papers on numerical methods. In the early fifties she took up plasticity again in a more general form. After the death of von Mises, Geiringer worked at Harvard under a grant from the Office of Naval Research to complete his work. In 1957 she published a new edition of his book *Probability, Statistics, and Truth.* Her work with G.S.S. Ludwig and von Mises, *Mathematical Theory of Compressible Fluid Flow,* appeared in 1958. The new edition of von Mises' *Mathematical Theory of Probability and Statistics,* with Geiringer's complementary material, was published in 1964. Geiringer wrote papers and lectured on probability during this period and wrote an article entitled "The Mathematical Theory of the Inelastic Continuum" with A.F. Freudenthal for the *Encyclopedia of Physics*.

Geiringer retired from Wheaton in 1959, but continued her research work at Harvard. Wheaton gave her an honorary degree in 1960. Geiringer was made Professor Emeritus by the University of Berlin in 1956, and was honored by the University of Vienna on the fiftieth anniversary of her graduation. On March 22, 1973, during a visit with her younger brother, Karl, a noted musicologist, in Santa Barbara, Geiringer died of influenzal pneumonia.

## SELECTED WRITINGS BY GEIRINGER

*Mathematical Foundations of the Theory of Isotropic Plastic Bodies* (title translated),
   Mémorial des Sciences Mathématiques, 1937
*Geometrical Foundations of Mechanics*, 1942

## FURTHER READING

### Books

*Notable American Woman: The Modern Period.*
   Cambridge, MA: Harvard University Press,
   1980, pp. 267–268.

### Periodicals

*Boston Sunday Globe* (March 25, 1973).
*New York Times* (July 19, 1953): 25; (March 24, 1973): 36.
Rees, Mina. "The Mathematical Sciences and World War II." *American Mathematical Monthly* (October 1980): 607–621.

—*Sketch by Sally M. Moite*

# Aleksandr Osipovich Gelfond
## 1906–1968(?)
### Russian algebraist, number theorist, and educator

Aleksandr Gelfond made significant contributions to the theory of transcendental numbers and the theory of interpolation and approximation of the functions of a complex variable. He established the transcendental character of any number of the form $a^b$, where $a$ is an algebraic number different from 0 or 1 and $b$ is any irrational algebraic number, which is now known as Gelfond's theorem.

*Aleksandr Osipovich Gelfond*

Gelfond was born in St. Petersburg (later Leningrad); his father was a physician who also dabbled in philosophy. Gelfond entered Moscow University in 1924 and completed his undergraduate degree in mathematics in 1927. He pursued postgraduate studies from 1927 to 1930 under the direction of A.J. Khintchine and V.V. Stepanov.

Gelfond's first teaching assignment was at the Moscow Technological College. He quickly won a more prestigious appointment at Moscow University, where he began teaching mathematics in 1931. He became a professor of mathematics in 1931, a position he held until his death. For several years, Gelfond served as the chairman of the mathematics department specializing in the theory of numbers. His enthusiasm for the history of mathematics was reflected not only by his own works on **Leonhard Euler**, but by incorporating a history of mathematics division into the theory department he chaired.

In 1933, Gelfond was also appointed to a post in the Soviet Academy of Sciences Mathematical Institute. He completed a doctorate in mathematics and physics in 1935 and was elected a corresponding member of the Academy of Sciences of the U.S.S.R. in 1939.

### Back to the Future

Gelfond found his greatest inspiration in the past. In 1748, Euler proposed that logarithms of rational number with rational bases are either rational or transcendental; in 1900, **David Hilbert** presented a statement of twenty-three fundamental open problems to challenge the mathematicians of the 20th century. For three decades, mathematicians were unable to trace a solution to the puzzle posed by Hilbert's seventh problem—the assumption that $a^b$ is transcendental if $a$ is any algebraic number other than 0 or 1 and $b$ is any irrational algebraic number.

In 1929, Gelfond established connections between the properties of an analytic function and the arithmetic nature of its values, publishing his first paper on the topic, "Sur les nombres transcendant," in 1929. He built on this discovery to unravel Hilbert's seventh riddle by using linear forms of exponential functions. Gelfond published the results of his work, "Sur le septieme probleme de Hilbert," in 1934. He continued his explorations, using his knowledge of functions to develop theorems related to rational integers, transcendental numbers (he was able to construct new classes in this area), mutual algebraic independence, and analytic theory.

Gelfond's interest in function theory was probably shaped by the Luzitania—an informal academic and social organization clustered around Nikolai Nikolaevich Luzin, a noted mathematician in the 1920s. Gelfond was a contemporary and colleague of **Nina Karlovna Bari**, a Luzin protegee; although Gelfond's name does not appear on the list of those who declared themselves Luzitanians. Luzin's prominence and the intellectual vigor of the students he attracted influenced the philosophy and direction of mathematics at the university. By 1930, the Luzitania movement sputtered and died, and Luzin left Moscow State for the Academy of Science's Steklov Institute.

## Politician Pragmatism

In 1936, during the dictatorship of Josef Stalin, Luzin was charged with ideological sabotage. Luzin's trial was abruptly and surprisingly canceled, but he was officially reprimanded and withdrew from academia. Luzin's fall demonstrated—in a way that could not be ignored—the inextricable interweaving of politics and academic achievement. Gelfond was permitted to pursue his studies in peace in part because of his political connections.

"He was a member of the Communist Party," wrote Ilya Piatetski–Shapiro, for whom Gelfond was an instructor, mentor and advisor. "His father was personally acquainted with Lenin . . . he said that his father and Lenin had disagreements in public life, but in private life they were friends. Being a member of the Communist Party, Gelfond felt that he had some influence. . . ." Such influence could not overcome the deep wave of anti–Semitism that swept over Russia after World War II. Despite Gelfond's recommendation, Piatetski–Shapiro, who was Jewish, was denied

admission to Moscow University's graduate school by the party committee of the mathematics department.

But Gelfond "was a very warm person, very humane and sensitive to me and to the other students," Piatetski–Shapiro wrote, and Gelfond was reluctant to let a promising student—winner of the Moscow Mathematical Society award for young mathematicians—languish. Although his sponsorship could have had dire implications for his own career, Gelfond persisted, and finally secured admission to the graduate program for Piatetski–Shapiro at the Moscow Pedagogical Institute.

Gelfond's most comprehensive publications were released in 1952. *Transtsendentnye i albegraicheskie chisla* provided an overview of his work in transcendental numbers, and his work on the theory of the functions of a complex variable is compiled in *Ischislenie knoechnyko raznostey*.

In 1968, Gelfond was named a corresponding member of the International Academy of the History of Science. He also served as chair of the scientific council of the Soviet Academy of Sciences Institute of the History of Science and technology, which refereed works on the history of physics and mathematics.

Gelfond's drive to expand the understanding of mathematics theory persisted to the day of his death. "When he died . . . I was present in the hospital," wrote Piatetski–Shapiro. "I remember he was trying to write some formula and tell me something which was clearly related to the zeta function. He could not because he was already paralyzed."

Gelfond died in Moscow; most sources list the year of his death as 1968, but Piatetski–Shapiro records it as 1966.

## SELECTED WRITINGS BY GELFOND

*Ischislenie konechnykh raznostey,* 1952; third edition, 1967

*Transtsendentynye i algegraicheskie chisla* (1952; translated and published as *Transcendental and Algebraic Numbers,* 1960)

(With Y.V. Linnik) *Elementarnye metody v teorii chisel,* 1962

### Periodicals

"Sur les nombers transcendants." *Comptes rendus hebdomadaires des seances de l'Academie des sciences,* 189 (1929): 1224–28.

## FURTHER READING

Bell, E.T. *Men of Mathematics.* New York: Simon and Schuster, 1965, p. 463.

Youschkevitch, A. P. "Aleksandr Osipovich Gelfond," in *Dictionary of Scientific Biography*. Volume V. Edited by Charles Coulston Gillispie. New York: Charles Scribner's Sons, 1973, pp. 342–43.

Zdravkovska, Smilka and Peter L. Duran, editors. *Golden Years of Moscow Mathematics*. Providence, RI: American Mathematical Society, 1993, pp. 200–203.

—*Sketch by A. Mullig*

# Geminus
## c. 130–c.70 B.C.
### Greek philosopher, astronomer, and mathematician

Geminus was born around 130 B.C., possibly on the Greek island of Rhodes. He is primarily known for his classification of mathematics, which Proclus recorded in his Commentary on **Euclid.** In addition to his efforts to define mathematics as a science, and to define science in general, Geminus wrote introductions to mathematics and astronomy. His astronomical work, *Introduction to Astronomy*, which surveys Greek astronomy from the standpoint of **Hipparchus,** is still extant, while the mathematics textbook, *Theory of Mathematics*, which Proclus consulted, survives only in fragments.

It is known that Geminus was influenced by the Stoic philosopher Poseidonius of Apamea, who opened a school of philosophy at Rhodes in 97 B.C. Poseidonius combined Stoic monism (which defines the world as a hierarchy, rising from inorganic to organic beings, and ruled by God, or Absolute Reason) with Plato's dualism (which divided the universe into two worlds: one mortal, the other immortal). Both facets of Poseidonius' doctrine evince a hierarchical conception, which seems evident in Geminus' classification of mathematics.

### Proposes New Definition and Classification of Mathematics

Since geometry was still the dominant mathematical science when Geminus wrote, his book on mathematics essentially focused on this discipline, also providing a historical review of Greek geometry. He discussed a number of geometric concepts including line, surface, and angle. His classification of mathematics, however, was an attempt to define the entire science. In accordance with the predominant Greek view of mathematics, inspired by Aristotle, that pure science (the quest for knowledge which justifies itself) is superior to applied science, Geminus maintained the fundamental dichotomy of "pure" and "impure" in his system.

Proclus tells us in his Commentary on Book I of **Euclid's** *Elements* that the terms Geminus uses to define pure and applied mathematics are, respectively, "intelligible" and "tangible." Of the classification itself, Proclus writes: "On one side [Geminus and others] place tangible things, or anything connected with them. No doubt they would call intelligible the subjects of contemplation that the soul elaborates within itself, cutting itself from thing material. They rank arithmetic and geometry as the two first and most important sectors of that form of mathematics that deal with that which is intelligible. With regard to the mathematics which covers things tangible, they designate six sectors: mechanics, astronomy, optics, geodesy, canonics, and logistics. On the other hand, for them, unlike others, tactics is not worthy to be considered a part of mathematics, even though it at times involves logistics, as in the enumeration of troops, and geodesy, as in the surveying and the division of land."

Geminus died around 70 B.C., but the place of his death is not known.

### FURTHER READING

Clagett, Marshall. *Greek Science in Antiquity*. New York: Abelard–Schuman, 1955.

Sarton, George. *From Homer to Omar Khayyam*. Volume 1: *Introduction to the History of Science*. Washington, D.C.: Carnegie Institution of Washington, 1953.

Vitrac, Bernard. "The Odyssey of Reason," in *From Five Fingers to Infinity: A Journey through the History of Mathematics*. Edited by Frank J. Swetz. Chicago: Open Court, 1994.

—*Sketch by Zoran Minderovic*

# Ruth Gentry
## 1862–1917
### American geometer

Ruth Gentry wrote her doctoral thesis, "On the Forms of Plane Quartic Curves," in 1896, before receiving her Ph.D. in Mathematics from Bryn Mawr College. Bryn Mawr was the first school to offer resident fellowships to women who aspired to earn a graduate degree. Gentry was one out of only 10 women to receive Ph.D.'s in mathematics during the 19th century. Her teaching career consisted of teaching at Vassar College and schools for girls and young women.

Prior to her graduate work at Bryn Mawr, Gentry earned a degree in mathematics from the University of Michigan. She was the second recipient of the Association of College Alumnae European Fellowship, as well as the first mathematician to ever receive the honor. From 1891 to 1892, Gentry used the fellowship to attend lectures at the University of Berlin; however, she was barred from enrollment. At that time, women were not admitted to German universities.

During her stay in Germany, Gentry wrote to **Christian Felix Klein** at the University of Göttingen, asking him whether or not she could be admitted to his lectures the following year. Klein replied that she could not. In 1893, however, Klein did suggest a trial program of admitting a few women into the mathematics program at the University of Göttingen. The program went forward, but Gentry was not selected to attend.

Throughout the 19th century myths were rampant about why women appeared to be inferior to men in science and mathematics. By the 1860s, people generally believed that the frontal lobe inside male brains, thought to be where intelligence is located, were larger than the female brain. However, discoveries in the 1870s dispelled that myth, and scientists brushed the information aside when it was revealed that women actually had the larger frontal lobe.

Perhaps because of such myths of that era, and because of basic cultural differences, Gentry, as a woman mathematician, struggled to be recognized for her work.

## SELECTED WRITINGS BY GENTRY

*On the forms of plane quartic curves,* 1896

## FURTHER READING

### Books

Dunham William. *The Mathematical Universe: An Alphabetical Journey through the Great Proofs, Problems, and Personalities.* New York, 1994.

Fenster, Della and Karen Parshall. "Women in the American Mathematical Research Community: 1891–1906," in *The History of Modern Mathematics* Volume. III. Edited by Eberhard Knobloch and David Rowe, pp. 229–261.

### Periodicals

Green, Judy and Jeanne LaDuke. "Women in the American Mathematical Community: The pre–1940 Ph.D.'s." *The Mathematical Intelligencer* 9, no.1 (1987): 11–22.

Kenschaft, Patricia. "The Students of Charlotte Angas Scott." *Mathematics in College* (Fall 1982): 16–20.

### Other

"Ruth Gentry." *Biographies of Women Mathematicians.* June 1997. http://www.scottlan.edu/lriddle/women/chronol.htm (July 22, 1997).

"Women in Mathematics Myths." http://www.telplex.bsu.edu/home/nshadle/web/myths.html. (June 2, 1997).

*—Sketch by Monica L. Stevens*

# Sophie Germain
## 1776–1831
### French number theorist

Sophie Germain's foundational work on Fermat's Last Theorem, a problem unsolved in mathematics into the late 20th century, stood unmatched for over one hundred years. Though published by a mentor of hers, **Adrien–Marie Legendre**, it is still referred to in textbooks as Germain's Theorem.

Germain worked alone, which was to her credit, yet contributed in a fundamental way to her limited development as a theorist. Her famed attempt to provide the mystery of Chladni figures with a pure mathematical model was made with no competition or collaboration. The three contests held by the Paris Académie Royale des Sciences from 1811 to 1816, regarding acoustics and elasticity of vibrating plates, never had more than one entry—hers. Each time she offered a new breakthrough: a fundamental hypothesis, an experimentally disprovable claim, and a treatment of curved and planar surfaces. However, even her final prizewinning paper was not published until after her death.

#### The Femme Savante

Marie–Sophie Germain was born April 1, 1776, in Paris to Ambroise–François Germain and Marie–Madeleine Gruguelu. Her father served in the States–General and later the Constituent Assembly during the tumultuous Revolutionary period. He was so middle class that nothing is known of his wife but her name. Their eldest and youngest daughters,

*Sophie Germain*

Marie–Madeleine and Angelique–Ambroise, were destined for marriage with professional men. However, when the fall of the Bastille in 1789 drove the Germains' sensitive middle daughter into hiding in the family library, Marie–Sophie's life path diverged from them all.

From the ages of 13 to 18 Sophie, as she was called to minimize confusion with the other Maries in her immediate family, absorbed herself in the study of pure mathematics. Inspired by reading the legend of **Archimedes,** purportedly slain while in the depths of geometric meditation by a Roman soldier, Germain sought the ultimate retreat from ugly political realities. In order to read **Leonhard Euler** and **Isaac Newton** in their professional languages, she taught herself Latin and Greek as well as geometry, algebra, and calculus. Despite her parents' most desperate measures, she always managed to sneak out at night and read by candlelight. Germain never formally attended any school or gained a degree during her entire life, but she was allowed to read lecture notes circulated in the École Polytechnique. She passed in her papers under the pseudonym "Le Blanc."

### Correspondence School

Another tactic Germain used was to strike up correspondences with such successful mathematicians as **Karl Gauss** and Legendre. She was welcomed as a marvel and used as a muse by the likes of **Jean B. Fourier** and **Augustin–Louis Cauchy**, but her contacts

did not develop into the sort of long–term apprenticeship that would have compensated for her lack of access to formal education and university–class libraries. Germain did become a celebrity once she dropped her pseudonym, however. She was the first woman not related to a member by marriage to attend Académie des Sciences meetings, and was also invited to sessions at the Institut de France—another first.

Some interpret Gauss' lack of intervention in Germain's education and eventual silence as a personal rejection of her. Yet this conclusion is not borne out by certain facts indicating Gauss took special notice. In 1810, Gauss was awarded one of his many accolades, a medal from the Institut de France. He refused the monetary component of this award, accepting instead an astronomical clock Germain and the institute's secretary bought for him with part of the prize. Gauss' biographer, G. Waldo Dunnington, reported that this pendulum clock was used by the great man for the rest of his life.

Gauss survived her, expressing at an 1837 celebration that he regretted Germain was not alive to receive an honorary doctorate with the others being feted that day. He alone had lobbied to make her the first such honored female in history. A hint of why Gauss valued her above the men who joined him in the Académie is expressed in a letter he sent to her in 1807, to thank her for intervening on his behalf with the invading French military. A taste for such subjects as mathematics and science is rare enough, he announced, but true intellectual rewards can only be reaped by those who delve into obscurities with a courage that matches their talents.

### No–Man's Land

Germain was such a rarity. She outshone even **Joseph–Louis Lagrange** by not only showing an interest in prime numbers and considering a few theorems, about which Lagrange had corresponded with Gauss, but already attempting a few proofs. It was this almost reckless attack of the most novel unsolved problems, so typical of her it is considered Germain's weak point by 20th century historians, that endeared her to Gauss.

Germain's one formal prize, the Institut de France's Gold Medal Prix Extraordinaire of 1816, was awarded to her on her third attempt, despite persistent weaknesses in her arguments. For this unremedied incompleteness, and the fact that she did not attend their public awards ceremony for fear of a scandal, this honor is still not considered fully legitimate. However, the labor and innovation Germain had brought to the subjects she tackled proved of invaluable aid and inspiration to colleagues and other mathematical professionals as late as 1908. In that year, L. E. Dickson, an algebraist, generalized Germain's Theorem to all prime numbers below

1,700, just another small step towards a complete proof of Fermat's Last Theorem.

Germain died childless and unmarried, of untreatable breast cancer on June 27, 1831 in Paris. The responsibility of preparing her writings for posterity was left to a nephew, Armand–Jacques Lherbette, the son of Germain's older sister. Her prescient ideas on the unity of all intellectual disciplines and equal importance of the arts and sciences, as well as her stature as a pioneer in women's history, are amply memorialized in the École Sophie Germain and the rue Germain in Paris. The house on the rue de Savoie in which she spent her last days was also designated a historical landmark.

## SELECTED WRITINGS BY GERMAIN

### Books

*Recherches sur la theorie des surfaces elastiques,* 1821
*Remarques sur la nature, les bornes et l'etendue de la question des surfaces elastiques,* 1826
*Oeuvres philosophiques de Sophie Germain,* (edited by H. Stupuy), 1879

### Periodicals

"Examen des principes qui peuvent conduire a la connaissance des lois de l'equilibre et du mouvement des solides elastiques." *Annales de chimie et de physique* 8 (1828): 123–131.
"Memoire sur la courbure des surfaces." *Journal für die reine und angewandte Mathematik* 7 (1831): 1–29.
"Memoire sur l'emploi de l'epaisseur dans la theorie des surfaces elastiques." *Journal de Mathematiques pures et appliques* 6 (1880) Supplement: S5–S64.

### Other

*Cinq lettres de Sophie Germain et C. F. Gauss.* Berlin: B. Boncompagni–Ludovici, 1880.

## FURTHER READING

### Books

Bucciarelli, Louis L. and Nancy Dworsky. *Sophie Germain: An Essay in the History of the Theory of Elasticity.* Dordrecht, Holland: D. Reidel Publishing Company, 1980.
Dunnington, G. Waldo. *Carl Friedrich Gauss: Titan of Science.* New York: Exposition Press, 1955, pp. 66–69, 93, 192.

Gray, Mary W. "Sophie Germain (1776–1831)," in *Women of Mathematics.* Edited by Louise S. Grinstein and Paul J. Campbell. New York: Greenwood Press, 1987, pp. 47–56.
Kramer, Edna E. "Sophie Germain," in *Dictionary of Scientific Biography.* Volume V. Edited by Charles Coulston Gillispie. New York: Charles Scribner's Sons, 1972, pp. 375–76.
Mozans, H. J. *Woman in Science.* New York: D. Appleton & Co., 1913, pp. 154–57.
Ogilvie, Marilyn Bailey. *Women in Science.* Cambridge, MA: MIT Press, 1986, pp. 16, 90–92.
Osen, Lynn M. "Sophie Germain." *Women in Mathematics.* Cambridge, MA: MIT Press, 1992, pp. 83–93.
Perl, Teri. *Math Equals: Biographies of Women Mathematicians.* Menlo Park, NJ: Addison–Wesley Publishing, 1978.

### Periodicals

Dalmedico, Amy Dahan. "Sophie Germain." *Scientific American* (December 1991): 117–122.
Dauben, Joseph W. "Reviews: Sophie Germain." *American Mathematical Monthly* 92 (1985): 64–70.
Gray, Mary. "Sophie Germain: A Bicentennial Appreciation." *Association for Women in Mathematics Newsletter* 6 (September–October 1976): 10–14.
Ladd–Franklin, Christine. "Sophie Germain: An Unknown Mathematician." *Century* 48 (1894): 946–49.
Sampson, J. H. "Sophie Germain and the Theory of Numbers." *Archive for History of Exact Science* 41 (1990–91): 157–61.

### Other

Caldwell, Chris K. "The Ten Largest Known Sophie Germain Primes." *The Largest Known Primes.* (1995–96). http://www.utm.edu/research/primes/largest.html#Sophie.
O'Connor, John J. and Edmund F. Robertson. "Sophie Germain." *MacTutor History of Mathematics Archives.* (December 1996). http://www-groups.dcs.st-and.ac.uk/~history/index.html
Swift, Amanda. "Sophie Germain." *Biographies of Women Mathematicians.* June 1997. http://www.scottlan.edu/lriddle/women/chronol.htm (July 22, 1997).

*—Sketch by Jennifer Kramer*

*Josiah Willard Gibbs*

# Josiah Willard Gibbs
## 1839–1903
### American mathematical chemical–physicist

J. Willard Gibbs is not as famous as the Europeans who discovered and lionized him. James Clerk Maxwell was the first and for a time nearly the only major scientist among his contemporaries to fully understand Gibbs' publications and what they implied. **Albert Einstein** called him "the greatest mind in American history." Gibbs' studies of thermodynamics and electromagnetics and discoveries in statistical mechanics made Einstein's later theories conceivable. He is also largely responsible for the field of physical chemistry, which impacted the steel and ammonia industries. Gibbs is known as the "father of vector analysis" for replacing **William Rowan Hamilton**'s quaternions in the field of mathematical physics. Thanks to him there are such ideas as the Gibbs phase rule, the Gibbs adsorption isotherm regarding surface tension, Gibbs free energy, and Gibbsian ensembles. Even two short letters to *Nature* in the late 1890s defined what is now known as the "Gibbs phenomenon" in the convergence of a Fourier series. Gibbs' deployment of probability set the stage for quantum mechanics to come about some decades after his death. For all of these achievements, he was elected to the Hall of Fame for Great Americans in 1950.

The Gibbs family was originally from Warwickshire, England, having emigrated to Boston in the 17th century. Josiah, born on February 11, 1839, bears the same name as his father but was not known as Josiah Willard Gibbs, Jr. The two men eventually differentiated themselves according to their use of initials. Gibbs' father, a professor of biblical or "sacred" literature at Yale University, went by J.W. Gibbs. His wife Mary Anna's maiden name was Van Cleve.

Josiah was the only boy in the family. Of his four sisters he would remain closest to Anna, who also never married, and Julia, who married Addison Van Name, a member of the Connecticut Academy that first published Josiah's articles. The Van Name home in New Haven, Connecticut, where Josiah would stay later in life until his death, was within walking distance of the house where he was born.

### A Practical Education

Josiah's childhood was marred by scarlet fever, which left him sensitive to illness in adulthood. However, his home life was otherwise supportive. It has been assumed that the young Gibbs' latent scientific talents were actually inherited from his mother, who was an amateur ornithologist. Her wit and charm were widely acknowledged, and her ingenuity extended to building dollhouses for her girls that included realistic plumbing and kitchen equipment. His father, who was considered a prodigy in his day, was an exemplary scholar and teacher in the humanities who received an honorary degree from Harvard after his retirement. Tragedy came later, as two sisters and both parents eventually died by the time Gibbs was a graduate student.

Gibbs began school at the age of nine, in a private boy's school known informally as "Mr. Farren's School." From there he transferred to Hopkins Grammar School, no more than half a block from the family home. At Yale, it seemed likely that Gibbs' would follow his father into philology. He was a highly decorated Latin student often chosen to give orations at university functions. Like his father, he earned a bachelor's degree at 19.

The American academy at that time valued and rewarded only applied science and mathematics, so when Gibbs' continued his studies at Yale he wrote a fairly pedestrian doctoral thesis on spur gear design in 1863. This made him one of three Ph.D. recipients that year, at the first American institution to offer the degree. His doctorate in particular was the first in engineering and the second in science ever conferred in the United States. After a short stint as a Latin tutor, he returned to his chosen field. Gibbs was awarded a patent for his redesigned railway car brakes in 1866. That same year he took his one major trip abroad. For three years he attended physics lectures at Paris, Berlin, and Heidelberg's universities given by the field's foremost practitioners. Gustav Kirchhoff and Hermann von Helmholtz were of particular influence. Luckily, Gibbs could subsist on monies inherited from his parents, because upon his return to Yale he was appointed a professor of mathematical physics at no pay. This has been explained as resulting from his lack of published works. Gibbs would keep that unpaid post until an offer from Johns Hopkins University in 1880 of three thousand dollars a year forced Yale authorities to counteroffer two thousand.

### A Late Bloomer

In 1873 Gibbs devised a geometrical representation of the surface activity of thermodynamically active substances, and wrote "Graphical Methods in the Thermodynamics of Fluids," his first publication. Although the publishers of *Transactions of the Connecticut Academy* did not fully understand this or his other papers, they raised money specifically to print his material, which was sometimes lengthy. Much has been made of the fact that he was by then 34 years old. Most mathematicians or scientists peak at a much younger age. However, Gibbs had apparently been developing his theories for quite a long time, and was only just beginning to articulate his discoveries. He built on principles previously set down by Helmholtz, Jules Joule, and Lord Kelvin, but whereas his predecessors had been specifically concerned with an immediate example like the heat engine, Gibbs preferred to keep it mathematically general. His diagrams treated entropy, temperature, and pressure, in their relations to volume, as coordinates. When he considered a three–dimensional surface, the coordinates he chose were entropy, volume, and energy.

By mathematically formulating the second law of thermodynamics regarding entropy and mechanical energy, Gibbs made thermodynamics scientifically viable. His phase rule is a simple looking formula: $f = n + 2 - r$. In that sequence $f$ represents the total degrees of freedom in temperature and pressure, $n$ the total of chemical elements in the object's makeup, and $r$ the number of phases the object may take over time—solid, liquid, or gas in any combination. This rule was central to his reconception of the thermodynamics of a complex of systems into a single system over time—the probability of the existence of any possible system in "phase–space" overall.

By doing so, and by applying the first and second laws of thermodynamics to complex substances, Gibbs set the theoretical basis for physical chemistry. This relatively new specialty deals with phenomena like hydrodynamics, and novel types of mathematical modeling like electrochemical simulations and genetic algorithms. Gibbs' "single system" method is now a fundamental part of statistical mechanics in the form of Gibbsian ensembles. These are defined as large numbers of thermodynamically equivalent macroscopic systems. By using these ensembles, laboratory and factory researchers save themselves the tedium and risk of trial and error experimentation when synthesizing new compounds or alloys.

"Gibbs free energy" refers to the likelihood of any one chemical reaction taking place, which takes into account both entropy—the disorder in a system—and enthalpy, its heat content. At least one biographer considers Gibbs' clarification of Rudolf Clausius' original definition of entropy, as it became ranked with other thermodynamic properties such as energy, temperature, and pressure. His ideas of chemical potential and free energy now take precedence in the conception of how chemical reactions take place.

### The American Lavoisier

Although he drew mixed reviews as a teacher, Gibbs was clearly concerned with his relationship with his students and with involving them in mathematics. In 1877 he founded the Math Club at Yale, the second of such informal groups there. He served as executive officer for ten years, and more than likely gave his first impressions of vector analysis and perhaps even the multiple algebra or matrix studies that gave birth to vectors. There are other indications that he used his classes as seminars.

For his vector analysis Gibbs drew from **Hermann Grassman** as well as William Hamilton. Even in its early stages he could use it to calculate the orbits of planets and comets such as Swift's, improving upon **Karl Gauss'** method, and also applied vectors to problems in crystallography. Gibbs accounted for most of the properties of light as an electromagnetic phenomenon according to Maxwell's theory, in purely theoretical terms, during the 1880s. In doing so, he succeeded in treating the relation between force and displacement waves in electricity in the same way as others had for mechanical and acoustic waves. All this was an outgrowth of Gibbs' courses from 1877 to 1880, including the first college–level course in vector analysis with concentration on electricity and magnetism and the first public usage of vector methods. His analyses have not had to be corrected since.

Gibbs' ideas could not be disseminated in Europe until they were translated, because no one on the Continent at that time followed American scientific or mathematical journals consistently. Consequently, both Helmholtz and Karl Planck unknowingly duplicated some of Gibbs' findings, and Jacobus Van't Hoff independently conceived of chemical thermodynamics. This situation changed when Friedrich Ostwald translated Gibbs into German in 1891. Ostwald was followed near the end of that decade by Henri Le Chatelier, who translated Gibbs into French. Edwin Wilson, who also wrote posthumous biographical commentaries on his professor, wrote a textbook published in 1901 entitled *Gibbs' Vector Analysis* that succeeded in reaching a larger audience. The book was edited from class notes used between 1881 and 1884.

Such efforts led to great fame for Gibbs near the last years of his life. After reading Gibbs in translation, one French scientist called him America's answer to Antoine Lavoisier. Gibbs was awarded honorary Doctor of Science degrees from Erlangen, Williams College, and Princeton University. Scientific and mathematical societies in Haarlem, Göttingen, Amsterdam, Manchester, and Berlin made him an honorary or foreign member. He was also given the Copley Medal by the Royal Society of London two years before he died. Before the advent of the Nobel Prize in 1901, the Copley Medal was the highest honor conferrable in the scientific world.

Gibbs became a member of the American Mathematical Society in March of 1903, one month before his death. He had also recently signed a contract to reprint his "Equilibrium" series with approximately 50 pages of additions. Because of his childhood illness, he had taken on a lifetime's regimen of mild outdoor activities such as long walks, horseback riding, and camping with close friends like fellow professor Andrew Phillips. His teaching schedule sometimes required "enforced rest" leaves of absence, but he never went against his doctor's advice. In fact, according to Wheeler's biography, Gibbs' health was excellent for more than 30 years before his final illness. However, an apparently mild illness could not be shaken off, and he died on April 28, 1903, the night before he was to resume his duties. Gibbs' last resting place, two blocks from his brother–in–law's house, is marked with a headstone identifying him only as a Yale professor.

## SELECTED WRITINGS BY GIBBS

### Books

*Elementary Principles in Statistical Mechanics Developed with Special Reference to the Rational Foundation of Thermodynamics,* 1981
*The Scientific Papers of J. Willard Gibbs.* Edited by H.A. Bumstead and R. G. Van Name, 1906
*The Collected Works of J. Willard Gibbs,* 1948

### Periodicals

"Graphical Methods in the Thermodynamics of Fluids." *Transactions of the Connecticut Academy* 2 (1873): 309–42.
"A Method of Geometrical Representation of the Thermodynamic of the Thermodynamic Properties of Substances by Means of Surfaces." *Transactions of the Connecticut Academy* 2 (1873): 382–404.

"On the Equilibrium of Heterogeneous Substances." *Transactions of the Connecticut Academy* 3 (1875–78).

## FURTHER READING

### Books

Asimov, Isaac. *Asimov's Biographical Encyclopedia of Science and Technology.* Second Revised Edition. Garden City, NY: Doubleday, 1982, pp. 485–86.
Eaves, Howard. *Great Moments in Mathematics (After 1650).* Washington, D.C.: Mathematical Association of America, Inc., 1980, pp.106–07.
Fien, Donald M. "Josiah Willard Gibbs," in *The Great Scientists.* Edited by Frank N. Magill. Danbury, CT: Grolier, 1989, pp. 117–23.
Grabiner, Judith V. "Mathematics in America: The First Hundred Years."
*The Bicentennial Tribute to American Mathematics 1776–1976.* Edited by Dalton Tarwater. Washington, D.C.: Mathematical Association of America, Inc., 1977.
Wheeler, L.P. *Josiah Willard Gibbs: The History of a Great Mind.* New Haven, CT: Yale University Press, 1951.

### Other

"Josiah Willard Gibbs." *MacTutor History of Mathematics Archive.* http://www–groups.dcs.st–and.ac.uk/~history/index/.html (July 1997).

*—Sketch by Jennifer Kramer*

# Albert Girard
## 1595(?)–1632
### French trigonomist

Few details are known about the life of Albert Girard outside of his contributions to the fields of trigonometry, geometry, arithmetic, and algebra. He also translated several important mathematical texts. Girard was born around 1595 in St. Mihiel, France, which at that time belonged to the duchy of Lorraine. The reason for this assumption is that he often added the word *Samielois* to his name, as did many printers from the town of St. Mihiel in the 17th century.

Girard was affiliated with the Reformed church and it is popularly assumed that his move to the Netherlands was brought on by opposition to Protestant religions in Lorraine. Girard was a student of **Willebrord Snell** at the University of Leiden in the Netherlands, as well as an engineer in the army of Frederick Henry of Nassau, the Prince of Orange. In addition, it is believed that he was also a professional musician.

Girard's contributions to mathematics were varied. He incorporated the use of the supplementary triangle in spherical trigonometry and simplified the concept of the plane polygon in geometry by defining three types of quadrilaterals, 11 types of pentagons, and 69 of 70 types of hexagons. He also developed a formula for determining the proper construction of two quadrilaterals with the sides of a convex quadrilateral inscribed in a circle. Girard was also the first mathematician to publicly state that the area of a spherical triangle is proportional to its spherical excess. In addition, Girard was the first mathematician to determine the geometric significance of negative numbers.

In the area of arithmetic, Girard determined the whole numbers that are the sums of two squares and revealed that certain numbers that cannot be decomposed into three squares can be decomposed into four squares. In algebra and the theory of numbers, he followed in the footsteps of **François Viète**, using his forerunners' "specious logistic" but referring to it as "literal algebra." He developed a clear rule for the extraction of the cube root of binomials, improving on a previous rule established by **Rafaello Bombelli**. Girard's rule was subsequently improved by **René Descartes** in 1640.

Girard also translated many significant mathematical texts, including Henry Hondius' 1625 treatise on fortifications from Flemish to French, and the works of Samuel Marolois and **Simon Stevin**. In addition, Girard developed a simplified version for demarking the cube root, which is still in use today.

For all his knowledge and research, however, Girard lived a poor life, as detailed in his posthumously published edition of the works of Stevin. He died young, driven out of his homeland, and unable to secure a patron. A note in a mathematical journal indicates that Girard died on December 8, 1632, and it is conjectured that he died in Leiden.

## SELECTED WRITINGS BY GIRARD

*Tables des sinus, tangentes, et sécantes selon le raid de 100,000 parties,* 1626, 1627; revised edition 1629
*Invention nouvelle en l'algèbre,* 1629, 1884

*Kurt Friedrich Gödel*

## FURTHER READING

### Books

Itard, Jean. "Albert Girard," in *Dictionary of Scientific Biography.* Volume V. Edited by Charles Coulston Gillispie. New York: Charles Scribner's Sons, 1970, pp. 408–10.
### Other

"Albert Girard." *MacTutor History of Mathematics Archive.* http://www–groups.dcs.st–and.ac.uk/~history/index.html (July 1997).

—*Sketch by Kristin Palm*

# Kurt Friedrich Gödel
## 1906-1978
### Austrian-born American logician

Kurt Friedrich Gödel was a mathematical logician who proved perhaps the most influential theorem of 20th-century mathematics—the incompleteness theorem. Although he was not prolific in his published research and did not cultivate a group of students to carry on his work, his results have shaped the development of logic and affected mathematics and philosophy, as well as other disciplines. The philosophy of mathematics has been forced to grapple with the significance of Gödel's results ever since they were announced. His work was as epoch-making as that of **Albert Einstein**, even if the ramifications have not been as visible to the general public. Gregory H. Moore, in *Dictionary of Scientific Biography*, related that in May of 1972 mathematician Oskar Morgenstern wrote that Einstein himself said that "Gödel's papers were the most important ones on relativity theory since his own [Einstein's] original paper appeared."

Gödel was born in Brünn, Moravia (now Brno, Czech Republic), on April 28, 1906, the younger son of Rudolf Gödel, who worked for a textile factory in Brunn, and Marianne Handschuh. Gödel had an older brother, Rudolf, who would study medicine and become a radiologist. The Gödels were part of the German-speaking minority in Brünn, which subsequently became one of the larger cities in the Czech Republic. The family had no allegiance to the nationalist sentiments around them, and all of Gödel's educational experience was in German-speaking surroundings. He was baptized a Lutheran and took religion more to heart than the rest of his family.

Gödel began his education in September, 1912, when he enrolled in a Lutheran school in Brünn. In the fall of 1916 he became a student in a gymnasium, where he remained until 1924. At that point he entered the University of Vienna, planning to major in physics. In 1926, influenced by one of his teachers in number theory, he changed to mathematics; he did, however, retain an interest in physics, which he expressed in a number of unpublished papers later in life. He also continued his studies in philosophy and was associated with the Vienna Circle, a gathering of philosophers of science that had great influence on the English-speaking philosophical community. Gödel never was one, however, to follow a party line, and he went his own way philosophically. He felt that his independence of thought contributed to his ability to find new directions in mathematical logic.

### Proves the Incompleteness Theorem

Gödel's father died in February of 1929, and shortly thereafter his mother and brother moved to Vienna. Gödel completed the work for his dissertation in the summer of that year. He received his doctorate in February of 1930 for his proof of what became known as the completeness theorem. The problem that Gödel had considered was the following: Euclidean geometry served as an example of a kind of branch of mathematics where all the results were derived from a few initial assumptions, called axioms. However, it was hard to tell whether any particular list of axioms would be enough to prove all the true statements about the objects of geometry. Gödel

showed in his dissertation that for a certain part of logic, a set of axioms could be found such that the consequences of the axioms would include all true statements of that part of logic. In other words, the collection of provable statements and the collection of true statements amounted to the same collection. This was a reassuring result for those who hoped to find a list of axioms that would work for all of mathematics.

In September of 1930, however, mathematical logic changed forever when Gödel announced his first incompleteness theorem. One of the great accomplishments of mathematical logic earlier in the century had been the work of two British mathematicians, **Alfred North Whitehead** and **Bertrand Russell**. Their three-volume work *Principia Mathematica* (Latin for "mathematical principles" and based on the title of a work by **Isaac Newton**), tried to derive all of mathematics from a collection of axioms. They examined some areas very thoroughly, and though few mathematicians bothered to read all the details, most were prepared to believe that Whitehead and Russell would be able to continue their project through the rest of mathematics.

Gödel's work was written up under the title "On Formally Undecidable Propositions of *Principia Mathematica* and Related Systems." In this paper, which was published in a German mathematical journal in 1931, Gödel introduced a new technique which enabled him to discuss logic using arithmetic. He translated statements in logic into statements involving only numbers, and he did this by assigning numerical values to symbols of logic. It had long been known that there were problems involved in self-reference; any statement that discussed itself, such as the statement "This statement is false," presented logical difficulties in determining whether it was true or false. The assumption of those who hoped to produce an axiomatization of all of mathematics was that it would be possible to avoid such self-referring statements.

Gödel's method of proof enabled him to introduce the technique of self-reference into the very foundations of mathematics; he showed that there were statements which were indisputably true but could not be proved by axiomatization. In other words, the collection of provable statements would not include all the true statements. Although the importance of Gödel's work in this area was not immediately recognized, it did not take long before those seeking to axiomatize mathematics realized that his theorem put an immovable roadblock in their path. The proof was not obvious to those who were not used to thinking in the terms that he introduced, but the technique of Gödel numbering rapidly became an indispensable part of the logician's tool kit.

Of the schools of mathematical philosophy most active at the time Gödel introduced his incompleteness theorem, at least two have not since enjoyed the same reputation. Logicism was the belief that all mathematics could be reduced to logic and thereby put on a firm foundation. Formalism claimed that certainty could be achieved for mathematics by establishing theorems about completeness. In the aftermath of Gödel's work, it was even suggested that his theorem showed that man was more than a machine, since a machine could only establish what was provable, whereas man could understand what was true, which went beyond what was provable. Many logicians would dispute this, but no philosophy of mathematics is imaginable which does not take account of Gödel's work on incompleteness.

### Moves to Princeton and Begins Work on Set Theory

Gödel was never a popular or successful teacher. His reserved personality led him to lecture more to the blackboard than to his audience. Fortunately, he was invited to join the Institute for Advanced Study at Princeton, which had opened in the fall of 1933, where he could work without teaching responsibilities. Despite the attractions of the working environment in Princeton, Gödel continued to return to Austria, and it was there that he lectured on his first major results in the new field to which he had turned attention, the theory of sets.

Set theory had been established as a branch of mathematics in the last half of the nineteenth century, although its development had been hindered by the discovery of a few paradoxes. As a result, many who studied the field felt it was important to produce an axiomatization that would prevent paradoxes from arising. The axiomatization which most mathematicians wanted was one which would capture the intuitions they had about the way sets behaved without necessarily committing them to points about which there was disagreement. Two of the statements about which there were disagreement were the axiom of choice and the continuum hypothesis. The axiom of choice said that for any family of sets there is always a function that picks one element out of each set; this was indisputable for finite collections of sets but was problematic when infinite collections of sets were introduced. The continuum hypothesis stated that, although it was known that there were more real (rational and irrational) numbers than whole numbers (integers), there were no infinite sets in size between the real numbers and the whole numbers.

Gödel's major contribution in set theory was the introduction of what are known as constructible sets. These objects formed a model for the standard axiomatization of set theory. As a result, if it could be shown that the axiom of choice and the continuum hypothesis applied to the constructible sets, then those disputed principles had to be at least consistent

with the standard axiomatization. Gödel successfully demonstrated both results, but this still left open the question of whether the two statements could be proved from the standard axiomatization. One of the major accomplishments of set theory in the second half of the century was the demonstration by **Paul Cohen** that neither the axiom of choice nor the continuum hypothesis could be proved from the standard axiomatization.

Gödel had suffered a nervous breakdown in 1934 which aggravated an early tendency to avoid society. He married Adele Porkert Nimbursky, a nightclub dancer, on September 20, 1938. He had met his wife when he was 21 years old, but his father had objected to the match, based on the difference in their social standing and the fact she had been married before. After his marriage, his domestic situation was something of a comfort in the face of the deteriorating political situation in Austria, especially after the union of Austria and Germany in 1938, when Adolf Hitler was in power. When he returned to Vienna from the United States in June of 1939, he received a letter informing him that he was known to move in "Jewish-liberal" circles, not an attractive feature to the Nazi regime. When he was assaulted by fascist students that year, he rapidly applied for a visa to the United States. It was a sign of his stature in the profession that at a time when so many were seeking to escape from Europe, Gödel's request was promptly granted. He never returned to Europe after his hasty departure.

Gödel was appointed an ordinary member of the Institute for Advanced Study in Princeton, where he would remain for the rest of his life. His closest friends were Einstein and Oskar Morgenstern, and he took frequent walks in Einstein's company. Einstein and Gödel were of opposing temperaments, but they could talk about physics and each respected the other's work. Morgenstern was a mathematical economist and one of the founders of the branch of mathematics known as game theory. Gödel and his wife were content with this small social circle, remaining outside the glare of publicity which often fell on Einstein.

After his arrival in Princeton, Gödel started to turn his attention more to philosophy. His mathematical accomplishments guaranteed his philosophical speculations a hearing, even if they ran counter to the dominant currents of thought at the time. Perhaps the most popular philosophical school then was naturalism—the attempt to ground mathematics and its language in terms of observable objects and events of the everyday world. Gödel, however, was a Platonist and he believed that mathematics was not grounded in the observable world. In two influential published articles, one dealing with **Bertrand Russell** and the other with the continuum hypothesis, Gödel argued that mathematical intuition was a special faculty which needed to be explored in its own right. Although the bulk of mathematical philosophers have not followed him, they have been obliged to take his arguments into account.

Although Gödel moved away from mathematics in his later years, he contributed occasionally to the field. One of his last mathematical articles, published twenty years before his death, dealt with the attempt to formalize the approach to mathematical philosophy known as intuitionism. Gödel himself was not partial to that approach, but his work had wide influence among the intuitionists. American mathematician **Paul Cohen** was also careful to bring his work on the axiom of choice and the continuum hypothesis to him for his approval.

In his years at the Institute for Advanced Study, awards and distinctions began to accumulate. In 1950, Gödel addressed the International Congress of Mathematicians and the next year received an honorary degree from Yale; in 1951, he also received the Einstein award and delivered the Gibbs lecture to the American Mathematical Society. Harvard gave him an honorary degree in 1952 and in 1975 he received the National Medal of Science. That same year he was scheduled to receive an honorary degree from Princeton, but ill health kept him from the ceremony. By contrast, Gödel refused honors from Austria, at least as long as he lived; however, the University of Vienna gave him an honorary doctorate posthumously.

Gödel had a distrust of medicine that amounted in his later years to paranoia. In late December of 1977 he was hospitalized and he died on January 14, 1978 of malnutrition, brought on by his refusal to eat because of his fear of poisoning. His wife survived him by three years; they had no children. Gödel's heirs were the mathematical community to which he left his work and the challenge of understanding the effects of his results. The year after his death Douglas Hofstadter's book *Gödel, Escher, Bach* became a bestseller, illustrating Gödel's ideas in terms of art and music.

## SELECTED WRITINGS BY GÖDEL

*Collected Works,* edited by Solomon Feferman and others, 4 volumes, 1986-

## FURTHER READING

### Books

Dawson, John W., Jr. *Logical Dilemmas.* A & K Peters, 1995.
Moore, Gregory H. "Kurt Gödel," in *Dictionary of Scientific Biography.* Volume XVII. Edited by Charles Coulson Gillespie. New York: Charles Scribner's Sons, 1990, pp. 348–357.
Hofstadter, Douglas R., *Gödel, Escher, Bach.* New York: Basic Books, 1979.
Nagel, Ernest, and J. R. Newman. *Gödel's Proof.* New York University Press, 1958.
Van Heijenoort, Jean, editor. *From Frege to Gödel.* Cambridge, MA: Harvard University Press, 1967.
Yourgrau, Palle. *The Disappearance of Time.* Cambridge, England: Cambridge University Press, 1991.

### Periodicals

Lucas, J. R. "Minds, Machines and Gödel," *Philosophy* (1961): 120–124.

*—Sketch by Thomas Drucker*

# Evelyn Boyd Granville
## 1924-
### African-American applied mathematician and educator

In 1949, Evelyn Boyd Granville at Yale University and **Marjorie Lee Browne** at the University of Michigan became the first African-American women to receive doctoral degrees in mathematics. Granville's specialty is complex analysis, and her career has encompassed both college teaching and applied work in the U.S. space program during its formative years.

Granville was born in Washington, D.C., on May 1, 1924. Her father, William Boyd, worked as a custodian in their apartment building. He did not stay with the family, however, and Granville was raised by her mother, Julia Walker Boyd, and her mother's twin sister, Louise Walker, both of whom worked as examiners for the U.S. Bureau of Engraving and Printing. Granville and her sister Doris, who was a year and a half older, often spent portions of their summers at the Linden, Virginia, farm of a family friend.

The public schools of Washington, D.C., were racially segregated when Granville attended them. However, Dunbar High School (from which she graduated as valedictorian) maintained high academic standards. Several of its faculty held degrees from top colleges, and they encouraged the students to pursue ambitious goals. With the encouragement of her family and teachers, Granville entered Smith College with a partial scholarship from Phi Delta Kappa, a national sorority for Black women. During the summers, she returned to Washington to work at

*Evelyn Boyd Granville*

the National Bureau of Standards. Granville majored in mathematics and physics, and in 1945 she graduated summa cum laude and was elected to Phi Beta Kappa.

Granville undertook graduate studies at Yale University, earning an M.A. in mathematics and physics in one year. Her doctoral studies were supported by an Atomic Energy Commission Predoctoral Fellowship and a Julius Rosenwald Fellowship that was offered to help promising Black Americans develop their research potential. Granville's research concentrated on functional analysis, and her advisor, Einar Hille, was a former president of the American Mathematical Society.

Granville spent a year doing postdoctoral research at New York University's Institute of Mathematics and Science. Apparently because of *de facto* housing discrimination, she was unable to find an apartment in New York City, so she moved in with a friend of her mother. Despite attending segregated schools, Granville had not encountered discrimination based on race or gender in her professional preparation. Only years later would she learn that her 1950 application for a teaching position at a college in New York City was turned down for such a reason. Biographer Patricia Kenschaft reported in *The American Mathematical Monthly* that, according to two different sources, Granville's application was rejected because of either her race or her gender. Whichever version was true, it is clear that an exemplary

academic record and a Ph.D. from a prestigious school, earned under a renowned mentor, were not enough to overcome such entrenched biases. In 1950, Granville accepted the position of associate professor at Fisk University, a historically-Black college in Nashville, Tennessee. She was a popular teacher, and at least two of her female students credited her with inspiring them to earn doctorates in mathematics in later years.

After two years of teaching, Granville went to work for the Diamond Ordnance Fuze Laboratories as an applied mathematician. Subsequently, from 1956 to 1960, she worked for IBM on the Project Vanguard and Project Mercury space programs, analyzing orbits and developing computer procedures. Her job included making "real-time" calculations during satellite launchings. "That was exciting, as I look back, to be a part of the space programs—a very small part—at the very beginning of U.S. involvement," Granville told Loretta Hall in an interview.

On a summer vacation in southern California, Granville met the Reverend Gamaliel Mansfield Collins, a minister in the Community Church. They were married in 1960, and made their home in Los Angeles. They had no children, although Collins' three children occasionally lived with them. In 1967, the marriage ended in divorce.

Upon moving to Los Angeles, Granville took a job at the Computation and Data Reduction Center of the U.S. Space Technology Laboratories, studying rocket trajectories and methods of orbit computation. In 1962, she became a research specialist at the North American Aviation Space and Information Systems Division, working on celestial mechanics, trajectory and orbit computation, numerical analysis, and digital computer techniques for the Apollo program. The following year, Granville returned to IBM as a senior mathematician.

In 1967, not wanting to leave Los Angeles during a restructuring phase at IBM, Granville took a teaching position at California State University in Los Angeles. Although she found the job enjoyable and rewarding, she was disappointed in the mathematics preparedness of her students. Granville began working to improve mathematics education at all levels. She taught an elementary school supplemental mathematics program part-time in 1968–1969 through the State of California Miller Mathematics Improvement Program. The following year, Granville directed a mathematics enrichment program that provided after-school classes for kindergarten through fifth grades, teaching grades two through five herself. She also taught at a National Science Foundation Institute for Secondary Teachers of Mathematics summer program at the University of Southern California in 1972. In 1975, Granville co-authored *Theory and*

*Application of Mathematics for Teachers,* which was used as a textbook at over 50 colleges.

In 1970, Granville married Edward V. Granville, a real estate broker. After her 1984 retirement from California State University in Los Angeles, they moved to a sixteen–acre farm in Texas, where they sold eggs produced by their 800 chickens. From 1985 to 1988, Granville taught mathematics and computer science at Texas College in Tyler. In 1990, she accepted an appointment to the Sam A. Lindsey Chair at the University of Texas at Tyler, and in subsequent years continued teaching there as a visiting professor. Smith College awarded Granville an honorary doctorate in 1989, making her the first Black woman mathematician to receive such an honor from an American institution.

Throughout her career, Granville has shared her energy with a variety of professional and public service organizations and boards. For example, she has been involved with the National Council of Teachers of Mathematics, the American Association of University Women, and the Psychology Examining Committee of the Board of Medical Examiners of the State of California.

When asked to summarize her major accomplishments, Granville told Hall, "First of all, showing that women can do mathematics." Then she added, "Being an African-American woman, letting people know that we have brains too."

## SELECTED WRITINGS BY GRANVILLE

(With Jason L. Frand) *Theory and Applications of Mathematics for Teachers.* Second Edition. 1978.

## FURTHER READING

### Books

Kenschaft, Patricia Clark. "Evelyn Boyd Granville." *Women of Mathematics: A Biobibliographic Sourcebook.* Edited by Louise S. Grinstein and Paul J. Campbell. Westport, CT: Greenwood Press, 1987, pp. 57–61.
Hine, Darlene Clark, editor. *Black Women in America.* Volume I. Brooklyn: Carlson, 1993, pp. 498–99.
Perl, Teri, and Joan M. Manning. *Women, Numbers and Dreams.* Santa Rosa, CA: National Women's History Project, 1982.

### Periodicals

Kenschaft, Patricia C., "Black Women in Mathematics in the United States." *The American Mathematical Monthly* (October 1981): 592–604.

### Other

"Evelyn Boyd Granville." *Biographies of Women Mathematicians.* June 1997. http://www.scottlan.edu/lriddle/women/chronol.htm (July 21, 1997).
Granville, Evelyn Boyd, interview with Loretta Hall conducted January 11, 1994.

—*Sketch by Loretta Hall*

# Hermann Günther Grassmann
## 1809(?)–1877
### German geometer

Hermann Günther Grassmann was a gifted German thinker whose work spanned the fields of mathematics and linguistics, theology, and botany. His decision to focus on mathematics came when he was 31, but he abandoned the field 20 years later when his formulation of a geometric calculus did not receive the recognition it deserved. Grassmann's conception of *n*–dimensional vector space and multi–linear algebra, laid out in his monumental work *Die Lineale Ausdehnungslehre,* were ahead of his time but had great impact once they were grasped by late 19th– and early 20th–century mathematicians.

Grassmann was born on April 15, 1809, in Stettin, Prussia (present–day Szczecin, Poland), the third of 12 children. His father, Justus Günther Grassmann, taught mathematics and physics at the local gymnasium and wrote several basic–level mathematics textbooks. The Grassmann family was a religious one: Justus had briefly served as a Protestant minister before becoming a teacher, and Grassmann's mother, Johanne Medenwald, was the daughter of a minister.

### Studies in Several Fields

Grassmann was schooled first by his mother and at a private academy before enrolling at the Stettin Gymnasium. He was a fine student, earning the second highest score on the final secondary school examination. During his three years at the University of Berlin, Grassmann focused his studies on theology and classical languages and literature. His work on mathematics and physics was done on his own once he returned to Stettin in 1830.

Over the next decade Grassmann took a series of examinations in order to secure a job in the scholastic community. He passed an examination in December of 1831 that allowed him to teach only at the elementary school level. The following spring, Grassmann took a position teaching at the Stettin Gymnasium. In 1834, he passed the first level of theological examinations administered by the local Lutheran church, but instead of pursuing a religious career, he took a job as senior master at the Gewerbeschule in Berlin. Grassmann changed jobs a year later to take a teaching post at the Otto Schule in Stettin. By 1840, he had completed his round of tests, passing the second–level theology examination and the mathematics examination that allowed him to teach at the secondary school level.

### Introduces Geometric Calculus

It was during this last mathematics examination that Grassmann first applied his geometric calculus to solve a problem on the theory of the tides. After this 1840 examination, Grassmann decided to devote his energy to mathematics, particularly to the development of the geometric calculus he had been working on since 1832. In 1844, *Die Lineale Ausdehnungslehre* was published, presenting Grassmann's geometric calculus as a combination of synthetic geometry's treatment of points (and not numbers) with analytic geometry's use of calculations. Grassmann introduced *n*–dimensional vector space and multi–linear algebra, concepts that paved the way for the creation of exterior algebra.

Despite containing profoundly revolutionary ideas, *Ausdehnungslehre* was largely ignored by Grassmann's contemporaries because of the work's abstract nature and unreadable style. With mathematicians such as **Julius Plücker** and August Ferdinand Möbius refusing to write a review about the book, the professional community generally disregarded the work. Grassmann used the concept of connectivity he established in *Ausdehnungslehre* in an 1845 paper in which he revised Ampere's fundamental law for the reciprocal effect of two infinitely small currents. Again, Grassmann's poor writing style obscured a scientifically important paper. When Grassmann applied for a position as a university professor in 1847, E. E. Kummer's critique of Grassmann's 1845 paper, which stated that it contained "commendably good material expressed in a deficient form," prevented Grassmann from landing the job.

Grassmann married Marie Therese Knappe on April 12, 1849, and the couple had 11 children, two of whom died as young children. Convinced of the importance of his geometric calculus, Grassmann revised his *Ausdehnungslehre* for republication in 1862. Although Grassmann tried a new approach in explaining his methodology, the second version met with the same reception as the first. By the mid–1860s, Grassmann was frustrated by the lack of recognition for his mathematical contributions, so he turned his full attention to his studies in linguistics and other sciences.

### Contributes to Linguistics and Other Fields

As early as 1849, Grassmann began studying Sanskrit, followed by Lithuanian, Russian, and older forms of Prussian and Persian. In 1854 he developed a theory about the tonal components of vowels, and his 1863 theory about aspirates and the sound shift in Germanic languages became the linguistic law that bears his name. By 1860, Grassmann had taken interest in the Hindu literary masterpiece, the *Rig–Veda*. Grassmann compiled a glossary and composed a translation of the *Rig–Veda* during the 1870s, finding the instant acclaim in linguistics that he was never awarded in mathematics.

Grassmann also tried his hand at a variety of other projects during his lifetime, including a study of colors in his 1853 *Zur Theorie der Farbenmischung* and the renaming, using German etymological roots, of plant species native to German–speaking areas in his 1870 *Deutsche Pflanzennamen*. Writing for a political newspaper in 1848, Grassmann penned a series of articles supporting a Germany united under constitutional monarchy. Grassmann even wrote folk songs in which he harmonized up to three voices.

### Returns to Mathematics

After his 1871 election to the Göttingen Academy of Sciences, Grassmann returned to mathematics. He published several papers before he died on September 26, 1877, of heart failure. The following year a third version of *Ausdehnungslehre*, prepared before his death, was published. An appreciation of Grassmann, appearing a year after his death in the *Schulprogramm* of the Stettin Marienstifts gymnasium, said of Grassmann, " . . . only a quite independent spirit could dare to break his own paths in mathematics, on which others followed him only after decades . . . ." Indeed, it was only after his death that mathematicians began drawing off Grassmann's *Ausdehnungslehre* and crediting his discoveries with the later development of linear matrix algebra.

## SELECTED WRITINGS BY GRASSMANN

*A New Branch of Mathematics: The "Ausdehnungslehre" of 1844 and Other Works by Hermann Grassmann.* Translated by Lloyd C. Kannenberg, 1995

## FURTHER READING

### Books

Burau, W. and C.J. Scriba. "Hermann Günther Grassmann," in *Dictionary of Scientific Biography*. Volume XV. Edited by Charles Coulston Gillisipie. New York: Charles Scribner's Sons, 1976, pp. 192–199.

Schubring, Gert, editor. *Hermann Günther Grassmann (1809–1877): Visionary Mathematician, Scientist and Neohumanist Scholar*. Dordrecht, Netherlands: Kluwer Academic Publishers, 1996.

### Periodicals

Fearnley–Sander, Desmond. "Hermann Grassmann and the Prehistory of Universal Algebra." *The American Mathematical Monthly* 89 (March 1982): 161–166.

————. "Hermann Grassmann and the Creation of Linear Algebra." *The American Mathematical Monthly* 86 (December 1979): 809–817.

*—Sketch by Bridget K. Hall*

# George Green
## 1793–1841
### English mathematical physicist

George Green essentially taught himself mathematics. An omnivorous reader, he frequented the Bromley House Subscription Library in Nottingham, and used the top of his father's mill in Sneinton, a suburb of Nottingham, as his study. His knowledge was gleaned from the French mathematicians, especially Denis Poisson. In 1828, Green published, by subscription, his best–known work, *An Essay on the Application of Mathematical Analysis to the Theories of Electricity and Magnetism*. Sources for his conclusions were based on Poisson's previous work on surface electricity and magnetism and the single–fluid theories of chemist and physicist Henry Cavendish. In the *Essay*, which analyzed electricity and magnetism from a mathematical standpoint, Green created the term, "potential," applying it as a central precept in electrical theory. As he used the term, it signified the sum of the masses of a system's particles; each mass divided by its distance from any given point. His paper also introduced concepts now known as Green's theorem and Green's functions. The theorem and functions are significant elements in Green's potential theory—the functions serve as an important tool for solving partial differential equations. Because this early work was available only to a limited and local group of scientists, it could have been lost to later scholars if it were not for the efforts of William Thomson (Lord Kelvin), who became acquainted with Green's work while an undergraduate at Cambridge University. His discovery and dissemination of Green's seminal work played an important role in furthering the development of electromagnetic research in the following decades. Thomson republished Green's *Essay* with an introduction in 1850. In spite of Green's mathematical talents, formal recognition of his accomplishments did not come until near the end of his life. In 1839, he was given, on the basis of his independent research, a Fellowship at Caius College at Cambridge.

Green was born in Nottingham, England, in July 1793. He was the son of a miller and baker. His formal education was cut short at an early age, when he left school to work in his father's bakery. However, his early interest and obvious skill in mathematics was evident to Robert Goodacre, the local schoolmaster, and he encouraged young George to continue his studies, albeit on a self–taught basis. The family mill at Sneinton was a successful business, and when Green's father died, the family business was sold. Green's inheritance made it possible for him to enter Cambridge in 1833 at the age of 40. (Green's involvement in the milling business over those years caused him to be known as the "miller–mathematician.") At Caius College, Green pursued a dual course of studies—those required by the University, and his own independent research. He received his undergraduate degree from Cambridge in 1837. His less–than–stellar academic standing as "fourth wrangler" was undoubtedly due to his lack of formal mathematical training and the demands made on his time by his private research.

### Writings Gain Him Perse Fellowship

Even before Green entered college, he possessed sufficient scientific credibility to present a paper to the Cambridge Philosophical Society. Read to those assembled on November 12, 1832, this second work was titled "The Laws of the Equilibrium of Fluids Analogous to the Electric Fluid." Another paper, entitled "Exterior and Interior Attractions of Ellipsoids of Variable Densities," was given at Cambridge the following May. In 1837, the year he received his degree, he presented to the Society a paper which explained the formula for computing the height of a wave. This paper, titled "On the Motion of Waves in a Variable Canal of Small Depth and Width," was followed in December of that year by "On the Reflexion and Refraction of Sound," and "On the Reflexion and Refraction of Light at the Common Surface of Two Non–Crystallized Media." In May 1839, he presented a paper on his theory of the conservation of mechanical energy, titled "On the

Propagation of Light in Crystallized Media." This paper (which, along with the one on light reflexion and refraction, build on the theories of **Augustin–Louis Cauchy,** the French mathematician) is viewed as second in importance to his work on the mathematical analysis of electricity and magnetism theory. The presentation of this series of papers gained him the Perse Fellowship in October 1839. As a Perse Fellow, Green never lectured and did not involve himself in academic life. A man of reserve, he was more comfortable tutoring and setting problems for examination, rather than teaching. By 1840, his health was failing and he returned to Nottingham. His death on May 31, 1841 cut short a life viewed by many as one full of remarkable accomplishments, although made in relative obscurity.

### Electrical Theories Influence Modern Manufacturing

Green's mathematical theories about electricity helped lay the foundation for modern practices of commerce and manufacturing. His influence on Lord Kelvin was generated by his first great work. Thanks to Kelvin's foresight in reprinting the forgotten work, Green's *Essay* went on to interest and engage the minds of other 19th–century scientists. James Clerk Maxwell is one example. The Scottish mathematician, who was ten years old when Green died, used Green's theory of potential in his own work, as did other prominent scientists specializing in electromagnetism. **Joseph Liouville,** a leading French mathematician; André Marie Ampère, a French contemporary; Michael Faraday, the physicist and chemist who established the connection between electricity and chemistry; and even later, Heinrich Hertz, the great German physicist, all built on the electromagnetic research of their peers and their predecessors. Thus, the self–taught "miller–mathematician" keeps prominent company with the great theorists of the industrial age.

## SELECTED WRITINGS BY GREEN

*An Essay on the Applications of Mathematical* Analysis to the Theories of Electricity and Magnetism, 1828. (Less than 100 copies printed, republished by William Thomson, 1850–1854.)
*Collected works,* edited by N. M. Ferrers, 1871

## FURTHER READING

### Books

*The Encyclopedia Americana,* international edition. Vol. 13. Danbury, CT: Grolier, Inc., 1995, p. 445.
Muir, Hazel, editor. *Larousse Dictionary of Scientists.* New York: Larousse, 1994, p. 218.
Porter, Roy, editor. *The Biographical Dictionary of Scientists.* New York: Oxford University Press, 1994, p. 288.
Taton, René, editor. *History of Science. Science in the Nineteenth Century.* New York: Basic Books, Inc., 1961, pp. 178–180.
Wallis, P.J. "George Green," in *Dictionary of Scientific Biography.* Volume XV, supplement I. Edited by Charles Coulston Gillispie. New York: Charles Scribners' Sons, 1978, pp. 199–201.

### Periodicals

"News and Views." *Nature* (October 25, 1947): 561.

*—Sketch by Jane Stewart Cook*

# James Gregory
## 1638–1675
### Scottish–born mathematician and astronomer

James Gregory's work laid the foundation for the development of calculus, and his work in astronomy and optics for astronomical observations influenced the works of **Isaac Newton.** He was born in Drumoak, Scotland, the son of John Gregory, a minister, and Janet Anderson Gregory. Gregory was a sickly child, and his mother guided his early education at home; she must have been an unusual woman for the 17th century, because she included geometry among the subjects she taught her son.

In 1651, Gregory left home for grammar school in Aberdeen, Scotland. He completed his preparatory work, then graduated from Aberdeen's Marischal College, where he focused his studies in mathematical optics and astronomy. Frustrated by the lack of scholarly opportunities in Aberdeen, Gregory traveled to London in 1662, where he met Robert Moray, an influential member of the Royal Society. In 1663, Gregory published *Optica promota*, a work that anticipated Newton's efforts in optics by suggesting the use concave mirrors in telescopes. He searched unsuccessfully to find a technician skillful enough to construct the prototype of such an instrument. Moray attempted to introduce Gregory to the Dutch mathematician **Christiaan Huygens** to help further the young man's studies, but was unsuccessful. Gregory then decided to pursue scientific studies in Italy with Stefano degli Angeli, and left London for Padua in 1664.

*James Gregory*

At the University of Padua, Gregory studied geometry, mechanics, and astronomy. In 1667 he published *Vera circuli et hyperbola quadratura*, in which he explored the nature of the area of circles and hyperbolas. *Geometriae pars universalis, inserviens quantitatum curvarum transmutationi & mensurae* followed in 1668, where Gregory introduced the concepts of convergent and divergent series. He also discussed the differences between algebraic and transcendental functions, and offered a series of expressions for trigonometric functions and a proof for the fundamental theorem of calculus.

## A Prophet in His Own Land

Gregory returned to London in 1668. His work in Italy and his contacts with the Italian scientific community initially earned him considerable notice, and he was quickly elected to the Royal Society. Gregory was named as the new chairman of mathematics at St. Andrew's College, Scotland, in 1668. Gregory's time was consumed by his duties for the next several years. "I am now much taken up and hath been . . . this winter bypast, both with my publik lectures, which I have twice a week, and resolving doubts . . . gentlemen and scholars proposeth to me," he wrote in 1671.

Despite the time devoted to teaching and academic administration, Gregory managed to maintain a voluminous correspondence with John Collins, who forwarded to Gregory copies and transcripts of material from such noted scholars as **Isaac Barrow**, René–Francois de Sluse, and Newton.

## The Price of Academic Freedom

Gregory waged war on the antiquated curriculum at St. Andrew's, but his efforts to incorporate contemporary science into the college's course of studies was resisted by the faculty and the college's governing board of regents. Gregory hoped to establish the first public observatory in Great Britain at St. Andrew's; in 1673, he journey to London to seek advice and obtain instruments for such a facility. Unsuccessful in his efforts to secure financial backing for the project, Gregory returned to St. Andrew's to find himself an academic outcast. A student rebellion against the established curriculum pushed the board of regents to action. Gregory and his radical ideas were the obvious scapegoat; servants were forbidden to wait on him and his salary was withheld. Colleagues and students were instructed to treat him as a pariah. "Scholars of most eminent rank were violently kept from me . . . the masters persuading them that . . . they were not able to endure mathematics," he wrote.

When Edinburgh University offered Gregory its newly endowed chairmanship of mathematics, he fled St. Andrew's. Sadly, within a year of the appointment, Gregory suffered a debilitating, blinding stroke while observing Jupiter through a telescope. He died a few days later, in October 1675.

Before his death, Gregory ceased publishing papers in pure mathematics. His private papers, however, are rich in theories, proofs, and questions that might have earned him wide acclaim during his lifetime, had the materials been issued. Gregory delved into the theory of equations and the location of their roots, and attempted to solve the general quintic. His letters to Collins include his work on quadrature and rectification of the logarithmic spiral, his independent discovery of the general binomial expansion, several trigonometrical series (including those for the natural and logarithmic tangent and secant), and a series solution to **Johannes Kepler**'s problem in which he outlined how the series could be applied to the roots of equations.

Much of the work Gregory pursued during the last years of his life is lost. In a short paper published in 1672, he proved that atmospheric height is logarithmically related to barometric pressure. Other surviving papers demonstrate that he deduced the elliptical integral expressing the time of vibration in a circular pendulum and pursued his work in theoretical astronomy.

Gregory's reluctance to publish limited his success during his lifetime. His work served as the springboard for the published works of others, including Newton and Gregory's nephew, David, who did not acknowledge their debt to Gregory. The scope of Gregory's work and the extent of its influence on the 17th century scientific community were little recognized until 1939, when some of his notes—scribbled in 1671 on the back of a letter from a bookseller—were published and scholarly curiosity in Gregory's work was awakened.

## SELECTED WRITINGS BY GREGORY

*Geometriae pars universalis, inserviens quantitatum curvarum transumtationi & mensurae,* 1668
*"An Account of a Controversy betwixt Stephano de Angelis and John Baptistat Riccioli," 1668
*Exercitationes gemoetricae,* 1668
*(Many of Gregory's papers were issued by the Royal Society of London in *The Philosophical Transactions of the Royal Society*)

## FURTHER READING

Daintith, John, Sarah Mitchell, and Elizabeth Tootill, editors. *A Biographical Encyclopedia of Scientists.* Volume 1. New York: Facts on File Inc., 1981, p.333.
Eves, Howard. *An Introduction to the History of Mathematics.* New York: CBS College Publishing, 1983, p.87.
Millar, David, Ian Millar, John Millar and Margaret Millar, editors. *The Cambridge Dictionary of Scientists.* Cambridge, England: Cambridge University Press, 1996, p.136.
Smith, David E. *History of Mathematics.* New York: Dover Publications, 1958, pp. 409–10, 413.
Whiteside, D.T. "James Gregory," in *Dictionary of Scientific Biography.* Volume V. Edited by Charles Coulston Gillispie. New York: Charles Scribner's Sons, 1973, pp.524–30.

—*Sketch by A. Mullig*

# Alexander Grothendieck
## 1928-
### French algebraic geometer and analyst

Alexander Grothendieck has had an influence on the mathematics of the second half of the 20th century well beyond the scale of his publications. Grothendieck started off as an especially prolific contributor to each of the areas to which he turned, including functional analysis, algebraic geometry, and

*Alexander Grothendieck*

category theory, only to move away from mathematics later in his career. As a result, he has had fewer students to carry on his research tradition than if he had followed a more orthodox path. Nevertheless, one of the chief activities of the mathematicians in several areas has been to recast their field in the terms introduced by Grothendieck.

Grothendieck's early years have been difficult to reconstruct, due to his reluctance to deal in ordinary reminiscences and his distrust of biographers. The generally accepted date and place of his birth are March 28, 1928, and Berlin, but the identity of his parents is less clear. At least one version that Grothendieck has given, as noted by Colin McLarty, indicates that his father was named Morris Shapiro and that he was sentenced to death for attempting to assassinate the Russian Czar in 1905. After Shapiro served a number of years in prison in Siberia, he was released by the Bolsheviks and went to Germany in 1922, about which time he met his future wife. Grothendieck was their first child and took his name from that of the governess who cared for him from 1929 to 1939. In the latter year, his mother took him to France, where he learned for the first time that he was Jewish by ancestry. His father died in the Auschwitz concentration camp and Grothendieck was saved thanks to the cooperation of Protestant and Catholic clergy in Le Chambon sur Lignon in southern France. His mother died in the 1960s.

The story becomes much clearer once Grothendieck entered the French higher education system after the war. He studied at the University of Montpellier and spent a year at the École Normale Supérieure, one of the leading traditional scientific universities in France. At this time France was undergoing a mathematical renaissance, thanks to the pedagogic efforts of the group known under the collective pseudonym of Nicolas Bourbaki. Among those who took part in the grand program of rewriting all of mathematics in the Bourbaki mode were Jean Dieudonné and Laurent Schwartz, both at the University of Nancy. Grothendieck went to work in the area of functional analysis with the two Bourbakists and rapidly produced material sufficient and appropriate for a thesis.

Grothendieck's first conspicuous success was in the area known as functional analysis. This mixed the traditional area of calculus with the more recent developments in topology, the field dealing in properties of geometric configurations, to be able to handle broad ranges of questions. The idea was to replace detailed and lengthy calculations with shorter and more insightful proofs. It is not surprising that such an area attracted Grothendieck, who did not feel that his greatest strength was in long, technical arguments. His contributions came in the area of reconsidering disciplines from new perspectives.

## Categorizes the Mathematical World

Grothendieck's most lasting influence came from his work in the area to which he now moved, algebraic geometry. This field had been in existence for many years and could be traced back to French mathematician and philosopher **René Descartes** in the 17th century. The idea of merging algebra and geometry to enhance the study of both received a new impetus with the accelerated development of abstract algebra in the late nineteenth century. There was a flourishing Italian school of algebraic geometry in the first half of the twentieth century, but it was effectively wiped out by the World War II. American mathematician Oscar Zariski carried on the Italian tradition in the United States, although he felt that he had added a good deal of algebraic sophistication.

Grothendieck was supported during his early investigations into algebraic geometry by the French national center for scientific research. This allowed him plenty of opportunity for travel and he spent part of the 1950s in Brazil and part in Kansas. Perhaps the most fruitful environment he found was at Harvard, where Zariski had settled. As his Harvard colleagues noted, Grothendieck was obsessed by mathematics and worked for many hours at a stretch in an unheated study, emerging with 3000-page manuscripts. On the strength of his energy and imagination, Grothendieck was able to revolutionize mathematics with his research.

One of the chief elements in Grothendieck's approach to mathematics involved the relatively recent field of category theory. Set theory had become an accepted part of the foundations of mathematics, but category theory sought to add a new idea to the basic notions of set and membership—the idea of function. Functions had long been used in mathematics, but category theory built them into the basis of the mathematical universe. One way of looking at the change was that mathematicians began to realize that what was important about the objects of mathematics was how they were connected by functions, not their composition out of basic elements.

Before the work of Grothendieck, category theory had been an active area of research but with limited applications. Grothendieck combined the ideas of category theory with the traditional studies of algebraic geometry to raise the latter to a new level of abstraction. The innovations introduced by Zariski in the previous generation shrank by comparison. As Zariski was quoted in *The Unreal Life of Oscar Zariski,* "After Grothendieck's great generalization of the field . . . what I myself had called abstract turned out to be a very, very concrete brand of mathematics."

In 1959 Grothendieck took a position with the Institut des Hautes Études Scientifiques (IHES), recently established in Paris upon the model of the Institute for Advanced Studies in Princeton, New Jersey. There Grothendieck had the chance to lecture on a regular basis on his work in algebraic geometry and to attract mathematicians from all over the world. Not surprisingly, in 1966 he received the Fields Medal from the International Mathematics Congress, the highest award that the mathematical community can convey. Among the attractions of his work was its applicability to extending a variety of theorems that had originally been established in narrow contexts. Questions about number fields that had required immense amounts of computation to answer could be replaced by conceptually simpler questions about algebraic varieties, and the answers would have wide domains of applicability.

This golden age for algebraic geometry came to an end in 1970. Grothendieck had never been comfortable with playing the role of the "great man" and felt that the adulation of students was not good for him as a human being or as a mathematician. He also moved in a radical direction politically and hoped to be able to galvanize the mathematical community into political action. As a result, he left the IHES and taught at other French universities, particularly Montpellier, from which he retired in 1988. In the meantime, his ideas about category theory continued to supply the fuel for other areas of mathematics, including the foundations. The idea of a topos, a particular kind of category especially useful for analyzing logic, was introduced by Grothendieck for purposes of algebraic geometry. The continued fertility of topos theory adds to the fields indebted to Grothendieck's work during his contributions to algebraic geometric issues.

Grothendieck's memoir, *Récoltes et Semailles,* discusses at length his views on a number of subjects, most of which are unrelated to mathematics. More representative of his career in mathematics is the three-volume set of papers gathered for his 60th-birthday *festschrift* and published in 1990. The range of contributors includes many of the names of leaders of the mathematical community. His vision of mathematics has led not just to individual results but to a new sense of the powers of the subject.

## SELECTED WRITINGS BY GROTHENDIECK

### Books

*Récoltes et semailles,* 1986

### Periodicals

"Sur quelques points d'algébre homologique." *Tohoku Journal of Mathematics* (1957): 119–221.

## FURTHER READING

### Books

Cartier, P., et al., editors. *The Grothendieck Festschrift.* Boston: Birkhauser, 1990.
Dieudonné, Jean. *A History of Algebraic and Differential Topology.* Boston: Birkhauser, 1989.
Parikh, Carol. *The Unreal Life of Oscar Zariski.* San Diego, CA: Academic Press, 1991.

### Periodicals

McLarty, Colin. "Category Theory in Real Time." *Philosophia Mathematica.* (1994): 36–44.

### Other

McLarty, Colin, notes for biography of Grothendieck.

*—Sketch by Thomas Drucker*

# H

*Jacques Hadamard*

# Jacques Hadamard
## 1865-1963
**French analyst**

Widely considered the preeminent French mathematician of the 20th century, Jacques Hadamard has made an impact on many fields of mathematics. Although an analyst and a student of theoretical calculus by training, he has influenced topology, number theory, and even psychology. His work on defining functions won him the Grand Prix of the Académie des Sciences early in his career, and his proof of the prime number theorem solidified his importance in the mathematical world. He wrote several textbooks on a variety of mathematical subjects, including one which explained a mathematician's thought processes. Hadamard was first and foremost a teacher, however, and he used his position to help both students and colleagues alike see the connections between seemingly unrelated fields.

Born in Versailles on December 8, 1865, Jacques-Salomon Hadamard was the son of two teachers. His mother, Claude-Marie Picard, taught piano, while his father, Amédée, taught Latin at a prominent Paris high school. In 1884, at the age of eighteen, Hadamard began studying at the École Normale Supérieure. His first teaching job was at a high school in Paris, the Lycée Buffon, in 1890. When he was not teaching, he worked on his doctoral dissertation, and the research he did during this period led to his first breakthrough in mathematics.

## Work on Taylor Series Wins Grand Prix

Hadamard's dissertation concerned determining the shape of a function and finding certain points on that function where division by zero was involved in the original equation. Such functions had previously been considered undefined and unsolvable, but Hadamard found a way to solve them using Taylor series. Published in 1892, his work was so revolutionary that the French Académie des Sciences immediately awarded him its highest honor, the Grand Prix. This was also the year Hadamard married Louise-Anna Trenel, with whom he would have five children. In 1892 Hadamard also accepted a position as lecturer at the Faculté des Sciences of Bordeaux, where he continued his work. Although his accomplishments in defining functions had been important to the mathematical community at large, for Hadamard it was just another step toward a larger goal. He wanted to find a proof of the prime number theorem. For years, some of the world's best mathematicians had attempted to find a formula for the density of prime numbers among all the whole numbers. Many had discovered estimates and close guesses, but no one had achieved accurate results.

In 1896 Hadamard (and independently **C.J. de la Vallee-Poussin**) proved that the number of primes less than $n$ is well approximately by $n/\log(n)$ when $n$ is a large integer. He used powerful methods from complex analysis. While his theory only works when the numbers used are sufficiently large, mathematicians generally only concern themselves with primes when such large numbers are involved. In 1949 **Atle Selberg** and **Paul Erdös** independently gave the first "elementary" proofs of the prime number theory not using modern complex analysis.

## Career Research Spans Many Topics

Following publication of the proof of the prime number theorem, Hadamard left Bordeaux for a lectureship at the Sorbonne in Paris. A return to the intellectual center of Paris also meant greater involvement with the mathematical community, in which Hadamard had earned a high place. While many mathematicians were content to specialize in a small area of mathematics, Hadamard saw the importance of finding connections between the various fields. He was openly critical of mathematicians who limited their work to their immediate subject. In 1902 he argued, for example, that the definitions Vito Volterra had offered for the calculus terms *continuity, derivative,* and *differential* were inadequate because they could not be generalized to other fields, especially the relatively new area of topology. Instead of merely criticizing Volterra, however, Hadamard applied himself to generalizing analysis so it would be more applicable to other fields. His creation and definition of the term *functional,* first put forth in 1903, is one result of this generalization. Though Hadamard had used standard analysis to come up with functionals, the application of the idea to topology was important to establishing the validity of that field.

Hadamard's work forming connections between topology and analysis was interrupted in 1904 by a debate over mathematical logic which raged through the mathematical community. Ernst Zermelo, a German mathematician, had proposed that given an infinite number of sets, it would be possible to select exactly one, definable item from each set. This proposal was called the axiom of choice. Zermelo argued that it was obvious and thus needed no proof, but many of the most prominent mathematicians of the time, including **Émile Borel, Jules Henri Poincaré**, and **Henri Lebesgue**, disputed it. As Morris Kline describes the controversy in *Mathematics: The Loss of Certainty:* "The nub of the criticism was that, unless a definite law specified which element was chosen from each set, no real choice had been made, so the new set was not really formed." Yet the axiom of choice was necessary to establish sections of abstract algebra, topology, and standard analysis. Hadamard supported Zermelo. He rejected the idea that the item taken from the set could necessarily be defined, yet he felt that any theory which allowed mathematics to progress should be accepted, with or without formal proof.

In 1908, Hadamard spoke at the Fourth International Congress of Mathematicians in Rome, where he met the famous German topologist L. E. J. Brouwer. They began a correspondence relating to the mathematical ideas of their time, and the exchange of letters was crucial to Brouwer. The German mathematician used Hadamard's ideas as a springboard to some of his most important topological discoveries. In 1909, Hadamard left the Sorbonne for a more prestigious appointment as professor at the École Centrale des Arts et Manufactures. He would remain there, teaching concurrently at the Collège de France after 1920, until his retirement at the age of 71.

In 1912, Hadamard's friend and colleague Jules Henri Poincaré died. Poincaré, like Hadamard, had been involved in several different fields of mathematics, and his work had greatly influenced Hadamard's interest in generalization. Saddened by the loss of this great mathematician, Hadamard devoted a great deal of his research time after Poincaré's death to writing biographical works of his friend. Hadamard did his last piece of major research in the field of calculus in 1932, when he addressed a problem posed by the French mathematician **Augustin- Louis Cauchy**. But even after his retirement in 1937, Hadamard continued to ponder some of the questions that had concerned him throughout his career. The old controversy over the axiom of choice became the basis of a new book on the importance of accepting intuition for the sake of mathematical progress. He published *The Psychology of Invention in the Mathematical Field* in 1945, at the age of 80, and it was widely considered an innovative attempt at understanding how mathematicians come up with their ideas. Some of the work on this book was done in the United States, where he was a visiting professor at Columbia University in New York in 1941. Unlike many European mathematicians, however, Hadamard did not stay in America. He returned home to France, living out the rest of his life quietly. He died in Paris on October 17, 1963, at the age of 97.

## SELECTED WRITINGS BY HADAMARD

*The Psychology of Invention in the Mathematical Field,* 1945
*Oeuvres de Jacques Hadamard,* 4 volumes, 1968

## FURTHER READING

### Books

Cajori, F. *A History of Mathematics.* New York: Chelsea Publishing Company, 1980.
Kline, Morris. *Mathematical Thought from Ancient to Modern Times.* New York: Oxford University Press, 1972.
Kline, Morris. *Mathematics: A Loss of Certainty.* New York: Oxford University Press, 1980.
Phillips, Esther, editor. *Studies in the History of Mathematics.* Washington, D.C.: Mathematical Association of America, 1987.

*—Sketch by Karen Sands*

*Edmond Halley*

# Edmond Halley
## 1656–1742
### English astronomer and geophysicist

Nearly everyone has heard of Halley's comet. The comet, whose periodicity was predicted by Edmond Halley, and whose return, 17 years after his death, ensured his posterity, was one of more than 24 comets whose movements were studied and tracked by this gifted astronomer. His theory that comets follow an elliptical path as they travel through space is one of his best-known achievements. In addition to his work with comets, Halley also is notable for discovering that stars are not fixed in space, but move about. He deduced this movement from the study of three bright stars: Arcturus, Procyon, and Sirius. Halley surmised that his theory would hold true for more distant, dimmer stars, but this was not proved until 150 years later, when more advanced instrumentation made this determination possible.

As a mathematician, Halley calculated the focal length of certain optical lenses, the trajectory of artillery shells, and prepared actuarial tables later used by insurance companies to determine life and annuity premiums and values. His actuarial work is recognized as an early form of social statistics. Halley published important papers on trade winds and monsoons, tides, and terrestrial magnetism. His work in this area lay the foundation for scientific geophysics. He was an officer and fellow in the Royal Society of London, was appointed the Savilian chair of geometry at Oxford University in 1704, and was made Astronomer Royal in 1720.

Edmond Halley was born outside London in the borough of Hackney. His father, Edmond, Sr., was a well-to-do merchant who owned soap-rendering and salting businesses in London, as well as various rental properties. His mother, Ann Robinson, died in 1672, when Halley was a young man about to enter college. The eldest in the family, Edmond had a sister, Katherine, and a brother, Humphrey. The Halley's retained their wealth in spite of suffering losses in the Great Fire of London in 1666, and young Edmond was given a good education. After being tutored at home, Halley was enrolled in St. Paul's School. When he was 17, he entered Queen's College at Oxford. He showed an early interest in astronomy, which his father encouraged by buying him the proper scientific instruments. Colin A. Ronan, in *Edmond Halley, Genius in Eclipse*, says that one of the instruments was a telescope 24 feet long (731 m), and another, a sextant 2 feet (60 m) in diameter.

During his formal study of astronomy at Queen's College, Halley began a correspondence with John Flamsteed, the Astronomer Royal at the Greenwich Observatory. Their initially positive relationship was to turn sour when Flamsteed began to regard him as a rival. Flamsteed went so far as to accuse Halley of plagiarizing the work of other astronomers. Flamsteed's attempts to undermine Halley's reputation were to continue until Flamsteed's death in 1719. In 1676, Halley traveled to the island of Saint Helena to map the stars of the southern hemisphere. He published this work as the *Catalogus Stellarum Australium* in 1678, when he was only 22 years old. He also created a planisphere of southern stars, which he dedicated to his king, Charles II. This accomplishment, which resulted in the first star charts made of the southern hemisphere, also earned Halley an M. A. degree from Oxford.

After publishing his work on the southern hemisphere stars, Halley visited Holland in 1679 to meet with the great observational astronomer of the time, Johannes Hevelius, and investigate his methods of star observation. Halley wrote in praise of the accuracy of Hevelius' work, directly contradicting the opinion of some other astronomers, one being Johannes Flamsteed. This dispute was to further estrange Halley and Flamsteed.

### Develops Theory on Movement of Comets

In 1680, Halley and Giovanni Cassini, director of the Paris Observatory, did observational work on a new comet that appeared. He went on to develop his theories on the motion and periodicity of comets. One of Halley's most significant findings was that multiple comet sightings of years past were of the same comet making a return. His calculations led him to predict that the comet of 1682 would return in 1758. It was this comet, which did make its return very close to when Halley predicted, that became known as Halley's comet.

Halley married Mary Tooke, the daughter of an auditor in the royal bank, in 1682. They had three children: Katherine, Margaret, and Edmond. His daughters survived him; his son, a naval surgeon, died a year before Halley. In 1684, Halley's father was found dead. His death was later ruled murder, but no one was ever charged with the crime. Halley's wife, of whom little is known, died in 1736.

### Begins Friendship and Collaboration with Newton

Halley's long and fruitful friendship with **Isaac Newton** began in 1684, when Halley paid a visit to Cambridge to discuss with Newton his views on planetary motion. This visit, according to Ronan, "was to have results so far–reaching that they can be said to have altered the whole course of physical science." Their collaboration over a long period concerned how bodies move through space and the way gravitational forces control them. Newton, building on **Johannes Kepler's** laws of planetary motion, had established mathematical proof for those laws. These theories—which Halley used to advance his own body of work—applied as well to movements of the moon and comets. With Halley's encouragement and financial assistance, Newton was able to publish his theories—his *Principia* is thought by many to be the greatest work ever produced in natural science.

Halley's appointment as Astronomer Royal in 1720 conferred on him full recognition of his work by his peers and by his monarch. In this important post, he continued, over a period of 18 years, his work observing and charting the movements of the stars, moon, and planets. He died at Greenwich on January 14, 1742, of a circulatory disorder at age 86.

## SELECTED WRITINGS BY HALLEY

### Books

*Catalogus stellarum Australium,* 1678
*Philosophical Collections,* 1679–1682
*Astronomiae cometicae synopsis,* 1705
*Correspondence and Papers of Edmond Halley.* Edited by E.F. MacPike, 1932

## FURTHER READING

### Books

Asimov, Isaac. *Asimov's Biographical Encyclopedia of Science and Technology.* Garden City, NY: Doubleday & Company, Inc., 1972.

*William Rowan Hamilton*

Magill, Frank N., editor. *The Great Scientists.* Danbury, CT: Grolier Educational Corporation, 1989, pp. 1–6.
Muir, Hazel, editor. *Larousse Dictionary of Scientists.* New York: Larousse, 1994, pp. 230–231.
Ronan, Colin A. *Edmond Halley: Genius in Eclipse.* Garden City, NY: Doubleday & Company, Inc., 1969.
————. "Edmund Halley," in *Dictionary of Scientific Biography.* Volume VI. Edited by Charles Coulston Gillispie. New York: Charles Scribners' Sons, 1972, pp. 67–72.

*—Sketch by Jane Stewart Cook*

# William Rowan Hamilton
## 1805–1865
### Irish algebraist

William Rowan Hamilton was an Irish mathematician and astronomer of the 19th century, considered by some to be near in intellect to **Isaac Newton**. He created a novel system of algebra for operating on complex numbers, coined the term "vector," founded vector analysis, developed icosian calculus, and made important contributions to the understanding of light and optics.

Born in Dublin at the stroke of midnight, between the 3rd and 4th of August 1805, Hamilton was the only son of Archibald Hamilton and Sarah Hutton. In 1808, possibly because of his family's difficult financial condition, Hamilton was sent to live with his uncle, the Reverend James Hamilton, head of a diocesan school in the village of Trim, 40 miles northwest of Dublin. Hamilton had little contact with his parents during childhood, but his four sisters lived with him intermittently, continuing into adulthood. His mother died in 1817 and his father in 1819. Hamilton began his education as soon as he arrived in Trim, quickly revealing himself as a child prodigy. His uncle was eager to pour as much knowledge into him as was possible and at the age of 10, Hamilton was supposedly schooled in 13 foreign languages. His mathematical education began around the same time, with a study of **Euclid**. Hamilton went on to read Newton's *Universal Arithmetic* and *Principia*, analytic geometry, calculus, and **Pierre Laplace**'s *Mecanique Celeste*, all independently. In 1824, Hamilton entered Trinity College in Dublin; that same year he published a paper correcting a mistake in Laplace's work. Hamilton was a top scholar, and the college awarded him two separate *optimes,* or "off-scale grades," for his performance on examinations. So rare was such an honor that no *optimes* had been awarded in the previous 20 years.

## Publishes Innovative Theory of Rays

While still a student, Hamilton wrote the first part of a pivotal treatise on optics and delivered a paper on the subject to the Royal Irish Academy. His work dealt with the patterns of light produced by reflection and refraction. Hamilton demonstrated that light travels by the path of least action and he used algebraic functions to express its path mathematically. He also predicted the phenomenon of conical refraction, an idea which was later proved experimentally by the astronomer Humphrey Lloyd. Hamilton's reputation in mathematics and astronomy became widespread. In 1827, Dr. Brinkley, a professor of astronomy, resigned from Trinity to become a bishop. Hamilton was unanimously voted in as Brinkley's replacement and agreed to take the professorship, despite the fact that he had never applied for the post. His new job included an appointment as Royal Astronomer of Ireland and the directorship of the Dunsink Observatory. As head of the observatory, Hamilton lived at Dunsink. In 1828, he completed his manuscript on light patterns, *The Theory of Systems of Rays*, which was then published by the Royal Irish Academy. Hamilton retained his position as Royal Astronomer and the overseer of Dunsink throughout his life. He performed his duties diligently, but his first love was always pure mathematics.

## Invents Algebra of Quaternions

In the early 1830s, Hamilton began searching for a way in which to interpret complex numbers geometrically. He started by developing algebraic rules for working with complex numbers of the class $a + ib,$ where $a$ and $b$ are real numbers and $i$ is imaginary. He characterized this class of complex numbers in relatively simple terms, as ordered pairs, or "couples" of real numbers $a$ and $b$. Hamilton represented these complex numbers geometrically, as line segments having both length and direction. He coined the term vector to describe the segments, and applied algebraic rules to their analysis. Hamilton showed that the sum of two complex numbers could be represented by a parallelogram. His initial work demonstrated that complex numbers could be useful tools for dealing with the concept of rotation in plane, or two-dimensional geometry.

Hamilton's *Theory of Algebraic Couples*, published in 1835 by the Royal Irish Academy, expounded on his application of algebra to complex numbers. In the following years, realizing its potential importance for three-dimensional geometry, Hamilton tried to develop an algebra for dealing with complex numbers as triples, that is, of the type $a + ib + jc,$ where $a$, $b$, and $c$ are real numbers and $i$ and $j$ are imaginary. On an October day in 1843, while walking along the Royal Canal in Dublin, Hamilton had the sudden insight that algebraic operations on such complex numbers required the imposition of a fourth dimension in geometric space. At that moment, he conceived of the idea of operating on quadruples, number groupings he later termed quaternions. With a knife, he carved the equation $i^2 = j^2 = k^2 = ijk = -1$ in a stone of the Brougham Bridge. This proved to be the formula for operating on complex numbers of the type $a + ib + jc + kd$, where $a, b, c,$ and $d$ are real numbers and $i, j,$ and $k$ are imaginary. Among the important realizations Hamilton made about the algebra of his quaternions was that multiplication is not commutative. That is, for example, $ij$ does not equal $ji$. This realization was especially important because ultimately it forced mathematicians to abandon the belief that the commutative law of multiplication was axiomatic. A plaque commemorating Hamilton's inspiration about quaternion algebra, is installed today at the site on the bridge where he carved out his discovery.

During the decade after Hamilton's initial discovery, he lectured widely on his theory of quaternions. He believed quaternion algebra would transform the field of mathematical physics and become as important historically as the invention of the calculus. Although his expectations were never realized, Hamilton's ideas did play a historical role in the development of matrix algebra and certainly, in the development of vector analysis. In 1853, Hamilton's *Lectures on Quaternions* was published. He devoted most of the remainder of his life to refining his quaternion theory and investigating its application. His *Elements of Quaternions* was published posthumously in 1866.

Outside of his study of quaternions, the only mathematical work Hamilton pursued in his latter years was that of the icosian calculus. In 1856 he developed this new calculus, based originally on the geometric properties of the icosahedron, a 20–sided solid. Hamilton's discoveries about the vertices and edges of icosahedrons and other solids provided groundwork for what became known in the 20th century as graph theory.

Hamilton was recognized throughout his life for his contributions to mathematics and astronomy. When he was 30 years old, he became an officer of the British Association for the Advancement of Science. In that same year he was knighted by the lord lieutenant of Ireland and received the Royal Medal of the Royal Society. In 1837, he was elected president of the Royal Irish Academy, and six years later the British Government awarded him a Civil List life pension of 200 pounds a year. Shortly before his death, Hamilton was voted the first foreign member of the National Academy of Sciences in the United States. This last honor, bestowed on him for his work on quaternions, is the one that he reportedly cherished the most.

Hamilton married Helen Marie Bayly, the daughter of a country parson, in 1833. An illness left her a semi–invalid for most of her life. Together they bore two sons, William Edwin, born in 1834, who squandered his father's money on various business schemes, and Archibald Henry Hamilton, born in 1835, who became a clergyman. A daughter, Helen Eliza Amelia Hamilton, was born in 1840. Hamilton's wife was known as a pious, but shy and timid woman, whose intellect did not match that of her husband. Hamilton's friends were critical of his marriage, though for his part, he behaved dutifully as a husband and never complained. Nonetheless, in the latter part of his life he indulged in romantic fantasies and a prolonged and secretive correspondence with a woman who had rejected him as a young man. In the summer of 1865, having suffered from occasional bouts of gout over the years, Hamilton took seriously ill and died on September 2, 1865, in Dublin. His wife survived him by four years.

Hamilton was good friends with **Augustus De Morgan** and physicist John Herschel, as well as poets Samuel Taylor Coleridge and William Wordsworth. He was known as a convivial and jovial man, though as he grew older he developed a severe drinking problem. In his private life, his habits were apparently slovenly, and after his death his library was found to be strewn not only with unpublished manuscripts and disheveled papers, but desiccated food and dirty dishes.

## SELECTED WRITINGS BY HAMILTON

*A Treatise on the Geometrical Representation of the Square Roots of Negative Quantities,* 1829
*Preliminary and Elementary Essay on Algebra as the Science of Pure Time,* 1835
*Theory of Algebraic Couples,* 1835
*Lectures on Quaternions,* 1853
*The Mathematical Papers of Sir William Rowan Hamilton,* 1931–1967

## FURTHER READING

### Books

Bell, E.T. *Men of Mathematics.* New York: Simon and Schuster, 1986, pp. 340–361.
Graves, Robert Perceval. *Life of Sir William Rowan Hamilton.* 3 volumes. Dublin: Hodges, Figgis and Company, 1882–1889.
Hankins, Thomas L. *Sir William Rowan Hamilton.* Baltimore: Johns Hopkins University Press, 1980.

*—Sketch by Leslie Reinherz*

# Godfrey Harold Hardy
## 1877-1947
### English pure mathematician

Godfrey Harold Hardy was one of the foremost mathematicians in England during the early part of the 20th century. He was primarily a pure mathematician, specializing in branches of mathematics that study the behavior of numbers (such as number theory and analysis). He also made important contributions to areas of applied mathematics, and is known for formulating the Hardy-Weinberg law of population genetics. He taught at both Cambridge and Oxford and published over 350 research papers, either alone or in collaboration with other mathematicians—most notably John Edensor Littlewood and **S. I. Ramanujan.**

Born on February 7, 1877, in Cranleigh, England, Hardy was the elder of two children of Isaac and Sophia Hall Hardy. Both his parents came from poor families and were unable to afford university education for themselves, but they were people with a taste for intellectual and cultural pursuits and had made a place for themselves as schoolteachers. Hardy's father was the geography and drawing master at Cranleigh School, where he also gave singing lessons, edited the school magazine, and played soccer. His mother

*Godfrey Harold Hardy*

taught piano lessons there and helped run a boarding house for the younger students. They took great pains to educate their children well, and both Hardy and his sister Gertrude inherited their parent's love for education and the intellect. A gifted student, Hardy displayed a special talent and interest for mathematics from a very young age. When he was just two, he was writing down numbers into the millions, a common sign of future numerical ability. Rather than attend regular classes in mathematics, he was coached by a private tutor, and he completed sixth form at Cranleigh when he was only thirteen—about five years younger than the usual age—ranking second in class. He then won a prestigious scholarship to attend Winchester College, a private secondary school where he spent six years before graduating in 1896, winning another scholarship to attend Trinity College at Cambridge University.

### Follows the Road to Mathematics

Hardy initially chose to attend Cambridge rather than Oxford because of its standing in mathematics, and Trinity College was the premier institution for the subject in England. During his first years at Cambridge, however, he very nearly gave up mathematics altogether, in disgust over the examination system then in existence. Mathematics students had to take the Tripos examination, which consisted of eight days of solving problems. Hardy disliked the system because, rather than gauging the ability and insight of the student, he believed it tested endurance and the ability to memorize formulae and equations. Special private coaches trained students for Tripos, while lecturers at the universities pursued their own mathematical research. Hardy considered Tripos an utter waste of time, and he tried to change his course of study to history. What kept him in the field was his professor, A. E. H. Love, who recognized Hardy's affinity for pure mathematics and recommended that he read a book by the French mathematician Camille Jordan.

Entitled *Cours d'analyse de l'Ecole Polytechnique,* Jordan's book kept Hardy in mathematics, and he persevered through Tripos, putting real mathematics aside for two years. In his autobiographical book, *A Mathematician's Apology* , Hardy wrote of his career after reading Jordan's book: "From that time onwards I was in my way a real mathematician, with sound mathematical ambitions and a genuine passion for mathematics." Despite his acute distaste for Tripos, Hardy ranked fourth in the first examination in 1898, and he scored the highest points in the second part of the examinations two years later. Upon his graduation in 1899, he was named a fellow of Trinity College at Cambridge.

As a fellow, Hardy was finally free to devote his time to pure mathematics, and he did so with great enthusiasm and fervor. Over the next ten years he produced several papers on number series that established his reputation as an analyst, and in 1908 he published a book, *A Course of Pure Mathematics.* This was the first mathematical textbook in the English language to explain rigorously the fundamental concepts of the subject. Until then, books and teachers had merely provided these formulae and moved on to using them in various practical applications. Continuing his interest in mathematical education, Hardy joined a panel and tried to reform Tripos as the first step—he hoped—to abolishing it altogether. Although this latter goal proved futile (Tripos is still in existence nearly a century later), the panel did succeed in eliminating the worst features of the system.

Also in 1908, Hardy made his only contribution to applied mathematics in the form of a letter to the American journal *Science.* Mendelian genetics being the subject of much debate at that time, an article that recently appeared in *The Proceedings of the Royal Society of Medicine* had disputed some of Mendel's theories of inheritance of various traits. In his letter, Hardy used simple algebraic principles to prove the error in the article, and he set down an equation that predicted the patterns of inheritance. In the same year, a German physician named Wilhelm Weinberg devised a similar mathematical method for prediction, and the principle was named the Hardy-Weinberg law in honor of them both. Widely used in the study of the genetic transmission of blood groups and i... ...ay as a fundamental prin... ...e of popula... ...

during the ye... ...1900 and 1910, Hardy himself felt t' ...d not do too much of value, and he said so i.. *A Mathematician's Apology*: "I wrote a great deal during the next ten years but very little of any importance; there are not more than four or five papers I can still remember with some satisfaction." He believed that his best work came later, out of his associations with John Edensor Littlewood and S. I Ramanujan.

## Cultivates Productive Partnerships

Hardy began his collaboration with the mathematician J. E. Littlewood in 1911. The partnership, which lasted for over 35 years and resulted in the publication of over one hundred papers, was described by C. P. Snow in his foreword to Hardy's *A Mathematician's Apology* as "the most famous collaboration in the history of mathematics. There has been nothing like it in any science or in any other field of creative activity." Some eight years younger than Hardy, Littlewood was a brilliant mathematician who had already made a name for himself in the mathematical community.

Not much is known about exactly how the two men worked together. At the height of their combined productivity, from 1920 to 1931, they were not even at the same university. Hardy had moved to Oxford by then, while Littlewood remained in Cambridge. According to Snow, Hardy always maintained that Littlewood was the better, more powerful mathematician, although at meetings and conferences it was always Hardy who presented the papers. Indeed, Littlewood seldom attended these meetings, and other mathematicians were known to have joked that they doubted whether he even existed.

Hardy's second collaboration began in 1913 with a letter from India. The writer of the letter, Srinivasa Iyengar Ramanujan, was then an unknown clerk in Madras, who had received no formal education or training in mathematics but who claimed to have made some original discoveries while working on his own. After a single paragraph of introduction, Ramanujan plunged into his mathematics, providing page after page (the letter was over ten pages long) of theorems and results written out neatly by hand. Hardy's first instinct was to disregard the letter as a crank, filled as it was with wild claims and bizarre theorems without offering any proofs to support them. Indeed, two other Cambridge mathematicians had done just that, having received similar letters from Ramanujan earlier.

But something about the letter, perhaps its very strangeness, also intrigued Hardy, and he decided it was worth a closer look. He invited Littlewood to join him. After three hours of perusing the papers, the two men decided that the work was that of a genius. Hardy then wrote back to Ramanujan asking for proofs for some of his results, but untrained as he was Ramanujan did not or could not furnish these. In fact, when he wrote again it was to present Hardy with even more results and theorems. To Hardy, who had reintroduced the concept of rigor in proof to England, Ramanujan's intuitive reasoning and unorthodox methods were very frustrating. He invited Ramanujan to England, where from 1914 to 1919 the two men worked together and published many important papers. Hardy personally trained Ramanujan in modern mathematics and analysis; as he did in his collaboration with Littlewood, he also wrote most of the papers and presented the talks. Ramanujan himself returned to India and died in 1920, having contracted tuberculosis in England, but Hardy continued to promote his work long after his death.

## Leaves Cambridge for Oxford and then Returns

Meanwhile, Hardy was growing disenchanted with life at Cambridge, and controversies surrounding World War I had much to do with this. In his foreword to *A Mathematician's Apology,* Snow describes the years from 1914 to 1918 as the "dark years" for Hardy. Most of his friends were away at the war. His work with Littlewood was also suffering, as the latter had gone away to serve as a second lieutenant in the army. In 1916, **Bertrand Russell**, the noted philosopher, mathematician, and pacifist, was dismissed from his lectureship at Trinity for his antiwar activities. Hardy was a close personal friend of Russell's; outraged at this dismissal, he fought bitterly with many of his mathematical colleagues. In 1918, the university dismissed yet another person for their antiwar views, this time a librarian, upsetting Hardy even more, and he actively opposed the firing. Snow writes in *A Mathematician's Apology* that "it was the work of Ramanujan which was Hardy's solace during the bitter college quarrels."

Adding to his discontent was the fact that his duties at Cambridge were becoming increasingly administrative, leaving him little time for research. In 1919, he moved to Oxford University as Savilian Professor of Geometry at New College. Here, he reached the pinnacle of his career, setting up a flourishing research school and enjoying the best years of his collaboration with Littlewood. His flamboyance, radical antiwar views, and outspokenness were appreciated at Oxford. Hardy had an exceptional gift for working well with other people, and besides Ramanujan and Littlewood he collaborated with many other leading mathematicians of the day. He also spent one year as an exchange professor at Princeton University. In 1931, he returned to Cambridge as Sadleirian Professor of Pure Mathematics.

He retired in 1942, after which he continued to live in his rooms at Trinity. Shortly before his death in 1947, the Royal Society awarded him their highest honor, the Copley Medal.

There was nothing in life Hardy cared about more than mathematics, but he did have other interests. During his early years at Cambridge, he was part of several social groups, including a secret intellectual society known as the Apostles. This society met weekly to discuss and debate philosophical issues, and over the years it boasted some brilliant minds among its membership. During Hardy's time the philosophers Bertrand Russell and G. E. Moore were members, and it was through this association that many of his closest friendships were fostered.

Hardy was intensely fond of sports, particularly cricket. He followed cricket matches and scores with great attention. Hardy was not above bringing his passion for cricket into the classroom, describing the quality of mathematical work he considered exceptionally good to be in the "Bradman" or "Hobbs" class. As both men were cricketeers, not mathematicians, such references were apt to confuse unsuspecting students. To the end of his days he remained passionately interested in cricket, and when he died he was listening to his sister read to him from a book on the history of Cambridge University cricket.

Hardy was also a talented writer. He was often called upon to write obituaries of famous mathematicians. In addition to numerous mathematical texts, he also wrote *Bertrand Russell and Trinity,* a recounting of the wartime controversy, and *A Mathematician's Apology,* a treatise describing his love for the subject. In this book he offered a justification for his choice of career: "I have never done anything 'useful'. . . . Judged by all practical standards the value of my mathematical life is nil." But he adds, "The case for my life is this: that I have added something more to knowledge, and helped others to add more."

## SELECTED WRITINGS BY HARDY

### Books

*A Course of Pure Mathematics,* 1908
(Editor) *Collected Papers of Srinivasa Ramanujan,* 1927
*A Mathematician's Apology,* 1940
*Bertrand Russell and Trinity* (reprint), 1970

### Periodicals

"Obituary, S. Ramanujan." *Nature* 105 (June 17, 1920): 495–495.

## FURTHER READING

### Books

Kanigel, Robert. *The Man Who Knew Infinity.* New York: Macmillan, 1991.

Snow, C. P. Foreword to *A Mathematician's Apology.* Cambridge, England: Cambridge University Press, 1940.

—*Sketch by Neeraja Sankaran*

---

# Thomas Harriot (also transliterated as Hariot)
## 1560(?)–1621
### English mathematician and scientist

Thomas Harriot, according to his epitaph, "cultivated all the sciences/ And excelled in all." He has two published works, one on algebra, the other about the first English colony in Virginia. However, Harriot's unpublished papers display accomplished work in algebra, spherical geometry, refraction of light, paths of projectiles, astronomy, and a phonetic alphabet.

Harriot entered St. Mary's Hall, Oxford University, on December 20, 1577. University records shows that he was 17 years old, from Oxfordshire, and his father was a commoner. Harriot studied science and mathematics, receiving a B.A. in 1580. Thomas Allen may have been his teacher. Harriot probably went to London and gave private lessons in mathematics.

### A Scientific Explorer

By 1584, Harriot was employed by Sir Walter Raleigh to train his seamen in navigation. For this, Harriot wrote a text, now lost, called the *Articon.* According to J. V. Pepper in *Thomas Harriot: Renaissance Scientist,* it gave instructions for determining the latitude from observations of the sun or stars, and corrections for a cross staff, for the position of the pole star, and for variation of the compass. Harriot gave methods for calculating distances along rhumb lines, the paths of ships on the globe, which cross meridians at a constant angle.

In 1585, Raleigh sent Harriot to the English colony of Virginia with Sir Richard Grenville. About 100 colonists were left in what is now North Carolina for a year. Harriot and the artist John White observed plants, animals and minerals, made maps, and visited villages. Harriot is thought to have studied the indigenous language beforehand. Harriot's short account, *A Briefe and True Report,* lists natural commodities, and commodities "for victuall and sustenance," then describes resources for building, methods of cultivation, and the "nature and manners of the people of the countrey." The report, published in

1588, tells what a colonist would need to know for survival in the New World.

Raleigh had an estate in Ireland, and Harriot was given a property, the Abbey of Molanna, which he kept for about ten years. During this time, Harriot stayed in Raleigh's employ, helping with his affairs in Ireland and England. In his will, Harriot asked that his papers relating to Raleigh and Ireland be destroyed. Harriot found a new patron, Henry Percy, the Earl of Northumberland and a friend of Raleigh's. Harriot joined Northumberland's household at Syon around 1595.

### Politics and Science

Raleigh lost favor with Queen Elizabeth in 1592. When James I became king in 1603, Raleigh was accused of plotting to overthrow him. Sentenced to death, Raleigh was pardoned but imprisoned in the Tower of London. At the trial, both Raleigh and Harriot were accused of atheism. In 1605, Northumberland was accused of participating in the Gunpowder Plot to unseat King James. Harriot was also arrested, but soon released. The Earl remained in the Tower of London until 1622. Harriot stayed at Syon House, conducting scientific investigations, visiting his patrons in prison, and helping with their affairs. After Raleigh's release for an unsuccessful voyage to Guyana in 1617, Raleigh was rearrested and executed in 1618.

Commissioned by Raleigh in the early 1590s, Harriot observed the parabolic path of projectiles. Subsequently, at Syon, he conducted chemical experiments and determined specific weights of materials. In mathematical navigation, Harriot calculated the areas of spherical triangles, demonstrating that the Mercator projection is conformal (preserves angles), and wrote on finite difference interpolation for numerical tables. He discovered **Willebrod Snell**'s sine law of refraction from observations of the bending of light in liquids and glass. Harriot also made telescopes of 30 times magnification, observed Halley's Comet in 1607, made a map of the moon in 1609, observed sunspots and calculated the speed of the sun's rotation (1610–1613), observed the moons of Jupiter and calculated their orbits (1610–1612), and observed the third comet of 1618.

Harriot invented the greater than and less than signs. For equations, he wrote all the terms on one side equal to zero, wrote an equation as a product of factors involving the solutions, gave a method of solving cubic equations, and recognized negative and imaginary solutions. He used lowercase letters for variables, wrote powers as repeated factors, used "+" and "–," and **Robert Recorde**'s equal sign. Harriot also wrote on binary numbers and the paradoxes of infinity.

In 1613 Harriot developed a cancer on his nostril, which caused his death on July 2, 1621 in London. In his will, he asked Nathaniel Torpeley to select some of his papers (10,000 pages) for publication. Walter Warner took over in 1627 and published Harriot's influential algebra text, *Artis Analyticae Praxis*, in 1631. Harriot's papers were lost, then discovered in 1784 at Petworth House, Sussex, by Baron von Zach. Oxford University refused to fund publication of the papers in 1802, however, deeming them of no scientific interest. However, using Harriot's observations of the 1607 comet, published by von Zach, a merchant apprentice, **Friedrich Wilhelm Bessel,** calculated the comet's orbit, and began his career as an astronomer and mathematician.

## SELECTED WRITINGS BY HARRIOT

*A Briefe and True Report of the New Found Land of Virginia* 1588; enlarged in 1590 as *Narrative of the First English Plantation of Virginia*

*Artis Analyticae Praxis, as Aequations Algebraicas Nova, Expedita, & Generali Methodo, Resolvendas,* 1631

## FURTHER READING

### Books

Cajori, Florian. *A History of Mathematical Notations.* LaSalle, IL: Open Court, 1952.

*Dictionary of National Biography.* London: Oxford University Press, 1937.

Earles, M. P. *The London Pharmacopoeia Perfected.* The Durham Thomas Harriot Seminar, Occasional Paper No 3. London: Chameleon Press, 1985.

Jaffe, Bernard. *Men of Science in America.* Revised Edition. New York: Simon & Schuster, 1958.

James, Robert C. *Mathematics Dictionary.* Fifth Edition. New York: Van Nostrand Reinhold, 1992.

Lohne, J. A. "Thomas Harriot," in *Dictionary of Scientific Biography.* Volume VI. Edited by Charles Coulston Gillispie. New York: Charles Scribner's Sons, 1970, pp. 124–29.

Rukeyser, Muriel. *The Traces of Thomas Hariot.* New York: Random House, 1971.

Sherman, William H. "Thomas Harriot." *Sixteenth–Century British Nondramatic Writers, Second Series.* Dictionary of Literary Biography, Volume 136. Detroit: Gale Research, 1994.

Shirley, John W., editor. *Thomas Harriot: Renaissance Scientist.* Oxford: Clarendon Press, 1974.

*Felix Hausdorff*

————, editor. *A Source Book for the Study of Thomas Harriot.* New York: Arno Press, 1981.

Shirley, John W. *Thomas Harriot: A Biography.* Oxford: Clarendon Press, 1983.

**Other**

Van Helden, Albert. "Thomas Harriot." http://www.es.rice.edu/ES/humsoc/Galileo/People/harriot.html.

Westfall, Richard S. "Thomas Harriot." http//www-groups.dcs.st-and.ac.uk/~history/Mathematicians/Harriot.html.

—*Sketch by Sally M. Moite*

# Felix Hausdorff
## 1868-1942
### German topologist

Felix Hausdorff laid the foundations of set theoretic topology, which has evolved into an elaborate discipline that interacts with nearly every other field of mathematics. He precisely developed such basic notions as limits, continuous maps, connectedness, and compactness, which have become fundamental in building many kinds of mathematical structures. One of Hausdorff's revolutionary ideas, spaces of non-integer dimension, plays an important role in various topics, including geometric measure theory, the theory of dynamical systems, and in the description of the popularized notion of fractals. He was also a philosopher and author.

Hausdorff was born on November 8, 1868, in Breslau, Germany, which is now Wrocaw, Poland. His mother was Johanna Tietz Hausdorff; his father, Louis Hausdorff, was a dry goods merchant. The family moved to Leipzig, Germany, in 1871. The young Hausdorff eventually attended Leipzig University, where he studied astronomy and mathematics, earning his Ph.D. in 1891. His early research concentrated in the areas of optics and astronomy. After graduation, Hausdorff volunteered to serve in the German infantry. He achieved the rank of vice-sergeant before removing himself from consideration for further promotion in 1894. Hausdorff was Jewish, and no acknowledged Jews had been commissioned as officers in the German military for nearly 15 years. In 1896, following his father's death, Hausdorff succeeded him as a partner in the publishing firm Hausdorff and Company, which produced the leading trade magazine for spinning, weaving, and dyeing. That same year, he was accepted as a lecturer at Leipzig University.

### The Literary Hausdorff

Hausdorff had a lively interest in the fine arts and in philosophy. An accomplished pianist, he occasionally composed songs. Like many others of his generation, Hausdorff was deeply influenced by the philosophy of Friedrich Nietzsche, though he maintained a critical distance to certain parts of Nietzsche's work.

The first of Hausdorff's four full-length literary works was published in 1897. He wrote under the pseudonym Dr. Paul Mongré so that he could express himself freely without jeopardizing his university position. The first book, *Sant' Ilario: Thoughts from the Landscape of Zarathustra,* was primarily a collection of aphorisms relating to Nietzsche's influential volume *Thus Spake Zarathustra.* It was published by the same company that had published Nietzsche's works and was even produced with a similar book cover.

*Chaos in Cosmic Choice,* the second book written under the name Mongré, dealt with relationships between space and time and was intended as a radical continuation of Immanuel Kant's criticism of traditional metaphysics. Hausdorff presented the same concepts in "The Space Problem," his 1903 inaugural lecture after being appointed as an associate professor at Leipzig University. Mongré's third major literary work was *Ecstasies,* a volume of sonnets and poems published in 1900. He also wrote *The Doctor's Honor,* a satirical play that was successfully produced in Hamburg and Berlin.

## The Mathematical Hausdorff

In 1897, Hausdorff began publishing papers on topics in mathematics, including non-Euclidean geometry, complex numbers, and probability. He became interested in **Georg Cantor's** work on set theory, and during the summer semester of 1901 he taught what may have been the first course on set theory to be presented in Germany. Also, about this time, **David Hilbert** was publishing work applying set theory to geometry; this work may have been the inspiration for Hausdorff's greatest mathematical accomplishment.

In 1910, Hausdorff accepted a position as associate professor at the University of Bonn. Although he had written one or two articles on set theory each year for two decades, he published nothing from 1910 until 1914; apparently this was a period of intense work on the creation of point set topology. Hausdorff moved to Greifswald in 1913 to become a professor at the university there, and the following year, he published his monumental *Grundzüge der Mengenlehre* ("Basic Features of Set Theory")

The *Grundzüge* was a comprehensive text dealing with set theory, point set topology (now more commonly called set theoretic topology), and real analysis. Although the book was written for students at the advanced undergraduate level, Hausdorff noted in the preface that the volume also offered new ideas and methods to his professional colleagues. By organizing point set theory with just the right choice of axioms, he so thoroughly revised the related existing work that his book became the foundation on which modern topology has been developed.

Topology generalizes concepts such as continuity and limits to sets other than real and complex numbers. A topological space is free of all imposed structure not relevant to the continuity of functions defined on it. While Hausdorff's definitions and axioms were so general that an unlimited variety of geometric interpretations was possible, he developed the Euclidean plane as a special case by adding appropriate postulates. As Carl B. Boyer wrote in *A History of Mathematics,* "Topology has emerged in the twentieth century as a subject that unifies almost the whole of mathematics, somewhat as philosophy seeks to coordinate all knowledge."

Because of its generality, topology gives rise to apparent paradoxes that violate intuition, two of which Hausdorff addressed in the *Grundzüge*. One involves the transfinite numbers developed by Cantor, in which there are different magnitudes of infinity, and an infinite, proper subset may have the same number of elements as its superset. The other,

now called the Hausdorff Paradox, shows that the surface of a sphere can be decomposed into three equal, nonintersecting sets so that the original sphere may be represented as the union of any two of them.

In 1919, Hausdorff introduced another revolutionary concept. He generalized the notion of dimension (e.g., a two–dimensional triangle or a three–dimensional cube) to include the possibility of objects with fractional dimensions. This has proven highly fruitful in various areas of mathematics, besides being popularized in the form of computer-generated fractal images.

Hausdorff returned to the University of Bonn in 1921, where he worked as a professor for the rest of his career. He was respected as the most capable mathematician in Bonn and as a professor whose lectures were well reasoned and clearly delivered. He taught until 1935, when he reached the mandatory retirement age of 67. Hausdorff continued to publish mathematical papers until 1938.

## The Personal Hausdorff

In 1899, Hausdorff married Charlotte Goldschmidt; his only child, a daughter named Lenore (usually called Nora), was born the following year. Although Charlotte came from a Jewish family, she had been baptized a Protestant Christian in 1896, and Lenore was similarly baptized. The family moved to Bonn in 1921, and the street on which they lived would be renamed Hausdorffstrasse in 1949.

The anti-Semitism that had blocked Hausdorff's promotion in the infantry and threatened to prevent his promotion at Leipzig University continued to plague him throughout his lifetime. For instance, a young professor whose appointment Hausdorff had supported in 1926 became openly anti-Semitic in 1933, repudiating any former contact with Jews and refusing to join the rest of the faculty in attending seminars given by Jewish mathematicians. Some of Hausdorff's Jewish friends left Germany to escape the persecution; others whose emigration was thwarted committed suicide.

Suicide was a topic often addressed by Nietzsche; consequently, it had been a subject for reflection by Hausdorff. *Zarathustra* advocated "voluntary death" as a consummation of life for the noble man. "Death and Return," an 1899 essay by Mongré, broached the subject in the form of a letter to a fictitious, depressed friend. In it, the author advised that "this final remedy really helps, that it does not [merely] plunge one into a futile expense for morphine or revolver cartridges." Apparently, Hausdorff viewed suicide as an effective impediment to the Nazis' strategy of destroying their victims' human dignity: in death, the individual had no future, but his past was indestructible.

The infamous November pogrom of 1938, in which 20,000 Jews were arrested and government-sanctioned attacks resulted in 25 million marks' worth of damage to hundreds of Jewish homes, shops, and synagogues, occurred the day after Hausdorff's seventieth birthday. Charlotte Hausdorff and her sister, Edith Pappenheim (who had come to live with them a few months earlier), tried to bolster Hausdorff's spirits. He continued to work on his mathematics, but he put his writings into storage rather than publishing them.

In mid–January of 1942, the Hausdorffs were ordered to report to an internment camp located at a former monastery; this would probably be followed by deportation to a concentration camp. After organizing their affairs and leaving property disposal and cremation instructions with trusted friends, Hausdorff, his wife, and her sister committed suicide on January 26, 1942, by taking an overdose of sedatives.

On January 25, 1980, a memorial plaque honoring Hausdorff was placed at the entrance of the Mathematical Institute at the University of Bonn. In 1992, an exhibition of photographs and personal, literary, and mathematical documents was held at the University of Bonn, commemorating the fiftieth anniversary of Hausdorff's death.

## SELECTED WRITINGS BY HAUSDORFF

*Set Theory,* second edition 1978

## FURTHER READING

### Books

Abbott, David, editor. *The Biographical Dictionary of Scientists: Mathematicians.* New York: P. Bedrick Books, 1985, pp. 68–69.

Boyer, Carl B. *A History of Mathematics.* New York: John Wiley & Sons, 1991, pp. 620–622.

Katêtov, M. "Felix Hausdorff," in *Dictionary of Scientific Biography.* Volume VI. Edited by Charles Coulston Gillispie. New York: Charles Scribner's Sons, 1974, pp. 176–177.

### Periodicals

Chowdhury, M. R. "Hausdorff." *The Mathematical Intelligencer* (Winter 1990): 4–5.

Shields, Allen. "Felix Hausdorff: *Grundzüge der Mengenlehre.*" *The Mathematical Intelligencer* (Winter 1989): 6–9.

### Other

Brieskorn, Dr. Egbert, letters to Loretta Hall, 1993-1994.

*Stephen W. Hawking*

*Felix Hausdorff—Paul Mongré—1868-1942* (in German). Catalog for the 1992 memorial colloquium and exhibition, University of Bonn.

—*Sketch by Loretta Hall*

# Stephen W. Hawking
## 1942-
### English cosmological physicist

Stephen William Hawking is an icon, thanks to his efforts to popularize science. However, it is less known that he applies cutting-edge mathematics, including geometric models and algebraic analyses, to the still speculative field of cosmological physics.

Hawking was born January 8, 1942, during the height of World War II. His mother was spirited off to Oxford to have her first child, but the family suffered no fatalities and soon returned to London. Hawking spent his early years in historic St. Albans, England, built in honor of a martyred Roman centurion. His grammar school of the same name was situated near an abbey and a cathedral that dates back to 303 A.D. Hawking's parents, Isobel and Frank, were both middle class and college educated, so Stephen and his three siblings grew up in a talkative, intellectual environment. His father was a research biologist who worked for the government. The elder Hawking expected young Stephen to follow him into medicine rather than bother with impractically abstract mathematics.

Hawking studied mathematics as an undergraduate at Oxford University and later went to Cambridge to pursue a doctorate. It was at Cambridge that he first developed signs of the condition known as "motor neuron disease" in England. Hawking was given a short life expectancy death at age 25—but his parents did not appreciate the lack of attention he got from his doctors and designed their own therapy program to help their son.

Hawking's career, once he won his battle with an understandable bout of depression, began to take off in 1965. Roger Penrose, whose own career would be intertwined with Hawking's, convinced physicists that a singularity was more than a theoretical conjecture. Black hole theory had begun.

### "I Tended to Think in Pictures"

As Hawking states in their 1996 book of debates, *The Nature of Space and Time,* Penrose took the messy world of general relativity and introduced novel models like spinors, and new ways to discover general properties without getting bogged down in partial differential equations. The possibility of conceiving of a complex space time came within a theorist's grasp. As Hawking admits in the film version of his most famous book, *A Brief History of Time,* he "tended to think in pictures" in reaction to his limited use of speech and writing as memory aids. The theoretical models Hawking invents are built with quartic geometry, essentially three–dimensional surfaces placed in the Gaussian or complex plane. This allows time to be part of the models in such a way that they double as histograms and wave functions.

Since theory outstrips technology in cosmological physics, the only proofs as to the existence, nature, and behavior of black holes are mathematical ones to date. Hawking's practical achievements so far all follow this pattern. In the late 1960s he developed a mathematical model of causality in curved space–time in order to show that "The Big Bang" must have coincided with a singularity. Beginning in the 1970s, Hawking proved that black holes must follow the second law of thermodynamics by giving off what is now called "Hawking radiation." In later years, Hawking collaborated on a proof of John Wheeler's "no hair" theorem for black holes, that only some of the properties of matter survive being sucked into a black hole, but not others.

At age 32, Hawking was already wheelchair bound when he became the youngest inductee into the

Royal Society of London. In 1980, he was inaugurated as Lucasian Professor of Mathematics at Cambridge, a post once held by **Isaac Newton.** By 1982, Hawking had been named a commander of the British Empire. The awards, honoraria, and special events devoted to Hawking over the past few decades are too numerous to mention, but many have helped to defray his sizable health care costs in addition to recognizing his work.

Others have been inspired to create new technologies to aid Hawking's work and improve his mobility and communications. The most famous is the electronic voice synthesizer and motorized wheelchair now identified with him. The Intel Corporation also developed an Internet ready system of devices not intended for the open market, which allows him to remotely access cellular phone networks and activate audiovisual equipment, as well as automatic doors and room lights.

### "You Can't Be Slightly Dead"

As perhaps the most famous living victim of amyotrophic lateral sclerosis (ALS, also known as Lou Gehrig's disease), Hawking has made contributions to the lives of the disabled, directly and indirectly. He lobbied to make the streets around his Cambridge office wheelchair accessible, and wrote *Computer Resources for People With Disabilities: A Guide to Exploring Today's Assistive Technology.*

Hawking continues to follow his interests, whether appearing on a television program or commercial, lecturing to students, or debunking popular ideas about quantum theory, parapsychology, or the demise of physics. As Hawking likes to say about the popular Schroedinger's Cat conundrum, "you can't be slightly dead any more than you can be slightly pregnant." Quasi-mystical paradoxes hold no interest for him. Aside from personal taste, there is a reason why he would not fear the consequences of the unified field theory in physics to which he aspires. The very mathematics Hawking applies to that goal resulted from the "end of math" when **Karl Gauss** finalized the number system at the turn of the 19th century.

Hawking's family entered the limelight with him. His first marriage was to Jane Wilde, whom he first courted in the early 1960s when they were both graduate students. Their three children, Lucy, Robert, and Timothy, were born between 1967 and 1979 as Hawking's reputation grew. After a difficult period when Hawking came to international attention while his physical situation grew more severe, including a near fatal illness and the loss of his voice, the marriage dissolved in the early 1990s. Hawking married his nurse, Elaine Mason, in 1995.

Early in 1997, Hawking reluctantly conceded to Kip Thorne and John Preskill that under improbably rare yet calculable circumstances "naked singularities" may appear in the universe. Their referee turned out to be a supercomputer at the University of Texas at Austin, programmed to run a series of mathematically projected simulations.

## SELECTED WRITINGS BY HAWKING

(With G. F. R. Ellis) *The Large Scale Structure of Space-Time,* 1973

(With G. W. Gibbons) *Supersymmetry and Its Applications.* Edited by P. K. Townsend. 1986

*A Brief History of Time: From the Big Bang to Black Holes,* 1988

*Three Hundred Years of Gravitation,* 1989

*Stephen Hawking's A Brief History of Time: A Reader's Companion,* 1992

*Black Holes and Baby Universes,* 1994

(With Roger Penrose) *The Nature of Space and Time,* 1996

### Other

*A Brief History of Time: An Interactive Adventure CD-ROM,* 1994

*The Cambridge Lectures: Life Works,* 1996 (Audio book)

## FURTHER READING

### Books

Abbott, David, general editor. *The Biographical Dictionary of Scientists and Mathematicians.* New York: Peter Bedrick Books, 1986, pp. 69–70.

Boslough, John. *Stephen Hawking's Universe.* New York: Avon Books, 1985.

Horgan, John. *The End of Science.* Reading, MA: Helix Books/Addison Wesley.

Lindley, David. *The End of Physics.* New York: BasicBooks/HarperCollins, 1993.

Overbye, Dennis. *Lonely Hearts of the Cosmos.* New York: HarperPerennial/HarperCollins, 1992.

White, Michael and John Gribbin. *Stephen Hawking: A Life in Science.* New York: Plume/Penguin Books, 1993.

### Periodicals

Browne, Malcolm W. "A Bet on a Cosmic Scale, and a Concession, Sort Of." *The New York Times* (February 12, 1997): A1.

### Other

Coulthard, Edmund. *The Making of A Brief History of Time.* London: Anglia Television Limited, 1992.

Morris, Errol. *A Brief History of Time.* London: Triton Pictures, 1992.

—*Sketch by Jennifer Kramer*

# Louise Schmir Hay
## 1935–1989
### French–born American logician and educator

Louise Hay's mathematical specialty was recursively enumerable sets, which are studied in mathematical logic and the theory of computation. She was head of the mathematics department at the University of Illinois at Chicago, and a founding member of the Association for Women in Mathematics.

Hay was born in Metz, France, on June 14, 1935. Her father, Samuel Szmir, who had emigrated to France from Poland, was in the clothing business. Her mother, Marjem Szafran Szmir, was also from Poland. Hay's mother died in 1938, and her father married Eva Sieradska Szmir late that year. There were three children in the family, Gaston, born in 1933, Hay, and Maurice, born in 1943. (Gaston Schmir became a professor of biochemistry at Yale University, and Maurice Schmir became an anesthesiologist.) The Szmirs were Jewish family and spent World War II from the Germans. In March 1944, Hay and her older brother managed to get to Switzerland, where they remained for the next year. They rejoined the others after the liberation of France, then emigrated to the United States in 1946, where the family name was changed to Schmir. Hay's father owned a delicatessen in New York from 1950 to 1965.

Hay attended William Taft High School in the Bronx, New York. She had no interest in mathematics until her tenth grade geometry teacher, David Rosenbaum, had her read a book about non–Euclidean geometry, which fascinated her. He also helped her get tutoring work. In her senior year, Hay won third prize in the Westinghouse Science Talent Search for a project in non–Euclidean geometry. She was also class valedictorian. The prize enabled her to get a scholarship to attend Swarthmore College, as well as summer jobs at the National Bureau of Standards. She also had a part time job at the Moore School of Electrical Engineering at the University of Pennsylvania. At the end of her junior year, she married John Hay, whose field was experimental psychology. Hay completed her B.A. at Swarthmore in 1956. She joined her husband at Cornell University, where she studied mathematical logic. After two years she left Cornell to go to Oberlin with her husband, and wrote a master's thesis, on infinite valued predicate calculus, for J.

Barkley Rosser. Hay earned a M.A. from Cornell in 1959. Because of some unexpected results in the thesis, it was deemed publishable and appeared in the *Journal of Symbolic Logic*.

Hay held a visiting teaching position at Oberlin. Next, she worked at the Cornell Aeronautical Laboratory in Buffalo, New York, for a year. The Hays moved to Massachusetts, where John Hay had a job at Smith College. Hay taught part–time at a junior college for one year, after which she taught for three years at Mt. Holyoke College. Again inspired by a book, Artin's *Geometric Algebra*, Hay attended seminars at Cornell in the summer of 1962, where her husband was participating in a research project. She taught at Mt. Holyoke that fall, and in the following year her first son, Bruce, (now a law professor) was born. A discussion with algebraist **Hannah Neumann** about raising children while pursuing a mathematical career convinced Hay to stay on at Cornell, where she and her husband had again gone for the summer, while her husband returned to Smith. Hay landed a research assistantship, took three courses, passed her preliminary examinations, worked on her dissertation for Anil Nerode, and gave birth to twins, Philip and Gordon (both sons studied mathematical finance) in 1964, completed her thesis in recursion theory and received her Ph.D. from Cornell in 1965. After staying home for a year, Hay taught as an assistant professor at Mt. Holyoke from 1965 to 1968, spending 1966 to 1967 at M.I.T. on an N.S.F. fellowship, and going "through at least 15 sitters." In 1968, she and John Hay divorced.

Hay went to the University of Illinois at Chicago as an associate professor in 1968. She married Richard Larson, a colleague in the mathematics department in 1970. In 1975, she was promoted to professor. In the late 1960s and the 1970s especially, Hay wrote a series of papers dealing with the classification of index sets of recursively enumerable sets, and she introduced the idea of a "weak jump." She enjoyed "the peculiar thrill of briefly knowing a sliver of mathematical truth that *nobody* else knows." Hay also wrote reviews of papers in recursion theory for the *Mathematical Reviews*.

Hay was involved in the Association for Women in Mathematics, giving one of their first invited addresses on a mathematical topic in January 1974. She also supported and encouraged women students at her school, and had two graduate students, Nancy Johnson and Jeanleah Mohrherr. Hay and her husband were awarded Fulbright fellowships to the Philippines in 1978. Hay's interests, according to her brother, Gaston Schmir, included "travel, good food, guitar playing, the feminist movement, and math education for women."

Hay became acting head of the mathematics department in 1979, and headed the department from 1980 until 1988. Personally warm and caring, she was a popular administrator, who democratized the department, made strong appointments, and saw the department name changed to include mathematics, statistics, and computer science. Hay served as secretary of the Association for Symbolic Logic from 1977 to 1982, and on the executive board of the Association for Women in Mathematics from 1980 to 1982.

Hay was diagnosed with cancer in 1974 and suffered a recurrence in 1988. She died on October 28, 1989 in Oak Park, Illinois, at the age of 54. A scholarship was established in Hay's honor at Oak Park/Forest River High School for senior women who intend to major in mathematics. In addition, the Association for Women in Mathematics gives an annual award named for Louise Hay. The award is given to a woman who has made an outstanding contribution to mathematics education.

## SELECTED WRITINGS BY HAY

### Periodicals

"Axiomatization of the Infinite–valued Predicate Calculus." *Journal of Symbolic Logic* (March 1963): 77–86.

"The Co–simple Isols." *Annals of Mathematics* 83, no. 2, (1966): 231–56.

"On the Recursion–Theoretic Complexity of Relative Succinctness of Representations of Languages." *Information and Control* 52 (1982): 2–7.

"How I Became a Mathematician (or How It Was in the Bad Old Days)." *Newsletter of the Association for Women in Mathematics* 19, no. 5, (1989): 3–4.

## FURTHER READING

### Books

Green, Judy and Jeanne Laduke. "Women in American Mathematics: A Century of Contributions." *A Century of Mathematics in America*, Part II. Edited by Peter Duncan. Providence, RI: American Mathematical Society, 1989, pp. 379–398.

Mathematical Society of Japan. *Encyclopedic Dictionary of Mathematics*. Second Edition. Edited by Kiyosi Ito. Cambridge, MA: MIT Press, 1987.

Soare, Robert I. *Recursively Enumerable Sets and Degrees*. Berlin: Springer–Verlag, 1987.

### Periodicals

Blum, Lenore. "A Brief History of the Association for Women in Mathematics: The Presidents' Perspectives." *Notices of the American Mathematical Society* (September 1991): 738–54.

Hughes, Rhonda. "Fond Remembrances of Louise Hay." *Newsletter of the Association for Women in Mathematics* (January–February 1990): 4–6.

"Louise S. Hay, Professor, 54." (Obituary) *The New York Times* (October 31, 1989): II, 10:5.

Soare, Robert I. "Louise Hay: 1935–1989." *Newsletter of the Association for Women in Mathematics* (January–February 1990): 3–4.

**Other**

American Mathematical Society. *Mathematical Reviews: MathSci Disc* (CD-ROM). Boston: SilverPlatter Information Systems.

Association for Women in Mathematics. "AWM Homepage." http://www.math.neu.edu/awm.

"Louise Hay." June 1997. *Biographies of Women Mathematicians.* http://www.scottlan.edu/lriddle/women/hay.htm (July 22, 1997).

Schmir, Gaston L. (brother of Louise Hay), interview with Sally M. Moite conducted April 11–April 28, 1997.

—*Sketch by Sally M. Moite*

# Ellen Amanda Hayes
## 1851–1930
### American mathematics educator

Ellen Amanda Hayes was born on September 23, 1851. Her maternal grandparents, originally from Granville, Massachusetts, founded the small town of Granville, Ohio, in 1805 and it was in their home that Hayes was born. Hayes' grandparents, as well as her parents, would set the stage for her love of learning, career, and political interests.

Hayes' father, Charles Coleman Hayes, made his living as a tanner after serving as an officer in the Civil War. Her mother, Ruth Rebecca (Wolcott), taught all six of her children to read, gave them a smattering of astronomy, and instructed them in botany, supplying them with the names plants in Latin. Both generations, parents and grandparents, believed in education without regard to gender. Hayes' mother had been trained as a teacher and graduated from the Granville Female Academy, a school that enjoyed the support of, and accepted as a trustee, Hayes' grandfather, Horace Wolcott. Although Hayes' father was uneducated, he too encouraged the education of his children.

Hayes left the home instruction supplied by her mother when she was seven and went to the Centerville school. That school had only one room for all levels of instruction and kept no grades. At age 16,

Hayes was herself a teacher at a country school, saving the money she earned to attend college. After entering Oberlin College in 1872 as a preparatory student, Hayes began her college career as a freshman in 1875. Her endeavors mainly centered on the fields of mathematics and science, but she also became well versed in English literature, Greek, Latin, and history. Her mother's introduction to astronomy must have left a lasting impression because Hayes spent time at the Leander McCormick Observatory at the University of Virginia in 1887–1888, where she studied the Minor Planet 267, confirming its definite orbit, and producing other important papers on Comet *a* and planetary conic curves.

After graduating from Oberlin with a bachelor of arts degree, Hayes spent a year as the principal of the women's department at Adrian College in Michigan. In 1879, she became a teacher of mathematics at Wellesley College. By 1888, she was a full professor and had assumed the role of chair of the department. In 1897, a department of applied mathematics was created at Wellesley and Hayes took the helm. Her responsibilities included giving instruction in seven levels of applied mathematics.

### A Controversial Woman

Although Hayes spent 37 years at Wellesley, the association was often far from congenial because of Hayes' view on education and politics. She was never silent or restrained about either subject. Hayes was adamant about females taking mathematics and science courses and highly critical of the school for allowing students to choose electives that would make it possible for them to evade these studies. Reforms concerning working conditions, politics, and the education of women was something Hayes worked toward all her life. Her views on and support of the union movement and workers rights caused her to receive threats and to be arrested. She closely studied the Russian Revolution of 1917 as it unfolded, writing and speaking openly about the situation. Although Hayes never affiliated herself with the Communist party and disagreed with many of its doctrines, her association with socialist causes did much to brand her a radical and incite serious criticism from Wellesley College. Upon her retirement from Wellesley, Hayes was denied the honorary position of Professor Emeritus usually bestowed on teachers for lengthy and faithful service.

### Legacy of a Life Well Spent

At the age of 72, Hayes began her own newspaper. The *Relay* was published monthly and was devoted to giving publicity to facts and movements that Hayes believed were not accurately presented in the mainstream press. Her description of the publication was that "the *Relay* plans to camp in a hut by the side of the road and to keep a lamp or two burning—in the hope of being a friend to wayfarers and especially to the limping Under Dog." Other books written after her retirement include *The Sycamore Trail Know?*, a book which asks readers to question the origin of their beliefs and superstitions, and study the nature of evidence. Most of Hayes' work was self-published.

Upon her death in 1930, Hayes' brain was donated to the Wilder Brain Collection at Cornell University. The epitaph assigned to her by her friends was her favorite quotation: "It is better to travel hopefully, than to arrive."

## SELECTED WRITINGS BY HAYES

### Books

*Lessons on Higher Algebra,* 1891, revised edition 1894
*Elementary Trigonometry,* 1896
*Algebra for High Schools and Colleges,* 1897
*Calculus with Applications: An Introduction to the Mathematical Treatment of Science,* 1900

### Periodicals

"Comet *a* 1904." *Science* 19 ( May 27, 1904): 833–34.

## FURTHER READING

### Books

Brown, Louise. *Ellen Hayes: Trail–Blazer.* West Park, NY: 1932.
Moskol, Ann. "Ellen Amanda Hayes," in *Women of Mathematics: A Biobibliographic Sourcebook.* Edited by Louise S. Grinstein and Paul J. Campbell. Westport, CT: Greenwood Press, 1987, pp. 62–66.

### Periodicals

Gordon, Geraldine. "Ellen Hayes: 1851–1930." *The Wellesley Magazine* (February 1931): 151–52.
Merrill, Helen A. "Ellen Hayes." *Scrapbook of the History Department of Mathematics* (1944): 41–46. Archives, Wellesley College.

### Other

"Ellen Amanda Hayes." *Biographies of Women Mathematicians.* June 1997. http://www.scotlan.edu/lriddle/women/chronol.htm (July 22, 1997).

—*Sketch by Kelley Reynolds Jacquez*

# Olive Clio Hazlett
## 1890–1974
### American algebraist and educator

Olive Clio Hazlett was one of the most active women working in mathematics prior to 1940. She is best known for her work in the area of linear algebra. The majority of her research was conducted in linear algebra and also in modular invariants, making important contributions in both areas.

Hazlett received her undergraduate degree from Radcliffe College in 1912. She began work on her Ph.D. at the University of Chicago in 1913, receiving her doctorate there in 1915. She conducted additional study and research work at Harvard University and in Rome, Zurich, and Göttingen as a Guggenheim fellow. Hazlett taught mathematics for more than 40 years. She began her career in 1916 at Bryn Mawr College in Pennsylvania, remaining there for two years before accepting a position as assistant professor at Mount Holyoke College in Massachusetts. Hazlett taught at Mount Holyoke for eight years, attaining the position of associate professor in 1924. In 1925, she took a position at the University of Illinois, enticed there by the excellent library facilities and the assurance of sufficient research time to develop her mathematical theories. She completed her career at the University of Illinois, retiring as emeritus associate professor in 1959.

### Career Advancement Denied

Although recognized for her outstanding and prolific accomplishments, Hazlett's career did not reflect her mathematical brilliance. As was common with many gifted women mathematicians of her era, she was denied advancement in her profession. She attained the level of associate professor, and remained at that level throughout her career. This meant Hazlett often was relegated to teaching introductory courses and undergraduates, long after her male peers had advanced to full professorships and graduate students. In spite of this, her name appeared in seven editions of *American Men and Women of Science.* Hazlett also took an active role in the professional associations of her profession, including the American Mathematical Society, and served as associate editor of the *Transactions of the American Mathematical Society* for 12 years, from 1923 to 1935. She was also a member of the Society's council from 1926 to 1928, and a member of the New York Academy of Sciences. The award of a Guggenheim Fellowship in 1928 allowed her to study in Europe for one year.

While at the University of Illinois, Hazlett suffered a series of mental breakdowns during the 1930s and 1940s. Margaret Rossiter, in her book *Women Scientists in America,* states that Hazlett never fully recovered from her illness. "Isolated and moderately successful but with aspirations of full equality, [she] denied the potential psychological dangers in [her] situation."

Hazlett died on March 11, 1974, in a Keene, New Hampshire, nursing home at age 83. She had never married, and lived out her life on what Rossiter termed a "pitiable pension," resulting from the low pay she had received throughout her career.

## FURTHER READING

### Books

Bailey, Martha J. *American Women in Science: A Biographical Dictionary.* Santa Barbara, California: ABC–CLIO, Inc., 1994, p. 159.

Rossiter, Margaret W. *Women Scientists in America: Struggles and Strategies to 1940.* Baltimore: The Johns Hopkins University Press, 1982.

### Periodicals

Obituary. *The New York Times.* March 12, 1974, p. 40:4.

### Other

Riddle, Larry. "Olive Clio Hazlett." *Biographies of Women Mathematicians.* June 1997. http:www/scottlan.edu./lriddle/women/chronol.htm (July 21, 1997).

—*Sketch by Jane Stewart Cook*

*Heinrich Eduard Heine*

# Heinrich Eduard Heine
## 1821–1881
### German number theorist

Heinrich Eduard Heine was one of the most productive mathematical writers in 19th century Germany, having published approximately 50 papers on a wide range of topics. His most notable achievements in mathematics were the formulation of the concept of uniform continuity and the Heine–Borel theorem.

Born in Berlin, Heine was the eighth child in a family of nine. His father, Karl, was a banker. Henriette Märtens, his mother, stayed home and tended to the children. Heine was privately schooled at home before enrolling in the Friedrichswerdersche Gymnasium, and in 1838 graduated from the Köllnische Gymnasuim in Berlin. He attended Göttingen University and often sat in on the lectures of **Karl Gauss** and Moritz Stern. In 1842, Heine received his Ph.D. in mathematics from Berlin University, where he was a student of **Peter Dirichlet**. He taught at Bonn University as a *privatdozent,* or unpaid lecturer, and as professor before moving to Halle University, where he remained throughout his career. Heine held the office of rector for the university in 1864–1865.

### The Heine–Borel Theorem

Heine is most noted for his work regarding the Heine–Borel Theorem, which is defined as the following by Carl Boyer in his *A History of Mathematics*: "If a closed set of points on a line can be covered by set intervals so that every point of the set is an interior point of at least one of the intervals, then there exists a finite number of intervals with this covering property." Heine formulated the notion of uniform continuity and proved its existence in continuous functions. What has been noted often is that Heine's was the essential discovery and **Émile Borel**'s reduction of uniform continuity to the covering property was secondary. Heine also studied Bessel functions, Lamé functions, and spherical functions, also known as Legendre polynomials, and in 1861 he published his most influential work, *Handbuch der Kugelfunctionen*, which was considered the authoritative text on spherical functions well into the turn of the 20th century.

In 1850, Heine married Sophie Wolff and had four children. He was an active member of the Prussian Academy of Sciences as well as a member of the Göttingen Gesellschaft der Wissenschaften. In 1875, Heine declined the offer to chair the mathematics department at Göttingen, but received the Gauss Medal in 1877. Heine died in 1881 in Halle.

## SELECTED WRITINGS BY HEINE

*Handbuch der Kugelfunctioen, Theorie und Anwendungen,* 1861; 2nd ed., 1878–1881

"Die Elemente der Funtionenlehre," in *Journal für die reine und angewand Mathematik* 74 (1872): 172–188.

(The majority of Heine's papers were published in the mathematical journal *Zeitschrift für die reine und angewandte Mathematik.*)

## FURTHER READING

Abbott, David, editor. *The Biographical Dictionary of Scientists: Mathematicians.* New York: Peter Bedrick Books, 1985, pp. 70–71.

Boyer, Carl. *A History of Mathematics.* Second edition. Revised by Uta C. Merzbach. New York: John Wiley & Sons, 1989, pp. 561, 563, 571, 618.

Freudenthal, Hans. "Heinrich Edward Heine," in *Dictionary of Scientific Biography.* Volume VI. Edited by Charles Coulston Gillispie. New York: Charles Scribner's Sons, 1978, p. 230.

—*Sketch by Tammy J. Bronson*

*Charles Hermite*

# Charles Hermite
## 1822–1901
### French analyst and algebraist

Charles Hermite was one of the founders of analytic number theory. This discipline uses the techniques of analysis (the calculus) to handle questions about positive whole numbers. Hermite is also remembered for having shown that one of the central constants of mathematics, e, the base of natural logarithms, belongs in the class of transcendental numbers.

The son of Ferdinand Hermite and Madeleine Lallemand, Hermite was born on Christmas Eve, 1822. His ancestry was both French and German and the town Dieuze, Hermite's birthplace, was at one time claimed by both France and Germany. Nevertheless, Hermite considered himself French all his life and became one of the mainstays of the French academic establishment.

Hermite attended the Collège Henri IV and proceeded from there to the Collège Louis–le–Grand, where he was taught mathematics by the same instructor who had supervised the work of the ill–fated French genius **Èvariste Galois**. When Hermite decided to continue his studies at the Ècole Polytechnique, he was admitted 68th in his class, thanks to his having neglect geometry. Throughout his life Hermite had a dislike for examinations and preferred to pick up material spontaneously rather than under the pressure of a deadline. Hermite enjoyed corresponding with the best mathematicians of Europe, including **Karl Jacobi**, and some of the material in Hermite's letters is remarkably sophisticated. In particular, Hermite generalized a result of **Niels Abel** that applied elliptic functions to the class of hyperelliptic functions as well.

Hermite's family life reflected his increasingly central position in the French mathematical establishment. His wife was the sister of the mathematician Joseph Bertrand, and one of his daughters married the eminent analyst Émile Picard (who was to edit Hermite's works after his father–in–law's death). From the time he became admissions examiner at the Ècole Polytechnique in 1848, Hermite devoted much of his effort to working with students at every level. In addition to Picard, his other distinguished students included **Henri Poincaré**, Camille Jordan, and Paul Painlevé. This record attests to his eagerness in welcoming students as colleagues. **Émile Borel** is said to have remarked that no one made people love mathematics so deeply as Hermite did.

On a professional level, Hermite accomplished some formidable tasks with the analytic apparatus which he had mastered. The solution of the general quadratic (second–degree) equation had been known since ancient times. Solutions to the cubic and quartic equations had been developed during the Italian Renaissance. When Galois showed that ordinary algebraic methods could not solve the general quintic (fifth–degree) equation, the subject appeared to have reached a dead end. Using once again the techniques of elliptic functions, Hermite showed that fifth–degree equations could be solved after all.

### The Transcendence of e

The single result for which Hermite is best known was the transcendence of e. A number is said to be algebraic if it is the solution of a polynomial equation with integer coefficients. For example, the square root of 2 is algebraic, since it is a solution of the equation $x^2-2=0$. A real number that is not algebraic is called transcendental. The French mathematician **Joseph Liouville** had shown that there were transcendental numbers but no familiar examples were known. Hermite was able to show that e could not be written as the solution of a polynomial equation and therefore had to be transcendental. His technique was used shortly thereafter to show that $\pi$ was also transcendental, although Hermite does not seem to have recognized just how useful his technique was.

After his appointment as professor analysis at both the Ècole Polytechnique and the Sorbonne, Hermite took to writing textbooks that were widely used and appreciated. Although he resigned his chair at the Ècole Polytechnique after only seven years in 1876, he remained at the Sorbonne for another 21 years. Hermite attached a great value to insight and did not include much rigor in his teaching of elementary material. If his papers suffer from a fault, it is the occasional tendency to allow the details to get in the way of the overall picture. A large number of his ideas were developed by others and the complex generalization of quadratic forms named for him proved to be central in the formulation of quantum mechanics.

At the time of his 70th birthday, the adulation Hermite received from across Europe attests to his reputation as an elder mathematical statesman. His interests were never narrow, and he was awarded an impressive collection of decorations both at home and abroad. Hermite died on January 14, 1901, leaving a solid basis of mathematics and an unmatched collection of students to carry on his work.

## SELECTED WRITINGS BY HERMITE

*Oeuvres*, 4 vols. Edited by Èmile Picard, 1905–17

## FURTHER READING

Bell, Eric. *Men of Mathematics*. New York: Simon and Schuster, 1937, pp. 448–465.
Freudenthal, Hans. "Charles Hermite," in *Dictionary of Scientific Biography*. Volume XI. Edited by Charles Coulston Gillispie. New York: Charles Scribner's Songs, 1973, pp. 306–309.
Prasad, Ganesh. *Some Great Mathematicians of the Nineteenth Century*. Volume 2. Benares, India: Benares Mathematical Society, 1934, pp. 34–59.

*—Sketch by Thomas Drucker*

# Hero of Alexandria
## c.65–125
### Greek geometer and engineer

Hero (or Heron) of Alexandria was a Greek mathematician and engineer whose major contributions to mathematics were Hero's formula and the first approximation in Greece of a number's square root. He also wrote a number of books on mathematics, including *Metrica*, is a treatise on geometry. While much of his writings in mathematics and mechanics stem from earlier authors, Hero is credited as being one of the earliest and most comprehensive and detailed recorders of ancient technology, especially through such works as *Penumatica* and *Automata*. Hero also designed many mechanical instruments, which earned him the name "the mechanic" or "the machine man."

Other than his writings, nothing is known about the life of Hero, who is sometimes referred to as Heron. The date of his birth is unknown, and estimates of when he lived vary by 150 years. However, Hero mentions an eclipse of the moon visible from Alexandria in his book *Dioptra*. Modern astronomers date this eclipse as having occurred in 62 A.D., thus providing the major clue as to the era when Hero lived. Scholars agree that Hero probably lived and worked in Alexandria. However, there is some question as to his nationality, which may have been Egyptian.

Despite the lack of historical records on Hero's life, the breadth of his writings on mathematics and mechanics leave little doubt that he was well educated. Hero was strongly influenced by the writings of Ctesibius of Alexandria and may even have been a student of the ancient mechanical engineer. His works draw on a wide range of sources, including, Greek, Latin, and Egyptian. Unlike most of his contemporaries, Hero makes no mention of working for a Roman patron.

Hero's writings in mathematics and especially in mechanics reveal that he was practical by nature, often using ingenious means to attain his goal, like his design for a steam engine, catapults for war, and various machines for lifting that used compound pulleys and winches. Hero was also precise in dictating the types of materials to be used to make the machine function properly. Interestingly, Hero designed several mechanical devices to simulate "temple miracles," including a device attached to the temple door which made a trumpet play when the door was opened.

## Compiles Encyclopedic Works on Geometry

Hero's background in mechanical engineering is clearly evident in his practical rather than theoretical approach to mathematics and geometry. Although credited with the first approximations of square roots and his famous formula for calculating the area of a triangle, Hero's primary contribution to geometry stem from a series of treatises in which he freely incorporated the writings and findings of others to compile a coherent body of work on the subject.

The most famous of these works is *Metrica*, which consists of three books focusing on the calculation of areas and volumes and their division. This book was lost until the last century, and scholars knew of its existence only through a 6 A.D. commentary by Eutocius. Then, in 1894, historian Paul Tannery discovered a fragment of the book in Paris. A completed copy was found by R. Schöne in Constantinople (Istanbul) in 1896. Book I of *Metrica* contains the famous Hero's formula, which he used to calculate the area of a triangle when the sides are given and provided surveyors of his day with a formula for determining the area of land lots. Like most of Hero's work, this formula may came from an earlier source. Although an ancient Islamic manuscript credit's **Archimedes** with developing the famous equation, no writings of Archimedes are known to contain the formula.

Hero's other works include *Definitions*, basically a catalogue of geometrical terms; *Geometrica*, an introduction to geometry, and *Sterometrica*, which focuses on solid geometry for spheres, cubes, pyramids, and other figures. He is also believed to have written a commentary on the famous Greek mathematician **Euclid**. Hero's emphasis on the practical use of geometry is evident in the types of problems he tackles. For example, he provides a method for calculating a theater's seating capacity and for determining the number of jars that could be stored in a ship. In the theoretical realm of mathematics, Hero is credited as the first Greek mathematician to use systematic geometrical terminology and symbols. Although the Babylonians had developed a formula for approximating square roots nearly 2,000 years before Hero, he was the first Greek to develop methods for finding approximate numerical square and cube roots.

## Advances the Principles of Mechanics

Hero's contributions to the field of mechanics are wide ranging, and he achieved considerable fame in his own day for some of his inventions. For the most part, these inventions focused on the practical, like keeping time with a water clock and developing a compressed air catapult for war. His most famous mechanical design was for the aeolipile, which used steam to rotate a sphere and has been compared to the modern–day jet engine.

Much of Hero's fame today lies in the fact that most of his treatises on mechanics have survived throughout the centuries. As a result, he is considered the definitive source on ancient Greek and Roman technology. His design for the aeolipile appear in his *Penumatica*, which focuses on designs for machines powered by compressed air, siphons, and steam pressure. Although the treatise contains some theoretical components, the most important aspect of the book is Hero's ability to combine different schools of mechanical thought into a cohesive treatise. In *Mechanica*, Hero focuses on basic mechanics and their application, including the levers, pulleys, screws, and various tools. Hero was also interested in mechanical "gadgets," primarily used to produce "miracles" in a religious context. He describes these devices in both *Pneumatica* and *Automata*, which includes designs for making miniature mechanized puppet theaters. A book on machines for lifting heavy objects, *Baroulkos*, is also lost.

The wide range of Hero's interests are evident in his other works, which include *Dioptra*, at treatise on surveying. In *Catoptrica*, Hero discusses mirrors, including the theory of refraction and the various kinds of mirrors, such as flat, concave, and complex. He also wrote lost works on topics like large and hand catapults (*Baelopoeica* and *Cheiroballistra*), time keeping devices, and vault construction.

Hero's stature as a historical figure in mathematics and mechanics has grown with the rapid advance of technology in the last two centuries. Hero was more than a mere chronicler of ancient devices and how they should be built, he also exhibit scrupulous attention as to their construction and the materials to be used. Although Hero never built many of his models, they presented interesting design problems that grew in importance with the industrial revolution.

Hero is not without his critics for a number of reasons. First and foremost, his works contain some notable errors, primarily in the area of mathematics. Hero is also known to have gathered much of his knowledge from previous writers. Despite these criticisms, Hero revealed an advanced understanding of harnessing power, especially in the form of wind and steam. As a scholar, he conducted comprehensive research and took a systematic approach to revealing the basic of many useful devices, including a coin–operated machine. As with his birth, the date of Hero's death is unknown.

## FURTHER READING

Gow, James. *A Short History of Greek Mathematics.* New York: Chelsea Publishing Company, 1968, pp. 276–286.

Dunham, William. *Journey through Genius: The Great Theorems of Mathematics.* New York, John Wiley & Sons, Inc., 1990, pp. 117–127.

Landels, J.G. *Engineering in the Ancient World.* Los Angeles: University of California Press, 1978, pp. 199–208.

Magill, Frank N., editor. *The Great Scientists 6.* Danbury, CT: Grolier Educational Corporation, 1989, pp. 49–53.

Van Der Waerden, L. Van. *Science Awakening.* New York: John Wiley & Sons, Inc., 1963, pp. 276–278

*—Sketch by David A. Petechuk*

*Caroline Lucretia Herschel*

# Caroline Lucretia Herschel
## 1750–1848
### German–born English astronomer

Caroline Herschel was born on March 16, 1750 in Hanover, Germany. She is noted for her scientific annals in astronomy more than for her mathematical knowledge. Yet, while her accomplishments were heralded in astronomy, Herschel deserves recognition in both fields. She never received formal mathematical training, which only serves to accent the dimension of her accomplishments and determination.

Herschel was a homely child who received no love and encouragement from her mother. Her father, Isaac Herschel, a musician in the Hanoverian Guard, on the other hand, encouraged her to obtain an education. She was, as he kept telling her, so homely and without money, no one would marry her until she was older and had more character. Herschel was a literate young woman, but did not receive a formal education. At age 17, her father died and she fell under her mother's domination.

Herschel led a harsh life until her brother William, who was eleven years her senior, empathized with her plight and invited her to live with him in Bath, England, where he was immersed in musical training and astronomy. Their mother refused to let her go until William promised to provide funds for her mother to retain a maid.

In August 1772, Herschel left for England. Over the next five years, her horizons expanded. A neighbor taught her cooking, marketing, and English. Unlike their parents, William encouraged his sister to be independent and enroll in voice lessons and the harpsichord. She soon became an integral part of William's musical performances at small gatherings.

In her spare time, she and William discussed astronomy, and her interest in the constellations grew. But as the sister of William, she needed to learn and incorporate English society into her schedule. Such activities seemed like nonsense to her staunch German upbringing, but she did learn with William's guidance, and soon began making appearances at the opera, theater, and concerts.

Herschel longed to be self–supporting, and at age 27 she was in demand as a soloist for oratorios. But William increasingly needed her efficient, meticulous talents in copying his astronomy catalogs, tables, and papers. She eventually drifted from her desires and devoted herself to his astronomy.

## Made Celestial Discoveries

Herschel assisted her brother in grinding and polishing his telescopes. He built a new six–foot telescope and began scanning the night skies. In 1781, William discovered Uranus, with, as he said, Caroline's devoted help. This discovery assured him recognition in British scientific circles. Originally, Uranus had been named "Georgium Sidus," after King George III. Thus, William was appointed to the position of court astronomer and was knighted. While such an appointment guaranteed financial security for William, Herschel was appointed William's assistant and given an annual stipend of 50 pounds. Herschel's appointment by the Court made her the first female in England honored with a government position.

Herschel focused on providing her brother with the support system he needed. She systematically collected data and trained herself in geometry, learned formulas and logarithmic tables, and gained an understanding of the relationship of sidereal time (time measured by means of the stars) to solar time. Her record–keeping was meticulous and systematic. The numerical calculations and reductions, which saved her brother precious time, were all done without error, and the volume of her work was enormous.

When Herschel was not engaged in other tasks, she too searched the night skies using a small Newtonian reflector. To her credit, in early 1783, Herschel discovered the Andromeda and Cetus nebulae, and by year's end, had discovered an additional 14 nebulae to those already catalogued. As a reward, William presented her with a new Newtonian sweeper of 27 inches, with a focal length of 30. Herschel was also the first woman to discover a comet, and between 1789 and 1797 she had discovered another seven comets.

Herschel calculated and catalogued nearly 2,500 nebulae. She also undertook the task of reorganizing John Flamsteed's *British Catalogue*, which listed nearly 3,000 stars. Herschel's listings were divided into one–degree zones in order for William to use a more systematic method of searching the skies.

Her brother married in 1788, causing her concern about having to share his home and affections. But her concerns were without merit, as she was warmly and graciously accepted by her new sister–in–law. The two women became friends and Lady Herschel was a great support to her.

On August 25, 1822, William died, leaving Herschel without support. She returned to Hanover after his death, still supported by the royal family. Herschel continued with her own work in the fields of

mathematics and astronomy. In 1825, she had donated the works of John Flamsteed to the Royal Academy of Göttingen.

Herschel never married, and spent the last years of her life in Hanover, organizing and cataloguing the works of her nephew, Sir John Herschel, William's son, who carried on his father's extensive work.

In 1828, at age 75, the Royal Astronomical Society voted Herschel a gold medal for her monumental works in science. At age 85, she was made honorary member of the Royal Astronomical Society; she was similarly honored by the Royal Irish Academy for her work in science. On her 96th birthday, she was awarded the gold medal of science by the King of Prussia.

On January 9, 1848, Herschel died at the age of 97. Her meticulous work aiding her famous brother was her legacy. While not credited with any original mathematical works, she applied her painstaking, meticulous skill to the advancement of human knowledge.

## FURTHER READING

### Books

Hoskin, Michael A. "Caroline Herschel," in *Dictionary of Scientific Biography*. Volume VI. Edited by Charles Coulston Gillispie. New York: Charles Scribners' Sons, 1981, pp. 322–23.

Osen, Lynn M. "Caroline Herschel." *Women in Mathematics*. Cambridge, MA: The MIT Press, 1974.

Schweighauser, Charles A., editor. *Astronomy from A to Z: A Dictionary of Celestial Objects and Ideas*. Springfield, IL: Sangamon State University, 1991.

### Other

Nysewander, Melissa. "Caroline Herschel." *Biographies of Women Mathematicians*. June 1997. http://www.scottlan.edu/lriddle/women/chronol.htm (July 21, 1997).

*—Sketch by Corrine Johnson*

---

# David Hilbert
## 1862-1943
### German number theorist

By the end of his career, David Hilbert was the best known mathematician in the world, as well as the

*David Hilbert*

most influential. His contributions did not merely affect but decisively altered the directions taken in many fields. In some ways, however, his career ended in disappointment; he had inherited one of the great mathematical centers for research and teaching, but from his retirement he had to watch its glory disappear under the ideological onslaught from the Nazi government. Nevertheless, the heritage of the contributions he made to mathematics, as well as the students he trained, has outlasted the disruptions of World War II.

Hilbert was born in Wehlau, near Königsberg, on January 23, 1862. His family was staunchly Protestant, although Hilbert himself was later to leave the church in which he was baptized. Otto Hilbert, his father, was a lawyer of social standing in the society around Königsberg, and his mother's family name was Erdtmann. The name "David" ran in the family—a fact Hilbert had subsequently to verify to the Nazi regime, which suspected that anyone with the name was of Jewish ancestry. Hilbert's early education was in Königsberg, which he would always consider his spiritual home.

In 1880, Hilbert entered the University of Königsberg, where he received his Ph.D. in 1885. By the next year he had become a privatdozent, and by 1892 Hilbert had been appointed to the equivalent of an assistant professorship at Königsberg, rising in the ranks to a professorship the next year. In 1895 he took a chair at Göttingen, where he remained until his

retirement. As this rapid progress attests, Hilbert knew enough about academic politics to advance through the complexities of the German system. In this he had the guidance of a mathematician with great political skills, named **Felix Klein** , who had devoted much of his life to building the University of Göttingen into the world's mathematical center.

### Researches Invariants and Contributes to Number Theory

Hilbert made his mathematical reputation on the strength of his research into invariant theory. The notion of an invariant had been created in the nineteenth century as an expression of something that remains the same under various sorts of transformations. As a simple instance, if all the coefficients in an equation are doubled, the solutions of the equation remain the same. A good deal of work had been done in classifying invariants and in trying to prove what sorts of invariants existed. The results were massive calculations, and books on invariant theory were made up of pages completely filled with symbols. Hilbert rendered most of that work obsolete by taking a path that did not require explicit calculation. Those who had been practicing invariant theory were taken aback by Hilbert's effrontery, and one of them described Hilbert's approach as "not mathematics, but theology." Invariant theory quickly disappeared from the center of mathematical interest, as Hilbert's work required some time to be absorbed. Only much later was the field reopened, as invariant theorists at last were ready to proceed from his calculations.

Perhaps the mathematician closest to Hilbert was **Hermann Minkowski**, two years younger than Hilbert but well-known at an even earlier age. At first, Hilbert's family did not approve of their friendship because Minkowski was the son of a Jewish rag merchant. Hilbert nonetheless kept in close contact with Minkowski, who had won a prize from the French Academy while still in his teens. Hilbert eventually managed to bring Minkowski to Göttingen.

In 1893 the German Mathematical Association appointed Hilbert and Minkowski to summarize the current state of the theory of numbers. Number theory was the oldest branch of mathematics, as it dealt with the properties of whole numbers. Much new work had been done by **Karl Friedrich Gauss**, and throughout the second half of the nineteenth century further progress had been made. The accessibility of the statement of problems in number theory made them attractive as an object of investigation, although the work of Leopold Kronecker, perhaps the biggest influence on Hilbert, was already well beyond what the nonmathematician could easily follow. Minkowski withdrew from the project, and in 1897 Hilbert submitted a report called *Der Zahlbericht* ("Number Report"). His work advanced the subject to a more technical level, one which has been maintained throughout the 20th century. Many of the results included still bear Hilbert's name, a tribute to the longevity of his influence.

## Poses Problems for the New Century

The next direction in which Hilbert pursued his research was somewhat unexpected. After all his work in algebra, he began to look at the foundations of geometry. **Euclid** had already laid the foundations more than 2,000 years before, but detailed examination of some of Euclid's proofs revealed gaps in his presentation; he had made assumptions that were neither explicit nor justified by what had been proven earlier. In addition to problems posed by these gaps, another source for a new approach to geometry was the discovery during the 19th century of non-euclidean geometries. These shared some axioms or assumptions with Euclid's system, but differed in other respects. For example, in Euclidean geometry the sum of the angles of a triangle was equal to 180 degrees, while in non-euclidean geometries the sum could be greater or less. One of the reasons that it had taken so long to develop non-euclidean geometries was a general disagreement over what was true about geometrical objects. Hilbert felt that the only way to make progress was to be entirely explicit about each proof and not to trust to unspoken assumptions.

The safest way to avoid these assumptions was to regard the terms of the subject as defined only by the axioms in which they were used. Mathematicians might think they know what a "line" means and may be tempted to use this mental image in trying to prove a fact about lines. But that mental image could easily add something to the notion of line beyond what is given in the axioms. Taken in this way, the axioms can be considered as a kind of definition for the terms used in them. As Hilbert noted, the question of which theorems followed from which axioms had to be unchanged if all the technical terms of the subject (like point, line, or plane) were replaced by words from some other area. It was the form of the axiom that mattered, not what the objects were. This brought Hilbert into conflict with **Gottlob Frege**, one of the founders of mathematical logic. The controversy between Hilbert and Frege involved issues about the philosophy of mathematics that remained central to the field for much of the 20th century. In general, it can be claimed that Hilbert's perspective has been more helpful in enabling mathematicians to pursue the foundations of geometry.

One of the highlights of Hilbert's career came in 1900, when he was invited to address the International Congress of Mathematicians in Paris. His talk consisted of the statement of twenty-three problems, which he challenged his peers to solve in the 20th century. Although not all of the problems have proved

to be of the same importance, by posing them Hilbert created an agenda that has been followed by many distinguished mathematicians. The first problem dealt with the question of how many real numbers there were, compared to the number of whole numbers. It was not resolved until 1963, when it was shown that the answer depends on which axioms are selected as the basis for the theory of sets. Hilbert's seventh problem dealt with the irrationality of certain real numbers, an area to which Hilbert himself had contributed. A number is rational if it is the ratio of two whole numbers, irrational if can not be so expressed, and transcendental if it is not the solution of a polynomial equation with whole number coefficients. Although Hilbert was not the first to prove that the numbers e and $\pi$ (the base of natural logarithms and the ratio of the circumference of a circle to its diameter, respectively) were transcendental, he had simplified the proofs considerably. **A. O. Gelfond** solved his seventh problem, by establishing that a whole class of numbers was transcendental. Hilbert's tenth problem, on the solubility of certain equations, required much progress in mathematical logic before it could be solved. Entire books and conferences have been devoted to the state of the solutions to Hilbert's problems.

In addition to the study of the foundations of geometry, Hilbert turned to mathematical analysis and left a decisive imprint on this field as well. The previous generation of mathematicians had found defects in one of the standard principles from earlier in the century. Hilbert showed that the principle could be preserved, and he proceeded from there to make great progress in the study of integral equations. Hilbert has been credited with the creation of functional analysis, and although there was more foundational work to be done after him, his brief involvement in the area had once again altered it irrevocably.

In the fall of 1910, the second Bolyai Prize was awarded to Hilbert as a confirmation of his mathematical stature. The best-known images of him come from the period surrounding World War I. His distinctive appearance, from Panama hat to bearded chin, and his sharp voice set the tone for mathematics in Germany and the world. During the war he refused to sign the "Declaration to the Cultural World," which claimed that Germany was innocent of alleged war crimes. He was also willing to put mathematics before nationality, and he included an obituary for a French mathematician in the journal *Mathematische Annalen* (the showpiece of German mathematics) during the war. These acts made him unpopular with German nationalists. In the same way, he also took pleasure in fighting the academic establishment over the rights of **Emmy Noether**, who was both a woman and a Jew. Of the 69 students who wrote their theses under Hilbert (an enormous number for any time), there were several women.

After a brief dalliance with theoretical physics (an area Hilbert felt too important to leave to the physicists but to which he made few lasting contributions), Hilbert returned to questions of the philosophy of mathematics that had arisen earlier during his work on geometry. He was eager to pursue a program that could result in the establishment of secure foundations for mathematics. While he was willing to grant some importance to finite mathematics, he felt that the infinite required special treatment. In his account, called formalism, he set out to prove the consistency of mathematics. This enterprise put him in conflict with the other philosophies of mathematics most frequently advocated. He gave expression to his views most notably in an address "On the Infinite" in 1925; he was challenged by many, including **L. E. J. Brouwer** and **Hermann Weyl**, but it was the incompleteness theorem of the young Austrian mathematician **Kurt Gödel** which threatened the entire program Hilbert was pursuing. Certain narrow interpretations of formalism were put to rest by Gödel's work, and some of Hilbert's views were included among these. On the other hand, Hilbert's account of the foundations of mathematics changed during his career, and a good part of the work he did in the area has survived within post-Gödel logic under the heading of proof theory.

Hilbert married Kathe Jerosch in 1892. While he was willing to be casual with regard to his appearance, his wife helped prevent at least some of his sartorial excesses. She also proved a source of strength to Hilbert in his disappointments, one of which was their only son Franz, who never lived up to his father's expectations and probably suffered from a mental disorder. The last years of Hilbert's life were also darkened by the advent of National Socialism and its dire effects on Germany's intellectual community. Hilbert was proud of receiving honorary citizenship from Königsberg in 1930, the year of his retirement from Göttingen, but nothing could assuage his grief over the sequence of losses that the university suffered from the departure of many of its leading minds. Hilbert turned 71 in 1933, the year the Nazis came to power, and it was too late for him to look for a new home. Many of his students had found academic homes abroad, and nothing could rebuild the university in the face of racial laws and hatred of the intellect. It is a measure of the state of German mathematics and the political atmosphere in Göttingen that at Hilbert's death on February 14, 1943, no more than a dozen people attended his funeral.

## SELECTED WRITINGS BY HILBERT

### Books

*The Foundations of Geometry*, 1902

*Gesammelte Abhandlungen* 3 volumes, 1932–1935

## FURTHER READING

### Books

Freudenthal, Hans. "David Hilbert," in *Dictionary of Scientific Biography*. Volume VI. Edited by Charles Coulston Gillispie. New York: Charles Scribner's Sons, 1970–1978, pp. 388–395.

*Mathematical Developments Arising from the Hilbert Problems*. Washington, D.C.: American Mathematical Society, 1979.

Reid, Constance. *Hilbert*. New York: Springer-Verlag, 1970.

Tiles, Mary. *Mathematics and the Image of Reason*. London: Routledge, 1991.

*—Sketch by Thomas Drucker*

## George William Hill
### 1838–1914
**American mathematical astronomer**

George Hill practiced mathematics only as it aided his astronomical research, but his methods and findings enriched both mathematical theory and the practice of celestial mechanics. He was the first to use infinite determinants in his calculations of "periodic orbits," a phrase and concept he initiated, and created his own detailed tables to describe the motions of Jupiter and Saturn. Hill also contributed to our knowledge of the three and four body problems with his painstaking studies of the effect of the moon's motion on other planets and vice versa. He served as vice-president of the American Mathematical Society in 1893–1894, during the organization's fifth anniversary, and was elected president the next year. Hill is considered the most famous mathematical graduate of Rutgers University in New Jersey, and his achievements are commemorated by the Hill Center on the Busch campus of the university.

George William Hill was born in New York City on March 3, 1838. His father, John William Hill, was an English-born artist who designed and produced engravings. His mother, the former Catherine Smith, was of French Huguenot extraction. Hill had at least one brother, with whom he kept close ties until his death. When Hill was eight years old, the family relocated to a farm in West Nyack, New York. He first attended school in the town of West Nyack, and for college he chose Rutgers in New Jersey. There, Hill came under the influence of Dr. Theodore

*George William Hill*

Strong. Dr. Strong was running the math department singlehandedly, as he had been since taking over the university's only endowed chair for the discipline in 1827. He had no great preference for natural philosophy, as physics was then called, but he emphasized the great classical works on celestial mechanics in his classes. Thus, Hill was exposed to the works of **Leonhard Euler**, Sylvestre Lacroix, **Pierre Simon Laplace**, **Joseph–Louis LaGrange** and **Adrien–Marie Legendre**. He graduated with an A.B. in 1859 and received his master's degree three years later. Hill published his first paper in 1859, while still an undergraduate at Rutgers. In 1861, his third paper, "On the Confirmation of the Earth," appeared in *Mathematical Monthly*, earning Hill a prize and the attention of the journal's editor J. D. Runkle.

### Distinguished Career

Hill became an assistant in the offices of the *American Ephemeris and Nautical Almanac*, headquartered in Cambridge, Massachusetts, in 1861. After a year or two, his employers allowed him to work from home in West Nyack. In 1874, Hill became a member of the National Academy of Sciences. Important studies of the moon soon followed. His 1877 paper, "On the Part of the Motion of the Lunar Perigee Which Is a Function of the Mean Motions of the Sun and Moon," depended upon the infinite determinant for its intricate yet superior calculations. Hill's second important paper in astronomy, "Re-

searches in the Lunar Theory," appeared in the premier issue of *American Journal of Mathematics* in 1878. This paper introduced both the periodic orbit and the surface of zero velocity, ideas that would soon influence **Jules Henri Poincaré**. Hill recorded his 1880 canoe trip from the Great Lakes to Hudson Bay in maps and photographs that were published soon after. As with all his excursions, Hill traveled alone.

In 1882, Hill undertook his most ambitious work at the urging of Simon Newcomb, director of the *American Ephemeris*. This study of Jupiter and Saturn would eventually take Hill a decade to complete. For this project he relocated with the office of the *Nautical Almanac* to Washington, D.C. to conduct observations. Hill had previously calculated the orbits of Jupiter and Saturn to 12 decimal places for the *Nautical Almanac*, but this project culminated in a new theory of why and how the two planets travel as they do. After completing the landmark study in 1892, Hill left the *Nautical Almanac*. Hill's work on the orbits of Jupiter and Saturn is widely considered to be one of the most significant 19th century contributions to mathematical astronomy.

Hill was appointed a lecturer in celestial mechanics at Columbia University in 1892. After a hiatus to assume the presidency of the American Mathematical Association from 1894–1896, he returned to Columbia to teach until his retirement in 1901. Hill refused to take a salary for his academic activities at Columbia. As he expressed it, this was less the result of largesse than convenience—he considered the fee too much trouble to collect.

Hill's work was greatly appreciated in England. The London Royal Astronomical Society made him a foreign associate and also gave him their gold medal in 1887. The University of Cambridge offered him his second honorary doctorate in 1892. (He received his first honorary doctorate from Rutgers in 1872.) By 1902, the Royal Society made him a fellow, and in 1909 they awarded him their Copley Medal, considered the highest British honor in the field of science. More than one country extended recognition, however. Hill also won the Paris Academy's Damoiscan Prize in 1898.

Hill cultivated few relationships, preferring to work and think alone, an outlook that fostered his reputation as a deeply innovative theorist. He never married and, in later life, he lived comfortably on his farm, indulging his passion for books and botany. His married brother lived nearby and Hill often visited him for meals.

Hill died on April 16, 1914. When his collected papers were reprinted in four volumes, they were prefaced by Hill's most famous adherent, Jules Henri Poincaré. Today the position of George W. Hill Professor of Mathematics and Physics at Rutgers is underwritten by the New Jersey Board of Governors.

Although his 1886 linear differential is now known as the Hill Equation, the fact that Hill was inducted into the Rutgers Alumni Hall of Fame in 1996 as an astronomer indicates that his application of mathematics to solve astronomical problems is what won him his place in history.

## SELECTED WRITINGS BY HILL

### Books

*The Collected Mathematical Works of George William Hill*, 4 volumes, 1905–07

### Periodicals

"Researches in the Lunar Theory." *American Journal of Mathematics* 1, no. 1 (1878).

## FURTHER READING

### Books

Brown, E.W. "George William Hill." *Dictionary of American Biography*. New York: Charles Scribner's Sons, 1928, pp. 32–3.

Elliott, Clark A., comp. *Biographical Index to American Science: The Seventeenth Century to 1920*. Westport, CT: Greenwood Press, 1990, p. 106.

*National Cyclopaedia of American Biography*. Volume 13. Reprint. Ann Arbor, MI: University Microfilms, 1967–71, p. 442.

Tarwater, Dalton, editor. *The Bicentennial Tribute to American Mathematics 1776–1976*. Washington, DC: Mathematical Association of America, 1977.

### Other

"George William Hill." *MacTutor History of Mathematics Archive*.
http://www–groups.dcs.st–and.ac.uk/~history/Mathematicians/index.html (July 1997).

Weibel, Charles. "A History of Mathematics at Rutgers." http://www.math.rutgers.edu/~weibel/history.html#hill (July 1997).

*—Sketch by Jennifer Kramer*

# Hipparchus of Rhodes
## c. 180–c. 125 B.C.
### Greek trigonometer and astronomer

Hipparchus of Rhodes was a renowned Greek astrologer whose mathematical computations to chart

*Hipparchus of Rhodes*

the sun, moon, and stars led to his being named the founder of trigonometry. Hipparchus is also considered the founder of Greek astronomy for his systematic approach to the discipline, which based it on a solid scientific foundation. In his book, the *Almagest*, the famous Greek astronomer and mathematician **Claudius Ptolemy** holds Hipparchus in the highest regard, referring to him as the "man who loved work and truth."

Details of Hipparchus' life remain lost. What little is known about him can be found in the writings of Strabo and Ptolemy. According to several ancient sources, Hipparchus was born in the city of Nicaea in a part of northwestern Asia Minor that is now Turkey. The era in which he lived has been determined primarily from the dates of his astronomical observations. In the *Almagest*, Ptolemy refers to Hipparchus making observations from Rhodes in Greece, indicating that he spent the latter part of his life there. According to some ancient references, Hipparchus also made astronomical observations from Alexandria, Egypt.

While Hipparchus the man remains a mystery, his reputation as a pioneer in astronomy and mathematics led to his image appearing on second– and third–century Nicaean coins showing a man contemplating a globe. The few surviving anecdotes concerning Hipparchus may be truth or legend. According to one anecdote, Hipparchus once caused a stir among fellow theater goers when he came in wearing a cloak to protect himself from a storm he had predicted. In another questionable anecdote handed down by Pliny, Hipparchus is credited with predicting eclipses of the sun and moon for 600 years.

## Transforms Greek Astronomy through Trigonometry

Like many ancient Greeks, Hipparchus' interest in mathematics stemmed from his devotion to astronomical studies. Pliny reports that Hipparchus became the first Western astronomer to record the observation of a nova, or new star, probably around 133 or 134 B.C. Hipparchus discovered the star in the constellation Scorpio during one of his nightly observations of the skies. The sighting marked a phenomenal event in his time since the Greeks believed that all stars were fixed, with their position and number firmly established. The discovery led Hipparchus to create a catalogue of stars so future generations of astronomers could accurately determine the appearance of new stars and record changes in their positions and brightness.

Before Hipparchus, Greek astronomy was based primarily on observation and a few geometrical models. In Hipparchus' time, it was believed that the stars were located on a single sphere and revolved around the Earth. To create his catalogue of stars, Hipparchus developed the idea that their positions and distance could be determined by taking three stars at a time and forming a spherical triangle. Since **Euclid**'s mathematical theorems focused on plane triangles, Hipparchus developed his own special theorems for working with spherical triangles. These efforts led him to create a table of chords, or straight lines in a circle. Using these chords, Hipparchus developed a method for finding general solutions for trigonometrical problems and, in effect, became the founder of trigonometry.

Hipparchus is also credited with introducing a new unit of measurement, the degree, which is still used to measure arcs and angles, and proposed the idea of dividing a circle into 360 degrees. He may also have been the first to use letters to indicate the points of a triangle. Substituting mathematical models for mechanical ones, Hipparchus extended the work of **Apollonius of Perga** on epicycles and eccentrics.

## Produces Ground Breaking Astronomical Measurements

In the realm of astronomy, Hipparchus' most noted achievement is the discovery of the precession of the equinoxes. By comparing early astronomers' measurement of the positions of stars with his own measurements, Hipparchus found that the stars seemed to have shifted systematically in the same direction. He went on to establish that this phenomenon was due to a shift in the position of the equinoxes.

Hipparchus' other accomplishments include accurately establishing the distance to the moon and calculating the length of the year to within 6.5 minutes of the modern year. He also improved upon or invented several observational instruments, including one for measuring the diameters of the sun and the moon. Hipparchus may also have invented the plane astrolabe and stereographic projection. In the area of geography, which was his other major field of study, Hipparchus advocated the use latitudes and longitudes.

Although Hipparchus was reported to have written several works on astronomy and geography, his only surviving work is the *Commentary on the Phaenomena of Eudoxus and Aratus*. This three–book treatise criticizes earlier astronomers' estimates of the position of stars and constellations. Other reported works by Hipparchus include *Geography, On the Length of the Year, On the Displacement of the Solstitial and Equinoctial Points*, and *On Bodies Carried down by Their Weight*. He also wrote *Against the Geography of Eratosthenes*, attacking the ancient Greek geographer's works, including **Eratosthenes'** treatise *On the Measurement of the Earth*.

Fortunately, Hipparchus' ideas were saved and reported by those who followed. Ptolemy credits Hipparchus with providing much of the foundation for the *Almagest*, which was long considered the "Bible of Astronomy." Unquestionably, Hipparchus was an open–minded seeker of truth who helped establish astronomy and mathematics as evolving disciplines. Ironically, Ptolemy's *Almagest* was such a comprehensive achievement that it nearly obscured Hipparchus' achievements. If Ptolemy had not acknowledged his debt to this pioneering thinker, Hipparchus may have been relegated to footnote status in the history of Greek science and thought.

## FURTHER READING

Magill, Frank N., editor. *The Great Scientists*. Volume 6. Danbury, CT: Grolier Educational Corporation, 1989, pp. 79–84.

Neugebauer, O. *A History of Ancient Mathematical Astronomy*. New York: Springer–Verlag, 1975, pp. 274–343.

Sarton, George. *A History of Science: Hellenistic Science and Culture in the Last Three Centuries B.C.* Cambridge, MA: Harvard University Press, 1959, pp. 284–288.

Terry, Leon. *The Mathmen*. New York: McGraw–Hill Book Company, 1964, pp. 197–212.

Toomer, G.J. "Hipparchus of Rhodes," in *Dictionary of Scientific Biography*. Volume XV. Edited by Charles Coulston Gillispie. New York: Charles Scribner's Sons, 1975, pp. 207–224.

*—Sketch by David A. Petechuk*

# Hippocrates of Chios
## c.470–c.410 B.C.
### Greek geometer

Hippocrates of Chios was a Greek merchant turned mathematician who wrote the first textbook on geometry. He is also noted for his efforts in elucidating the properties of circles and the quadrature of the lune. Despite turning to mathematics later in life, Hippocrates, who was also interested in astronomy, has been called the greatest mathematician of the fifth century B.C.

Hippocrates was born on the island of Chios but little else is known about his life. He is not to be confused with the more famous Hippocrates of Cos, who was born in the same century and is known as the father of medicine. According to information passed down by Aristotle, Proclus, Simplicius, and others, Hippocrates' first calling was as a merchant. However, unlike the famous Greek mathematician **Thales**, who made a fortune in commerce, Hippocrates' endeavors in trade led to his financial ruin.

It is speculated that Hippocrates was a thriving merchant until an unfortunate turn of events. According to one legend, he lost most of his wares when attacked by Athenian pirates near Byzantium. Another version points to dishonest customs men who threatened him with imprisonment and then bilked him of almost everything he owned. Hippocrates is said to have traveled to Athens to recoup his losses in a court of law. Required to stay in Athens for an extended period of time, Hippocrates began to attend lectures on mathematics and philosophy. He eventually became proficient enough in mathematics to open his own school. Although Aristotle characterized Hippocrates as a "competent geometer," he also—perhaps unfairly—said Hippocrates lost his fortune because he was "stupid and lacking in sense."

### Writes First Account of Elementary Mathematics

Hippocrates is believed to have been greatly influenced by the Pythagorean school of mathematics, named after the famed Greek mathematician and philosopher **Pythagoras** . Whether he came under this influence in his home of Chios, which is close to Samos where Pythagoras was born, or in Athens is debatable. Hippocrates' concept of proportion and his astronomical theories are both related to the Pythagorean school of thought.

Fortunately, Hippocrates' misfortunes in commerce had a silver lining. Although the Pythagoreans believed it was taboo to earn money from their knowledge, Hippocrates was reportedly allowed to establish a school in Athens because of his financial troubles. Hippocrates went on to write the first mathematical textbook, called the *Elements of Geometry*. This work precedes the better known *Elements* written by **Euclid** more than a century later.

Although Hippocrates' book is lost, it had a profound influence on the mathematicians who followed him. Through his pioneering book, Hippocrates was the first to develop geometrical theorems from axioms and postulates in a scientifically precise and logical manner. His book may also have contained the first written accounts of Pythagorean mathematics since the Pythagoreans themselves did not believe in written texts. Although Euclid's book was far superior in its approach to geometry and went on to become the most famous textbook of all time, Euclid most certainly based some of his work on that of his predecessor, including much of what appears in Books I and II of his *Elements.*

### Addresses Ancient Problems in Mathematics

One of the most famous problems faced by ancient Greek mathematicians was doubling the cube, also called the Delian Problem. According to one legend, the Delian Problem arose when a Greek concluded that a typhoid plague was a scourge sent by the god Apollo, who was displeased with his altar. Apollo ordered a second altar built in his honor that would be double in size but have the same cubical form as the first altar. Mistakenly, the Athenians thought the problem was solved simply by doubling each of the old altar's edges. As the legend goes, the plague continued and the problem of doubling the cube became a preeminent mathematical problem in ancient Greece.

In his attempts to solve the problem of doubling the cube, Hippocrates used the method of reduction. Although Plato developed a method of reduction for philosophical problems, Proclus credits Hippocrates as the first to use such an approach in geometry. Basically, this method operated by altering a difficult problem into a simpler form, solving this simpler form of the problem, and then attempting to apply the solution to the more difficult problem. In the case of the Delian Problem, Hippocrates proposed that a cube could be doubled by finding the two mean proportionals (geometric means) between two given lines or between a number and its double. While Hippocrates never completely solved the problem of doubling the cube, others followed up on his directive and went on to develop several solutions to this ancient geometric puzzle.

Besides his book, Hippocrates' most noteworthy contribution to ancient mathematics was his quadrature of the lune, a figure bounded by two crescent–shaped arcs of unequal radii. Hippocrates' interest in the quadrature of the lune probably stemmed from his attempt to solve another popular problem of ancient Greece, namely the squaring of the circle. Hippocrates' based his work on the theorem that the areas of two circles are the same as the ratio of the squares of their diameter, or radii. According to some accounts, Hippocrates falsely claimed that his work on the quadratures of lunes led him to discover how to square a circle.

Among Hippocrates' other contributions to mathematics were geometrical solutions to quadratic equations and an early method of integration. Through his theorems on circles, Hippocrates may have also introduced the indirect method of proof to mathematics, also known as the *reductio ad absurdum*. In essence, this approach first assumes the opposite of what is wanted to be proved is true. By proving the opposite to be false, the alternative is then considered true.

### Develops Theories for Comets and the Galaxy

Like many of his contemporaries, Hippocrates was also enamored by the heavens. Chios had long been a center of astronomical studies, and Hippocrates is believed to have formed his own theories concerning comets and the galaxy. In keeping with the Pythagorean view of the heavens, Hippocrates believed that there was only one comet, which was a planet that appeared at long intervals. He added the belief that the comet's tail was a type of mirage caused by the comet taking up moisture when it neared the sun. Some ancient commentators also say Hippocrates created a similar theory to explain the appearance of the galaxy.

Since Hippocrates turned to math and astronomy rather later in his life, it is noteworthy that he apparently attained great renown in a relatively short period of time. Much of his work is known only through commentaries by later mathematicians and historians, like Simplicius, who based his work on the *History of Geometry* by Eudemus. As a result, certain claims pertaining to Hippocrates are difficult to substantiate, including that he was the first to use letters in geometric figures and that he established the technical meaning of the word "power," which is now used in algebra. However, his reputation as an excellent geometer who influenced the course of mathematical thought is well–founded. Nothing is known about his death.

*Grace Hopper*

## FURTHER READING

Boyer, Carl B. *A History of Mathematics*. Princeton: NJ: Princeton University Press, 1985, pp. 71–75.

Bulmer–Thomas, Ivor. "Hippocrates in Chios," in *Dictionary of Scientific Biography*. Volume VI. Edited by Charles Coulston Gillispie. New York: Charles Scribners' Sons, 1975, pp. 410–418.

Dunham, William. *Journey through Genius: The Great Theorems of Mathematics*. New York: John Wiley & Sons, 1990, pp. 10–11, 26.

Terry, Leon. *The Mathmen*. New York: McGraw–Hill Book Company, 1964, pp. 123–136.

—*Sketch by David A. Petechuk*

# Grace Hopper
## 1906-1992
### American computer scientist

Grace Hopper, who rose through Navy ranks to become a rear admiral at age 82, is best known for her contribution to the design and development of the COBOL programming language for business applications. Her professional life spanned the growth of modern computer science, from her work as a young Navy lieutenant programming an early calculating machine to her creation of sophisticated software for microcomputers. She was an influential force and a legendary figure in the development of programming languages. In 1991, President George Bush presented her with the National Medal of Technology "for her pioneering accomplishments" in the field of data processing.

Admiral Hopper was born Grace Brewster Murray on December 9, 1906, in New York City. She was the first child of Marry Campbell Van Horne Murray and Walter Fletcher Murray. Encouraged by her parents to develop her natural mechanical abilities, she disassembled and examined gadgets around the home, and she excelled at mathematics in school. Her grandfather had been a senior civil engineer for New York City who inspired her strong interest in geometry and mathematics.

At Vassar College, Hopper indulged her mathematical interests, and also took courses in physics and engineering. She graduated in 1928, then attended Yale, where she received a master's degree in 1930 and a doctorate in 1934. These were rare achievements, especially for a woman. As Robert Slater points out in *Portraits in Silicon,* U.S. doctorates in mathematics numbered only 1,279 between 1862 and 1934. Despite bleak prospects for female mathematicians in teaching beyond the high school level, Vassar College hired her first as an instructor, then as a professor of mathematics. Hopper taught at Vassar until the beginning of World War II. She lived with her husband, Vincent Foster Hopper, whom she had married in 1930. They were divorced in 1945 and had no children.

### Begins Computer Work in Navy

In 1943, Hopper joined the U.S. Naval Reserve, attending midshipman's school and obtaining a commission as a lieutenant in 1944. She was immediately assigned to the Bureau of Ships Computation Project at Harvard. The project, directed by **Howard Aiken** , was her first introduction to Aiken's task, which was to devise a machine that would assist the Navy in making rapid, difficult computations for such projects as laying a mine field. In other words, Aiken was in the process of building and programming America's first programmable digital computer—the Mark I.

For Hopper, the experience was both disconcerting and instructive. Without any background in computing, she was handed a code book and asked to begin computations. With the help of two ensigns assigned to the project and a sudden plunge into the works of computer pioneer **Charles Babbage**, Hopper began a crash course on the current state of computation by way of what Aiken called "a computing engine."

The Mark I was the first digital computer to be programmed sequentially. Thus, Hopper experienced first-hand the complexities and frustration that have always been the hallmark of the programming field. The exacting code of machine language could be easily misread or incorrectly written. To reduce the number of programming errors, Hopper and her colleagues collected programs that were free of error and generated a catalogue of subroutines that could be used to develop new programs. By this time, the Mark II had been built. Aiken's team used the two computers side by side, effectively achieving an early instance of multiprocessing.

By the end of the war, Hopper had become enamored of Navy life, but her age—a mere 40 years—precluded a transfer from the WAVES into the regular Navy. She remained in the Navy Reserves and stayed on at the Harvard Computational Laboratory as a research fellow, where she continued her work on the Mark computer series. The problem of computer errors continued to plague the Mark team. One day, noticing that the computer had failed, Hopper and her colleagues discovered a moth in a faulty relay. The insect was removed and fixed to the page of a logbook as the "first actual bug found." The words "bug" and "debugging," now familiar terms in computer vocabulary, are attributed to Hopper. In 1949, she left Harvard to take up the position of senior mathematician in a start-up company, the Eckert-Mauchly Computer Corporation. Begun in 1946 by J. Presper Eckert and John Mauchly, the company had by 1949 developed the Binary Automatic Computer, or BINAC, and was in the process of introducing the first Universal Automatic Computer, or UNIVAC. The Eckert-Mauchly UNIVAC, which recorded information on high-speed magnetic tape rather than on punched cards, was an immediate success. The company was later bought by Sperry Corporation. Hopper stayed with the organization and in 1952 became the systems engineer and director of automatic programming for the UNIVAC Division of Sperry, a post she held until 1964.

Hopper's association with UNIVAC resulted in several important advances in the field of programming. Still aware of the constant problems caused by programming errors, Hopper developed an innovative program that would translate the programmer's language into machine language. This first compiler, called "A-O," allowed the programmer to write in a higher-level symbolic language, without having to worry about the tedious binary language of endless numbers that were needed to communicate with the machine itself.

One of the challenges Hopper had to meet in her work on the compiler was that of how to achieve "forward jumps" in a program that had yet to be written. In *Grace Hopper, Navy Admiral and Computer Pioneer,* Charlene Billings explains that Hopper used a strategy from her schooldays—the forward pass in basketball. Forbidden under the rules for women's basketball to dribble more than once, one teammate would routinely pass the basketball down the court to another, then run down the court herself and be in a position to receive the ball and make the basket. Hopper defined what she called a "neutral corner" as a little segment at the end of the computer memory which allowed her a safe space in which to "jump forward" from a given routine, and flag the operation with a message. As each routine was run, it scouted for messages and jumped back and forth, essentially running in a single pass.

During the early 1950s, Hopper began to write articles and deliver papers on her programming innovations. Her first publication, "A Manual of Operation for the Automatic Sequence Controlled Calculator," detailed her initial work on Mark I. "The Education of a Computer," offered in 1952 at a conference of the Association of Computing Machinery, outlined many ideas on software. An article appearing in a 1953 issue of *Computers and Automation,* "Compiling Routines," laid out principles of compiling. In addition to numerous articles and papers, Hopper published a book on computing entitled *Understanding Computers,* with Steven Mandrell.

## The Development of COBOL

Having demonstrated that computers are programmable and capable not only of doing arithmetic, but manipulating symbols as well, Hopper worked steadily to improve the design and effectiveness of programming languages. In 1957, she and her staff at UNIVAC created Flow-matic, the first program using English language words. Flow-matic was later incorporated into COBOL, and, according to Jean E. Sammet, constituted Hopper's most direct and vital contribution to COBOL.

The story of COBOL's development illustrated Hopper's wide-reaching influence in the field of programming. IBM had developed FORTRAN, the densely mathematical programming language best suited to scientists. But no comparable language existed for business, despite the clear advantages that computers offered in the area of information processing.

By 1959, it was obvious that a standard programming language was necessary for the business community. Flow-matic was an obvious prototype for a business programming language. At that time, however, IBM and Honeywell were developing their own competing programs. Without cooperative effort, the possibility of a standard language to be used throughout the business world was slim. Hopper, who campaigned for standardization of computers and programming throughout her life, arguing that the lack of standardization created vast inefficiency and waste, was disturbed by this prospect.

The problem was how to achieve a common business language without running afoul of anti-trust laws. In April 1959, a small group of academics and representatives of the computer industry, Hopper among them, met to discuss a standard programming language specifically tailored for the business community. They proposed contacting the Defense Department, which contracted heavily with the business industry, to coordinate a plan, and in May a larger group met with Charles Phillips. The result was the formation of several committees charged with overseeing the design and development of the language that would eventually be known as COBOL—an acronym for "Common Business Oriented Language." Hopper served as a technical advisor to the Executive Committee.

The unique and far-ranging aspects of COBOL included its readability and its portability. Whereas IBM's FORTRAN used a highly condensed, mathematical code, COBOL used common English language words. COBOL was written for use on different computers and intended to be independent of any one computer company. Hopper championed the use of COBOL in her own work at Sperry, bringing to fruition a COBOL compiler concurrently with RCA in what was dubbed the "Computer Translating Race." Both companies successfully demonstrated their compilers in late 1960.

Hopper was elected a fellow of the Institute of Electrical and Electronics Engineers (IEEE) in 1962 and of the American Association for the Advancement of Science (AAAS) in 1963. She was awarded the Society of Women Engineers Achievement Award in 1964. She continued her work with Sperry, and in 1964 was appointed staff scientist of systems programming, in the UNIVAC Division.

## Returns to Navy Life

While at Sperry, Hopper remained active in the Navy Reserves, retiring with great reluctance in 1966. But only seven months later, she was asked to direct the standardization of high level languages in the Navy. She returned to active duty in 1967 and was exempted from mandatory retirement at age of 61. She served in the Navy until age 71.

Although she continued to work at Sperry Corporation until 1971, her activities with the Navy brought her increasing recognition as a spokesperson for the usefulness of computers. In 1969, she was named

"Man of the Year" by the Data Processing Management Association. In the next two decades, she would garner numerous awards and honorary degrees, including election as a fellow of the Association of Computer Programmers and Analysts (1972), election to membership in the National Academy of Engineering (1973), election as a distinguished fellow of the British Computer Society (1973), the Navy Meritorious Service Medal (1980), induction into the Engineering and Science Hall of Fame (1984) and the Navy Distinguished Service Medal (1986). She lectured widely and took on vested interests in the computer industry, pushing for greater standardization and compatibility in programming and hardware.

Hopper's years with the Navy brought steady promotions. She became captain on the retired list of the Naval Reserve in 1973 and commodore in 1983. In 1985 she earned the rank of rear admiral before retiring in 1986. But her professional life did not end there. She became a senior consultant for the Digital Equipment Corporation immediately after leaving the Navy and worked there until her death, on January 1, 1992. In its obituary, the *New York Times* noted that "[Like] another Navy figure, Admiral Rickover, Admiral Hopper was known for her combative personality and her unorthodox approach." Unlike many of her colleagues in the early days of computers, Hopper believed in making computers and programming languages increasingly available and accessible to nonspecialists.

## SELECTED WRITINGS BY HOPPER

### Books

(With Steven Mandrell) *Understanding Computers,* 1984

### Periodicals

"A Manual of Operation for the Automatic Sequence Controlled Calculator." *Annals of the Harvard Computation Laboratory* 1 (1946).
(With John W. Mauchly) "Influence of Programming Techniques on the Design of Computers." *Proc. IRE* 41 (October 1953).
"The Education of a Computer." *Annals of the History of Computing* 9 (1988).

## FURTHER READING

### Books

Billings, Charlene W. *Grace Hopper, Navy Admiral and Computer Pioneer.* . Springfield, NJ: Enslow, 1989.

King, Amy C. and Tina Schalch. "Grace Hopper," in *Women in Mathematics.* Edited by Louise S. Grinstein and Paul J. Campbell. Westport, CT: Greenwood Press, 1987, pp. 67–73.
Slater, Robert. *Portraits in Silicon.* Cambridge, MA: MIT Press, 1987.

### Periodicals

*New York Times.* (January 3, 1992).
Sammet, Jean E. "Farewell to Grace Hopper— End of an Era!" *Communications of the AMC.* (April 1992).

### Other

Norman, Rebecca. "Grace Hopper." *Biographies of Women Mathematicians.* http://www.scottlan.edu/lriddle/women/chronol.htm. (July 22, 1997.)

*—Sketch by Katherine Williams*

---

# Christiaan Huygens
## 1629–1695
### Dutch astronomer and mathematical physicist

Christiaan Huygens is best known for his work in astronomy and physics, but he also did important work in mathematics. Although he advanced no new theories, Huygens made improvements to existing methods of calculation and applied them to solving problems in the natural sciences. Had he not been diverted from mathematics by astronomy and physics, he might have be one of history's greatest mathematicians. However, the modern world might be lacking the pendulum clock and the wave theory of light, Huygens' greatest accomplishments.

Born April 14, 1629, at The Hague in the Netherlands, Christiaan Huygens was the son of a family of diplomats. His grandfather, also named Christiaan, had been secretary to William the Silent and Prince Maurice, and his father, Constantjin, had been in service to Prince Frederic Henry and the House of Orange his whole life. Young Christiaan was expected to follow the family path (his brother Constantjin, like his father, served the House of Orange), but he was far too interested in the sciences.

Huygens studied law and mathematics at the University of Leiden and the Collegium Arausicum at Breda. There, he studied classical mathematics and the more modern techniques of **René Descartes** and others. At the end of these formal studies, Huygens

*Christiaan Huygens*

was able to live at home from 1650 to 1666, thanks to support from his father. Those sixteen years at home were the most fruitful of his scientific career.

During this period, Huygens made a number of trips to Paris and London. On his first journey, to Paris in September 1655, he met the men who would become the core of the Académie Royale des Sciences, which was founded in 1666 (Huygens soon became a member early). He traveled to Paris again and stayed from October 1660 to March 1661; he then spent two months in London, attending meetings at Gresham College and meeting some of England's great thinkers. Huygens eventually settled in Paris and lived there from 1666 to 1680, but political unrest and ill health made it essential for him to return to The Hague, where he died in 1695.

### Takes the Cartesian Approach

Huygens first studied mathematics, concentrating on determinations of quadratures and cubatures. He published *Theoremata de quadratura hyperboles, ellipses et ciculi*; in it, he related the quadriture and the center of gravity of these various shapes. His next book, *De circuili magnitudine inventa*, appeared in 1654.

After Huygens heard about **Blaise Pascal**'s work in probability, he began to investigate it. He published *Tractatus de ratiociniis in alea ludo* (a book about the theory of probabilty) in 1657, and applied

the theory to calculating life expectancy. It remained the only book on the subject until the 18th century.

Huygens was known as a Cartesian—that is, he followed the ideas of René Descartes. As such, he did not believe in the action of forces as **Isaac Newton** proposed, believing instead that there could be a mechanical explanation for everything. Huygens' great gift to others was demonstrating how mathematics could be applied to natural problems, and his work, *Holorogium oscillatorium*, shows the strength of this approach.

### His Two Major Achievements

Huygen's first major achievement was the development of the pendulum clock. Years before, **Galileo** had observed the movement of a lamp in a church and realized that the time of each swing was nearly the same, no matter the extent of the swing. Despite his brilliance, Galileo never managed to develop a working pendulum, which would have led to more accurate timekeeping (of increasing importance in this era of scientific revolutions). Huygens realized that the circular arcing swings that Galileo had observed were not identical; rather, they were nearly so. For a pendulum to move in swings of equal time, its movement would have to follow a curve known as a cycloid. Huygens designed a clock with such a pendulum that had weights attached near its fulcrum that would make it move in this nearly circular, cyloidian arc; he then attached the pendulum to a clock's works and used a system of falling weights that would keep the pendulum moving despite friction and air resistance. The mechanisms of the modern grandfather clock have changed little from Huygens' first one, which he presented to the Dutch estates general. He described the clock in his work *Horologium*, published in 1658.

More important than this invention was Huygens' development of the wave theory of light. At the time, it was believed only particles that traveled in a straight line could cast shadows as light did; Newton had advanced the idea of light consisting of such straight–traveling particles. Because of the examples of water waves, it was thought that waves bent around objects. In an effort to correct Newton, Huygens proved that in some cases waves could indeed travel in straight lines. He believed that light was a series of shock waves that disturbed particles existing in the "ether." When struck by the light, these closely packed particles would move and form new wave fronts. Where these wave fronts overlapped, there was light—a concept known as Huygen's principle. Although his theory allowed him to explain refraction and reflection, and predicted that light would more slowly in a denser medium such as water, it did not explain polarization. Newton's theory remained at the forefront; Huygens' idea laid dormant until it was rediscovered and improved on by Thomas Young in the 19th century.

### Work in Optics and Astronomy

Huygens also worked in optics, creating fine telescopes that led to important astronomical discoveries. Thanks to his improved instrument, he was able to correct Galileo's error that stated Saturn was a triple planet. Galileo's telescope did not have sufficient resolution; Saturn's rings appeared to him as two planets snuggled up to the main body. Huygens was able to discern a single ring around the planet.

Huygens also was the first to guess at the distances of the stars. Assuming that Sirius was as bright as the sun, he calculated it to be 2.5 trillion miles from Earth. We now know that Sirius is far brighter than the sun and its true distance is 20 times Huygens' approximation. Huygens also was the first to notice surface markings on Mars and the largest of Saturn's moons, Titan.

## SELECTED WRITINGS BY HUYGENS

Huygens' works were collected by the Society of Sciences of Holland, a 60–year project that began in 1885. Titled *Ouevres complètes de Chrisitaan Huygens publiees par la Socitété Hollandais des Sciences*, the set consists of 22 volumes. The first 10 comprise his correspondence, the remainder, his scholarly writings. Volume 22 includes a biography of the scientist. The papers cited in this essay are all included in this collection.

## FURTHER READING

### Books

Abbott, David. *The Biographical Dictionary of Scientists: Astronomers*. New York: Bendrick Books, 1984.

Asimov, Isaac. *Asimov's Biographical Encylopedia of Science and Technology*. Second Revised Edition. Garden City, NY: Doubleday, 1982.

Bell, A.E. *Christiaan Huygens and the Development of Science in the 17th Century*. London, 1947.

*Biographical Encyclopedia of Scientists*. Second Edition. Volume 1. Bristol and Philadelphia: Institute of Physics Publishing, 1994.

Bos, H.J.M. "Christiaan Huygens," in *Dictionary of Scientific Biography*. Volume VI. Edited by Charles Coulston Gillispie. New York: Charles Scribner's Sons, 1972, pp. 597–613.

*Hypatia of Alexandria*

Turner, Ronald, and Steve Goulden, editors. *Great Engineers and Pioneers in Technology*. Volume 1. New York: St. Martin's Press, 1981.

*—Sketch by Fran Hodgkins*

# Hypatia of Alexandria
## 370(?)–415
### Greek geometer, astronomer and philosopher

Hypatia, the earliest known woman mathematician, wrote commentaries on several classic works of mathematics. The daughter of a mathematician, she was trained in mathematics and philosophy and became head of the Neoplatonic school at Alexandria, where she taught philosophical doctrines dating back to Plato's Academy. Hypatia was a respected teacher and influential citizen of Alexandria, greatly admired for her knowledge as well as for her decorum and dignity. Although Hypatia's original work has not survived, she is known from the letters of her student Synesius of Cyrene. She is also mentioned in the fifth–century *Ecclesiastical History* of Socrates Scholasticus, and in the tenth–century *Lexicon* of Suda (or Suidas). Information about Hypatia is fragmentary and oblique, fact and fiction have mingled, and her life has become the stuff of legend, inconsistencies, and conflicting opinions.

Hypatia was born in Alexandria, Egypt; the year is generally thought to be 370, although some scholars argue for an earlier date, 355. Founded on the Nile River by Alexander the Great in 332 B.C., Alexandria had been the center of scholarly attainment in science, and during Hypatia's time was the third largest city in the Roman Empire. Hypatia's father, **Theon**, was a member of the Museum, a place of residence, study, and teaching similar to a modern university. A mathematician and astronomer, Theon had predicted eclipses of the sun and the moon which were observed in Alexandria, and his scholarship included commentaries on **Euclid** and **Claudius Ptolemy**. Hypatia was taught by Theon, collaborated with him, and did independent work. Whereas Theon also produced poetic work and a treatise on the interpretation of omens, Hypatia's works seem to have been strictly mathematical.

### Admired for Her Knowledge and Deportment

Hypatia was recognized as a gifted scholar and eloquent teacher, and by 390 her circle of influence was well–established. By 400, she was head of the Neoplatonic school, for which she received a salary. Socrates Scholasticus, the Byzantine church historian, wrote that Hypatia was so learned in literature and science that she exceeded all contemporary philosophers. Philostorgius, another historian, noted that she surpassed her father in mathematics, and especially in astronomy. From Synesius' letters to and about her, it is clear Hypatia had extensive knowledge of Greek literature. Her students were aristocratic young men, both pagan and Christian, who rose to occupy influential civil and ecclesiastical positions. They came from elsewhere in Egypt, and from as far away as Cyrene, Syria, and Constantinople to study privately with Hypatia in her home. They were united through intellectual pursuits and considered Hypatia their "divine guide" into the realm of philosophical and cosmic mysteries, which included mathematics. Hypatia combined the principles of free thinking with the ideal of pure living. She was known for her prudence, moderation and self–control, for her ease of manner, and for her beauty. She chose to remain a virgin and to devote her life to pursuit of knowledge and the philosophical ideal. According to an account in Suda, which may be apocryphal, when one of her students fell in love with her, she threw at him a rag that was the equivalent of a sanitary napkin, saying "You are in love with this, not with [the Platonic ideal of] the Beautiful."

By wearing a tribon, the characteristic rough white robe of the philosopher, Hypatia indicated that she did not wish to be treated as a woman. She traveled freely about the city in her chariot, instructed her students in Platonic and Aristotelian philosophy, visited and lectured at public and scientific institutions. She had exerted political influence and may have held a political position. In one of his letters to her, Synesius asks Hypatia to intervene with her powerful friends to restore the property of two young men. Throughout his life Synesius remained devoted to Hypatia, praised her erudition, and asked for advice on his own writings. He must have visited her the several times he visited Alexandria, including in 410, when he was consecrated as Bishop of Ptolemais by Theophilus, the patriarch of Alexandria.

### Works on Commentaries and Scientific Instruments

Although several of Theon's mathematical and astronomical works have survived, Hypatia's have not. It is known that Hypatia wrote a treatise titled *Astronomical Canon*, presumably on the movements of the planets, and a commentary on the algebraic work of Diophantus of Alexandria, which contains the beginnings of number theory. Diophantus, who lived in the third century A.D., is quoted by Theon, and some scholars believe that the survival of most of Diophantus' original thirteen books of the *Arithmetica* is due to the quality of Hypatia's work. The surviving texts, including six in Greek and four translated into Arabic, contain notes, remarks, and interpolations that may come from Hypatia's commentary. Hypatia also wrote *On the Conics of Apollonius*, in which she elaborated on **Apollonius'** third–century B.C. theory of conic sections.

In collaboration with Theon, Hypatia also worked on Ptolemy's *Almagest*, the second–century work which brought together disparate works of early Greeks in 13 volumes and served as the standard reference on astronomy for more than 1,000 years. In the *Almagest*, Ptolemy introduced a method of classifying stars, and used Apollonius' mathematics to construct a masterful (though incorrect) theory of epicycles to explain the movement of the sun, moon and planets in a geocentric system. Hypatia may have corrected not only her father's commentary but also the text of *Almagest* itself, and may also have prepared a new edition of Ptolemy's *Handy Tables*, which appears in the work of Hesychius under the title *The Astronomical Canon*.

Synesius' letters reveal Hypatia's interest in scientific instruments. In one instance he asks her to have a hydroscope (an instrument for measuring the specific gravity of a liquid) made for him. In another, he consults her about the construction of an astrolabe, an instrument used to measure the position of the stars and planets.

### Meets a Violent Death

In Hypatia's time, Christianity became the official religion of the Roman empire, and Greek temples were converted to Christian churches. In 411, Cyril succeeded Theophilus as bishop of Alexandria. One of his actions, following Jewish–Christian riots, was to expel Jews from the city. Orestes, the civil governor, disapproved of Cyril's actions and the growing encroachment of the Christian church on civil authority. Cyril roused negative sentiment toward Orestes, and Orestes was attacked by five hundred Nitrian monks, who lived in monasteries outside the city. The monk Ammonius threw a stone that wounded Orestes. Intervention by the populace saved Orestes, who then ordered Ammonius tortured to the extent that he died. Cyril applauded Ammonius' actions as admirable.

Hypatia fell victim to these political hostilities. She was a close associate of Orestes, and undoubtedly was defamed by Cyril. Admiration for her turned to resentment, and she was perceived as an obstacle to the conciliation of Orestes and Cyril. In March of 415, during Lent, as Hypatia rode in her chariot through the streets of Alexandria, she was attacked upon by a fanatical mob of antipagan Christians. The mob dragged Hypatia into the Caesareum, then a Christian church, where she was stripped naked and murdered. According to ancient accounts, Hypatia's flesh was stripped from her bones, her body mutilated and scattered throughout the streets, then burned piecemeal at a place called Cinaron.

Following Hypatia's murder many of her students migrated to Athens, where they contributed to the Athenian school, which in 420 acquired a considerable reputation in mathematics. The Neoplatonic school at Alexandria continued until the Arab invasion of 642. The books in the library at Alexandria were subsequently used as fuel for the city's baths, where they lasted six months. Hypatia's works were probably among them.

## FURTHER READING

### Books

Bregman, Jay. *Synesius of Cyrene: Philosopher–Bishop.* Berkeley: University of California Press, 1982.

Dzielska, Maria. *Hypatia of Alexandria.* Translated by F. Lyra. Cambridge, MA: Harvard University Press, 1995.

Fitzgerald, Augustine. *The Letters of Synesius of Cyrene.* London: Oxford University Press, 1926.

Lefkowitz, Mary R. *Women in Greek Myth.* Baltimore: The Johns Hopkins University Press, 1986.

Snyder, Jane McIntosh. *The Woman and the Lyre: Women Writers in Classical Greece and Rome.* Carbondale: Southern Illinois University Press, 1989, pp. 113–121.

Socrates (Scholasticus). *Ecclesiastical History.* London: Samuel Bagster and Sons, 1844, pp. 480–483.

Waithe, Mary Ellen. "Finding Bits and Pieces of Hypatia," in *Hypatia's Daughters: Fifteen Hundred Years of Women Philosophers.* Edited by Linda Lopez McAlister. Bloomington: Indiana Univ. Press, 1996, pp. 4–15.

### Periodicals

Rist, J. M. "Hypatia." *Phoenix* (Autumn 1965): 214–225.

### Other

Adair, Ginny. "Hypatia." *Biographies of Women Mathematicians.* June 1997. http://www.scottlan.edu/lriddle/women/chronol.htm (July 22, 1997).

*—Sketch by Jill Carpenter*

*Karl Gustav Jacob Jacobi*

# Karl Gustav Jacob Jacobi
## 1804–1851
### German mathematical physicist

As the impact of the American and French Revolutions was felt across Europe, a social atmosphere arose that encouraged ground breaking work in mathematics. Karl Gustav Jacob Jacobi, who attracted early attention from luminaries such as **Adrien–Marie Legendre** and **Karl Gauss**, appeared alongside and sometimes worked with a handful of innovative contemporaries like **William Hamilton, Augustin–Louis Cauchy, Peter Dirichle**t, and **Niels Abel**. Jacobi's own contributions range across older subjects in math such as number theory and newer fields like analysis. Two mathematical terms he devised now bear Jacobi's name. He electrified the teaching profession with an unprecedented practice of opening up his theoretical notes to his students, thereby inventing the research seminar now common in universities. While the political instabilities of post–revolutionary times sometimes threatened Jacobi's livelihood, his sheer genius was always enough to attract a new protector to sponsor his continued work. Jacobi's mathematics had an immediate effect on the classical mechanics of **Isaac Newton, Pierre Laplace** and **Joseph Lagrange**, and later on the quantum mechanics and relativity theories of the 20th century.

Jacobi, whose first name is sometimes spelled "Carl," was born in Potsdam, Prussia (Germany) into a wealthy and well–educated Jewish family on December 10, 1804. He was one of two boys who would both gain a measure of fame during their lifetimes. Karl's older brother, Moritz, would later be celebrated—and even further on dismissed—as the founder of an experimental concept in electricity known as "galvanoplastics." The other children in the family were Eduard and Therese. All that is generally known of their mother is her own family name, Lehmann, and that her brother tutored Karl until he was 12. His father, Simon Jacobi, was a banker in Berlin.

### An Independent Child

Young Jacobi was a prodigy. After being home schooled in the classics and in mathematics, he entered Potsdam Gymnasium in 1816. He promptly rose to the top of his class within a few months, proving he was ready for university training. He was still only 12 years old. Early entry to higher education was forbidden by the authorities, so the Gymnasium kept him until the legal age of sixteen. The rector of the school described him as "a universal mind," expressing high hopes for his future.

The University of Berlin had to make way for Jacobi, who rebelled against his teacher, Heinrich Bauer, and preferred to read **Leonhard Euler** and Lagrange on his own. Jacobi graduated within a year with top marks for classical languages, history, and mathematics. As a graduate student Jacobi majored in philology, since he had no peer at Berlin in mathematics. Instead, he continued to read on his own and correspond with Gauss at Göttingen. Jacobi qualified for a teaching position at age 19, which he took without pay the next year, after completing his doctorate on partial fractions. Jacobi achieved all this as a Jew, but for perhaps more than one reason he converted to Christianity at age 20.

### "Only Cabbages Have No Nerves"

During his student days Jacobi continued to write to his uncle and former teacher, and corresponded with other mathematicians, a habit he continued throughout his career. To one who complained of the physical costs of intellectual labor, Jacobi replied, "Only cabbages have no nerves." How do they benefit, he asked rhetorically, from such well–being? This riposte shows Jacobi's lack of concern for his own health, which may well have contributed to his early death from a combination of illnesses. Some of his vacations were forced, whenever his schedule threatened to wear him down.

Jacobi came into his own as a lecturer. After six months at Berlin he transferred to the University of Königsberg on the recommendation of Legendre, who was impressed with Jacobi's additions to his own pioneering work in elliptic integrals. Jacobi combined new mathematical ventures and classwork to show his new inventions as they took shape. More importantly, he presented himself as one who knew little and desired to know more, inspiring his students to forge ahead according to his example. You do not have to "meet" all subjects in math, he argued by analogy, before "marrying" one.

Jacobi exclusively studied (at least for a while) the work of Legendre on elliptic integrals. Independently of Abel in Norway, Jacobi fully developed the new area of elliptic functions and introduced the notation used today for these functions. He investigated hyperelliptic integrals, making important discoveries about the generalizations of elliptic functions that were later to become known as Abelian functions. Because of his work in mathmatical physics, the ellipsoids of equilibrium for rotating liquid masses are known as Jacobi ellipsoids. Jacobi's work in functions inspired Cauchy and also **Joseph Liouville**. Other interests included determinants, especially those used in relation to partial differential equations and called the Jacobian determinants. Jacobi published three papers on determinants in 1841. For the first time he gave an algorithmic definition of the determinant that applied to cases when the entries were either numbers or functions. These papers helped to make the idea of a determinant more widely known. Partial differential equations came into play in dynamics, a subject which interested Jacobi. Here, he parlayed the findings of Hamilton into results later applied to quantum mechanics. To classical mechanics he contributed work on the three–body problem and other dynamical problems. In number theory, Jacobi proved an assertion of **Pierre de Fermat**'s regarding the expression of integers as sums of squares.

Simon Jacobi died in 1832, and the family was able to live on his bequest for another eight years. In 1840, however, financial troubles forced them into bankruptcy. These difficulties likely contributed to Jacobi's subsequent collapse from overwork. Jacobi relinquished his chair of Ordinary Professor of Mathematics at Königsberg in 1842. He thereafter subsisted on a pension granted by Frederick William IV from the Prussian government, staying in Italy for a time due to ill health. For personal reasons, Jacobi ran for local office in Berlin as a liberal in 1848, leading to a temporary suspension of his Prussian funding, a great sorrow to Jacobi's wife and seven children. However, he was soon back in favor. He taught intermittently at Berlin meanwhile, but died February 18, 1851. Suffering from diabetes, he developed a fatal case of smallpox, contracted after a bout of influenza. Jacobi was memorialized by his friend and colleague **Lejeune Dirichlet** in 1852 with a special lecture given at the Berlin Academy of Sciences. In it, Dirichlet called Jacobi the greatest Academy member since Lagrange.

## SELECTED WRITINGS BY JACOBI

*Fundamenta nova theoriae functionum ellipticarum,* 1829
*Canon arithmeticus,* 1839
*Mathematical Werke,* 3 Volumes. 1846–71

## FURTHER READING

### Books

Abbott, David, editor. *The Biographical Dictionary of Scientists.* New York: Peter Bedrick Books, 1986, p. 74.

Asimov, Isaac. *Asimov's Biographical Encyclopedia of Science and Technology.* Second Revised Edition. Garden City, NY: Doubleday, 1982, pp. 356–57.

Bell, E.T. "The Great Algorist." *Men of Mathematics.* New York: Simon & Schuster, 1937, pp. 327–339.

Eaves, Howard. *An Introduction to the History of Mathematics.* Fourth Edition. Holt, Rinehart & Winston, 1976, pp. 380–81.

Hawkins, T. "Jacobi and the birth of Lie's theory of groups," in *Amphora.* Edited by S.S. Demidov et al. Boston/Berlin: Basel, 1992, pp. 289–313.

Königsberger, Leo. *C.G.J. Jacobi.* 1904.

Porter, Roy, editor. *The Biographical Dictionary of Scientists.* Second Edition. New York: Oxford University Press, 1994.

Williams, Trevor I., editor. *A Biographical Dictionary of Scientists.* Third Edition. New York: John Wiley & Sons, 1982, p. 277.

### Other

O'Connor, John J. and Edmund F. Robertson. "Karl Gustav Jacob Jacobi." *MacTutor History of Mathematics Archive.*

http://www–groups.dcs.st–and.ac.uk/~history/Mathe-maticians/Jacobi.html (July 1997).

*—Sketch by Jennifer Kramer*

# Sof'ja Aleksandrovna Janovskaja (also transliterated as Sofia Yanovskaya)
## 1896–1966
### Russian logician

Sof'ja Aleksandrovna Janovskaja made her mark in the mathematical community not for what she discovered but what she promoted: the legitimacy of mathematical logic as an independent and worthy discipline. A Bolshevik during Russia's civil war (1918–1921), Janovskaja believed mathematical logic was a science with real–world applications, distinctly different from the philosophical idealism that she considered an exclusively bourgeoisie concept.

Janovskaja was born on January 31, 1896, in Pruzhany, Poland (now part of Belarus). Although scholars know little about Janovskaja's early life, some evidence suggests that the Janovskajas were native Poles, perhaps belonging to the local gentry. When Janovskaja was just a few years old, the family moved to Odessa, where she was educated in the classics and mathematics. In 1915, Janovskaja enrolled in the Higher School for Women in Odessa, where she studied until the 1917 Revolution disrupted life throughout Russia.

Janovskaja began aiding anti–royalist political prisoners in 1917 as a member of the underground Red Cross, and in November of 1918 she joined the Bolshevik faction of the Russian Communist Party—a risky move, as the party remained illegal in Odessa until late the following year. In 1919 she served as a political commissar in the Red Army and edited the *Kommunist*, Odessa's daily political newspaper that was printed out of the city's catacombs. Janovskaja was a worker in the Odessa Regional Party until 1923, when she decided her duty as a party member would be better served by using the sciences to support the tenets of the revolution.

## Pursues Career in Mathematics

In the early years of the Soviet Union, the principles of the revolution gave way for the advancement of women in professional—especially scientif-ic—fields. Talented and dedicated, Janovskaja quickly established herself in the Soviet mathematical community. From 1924 to 1929, she studied at the Institute of Red Professors in Moscow and took seminars at Moscow State University. She began directing a seminar on mathematical methodology at Moscow State University in 1925, and she officially joined the faculty a year later. By 1931, she was a professor, and she earned her doctorate from the Mechanical–Mathematical Faculty of Moscow State University in 1935.

With the breakout of World War II, Janovskaja was evacuated to Perm, where she taught at Perm University from 1941 to 1943. She returned to Moscow in 1943 to take the post as director of Moscow State University's seminar on mathematical logic, and three years later she became the first faculty member to teach mathematical logic in the philosophy department.

## Joins the Debate on Mathematical Logic

For years, Western dialectical philosophers dismissed mathematical logic as being idealist—that is, based on preconceived (or *a priori*), abstract notions of number. Such a logic, these philosophers charged, lacked any material applications, and therefore the pursuit of mathematical logic would not further the proletarian cause. Because the concepts used in mathematical logic were said to be predetermined in an ideal realm removed from the material world, mathematical logic appeared to lack the Marxist–Leninist worldview derived from historical experience.

Janovskaja disagreed with these dialecticians about the nature of mathematical logic. The notion of numbers, Janovskaja contended, came from observing groups of things in the material world, and from those many experiences arriving at the general idea of numbers. Other concepts and rules used in mathematical logic were similarly induced from the collective human experience in the material world. Janovskaja saw parallels between the laws governing logic and the Communist Party's monistic philosophy; her writings frequently include the statement by Lenin that all laws are the result of billions of experiences. Janovskaja not only promoted mathematical logic as a pure science, she became a mathematical historian, studying how mathematical methods had evolved throughout time. Janovskaja believed the value of both pursuits would be in their applicability to real–world problems.

## Contributions as Educator

Janovskaja produced no original works, but her translations and commentaries made the works of **René Descartes**, Georg Hegel, and Karl Marx accessible to Russian students. She wrote several lucid articles for the *Great Soviet Encyclopedia*, explaining formalism, logistics, and mathematical paradoxes in simple language accessible to most readers. Still, scholars consider Janovskaja's greatest contributions to be her two journal articles on the history of mathematical logic in the Soviet Union: "The Foundations of Mathematics and Mathematical Logic " (1948) and "Mathematical Logic and Foundations of Mathematics" (1959).

In the classroom Janovskaja did not prove abstract theorems. She preferred to explore actual problems such as the hangman's paradox, in which a man sentenced to death must use his final request to put the hangman in a logical bind. Janovskaja taught at least two courses each school year, incorporating new ideas and material in such a way that she never taught the same course twice. According to Boris A. Kushner, a student who knew Janovskaja at Moscow State University during the 1960s, students found Janovskaja's style engaging and the faculty were not incensed by her lack of original work. "Her whole personality, kind, open and deep, the tremendous and dangerous war she conducted against demagogic dialecticians—all that commanded respect," Kushner wrote in his remembrance of the teacher in *Modern Logic*.

Janovskaja was recognized for her years of contribution to the field in 1951, when she received the prestigious Order of Lenin award. Her efforts to create a department of mathematical logic at Moscow State University were successful, and she was named the department's first chair on March 31, 1959. She died on October 24, 1966.

In many ways, Janovskaja is considered more of a philosopher than a mathematician. Her work is more concerned with the nature of problems and the methods to be used than with the answers themselves. In her 1963 paper, "On Philosophical Questions of Mathematical Logic," Janovskaja listed logical concepts and problems and described how they were philosophical in nature. "These questions have not been listed in order to offer any answers at all. I do not know sufficiently the questions to allow myself to do so."

## SELECTED WRITINGS BY JANOVSKAJA

"Foundations of mathematics and mathematical logic" *Matematika v SSSR za tridcat let 1917–1947* (1948): 11–45.

"Mathematical logic and foundations of mathematics" *Matematika v SSSR za sorok let 1917–1957* (1959): 13–120.

## FURTHER READING

### Books

Anellis, Irving H. "Sof'ja Janovskaja," in *Women in Mathematics*. Edited by Louise S.Grinstein and Paul J. Campbell. New York: Greenwood Press, 1987, pp. 80–85.

### Periodicals

Anellis, Irving H. "The Heritage of S.A. Janovskaja." *History and Philosophy of Logic* 8 (1987), 45–56.

Bashmakova, I.G., et. al. "Sofia Aleksandrovna Yanovskaya." *Russian Mathematical Surveys* 21 (May–June 1966), 213–221.

Bochenski, J.M. "S.A. Janovskaja." *Studies in Soviet Thought* 13 (1973), 1–10.

Kushner, Boris A. "Sof'ja Aleksandrovna Janovskaja: A Few Reminiscences." *Modern Logic* 6 (January 1996): 67–72.

### Other

"Sof'ja Janovskaja." *Biographies of Women Mathematicians*. Http://www.scottlan.edu/lriddle/women/chronol.htm (July 22, 1997).

*—Sketch by Bridget K. Hall*

# William Jones
## 1645–1749
### Welsh–born English geometer and educator

William Jones is noted as the person who introduced π as the symbol for the ratio of the circumference of a circle to its diameter; otherwise, his major contribution seems to have been as a correspondent and manuscript collector for other notable mathematicians of his era, particularly Sir **Isaac Newton**.

Jones was born in Anglesey, Wales, to John George, a farmer, and Elizabeth Rowland. As was the Welsh custom, Jones translated his father's first name into a surname. As a schoolboy, Jones' quickness caught the attention of a local landowner, Bulksley of Baron Hill. In 17th century Britain, the patronage of the gentry was crucial for a farmer's son who wished to advance; thanks to Bulksley's interest, Jones obtained a position with a countinghouse—today's equivalent of an accounting firm—in London. As a representative of the firm, Jones had the opportunity to travel; on one trip to the West Indies, he taught mathematics aboard a man–of–war and discovered his own talent for teaching.

When he returned to London, Jones set himself up as a mathematics tutor. He secured teaching positions with some of England's great families; one of his pupils, Philip Yorke, who became the first Earl of Hardwicke, was eventually named Lord Chancellor. The earl apparently remembered his mathematics tutor with respect and affection, as he invited Jones to travel with him on his official rounds as Lord Chancellor and appointed Jones as "secretary of peace."

Jones also tutored Thomas Parker, who later became the first earl of Macclesfield; the earl retained Jones as a tutor to his son George, who later became president of the Royal Society of London. As the family tutor, Jones lived in comfort at Shirburn Castle, Tetsworth, Oxfordshire, with the family. It was there he met Maria Nix, daughter of a London cabinetmaker, who became his wife and the mother of his two sons and daughter.

In 1702, Jones published *A New Compendium of the Whole Art of Navigation*, a how–to manual that explained the application of mathematics and astronomy to sailing. His second book in 1706, *Synopsis palmariorum matheseos*, published in 1706, attracted the attention of Sir Isaac Newton. A primer for beginning mathematics students, *Synopsis* included an overview of recent advances in mathematics. It was in this work he introduced pi ($\pi$) for the ratio of the circumference of a circle to its diameter.

Newton's appreciation of Jones' *Synopsis* was the starting point of voluminous correspondence between the two. In 1708, Jones acquired a transcript of Newton's 1669 *De analysis* as part of the collected papers of John Collins; in 1711, Newton permitted Jones to print his works *De analysis per aequationes numero terminorum infinitas* and *Methodus differntialis* and other tracts. Jones issued the collected works as *Analysis per quantitatum series, fluxiones ac differentias: cum enumeratione linearum tertii ordinis.*

In the same year, Jones was appointed a member of the committee set up by the Royal Society to probe the invention of calculus. Along with John Machin and **Edmund Halley**, he was responsible for the printed report. In 1712, he was elected a fellow of the Royal Society and eventually served as its vice president.

When Jones died in 1749, he left a huge collection of manuscripts and letters. Notes and journals discovered after his death indicate that he intended to compile and publish an extensive work on mathematics, but Jones' papers became so intermixed with the works of others that the task of separating Jones' notes from other manuscripts was not completed until the 1970s.

## SELECTED WRITINGS BY JONES

*A New Compendium of the Whole Art of Navigation,* 1702

*Snyopsis palmariorum matheseos, or a New Introduction to the Mathematics,* 1706

## FURTHER READING

Baron, M.E. "William Jones," in *Dictionary of Scientific Biography.* Volume VII. Edited by Charles Coulston Gillispie. New York: Charles Scribner's Sons, 1973, pp. 162–63.

Eves, Howard. *An Introduction to the History of Mathematics.* New York: CBS College Publishing, 1983, p. 87.

Hoffman, Joseph E. *The History of Mathematics to 1800.* Volume II. Totowa, N.J.: Littlefield, Adams and Co., 1967.

Smith, David E. *History of Mathematics.* Volume II. New York: Dover Publications Inc., 1958, p. 312.

*—Sketch by A. Mullig*

*Tosio Kato*

# Tosio Kato
## 1917-

**Japanese-born American mathematical physicist**

Tosio Kato, whose career in mathematics ranged over more than 40 years, made major contributions to the field of mathematical physics. A prolific writer, he produced hundreds of published articles during his career. His most important research, on perturbation theory, won him acclaim and awards in both his native country of Japan and in the United States, where he spent most of his career.

Kato was born on August 25, 1917, in Tochigiken, Japan, the son of Shoji and Shin (Sakamoto) Kato. He attended the University of Tokyo, where he received a bachelor's degree in 1941; he would receive a doctor of science degree from the university ten years later. In 1943, he began teaching at the University of Tokyo, and due to the stability this position provided he was able to marry Mizue Suzuki the following year.

### Discovers Key Ideas behind Perturbation Theory

Even before his appointment to full professorship at the university, which he achieved upon receiving his doctorate in 1951, Kato had begun to publish the beginnings of his research in perturbation theory. Perturbation theory is the study of a system which deviates slightly from a less complex, ideal system. This is an important field of research because most systems that mathematicians and physicists study are not ideal. Kato examined only the perturbation theory which relates to linear operators (functions). The groundbreaking work in the field had been accomplished by John Rayleigh and Erwin Schrödinger in the 1920s.

Kato's contributions to perturbation theory were threefold. First, he laid the mathematical foundation for the theory, applying ideas in modern analysis and function theory. Second, he established the selfadjointness of Schrödinger operators—in other words, he showed that Schrödinger operators are symmetric. This was significant because these operators are a fundamental tool of quantum physics and knowledge of their symmetry makes their manipulation much simpler. Finally, he began the study of the spectral properties of the operators, which describes the variety of simple effects which the operators can have when applied to specific elements of a given set. This is important as it allows the operators to be described in terms of many simple effects which can be combined into larger ones. The term *spectral* is related to the way in which a complicated operator splits into distinct effects, as light can be split into distinct colors.

The culmination of his work was his definitive book on the subject, *Perturbation Theory for Linear Operators,* which he began writing in Japan and completed in the United States, after accepting a professorship at the University of California at Berkeley in 1962. Although the book did not appear until 1966, his home country had already recognized him as a leading researcher in his field, presenting him with the Asahi Award in 1960.

In the United States, Kato continued his research into perturbation theory and used functional analysis to solve problems in hydrodynamics and evolution equations. Even though Kato was well known in the American mathematical community by his hundreds of published articles, acknowledgement of the importence of his work came late in the United States. It was not until 1980 that the American Mathematical Society and the Society for Industrial and Applied

Mathematics jointly awarded him the Norbert Wiener Prize for applied mathematics. In Japan, Kato continued to receive recognition. On the occasion of his retirement from the University of California in 1989, the University of Tokyo held a conference in his honor. The conference, entitled "The International Conference on Functional Analysis in Honor of Professor Tosio Kato," paid homage to Kato's many contributions in the field of mathematical physics.

## SELECTED WRITINGS BY KATO

### Books

*Perturbation Theory for Linear Operators,* 1966
*A Short Introduction to Perturbation Theory for Linear Operators,* 1982

### Periodicals

"On the Convergence of the Perturbation Method, I, II." *Progressive Theories of Physics* 4 (1949): 514–523.

## FURTHER READING

### Books

Fujita, H., T. Ikebe, and S. T. Kuroda, editors. *Functional-Analytic Methods for Partial Differential Equations.* New York: Springer-Verlag, 1980.

### Periodicals

*Notices of the American Mathematical Society* 27 (October 1980): 528–529.

—*Sketch by Karen Sands*

# Linda Keen
## 1940–

### American geometric analyst

Linda Keen devotes her time and energies to some of the hottest topics of the day. Her work in complex analysis and dynamical systems deals with the mathematics responsible for the vibrant graphics seen in science shows, fractal art, and lifelike computer animations. For more than 30 years, her research has been funded by grants and fellowships from the National Science Foundation (NSF). Keen has helped evaluate other postdoctoral fellowships for the NSF as well as for NATO. She is also active in the mathemati-

*Linda Keen*

cal community, through her professional participation on various editorial boards and steering committees. Her influence has been felt at local and national levels regarding such pressing issues as women and minority involvement, librarianship, project funding, educational and test standards, research goals, and professional ethics. Among her most visible posts have been the presidency of the Association for Women in Mathematics (1985–1986), and the vice–presidency of the American Mathematical Society (1992–1995).

Linda Keen was born Linda Goldway in New York City on August 9, 1940. Her father was an English teacher who did not take his daughter's interest in the comparatively obscure language of mathematics personally. In fact, he encouraged her to study at a local magnet school, the Bronx High School of Science. She would stay in New York throughout her early academic career, earning a B.S. in 1960 from City College, an M.S. from New York University (NYU) two years later, and a Ph.D. from the Courant Institute of Mathematical Sciences in 1964.

### "The Children's Lunch"

Keen was fortunate to have as her Ph.D. advisor the celebrated Lipman Bers. "Lipa," as he was called, was a political refugee throughout the 1930s, coming to the U.S. from Latvia via Prague. At NYU, he was a colleague and friend as much as an authority on complex analysis. In 1964, it was still quite rare for a male mathematician to be even tolerant of women.

Yet, as Bers said himself in an interview with Donald Albers and Constance Reid, it never occurred to him that women could be intellectually inferior to men. While studying with Bers, Keen focused on the analytic aspects of Riemann surfaces.

Keen's obituary for Bers highlighted his lack of pretense and approachability. Classes were held Friday afternoons, so Bers extended the time with his students to include lunch—the "children's lunch" as he called it. Keen, who was almost always the youngest at the table, would be given the check to divide in her head. Once, when the "children" all insisted that the oldest person take the chore, Bers was unprepared and his calculations were incorrect.

### Committee on Professional Ethics

Bers' commitment to political activism and human rights, resulting from his years of living beneath the shadow of dictatorships, seems to have influenced Keen as well. From 1992 to 1996, she chaired the special advisory committee to write Ethical Guidelines and Procedures for the American Mathematical Society (AMS) Committee on Professional Ethics (COPE). The document developed by the special advisory committee provides professional mathematicians with guidance about ethical issues including giving credit for new findings, refereeing papers responsibly, protecting "whistle blowers," and social responsibility.

Among the issues with which the committee's report grappled are those arising from work in industry and with the government, as well as standards of conduct within professional organizations such as the AMS. Protecting confidentiality, anonymity, and privileged information is given top priority. As the guidelines state, "Freedom to publish must sometimes yield to security concerns, but mathematicians should resist excessive secrecy demands whether by government or private concerns." Those AMS members who advise graduate students are now expected to paint a realistic picture of employment prospects, and not to exploit their students by giving them heavy workloads at low pay. The guidelines also include a standard nondiscrimination policy. The special advisory committee's proposed guidelines were ratified in 1995 by a 25 to 3 vote.

Keen has served her profession in similar capacities a number of times. She began her involvement with COPE in 1986, and became a member of various policy boards for the AMS throughout the 1990s. Keen has also worked with the International Mathematics Union. She was of a member of the panels charged with evaluating the mathematics departments of the State University of New York–Potsdam and Rutgers University–Newark, and the minority program at the University of Minnesota. Keen was also a charter member of the Mayor's Commission for Science and Technology of the City of New York, serving on this commission from 1984–1985.

Keen's professional career has taken her to various institutions in her home state, including Hunter College and the City University of New York. When Lehman College, formerly the Bronx campus of Hunter, became independent in 1968, Keen remained on the faculty. She was promoted to full professor at Lehman in 1974 and presently holds a dual appointment in the Graduate Center Doctoral Faculties in Computer Science and in Mathematics. Keen has also held visiting professorships at the University of California at Berkeley, Columbia University, Boston University, Princeton, and MIT, as well as at mathematical institutions in several foreign countries, including Germany, Brazil, Denmark, Great Britain, and China. Keen's editorial services are equally international. Currently she serves on editorial boards for the *Journal of Geometric Analysis* and the *Annales* of the Finnish Academy of Sciences.

Throughout her career, Keen has preferred working collaboratively with other mathematicians to working alone. During the 1980s she worked with Caroline Series on the geometric aspects of Riemann surfaces, and, more recently, she contributed to the field of dynamical systems in cooperation with Paul Blanchard, Robert Devaney, and Lisa Goldberg. As Keen puts it, "I am basically a social person and enjoy people." She currently counts her husband and two children as her chief supporters, worthy successors in this regard to her father and Lipman Bers.

## SELECTED WRITINGS BY KEEN

### Books

*The Legacy of Sonya Kovalevskaya: Proceedings of a Symposium,* 1987
(Edited with R. Devaney) *Chaos and Fractals: The Mathematics Behind the Computer Graphics,* 1989
(Edited with J. Dodziuk) *Lipa's Legacy,* (in publication)

### Periodicals

"Lipman Bers (1914–1993)." *AWM Newsletter* 24 (1994): 5–7.

## FURTHER READING

### Books

Albers, D.J., G.L. Alexanderson, and C. Reid, editors. *More Mathematical People: Contemporary Conversations.* Boston: Harcourt Brace Jovanovich, 1990.

*Johannes Kepler*

*American Men & Women of Science.* Nineteenth edition. New Providence, NJ: Bowker, 1994, p. 262.

**Other**

"Linda G. Keen." http://www.math.neu.edu/awm/NoetherBrochure/Keen93.html

"Linda Keen." *Biographies of Women Mathematicians.* June 1997. http://www.scottlan.edu/lriddle/women/chronol.htm (July 1997).

—*Sketch by Jennifer Kramer*

# Johannes Kepler
## 1571–1630
### German astronomer and mathematician

Johannes Kepler is best known for his discovery of the three laws of planetary motion, demonstrating that the universe operates according to fixed, natural laws. A disciple of the great astronomer Copernicus, Kepler refined and advanced Copernican theory. After the death of Tycho Brahe, a leading astronomer of the 16th century, Kepler was appointed Imperial Mathematician in Prague by Emperor Rudolph II. His many published works describe his search for the mathematical accordances on which the universe is structured, and his study laid the foundation for the understanding of the cosmos, out of which grew the science of modern astronomy. **Isaac Newton**'s work in universal gravitation was influenced by Kepler's discoveries and the standards he set for scientific inquiry. Kepler's research in light refraction resulted in two important books on the subject of optics and improvements in the astronomic telescope. As a mathematician, his efforts to complete Brahe's planetary tables were furthered by his discovery of the usefulness of logarithms in making computations. Although logarithm formulas and tables were published by **John Napier**, a Scottish mathematician, in 1614, Kepler used a modified system of his own creation, and these early computations prefigure the invention of calculus. Kepler's published works were numerous, and of great influence to those astronomers who followed him. Some of his most prominent writings were: *Mysterium cosmographicum, Astronomiae pars optica, Astronomia nova, Dioptrice, Harmonice mundi, Epitome astronomiae Copernicanae, Tabulae Rudolphinae,* and *Ephemerides novae.*

Kepler was born December 27, 1571, in Weil der Stadt, Germany. He was the eldest son of Heinrich and Katharina (Guldenmann) Kepler. His grandfather was the local mayor. Although the family was of the Lutheran faith, Kepler's father became a mercenary soldier in the service of the Catholic Duke of Alva, fighting a Protestant reform movement in the Netherlands when Johannes was a small boy. In 1588, he abandoned his family. All accounts describe his mother as a difficult, quarrelsome woman, and in later life, Kepler was obliged to defend her against charges of witchcraft. Katharina influenced her son in a more positive way, when as a small boy of six, he was taken by her to observe the comet of 1577. Kepler married twice; first to Barbara Müller (April 27, 1597), with whom he had five children, two of whom died young. His first wife died in 1611 of typhus. On October 30, 1613, he married Susanna Reuttinger. They had seven children, five of whom died in infancy.

Kepler's elementary schooling began at Leonberg, when he entered a church school. His religious nature was apparent at a young age, and he began to prepare for a life in the church. After two years at a monastery school in Adelberg, he enrolled at Maulbronn, preparing for the University of Tübingen. He entered the University in 1589. There, Michael Maestlin, the astronomy professor at Tübingen, was to open the Copernican door that led to Kepler's future.

### Teaching Duties Combine Mathematics, Astronomy, and Astrology

A brilliant student, Kepler was seen as destined for great accomplishments. In August 1591, he received a master's degree from the University and began his theological studies. However, his plans to enter the ministry were forever altered by the death of Georgius Stadius, teacher of mathematics at the Lutheran high school in Graz. Searching for a replacement, the school turned to the University for a recommendation. Kepler was nominated, and accepted the post. His duties were to teach mathematics and oversee the annual publication of astrology almanacs—weather forecasts and other predictions based on astrological rules. It may seem ironic that a learned scientist would consent to dabble in the astrological arts. But, although he regarded astrology as popular superstition and at best a lesser form of astronomy, Kepler's observation of and profound feeling for cosmic harmonies made him tolerant of astrology's attraction as a guiding force for the common individual.

### Cosmic Insight Brings New Discoveries

Kepler's teaching at Graz was not very fulfilling to him. But inspiration struck one day as he pondered such questions as: Why are there only six planets? How can their distance from the sun be determined? Why do planets farthest from the sun move the slowest? He theorized that these questions could be answered by relating their placement and movement to the five regular solids of Euclidean geometry. This insight resulted in his first major publication, *Mysterium cosmographicum*, in 1596. In it, Kepler describes a mathematical relationship between a planet's distance from the sun and its orbit periodicity. From this intriguing beginning was to evolve Kepler's laws of planetary motion.

Seeking to disseminate his findings, Kepler sent copies of *Mysterium cosmographicum* to other European scientists. Tycho Brahe, a leading observational astronomer, was impressed by Kepler's mathematical theories (if not their Copernican basis) and invited him to come to Prague to work with him. In 1598, Kepler and all other Protestant teachers were forced by the Catholic authorities to leave Graz on short notice. Newly married, and in need of work, he accepted Brahe's invitation. Although his wife was from a wealthy family, her inheritance was tied to property in Graz. The Keplers arrived in Prague in 1600 without assets and were to be plagued with financial problems throughout their marriage.

### The Years with Brahe

Kepler's first assignment was to study the orbit of Mars. This work was to occupy him for eight years, and resulted in the discovery of his first two laws of planetary motion. The first law shows that every planet's orbit is an ellipse with the sun at one focus. This was revolutionary, because until Kepler proved otherwise, it was believed that a planet's orbit was circular. The second law shows that the line from the sun to a planet sweeps across equal areas in equal time. Kepler's great work, *Astronomia nova*, published in 1610, presents these findings. During the years spent on the Mars orbital study, Kepler also began his work in optic research. This interest came about when he built a pinhole camera to observe a solar eclipse in 1600. His optic discoveries formed the basis of modern geometrical optics. In 1604, Kepler published his theories in *Astronomiae pars optica*. In 1611, his *Dioptrice* applied those optical laws to the telescope.

Brahe had died in 1601 and Kepler assumed the position of Imperial Mathematician and the obligation to complete and publish his mentor's tables of planetary motion. Although their working relationship had many times been strained, Kepler took on this task. Called the *Tabulae Rudolphinae* after his patron, the Emperor Rudolph, the work was eventually published in 1627.

### The Third Law and the Writing of Astronomical Epics

After the death of his first wife in 1611, Kepler took a position as a mathematics teacher in Linz, Austria. He remarried, and was to remain in Linz for fourteen years, during which he produced *Harmonice mundi*, his theories on the harmonies of the universe. This book contained Kepler's third law: that the square of a planet's period measured in years equals the cube of its distance from the sun measured in astronomical units. His textbook series of seven books, *Epitome astronomiae Copernicanae*, described all that was then known of heliocentric astronomy, and included Kepler's three laws. During this time, he also produced a work called *Stereometria doliorum vinariorum*, stemming from his efforts to measure the volume in a wine cask. This work is considered to be a preliminary basis of calculus.

Kepler left Linz during the Counter Reformation and settled his family in Regensburg. After the publication of the *Tabulae Rudolphinae*, he began once again to look for employment. In 1628, Kepler was promised a post by the king of Bohemia, Ferdinand III, in the newly–acquired duchy of Sagan. It was there that he published the last of his works, *Emphemerides pars II* and *Emphemerides pars III* (astronomical charts and weather observations); and a curious book of science fiction, *Somnium seu astronomia lunari*, about a voyage to the moon.

Financial hardship (he did not receive the promised salary from King Ferdinand) caused Kepler to return to Regensburg, bringing with him all his books and research papers of a lifetime of work. There, he became ill with a fever and died on November 15, 1630. He was buried in the Regensburg Protestant cemetery.

Kepler has been called a mathematical mysticist. His deep religious beliefs led him to search for harmony between humanity and the mathematical principles by which the universe was ordered.

## SELECTED WRITINGS BY KEPLER

*Epitome astronomiae Copernicanae,* Books IV and V. Edited by Charles Glenn Wallis. In *Great Books of the Western World, XVI,* 1952.

*Joannis Kepleri astronomi opera omina,* 8 volumes. Edited by Christian Frisch, 1858–1871; reprinted 1971. (In German; includes all of Kepler's major printed works and extensive excerpts from his correspondence.)

*Johannes Kepler Gesammelte Werke.* Edited by Caspar and Franz Hammer, 1937 (Contains major works, theological writings, and correspondence.)

## FURTHER READING

### Books

Caspar, Max. *Johannes Kepler.* Translated by C. Doris Hellman. New York: Abelard–Schuman, 1959.

Considine, Douglas M., editor. *Van Nostrand's Scientific Encyclopedia.* Eighth edition. New York: Van Nostrand Reinhold, 1995, pp. 1812–1813.

Daintith, John, editor. *Biographical Encyclopedia of Scientists.* Volume I. New York: Facts on File, Inc., 1981, pp. 434–436.

Gingerich, Owen. "Johannes Kepler," in *Dictionary of Scientific Biography.* Volume VII. Edited by Charles Coulston Gillispie. New York: Charles Scribner's Sons, 1973, pp. 289–312.

Illingworth, Valerie, editor. *The Facts on File Dictionary of Astronomy.* Third edition. United Kingdom: Harper Collins Publishers Ltd., 1994, pp 240–241.

Koestler, Arthur. *The Watershed: A Biography of Johannes Kepler.* Garden City, NY: Anchor Books, 1960.

Magill, Frank N., editor. *The Great Scientists.* Danbury, CT: Grolier Educational Corp., 1989, pp. 199–206.

Muir, Hazel, editor. *Larousse Dictionary of Scientists.* New York: Larousse, 1994, pp. 283–284.

—*Sketch by Jane Stewart Cook*

*Omar Khayyan*

# Omar Khayyam
## c.1048–c.1131
### Persian mathematician, poet, astronomer, and philosopher

Known in the West as Omar Khayyam, Ghiyāth al–Dīn Abu' l–Fath 'Umar ibn Ibrāhīm al–Nīsābūrī al–Khayyāmī was born and died in Nīshāpūr, Khorāsān, Persia (now Iran). Khayyam's given name was 'Umar, while "al–Khayyāmī" means tent–maker, which may have been the family trade. He was one of the most brilliant figures of Islamic civilization. His passionate and thought–provoking *Ruba'iyat* ("Quatrains"), in the West far better known than his extraordinary work as a mathematician, is a much–anthologized verse collection that has been praised as one of the treasures of world literature. As a mathematician, Khayyam is noted for his work in cubic equations.

### Works as Astronomer in Isfahan

Around 1070, Khayyam traveled to Samarkand, subsequently proceeding to Isfahan, upon the invitation of the Seljuk sultan Jalal–al–Din Malik–shah. Employed by the sultan as the court astronomer, Khayyam supervised a team of royal astronomers whose task it was to compile astronomical tables. Among Khayyam's accomplishments in Isfahan was a projected calendar reform. Khayyam's calendar is more accurate than the Gregorian calendar, accumulating an extra day in 3,770 years (as opposed to 3,333 years in the Gregorian calendar currently in use). His calculation of the length of the year—365.24219858156 days—is more precise than the Gregorian 365.2425 days. During this period, Khayyam also wrote commentaries on **Euclid**, as well as philosophical treatises on such fundamental questions as being, existence, and universal science.

In 1092, Khayyam found himself in a difficult situation following the sultan's assassination. He not only lost a patron but acquired powerful enemies. The observatory lost its funding, the calendar reform was halted, and Khayyam, who may have not been religious in a traditional sense, had to defend himself against charges of atheism. Nevertheless, Khayyam stayed in Isfahan until 1118, working hard to convince the late sultan's successors to provide support for science and education. In Merv, the new Seljuk capital, Khayyam continued his scientific and literary work. He eventually returned to Nishapur toward the end of his life.

### Breaks New Ground in Algebra

Khayyam's great contribution to algebra, described in his book *Algebra*, is his method of solving cubic equations. In his approach to quadratic equations, described in his *Elements,* **Euclid** relied on geometry. In fact, Euclid used line segments to indicate magnitudes which modern algebra represents by letters. In other words, Greek algebra was geometric, i.e., mathematicians used geometry to map algebraic problems. For example, the quadratic equation $x^2 - ax + b = 0$, transposed as $x^2 + b = ax$, would be translated into the geometrical problem of creating a square ($x^2$) so that the sum of $x^2$ and the given area $b$ equals the area of the rectangle $xa$, $a$ being a given side. Continuing in the tradition of Greek algebra, Khayyam successfully applied the geometric method for the solution of quadratic equations to cubic equations. In doing so, he had to imagine cubes and parallelepipeds instead of squares and other rectangles. However, cubic equations required geometry that cannot be done by straightedge and compass only, and Khayyam, drawing on Greek sources, arrived at a solution by using conic sections.

Although relying on geometry to solve equations, Khayyam used numbers, instead of line segments, exemplifying, as Carl B. Boyer has written, the seemingly clairvoyant efforts of Arabic mathematics "to close the gap between numerical and geometric algebra," a problem which will occupy mathematicians for several centuries. According to historians, Khayyam's work with cubic equations is one the great accomplishments of Arabic mathematics.

### Work on Euclid Leads to New Insights

In his book *Commentaries on the Difficulties in the Postulates of Euclid's Elements*, Khayyam attempted to prove the "parallel postulate," which states that, given a line $a$, only one line parallel to $a$ can be drawn through a point which is not on $a$. Euclid's fifth postulate, as it is known, is not, despite its apparent simplicity and transparence, easy to prove, and has therefore posed a great challenge to mathematicians. The fifth postulate is, as mathematician and musicologist Edward Rothstein calls it, "a troublesome axiom," which indeed explains why many generation of mathematicians tried to prove it. As Rothstein points out, this axiom is troublesome because it cannot be derived from the other axioms of the Euclidean systems. And it can be contradicted, by imagining an infinity of lines drawn through a point (not on line $a$) which are parallel to $a$. This new situation is possible in a new geometric system. According to Rothstein, "Properly reinterpreted, the new axiomatic system actually creates a different geometric universe, one that is called 'non–Euclidean'."

Unsatisfied by previous efforts to prove the fifth postulate, Khayyam attacked the problem by constructing a quadrilateral with two equal sides that are perpendicular to their base. Accepting that the quadrilateral's upper angles are by definition equal, Khayyam considered three hypotheses, namely, that these angles are right, acute, or obtuse. The second and third hypotheses had to be rejected, because, in accordance with the fifth postulate, converging lines will intersect. Although Khayyam dismissed the problematic hypotheses, he entered a theoretical universe in which, centuries later, non–Euclidean geometry was accepted as possible.

In his commentary on Euclid, Khayyam also discusses the theory of proportions. According to Euclid, the validity of a ratio, for example, $a : b = c : d$, can only be established if the alphabetical symbols stand for commensurable numbers, i.e., numbers that can be divided into units. Thus, the Euclidean definition accepts only rational numbers, (numbers that are integers or a quotient of two integers). Khayyam found Euclid's definition too narrow, suggesting instead that the idea of proportion be expressed purely numerically, without reference to the notion of unit. Khayyam's conception of proportion implies that numbers, but not only rational numbers, can represent proportion. Thus, by examining and expanding the Euclidean concept of viable number, Khayyam anticipated a line of theoretical work which culminated in the definition of the real number concept by **Julius Dedekind** in the 19th century.

### Applies Theory of Proportion to Music

Possessing a truly encyclopedic mind, Khayyam also applied his mathematical insights to musical theory. In his "Discussion on Genera Contained in a Fourth," he tackled the problem, already studies by Greek mathematicians, of dividing the interval of a fourth, which **Pythagoras** defined as the ratio of 3 : 4 (the integers represent string lengths) into three intervals which would basically fit into tonal systems whose building–blocks are whole tones and half–tones (diatonic) or only half–tones (chromatic). What made the problem interesting was not just a mathematical challenge, but also the fact that the Greek scale was *enharmonic*, unlike the modern scale of equally–spaced 12 half–tones constituting an octave. This means that, depending on the tuning of a string instrument, (c-sharp and d-flat), which on a piano would be represented by the same key, would not necessarily have the same pitch. Khayyam divided the fourth in 22 ways, three being new.

### Poet and Philosopher

To readers of poetry Khayyam is known as the author of the *Ruba'iyat*, widely praised as one of the great treasures of world poetry. A *ruba'i* (plural: *ruba'iyat* ) is a two–line stanza. As each line consists of two hemistiches, the stanza has four hemistiches, which explains the term *ruba'i*—the Arabic word meaning *foursome*. The *Ruba'iyat*, which consist of over 200 stanzas attributed to Khayyam, address such ultimate questions as death, eternity, reality, time, as well as the paradox of human existence. Rich in imagery and profound wisdom. Khayyam's poetry has, through numerous translation, found enthusiastic readers all over the world, and inspired almost cultic devotion in England.

Among the numerous philosophical, scientific, and literary discussions generated by Khayyam's poetry throughout the centuries possibly the most fascinating, and inconclusive, one is the question concerning his possible adherence to Sufism, the venerable mystical tradition which developed within Islam. While some scholars have found no traces no Sufism in the works of the mathematician–poet, writers such as Idries Shah have identified the timeless quality of Khayyam's poetry as typical of the Sufi world view. However, a scientist's or poet's writings, as commentators often realize, may not reveal his (or her) deepest beliefs, and this probably, as B. W. Robinson has remarked, explains why Khayyam eludes attempts at classification, scholars offering no "reliable gauge of Omar Khayyam's own position in relation to the life of the spirit." Robinson concludes: "Life is lived very publicly in the East. As a result, though the heart is much talked of, men have learnt how to conceal what is really in it and, while verses might beguile a moment of leisure among friends and win applause for the poet's skill, they very rarely, as any student of Persian lyrical poetry knows, reveal precisely what the poet believed or did not believe."

## SELECTED WRITINGS BY KHAYYAM

*The Algebra of Omar Khayyam,* 1931
*The Ruba'iyat of Omar Khayyam*, translated by Peter Avery and John Heath–Stubbs, 1981

## FURTHER READING

Boyer, Carl B. *A History of Mathematics.* Second edition. Revised by Uta C. Merzbach. New York: Wiley, 1991.

Dantzig, Tobias. *Number: The Language.* Fourth revised edition. Garden City, NY: Doubleday, 1954.

Eves, Howard. "Omar Khayyam's Solution of Cubic Equations," in *From Five Fingers to Eternity: A Journey through the History of Mathematics.* Edited by Frank J. Swetz. Chicago: Open Court, 1994, pp. 302–03.

Joseph. George Gheverghese. *The Crest of the Peacock: Non–European Roots of Mathematics.* London: I. B. Tauris, 1991.

Kline, Morris. *Mathematical Thought from Ancient to Modern Times.* New York: Oxford University Press, 1972.

Robinson, B. W. "Appendix 3," in *The Ruba'iyat of Omar Khayyam.* Translated by Peter Avery and John Heath–Stubbs. London: Penguin Books, 1981.

Rothstein, Edward. *Emblems of Mind: The Inner Life of Music and Mathematics.* New York: Avon Books, 1995.

Sarton, George. *Introduction to the History of Science.* Volume I: *From Homer to Omar Khayyam.* Washington, DC: William and Wilkins, 1927.

Shah, Idries. *The Sufis.* New York: Doubleday, 1964.

Struik, D. J. "Omar Khayyam, Mathematician," in *From Five Fingers to Infinity: A Journey through the History of Mathematics.* Edited by Frank J. Swetz. Chicago: Open Court, 1994, pp. 297–301.

—*Sketch by Zoran Minderovic*

# al–Khwārizmī (also known as Alchorizmi and Algorismus) 780(?)–850(?)

**Arab algebraist and astronomer**

Al–Khwārizmī was a noted Arab mathematician and astronomer who studied and worked during the first half of the ninth century. Although his work was not as significant as that of other Arab scholars of the time, he was a prolific author. Several of his writings about mathematics endured until the 12th century, when they were translated into Latin and began to influence the Hindu system of numbers (later erroneously called Arabic numbers). Al-Khwārizmī was also one of the early developers of the mathematical process used to solve for unknown quantities. One of the terms he used for this process was *al–jabr*, which eventually became the word "algebra" that we use today.

Al-Khwārizmī 's full name was Abū Ja'far Muhammad ibn Mū sā al-Khwā rizmī. Some historians add al-Qutrubbullī to his name, meaning he came from the region of Qutrubbull, located between the Tigris and Euphrates Rivers in what is now known as Iraq. The name al-Khwārizmī probably refers to his ancestors, who came from the district of Khwarizm south of the Aral Sea in central Asia. His exact birth date is unknown, but it is believed to be before 800, and possibly as early as 780. The names of his parents are also unknown, as are the details of his childhood and education.

## Embarks on a Career of Scholarly Studies and Writings

All that is known about al-Khwārizmī is that by the time he was an adult, he had become a scholar in astronomy and mathematics. During the reign of the Caliph al-Ma'mūn, in the period of 813–833, al-Khwārizmī became a member of the *Dār al–Hikma*, or House of Wisdom, in Baghdad. The Caliph was a patron of learning and scientific investigation, and the members of the Dār al-Hikma were among the most noted scholars of the day. Soon after he joined the Dār al-Hikma, al-Khwārizmī began writing on several topics. The exact dates and order of composition of each work are not certain, but it is believed that one of his first books was on geography, which listed the longitudes and latitudes of locations within the known world of the Mediterranean and Near East. These lists were based on previous work done by the Greeks, but included many significant revisions. Al-Khwārizmī's other writings included an astronomy text in which he revised work done by Indian astronomers; two works about the astrolabe, an instrument often used to make astronomical measurements; a book on sundials; an examination of the Jewish calendar; and a chronology of events that occurred in the Moslem world during the early part of the ninth century.

## Writes Two Important Works on Mathematics

In addition to his other writings, al-Khwārizmī wrote two important works on mathematics. Both are believed to have been written in the period 825–830,

although this is not certain. The first, entitled *al-Kitāb al-mukhtasar fihisāb al-jabr wa'l-muqābala*, is translated as "The Compendious Book on Calculation by Completion and Balancing." It was a handbook of practical mathematics whose purpose, according to the English translation by Frederic Rosen in 1831, was to provide "what is easiest and most useful in arithmetic, such as men constantly require in cases of inheritance, legacies, partition, lawsuits, and trade, and in all their dealings with one another, or where the measuring of lands, the digging of canals, geometrical computations, and other objects of various sorts and kinds are concerned." This book consisted of three parts. The first part, and the most significant for later mathematicians, was the development of solutions to problems where there was an unknown quantity. In presenting these solutions, al-Khwārizmī introduced two operations. One, called *al-jabr*, is translated as "restoration" or "completion" and consisted of moving known or unknown quantities of the same power to one side or the other of the equation to eliminate negative quantities perhaps involving the unknown. For example, $x=10-3x$ would be "transformed" to $4x=10$. The second operation, called *wa'l-muqābala*, is translated as "balancing" and involved reducing (subtracting) positive quantities of the same power on both sides of the equation. Like all Arab mathematicians of the time, al-Khwārizmī did not use symbols for unknowns or for the equal sign, but instead wrote all his solutions in words. For example, he used the Arabic word *shay'*, meaning "thing" or "something," for an unknown, and the term "it becomes" for equals.

Although al-Khwārizmī is often credited with being the first to work in this area, his solutions only involved unknowns of the first and second powers and were fairly elementary compared to the works of earlier Greek and Indian mathematicians. His real claim to fame came later, when Arab mathematical writings were translated into Latin. At that time, the term *al-jabr wa'l-muqābala* ("completion and balancing"), found in the title of al-Khwārizmī's book, was applied to all writings dealing with the mathematics of solving for unknown quantities. This was often shortened to *al-jabr* and it became the name of the mathematical branch we know today as algebra.

Al-Khwārizmī's second book on mathematics explained the use of Hindu numbers and the concept of place-value system. The original version of this work in Arabic has been lost, but the title may have been translated as "Treatise on Calculation with the Hindu Numerals." It deals with the four basic arithmetic operations—addition, subtraction, multiplication, and division—as well as with fractions and square roots. The concept of using separate symbols for numbers was new to the Arabs, who had developed an alphabetic numbering system similar to the Greek (and later Roman) system of using letters for numbers. More important was the concept of place-value, which had been developed in India and brought to the attention of Arab mathematicians in the early ninth century. Instead of using letters or tedious graphic representations for numbers larger that nine, the Hindu system was based on the value ten and assigned values to the position of numerals when grouped together. Thus, the number "10" meant "one unit of ten, being one more than nine, plus no other units," and the number "87" meant "eight units of ten plus seven other units."

Although al-Khwārizmī found the Hindu numbering system superior to others and recommended its use to mathematicians and merchants everywhere, the system was not put into general usage until several centuries later. At that time, partially thanks to al-Khwārizmī's book on the subject, the numerals and counting system were labeled Arabic numbers and slowly gained wide acceptance in Europe.

Al-Khwārizmī continued his work as a scholar until the middle of the ninth century. According to one historian, he was one of several astronomers called to the sickbed of the Caliph al-Wathiq in 847 to cast his horoscope. They predicted he would live another 50 years based on the positions of the stars, and were astounded when he died ten years later. The exact date of al-Khwārizmī's death is not known, but may have been as late as 850.

### His Works Influence Later Arab and European Mathematicians

The significance of al-Khwārizmī's work in mathematics rests on his writings. His mathematical handbook, including the portions on algebra, continued to be used by students and scholars and helped form the basis for later advances in algebra by Arab mathematicians during the next three hundred years. The handbook was translated into Latin in the 12th century by Robert of Chester and again by Gerard of Cremona. It had a significant influence on the development of algebra in Europe, and some of the terms used were literal translations of al-Khwārizmī's original terms.

Al-Khwārizmī's book on the Hindu number system achieved only limited use among Arab mathematicians, but was of great importance after it was translated into Latin in the early 12th century. The Latin translation quickly stirred interest in the new system as accounting clerks for the major merchants and shipping firms of Europe found they could do their accounts faster than with other systems.

### SELECTED WRITINGS BY AL-KHWĀRIZMĪ

*al-Kitab al-mukhatasar fi hisab al-jabr wa'l-muqabala* ("The Compendious Book on Calculation by Completion and Balancing"), date unknown

*Christian Felix Klein*

"Treatise on Calculation with the Hindu Numerals," date unknown

### FURTHER READING

#### Books

American Council of Learned Societies. *Biographical Dictionary of Mathematicians*. New York: Charles Scribner's Sons, 1991, pp. 1246–1253.
Bergamini, David. *Mathematics*. New York: Time Incorporated, 1963, pp. 17, 67.

—*Sketch by Chris Cavette*

# Christian Felix Klein
## 1849–1925
### German analyst and geometer

Felix Klein is arguably one of the most influential mathematicians of the 19th century. He is best known for building the mathematical community at the University of Göttingen which became a model for research facilities in mathematics world wide.

Christian Felix Klein was born on November 25, 1849 in Dusseldorf, the son of an official in the local finance department. Klein graduated from Gymnasium (the German equivalent of an academic high school) in Dusseldorf and began studying at the University of Bonn in 1865. At Bonn, he fell under the influence of **Julius Plücker**, one of the best-known geometers of the century. Plücker had moved the center of his interest to physics, and it had been in physics that Klein originally wanted to work, but Plücker returned to his original interest in geometry and took Klein with him. After Plücker's death in 1867, Klein became responsible for finishing a manuscript of Plücker's, which gave him an early introduction to the scholarly community and, in particular, to Alfred Clebsch, another prominent geometer of the time.

After receiving his doctorate in 1868 Klein spent a year traveling between Göttingen, Berlin, and Paris. Of the three, he enjoyed Göttingen immensely, did not like Berlin, and had to leave Paris ahead of schedule because of the outbreak of the Franco-Prussian War. Some of his travels were spent with the young Norwegian mathematician **Marius Sophus Lie**, whose ideas on geometry and analysis were much in common with Klein's. Klein's patriotism led him to enlist as a medical orderly during the war, but before the year was over he had returned to Dusseldorf, suffering from typhoid fever. The next year Klein qualified as a lecturer at Göttingen, but the following year he accepted a chair at the University of Erlangen. The complexities of academic promotion within the German university system at the time frequently required moving about from one university to another, merely for the sake of promotion within the original university.

## Imposes Order on Geometry

It was the custom for a new professor to deliver an inaugural address at a German university, and in 1872 Klein followed suit at Erlangen. At the time, it was difficult to speak of one geometry, as recent developments had led to a collection of geometries whose relation to one another was unclear. There was the familiar Euclidean geometry, based on the ordinary axioms including the parallel postulate (which stipulated that there was exactly one parallel to a line through a point not on that line). There were at least two non-Euclidean geometries, one denying the existence of any parallels through a point not on a line, the other allowing the existence of an infinite number of parallels. Finally, projective geometry, which had been known since the 17th century, had been given a more quantitative turn in the work of **Arthur Cayley**, among others.

As outlined by Klein, geometry is the study of the properties of figures preserved under the transformations in a certain group. Which group of transformations one started with determined the geometry in which one was working. For example, if the transformations were limited to rigid motions, then one had Euclidean geometry. If projections were allowed, then one had projective geometry. If an even wider class were included, then one could end up with topology. This view (called the Erlangen program) has infused the spirit, not just of geometry, but of mathematics as a whole ever since.

Also, in 1872 Klein took over editing *Mathematische Annalen* after the death of Clebsch. Under his editorship this was the leading mathematical journal in the world and it was to remain so until World War II. By 1875, Klein had left Erlangen for the Technische Hochschule in Munich and then in 1880 he went to the University of Leipzig. In 1884 he was invited to take the place of **James Joseph Sylvester** at Johns Hopkins University in Baltimore, but he declined. He did make several visits to the United States subsequently, where both his personal influence and those of his students were strong. Finally, in 1886, Klein achieved the goal of a chair at Göttingen.

## Builds a Home for Mathematics

Two factors in particular led Klein to successfully create a mathematical center at Göttingen. One was personal, as he was married to Anne Hegel, a descendant of the German philosopher Georg Wilhelm Friedrich Hegel. Her striking beauty may have been a draw even for those who were not yet convinced of the mathematical attractions of her husband. In the course of their married life the Kleins had one son and three daughters.

The other factor was not so pleasant. One of the subjects on which Klein had been working while at Leipzig were automorphic functions, transformations of the complex plane into itself that satisfied certain conditions. Unfortunately, for Klein the year 1884 turned into a competition with the younger French mathematician **Henri Poincaré** seeking fundamental results. Although Klein's work during this period was of a high quality, he felt that he had not lived up to expectations and suffered a nervous breakdown.

Thereafter, Klein immersed himself in creating a major mathematical center at Göttingen. The mathematical discussions did not stop with the classroom walls, but continued at the Kleins' home or on walks into the woods around Göttingen. One feature of the institute was a room filled with geometrical models to help with visualization. The presence of such a room was a reminder of Klein's antipathy to the abstract style of analysis favored by **Karl Weierstrass** at Berlin. Klein wanted his mathematics to have intuitive content, which explains why he was anathema to Weierstrass. Klein attracted many of the leading German mathematicians to Göttingen, the most outstanding being **David Hilbert**. Göttingen's creative atmosphere encouraged the presence of women in the lecture hall and foreign visitors.

At the time of Klein's retirement shortly before the outbreak of World War I, he could take pride in having brought together a mathematical research community the like of which the world had never seen. In 1912, he received the Copley medal of the Royal Society, one of just many honors. His last years were saddened by the death of his son on the battlefield during the war, and he died on January 22, 1925.

Within ten years of his death the Nazi government had undertaken the dismantling of the research community in Göttingen. When the Institute for Advanced Studies was founded at Princeton in the 1930s, it modeled itself after Göttingen. The dream which Klein had brought into reality lived on.

## SELECTED WRITINGS BY KLEIN

*The Evanston Colloquium*, 1911
*Gesammelte Mathematische Abhandlungen*,
    1921–1923
*Elementary Mathematics from an Advanced Standpoint*. Translated by E.R. Hedrick and C.A. Noble, 1939

## FURTHER READING

Burau, Werner, and Bruno Schoeneberg. "Felix Klein," in *Dictionary of Scientific Biography*. Volume VII. Edited by Charles Coulston Gillispie. New York: Charles Scribner's Sons, 1973, pp. 396–400.
Reid, Constance. *Hilbert*. New York: Springer–Verlag, 1970.
Yaglom, I.H. *Felix Klein and Sophus Lie*. Translated by Sergei Sossinsky. Boston: Birkhauser, 1988.

—*Sketch by Thomas Drucker*

# Donald E. Knuth
## 1938–

### American computer scientist and mathematician

It is not often in the world of scientific discovery where one man has contributed a body of knowledge to a topic so far in excess of others in the field. Donald Knuth has done for computer science as

*Donald E. Knuth*

Albert Einstein did for physics with his theories of relativity. Knuth was born January 10, 1938 in Milwaukee, Wisconsin, to Ervin Henry Knuth and Louise Marie Bohning. His father had the distinction of being the first college graduate in the Knuth family and started his professional career as a grade–school teacher. Later he was employed as a bookkeeping teacher in a private Lutheran high school. Knuth's father also instilled an appreciation for music, playing the church organ for Sunday services. His talent for mathematics, education and music left a legacy for his son. Today, Knuth's talent and accomplishments have inspired and improved the entire world.

### Knuth Learns Music, English and Math

Early on, Knuth realized that most of the schoolwork required of him from middle school to college consisted mostly of writing and mathematics. Understanding this, he mastered both disciplines. Since his father was an educator in a Lutheran high school, Knuth attended Lutheran schools. These schools were very strong in English education, particularly in sentence structure and grammar. Knuth remembers that in the 7th and 8th grades, one of his favorite activities was diagraming sentences with classmates. Although they could diagram all the sentences in their English books, they always had trouble with sentences that were seen in everyday life—on billboards, posters, advertisements—but most particularly those found in the church hymnals. The students worked very hard on these problems, and although Knuth never says if they figured them out, he does say those same students breezed through English class when they moved on into high school.

One of Knuth's fond memories is the Ziegler's Candies contest. The manufacturers of the Ziegler's candy bar sponsored a contest where contestants had to see how many words they could find in the letters in "Ziegler's Giant Bar." In the 8th grade, Knuth knew he had a knack for problems like this, so he entered. He told his parents he had a stomachache and for two weeks he stayed home "sick"—all the while poring over an unabridged dictionary finding as many words as possible. Without using the apostrophe, Knuth's list contained about 4,500 words; there were only 2,500 words on Ziegler's master list. The grand prize was a television set for the school and enough Ziegler candy bars to feed the entire student body."

In high school, Knuth entered the Westinghouse Science Talent Search with his proposal "The Potrzebie System of Weights and Measures." This was an imaginary, but very clearly defined, system that would replace our units of ounces, pounds, inches, and feet. In a style that is typical of Knuth, he defined his units, including the "potrzebie" (the thickness of *MAD Magazine* issue 26) and the "whatmeworry" (the basic unit of power). Not only did this proposal win an honorable mention in the Westinghouse contest, he also earned $25 from *MAD Magazine* when they published the paper.

Knuth also did considerable work in high school learning about graphs. He would take a function of several variables, fix all but one, then vary the last one to see the changes in the shape of the graph. Of course, this was before the advent of graphing calculators so each of his graphs were computed and plotted by hand.

Knuth graduated with the all–time record for grades at his high school. By the time he left high school Knuth was an accomplished mathematician, writer, and musician but was undecided about what he wanted to study in college. He said that his choice was determined by the school he chose. Due to scholarship considerations, he chose to attend the Case Institute of Technology, majoring in physics.

## Knuth Meets the Computer

It was in college that Knuth was first introduced to an IBM 650 computer. This was one of the first mainframe computers made by the computer system pioneer. He obtained a copy of the manual and studied it cover to cover. In an interview with Donald Albers, Knuth said, "the manual we got from IBM would show examples of programs and I knew I could do . . . better than that. So I thought I might have

some talent." Little did he realize how prophetic that statement was. He wrote programs on the old IBM machine that would teach it how to play tic–tac–toe and performing prime number factorization.

In 1958, Knuth developed a basketball player analysis for the college team, so impressing the coach that he used it, claiming that it helped the team win a league championship. *Newsweek* wrote an article about the program and IBM used a publicity photo of Knuth standing next to the machine.

In college, Knuth always had a fear that he was not good enough to succeed, so he took on a tremendous amount of unassigned, ungraded work to solidify his knowledge of the subject. Although he started at Case as a physics major, Knuth's interests turned toward mathematics as a sophomore. While taking a course in abstract mathematics, the professor assigned a problem—finding the correct solution would earn the solver an automatic "A" in the class. Knuth found himself having some extra time, and by what he calls "a stroke of luck" was able to solve it. He turned it in, the professor gave him the "A" and he cut the rest of the class for the semester. Knuth's conscience caught up with him, so when the class was offered the following year, he worked as the class grader. His difficulty with the physics lab classes that were required for his major finally caused him to switch to mathematics. Knuth's work at the university was so distinguished, that upon receiving his Bachelor of Science degree in 1960, the faculty made the unprecedented move of awarding him a concurrent Master's degree in mathematics.

After graduation, Knuth moved to California and enrolled at the California Institute of Technology. He was awarded in doctorate in mathematics from CalTech in 1963 and joined the faculty as an assistant professor of math. Computer science was in his blood though, and he continued his study through his consulting work with the Burroughs Corporation while working on his doctorate. In the early 1960s, language compiler theory was not well developed. In fact, most computer scientists had programmed computers using a very primitive assembly language. Language compilers are programs that convert a written program using words and structured syntax into the instruction codes that are understood by the computer hardware.

## The Art of Computer Programming

Knuth had a fascination with compilers and compiler theory and in early 1962, while still in graduate school, the publisher Addison–Wesley asked him to write a book about compilers. His project began that summer and after four years of writing and expanding the book's focus, his manuscript had grown to over 3,000 pages. His finest work, *The Art of Computer Programming* was the premier textbook through the 1970s and remains an invaluable reference and teaching tool for computer programmers. As of 1995, three of the seven volumes have been published and his fourth volume, about combinatorial algorithms, is in progress.

Knuth's development of computer science has resulted in the discovery and establishment of many fundamental rules and ideas. Included in this list is the concept of "lookahead," where a compiler will look ahead a few words to decide on a grammatical context for the prior words. He also expanded the concept of the "attribute grammar" first proposed by Backus and Naur. The idea of inherited attributes forms the basis for the object–oriented programming techniques that are so prevalent in the computer programming industry in the 1990s. But some of Knuth's most useful work has been his extensive exploration of different computer algorithms and their efficiency. He dedicated an entire volume of *The Art of Computer Programming* just to the study of algorithms.

Another interest of Knuth's has been typography. He wrote a paper called "The Letter S" in which he studied the mathematical shape of the letter throughout history and what equations lead to the most aesthetically pleasing letter. Always one to do things one thing at a time, Knuth put his other projects on hold to develop two computer languages dealing with typography. His first, called TEX, is a typesetting program; the second, METAFONT develops the shapes of the letters.

Knuth held a position as professor at Caltech until 1968, then he was offered a position at Stanford University. He continues as a professor emeritus at Stanford. In addition, Knuth has received numerous honorary degrees from many universities, including the University of Pennsylvania, State University at Stony Brook, Grinnell College, Concordia University in Montreal, the University of Paris, the Royal Institute of Technology in Stockholm, and at St. Petersburg's University in Russia.

Knuth won the Steele Prize, a high honor for computer scientists, for *The Art of Computer Programming* in 1987. He is a fellow of the Guggenheim Foundation and recipient of the National Medal of Science, presented by President Jimmy Carter in 1979. Knuth was presented with the Alan M. Turing award in 1974 and the Computer Pioneer award in 1982. Knuth was married Jill Carter in 1961 and have two children, John and Jennifer. He lives near Stanford University and continues to enjoy another of his loves: playing the pipe organ (which he designed).

Knuth's approach to the problems he has tackled in mathematics and computer science are summed up in a quotation from Shasha and Lazere, "It's not true that necessity is the only mother of invention. The other part is that a person has to have the right

background for the problem . . . The ones I solve, I say, 'Oh, man, I have a unique background that might let me solve it—it's my destiny, my responsibility'."

## SELECTED WRITINGS BY KNUTH

*The Art of Computer Programming,* 1968
*Surreal Numbers: How Two Ex–Students Turned on to Pure Mathematics and Found Happiness, A Mathenmatical Novelette,* 1974
*Computer Modern Typefaces,* 1986
*Computers and Typesetting,* 1986
*METAFONTbook,* 1986
*The TEXbook,* 1986
*Axioms and Hulls,* 1992
*Literate Programming,* 1992
*Stable Marriage and Its Relation to Other Combinatorial Problems: An Introduction to the Mathematical Analysis of Algorithms,* 1997

### Further Reading

Albers, Donald J. and Lynn A. Steen. *Mathematical People.* Boston: Birkhauser, 1985, pp. 183–203.
Shasha, Dennis E. and Cathy A. Lazere. *Out of Their Minds: The Lives and Discoveries of 15 Great Computer Scientists.* New York: Springer–Verlag, 1995, pp 89–101.

*—Sketch by Roger Jaffe*

# Pelageya Yakovlevna Polubarinova Kochina
## 1899-
### Russian applied mathematician

During Pelageya Polubarinova-Kochina's remarkable career of over 70 years, she played a major role in the worldwide development of the theory of hydrodynamics through the application of complex functions and differential equations.

Kochina was born in Astrakhan, Russia, on May 13, 1899. Her mother was Anisiya Panteleimonovna, and her father was an accountant named Yakov Stepanovich Polubarinov. Kochina had an older brother and a younger sister and brother. During her school years, the family moved to St. Petersburg (which was renamed Petrograd in 1914) to obtain the best possible education for the children. After Kochina graduated from the Pokrovskii Women's Gymnasium in 1916, she began taking courses in the

Bestudzevskii women's program, which was incorporated into the University of Petrograd (later Leningrad) following the revolution of 1917.

After her father died in 1918, Kochina worked at the Main Physics (later Geophysics) Laboratory to support her mother and younger siblings while she pursued her education. Her sister died of tuberculosis; Kochina also developed the disease, but managed to graduate in 1921 with a degree in pure mathematics from Petrograd University. She continued working at the Main Geophysics Laboratory, in the Division of Theoretical Meteorology, under the direction of A. A. Friedmann, who sparked her interest in hydrodynamics.

Russia's experience in World War I had exposed the country's industrial deficiencies, and the new Soviet government expanded research efforts to apply mathematics to technological problems. Kochina excelled in this endeavor, as did Nikolai Evgrafovich Kochin, a colleague who attended night classes at Petrograd University. The two young people shared more than a professional interest; after three years of working and vacationing together, they married in 1925. Philosophically, they espoused their country's post-revolutionary attitudes; their wedding was a simple matter of registering their marriage at a Leningrad office and then taking their witnesses to tea.

Kochina quit her job at the Main Geophysics Laboratory to raise her daughters, Ira and Nina, although she remained professionally active. During the decade prior to her appointment as a professor at Leningrad University in 1934, she taught at a workers' high school, at the Institute of Transportation, and at the Institute of Civil Aviation. She also served as a deputy in the Leningrad city soviet (legislature).

In 1935, Kochina's husband became head of the mechanics division of the Steklov Mathematics Institute, and the family moved to Moscow. Kochina turned her attention from teaching to research, becoming a senior researcher in Kochin's division. She also served in the Moscow soviet and eventually became a deputy in the Supreme Soviet of the Russian Republic. In 1939, Kochina's husband was named an Academician (full member) of the Academy of Sciences, and his division became part of the Academy's new Institute of Mechanics. In 1940, while working at the Institute, Kochina completed her dissertation on theoretical aspects of filtration and was awarded the degree of doctor of physical and mathematical sciences.

During World War II, Kochina and her daughters were evacuated 450 miles (724 km) east to the city of Kazan, while her husband stayed in Moscow doing military research. They returned after the Soviet army's 1943 victories at Stalingrad and Kursk. Kochin became ill and died on December 31, 1944;

Kochina finished delivering his course of lectures on the theory of interrupted currents.

In addition to working at the Academy of Sciences, Kochina lectured on her research activities, teaching at the Hydrometeorological and Aircraft Building Institute and at the University of Moscow's Aviation Industry Academy. In 1946, she was named a corresponding member of the USSR Academy of Sciences and awarded the State Prize of the Soviet Union. Two years later, she became director of the Institute of Mechanics' division of hydromechanics, which focused on filtration problems.

In 1958, Kochina was named an Academician of the USSR Academy of Sciences and was asked to help create a Siberian branch of that institution. The following year, at the age of 60, she left Moscow for a decade of work in Siberia. During that time, she held the positions of department director at the Hydrodynamics Institute and head of the department of theoretical mechanics at the University of Novosibirsk. She returned to Moscow in 1970 to direct the section for mathematical methods of mechanics at the USSR Academy of Sciences' Institute for Problems in Mechanics.

Although Kochina's training was in pure mathematics, her professional life was dedicated to finding solutions for practical problems in hydrodynamics. In 1952, she wrote *Theory of Ground Water Movement*; Roger De Wiest's English translation notes that "in this book, reference is made to over thirty of her original and significant contributions on the hydromechanics of porous media (groundwater and oil flow)." One major accomplishment was her development of a very general method for solving two-dimensional problems on the steady seepage of subsurface water in homogenous subsoils, which has important applications in the design of dam foundations. She obtained significant results in the theory of tides and free-flowing currents, and she solved problems relating to soil drainage and salt accumulation during her work on irrigation and hydroelectric projects. The noted topologist **Pavel S. Aleksandrov** praised her "elegant" solution to the problem of describing the location of the boundary between an oil-bearing domain and surrounding water as oil is removed by wells. She proposed the inverse problem of the theory of filtration— that is, to determine the domain of groundwater flow given a proposed construction such as a dam; since her pioneering work, the topic has been widely researched by others.

In addition to technical topics in mathematics and hydrology, Kochina was also fascinated with the history of mathematics and mechanics. She wrote the first extensive studies of **Sofia Kovalevskaia's** life and work and published descriptions of the scientific legacies of **Karl Weierstrass** and A. A. Friedmann. She also authored two biographies of her husband—

one in 1950 during the Stalin era and one in the post-Stalinist atmosphere of 1970.

On her 70th birthday, Kochina was named a Hero of Socialist Labor. She has actively participated in women's movements for peace, and on her eightieth birthday she was awarded the order of the Friendship of Nations. In 1994, Kochina delivered the opening remarks at an international conference on complex analysis and free boundary problems held in St. Petersburg, Russia, in honor of her 95th birthday.

## SELECTED WRITINGS BY KOCHINA

*Theory of Ground Water Movement.* Translated by Roger De Wiest, 1962

## FURTHER READING

### Books

Phillips, George W. "Pelageya Yakovlevna Polubarinova-Kochina," in *Women of Mathematics: A Biobibliographic Sourcebook.* Edited by Louise S. Grinstein and Paul J. Campbell. Westport, CT: Greenwood Press, 1987, pp. 95–102.

### Periodicals

Aleksandrov, P. S., et al. "Pelageya Yakovlevna Kochina: On Her 80th Birthday." *Association for Women in Mathematics Newsletter* (January-February 1982): 9–12.
"Pelageya Polubarinova-Kochina." *Biographies of Women Mathematicians.* June 1997. http://www.scottlan.edu/lriddle/women/chronol.htm (July 22, 1997).

—*Sketch by Loretta Hall*

# Kunihiko Kodaira
## 1915-
### Japanese analyst

The high reputation of Japanese mathematics owes a great deal to the work of the analyst Kunihiko Kodaira, both for his specific contributions and for his interest in education. Through his pioneering research in algebraic varieties, harmonic integrals, and complex manifolds, Kodaira became the first Japanese mathematician ever awarded the Fields

*Kunihiko Kodaira*

Medal, as well as being frequently honored in his own country. He has provided the entire mathematical community with a legacy of research papers on a variety of subjects and textbooks for all ages.

Born March 16, 1915, in Tokyo to Gonichi and Ichi (Kanai) Kodaira, Kodaira grew up in Japan's biggest city and remained there when he began his university education in 1935. At the University of Tokyo, he maintained a wide variety of interests in various mathematical fields and published his first research paper, written in German, a year before he took his first undergraduate degree. His early influences were the works of M. H. Stone, **John von Neumann**, W. V. D. Hodge, **André Weil** and, most importantly, **Hermann Weyl**, whose work on Riemann surfaces was very well known. Kodaira was fascinated by algebraic geometry, in spite of his analytic bent, and he found the impetus for much of his work in the book *Algebraic Surfaces* by the Italian geometer Oscar Zariski. In 1938, Kodaira earned his undergraduate degree in mathematics; three years later, he earned a second in theoretical physics.

By the time Kodaira began his Ph.D. studies in 1941, World War II had isolated Japan from the rest of the world. Much of the work he did during this period was in ignorance of some of the most recent and important advances in his field, yet Kodaira still managed to solve some difficult problems. In fact, the major ideas that would form his doctoral thesis, his first important work, were present in a paper written in the depths of the war. The end of the war and peace with the Americans afforded Kodaira new opportunities to continue his research. In 1943, he married Sei Iyanaga; they have four children.

## Advances Career in America

Kodaira worked on his dissertation while teaching at the University of Tokyo, where he had been appointed associate professor in 1944. His thesis covered the relation of harmonic fields to Riemann manifolds, a type of mathematical surface whose definition Kodaira would help establish. In mathematics something is defined as harmonic if, by examining only the boundary of the surface, the interior of that surface can be described. Kodaira wanted to use this idea to uncover the makeup of the Riemann manifold; his thesis was the beginning of this work.

He received his doctorate in 1949; during the same year his dissertation was published in an international mathematical journal, which Hermann Weyl read. Impressed with the article, Weyl became interested in the Japanese mathematician, and in 1949 he invited Kodaira to join him at Princeton's Institute for Advance Studies (IAS). Although Kodaira did not relish leaving his homeland, he was deeply honored by the invitation. Many of the world's top mathematicians were at the IAS, and Kodaira felt his best opportunities were there, so he made up his mind to bring his young family to the United States.

Through various collaborations with IAS mathematicians such as W. L. Chow F. E. P. Hirzebruch, and especially D. C. Spencer, Kodaira continued his work on manifolds. He discovered that, by using a type of integral known as a harmonic integral, he could more completely define the Riemann manifolds, and he sought to generalize this knowledge. Complex manifolds, which are those involving complex numbers, form the basis of much of modern calculus; however, at the time, little was understood about their properties because many were not defined. Therefore, Kodaira's studies were critical to the advancement of the field of calculus. After a great deal of work on Riemann manifolds, he turned to another type of manifold called the Kählerian manifolds, where he would produce his most spectacular results.

Kodaira wanted to prove that the Kählerian manifolds, like the Riemann manifolds, were analytic in nature—in other words, that calculus could be used to solve or define them. He began to examine a small subset of the Kählerian manifolds known as the Hodge manifolds. Using a theorem he had created earlier, called the vanishing theorem, he successfully proved the existence of meromorphic functions, a type of analytic function, on the Hodge manifolds. These meromorphic functions could be solved using algebraic varieties, a set of points that satisfy certain polynomial equations, thus making the Hodge manifolds analytic. By extension, then, in a theorem Kodaira labelled the embedding theorem, he proved that if the Hodge manifolds were analytic, then all Kählerian manifolds were analytic as well.

Because each manifold is different and because they are so crucial to the theory of modern calculus, any theory or set of theories that can classify an entire group of them constitutes a major advance. For his work on algebraic varieties and Kählerian manifolds, Kodaira received the Fields Medal in 1954, which was presented to him in Amsterdam by Kodaira's friend and mentor, Hermann Weyl. Three years later, his native country followed with two of their most prestigious awards: the Japan Academy Prize and the Cultural Medal, given by the government of Japan.

Kodaira's research continued in the midst of all the attention he received. Between 1953 and 1960, he further examined complex manifolds with D. C. Spencer, using the idea of deformations to refine his definitions. A deformation is a function which twists or bends a surface without tearing it; this function is important for both theoretical and practical reasons. Theoretically, deformations are one of the fundamental ideas of topology; practically, they are applied in mechanical engineering to describe the bending of metals. Therefore, Kodaira was applying an extremely practical and concrete concept to a very abstract theory of surfaces.

Much in demand because of his high reputation, Kodaira spent the next several years as visiting professor at prestigious American universities. He remained at the IAS and Princeton until 1961, then spent a year at Harvard, and two years at Johns Hopkins; he taught at Stanford from 1965 to 1967. In 1967, he returned to Japan, accepting a position as full professor at the University of Tokyo.

## Focuses on Education

Although Kodaira continued his research after his return to Japan, he became increasingly interested in the teaching of mathematics. In 1971, he collaborated with James Morrow on a textbook based on his research into complex manifolds, the first of several such works. In 1975, he took on additional teaching responsibilities at Gakushuin University in Tokyo. Also in that year, Kodaira's collected works appeared and he was granted emeritus status at the University of Tokyo. As an additional honor in a long list, he won the Fujihara Foundation of Science Prize for his theory of complex manifolds.

In the early 1980s, Kodaira joined a government-sponsored project to produce mathematics textbooks for students from grades seven to eleven. Kodaira produced the compulsory curriculum for grades seven through nine. These texts provided a weighty mathematics background with the intent of preparing Japanese students for any career. Kodaira's texts appeared in 1984 and were later published in translation in the United States in 1992 as an example of the high quality of foreign texts.

Just before his retirement from teaching in 1985, the Wolf Foundation of Israel awarded Kodaira their mathematics prize for his contributions to the field of complex manifolds. Even after retirement, Kodaira remained interested in mathematics, and he published another work which summarized his theory of complex manifolds in 1986.

## SELECTED WRITINGS BY KODAIRA

### Books

(With James Morrow) *Complex Manifolds,* 1971
*Collected Works,* 3 volumes, 1975
*Introduction to Complex Analysis,* 1978
*Complex Manifolds and Deformation of Complex
    Structures,* 1986
*Japanese Grade 7, 8, 9 Mathematics,* 3 volumes,
    1992

## FURTHER READING

### Books

Baily, W. L., and T. Shioda, editors. *Complex Analysis and Algebraic Geometry.* Cambridge, UK: Cambridge University Press, 1977.
Spencer, D. C., and S. Iyanaga. *Global Analysis: Papers in Honor of K. Kodaira.* Princeton, NJ: Princeton University Press, 1969.

### Periodicals

Weyl, H. "On the Work of Kunihiko Kodaira." Paper presented at the 1954 International Congress of Mathematicians in Amsterdam.

*—Sketch by Karen Sands*

*Andrei Nikolaevich Kolmogorov*

# Andrei Nikolaevich Kolmogorov
## 1903-1987
### Russian probabilist and educator

Andrei Nikolaevich Kolmogorov made major contributions to almost all areas of mathematics and many fields of science and is considered one of the 20th century's most eminent mathematicians. He was the founder of modern probability theory, having formulated its axiomatic foundations and developed many of its mathematical tools. Kolmogorov also helped make advances in many applied sciences, from physics to linguistics. A great teacher, he did much to keep the Soviet Union in the forefront of research in theoretical and applied mathematics and was responsible for reforms in mathematics education at the elementary and high–school levels.

Kolmogorov was born in the town of Tambov in central Russia on April 25, 1903. His father, Nikolai Kataev, became a professional agriculturalist and was killed during World War I. His mother, Mariya Yakovlevna Kolmogorova, was not formally married to his father and died during his birth. Her sister, Vera Yakovlevna Kolmogorova, adopted and raised the boy in the family's home village of Tunoshna. As a child, young Kolmogorov and his friends attended a school run by his two aunts. At the age of five, he made his first mathematical discovery by noticing the pattern that $1=1^2$, $1+3=2^2$, $1+3+5=3^2$, etc.

In 1920, at the age of 17, Kolmogorov enrolled in Moscow University. To help support himself while he attended the university, he worked as a secondary school teacher. He took an active role in the school, and he is said to have been more proud of that work than of the honors he garnered for his own academic progress. Within two years, Kolmogorov had completed a study in the theory of operations on sets, which was eventually published in 1928. A second project he also completed in 1922 brought immediate recognition: He formulated the first known example of an integrable function with a Fourier series that diverged almost everywhere (he soon extended that result to *everywhere*). The international mathematics community took notice of the bright 19–year–old. During his years as a university student, he published 18 mathematical papers including the strong law of large numbers, generalizations of calculus operations, and discourses in intuitionistic logic. In 1925, Kolmogorov received a doctoral degree from the department of physics and mathematics and became a research associate at Moscow University. At the age of 28, he was made a full professor of mathematics; two years later, in 1933, he was appointed director of the university's Institute of Mathematics. In 1942, Kolmogorov married Anna Dmitrievna Egorova.

## Formulates Axiomatic Basis of Modern Probability Theory

While he was still a research associate, Kolmogorov published a paper, "General Theory of Measure and Probability Theory," in which he gave an axiomatic representation of some aspects of probability theory on the basis of measure theory. His work in this area, which a younger colleague once called the "New Testament" of mathematics, was fully described in a monograph that was published in 1933. The paper was translated into English and published in 1950 as *Foundations of the Theory of Probability*. Kolmogorov's contribution to probability theory has been compared to **Euclid**'s role in establishing the basis of geometry. He also made major contributions to the understanding of stochastic processes (involving random variables), and he advanced the knowledge of chains of linked probabilities.

Kolmogorov developed many applications of probability theory, creating a powerful technique for using probability to make observa- tion–based predictions in the face of randomness and researching statistical inspection methods for mass production. One of the applications that Kolmogorov developed is known as reaction–diffusion theory, which deals with the manner in which an event spreads through a given population. It is now used to study the dispersion of epidemics, cultural changes, effects of advertising, and a variety of other situations in such fields as biology and chemistry. He was largely responsible for demonstrating the applicability of the emerging fields of probability and statistics on substantial problems in science and engineering. In fact, Kolmogorov contributed to the war effort in the 1940s by solving statistical problems relating to artillery fire and by applying stochastic theory to suggest the most effective placement of barrage balloons for protecting Moscow from Nazi bombing assaults.

## Leaves His Mark on Many Fields

An appointment as Chair of Theory of Probability at Moscow University in 1937 served as official recognition of Kolmogorov's achievements in probability theory. Communication with the West was sporadic, however, and it was not until the late 1950s that Western mathematicians discovered that Kolmogorov had already determined the nature of many issues in probability theory that they were still working to discover. In 1939, at the age of 36, Kolmogorov became one of the youngest full members elected to the Soviet Academy of Sciences. He was later appointed academician–secretary of the Academy's department of physical and mathematical sciences. These honors were in recognition not only of his work in probability theory but of his contributions to other areas of theoretical and applied mathematics.

Kolmogorov made significant contributions to set theory, measure theory, integration theory, topology, functional analysis, constructive logic, differential equations, and the theory of approximation of functions. Among his many accomplishments in applied fields, Kolmogorov developed important results in such areas as biological statistics, econometrics, mathematical linguistics, and the theory of fluid turbulence. His work in fluid turbulence was so profound that in 1946 he was chosen to head the Turbulence Laboratory at the Academy Institute of Theoretical Geophysics. He continued to work in this field for many years, sailing around the world in 1970–1972 to study ocean turbulence. He helped construct the Kolmogorov–Arnold–Moser (KAM) theorem, which is used to analyze stability in dynamic systems. Kolmogorov introduced the concept of entropy (a theoretical measure of unavailable energy in a thermodynamic system) as a measure of disorder. In the first detailed solution of the three-body problem, which had been proposed by **Isaac Newton,** Kolmogorov analyzed the interactions of two celestial bodies orbiting in the same plane, one with an elliptical orbit and the other with a circular path.

In an obituary published in *Physics Today,* V. I. Arnold of the Steklov Mathematical Institute wrote that "Kolmogorov considered his most difficult achievements to be his work from 1955 to 1957 on the 13th **[David] Hilbert** problem." The problem involves finding a way to represent a function of many variables in terms of a combination of functions having fewer variables.

In a remembrance of Kolmogorov published in *Statistical Science,* A. N. Shiryaev wrote that "One sensed that he had continuously intensive brain activity." His active intellect led him to investigate questions in a wide range of mathematical fields as well as a number of applied subjects including meteorology, hydrodynamics, celestial mechanics, genetics, history, and linguistics. Shiryaev wrote, "According to his own words, Kolmogorov had a lively interest in a problem only until it became clear what the answer should be. As soon as the picture became clear he tried to avoid writing down the results and proofs; he would look for someone else to take over." Indeed, by the time Kolmogorov solved a problem, he would have identified other topics to investigate. Shiryaev quoted the mathematician A. Ya. Khinchin as saying, "The most important and most fascinating feature of [Kolmogorov] as a mathematician is the wealth of his ideas. Each sentence of his about any work could become the basis for a Ph.D. dissertation." Also speaking at a 50th birthday tribute to Kolmogorov, **Aleksandr Gelfond** said, "The fact that mathematics is viewed as a unified discipline is due to a large extent to Kolmogorov."

Kolmogorov was also actively interested in mathematical education in the U.S.S.R., working as the chairman of the Academy of Sciences Commission of Mathematical Education. He played a pivotal role in overhauling the teaching of mathematics during the 1960s, and his leadership in mathematics education for secondary schools and universities helped move the U.S.S.R. to the forefront of mathematics internationally during the following decades. In fact, being of the opinion that no mathematician could possibly do meaningful research after the age of 60, Kolmogorov retired in 1963 and spent the following 20 years teaching high school. During his final years, he compiled his collected works; an English version was published in the United States in 1991. He died in Moscow on October 20, 1987, at the age of 84.

## SELECTED WRITINGS BY KOLMOGOROV

*Foundations of the Theory of Probability,* 1950
*Selected Works of A. N. Kolmogorov.* 2 volumes. 1991, 1992

## FURTHER READING

### Books

"Andrey Nikolayevich Kolmogorov." *The New Encyclopedia Britannica.* 15th Edition Micropaedia, Volume 6. Chicago: Encyclopedia Britannica, Inc., 1997, pp. 940–41.
Boyer, C., and U. Merzbach. *A History of Mathematics.* New York: John Wiley & Sons, 1990.

### Periodicals

Arnold, V. I. "A. N. Kolmogorov" (obituary). *Physics Today* 42 (October 1989): 148, 150.
Chentsov, Nikolai N. "The Unfathomable Influence of Kolmogorov." *The Annals of Statistics* 18 (September 1990): 987–95.

*Sonya Vasilievna Kovalevskaya*

Kendall, David G. "Kolmogorov As I Remember Him." *Statistical Science* 6 (August 1991): 303–12.
Shiryaev, A. N. "Everything About Kolmogorov Was Unusual. . . ." *Statistical Science* 6 (August 1991): 313–18.

*—Sketch by Maureen L. Tan and Loretta Hall*

# Sonya Vasilievna Kovalevskaya (also transliterated as Sofia Vasilevna Kovalevskaia) 1850–1891
### Russian mathematician and educator

Sonya Vasilievna Kovalevskaya has been applauded by some as the most astounding mathematical genius to surface among women in the last two centuries, and one of the first women to make contributions of high quality to the field.

The middle of three children, Kovalevskaya was born on January 15, 1850, in Moscow. Her father, Vasilli Korvin-Krukovski, was an Artillery General in the Russian army. He was a educated and disciplined man who was fluent in English and French, and was a stern but benevolent parent. Her mother, Elizaveta Fyodorovna Schubert, came from a family of German scholars who had emigrated to Russia in the mid–1700s. Kovalevskaya's grandfather, Fyodor Fyodorovich Schubert, and great–grandfather, Fyodor Ivanovich Schubert, were noted mathematicians.

A singular incident in Kovalevskaya's childhood seems to have been a portent of her devotion to the study of mathematics. While living at the family estate, she came upon a room where the wallpaper consisted of sheets of Mikhail Ostrogradsky's lithographed lectures on the differential and the integral calculus. The child spent hours trying to decipher the formulae. Years later, at the age of 15, she astonished her tutor with how quickly she grasped and assimilated the conceptions of differential calculus. Kovalevskaya wrote in her memoirs that she "vividly remembered the pages of Ostrogradsky . . . and the conception of space seemed to have been familiar to me for a long time."

It was also rather exceptional that her father allowed Kovalevskaya to study with a tutor at all. She described her father as one who "harbored a strong prejudice against learned women." It has been suggested that the best explanation is that Kovalevskaya's own fierce determination was the catalyst for changing her father's mind. Once the tutelage was over, however, she faced an uncertain future for obtaining advanced education. She knew her father would never agree to sending her to a university. During that time, women were not allowed to attend the universities in Russia and most fathers, including Kovalevskaya's, were unwilling to give consent to daughters to study abroad. Again, Kovalevskaya's determination was stronger than her father's will.

The device she used to get her way was a popular one at the time; she began searching for a husband. The type of husband she was looking for had to agree to sign papers allowing Kovalevskaya to travel, live apart from him, and pursue an education. The agreement also came with the understanding that the marriage was a platonic one, without the marital rights usually afforded a husband. She found such a man in Vladimir Kovalensky, who made his living translating and publishing books while pursuing a degree in paleontology. Along with his high intellect, Kovalensky also distinguished himself by supporting liberal causes.

The resistance by Kovalevskaya's family to the marriage was anticipated and overcome by using the same sort of guile that had created the situation. She sent notes to a number of distinguished family friends happily announcing her impending marriage, thereby forcing her father to either give public approval or publicly admit that his daughter was rebellious. To make certain her father would not renege on the announcement, Kovalevskaya ensconced herself in Kovalensky's apartment, refusing to leave, until she felt secure that the marriage would indeed take place. The couple was married in September 1868.

### Studies with the Masters

In early 1869 the newlyweds left Russia and settled in Heidelberg, Germany. This was where Kovalevskaya was to fulfill her dream of a higher education. Because she was a woman, the officials at the University of Heidelberg demanded that she secure the written permission of each of her professors before full admittance was granted. She undertook a class schedule of 22 hours per week, 16 of which were spent studying mathematics with Paul Du Bois–Reymond and Leo Köenigsberger, both of whom were students of the renowned mathematician **Karl Weierstrass**. After three successful semesters of study at the university, Kovalevskaya left for Berlin, seeking out Weierstrass. Their initial meeting marked a personal as well as professional relationship that lasted a lifetime.

Kovelevskaya did not arrive in Berlin unannounced, however. The praises of her professors at Heidelberg preceded her, and this did much to persuade Weisertrass' decision to become Kovelevskaya's mentor. In addition, Weierstrass had written a paper where he gave credit to Kovalevskaya's grandfather, Fyodor Schubert, for a mathematical maxim eight years before meeting Kovelevskaya. Unfortunately, winning Weierstrass' acceptance was not the only obstacle to the continuation of her studies—the university forbade women from attending Weierstrass' formal lectures. The obstacle was removed when Weierstrass agreed to teach Kovalevskaya privately twice a week, giving her the same courses as his regular university students.

In the beginning, Weierstrass never imagined that Kovalevskaya would want a formal degree in mathematics, believing that a married woman would have no use for one. In the fall of 1872 however, Kovalevskaya confided the truth about her marriage and he began to steer her toward work on a dissertation. By the spring of 1874, Kovalevskaya had written three doctoral dissertations, each of them in Weierstrass' opinion worthy of a degree, and one so outstanding that both were confident of forthcoming recognition. They were not disappointed. Weierstrass submitted Kovalevskaya's work to the University of Göttingen and she was awarded her doctoral *summa cum laude* in the fall of 1874, becoming the first woman to earn a doctorate in mathematics.

### Confronted Detours Along the Way

Elated but exhausted by her labors, Kovalevskaya and her husband returned to Russia to relax with friends and family. Both were also hoping to secure positions in the academic world, but for a combination of reasons neither was welcomed to university posts. Kovalevskaya found herself discriminated against because of her gender, and Kovalensky's liberal activities spawned suspicion among Russian academics. Kovalevskaya and her husband decided to consummate their relationship, and Kovalevskaya's only child, Sofia, was born in 1878. For the next five years, Kovalevskaya and her husband put aside their respective fields of study and concentrated on trying to make a living at various commercial endeavors.

During this time it became apparent that Kovalevskaya was as gifted at writing as she was at mathematics and for a time her heart was torn between the two pursuits. The fiction she produced, including the novella *Vera Barantzova*, were met with acclaim and translated into several foreign languages. Meanwhile, Kovalensky was involved in questionable financial dealings. Faced with prosecution from charges of mishandling stock, Kovalensky committed suicide in April of 1883.

### Recognition at Last

Kovalevskaya returned to her study of mathematics and through the efforts of a friend and fellow student of Weierstrass, Gosta Mittag-Leffler, Kovalevskaya was offered a position at Stockholm University as a *privatdozent* (a licensed lecturer who could receive payment from students but not from the university) in 1884. Five years later, Kovalevskaya became the first female mathematician to hold a chair at a European university. This appointment was accompanied by the editorship of the journal *Acta Mathematica*, where she came in contact with the leading European mathematicians of the day. In 1888, Kovalevskaya's paper on the study of the motion of a rigid body received the Prix Bordin, given by the French Académie Royale des Sciences.

Kovalevskaya died in 1891 of influenza when she was only 41 years old, at the height of her mathematical career. Although she published only ten papers during her lifetime, Kovalevskaya's work has withstood the test of time. The research that won her the Prix Bordin is now known as the Kovelevskaya top, and her doctoral dissertation on partial differential equations lives on as the Cauchy-Kovelevskaya Theorem.

### SELECTED WRITINGS BY KOVALEVSKAYA

*Scientific Works,* 1948

### FURTHER READING

#### Books

Bell, Eric Temple. "Master and Pupil: Weierstrass and Sonya Kovalevskaya." *Men of Mathematics.* New York: Simon and Schuster 1937, pp. 423–429.

Cooke, Roger. *The Mathematics of Sonya Kovalevskaya.* New York: Springer–Verlag, 1984.

Kennedy, Don H. *Little Sparrow.* Athens: Ohio University Press, 1983.

Koblitz, A.H. *A Convergence of Lives: Sophia Kovaleskaia, Scientist, Writer, Revolutionary.* Boston: Birkhäuser, 1983.

Koblitz, Ann Hibner. "Sonya Kovalevskaya," in *Women for Mathematics: A Biobibliographic Sourcebook.* Edited by Louise S. Grinstein and Paul J. Campbell. Westport, CT: Greenwood Press, 1987, pp. 103–113.

Osen, Lynn M. "Sonya Corvin–Krukovsky Kovalevsky." *Women in Mathematics.* Cambridge, MA: MIT Press, 1979, pp. 117–140.

#### Periodicals

Tabor, Michael. "Modern Dynamics and Classical Analysis." *Nature* 310 (July 26, 1984): 277–282.

#### Other

Wilson, Becky. "Sonya Kovalesvskya." *Biographies of Women Mathematicians.* June 1977. http://www.scottlan.edu/lriddle/women/chronol.htm (July 22, 1997).

*—Sketch by Kelley Reynolds–Jacquez*

# Cecilia Krieger
## 1894–1974
### Polish–born Canadian topologist

Cecilia Krieger is best known as the woman who translated the work of **Wacław Sierpiński** from Polish to English, returning to her native land to see him in 1931 while working on the project. She was also awarded the third mathematics doctorate ever in Canada, and the third woman in the world to earn a Ph.D. Though she never rose above the rank of assistant professor, Krieger taught for over thirty years and is credited with urging **Cathleen Synge Morawetz** to take up a career in mathematics. Her studies in set theory and topology as originated by

**Georg Cantor** reach into the heart of what we now know as the interdimensional geometry of fractals.

Cypra Cecilia Krieger was born in Poland on an unknown date in 1894. She was one of five children born to a Jewish merchant named Moses and his wife Sarah. There were three daughters; Cecilia as she was called, Regina, and Rae. The sons were Samuel, who would later sponsor Cecilia's immigration to Canada in 1920, and Nathan. After early schooling in Poland, Krieger studied mathematical physics at the University of Vienna for a year. When she arrived in Toronto, however, she did not know a word of English.

### "It Is Really Easy"

Much was made in the local Canadian press about the fact that only four years later Krieger took her B.A. at the University of Toronto, supporting herself by working at the Muskoka Inn in the lake district. Her master's degree in modern elliptical functions, number and set theory, and the minimum principles of mechanics followed one year later. During the years 1928 and 1929, Krieger was a Ph.D. instructor in math and physics. She became the first woman in Canada to take a Ph.D. the same year she was promoted to lecturer, in 1930. Her advisor, W.J. Webber, was chiefly responsible for the growth and development of graduate studies in his department during this time.

Krieger had already begun publishing in 1928, but her 1934 translation of Sierpiński's second volume on topology, the first into English, was earned her recognition. When asked how difficult it had been to come to Canada, learn English and obtain her degrees in such a short period of time, Krieger demurred, prompting the reporter to declare her "a miracle of modesty." She replied that other languages are harder to learn than English. Krieger's English version of *General Topology* also included a 30–page outline of volume one in Sierpiński's series "The Theory of Aggregates," called *Transfinite Numbers*. The full–length volume one only existed in Polish and French at that point.

Unfortunately her father did not live to see this triumph, having died at age 70 in 1929 from complications following a streetcar accident.

### Touched Twice by the Holocaust

Following the death of her father in 1929, Krieger did her best to help her family, supporting her unmarried sisters throughout the 1930s and the early 1940s. Krieger also "adopted" the family of Alex Rosenberg, immigrant Jewish refugees, during World War II. After 12 years at the University College, in 1942, she received her last promotion to assistant professor. Krieger averaged a course load of 13 classes a week, ranging from six to 75 students per class.

Krieger married late in life to a Holocaust survivor, the physician Zygmund Dunaij, in 1953. She continued to teach, but also spent some time in the United States and was good friends with the Synge family in New York. However, this period is not widely documented. Her husband died in 1968. Krieger then taught in a private boy's preparatory school in upper Canada from 1969 until her death at age 80.

After spending her early life during a formative period for university mathematics, being female was just one of many firsts in the field for Krieger. When asked about women's prospects for success in mathematics, she had declined to generalize. "It depends upon the individual," she concluded.

## SELECTED WRITINGS BY KRIEGER

### Books

(Translator) *General Topology*, by Wacław Sierpiński, 1934

### Periodicals

"On the Summability of Trigonometric Series with Localized Properties." *Royal Society of Canada Transactions* 22, no. 3 (1928): 139–147.

"On Fourier Constants and Convergence Factors of Double Fourier Series." *Royal Society of Canada Transactions* Series 3 24, section 3 (May 1930): 161–196.

## FURTHER READING

### Books

Anand, Kailash K. "Cypra Cecilia Krieger and the Human Side of Mathematics." *Despite the Odds: Essays on Canadian Women and Science.* Edited by Marianne Ainley. Montreal, P.Q., Canada: Vehicle Press, 1990, pp. 248–251.

"Cathleen S. Morawetz." *More Mathematical People.* Edited by D.J. Albers, G.L. Alexanderson and C. Reid, pp. 221–238.

### Periodicals

*The Toronto Monthly* and other Toronto periodicals, 1924–1934, in the collection of the Thomas Fisher Rare Book Library, University of Toronto Archives.

### Other

"Cecilia Krieger." *Biographies of Women Mathematicians.*

http://www.scottlan.edu/lriddle/women/chronol.htm
(July 20, 1997).

—*Sketch by Jennifer Kramer*

# Krystyna Kuperberg
## 1944–
### Polish topologist

Krystyna Kuperberg is a researcher and educator best known for disproving the famous Seifert conjecture in topology. Her counterexample, first announced in the mid–1990s, was termed a "small miracle" of geometry by Ian Stewart. It was quickly generalized and should prove central to the continued development of dynamic systems theory, by way of the vector fields used to study physical and statistical phenomena.

Kuperberg was born Krystyna M. Trybulec in Tarnow, a city in southern Poland, on July 17, 1944. Her parents, Jan W. and Barbara H. (Kurlus) Trybulec, were both trained pharmacists. Her brother, Ardrzej, also became a mathematician. After receiving a master's degree from Warsaw University in 1966, Kuperberg had to wait until settling in the United States to earn her Ph.D. This was awarded by Rice University in 1974. Upon graduating she accepted her first post at Auburn University. Kuperberg remains a member of the faculty at Auburn, and has been a full professor there since 1984. She has also held visiting positions at Oklahoma State University, the Courant Institute of Mathematical Sciences in New York, the Mathematical Sciences Research Institute at Berkeley, and l'Universite de Paris–Sud, Centre d'Orsay.

In 1974, the first counterexample to the famous Seifert conjecture was found by P. A. Schweitzer. His "plug" was devised to cancel out any circular orbit, but it broke down to two minimal sets. Kuperberg, who had begun publishing papers in 1971, was already interested in dynamical systems. She resolved a conjecture about fixed points in 1981 with Coke Reed, and built upon the methods used in this work to find a new kind of counterexample with only one minimal set. What she eventually found served to disprove the Siefert conjecture for all three–dimensional manifolds.

### Hairy Donuts

The Seifert conjecture is a higher–dimensional extension of the "hairy billiard ball" theorem for the two–dimensional surface of a sphere. The idea that you cannot comb down all the hairs on a fully hairy ball without getting a cowlick is really a geometric statement about a dynamical system. The one–dimensional version, the circle or "1–sphere," is "combable," allowing for a smooth vector field. The 2–sphere is combable because it contains at least one "bald spot" consisting of a fixed point around which the trajectories can flow. The fact that there is always some place on Earth where the wind is not blowing is a real–world example of this "bald spot."

More complex surfaces proved more difficult to analyze. In the case of a torus–shaped vector field or "hairy donut," for instance, whether a trajectory is fixed or not depends upon how it advances along the circumference of the torus as it flows. This explains why it was impossible to prove a simple conjecture about one of the three–dimensional shapes for more than 40 years. In 1950, Herbert Seifert had proposed that in the three–dimensional case any smooth vector field will have at least one "closed" or periodic orbit. It was already known that 3–spheres did not have any "bald spots," but it seemed reasonable enough to think that they would have at least one closed orbit.

Kuperberg disproved this conjecture in 1993 by constructing a smooth vector field with no closed orbits, and her construction applies not only to 3–spheres, but to all three–dimensional manifolds. To do this she used a Wilson plug, a kind of topological tool, to break up any closed orbits that might be present. This plug is a three–dimensional shape that traps the trajectories of one or more formerly closed orbits inside itself. The trick is to apply the plug without creating any new closed orbits. To accomplish this, Kuperberg modified the plug so that it "eats its own tail" like a snake. Thus, the trajectories that enter get trapped in an infinite spiral and no new closed orbits can be formed. In addition to disproving the Siefert conjecture, Kuperberg's construction produces a "minimal set" that may be of an entirely new kind, according to John Mather, a dynamical systems theorist at Princeton. Since minimal sets are basic components of dynamical systems, Kuperberg's plug may help mathematicians better understand the range of things that can happen in these systems.

Since her success in disproving the Siefert conjecture, Kuperberg has been especially in demand as a speaker at events devoted to topology or dynamical systems, and at honorary symposia worldwide. She delivered the MSRI–Evans lecture at Berkeley in 1994, and addressed the American Mathematical Society and Mathematical Association of America meetings in 1995 and 1996.

Kuperberg's husband and frequent collaborator, Włodzimierz, received his Ph.D. in mathematics from Warsaw University. He is also a professor at Auburn University. Krystyna and Włodzimierz were married in Poland and lived there until 1969. Their son, Greg, born in Poland in 1967, is also a mathematician; he received a Ph.D. in mathematics from the University of California at Berkeley. Their daughter, Anna, born two years later in Sweden, holds a M.F.A. from the San Francisco Art Institute.

In addition to awards from Auburn University for her research and professorship, and National Science Foundation grant support, Kuperberg also won the Alfred Jurzykowski Foundation Award in 1995 from the Kosciuszko Foundation in New York. In 1996, Kuperberg was elected to the American Mathematical Society council as a member at large for a three–year term. She also currently edits the *Electronic Research Announcement* of the American Mathematical Society.

## SELECTED WRITINGS BY KUPERBERG

### Books

*Collected Works of Witold Hurewicz*, 1995
(With H. Cook, W.T. Ingram, A. Lelek, and P. Minc) *Continua with the Houston Problem Book*, 1995

### Periodicals

"A Smooth Counterexample to the Seifert Conjecture." *Annals of Mathematics* 140 (1994): 723–32.
(With W. Kuperberg) "Generalized Counterexamples to the Seifert Conjecture." *Annals of Mathematics* 144 (1996): 239–68.

## FURTHER READING

### Periodicals

Cipra, Barry. "(Vector) Field of Dreams." *What's Happening in the Mathematical Sciences* 2 (1994): 47–51.
Stewart, Ian. "Hairy Balls in Higher Dimensions." *New Scientist* (November 13, 1993): 18.

### Other

"Krystyna Kuperberg." *Biographies of Women Mathematicians*. June 1997. http://www.agnesscott.edu/lriddle/women/chronol.html (July 1997).

—*Sketch by Jennifer Kramer*

*Thomas Eugene Kurtz*

# Thomas Eugene Kurtz
## 1928-
### American statistician and computer scientist

Thomas Eugene Kurtz, cofounder of True BASIC, Inc., was a professor of mathematics and computer science at Dartmouth College for 37 years. During that time, he and John G. Kemeny, with whom he collaborated on many projects, designed and developed the Dartmouth Time Sharing System (DTSS) and the computer programming language, Beginner's All-purpose Symbolic Instruction Code, or BASIC. For those accomplishments, Kurtz and Kemeny received the first Pioneer's Day award from the American Federation of Information Processing Society in 1974.

Kurtz was born on February 22, 1928, in Oak Park, Illinois, to Oscar Christ Kurtz, who worked in various capacities at the International Lion's Club headquarters, and Helen Bell Kurtz. Interested in science from his youth, Kurtz entered Knox College in Galesburg, Illinois, with the intention of majoring in physics. He also took all of the mathematics courses available. Following the suggestion of an adviser to consider a career in statistics, which would allow him the opportunity to apply his mathematical skills to many different scientific problems, Kurtz switched majors in his senior year and graduated in 1950 with a B.A. in mathematics.

Kurtz earned his graduate education at Princeton University, where his interest in computing was forged by Forman Acton, a professor of engineering. Acton made it possible for him to spend the summer of 1951 at the Institute of Numerical Analysis, a branch of the National Bureau of Standards located on the University of California at Los Angeles (UCLA) campus. There, in addition to attending lectures on computing, Kurtz interacted with a number of the early computer pioneers, many of whom frequented UCLA during the summer.

From 1952 to 1956, Kurtz served as a research assistant in the Analytical Research Group at Princeton, where he wrote programs to help solve classified research problems, such as those concerned with the effectiveness of air-to-air rocket salvos. The programs were run on an IBM Card Programmed Calculator, and occasionally his job involved tending the machine throughout the night, transferring cards from the output bin back to the input hopper.

Upon graduating from Princeton in 1956 with a Ph.D. in mathematical statistics, Kurtz was recruited by John G. Kemeny, who was chair of Dartmouth's mathematics department. Though Kemeny had previously taught at Princeton until 1953, and had even lived a short distance away from Kurtz at one point, the two scientists had not met before Kurtz was recruited. One of Kurtz's first assignments was as liaison to the New England Regional Computer Center, which had been established at the Massachusetts Institute of Technology (MIT) with funding from IBM and had the provision that educational institutions in the northeast could have access to its facilities. Kurtz spent August of 1956 at MIT, learning assembly language programming—the language that the machine understands—for the center's IBM 704, which was the first commercially available machine with a magnetic core memory.

### Develops Dartmouth Time Sharing System

In 1959 Dartmouth finally purchased its own computer, an LGP–30, and Kurtz was appointed director of computing. Initially, the computer was used by just a small fraction of the Dartmouth student body and faculty, but Kurtz felt that all students should be able to use the computing facilities. One of the drawbacks to the widespread use of computers in the late 1950s was that users could not reserve time on a given machine, but had to submit their programs to be processed. The computer would run each request in the order it was received and then store the result. Such "batch processing" meant that users had to wait as much as a day or more to see their results, so that debugging a program could turn into a lengthy and frustrating process. "Time sharing," which allows many people to use a computer simultaneously by having the computer work on each person's problem

for short periods of time, avoided the delay, but general-purpose time sharing systems were not available.

In February of 1964, Kurtz and Kemeny began developing a time sharing computer system with the General Electric Corporation. Completed in June, the Dartmouth Time Sharing System, one of the first general-purpose systems of its kind, was comprised in part of one GE–235, which served as the central processors, as well a GE Datanet–30, which handled communications with terminals all over campus. The goal of the project was to make access to computing as simple as checking out a book in the college library. It gave all Dartmouth students, as well as students from area colleges and schools, access to the computer whenever they wished, without the bureaucratic obstacles of forms, permission, and restricted hours. To ensure that this democratic approach worked in practice, the computer gave precedence to small jobs (typically student's programs) as opposed to large ones (typically those submitted by the faculty).

### Develops BASIC computer programming language

Having removed one of the primary barriers to computer use, Kurtz and Kemeny went on to simplify the user interface, so that a student could essentially learn enough to use the system in an hour or less. But writing programs in the computer languages then in use was a more challenging task. Though Kurtz initially tried to simplify certain existing languages, namely ALGOL and FORTRAN, Kurtz and Kemeny decided that a new, simplified programming language was needed. The resulting programming language was called Beginner's All-purpose Symbolic Instruction Code, or BASIC, and has become the most widely used language in the world. In 1983, Kurtz, Kemeny and several others formed True BASIC, Inc., with the intention of creating a personal computer version of BASIC for educational purposes.

Between stints as professor of mathematics and computing at Dartmouth, Kurtz served as director of the Kiewit Computation Center from 1966 to 1975, director of the Office of Academic Computing from 1975 to 1978, and vice-chair, 1979 to 1983, and chair, 1983 to 1988, of the Program in Computer and Information Science. Over the years, Kurtz served as principal investigator for various projects supported by the National Science Foundation to promote the use of computers in education. He also participated in many other activities related to the use of computing in teaching, including the Pierce Panel of the President's Scientific Advisory Committee.

From 1974 to 1984, Kurtz chaired a committee of the American National Standards Institute, devoted to developing a national standard for BASIC, which by then existed in many incompatible forms. He served from 1987 to 1994 as convener of an

International Standards Organization working group, concerned with developing an international standard for BASIC. Kurtz received an honorary degree from Knox College in 1985, was recognized as a Computer Pioneer by the Institute of Electrical and Electronics Engineers (IEEE) in 1991, and was made a Fellow of the Association for Computer Machinery (ACM) in 1994.

## SELECTED WRITINGS BY KURTZ

*Basic Statistics,* 1963

(With J. G. Kemeny) *Basic Programming,* 1967
"BASIC." In *History of Programming Languages,* edited by Richard L. Wexelblat, pp. 515–49
(With J. G. Kemeny) *Back to BASIC,* 1985
(With J. G. Kemeny and J. L. Snell) *Computing for a Course in Finite Mathematics,* 1985
(With J. G. Kemeny) *Structured BASIC Programming,* 1987

*—Sketch by Rowan L. Dordick*

*Christine Ladd-Franklin*

# Christine Ladd-Franklin
## 1847-1930
### American logician and psychologist

Christine Ladd-Franklin made fundamental contributions to the scientific understanding of color vision and to syllogistic reasoning (deductive reasoning) and symbolic notation in logic. Although official policies of her era excluded women from advanced studies and academic positions at major universities, Ladd-Franklin studied logic and mathematics at Johns Hopkins University, researched color vision at universities in Göttingen and Berlin, and went on to lecture in psychology and logic at Johns Hopkins and Columbia University. Throughout her career she was an outspoken and effective campaigner for opening graduate programs and academic employment to women.

Christine Ladd was born in Windsor, Connecticut, on December 1, 1847. Her parents were Eliphalet Ladd, a New York merchant, and Augusta (Niles) Ladd. Her relatives included William Ladd, who founded the American Peace Society, and John Milton Niles, a former postmaster-general of the United States. Ladd-Franklin grew up in Connecticut and New York. At the age of twelve, after the death of her mother, Ladd-Franklin went to stay with her father's family in Portsmouth, New Hampshire. She graduated as valedictorian of her class from Wesleyan Academy in Wilbraham, Massachusetts, in 1865 and attended Vassar College, where she studied mathematics. Ladd-Franklin received her A.B. from Vassar in 1869 and spent the next nine years teaching secondary school. During this period, she wrote articles on mathematics for the *Educational Times,* an English publication.

When Ladd-Franklin sought to attend lectures in mathematics at the recently established Johns Hopkins University, she was admitted by the English mathematician **J. J. Sylvester**, who knew of her work. She also attended the lectures of logician **Charles Sanders Peirce** and of mathematics professor William Story. Ladd-Franklin studied at Johns Hopkins from 1878 to 1882. Sylvester persuaded the mathematics department to grant Ladd a $500 annual fellowship, which was renewed for three years. Her thesis, entitled "The Algebra of Logic," was published in 1883 in Peirce's *Studies in Logic by Members of the Johns Hopkins University*. In this work, Ladd proposed that logical statements could be analyzed more easily for validity when presented in the form of "inconsistent triads," which she later called "antilogisms," than when expressed as classical syllogisms. An antilogism comprises "three statements that are together incompatible." One example given by Ladd-Franklin in "The Antilogism," ( *Mind,* 1928), is "It is impossible that any of these measures should be idiotic, for none of them is unnecessary, and nothing that is necessary is idiotic." Her work was praised in its time, and Eugene Shen wrote in *Mind,* 1927, "No scheme in logic is more beautiful than that based on the eight propositions of Dr. Ladd-Franklin."

### An Authority on Color Vision

Ladd-Franklin turned to investigations of vision in the 1880s and began publishing articles on this subject in 1887. During a visit to Europe in 1891 and 1892, she studied Ewald Herwig's theory of color perception with G. E. Müller in Göttingen and did experiments in Müller's laboratory. She also attended lectures by the mathematician **Felix Klein**. Ladd-Franklin went on to visit Berlin, where she worked in the laboratory of Hermann von Helmholtz and attended lectures by Arthur König on Helmholtz's theory of color vision. Herwig believed that color perception arises from three opposing pairs of basic colors, while Helmholtz maintained that all the colors the eye sees can be generated from three basic colors-red, green, and blue. Ladd-Franklin synthesized these ideas, proposing her own color theory, which she presented to the International Congress of Psychology in London in 1892. She claimed that color vision had evolved from light (white) sensitivity by the addition of differentiation between yellow and blue light, followed by the separation of the yellow sensitivity into the perception of red and green. Consequently, yellow and white as well as blue, red and green, are perceived as basic colors. After a period of controversy, Ladd-Franklin's ideas were accepted by psychologists for many years.

An associate editor of *Baldwin's Dictionary of Philosophy and Psychology* in 1901 and 1902, Ladd-Franklin resumed lecturing in logic and psychology at Johns Hopkins from 1904 to 1909. In 1914 she became a lecturer at Columbia, where she continued teaching until 1927, when she was nearly 80 years old. In the late 1920s, she investigated the visual phenomenon of "blue arcs," which she believed showed that active nerve fibers emit a faint light. A collection of her major works on vision was published as *Colour and Colour Theories* in 1929.

### Opening Closed Doors

Ladd-Franklin's professional career was shaped by the restrictions placed on women scientists and scholars in the late 19th and early 20th centuries. She studied mathematics instead of physics because university laboratories did not admit women, and mathematics did not require laboratory work. Like other women, Ladd-Franklin was a "special student" at Johns Hopkins, a status outside normal admissions. When she completed her graduate work in 1882, the university would not grant her a degree. The situation was similar during her visit to Germany, where Ladd-Franklin was only allowed auditor status, and Müller delivered lectures to her privately. Even Ladd-Franklin's teaching positions at Johns Hopkins and Columbia were temporary, not permanent, appointments. One of the greatest disappointments of her career was suffered in 1914 when a leading group of experimental psychologists refused her request to attend their meeting on color vision at Columbia. Ladd-Franklin, a leading authority on the subject, gained admittance only by having one of the members take her as his guest. Johns Hopkins awarded its first doctorate to a woman in 1893, and officially began to admit women in 1907. In 1926, the school awarded Ladd-Franklin the Ph.D. in mathematics that she had earned in 1882.

Ladd-Franklin and other women in academic fields devised strategies to open American doctoral programs to women. The Association of Collegiate Alumnae, predecessor of the American Association of University Women, was formed in 1881. Ladd-Franklin proposed that the ACA start a fellowship for study overseas. The $500 fellowship that was established in 1890 helped American women gain entrance to lectures and later to earn doctoral degrees at German universities. Once women had been admitted abroad, it became easier to persuade American graduate schools to accept them also. From 1900 to 1917, Ladd-Franklin administered the Sarah Berliner fellowship, which supported new women doctorates in research. She hoped to persuade graduate schools to take the fellows into their faculty, since there was no cost to the school. This program did help women establish academic careers, but most worked at women's colleges. The scholarship did not fulfill Ladd-Franklin's goal of placing women in academic positions at coeducational schools.

Christine Ladd married Fabian Franklin, a member of the mathematics faculty at Johns Hopkins, on August 24, 1882. The couple had two children, of whom only one, Margaret, survived into adulthood. In 1895 Franklin became a journalist, and the family later moved to New York when he became an associate editor of the *New York Evening Post*. In addition to her scholarly work and articles about women's education, Ladd-Franklin published opinions on many subjects. In letters to the editors of the *New York Times* during World War I, she objected to tight collars for soldiers and advocated calling citizens of the United States "Usonians." Ladd-Franklin died of pneumonia in New York City on March 5, 1930.

## SELECTED WRITINGS BY LADD-FRANKLIN

### Books

"On the Algebra of Logic," in *Studies in Logic by Members of the Johns Hopkins University.* Edited by C. S. Peirce.
*Colour and Colour Theories* 1929

### Periodicals

"Some Proposed Reforms in Common Logic." *Mind* 15 (1890): 75–88.
"Women and Letters." *New York Times* (December 13, 1921): 18.
"Women and Economics." *New York Times* (May 28, 1924): 22.
"The Antilogism." *Mind* 37 (1928): 532–34.

## FURTHER READING

### Books

Ogilvie, Marilyn Bailey. "Christine Ladd-Franklin," in *Women in Science*. Cambridge, MA: Massachusetts Institute of Technology, 1986, pp. 116–17.

Rossiter, Margaret W. *Women Scientists in America.* Baltimore, MD: Johns Hopkins University Press, 1982.

### Periodicals

Church, Alonzo. "A Bibliography of Symbolic Logic." *Journal of Symbolic Logic* (December 1936): 138.

"Dr. Ladd-Franklin, Educator, 82, Dies." *New York Times* (March 6, 1930): 23.

Shen, Eugene. "The Ladd-Franklin Formula in Logic: The Antilogism." *Mind* 37 (1927): 54–60.

"To Restore Ideal at Johns Hopkins." *New York Times* (February 23, 1926): 12.

Venn, J. "Studies in Logic." *Mind* 8 (1883): 594–603.

### Other

Phan, An. "Christine Ladd-Franklin." *Biographies of Women Mathematicians.* http://www.scottlan.edu/lriddle/women/chronol.htm. (July 22, 1997.)

*—Sketch by Sally M. Moite*

*Comte Joseph–Louis Lagrange*

# Joseph–Louis Lagrange
## 1736–1813

**Italian–born French algebraist and number theorist**

Comte Joseph–Louis Lagrange is considered by many historians to be the foremost mathematician of 18th century Europe. He invented the calculus of variations, laid the foundation for modern mechanics, and made major contributions to the fields of algebra and number theory. He is also credited with establishing the standard of the metric system, now in widespread use throughout the world. Lagrange is highly regarded both for the originality of his work and the rigor and generality of his mathematical proofs.

Lagrange was born in Turin, Italy, on January 25, 1736. His father, a Frenchman, served as Treasurer of War in Sardinia, and his mother, Marie–Therese Gros, was the daughter of a wealthy Italian physician. Lagrange was the youngest of 11 children and the only one to survive past infancy. His early schooling focused on the classics, and although he read the works of **Euclid** and **Archimedes**, Lagrange displayed little initial interest in mathematics. It was not until he came across an article about calculus and its

superiority over ancient Greek geometry that his interest was kindled. Its author, English astronomer and mathematician **Edmund Halley**, was cited years later by Lagrange as the individual who had most influenced his decision to pursue mathematics. By the time Lagrange was 16 years old, he had mastered so much of the subject that he was appointed as a professor of mathematics at the Artillery School in Turin.

### Invents the Calculus of Variations

At the age of 19, Lagrange began working on a solution to several isoperimetric problems that were then under discussion among the leading European mathematicians of the era. In the course of deriving his proofs, Lagrange developed what was to become his most important intellectual achievement, the creation of the calculus of variations. In 1756, he sent a letter describing his work to **Leonhard Euler**, then the director of the mathematics division at the Berlin Academy of Sciences. Euler praised and encouraged Lagrange, realizing at once the significance of his results, and the two mathematicians began a long correspondence. By 1759, through Euler's influence, Lagrange was elected to the Berlin Academy.

Lagrange continued teaching in Turin. Pulling together his most talented students, he organized a research society that evolved into the Turin Academy of Sciences. In 1759, the Academy published its first volume of memoirs. Lagrange contributed three papers, one introducing his calculus of variations, another on the application of differential calculus to the field of probability, and a third regarding the theory of sound. In this last paper, he provided a mathematical description of string vibration, using a partial differential equation. His result settled an ongoing controversy about the subject between Euler and French mathematician **Jean le Rond d'Alembert**, favoring Euler's position.

### Captures Awards for Work on Celestial Mechanics

Throughout 18th century Europe, the learned academies encouraged research in celestial mechanics, frequently offering the incentive of a prize, since knowledge in this area was valuable to navigation. In 1764, Lagrange entered a competition sponsored by the French Académie Royale des Sciences to determine the gravitational forces that caused the moon to present a relatively consistent face to earth. For his calculations he received the Grand Prize. Two years later Lagrange again won the Grand Prize from the Académie, this time for deriving a partial solution to a more complicated gravitational problem involving the planet Jupiter, its four then-known satellites, and the sun. That same year, Lagrange received an invitation from King Frederick of Prussia to become director of mathematics at the Berlin Academy, replacing Euler, who had departed for St. Petersburg. Accepting the appointment in November of 1766, Lagrange entered a prolific period, composing memoirs nearly every month on subjects ranging from probability to the theory of equations.

In 1767, Lagrange published *On the Solution of Numerical Equations*, a treatise in which he explored universal methods for reducing equations from higher to lower degree. This work would help set the stage for the development of modern algebra. Lagrange also made early contributions to number theory, solving several of Fermat's theorems. He continued his investigation of gravitational interactions among planetary bodies, winning, in 1772, his third Grand Prize from the French Académie for a memoir on attractions among the sun, moon, and Earth. In 1774 and in 1778 he again won Grand Prizes, first for work related to lunar movement, then for a study of the perturbations of comets.

### Transforms Field of Mechanical Analysis

During his tenure at the Berlin Academy, Lagrange worked steadily on the topic of mechanical analysis, employing his calculus of variations. Through his efforts, the study of fluid and solid mechanics was to be unified. His *Mecanique Analytique*, finally published in 1788, applied calculus to the mechanics of rigid bodies and contained the general equations for describing motion in mechanical systems. This work is universally considered his

most important masterpiece. Ironically, Lagrange lost interest in the subject, and in mathematics in general, in the years before its publication. He began suffering from depression around 1780. When Frederick the Great died in 1786 an indifference toward science and a resentment toward foreigners arose in Berlin. Lagrange sought and obtained a position with the French Académie, where he was well-received. In Paris, his depression worsened. He was known to stare out the window for long periods of time and he hardly spoke. In letters to his friends and colleagues he wrote that mathematics was no longer important.

It was not until after the French Revolution that Lagrange finally began to emerge from his apathy and end his ambivalence toward mathematics. The monarchy had fallen and the Académie lost its royal patronage. Many of Lagrange's colleagues were beheaded, but Lagrange himself had managed to remain a politically neutral figure. He was granted a pension by the revolutionists and eventually appointed to a government committee that was charged with establishing standards for weights and measures. It was in this position that he persuaded the French to adopt the metric system. In 1795, Lagrange became a professor of mathematics at the newly formed École Normale and when it closed two years later, he assumed a professorship at the École Polytechnique. In an attempt to clarify the topic of calculus for his students he wrote *Theory of Analytic Functions* in 1797 and *Lessons on the Calculus of Functions* in 1801. In these texts Lagrange attempted to reduce calculus to an algebraic system and was unsuccessful. Still, the works proved valuable as a catalyst to 19th century mathematicians who would refine his ideas and develop a more coherent calculus.

When Lagrange reached his seventies, he began revising and extending his *Mecanique Analytique*, completing a second edition. The long hours of work diminished his strength and energy and he suffered increasingly from fainting spells. On April 10, 1813, Lagrange died. His body was brought to rest in the Pantheon, as a tribute to the contributions he made to France.

Lagrange was known for his gentle demeanor and his diplomatic skills. At the Berlin Academy, he fell into favor with Frederick the Great, who had been highly critical of Lagrange's predecessor, Leonhard Euler. When Lagrange arrived at the French Académie he was doted upon by Queen Marie-Antoinette, yet he also managed to remain on good terms with leaders of the French Revolution. Napoleon Bonaparte, who made Lagrange a Senator, a Count of the Empire, and a Grand Officer of the Legion of Honor, consulted him frequently on philosophical and technical matters. Shortly after accepting his appointment at the Berlin Academy in 1766, Lagrange married a young woman to whom he was related. In his correspondence with his friend and colleague d'Alembert, he referred to the marriage as "inconsequential" and one of convenience. Nonetheless, when his wife took ill and died a few years later, Lagrange was reportedly heartbroken. He did not marry again until he was living in Paris and suffering from his depression. Then, at the age of 56 he took as his second wife the teenage daughter of his friend, astronomer Pierre–Charles Lemonnier. His new bride was devoted to him, and it is said she helped Lagrange regain his interest in mathematics. Neither marriage resulted in children.

## FURTHER READING

### Books

Bell, E.T. *Men of Mathematics*. New York: Simon and Schuster, 1986, pp. 153–171.

Cajori, Florian. *A History of Mathematics*. New York: Chelsea Publishing Co., 1991.

Eves, Howard. *An Introduction to the History of Mathematics*. New York: Holt, Rinehart and Winston, 1960.

*—Sketch by Leslie Reinherz*

# Johann Heinrich Lambert
## 1728–1777
### German statistician, geometer, and analyst

Johann Heinrich Lambert stood largely outside the academic environment of his time. Although his ideas did not become popular until the next century and some of his other work were lost for many years, his scientific investigations gave rise to questions about how to handle data, from which he developed some basic ideas of statistics.

Lambert was born on August 25, 1728 in the town of Mulhouse, at that time a free city allied with Switzerland. Lambert's father, Lukas, was a poor tailor and his mother was named Elisabeth Schmerber. At the age of 12 Lambert left school in order to help his father. Within the next few years Lambert went through a variety of professions, including clerk at an ironworks and (at the age of 17) secretary to the editor of a Basel newspaper. Lambert's father died in 1747 and the next year Lambert was hired by the von Sails family as a tutor for their children. From 1752 until his death, Lambert kept a diary which furnishes a guide to the direction of his thoughts and the subjects which he was investigating.

From 1756 through 1758 Lambert traveled throughout Europe with his pupils. It may have been intended as an educational experience for the children, but Lambert used it as an opportunity to connect with various scientific societies in the towns through which they traveled. He was elected a corresponding member of the Learned Society at Göttingen but by the end of the trip he was looking for a permanent scientific position. After an opportunity to organize a Bavarian Academy of Sciences failed, Lambert was offered a position at St. Petersburg. He was reluctant to follow some of his countrymen to the Russian court, however, and preferred to wait until something arose at the Prussian Academy instead. He was proposed for membership in 1761 and began receiving a salary in 1765.

During his time as an employee of the Prussian Academy, Lambert's wrote more than 150 papers. He based his investigations from some of the same mathematical perspectives as **Gottfried Leibniz** and followed some of his philosophical roads as well. One of Leibniz's goals had been to create an ideal language for carrying out reasoning and Lambert imitated his predecessor with no greater success. It was not until the work of **Gottlob Frege** in the 19th century that technical developments allowed for the creation of a language useful for imitating reasoning.

Lambert's best–known work in mathematics was connected with an ancient geometrical problem, the squaring of a circle. In an effort to construct by geometrical means a square the same size as a circle, it was crucial to know as much about the number $\pi$, the ratio of the circumference of a circle to its diameter, as possible. One fundamental question was whether $\pi$ was a rational number (that is, could be expressed as the ratio of two whole numbers) or not. Certain irrational numbers like the square root of two were well known, but $\pi$ was not similar to those irrationals. Finally, Lambert was able to prove that $\pi$ was not a rational number by looking at a continued fraction expanion. A continued fraction is one with a denominator within a denominator and so forth, leading to an expression that is cumbersome to express typographically but which can be used to stand for an infinite process. The same technique could also be employed to establish the irrationality of $e$, the base of natural logarithms.

Another area in which Lambert investigated is what was to become known as non–Euclidean geometry. In Euclidean geometry, there is only one line parallel to a given line through a point not on the line. By altering that condition, one can end up with geometries with different properties. Lambert was attempting to prove the parallel postulate of **Euclid** and stumbled on these non–Euclidean consequences in the process. Unfortunately, his work was not discovered until after non–Euclidean geometry had been more thoroughly explored.

### Creates a Theory of Errors

Perhaps the work with the longest–term consequences which Lambert published was a general theory of errors. In a work on the measurement of light he discussed the problem of determining the probability distribution for errors and formulated a method analogous to that of maximum likelihood as developed by statisticians more than a century later. He viewed weather as though it were produced by an infinite number of unknown causes, like the outcomes of a game of chance, and argued that if positive and negative errors are equally possible, then they will occur with equal frequency. As the importance of errors in measurement became more widely recognized, Lambert's work on the theory of errors was applied by **Karl Gauss** and others.

Lambert's work on map projections was among the first to give the theory of the subject, although the details of the formulation were improved by others. He allowed for the possibility of probabilities that did not simply add up, thereby imitating the structure of what has become known as "belief functions." In short, Lambert made contributions to many sciences as well as to mathematics and statistics. What is discouraging is to realize how much of Lambert's work disappeared, only to be rediscovered much later. At least his work on the irrationality of $\pi$ and e became part of a living tradition leading to the proof of their transcendence as executed in the 19th century. Lambert died in Berlin on September 25, 1777.

### SELECTED WRITINGS BY LAMBERT

*Opera Mathematica.* Edited by Andreas Speiser, 1946–48

### FURTHER READING

#### Books

Hacking, Ian. *The Emergence of Probability.* Cambridge, UK: Cambridge University Press, 1975.

Scriba, Christoph J. "Johann Heinrich Lambert," in *Dictionary of Scientific Biography.* Volume VII. Edited by Charles Coulston Gillispie. New York: Charles Scribner's Sons, 1973, pp. 595–600.

#### Periodicals

Sheynin, O.B. "J.H. Lambert's Work on Probability." *Archive for the History of Exact Science* (1971): 244–256.

—*Sketch by Thomas Drucker*

# Edmund Landau
## 1877–1938
### German number theorist

Edmund Landau profoundly influenced the development of number theory. His primary research focused on analytic number theory, especially the distribution of prime numbers and prime ideals. An extremely productive author of at least 250 publications, Landau's writings had a distinct style. His prose was carefully crafted, highlighted by lucid, comprehensive argumentation and a thorough explanation of the background knowledge required to understand it. Landau's writing style became more succinct over the course of his career. He was forced to retire from teaching at the behest of Nazi anti–Semitic policies.

Born in Berlin on February 14, 1877, Landau was the son of Leopold, a gynecologist, and Johanna (Jacoby) Landau. Johanna Landau came from a wealthy family with whom the Landaus lived in an affluent section of Berlin. Although Leopold Landau was an assimilated Jew and a German patriot, in 1872 he helped found an Judaism academy in Berlin. Landau himself studied in Berlin at the *Französische Gymnasium* (French Lycée), graduating two years early at age 16. He promptly began studying at Berlin University. Landau had published twice before receiving his Ph.D; both pieces explored chess related mathematical problems.

Under the tutelage of Georg Frobenius, Landau was awarded his doctorate at Berlin University in 1899 at the age of 22 years old. His dissertation dealt with what became his life's work: number theory. Landau began teaching at Berlin in 1901, when he earned the advanced degree which allowed him to teach mathematics. He proved to be a popular lecturer at the university because of his personal excitement of the carefully prepared material he presented to his students.

Landau's first major accomplishment as a mathematician came in 1903, when he simplified and improved upon the proof for the prime number theorem conjectured by **Karl Gauss** in 1796, and demonstrated independently by **Jacques Hadamard** and **C.J. de la Vallée–Poussin** in 1896. In Landau's proof, the theorem's application extended to algebraic number fields, specifically to the distribution of ideal primes within them.

Landau married Marianne Ehrlich (daughter of Paul Ehrlich, a friend of Landau's father, who won the 1908 Nobel prize in medicine or physiology) in 1905 at Frankfurt–am–Main, and fathered two daughters and two sons (one of whom died before age five). He served as a professor of mathematics at Berlin until 1909.

Landau published his first major work in 1909, the two–volume *Handbuch der Lehre von der Vertiolung der Prizahalen*. The volumes were the first orderly discussion of analytic number theory, and were used for many years in universities as a research and teaching tool. Landau's texts are still considered important documents in the history of mathematics.

In the same year, Landau became a full professor at the University of Göttingen. Although the faculty at Berlin tried twice to keep Landau on staff, the government wanted to make Göttingen a center of German mathematical learning. They succeeded in their objective, and Landau stayed there until 1934. In 1913, Landau even declined an offer from a university in Heidelberg for a chair position. Although he was still a charismatic, inspiring teacher by the 1920s he was criticized for his rigid, almost perfectionistic lecture style. A demanding lecturer, he insisted that one of his assistants sit through his presentations so any errors could be immediately corrected.

Landau continued his father's support of Jewish institutions. In 1925, he gave a lecture on mathematics in Hebrew at the Hebrew University in Jerusalem, an institution Landau heartily embraced. His activities there continued when he took a sabbatical from Göttingen and taught a few mathematics classes in 1927–28. Landau even contemplated staying in Jerusalem at one point.

Landau also published another important treatise in 1927, the three volumes of *Vorlesungen über Zahlentheorie*. In these texts, Landau brought together the various branches of number theory in one comprehensive text. He throughly explored each branch from its origins to the then–current state of research. Two years later, the widely respected Landau received a honorary doctorate of philosophy from the University of Oslo in Norway. The next year, Landau published another landmark book, entitled *Grundlagen der Analysis*. Beginning with **Giuseppe Peano**'s axioms for natural numbers, this volume presented arithmetic in four forms of numbers: whole, rational, irrational, and complex.

The Nazi Party and their policies of discrimination against Jews led to a premature end to Landau's academic career. In late 1933, he was forced to cease teaching at Göttingen, although he was one of the last Jewish professors to be purged from that institution. While technically not subject to the 1933 non–Aryan clause attached to Nazi civil servant laws, all Jewish mathematical professors were forced to leave Göttingen. Landau stayed on through the summer and fall terms of 1933, but he could only teach classes through assistants. Landau would sit in the back of every class, ready to teach at any moment if his ban was raised.

On November 2, 1933, Landau attempted to resume teaching his class. The students, alerted to this impropriety in advance, boycotted his lecture. SS Guards were stationed at the entrance in case a student did not want to boycott; only one got in. When it was clear he would not be allowed to lecture, Landau returned to his office. The boycotting students explained by letter that they no longer wanted to be taught by a Jew and be indoctrinated in his mode of thought.

In 1934, Landau was given his retirement leave, and he and his family moved back to Berlin. Although he never taught in Germany again, he did lecture out of the country at universities such as Cambridge in 1935 and Brussels in 1937. Landau died in Berlin of natural causes on February 19, 1938, and was buried in the Berlin–Weissensee Jewish cemetery.

## SELECTED WRITINGS BY LANDAU

*Differential and Integral Calculus*. Third Edition. Translated by Melvin Hausner and Martin Davis, 1965
*Elementary Number Theory*. Translated by Jacob E. Goodman, 1968
*Foundations of Analysis: The Arithmetic of Whole, Rational, Irrational, and Complex Numbers*. Third edition. Translated by F. Steinhardt, 1966

## FURTHER READING

### Books

Schoeneberg, Bruno. "Edmund Georg Hermann," in *Dictionary of Scientific Biography*. Volume VII. Edited by Charles Coulston Gillispie. New York: Charles Scribner's Sons, 1970, pp. 615–16.

### Periodicals

Chowdhury, M.R. "Landau and Teichmuller." *The Mathmatical Intelligencer* (Spring 1995): 12–14.
Norbert, Schappacher. "Edmund Landau's Göttingen: From the Life and Death of a Great Mathematical Center." *The Mathematical Intelligencer* (Fall 1991): 12–18.

—*Sketch by Annette Petruso*

*Robert P. Langlands*

# Robert P. Langlands
## 1936–

### Canadian–American mathematical physicist

Robert Langlands' subspecialities are group representations, number theory, and automorphic forms. His earliest theories faced skepticism, but he eventually persuaded fellow mathematicians of the possibility of links between algebra and analysis, promising what some have termed a "unifying principle" for all mathematics. More than 25 years ago, Langlands saw that seemingly disparate fields of mathematics are related in often unexpected ways. For example, one of his conjectures, ostensibly dealing with Lie groups, also by implication involves number theory and algebraic geometry.

In 1982, Langlands was awarded the American Mathematical Society Cole Prize for his work in automorphic forms. He received the first National Academy of Sciences Award in Mathematics in 1988 for the "extraordinary vision that has brought the theory of group representations into a revolutionary new relationship with the theory of automorphic forms and number theory." He shared the 1995 Wolf Prize with **Andrew Wiles**, again for work involving group representations, number theory, and automorphic forms. Without Langlands earlier work, Wiles would not have been able to accomplish his recent solution of Fermat's Last Theorem. However,

not all of Langlands' work is in pure mathematics. His activities in the realm of applied mathematics, often lauded for a spirited use of metaphors, directly grapple with the common pitfalls that await those who use pure mathematical models to guide physics experiments.

Robert Phelan Langlands was born in New Westminster, British Columbia, Canada on October 6, 1936, to Robert and Kathleen (Phelan) Langlands. Little is known of his early life. Though he eventually earned his Ph.D. (1960) at Yale University in the United States, he received his B.A. (1957) and M.A. (1958) from the University of British Columbia. He married Charlotte Lorraine Cheverie on August 13, 1956, and they eventually had four children—William, Sarah, Robert, and Thomasin.

Langlands joined the faculty of Princeton University as an instructor. He worked his way up to associate professor at Princeton before returning to Yale as a full professor in 1967. While at Princeton, he began his relationship with the nearby Institute for Advanced Study (IAS), where he eventually became a professor when he left Yale in 1972. During this time, Langlands was also busy writing and publishing nearly 30 papers that formulated a set of still unsolved problems and still unproven conjectures now loosely termed "the Langlands program."

### "A Driving Force"

Langlands has always been welcoming to other researchers. In 1977, he invited **Pierre Deligné**, who was at the time working on several elements of the Langlands program, to the U.S. to organize a conference with him. Li Guo, now with Rutgers University, did postdoctoral research during 1995 with Langlands as a mentor. Though the program had run out of funds and there was officially no room for Guo, Langlands personally intervened so the young mathematician could work at the Institute for Advanced Study for half a year. For these and many other reasons, he has been termed "a driving force in mathematics" by Enrico Bombieri, an IAS School of Mathematics colleague.

Langlands remains a professor at the Institute for Advanced Study. To celebrate his 60th birthday, the Institute organized the Langlands Conference on automorphic forms, geometry, and analysis in 1996. Langlands possesses dual American and Canadian citizenship, and holds memberships in both the American Mathematical Society and the Canadian Mathematical Society.

### SELECTED WRITINGS BY LANGLANDS

#### Books

"Einstein Series, the Trace Formula, and the Modern Theory of Automorphic Forms," in *Number Theory, Trace Formulas, and Discrete Groups.* Edited by K.R. Aubert, et al, 1988

(With D. Ramakrishnan) *The Zeta Functions of Picard Modular Surfaces*, 1992

"Representation Theory: Its Rise and Its Role in Number Theory," in *Proceedings of the Gibbs Symposium, AMS*, 1990

## Periodicals

"The Factorization of a Polynomial Defined by Partitions." *Comm. Math. Physics* 124 (1989): 251–284.

"Some Holomorphic Semi–Groups." *Proceedings of the National Academy of Sciences* 46 (1960): 361–363.

(With Ph. Pouliot, and Y. Saint–Aubin.) "Conformal Invariance in Two–Dimensional Percolation." *Bulletin of the American Mathematical Society* 30 (1994): 1–61.

## FURTHER READING

### Books

*American Men & Women of Science.* Nineteenth edition. New Providence, NJ: Bowker, 1994, p. 627.

McMurray, Emily J., editor. *Notable Twentieth Century Scientists.* Detroit, MI: Gale Research, 1995.

*Who's Who in Science and Engineering.* New Providence, NJ: Reed Publishing, 1994, p. 504.

### Periodicals

"Langlands Conference: Robert Langlands's 60th Birthday." *Institute for Advanced Studies Newsletter* (Fall 1996): 1–2.

### Other

Guo, Li., in a telephone interview with Jennifer Kramer, conducted July 14, 1997.

—*Sketch by Jennifer Kramer*

# Pierre Simon Laplace
## 1749–1827
### French mathematical physicist, statistician, and astronomer

Pierre Simon Laplace's work in both celestial mechanics and probability represent pinnacles of intellectual achievement. Laplace extended the work

*Pierre Simon Laplace*

of **Isaac Newton,** explaining variations in the planetary orbits and establishing the stability of the solar system. In addition, Laplace's nebular hypothesis used natural law to explain the origins of the solar system, providing a new cosmogony. His work in celestial mechanics led him into the area of statistical inference and probability, to which he made substantial contributions. In physics, Laplace's theory of the potential enabled advances in the understanding of electromagnetism and other phenomena. He also predicted the existence of black holes.

Laplace was born on March 23, 1749, in Beaumont–en–Auge, Normandy, France, to Pierre and Marie–Anne Sochon Laplace. His father was in the cider business and an official of the local parish; his mother's family were well–to–do farmers from Tourgéville. Laplace had one sister, Marie–Anne, born in 1745. From age 7 to 17 he was a day student at the Benedictine school at Beaumont–en–Auge, where his paternal uncle, Louis Laplace, an unordained abbé, was a teacher. Laplace was recognized for his intelligence, prodigious memory, and skill at argument.

### Wins Patronage of d'Alembert

In 1766, Laplace entered the University of Caen for theological training. His mathematical interests and talents were readily apparent, however, and he was encouraged in mathematics by his teachers, Christophe Gadbled and Pierre Le Canu. In 1768, at the age of 19, Laplace went to Paris to meet the mathematician **Jean d'Alembert,** carrying a letter of recommendation from Le Canu. Legend has it that d'Alembert twice gave Laplace mathematical problems and Laplace solved them overnight. In any case, d'Alembert, who was permanent secretary of the Paris Académie Royale des Sciences, was sufficiently impressed with Laplace's abilities that he used his influence to secure a job for him at the Military School, where Laplace taught elementary mathematics to young cadets. Several years later, in 1785, 16–year–old Napoleon Bonaparte attended the school and Laplace examined and passed him.

Settled in Paris, Laplace began to submit mathematics papers to the Académie. Nicolas Condorcet, the secretary, wrote that never had the Académie received in so short a time so many important papers on such varied and difficult topics. Laplace was proposed for membership in 1771, and was elected to the Académie on March 31, 1773, at age 24.

### Major Works in Celestial Mechanics and Probability

Laplace's first contributions involved adapting integral calculus to the solution of difference equations. He then addressed problems in mathematical astronomy. He explained the observed shrinking orbit of Jupiter and the expanding orbit of Saturn, demonstrating that the orbital eccentricies are self–correcting and that mean motions of the planets are invariable. Laplace also addressed and explained the acceleration of the moon around the Earth, cometary orbits, and the perturbations produced in the motion of the planets by their satellites.

From 1799 to 1825, Laplace produced his monumental five–volume *Traité de mécanique céleste (Treatise on Celestial Mechanics).* In it, he completed Newton's work and extended **Joseph Louis Lagrange**'s planetary work. Although Newton had believed that divine intervention would be necessary periodically to "reset" the solar system, Laplace showed that Newton's law of universal gravitation implied its long–term stability. When Napoleon observed that Laplace's voluminous treatise did not mention God as creator of the universe, Laplace replied, "Sir, I do not need that hypothesis."

In 1796, Laplace published *Exposition du système du monde,* in which he stated in an extended footnote the idea that the solar system condensed from a rotating cloud of gas. This was the nebular hypothesis originally proposed by the philosopher Immanuel Kant without mathematical elaboration. In his analysis of the gravitational field surrounding a sphere, Laplace's equations predicted the concept of the black hole, whose characteristics were deduced much later in **Albert Einstein**'s general theory of relativity.

With Antoine Laviosier, his colleague at the Académie, Laplace worked on several problems in physics, including thermal conductivity and capillary action. He is best known to physicists for his theory of the potential, which is useful in studying gravity as well as electromagnetic interactions, acoustics, and hydrodynamics.

Laplace contributed greatly to the field of probability and statistical inference. In 1774, using principles similar to those developed by **Thomas Bayes**, he derived essentially the same result involving integrals for determining probability, given empirical evidence. His *Théorie analytique des probabilités* ("Analytic Theory of Probability"), published in 1812 and expanded in 1814 with his *Philosophical Essay,* summarized all the materials known at that time in the area of probability, including the theory of games, geometrical probabilities, the theory of least squares, and solutions of differential equations. In his probability work Laplace introduced the idea of the Laplace transform, a simple and elegant mathematical technique for solving integral equations. Some of Laplace's contributions to probability were derived from questions in astronomy, for example, the central limit theorem which applied to the inclination of the orbits of comets. Laplace believed that through probability mathematics could be brought to bear on the social sciences, and suggested applications in insurance, demographics, decision theory, and the credibility of witnesses. In 1786, he published a study of the vital statistics of Paris, using probability techniques to estimate the population of France.

### Creation of the Metric System

In May 1790 the revolutionary government passed a law requiring standardization of weights and measures in France. The Académie Royale de Sciences was charged with making recommendations, and Laplace, along with Lagrange and **Gaspard Monge**, served on the committee appointed to consider the issue. They made recommendations for units of length, area, volume, and mass, with decimal subdivisions and multiples. They also devised decimal systems for money, angles, and the calendar. The basic unit of length as it was defined was named the "meter" at Laplace's suggestion. The decimalization of angles and the calendar lasted only a few years, but the other metric units were gradually accepted around the world.

In the late 18th century and into the 19th century, Laplace dominated the Académie Royale des Sciences, imposing his scientific preferences and deterministic ideology on younger colleagues. He presided over the Bureau des Longitudes, which addressed the needs of astronomy and navigation, taught at the École Normale when it was opened briefly in 1795, and in 1800 was instrumental in creating the governing body for the École Polytechnique (founded in 1794), where he served as a graduation examiner. Laplace was known for the "rapidity" of his teaching, and in his writing he became known for his use of the phrase "it is easy to see," by which he skipped steps in his explanations, confounding some of his later readers and translators.

Laplace's prominence in science brought him conspicuously into the political arena, where he served on numerous blue-ribbon commissions, including a commission that investigated hospital care. In 1799, Napoleon appointed him minister of the interior, but dismissed him after six weeks to place his brother in the position and appointed Laplace instead to the senate, where he eventually became chancellor. Later, Napoleon made Laplace a count of the empire and conferred upon him France's highest honors, the Grand Cross of the Legion of Honor and the Order of the Reunion. At one time, however, Napoleon noted that Laplace was a "mathematician of the first rank," but a mediocre administrator. As a member of the senate, Laplace voted against Napoleon in 1814, supporting Louis XVIII instead. After the restoration of the Bourbon monarchy in 1815, Laplace was rewarded with the title of marquis and was appointed president of the committee to oversee the reorganization of the École Polytechnique. His political opportunism allowed him to prosper and to continue his scientific work.

### Last Years at Arcueil

Laplace retired to his country estate in Arcueil, near Paris, where he and Claude Berthollet, a chemist and physician who also boasted a distinguished career in the service of science and France, formed an informal school. The discussions of the Société d'Arcueil were published between 1807 and 1817.

Although Laplace has been criticized for his arrogance, unreliability in giving credit to others, and wavering allegiances, he is considered to be France's most illustrious scientist in its golden age, and one of the most influential scientists of all time. Laplace had an acute mind and was generally healthy and vigorous until the end. After a short illness, he died on March 5, 1827, at Arcueil, just short of his 78th birthday. At his funeral, he was eulogized by many, including his student Siméon-Denis Poisson, who called him "the Newton of France." Laplace was buried at Pére Lachaise. In 1878, the monument erected to him was moved to Beaumont-en-Auge, and his remains were transferred to the small village of St. Julien de Mailloc. Laplace's papers were destroyed by a fire at the château of Mailloc in 1925, which was then owned by his great-great grandson, the comte de Colbert-Laplace.

## FURTHER READING

Anglin, W. S. *Mathematics: A Concise History and Philosophy*. New York: Springer-Verlag, 1994, p.188.

Bell, E. T. *Men of Mathematics*. New York: Simon & Schuster, 1937, pp. 172–182.

Grattan-Guinness, I. "Pierre Simon Laplace," in *Dictionary of Scientific Biography*. Vol. XV, Supplement I, Topical Essays. Edited by Charles Coulston Gillispie. New York: Charles Scribner's Sons, 1978, pp. 273–403.

—, editor. *Companion Encyclopedia of the History and Philosophy of the Mathematical Sciences*. Two volumes. London: Routledge, 1994.

Grattan-Guinness, I. and J. R. Ravetz. *Joseph Fourier: 1768–1830*. Cambridge, MA: The MIT Press, 1972.

Hahn, Roger. *Laplace as a Newtonian Scientist*. Los Angeles: University of California William Andrews Clark Memorial Library, April 8, 1967. 26 pp.

Motz, Lloyd and Jefferson Hane Weaver. *The Story of Mathematics*. New York: Plenum Press, 1993, pp. 162–167.

Numbers, Ronald L. *Creation by Natural Law: Laplace's Nebular Hypothesis in American Thought*. Seattle: University of Washington Press, 1977.

Stigler, Stephen M. *The History of Statistics: The Measurement of Uncertainty Before 1900*. Cambridge, MA: Harvard University Press, 1986.

*—Sketch by Jill Carpenter*

# Edna Ernestine Kramer Lassar
## 1902-1984
### American mathematics educator and author

Edna Ernestine Kramer Lassar was a mathematics professor who made her mark by writing books about math concepts for general readers. *The Nature and Growth of Modern Mathematics*, first published in 1970 under the name Edna E. Kramer, was her crowning achievement. In a review in *Science* magazine, Donald J. Dessart concluded that the volume "richly deserves a place on any mathematical bookshelf."

Lassar was born on May 11, 1902, in New York City. She was the eldest child of Joseph Kramer and Sabine Elowitch Kramer, Jewish immigrants from

Rima–Sombad, Austria–Hungary (now Czechoslovakia). Lassar's father worked mainly as a salesman of men's clothing. However, both of her parents showed a strong intellectual bent; Joseph was interested in political science, and Sabine enjoyed opera. Education was highly valued in their home. Lassar was named for an uncle, Edward Elowitch, who had died at age 19 shortly before her birth. This uncle had been a gifted math student, and Lassar later explained that she was motivated as a child to excel in the subject partly in his honor.

In 1922, Lassar received a B.A. degree in mathematics from Hunter College in New York City. After graduation she began teaching high school math, first at DeWitt Clinton High School in the Bronx, and then at Wadleigh High School in Manhattan. At the same time, Lassar attended graduate school at Columbia University, where she received a Ph.D. degree in 1930. She was only the third woman to earn a doctorate in pure mathematics from that institution. Her advisor at Columbia was Edward Kasner, who had published papers on polygenic functions of one complex variable. For her dissertation, Lassar developed an analogous theory of polygenic functions of the dual variable.

### Builds Reputation as Teacher and Writer

Lassar had originally hoped to obtain employment as a mathematical researcher, but such positions were scarce during the Depression of the 1930s. She gravitated instead toward teaching. In 1929, Lassar became the first female instructor of mathematics at the New Jersey State Teachers College in Montclair. Five years later she returned to the New York City public school system, taking a job at Thomas Jefferson High School in Brooklyn, thereby doubling her salary. Her first book, *A First Course in Educational Statistics*, appeared in 1935. It was followed by another educational volume, *Mathematics Takes Wings: An Aviation Supplement to Secondary Mathematics*.

On July 2, 1935, Lassar married Benedict Taxier Lassar, a high school French teacher and guidance counselor who later became a clinical psychologist. The couple shared a passion for travel, which they indulged in trips across the United States, Canada, and the Near and Far East. Lassar's husband was always supportive of her career, helping with library research and manuscript typing. Two of her books were dedicated to him.

During World War II Lassar had found an opportunity to apply her academic knowledge of statistics to practical use. While still teaching high school, she also worked part–time at Columbia University as a statistical consultant to the school's Division of War Research. Her work there included probabilistic analyses of the war in Japan. In 1948,

Lassar became an instructor at the New York Polytechnic Institute. She remained at the institute until her retirement in 1965.

Lassar collected real–life examples, historical tidbits, and enrichment activities to enliven her math lessons. By 1951, her collection had grown into a book, *The Main Stream of Mathematics*. This work explores mathematical history and concepts from ancient times to the early twentieth century. The theme was later expanded in *The Nature and Growth of Modern Mathematics*, a comprehensive 30–chapter work that emphasizes twentieth–century math through the 1960s. This book was widely praised for its lively style and clear explanations. It was chosen as a Science Book of the Month Club offering, and *Kirkus Reviews* lauded Lassar for "the remarkable lucidity of her expression and her attack."

Lassar suffered from Parkinson's disease for the last decade of her life. She died of pneumonia on July 9, 1984, at her home in New York City. One of Lassar's special interests had been the historical achievements of women in mathematics. Her unique blend of talent in both math and language assured her own place in the history of the field.

## SELECTED WRITINGS BY LASSAR

### Books

*A First Course in Educational Statistics*, 1935
*Mathematics Takes Wings: An Aviation Supplement to Secondary Mathematics*, 1942
*The Main Stream of Mathematics*, 1951
*The Nature and Growth of Modern Mathematics*, 1970

## FURTHER READING

### Books

Lipsey, Sally Irene. "Edna Ernestine Kramer Lassar," in *Women of Mathematics: A Biobibliographic Sourcebook*. Edited by Louise S. Grinstein and Paul J. Campbell. New York: Greenwood Press, 1987, pp. 114–120.

### Periodicals

Dessart, Donald J. "Tracing Mathematical Concepts." *Science* (October 23, 1970): 432.
"Dr. Edna Kramer–Lassar, 82, Ex–Professor of Mathematics." *New York Times* (July 25, 1984): D23.
Review of *The Nature and Growth of Modern Mathematics*. *Kirkus Reviews* (January 15, 1970): 91.

## Other

Long, Jennifer. "Edna Ernestine Kramer Lassar." *Biographies of Women Mathematicians.* June 1997. http://www.scottlan.edu/lriddle/women/chronol.htm (July 21, 1997).

        *—Sketch by Linda Wasmer Smith*

# Henri Lebesgue
## 1875-1941
### French mathematician

In the first decade of the 20th century, Henri Lebesgue developed a new approach to integral calculus in order to overcome the restrictions of previous theories. At that time, integration was used to calculate the area under a curve, but if the curve was discontinuous the theory was difficult to apply and left some questions unanswered. Lebesgue's theory of integration circumvented the problems caused by these discontinuities and was compatible with other basic mathematical operations.

Henri Léon Lebesgue was born in Beauvais, France, on June 28, 1875. His father was a typographical worker, and his mother was an elementary school teacher. He entered the École Normale Supérieure in 1894 and quickly demonstrated mathematical talent along with an irreverent attitude that gave him the tendency to ignore subjects that did not interest him. For instance, he passed his chemistry course only by mumbling his answers to the examiner who was hard of hearing. Even in mathematics he graduated third in his class in 1879. Nevertheless, his questioning of the traditional methods of mathematics was the basis for his reexamination of the concepts of length, area, and volume. He stayed on to work in the library for two years after his graduation in 1879.

Lebesgue inherited the solid foundation for the theory of calculus that was laid by the mathematical giants of the 19th century. **Karl Friedrich Gauss**, **Augustin-Louis Cauchy** , **Niels Henrik Abel** , and others of this period had introduced rigorous definitions of convergence, limit, and continuity. They had also formulated a precise definition of the integral, one of the two central concepts of calculus. Just as addition gives the total of a finite set of numbers, the integral pertains to the limiting case of the sum of a quantity that varies at every one of an infinite set of points. Such a quantity is described mathematically by a function, and integration of a function can represent the area bounded by a curve, the total work done by a variable force, the distance a planet travels in its elliptical orbit around the Sun, among other

possibilities. Cauchy's definition of the integral applied to functions that were continuous, that is, curves without any jumps. It could also handle a finite number of discontinuities, points where jumps occurred.

In 1854 **Georg Riemann** introduced an extension of the concept of integration which found its way into most calculus books. Unfortunately, the Riemann integral was unsatisfactory for dealing with some sequences of functions. Even if the sequence of functions approached a limit and each function in the sequence was continuous, the limit might not be a function that could be handled by Riemann integration. The problem was to find a definition for integration that would be compatible with taking the limit of a sequence of functions.

### Builds Upon Previously Established Theories

Lebesgue was influenced by the work of René Baire (another recent graduate of the École Normale Supérieure) and **Émile Borel** . By 1898, when Lebesgue published his first results, Baire had formulated an insightful theory of discontinuous functions. In that same year Borel published a theory of measure that generalized the concepts of area to new types of regions obtained as limits.

Lebesgue taught at the Lycée Central in Nancy from 1899 to 1902. During that time he developed the ideas for his doctoral thesis at the Sorbonne. The work, "Intégrale, longueur, aire," extended Borel's theory of measure, defined the integral geometrically and analytically, and established nearly all the basic properties of integration. J. C. Burkill notes in the obituary of Lebesgue for the *Journal of the London Mathematical Society:* "It cannot be doubted that this dissertation is one of the finest which any mathematician has ever written."

Lebesgue's consideration of discontinuous curves and nonsmooth surfaces was shocking to some of his contemporaries. Camille Jordan cautioned Lebesgue that he should not expect other scholars to appreciate his work. Fortunately, the usefulness of his new ideas quickly overcame any resistance, and Lebesgue received a university appointment as maître des conférences at Rennes in 1902. During his first year he gave lectures for the Cours Peccot at the Collège de France on his new integral and the next year on its application to trigonometric series. Lebesgue published these lectures in the series of tracts edited by Borel and was the first author in this series of monographs other than Borel himself. He gave a thorough exposition of the historical background of the problems leading up to the properties an integral should satisfy, including the compatibility with the limit of a sequence of functions.

In 1906 Lebesgue left Rennes to become chargé de cours for the faculty of sciences, and later a professor, at Poitiers. In 1910 he was maître des conférences at the Sorbonne, and in 1921 he became professor at the Collège de France. Among his many prizes and honors was his election to the Académie des Sciences in 1922. By that time Lebesgue had nearly 90 publications on measure theory, integration, geometry, and related topics. Although his ideas were ignored at the great centers of mathematics such as Göttingen, his integral was presented to undergraduates at Rice Institute as early as 1914, and served as an inspiration to the founders of the Polish schools of mathematics at Lvov and Warsaw in 1919.

During the last 20 years of his life, Lebesgue's work became widely known, and his approach to integration evolved as a standard tool of analysis. Lebesgue himself began to concentrate more on the historical and pedagogical issues associated with his work. He believed that mathematicians should work from the problems that motivate theory and resist being bound to tradition. He felt that mathematical education should follow this same principle. He freely used the words "deception" and "hypocrisy" to describe the lack of connection between students' natural intuition of numbers and geometry and the manner in which these subjects were taught. In "Sur le mesure des grandeurs," Lebesgue complained about teaching of mathematics: "An infinite amount of talent has been expended on little perfections of detail. We must now attempt an overhaul of the whole structure."

Lebesgue died in Paris on July 26, 1941. Even during the last months of his terminal illness, he continued his course on geometrical constructions at the Collège de France and dictated a book on conic sections. He was survived by his wife, mother, a son, and a daughter. His view of mathematics can be summed up in the concluding words of "Sur le développement de la notion d'intégrale": "A generalization made not for the vain pleasure of generalizing but in order to solve previously existing problems is always a fruitful generalization. This is proved abundantly by the variety of applications of the ideas that we have just examined."

## SELECTED WRITINGS BY LEBESGUE

### Books

*Leçons sur l'intégration et la recherche des fonctions primitives,* 1904
*Leçons sur les séries trigonométriques,* 1906
*Notice sur les travaux scientifiques de M. Henri Lebesgue,* 1922
*Measure and the Integral,* 1966

### Periodicals

"Intégrale, longueur, aire." *Annali di Mathematica* 7 (1902): 231–359.
"Sur le développement de la notion d'intégrale." *Matematiska Tidskrift* (1926): 54–74.
"Sur le mesure des grandeurs." *l'Enseignement Mathématique* 31 (1931).

## FURTHER READING

### Books

Hawkins, Thomas. *Lebesgue's Theory of Integration.* Madison: University of Wisconsin Press, 1970.

### Periodicals

Burkill, J. C. "Henri Lebesgue." *Journal of the London Mathematical Society* 19 (1944): 56–64.
Denjoy, A., L. Felix, and P. Montel. "Henri Lebesgue, le savant, le professeur, l'homme." *l'Enseignement Mathematique* 1957.
Kac, Mark. "Henri Lebesgue et l'Ecole Mathematique Polonaise: apercu et souvenirs." *l'Enseignement Mathematique* 21 (1975): 111–114.

*—Sketch by Robert Messer and Tom Chen*

# Adrien–Marie Legendre
## 1752–1833
### French geometer, number theorist, and elliptic function theorist

Adrien–Marie Legendre is best known as the author of *Éléments de géométrie*, a simplification of **Euclid**'s *Elements*. Published in 1794, it went through numerous editions and translations, and served as a standard geometry text for the next hundred years. Legendre made significant contributions to several other fields, however, including number theory and celestial mechanics. In addition, he is generally regarded as the founder of the theory of elliptic functions.

Legendre was born in Paris on September 18, 1752. Although little is known of his early life, Legendre apparently came from a well–to–do family. He studied at the Collége Mazarin in Paris, where he received an unusually progressive education in mathematics. One of his professors there was Abbé Joseph–François Marie, a highly regarded mathematician. At age 18, Legendre successfully defended his theses in mathematics and physics, and at age 22 he issued his first publication, a treatise on mechanics.

Legendre's modest family fortune was sufficient to support his research. Nevertheless, in 1775 he accepted a position at the École Militaire in Paris. He taught there until 1780, but he had yet to make an impact on the scientific world. Then in 1782, he won a prize awarded by the Berlin Academy for an essay on the path of projectiles taking the resistance of air into account. Legendre's winning essay, which was published in Berlin, attracted the attention of **Joseph–Louis LaGrange,** one of the finest mathematicians of the day. LaGrange, in turn, sought more information on the young author from another great French mathematician, **Pierre Simon Laplace.**

In January 1783, Legendre presented his first major paper before the French Académie Royale des Sciences. This paper dealt with the attraction of planetary spheroids (objects that have a sphere–like shape, but are not perfectly round). He also submitted to Laplace essays on various topics, including the properties of continued fractions and the rotation of bodies subject to no accelerating force. As a result of these efforts, Legendre was elected to the Académie on March 30, 1783.

### Celestial Mechanics and Elliptic Functions

As Legendre strove to make his work better known, his career continued to flourish. In 1784 he presented another paper on celestial mechanics. This one dealt with the form of equilibrium of a sphere–like mass of rotating liquid. The paper introduced the famous "Legendre polynomials," which are solutions to a particular kind of differential equation that still plays an important role in applied mathematics. Legendre's interest in celestial mechanics eventually led to two further papers, one on the attraction of certain ellipsoids, and the other on the form and density of fluid planets.

A second line of productive research by Legendre involved number theory, and a third comprised elliptic functions. Legendre began both investigations in the mid–1780s, although it was not until later that he made his most significant contributions. Legendre's first paper on number theory dates from 1785. Among other things, it set forth the law of quadratic reciprocity, by which any two odd prime numbers can be related. His first publication on elliptic functions came the following year, which addressed the integration of elliptic curves.

In 1787 Legendre was assigned to work on a project in geodesy (a branch of applied mathematics dealing with the measurement of the Earth). This project was a joint undertaking of observatories in Paris and Greenwich, England. The most important result was a theorem that stated how a spherical triangle may be treated as a plane, provided certain corrections are made to the angles.

### Obstacles Along the Path to Success

By the close of the 1780s, however, Legendre's scientific progress was being impeded by the French Revolution. He was particularly affected by the suppression of the French Académie Royale des Sciences, which forced him to publish some of his research himself. Legendre's small fortune dwindled, and he had to seek work to support himself and his new bride, Marguerite Couhin. Beginning in 1791, Legendre served on several public commissions, including one that converted the measurement of angles to the decimal system. In 1794, he became a professor of mathematics at the short–lived Institut de Marat in Paris. Legendre succeeded Laplace as the examiner in mathematics of students assigned to the artillery in 1799, a position he held until 1815.

During this same period, Legendre published the popular text *Éléments de géométrie*, which dominated geometry instruction in many parts of the world for the next century. Later editions of the book also contained the elements of trigonometry, as well as proofs of the irrationality of $\pi$ and $\pi^2$. In addition, an appendix on the theory of parallel lines was issued in 1803. Although this book was among Legendre's more mundane achievements, it is one for which he is still remembered.

Another obstacle to Legendre's success was the jealousy of Laplace. In *A Short Account of the History of Mathematics*, W.W. Rouse Ball describes the situation this way: "The influence of Laplace was steadily exerted against [Legendre's] obtaining office or public recognition, and Legendre, who was a timid student, accepted the obscurity to which the hostility of his colleague condemned him."

### Major Publications Cap a Career

These distractions may have slowed Legendre's progress somewhat, but they could not completely stifle his creativity. In 1798, he published the first edition of *Essai sur la théorie des nombres*, in which he further explored number theory. Among other topics, Legendre returned to the law of quadratic reciprocity, which he had stated 13 years before. (**Karl Gauss** offered the first rigorous demonstration of this law in 1801.) Appendices were added to the book in 1816 and 1825. The book's third edition, issued as two volumes in 1830, is considered a standard text in the field.

In 1811, Legendre published the first volume of another major work, *Exercices de calcul intégral*, much of which dealt with elliptic functions. Second and third volumes followed by 1817. However, Legendre saved his most influential work for last. From 1825 through 1828, he published three volumes of *Traité des fonctions elliptiques*, in which he expanded upon his elliptic function research. Unfortunately, Legendre treated this subject merely as a problem in integral calculus, failing to realize that it might be considered as a higher trigonometry and, thus, a distinct branch of analysis. The discovery of superior methods for handling elliptic functions were left to a younger generation of mathematicians, including **Karl Gustav Jacob Jacobi** and **Niels Henrik Abel.**

Legendre recognized the superiority of Jacobi's and Abel's methods at once. In the final years before his death, Legendre issued three supplements to the third volume of *Fonctions elliptiques*. In these supplements, he discussed the younger mathematicians' ideas. As Ball notes in *A Short Account of the History of Mathematics*: "Almost the last act of [Legendre's] life was to recommend those discoveries which he knew would consign his own labours to comparative oblivion."

Legendre's contributions did not go entirely unheralded. He was made a member of the French Legion of Honor and was granted the title of Chevalier de l'Empire. This was a minor honor compared to the title of count, however, which was bestowed upon Laplace and LaGrange. When LaGrange died in 1813, his post at the Bureau des Longitudes was given to Legendre, who held it for the rest of his life.

Legendre died in Paris on January 9, 1833, after a long, painful illness. He and his wife had never had children. Legendre's widow made a cult of his memory, carefully preserving his belongings. Upon her death in 1856, she left to the village of Auteuil (now part of Paris) the last country house where the couple had lived.

## SELECTED WRITINGS BY LEGENDRE

*Elements of Geometry and Trigonometry*. Revised and edited by Charles Davies, 1859

## FURTHER READING

### Books

Ball, W.W. Rouse. *A Short Account of the History of Mathematics*. New York: Dover Publications, 1960, pp. 421–425.

Itard, Jean. "Adrien–Marie Legendre." *Biographical Dictionary of Mathematicians*. Volume 3. New York: Charles Scribner's Sons, 1991, pp. 1477–1486.

Porter, Roy, editor. *The Biographical Dictionary of Scientists*. Second edition. New York: Oxford University Press, 1994, pp. 421–422.

*—Sketch by Linda Wasmer Smith*

*Gottfried Wilhelm von Leibniz*

# Gottfried Wilhelm von Leibniz
## 1646–1716
### German logician and philosopher

Gottfried Wilhelm Leibniz' restless intellect ranged from mathematics, physics and engineering to politics, linguistics, economics, law and religion. He is best known for his contributions in metaphysics and logic, and for his invention, independent of **Isaac Newton**, of differential and integral calculus, which is an indispensable mathematical tool. Leibniz received his doctoral degree when he was 19 years old, declined an early offer of an academic career, and became a lifelong employee of noblemen. Although he had met and corresponded with many influential and learned men, his death was marked by indifference. A nationalistic debate over priority in developing the calculus during Leibniz's lifetime resulted in a century–long rift between English and continental mathematics.

Leibniz was born in Leipzig, Saxony (now Germany), on July 1, 1646, four years after the birth of Newton. His father, Friedrich Leibnütz, was a lawyer and professor of moral philosophy at the University of Leipzig; Gottfried's mother, Catherina Schmuck, was Friedrich's third wife. Both sides of the family enjoyed social standing and scholarly reputations. Leibniz (who changed the spelling of his name) had a half–brother, Johann Friedrich; a half–sister, Anna

Rosina; and a sister, Anna Catherina, whose son, Friedrich Simon Löffler, became his sole heir. His father died when he was six, but young Gottfried had already begun to demonstrate a passion for knowledge and omnivorous reading. He studied his father's library of classic, philosophical and religious works, and his school syllabus included German literature and history, Latin, Greek, theology and logic. By age 12, Leibniz read Latin and was adept at writing Latin verse. He began to formulate his own ideas in logic, among them the ideas of an alphabet of human thought and a universal encyclopedia.

### Receives Doctoral Degree in Law

In 1661, at age 15, Leibniz entered the University of Leipzig, where he studied with philosophy professor Jakob Thomasius. He received his bachelor's degree in 1663, then spent the summer in Jena studying with the mathematician Erhard Weigel, who introduced him to elementary algebra and Euclidean geometry. After receiving his master's degree in 1664, Leibniz wrote a dissertation for the Doctor of Law degree, but because of his youth, the university refused to award it to him. Leibniz subsequently entered the University of Altdorf in Nuremburg, where his dissertation was accepted and his degree was awarded in 1666. The university then offered him a professorial position, but he declined.

Leibniz's first job was in Nuremburg, as secretary of a society of intellectuals interested in alchemy. By chance he met Baron Johann Christian von Boineburg, a former Chief Minister of the Elector of Mainz. Boinburg became Leibniz' patron, and helped him obtain a position as assistant to the Elector's legal adviser; later Leibniz was promoted to Assessor in the Court of Appeals and to a diplomatic position. Leibniz spent the next five years in Mainz and Frankfurt. In addition to writing legal and position papers for the Elector, he wrote papers on religious subjects relating to the reunion of the Protestant and Catholic churches, and continued to develop his philosophy, to which logic was central. Leibniz began a voluminous correspondence with hundreds of people on every conceivable topic. More than 15,000 of his letters survive.

### Leibniz in Paris

France had been encroaching on the Rhineland, and in the winter of 1671–1672, Leibniz devised a plan to distract the French by encouraging them to conquer Egypt and build a canal across the isthmus of Suez. The plan eventually came to nothing, but it allowed Leibniz to accompany a diplomatic mission to Paris, meet prominent philosophers and scholars, and immerse himself in Parisian salons. This was the time of Leibniz' greatest advances in mathematics, and he formed a lifelong friendship with **Christiaan Huygens**, who was in the employ of the Académie Royale des Sciences. He was also able to obtain and copy unpublished manuscripts of **Blaise Pascal** and **René Descartes**, and he devised a calculating machine that could add, subtract, divide, multiply and extract roots.

In 1673, Leibniz visited London on a diplomatic mission, and at a meeting of the Royal Society of London displayed a model of his calculating machine. He met Society secretary Heinrich Oldenburg, with whom he had corresponded, and was elected to membership in the Society. He also talked with the chemist Robert Boyle, the microscopist Robert Hooke, and the mathematician John Pell. The latter pointed out the gaps in Leibniz' mathematical knowledge. Leibniz left London in the grip of mathematical ideas, and returned to Paris to study higher geometry under Huygens, beginning the work that led him to the discovery of differential and integral calculus. In 1673 he developed his general method of tangents.

In the meantime, both Leibniz' patron and the Elector had died and Leibniz' future with his successor was uncertain. He wanted to remain in Paris, and hoped that his reputation might win him a paid position at the Académie Royale des Sciences or the professorship vacated by the death of Roberval at the Collége Royal but was disappointed. Leibniz visited London again briefly, and in 1676 left Paris for Hanover to act as advisor to the Duke of Brunswick and take charge of the ducal library. En route to Hanover, where he would spend most of the next forty years, Leibniz stopped in Holland for discussions with mathematician Jan Hudde and microscopists Jan Swammerdam and Antoni van Leeuwenhoek, and at The Hague to talk with the philosopher Spinoza.

### Introduces the Calculus

In 1684, Leibniz' brief paper, "A New Method for Maxima and Minima, as well as Tangents, which is impeded neither by Fractional nor Irrational Quantities, and A remarkable Type of Calculus for This" appeared in *Acta Eruditorum*. Leibniz used the term "calculus" to mean rules. This paper addressed differential calculus; he introduced integral calculus in 1686 in the same publication. Newton's similar work was published in 1689; however, because of its approach, it was not immediately obvious that it was the same as Leibniz'. Newton had used a geometrical approach; Leibniz' was algebraic. Early in the 1700s, supporters of Newton in Great Britain and of Leibniz on the Continent began to argue about the merits of the two systems, and about the priority of their discovery. Newton and Leibniz were eventually drawn into the animosities, and the consequences lasted for more than a century. For a hundred years, Continental mathematicians using Leibniz' methods and superior notation moved ahead in the theory of the calculus, while the English were held back by Newton's more cumbersome method. It is clear that Newton and Leibniz developed calculus independently; however, Leibniz has provided its notation and its name.

Leibniz was a versatile and prolific contributor to mathematics. In *On the Secrets of Geometry and Analysis of Indivisible and Infinite Quantities* (1686) Leibniz first used the integral sign. In a 1693 letter, Leibniz used multiple indices to state the result of three linear equations. He also worked on the problem of elimination in the general theory of equations and laid the foundations of the theory of determinants. In another letter, he suggested expanding cube roots into infinite series. He introduced terminology, including the term "function," borrowed by Bernoulli.

In 1702, Leibniz published *A Justification of the Calculus of the Infinitely Small*, in which he attempted to justify his algorithms for differentation and integration. Between the years 1702–1703 he integrated rational fractions in trigonometric and logarithmic functions, produced a theory of special curves, and stated equations important in navigation. Leibniz was one of the first to work out the properties of the binary number system; he anticipated the central concerns of modern computer science, bringing human reasoning under mathematical law.

### Final Years

In 1686, Leibniz began a study of the genealogy of the House of Brunswick that would occupy the remainder of his life. He began his research with a three-year trip to Bavaria, Austria, and Italy, along the way meeting with scholars and scientists and writing treatises on other topics. In 1696 he was promoted to Privy Councillor at Brunswick-Wolfenbüttel. He had promoted the establishment of scientific academies in several countries, and in 1700, through the influence of the Electress Sophie Charlotte, the Berlin Academy was founded with Leibniz as president. He was also elected an external member of the Paris Académie Royale des Sciences in 1700.

As a young man, Leibniz had a reputation as a savant, a wit, and an elegant courtier. In his last years he was increasingly disregarded. On November 14, 1716, after a week in bed suffering from arthritis, gout and colic, Leibniz died in Hanover in the presence of his secretary, Johann Georg Eckhart. His funeral and burial took place on December 14 in Neustüdter Church. No one from the Court attended his funeral, and his grave went unmarked for 50 years. An elegy was read at the Académie Royale des Sciences a year after Leibniz' death, but neither the Royal Society of London nor the Berlin Academy published an obituary.

### FURTHER READING

Aiton, E. J. *Leibniz: A Biography*. Bristol: Adam Hilger Ltd., 1985.

Broad, C. D. *Leibniz: An Introduction*. Edited by C. Lewy. London: Cambridge University Press, 1975.

Dunham, William. *The Mathematical Universe: An Alphabetical Journey through the Great Proofs, Problems, and Personalities*. New York: John Wiley & Sons, 1994, pp. 143–158.

Hofmann, Joseph E. *Leibniz in Paris 1672–1676: His Growth to Mathematical Maturity*. London: Cambridge University Press, 1974.

Katz, Victor J. *A History of Mathematics: An Introduction*. New York: HarperCollins, 1993, pp. 472–482.

Motz, Lloyd and Jefferson Hane Weaver. *The Story of Mathematics*. New York: Plenum Press, 1993, pp. 141–150.

Ross, G. MacDonald. *Leibniz*. Oxford: Oxford University Press, 1984.

Schrader, Dorothy V. "The Newton–Leibniz Controversy Concerning the Discovery of the Calculus," in *From Five Fingers to Infinity*. Edited by Frank J. Swetz. Chicago: Open Court, 1994, pp. 509–520.

Stillwell, John. *Mathematics and Its History*. New York: Springer–Verlag, 1989, pp. 110–117.

Styazhkin, N.E. *History of Mathematical Logic from Leibniz to Peano*. Cambridge, MA: The M.I.T. Press.

*—Sketch by Jill Carpenter*

# Tullio Levi–Civita
## 1873–1941
### Italian geometer and theoretical physicist

Tullio Levi–Civita's most crucial contribution to mathematics lay in his development of tensor calculus (originally called absolute differential calculus). This new calculus had a widespread effect. Although **Albert Einstein**'s theory of relativity depended on its development, it had many mathematical applications. Related to his tensor calculus, Levi–Civita developed theorems about the curvature of spaces and Riemannian geometry's covariant differentiation. Levi–Civita also propagated work in pure geometry, hydrodynamics, engineering, celestial and analytical mechanics. In addition, Levi-Civita was a prolific author of more than 200 publications as well as a gifted and well-liked teacher.

Levi-Civita was born on March 29, 1873, in Padua, Italy. He was the son of Giacomo Levi-Civita, a lawyer and senator. Educated in his hometown, Levi-Civita was an outstanding student in secondary school there. He also studied at the University of Padua from 1891 to 1895, under the tuteledge of Gregario Ricci Curbastro. Levi-Civita received his BA in 1894, and published while an undergraduate. He began his teaching career at a Pavia-based teacher-training college in 1895. A year later, he published his first piece dealing with what would be his most important contribution to mathematics: tensor calculus. Entitled "Sulle trasformazioni delle equazioni dinamiche," Levi-Civita stretched the work of his mentor, Ricci Curbasto, by using Curbastro differential geometry methods in what became known as absolute differential calculus.

## Begins a Teaching Career

Levi-Civita began teaching at the University of Padua in 1898 and he served as the chair of Mechanics. Around 1900, Levi-Civita and Ricci Curbasto jointly published the progress they had made in the development of absolute differential calculus in the *Méthodes de calcul differéntiel absolu et leurs applications* ("Methods of the Absolute Differential Calculus and Their Applications"). They had been working on this calculus together for several years. As developed by them, absolute differential calculus was an unparalled breakthrough in part because it was applicable to many fields. It could be applied to three key mathematical spaces: both Euclidean and non-Euclidean, as well as Riemannian curved spaces. Albert Einstein used absolute differential calculus in formulating his theory of general relativity 15 years later. For several years after Einstein's discovery, however, Levi-Civita was uncertain about parts of the theory of relativity, but he eventually embraced it.

In 1902, Levi-Civita became a professor of rational mechanics at Padua, a position he held until 1918. Levi-Civita's contributions to science were not limited to absolute differential calculus. Two subjects in which he showed extreme interest in and did the most work on while at Padua were celestial mechanics and hydrodynamics. From 1903-1916, Levi-Civita studied various aspects of celestial mechanics. He was especially concerned with the three body problem (how three bodies move when considered as mass centers and under Newton's mutual attraction theories). He achieved results on the problem from 1914-1916, but graciously admitted another scholar, Karl F. Sundmann, had reached the same conclusions in a roundabout manner. Levi-Civita also published relevant work on hydrodynamics, a subject of which

he was particularly fond, from 1906 on. He focused on concepts such as how an immersed solid's translational movements relate to the resistance of a liquid.

## Absolute Differential Calculus Becomes Tensor Calculus

Levi-Civita married one of his students, Libera Trevisani, in 1914. Before leaving Padua with his wife in 1917, Levi-Civita published his most important solo work in absolute differential calculus. In it, he promulgated the idea of parallel displacement and its place in curved space. It was this breakthrough that helped Einstein with the mathematical models of relativity. Parallel displacement had relevancy in other areas as well; in pure mathematics, it affected the growth of the theory of modern differential as applied to generalized spaces in topology and the geometry of paths. The scholarly discourse that rose from Levi-Civita's absolute differential calculus influenced its evolution into tensor calculus. Tensor calculus is used by mathematicians in the derivation of gravitation and electromagnetism unified theories.

Levi-Civita left Padua in 1918 to become a professor of higher analysis at University of Rome. In 1920, he became a Rational Mechanics Professor, a post he served at until he was forced to retire in 1938. Levi-Civita's work did not go unnoticed by his international peers. In 1922, Levi-Civita won the Sylvestor Medal from the Royal Society of London, which elected him a foreign member in 1930.

In Rome, Levi-Civita published a series of important works. He wrote *Lezioni di meccanica razionale* ("Lessons in Rational Mechanics") in 3 volumes, from 1923-27. In 1924, he produced *Questioni di meccanica classica e relativistica* ("Questions of Classical and Relativistic Mechanics"). A year later, his *Lezioni di calcolo differenziale assoluto* ("The Absolute Differential Calculus") appeared. He also published more on hydrodynamics and his theory of canal waves in the same year. Levi-Civita returned to questions raised by Einstein in 1929, when he published *A Simplified Presentation of Einstein's Unified Field Equations*. During the 1920s, Levi-Civita also was inspired by the rise of atomic physics to explore related mathematical problems.

An outspoken opponent of fascism, Levi-Civita was forced out of his professorship and into retirement in 1938 because of anti-Semitic laws. His membership in scientific societies of Italy was also revoked. Soon after his forced removal, Levi-Civita's health began to decline and he suffered from heart problems. A stroke ultimately caused his death in Rome on December 29, 1941.

## SELECTED WRITINGS BY LEVI-CIVITA

*The Absolute Differential Calculus.* Edited by Entrio Persico and translated by Miss M. Long, 1977

*The N–body Problem in General Relativity,* 1964

## FURTHER READING

### Books

Abbott, David Abbott, editor. *The Biographical Dictionary of Scientists: Mathematicians*, New York: Peter Bedrick, 1986, p. 85.

Gliozzi, Mario. "Tullio Levi–Civita," in *Dictionary of Scientific Biography*. Volume VIII. Edited by, Charles Coulston Gillispie. New York: Charles Scribner's Sons, 1970, pp. 284–85.

*—Sketch by Annette Petruso*

# Marius Sophus Lie
## 1842–1899
### Norwegian geometer

Marius Sophus Lie was one of the first prominent Norwegian scientists and among the last of the great 19th-century mathematicians. His main contribution was his theory of groups. Lie groups and Lie algebras are fundamental tools in many parts of 20th century mathematics, from the theory of differential equations to the understanding of elementary particle physics. Although he was an isolated academic who generally lacked regular contact with colleagues or interested students, Lie produced his finest work in collaboration with **Felix Klein** and later Friedrich Engel.

Lie was born on December 17, 1842, in Nordfjordeide, Norway, the youngest of six children. His father, Johann Herman Lie, was a Lutheran pastor. Lie's education was standard: he first studied in Moss, moving onto Kristiania (present–day Oslo) to study at Nissen's Private Latin School from 1857 to 1859. For the next six years, Lie studied mathematics and science at Kristiania University, graduating without distinction in 1865.

### Chooses a Career in Mathematics

Lie tutored other students for the next few years and pursued his own interests in astronomy and mechanics, but he did not find a professional field that appealed to him until 1868, when he was introduced to writings by **Jean–Victor Poncelet** and **Julius Plücker**. Lie was particularly intrigued by Plücker's proposal of a new kind of geometry that would use lines and curves instead of points as the elements of a given space.

After publishing his first paper in 1869, Lie was awarded a scholarship to study in Berlin, where he met Felix Klein, another mathematician indebted to Plücker's ideas. Lie and Klein spent the following summer in Paris, publishing several papers together, and it was here that Lie developed his idea of contact transformations.

Midway through the summer of 1870, the Franco–Prussian war broke out. En route to Italy, Lie was arrested by the French on charges that he was a Prussian spy. The arresting officers alleged that Lie's mathematical notes were in fact coded messages. Through the intervention of J.G. Darboux, a French mathematician he had worked with earlier that summer, Lie was released after spending a month in jail.

### Discovers Integration Theory and Lie Groups

Lie returned to Kristiania the following year to teach at two of his alma maters: the university and Nissen's Private Latin School. At this time, Lie formulated his integration theory of partial differential equations, sharing his results with Adolph Mayer, who was concurrently working on the same method. Lie was not recognized for this development at the time, perhaps because he did not write about his discovery using the accepted contemporary analytical language.

Lie earned his Ph.D. in 1872 and married Anna Sophie Birch two years later, having two sons and a daughter by her. During the 1870s, Lie worked on transformation groups, which he called finite continuous groups. These groups, later called Lie groups, possessed a fixed number of parameters but could be differentiated in any desired order. Lie applied his theory of transformation groups to show that a majority of the known methods of intergration could be introduced all together by means of group theory. He also used transformation groups to help classify ordinary differential equations and to give a unified method of solution using group–theoretic considerations. Lie later used his finite continuous groups in 1890 to identify the defects in Hermann von Helmholtz's application of group theory to the foundations of geometry.

While Lie was fond of taking hikes and admiring the Norwegian landscape, he found that Kristiania offered him little contact with other mathematicians or interested students. In 1886, his friend Klein, with whom he had maintained correspondence over the years, suggested that Lie take his place as professor of geometry at Leipzig University. Lie accepted, and his 12 years at Leipzig proved to be his most prolific.

### Writes Definitive Work on Transformation Groups

Beginning in 1884, Lie teamed up with Friedrich Engel, a student recommended by Klein and Mayer, and together they produced a three–volume work on transformation groups. *Theorie der Transformationsgruppen*, published between 1888 and 1893, owes much of its completion to Engel, as Lie failed to finish most of his other works on his own. It was during this time that Lie began suffering an acute nervous breakdown, diagnosed by doctors at the time as "neurasthenia." Although he was treated in a mental hospital in 1890, Lie remained severely depressed. In the third part of the *Theorie der Transformationsgruppen*, Lie wrote that "I am no pupil of Klein, nor is the opposite the case, although this might be closer to the truth." Klein was deeply hurt by the remark but still welcomed Lie into his home after the incident.

In 1898, Lie was awarded the first International Lobachevsky prize, and in September of that year he returned to teach at Kristiania University. The university had created a special mathematics chair to lure back the Norwegian native, but Lie occupied it for only a short while. He died on February 18, 1899, at the age of 56, of pernicious anemia.

Lie's influence continued well after his death, with mathematicians all over Europe continuing to work on Lie groups. Wilhelm Killing began classifying Lie groups during Lie's lifetime, only to garner Lie's criticism when his work produced errors. Killing's final work, which **Elie Joseph Cartan** later revised and based much of his work on, was still an important contribution. **Hermann Weyl** breathed new life into Lie's groups in his papers from 1922 and 1923, and subsequent generalizations of Lie's groups gave them a greater role in  quantum physics and quantum mechanics.

## SELECTED WRITINGS BY LIE

*Lie Groups: History, Frontiers, and Applications.* Translated by Michael Ackerman, 1975
*[1880 transformation group paper].* Translated by Michael Ackerman, 1975
*[1884 differential invariant paper].* Translated by Michael Ackerman, 1976

## FURTHER READING

### Books

Freudenthal, Hans. "Marius Sophus Lie," in *Dictionary of Scientific Biography*. Volume VIII. Edited by Charles Coulston Gillispie. New York: Charles Scribner's Sons, 1976, pp. 323–327.
Yaglom, I.M. *Felix Klein and Sophus Lie.* Translated by Sergei Sossinsky, edited by Hardy Grant and Abe Shenitzer. Boston: Birkhauser, 1988.

### Periodicals

Baas, N.A. "Sophus Lie." *Mathematical Intelligencer* 16 (Winter 1994): 16–19.
Ibragimov, N.H. "Sophus Lie and Harmony in Mathematical Physics, on the 150th Anniversary of His Birth." *Mathematical Intelligencer* 16 (Winter 1994): 20–28.

*—Sketch by Bridget K. Hall*

# Chia-Chiao Lin
## 1916-
### Chinese-born American applied mathematician and astronomer

Born in Fukien, China, on July 7, 1916, the son of Kai and Y. T. Lin, Chia-Chiao Lin pursued his undergraduate studies at the National Tsing Hua University in China, receiving a B.Sc. in 1937. He was awarded a master's degree in applied mathematics from the University of Toronto in 1941 and a Ph.D. in aeronautics from the California Institute of Technology in 1944. A former student of Theodore von Kármán whose varied background includes mathematics, aeronautics, and fluid mechanics, Lin's work has contributed to multiple disciplines within the scientific community, as well as to government and industry. His importance lies in his use of mathematical modeling to create new formal tools for theoretical investigation in a number of sciences, including meteorology, oceanography, astrophysics, chemical engineering, and planetary sciences.

Following early work on fluid mechanics, Lin turned to a concentration on the hydrodynamics of superfluid helium and, later, on astrophysics. It was his work in this latter field that led to his development of the density wave theory of the spiral structure of galaxies, which provided an answer to one of the most long-standing puzzles in astronomy. This theory attempts to explain the formation of galaxies, their shapes (elliptical, normal spiral, barred spiral, etc.), and their luminosity.

Scientists had long recognized that the shapes of galaxies appeared to have a certain regularity—a regularity that could be governed by wave phenomena. The density wave theory was developed to explain these patterns and to analyze their dynamical implications; such implications include star formation, which the theory explains as being triggered by galactic shocks induced by a low-amplitude density wave pattern. The density wave theory also classifies spiral galaxies by associating specific shapes with particular wave patterns, or "modes'; according to Lin, normal spiral modes and barred spiral modes correspond to normal spiral galaxies and transition barred galaxies and some barred spiral galaxies. Lin's theory has since been confirmed by a wealth of observational data and has provided a model for applying theoretical mathematics and physics—often in conjunction with computer science—to other disciplines.

Lin has been instrumental in the exchange of information between scientists in the United States and China and has organized trips by Chinese scientists to the United States. He has served as an advisor to the Chinese government on issues involving education and as an educational consultant for applied mathematics groups. He is a member of the National Academy of Science's Committee on Support of Research in the Mathematical Sciences; he served as the president of the Society for Industrial and Applied Mathematics and the chairman of the Committee on Applied Mathematics of the American Mathematical Society.

After a two-year stint as assistant and later associate professor of applied mathematics at Brown University, Lin took a position at Massachusetts Institute of Technology as professor of applied mathematics in 1947; he is now professor emeritus. He has also worked at Jet Propulsion Laboratories, and has served as a consultant to a number of industries. Lin married Shou-Ling Liang in 1946; they have one daughter.

## SELECTED WRITINGS BY LIN

### Books

*The Theory of Hydrodynamic Stability,* 1955
*Selected papers of C. C. Lin* (two volumes), 1987

### Periodicals

(With W. H. Reid) "Turbulent Flow, Theoretical Aspects." *Handbuch der Physik-Encyclopedia of Physics* 8 (1963).
(With Frank H. Shu) "On the Spiral Structure of Disk Galaxies." *Astrophysical Journal* number 140 (1964): 646–655.
(With Shu) "On the Spiral Structure of Galaxies II: Outline of a Theory of Density Waves." *Proceedings of the National Academy of Sciences* number 55 (1966): 229–234.

—*Sketch by Michael Sims*

# Carl Louis Ferdinand von Lindemann
## 1852–1939
### German analyst and geometer

The classic problem of squaring the circle had intrigued mathematicians since the time of **Euclid.** Only in 1882, however, when Ferdinand Lindemann proved that π is a transcendental number, was this problem finally resolved. While Lindemann is best known for this one result, he also played an important role in the development of mathematics in Germany during the turn of the 20th century.

Lindemann was born in Hanover, Germany, on April 12, 1852. His father was a teacher of modern languages and later a manager of a gas works while his mother was the daughter of a famous teacher of classical languages, so it is not surprising that their son finished first in his class upon graduating from his gymnasium in 1870. France and Germany had recently gone to war, but Lindemann's poor health prevented him from being called into the army. Instead, he enrolled at the University of Göttingen to study mathematics.

Göttingen attracted many of Europe's leading mathematicians. During his time there Lindemann attended lectures by Alfred Clebsch on analytic spatial geometry, algebraic curves, elliptic functions, and the theory of algebraic forms. He also met **Felix Klein,** who was then a lecturer at the university. In 1872 Klein became a full professor at the University of Erlangen; Lindemann joined him as Klein's second Ph.D. student, receiving his degree in 1873 with a thesis on non–Euclidean geometry and its connection with mechanics. In addition, after Clebsch's sudden death in 1872 and with Klein's encouragement, Lindemann edited and revised Clebsch's geometry lectures which he published as a textbook in 1876. The Clebsch–Lindemann text won wide acclaim and was used for several decades.

### Proves π Is Transcendental

Lindemann spent part of the 1876–77 academic year in Paris, where he began a long friendship with **Charles Hermite.** Because of the success of the Clebsch–Lindemann text, he was introduced to many of the leading French mathematicians. He returned from Paris to become associate professor at the University of Freiburg after a promised position at the University of Würzburg never materialized. During his six years at Freiburg, Lindemann published several minor papers on special functions and Fourier series, and also wrote a paper on the vibration of strings, inspired by the recent invention of the microphone. But his main success came with his work on the number π. During Lindemann's visit to Paris in 1876, Hermite had shown him his proof that the number *e* is transcendental, that is, that *e* is not the root of any polynomial with integer coefficients. Building upon his friend's earlier work, Lindemann finally succeeded in 1882 in proving that π is also transcendental. He sent his paper "Über die Zahl π" (Concerning the number π) to Klein for publication in the *Mathematische Annalen*. Klein sent the paper to **Georg Cantor**, who could find no errors, and who passed the paper on to **Karl Weierstrass** in Berlin for final verification of the proof. With Lindemann's permission, on June 22, 1882, Weierstrass presented the result to the Berlin Academy of Sciences to great acclaim.

The problem of squaring the circle, that is, constructing a square with the same area as that of a given circle, fascinated mathematicians for more than two thousand years. A solution had been found by **Dinostratus** around 350 B.C., but no one had ever been able to find a solution using just the classical Euclidean tools of the straightedge and compass. Mathematicians knew that if a number was transcendental, and hence not algebraic, then no line of that length could be constructed using these tools. Lindemann's proof that π was transcendental, and hence unconstructible, finally established unequivocally that the squaring of the circle was impossible by means of straightedge and compass alone.

### Teacher, Advisor, and Adminstrator

With the fame of his work on π freshly behind him, Lindemann accepted an appointment as full professor at the University of Königsberg in 1883. After ten years, he moved one final time to take a chair in mathematics at the University of Munich. He never again published a paper to rival the importance of his work on π. Nevertheless, Lindemann had a successful career as a teacher, an advisor of students, and an administrator. He supervised more than 60 German and foreign Ph.D. students, including **Hermann Minkowski** and **David Hilbert**. During his years in Munich, Lindemann served as dean of the arts and sciences, as rector of the university (an elected position comparable to that of president), and for 25 years as the director of the university's administrative committee. For several years Lindemann was also a confidential advisor to the king's court. In 1918, he received the Knight's Cross of the Order of the Bavarian Crown, an honor that granted nobility and the right to be known as Ferdinand Ritter von Lindemann.

In 1887, Lindemann married Lisbeth Küssner, a successful actress from Königsberg. They had two children, both born in Königberg, a son in 1889 and a daughter in 1891. Their son died tragically at the age of 22 during a mountain climbing accident in the Alps. Lisbeth apparently had mathematical as well as acting talents as she collaborated with her husband in translating and revising some of the works of the French mathematician **Henri Poincaré**. Lindemann died on March 6, 1939, three years after his wife. In his article on the man who discovered the transcendence of π, Fritsch writes that "he still published mathematical papers and thought about problems up to the day before his death."

## SELECTED WRITINGS BY LINDEMANN

"Alfred Clebsch: 'Vorlusungen über Geometrie," bearbeitet and herausgegeben von Dr. Ferdinand Lindemann, 1876
"Über die Zahl π," in *Mathematische Annalen* 20 (1882): 212–25.
"Wissenschaft und Hypothese" (translation of Henri Poincaré's "La Science et l'hypothèse"), 1904

## FURTHER READING

Dunham, William. *Journal through Genius: The Great Theorems of Mathematics*. New York:John Wiley & Sons, Inc., 1990.
Fritsch, R. "The Transcendence of π Has Been Known for about a Century—but Who Was the Man Who Discovered It?" *Results in Mathematics* 7 (1994): 164–83.
Wussing, H. "Carl Louis Ferdinand Lindemann," in *Dictionary of Scientfic Biography*. Volume VIII. Edited by Charles Coulston Gillispie. New York: Charles Scribner's Sons, 1974, pp. 367–68.

—*Sketch by Larry Riddle*

# Joseph Liouville
## 1809–1882
### French number theorist

Although Joseph Liouville's primary contribution to mathematics was the first proof of the existence of transcendental numbers (real numbers that are not roots of polynomials with integer coefficients), he had a wide range of mathematical interests. Liouville's publishing and teaching activities were vital to French mathematics during the 19th century. He made critical contributions to number theory, differential geometry, celestial mechanics, and rational mechanics. An enthusiastic lecturer, Liouville held numerous teaching positions, usually two or more at a time. Simultaneously, he was a prolific author, publishing 400 pieces in his lifetime, including over 200 on number theory alone.

Liouville was born in St. Omer, Pas–de–Calais, France, on March 24, 1809. He was the second son of an army captain, Claude–Joseph Liouville, and his wife, Thérèse (nee Balland). Liouville received his early education in Commery and Toul, before being accepted at the l'École Polytechnique. He began studying there in 1825, when he was 16 years of age, and left that institution two years later to enter l'École des Ponts et Chaussées. Liouville switched schools because of changing interests; he wanted to be an engineer. When he graduated in 1830, he was offered an engineering position, but by then he had decided he wanted to study mathematics full–time in the French center for mathematics, Paris.

Even before Liouville graduated from l'École des Ponts et Chaussées, he began to publish articles as early as 1828. He produced articles and notes on electricity and heat in scholarly journals. This early activity marked the beginning of a lifetime of fruitful scholarship. Liouville's primary publishing phase lasted until 1857. In that time, he had 100 or more treatises printed on mathematical analysis, as well as such topics as geometry, physics, algebra, and number theory.

Several years after graduation, in 1830, Liouville married his maternal cousin, Marie–Louise Balland. They had a family of three daughters and one son. To support his family, Liouville took on as many teaching positions as possible in secondary schools, sometimes teaching 34 hours or more a week. His publishing activities combined with his teaching experience led to his post as a *répétiteur* at the l'École Polytechnique at 1831. Liouville became a lecturer at l'École Centrale des Arts et Manufactures (an engineering school) in 1833. His enthusiasm for teaching led him to earn his doctorate in 1836 so he could hold professorships at the university level.

In this same year, 1836, Liouville filled a gaping hole in French mathematics academia when he founded the *Journal de mathématiques pures et appliquées* (also known as the *Journal de Liouville* or *Liouville's Journal*). There had been no forum for publishing French mathematical papers for the five years previous to his founding of the *Journal*. Although Liouville had no editorial experience, he edited this important publication for almost 40 years. He used his editorial power judiciously, publishing the best of the contemporary greats and helping young mathematicians get their first works in print. Liouville relinquished control of the *Journal* in 1874 when Henry Résal took over the editorship.

At the same time his teaching and publishing activities bloomed, Liouville experienced the most fruitful research period of his career. From 1832 to 1833, he concentrated on algebraic functions, specifically looking at integrals and their analytic behavior. This work led to his 1844 discovery of the proof of the existence of transcendental numbers, one of his most influential contributions to mathematics. In 1836–1837, he published influential papers with fellow mathematician Charles–François Stürm in the *Journal de mathématiques*. They delineated what became known as the Stürm–Liouville theory, based on Liouville's methodology for boundary–value problems. This theory became important in physics and integral equation theory. In analysis, Liouville is well known for proving that a bounded entire analytic function on the complex plane must be a constant function. The fundamental theorem of algebra follows as a simple corollary of Liouville's theorem.

Liouville continued to hold multiple teaching positions throughout his career. He resigned from l'École Centrale des Arts et Manufactures in 1838 when he became the professor and chair of Analysis and Mechanics at l'École Polytechnique. The previous year, 1837, Liouville began teaching at the Collège de France as an assistant for a professor there. He resigned in 1843 in protest of the Collège's choice for mathematics chair, Count Libri–Carrucci.

Liouville's contributions to French mathematics did not go unnoticed by his peers. In 1839, he became a member of the French Académie Royale des Sciences. A year later he became a member of Bureau des Longitudes, and served as its director at one point.

In 1848, Liouville took an unexpected turn into politics after the French Revolution of 1848. He won a seat on the constituent assembly as a moderate republican, but lost in an election for a position on the Legislative Assembly. This loss marked the end of his brief foray into elective politics.

Liouville resigned from l'École Polytechnique in 1851 and returned to the Collège de France. Libri–Carrucci had left France and Liouville was appointed to the mathematics chair in his place. Liouville taught at the Collège nearly continuously until 1879 or until his death in 1882 (depending on the source). He still maintained multiple teaching posts late in his career. He served as a professor of rational mechanics at the Sorbonne from 1857 to 1874, although he often used substitute teachers as his health began to decline. After his appointment at the Sorbonne, Liouville's research focused almost exclusively on two specialized topics in number theory.

Liouville's final years were marked by pain and suffering. From 1876 until his death, Liouville suffered from intense insomnia and gout. His wife and only son both died in 1880. Liouville himself died in Paris on September 8, 1882.

## SELECTED WRITINGS BY LIOUVILLE

(With Joseph Ritt) *Integration in Finite Terms;
Liouville's Theory of Elementary Methods,*
1948

## FURTHER READING

### Books

Abbott, David. *Biographical Encyclopedia of Scientists: Mathematicians.* New York: Peter Bedrick Books, 1986, p. 87.

Asimov, Issac. *Asimov Biographical Encyclopedia of Science and Technology.* Second revised edition. New York: Doubleday & Co., Inc., 1982, pp. 368–69.

Daintith, John, Sarah Mitchell, and Elizabeth Toothill, editors. *A Biographical Encyclopedia of Scientists.* Volume II. New York: Facts on File, Inc., 1981, p. 486.

Gascognie, Robert, editor. *A Chronological History of Science: 1450–1900.* New York: Garland Publishing, 1987, p. 339.

Taton, René. "Joseph Liouville," in *Dictionary of Scientific Biography.* Volume VIII. Edited Charles Coulston Gillispie. New York: Charles Scribner's Sons, 1970, pp. 381–87.

### Periodicals

Dieudonné, J. Review of *Joseph Liouville 1809–1882: Master of Pure and Applied Mathematics,* by Jesper Lützen. *The Mathematical Intelligencer* (Winter 1992): 71–73.

*—Sketch by Annette Petruso*

# Elizaveta Fedorovna Litvinova
## 1845–1919(?)
### Russian mathematics educator and biographer

Elizaveta Fedorovna Ivashkina Litvinova, a daughter of a landowning family in Russia's Tula region, was born in 1845. At a time when education for girls was generally limited to housekeeping and etiquette, Litvinova attended the Marinskaia in St. Petersburg, a progressive all–girls academy that offered some basic academic instruction. Still, the course of study available at the Marinskaia was far less comprehensive than the academic programs available at institutions for boys.

As a young woman living in St. Petersburg during the 1860s, Litvinova discovered a radical group that based its philosophy on nihilism. Members of this cultural movement believed that the natural sciences would provide the foundation for peaceful social change. For an intellectually vigorous young woman like Litvinova, such a philosophy was irresistible. She joined nihilist discussion groups and wrote revolutionists' poetry. And, as a nihilist, Litvinova embraced her duty to pursue advanced studies in the natural sciences.

Her family adamantly refused to support her in her efforts to earn a place in a university. Litvinova also had to work, furiously and independently, to make up the deficiencies in her academy training so she could pass the certification examination for graduates of male academies, a requirement for university admission. Some sympathetic university professors held seminars for women in the homes of wealthy nihilist supporters. Litvinova studied mathematics with A. N. Strannoliubskii, an academician who donated his spare time to helping women qualify for university admission and who counted **Sofia Kovalevskaia**, who would become Russia's most famous female mathematician, as one of his protegees.

In 1865, Russian women began to emigrate to institutions in Geneva, Bern, Heidelberg and Paris, which were opening their doors to women. The University of Zurich became a magnet for Russian women pursuing advanced degrees. As others flocked to Switzerland, Litvinova was barred from following them by an obstacle of her own making. In 1866, she married. Little is known about her husband, Dr. Litvinov (even his first name is not recorded) except for his objection to his wife's academic aspirations. Russian women needed the approval of a father or husband to obtain passports, and Dr. Litvinov consistently refused to give his wife permission to leave Russia. By 1870, Litvinova had passed the required competency examination. Still, she languished in Russia; then, as she wrote in her diary, "fate itself" intervened. She was somehow freed of her husband—whether he died, disappeared, or divorced her, Litvinova never disclosed it. She managed to obtain a passport, and in 1872 she left Russia for Zurich.

### A Colony of Scholars in Zurich

Litvinova arrived in Zurich as a typical college student—short on money but rich with enthusiasm. But the Swiss viewed the colony of Russian female students with suspicion. They were cynical about the morals of women who had left husbands and families to pursue education, and they deplored the perceived assault on femininity by this cohort of Russian women who were invading the natural sciences. Instead of studying at the university, where the growing number of women students inspired some tolerance, Litvinova enrolled at the Polytechnic Institute. In many of her lectures, Litvinova was the only woman among hundreds of males; later, she wrote that she feared to raise her eyes because the men would think her a woman of loose morals.

Despite such social pressures, Litvinova was happy in Zurich. Her studies absorbed her, and some of her professors were kind. She studied with the French professor Mequet; she not only attended lectures by mathematical analyst Hermann Schwarz, but was occasionally invited to share family tea in Schwarz's home.

Litvinova moved to Zurich with a four–year plan for completing her baccalaureate. But for some time, the autocratic Russian government had eyed the female Zurich colony with distrust. These women had not only defied Russian social conventions—they tended to embrace unconventional, even revolutionary political philosophies. In 1873, the tsar issued an edict that required all women studying abroad to return to Russia by January 1, 1874. Those who ignored the tsar's *ukase* would be banned from entering Russian universities (when and if they began admitting women) and would be barred from sitting for all licensing and civil service exams. Litvinova refused to obey. She remained in Zurich and completed her baccalaureate degree in 1876, then earned a doctorate from Bern University in 1878. She could not know at the time that her dissertation in function theory would be the only mathematical paper she would publish.

### The Price of Defiance

Following her return to Russia in 1878, Litvinova immediately faced the consequences of her rebellion. The doctorate, the Swiss teaching certification, and the stack of glowing recommendations from her Zurich professors meant nothing. She was prohibited from sitting for teachers' licensing exams and she was banned from full–time posts in any state–licensed institution. Litvinova was reduced to teaching in the lower classes of a women's academy. As an unlicenced teacher, she had no rights to a pension, vacation, or salary. She was paid by the hour and supplemented her meager earnings by writing popular accounts of the lives of mathematicians and philosophers, including a biography of Sofia Kovalevskaia.

After nine years, Litvinova finally won the right to teach upper level courses, becoming the first woman in Russia to teach the equivalent of high school mathematics. Still, she was denied the rights, privileges, and compensation afforded men at the same teaching rank. Her interest in teaching itself grew. Litvinova began using word problems in her classes and emphasized alternative approaches to proofs. She encouraged her students to use mathematics as a guide to logical thinking, making generaliza-

tions, and identifying underlying principles by studying groups of individual cases. During her 35–year teaching career, she managed to publish more than 70 papers on the philosophy and method of teaching mathematics.

Little is known about Litvinova's life after she retired from teaching. It is believed she lived with a sister in the country and died in 1919, but the date of her death has never been confirmed.

## SELECTED WRITINGS BY LITVINOVA

### Books

*Lösung einer Abbildungsaufgabe*, 1879 (doctoral thesis).
*Rulers and Thinkers: Biographical Essays*, 1897
*S. V. Kovalevskaia (Woman–Mathematician): Her Life and Scientific Work*, 1894

### Periodicals

"'Little One' (From the Life of the Zurich Women Students)," in *Pervyi Zhenskii Kalendar'na 1912.* (1912): 112–116.

## FURTHER READING

Koblitz, Ann Hibbner. "Elizaveta Fedorovna Litvinova," in *Women of Mathematics: A Biobibliographic Sourcebook.* Edited by Louise S. Grinstein and Paul J. Campbell. Westport, CT: Greenwood Press, Inc., 1987, pp. 129–134.

————. *A Convergence of Lives: Sofia Kovalevskaia: Scientist, Writer, Revolutionary.* Boston: Birkhauser Boston Inc., 1983.

————. "Elizaveta Fedorovna Litvinova (1845–1919): Russian Mathematician and Pedagogue." *Association for Women in Mathematics Newsletter* 14, no. 1 (January–February 1984): 13–17.

### Other

"Elizaveta Fedorovna Litvinova." *Biographies of Women Mathematicians.* June 1997. http://www.scottlan.edu/lriddle/women/chronol.htm (July 22, 1997).

*—Sketch by A. Mullig*

# Nikolai Ivanovich Lobachevsky
## 1792–1856
### Russian geometer and educator

Nikolai Ivanovich Lobachevsky is the first mathematician to publicly publish a system of non–Euclidean geometry. Although **Karl Friedrich Gauss** preceded him in the late 18th century and **János Bolyai** had devised a similar (though less analytical) conclusions around the same time, Lobachevsky showed that Euclid's Fifth postulate (also known as the Parallel postulate) could not be proved on the basis of the other postulates, and in turn created a new way of looking at geometry and geometric problems. Most of Lobachevsky's contemporaries scoffed at his conclusions, and he only became credited with his discoveries after his death. In fact, Lobachevsky sought credibility by publishing in different languages, but only a few of his colleagues supported his findings, including Gauss. Lobachevsky also did relevant research in other areas, including infinite series theory, integral calculus, probability, and the approximation of roots of algebraic equations.

Lobachevsky was born on December 1, 1792, in Nizhny Novgorod (known as Gorky from 1932 to 1990), Russia. His father, Ivan Maksimovich Lobachevsky, was a peasant of Polish descent who worked as a clerk in a provincial land–surveying office. His mother was named Praskovia Aleksandrovna Lobachevskaya. Lobachevsky's father died when he was about six or seven, depending on the source, and his mother took him and his two brothers, Alexander and Alexei, to Kazan where he spent the rest of his life. He attended the local Gymnasium on scholarship, then entered the University of Kazan when he was 14. Lobachevsky began his higher education in medicine, but when he began to study mathematics with Johann Bartels (a friend of Gauss), Lobachevsky switched majors. He earned his Master's degree in both mathematics and physics in 1811 or 1812.

### Serves University of Kazan in Many Ways

Soon after graduation, in 1814, Lobachevsky joined Kazan's faculty as a lecturer. Later that year, he was promoted to associate professor of mathematics, then became an extraordinary professor in 1816. In these early years, Lobachevsky attempted to deduce the Fifth postulate as a theorem, but he found that he could not make it work. He then started critically analyzing various versions of the postulate.

As Lobachevsky came to his earthshattering non–Euclidean conclusions, he also played a huge role in the development of his University. In 1820, he began administrative duties at Kazan. He served as dean of the mathematics and physics' faculty twice, from 1820 to 1821, then again in 1823 to 1825. In 1822, he became a member of the committee that supervised new building construction on campus and studied architecture to better understand the duties involved. He became the chair of this committee in 1825. Lobachevsky then became the University librarian from 1825 to 1835, and was elected University Rector (equivalent to president) in 1827, a post he held until 1846. His skills in administration did much to improve Kazan, taking it from a chaotic state, and in addition to restoring innovation, raised both academic standards and the faculty's equanimity. Lobachevsky also founded an academic journal *Uchenye zapiski* ("Scientific Memoirs").

In addition to these many administrative duties, Lobachevsky held a full professorship at Kazan from 1823 to 1846 and still found time to accomplish important mathematical progresses. He wrote *Geometriya* in 1823, in which he outlined his initial ideas that developed into his non–Euclidean geometry. This book remained unpublished as written until 1909, another example of his contemporaries' disregard of his ideas. Still, Lobachevsky described his ideas in a lecture for the Kazan faculty in 1826.

### Publishes "Imaginary Geometry" Findings

In 1829, the first complete published account of non–Euclidean geometry appeared in a Kazan–based publication called the *Kazan Messenger* after the leading scientific journals in Russia declined to print it. In 1837 he published his article "Géométrie imaginaire" ("Imaginary Geometry"), and the most important and full version of his new geometry was published in 1840 as *Geometrische Untersuchungen zur Theorie der Parellellinien* ("Geometric Investigations of the Theory of Parallel Links"). In these articles, Lobachevsky demonstrated that logical possibility of non–Euclidean geometry, which he called "imaginary geometry," as an analogy to imaginary numbers. In particular, Lobachevsky proved that Euclid's Fifth postulate was not a deducible result of the rest of Euclid's postulates.

Lobachevsky did not marry until he was 40 years old in 1832. He married the daughter of an aristocrat, Lady Varvara Aleksivna Moisieva. It was a generally unhappy marriage, and, although she came into the marriage with wealth, their economic situation deteriorated over the rest of Lobachevsky's life.

Lobachevsky also published in other areas of mathematics. In 1834, Lobachevsky devised a way to approximate the roots of algebraic equations. He also published a paper called "Algebra ili ischislenie konechnykh" ("Algebra, or Calculus of Finites"), which concerned the theory of infinite series, and a paper on the convergence of trigonometric series in which he proposed a general definition of a function similar to that later suggested by **Peter Gustav Lejeune Dirichlet.**

Though much of his work was dismissed in his lifetime, Lobachevsky was given a heredity nobility in 1837. Perhaps this honor was bestowed on him in part for such actions as personally saving lives during the cholera epidemic of 1830.

### Forced from the University of Kazan

Despite his valiant actions as administrator, teacher, and citizen of the university, Lobachevsky was relieved of his professorship in 1846. Some sources suggest this event occurred for political reasons. In that year, Lobachevsky became a government official, working in the Kazan educational district as an assistant trustee (or assistant guardian), where he remained until 1855 when he left because of his deteriorating health. Lobachevsky was not limited to his many university activities. He was a longtime member of the Kazan Economic Society, and was interested in agriculture as a hobby, which tied in with the society.

Lobachevsky's last publication of note, *Pangéométrie* (1855–56) was dictated by him in both French and Russian. (He spent the last years of his life blind or nearly so, because of cataracts.) The title means Pangeometry because he rightly thought non–Euclidean geometry had universal characteristics and applications. The text sums up his work in this field.

### Receives Acclaim Only After Death

Lobachevsky died in Kazan, on February 24, 1856. After his death, his work continued to be reprinted and translated elsewhere, which helped to spread his work and his reputation. Finally, Lobachevsky was given his due and put in a canon with other scientific greats. In the beginning of a book on his life and work, the author quotes W.K. Clifford: "What Vesalius was to Galen, what Copernicus was to Ptolemy, that was Lobachevsky to Euclid. There is, indeed, a somewhat instructive parallel between the last two cases. Copernicus and Lobachevsky were both of Slavic origin. Each of them has brought about a revolution in scientific ideas so great that it can only be compared with that wrought by the other. And the reason of the transcendent importance of these two changes is that they are changes in the conception of the Cosmos."

## SELECTED WRITINGS BY LOBACHEVSKY

*Geometrical Researches on the Theory of Parallels.* Translated by George Bruce Halstead, 1914

## FURTHER READING

Abbott. David, editor. *The Biographical Dictionary of Scientists: Mathematicians.* New York: Peter Bedrick Books, 1986, p. 86.

Asimov, Issac. *Asimov's Biographical Encyclopedia of Science and Technology.* Second revised edition. New York: Doubleday & Company, Inc., 1982, pp. 325–26.

Daintith, John, Sarah Mitchell, and Elizabeth Tootill, editors. *A Biographical Encyclopedia of Scientists.* New York: Facts on File, 1981, pp. 500–01.

Kagan, V. F. *N. Lobachevsky and his Contribution to Science.* Moscow: Foreign Languages Publishing House, 1957.

Rosenfeld, B.A. "Nikolai Ivanovich Lobachevsky," in *Dictionary of Scientific Biography.* Volume VIII. Edited by Charles Coulston Gillispie. New York: Charles Scribner's Sons, 1973, 428–35.

*—Sketch by Annette Petruso*

# Sheila Scott Macintyre
## 1910–1960
### Scottish analyst and educator

Although Sheila Scott Macintyre's career ended prematurely by cancer, she successfully juggled professional and family responsibilities while continuing to publish in her field. Her success in academics reflected the growing tolerance in Western society towards women entering higher education and specializing in previously male–dominated subjects. Macintyre's relationship with her husband, Archibald James Macintyre, was professional as well as personal, a successful role model for working couples of today. They worked on joint papers, served the British war effort together, and taught at the same universities, moving as a family from one post to the next. Macintyre helped produce a bilingual mathematics dictionary that went through two editions.

### An Only Child

Helen Myers Meldrum and James Alexander Scott, both natives of Scotland, had their only child on April 23, 1910. Macintyre was first sent to school at Trinity Academy in Edinburgh, where her father would serve as rector from 1925 to 1942. In 1926, she entered what was then known as Edinburgh Ladies' College, and in two years she became a "dux," or valedictorian, in mathematics and in her studies overall. This attracted the attention of the faculty of the University of Edinburgh, who granted Macintyre two bursaries. She was also awarded the Bruce of Grangehill Mathematical Scholarship.

Macintyre's M.A. was granted in 1932 with first class honors in mathematics and natural philosophy, and Edinburgh's faculty encouraged her to enter Cambridge University to pursue another B.A. of higher esteem. By the early 1930s women were allowed to be "wranglers," or top–rung math students, at Girton, the women's college of Cambridge. In two years Macintyre placed as a wrangler. This qualified her for a year–long research project, supervised by **Mary Lucy Cartwright**, who specialized in integral functions. The year bore fruit with Scott's first publication at the age of 25.

Another year of research followed, which was not equally satisfying. The next five years Scott spent teaching at girl's schools in Scotland, among them St.

Leonard's, in the town of St. Andrews, and James Allen's School for Girls.

### New and Original Problems

Archibald James Macintyre had earned a Ph.D. at Cambridge, but he did not meet his future wife until 1933, when they were introduced by a fellow academic in Scotland. After a lengthy courtship, they married on December 27, 1940. By this time, World War II took precedence for all citizens of Great Britain. The Macintyres worked in the same department at Aberdeen and taught courses sanctioned by the War Office and the Air Ministry. Macintyre took a year off between 1943 and 1944 for the birth of her first child, Alister William, before resuming her duties. She was retained for a permanent post as assistant lecturer at that time and simultaneously launched a thesis project. Her second son, Douglas Scott, was delivered almost the same day as her doctorate on the Whittaker constant. Unfortunately, this child succumbed to enteritis just before his third birthday.

A year later the Macintyres' daughter, Susan Elizabeth, was born. Over the next five years Macintyre raised their two children, continued to teach, and published a handful of papers on various problems related to the theory of functions of a complex variable. At least one commentator noted her ability to find "new and original problems" as well as her knack for refining older techniques and existing proofs. Macintyre also joined up with a member of the German department to produce a bilingual dictionary of mathematical terms, published in 1955.

The year 1958 seemed to mark a new phase of the Macintyres' careers. Long active in the Edinburgh Mathematical Society, Macintyre was elected a full Fellow of the Royal Society of Edinburgh. She published her last paper, on Abel's series. Macintyre and her family, with her father, joined Archibald at the University of Cincinnati. The couple served as visiting research professors there for a few years, where she was particularly successful and popular with her students. Macintyre died at age 50 on March 21, 1960.

### SELECTED WRITINGS BY MACINTYRE

#### Books

(With Edith Witte) *Mathematical Vocabulary (German–English)*. Second Edition, 1966

### Periodicals

"On the Asymptotic Periods of Integral Functions." *Proceedings of the Cambridge Philosophical Society* 31 (1935): pp. 543–554.

"A Functional Inequality." *Journal of the London Mathematical Society* 23 (1948): pp. 202–209.

(With A.J. Macintyre) "Theorems on the Convergence and Asymptotic Validity of Abel's Series." *Proceedings of the Royal Society of Edinburgh* A63 (1952): pp. 222–231.

## FURTHER READING

### Books

Fasanelli, Florence D. "Sheila Scott Macintyre," in *Women of Mathematics*. Edited by Louise S. Grinstein and Paul J. Campbell. Westport, CT: Greenwood Press, 1987, pp. 140–143.

*National Cyclopaedia of American Biography.* Volume 48. New York: James T. White & Co., 1965.

### Periodicals

Cartwright, Mary L. "Sheila Scott Macintyre." *Journal of the London Mathematical Society* 36 (1961): pp. 254–256.

Cossar, J. "Sheila Scott Macintyre." *Edinburgh Mathematical Notes* 43 (1960): p. 19.

Wright, E.M. "Sheila Scott Macintyre." *Year Book of the Royal Society of Edinburgh* (1961): pp. 21–23.

### Other

"Sheila Scott Macintyre." *MacTutor History of Mathematics Archive.*

http://www–groups.dcs.st–and.ac.uk/~history/Mathematicians/ Macintyre.html (July 1997).

"Shelia Scott Macintyre." *Biographies of Women Mathematicians.* June 1997. http://www/scottlan.edu/lriddle/women/chronol.htm (July 20, 1997).

—*Sketch by Jennifer Kramer*

*Colin Maclaurin*

# Colin Maclaurin
## 1698–1746
### Scottish geometer and physicist

Colin Maclaurin was one of Europe's foremost mathematicians during the 1700s. He was the first to provide systematic proof of **Isaac Newton**'s theorems.

Some of his noted accomplishments include explanations of the properties of conics and the theory of tides. Maclaurin was also a brilliant mathematician in his own right, who solved many problems in geometry and applied physics. Besides being an esteemed mathematician, Maclaurin was a creative inventor who loved to devise mechanical appliances. He was skilled in astronomy, mapmaking and sometimes spent his spare time acting as an actuary for insurance companies.

Maclaurin was born in Kilmodan, Scotland, in 1698, the son of a minister named John Maclaurin, a man of great learning. Unfortunately, his father died when Colin was just six years old. When his mother died nine years later, Maclaurin moved in with his uncle, Daniel Maclaurin.

Maclaurin's eldest brother, John, studied for the ministry and became a noted religious expert. Following his brother's lead, Maclaurin studied divinity at the University of Glasgow for a year. While at Glasgow, he met Robert Simson, professor of mathematics, who inspired Maclaurin's interest in geometry, especially the geometry of ancient mathematicians such as **Euclid**.

Maclaurin became interested in Newton's theories early on in his career. In 1715, he presented his thesis "On the Power of Gravity," which demonstrated real distinction, even though he was only in his teens when he defended it and earned a master of arts degree. The thesis also won him an appointment as a professor in mathematics at Marischal College in Aberdeen a year later.

In 1719, Maclaurin made two trips to London where he met some of the renowned scientists of the day, including Newton and Martin Folkes, who later became the president of the Royal Society of London. In 1720, Maclaurin published one of his signature works, *Geometrica Organica, sive descriptio linearum curvarum universalis*, which explained higher plane curves and conics. It proved many of the theories that Newton had proposed as well as solving other important problems in geometry. Maclaurin, for instance, showed that the cubic and the quartic could be represented by rotating these angles around their vertices. Newton had demonstrated similar properties for the conic sections.

In 1722, Maclaurin left Marischal to take on the tutoring of the son of Lord Polwarth, a powerful British diplomat. The two traveled through France, where Maclaurin produced another masterpiece *On the Percussion of Bodies*. For this work, the French Académie Royale des Sciences presented him with its prize in 1724.

When Polwarth's son died, Maclaurin hoped to reclaim his teaching position at Marischal, but during his three-year absence, it had been declared vacant and filled. When the chair of professor of mathematics at the University of Edinburgh fell vacant in 1725, he was appointed there. The prodigious Maclaurin then began lecturing on some of his favorite topics: the theories of Euclid, conics, astronomy, trigonometry, and Newton's *Principia*.

Maclaurin moved in Scotland's most inner circles. He was a fellow of the Royal Society and the Philosophical Society, where he acted as secretary. In 1733, he married Anne Stewart, the daughter of the solicitor general in Scotland. They had seven children.

In 1740, Maclaurin attracted international notice when he submitted an essay called "On the Tides" for the Académie Royale des Sciences prize. In the essay, Maclaurin explained the theory of tides, based on Newton's *Principia*, and defined the tides of the sea as an ellipsoid revolving around an inner point. Maclaurin shared the prize for his theory of tides with the mathematicians **Leonhard Euler** and **Daniel Bernoulli**. They all provided proof of Newton's theories about the movement of the ocean. The mathematician **Alexis Clairaut** was so taken with Maclaurin's success in explaining tides that he began probing the mystery of the Earth's shape with geometry.

Maclaurin's *Treatise of Fluxions*, published in 1742, was another of his major works. The treatise was written as a reply to **George Berkeley**'s 1734 publication *The Analyst, A Letter Addressed to an Infidel Mathematician*, in which Berkeley criticized Newton's theory of fluxions as "ghosts of departed quantities." Other mathematicians had also decried Newton's methods for their lack of systematic foundation. In his two-volume work, Maclaurin explained Newton's theories in detail and also solved other great quandaries of mathematics. The book contains descriptions of the infinite series as well as much praised work on the curves of quickest descent.

Though the *Treatise of Fluxions* was considered noteworthy at the time, it had little influence on international mathematics. The book persuaded British scientists to continue with the geometrical methods of Newton rather than the analytical calculus just being devised in other nations in Europe. As a result, Britain was left far behind in the development of mathematics during the late 1700s.

Maclaurin worked ceaselessly to defend Edinburgh during an attack by Jacobites in the rebellion of 1745. While planning and erecting the defenses of the city, however, he became so exhausted that he fell physically ill. When the city surrendered to the Jacobites, Maclaurin fled to England. A year later he returned to Edinburgh, where he died at age 48.

Until his death in 1746, Maclaurin remained the consummate mathematician and defender of Newtonian theories. A few hours before his death he dictated his last writing on Newton's work, which firmly set forth Maclaurin's belief in life after death. His *Treatise on Algebra* and *An Account of Sir Isaac Newton's Philosophical Discoveries*, an incomplete work, were published posthumously.

## SELECTED WRITINGS BY MACLAURIN

*Geometrica Organica, sive descriptio linearum curvarum universalis*, 1720
*The Treatise of Fluxions*, 2 volumes, 1742
*A Treatise of Algebra*, 1748
*An Account of Sir Isaac Newton's Philosophical Discoveries*, 1748

## FURTHER READING

Mooney, J. "Colin Maclaurin and Glendaruel." *The Mathematical Intelligencer* 16 (1994): 48–49.
Turnbull, H.W. "Colin Maclaurin." *American Mathematical Monthly* 54, no. 6 (1947).

*—Sketch by Barbara Boughton*

*Robert MacPherson*

# Robert MacPherson
## 1944–

### American algebraic topologist

MacPherson's mathematical writings cover a number of topics related to the relatively new field of topology, including combinatorics and group theory. His areas of interest also include catastrophe theory and random numbers. MacPherson has received numerous grants from the National Science Foundation, as well as honorary doctorates from the Universite de Lille and Brown University. In 1992, he was honored with the National Academy of Sciences award in mathematics.

Robert MacPherson is dedicated to teaching, but his highest profile in the mathematics community comes through his international outreach activities, especially within Russia and the former Soviet satellite nations. He has served on a number of committees devoted to aiding mathematicians caught in the collapse of the U.S.S.R., even raising and distributing much needed cash personally. MacPherson's most successful book, *Stratified Morse Theory*, has been translated into Russian for publication by Mir Press of Moscow.

Robert Duncan MacPherson was born in Lakewood, Ohio, to Herbert G. and Jeanette (née Wolfenden) MacPherson on May 25, 1944. He graduated from Swarthmore College in 1966, and his B.A. was

conferred with highest honors. Upon being granted a Ph.D. from Harvard University in 1970, MacPherson joined the faculty of Brown University. He remained at Brown for 17 years, holding a number of positions including Florence Pirce Grant University Professorship (1985–1987). From 1987 to 1994, MacPherson was a full–professor at MIT, and in 1994 he joined the staff of the Institute for Advanced Study (IAS) in Princeton, New Jersey.

Because the IAS is not a teaching facility, MacPherson accepted a concurrent, non–paying post at nearby Princeton University in order to maintain interaction with students. Research activities with the more than 20 Ph.D. candidates he has advised have often informed his presentations, if a recent talk on oriented matroids and topology is any indication. In the abstract, MacPherson notes that the talk is "mainly a report on work of Laura Anderson, Eric Babson, and Jim Davis," two of whom were his graduate students.

### Foreign Aid

During the spring of 1980, MacPherson worked at the Steklov Institute of Mathematics in Moscow as part of a National Academy of Sciences exchange program. Since then he has become involved with a number of joint international planning and advisory committees devoted to helping mathematicians weather the political and financial upheavals of post–Communism. MacPherson currently serves on the governing board of the Moscow Mathematical Institute and chairs the American Mathematics Society's Former Soviet Union Aid Fund Advisory Committee.

As noted in a 1995 issue of the *Notices of the AMS*, MacPherson was largely responsible for bringing attention to the financial plight of mathematicians in the former Soviet Union. In response, AMS established an aid program, to which members contributed. George Soros arranged for matching grants from the International Science Foundation and the Sloan Foundation, resulting in aid for 400 students and working mathematicians ranging from $50 to $80 per month.

As MacPherson remembers it, this activity sometimes involved a bit of skulduggery. When no quick, inexpensive way could be found to transfer the money they had raised to the intended recipients, MacPherson and Tim Goggins of the AMS carried $25,000 into Russia secretly. After smuggling the money past the border guards, the two men distributed it at meetings in Russia's major cities, an almost unheard of example of direct financial aid.

In 1997, MacPherson was elected to the chair of the mathematics committee of the National Research Council, a group that advises the United States government on scientific matters. He serves on the editorial boards of *Compositio Mathematica* and *Advances in Mathematics*, is an editor of the *Annals of Mathematics* and associate editor of *Selecta Mathematica*. MacPherson is also active in GLBMATH, an organization for gay and lesbian mathematicians. As he recently pointed out, "There are very few openly gay scientists, and young gays have no role models." In part through his GLBMATH activities, MacPherson hopes to change this situation.

## SELECTED WRITINGS BY MACPHERSON

### Books

(With W. Borho and J.L. Brylinski) "Primitive Ideals and Cone Bundles," in *Progress in Mathematics 78*, 1989

(With M. Goresky) *Stratified Morse Theory*, 1988

(With M. McConnell) "Projective Geometry and Modular Varieties," in *Algebraic Analysis, Geometry, and Number Theory*, 1989

### Periodicals

(With W. Fulton) "A Compactification of Configuration Space." *Annals of Mathematics* 139 (1994): 183–225.

(With W. Fulton) "Intersection Theory on Spherical Varieties." *Jouranl of Algebraic Geometry* 4 (1995): 181–93.

(With S. Gelfand and K. Vilonen) "Perverse Sheaves and Quivers." *Duke Math. Journal* 83 (1996): 621–643.

## FURTHER READING

### Books

*Who's Who in Science and Engineering*. New Providence, NJ: Reed Publishing, 1994, p. 549.

### Other

MacPherson, Robert, in an electronic mail interview with Jennifer Kramer, conducted July 14, 1997.

—*Sketch by Jennifer Kramer*

# Ada Isabel Maddison
## 1869–1950
### English algebraist and educator

In spite of her accomplishments and education, Ada Isabel Maddison remained a shy woman through-

out her life. In fact, while serving as an assistant at Bryn Mawr to President M. Carey Thomas in 1913, she received a note from Thomas invoking her to "[Speak] distinctly. When you get embarrassed, your voice gets lower and lower. I am sure the Faculty thinks it is shyness, as it is. You must conquer it."

Maddison was born on April 13, 1869, in Cumberland, England. Her parents were John and Mary Maddison. She took college preparatory courses at Miss Tallies School in Cardiff, South Wales, then entered the University of South Wales where she studied from 1885 to 1889. After leaving the University of South Wales she attended Girton College, Cambridge, for another three years. While at Girton she met and befriended **Grace Chisholm Young**, the first woman to receive a doctorate in Germany.

During their first year at Girton, both women attended a lecture given by **Arthur Cayley**, a mathematician who played a central role in founding the modern British school of pure mathematics and author of more than 900 published articles addressing almost every aspect of modern mathematics. Later, while at Bryn Mawr attending graduate lectures, Maddison studied Cayley's papers on modern algebra.

Maddison succeeded at passing the examinations of the Honour School at Oxford in 1892, and in the same year passed the Cambridge Mathematical Tripos Examination, first class. The examination was equivalent to the highest class of honors at Cambridge, 27th Wrangler, but the accomplishment did little to secure a degree since Maddison was not allowed to receive one.

Between 1892 and 1893 Maddison attended lectures given by Dr. **Charlotte Angas Scott** at Bryn Mawr. During the same time, she began her investigation into the singular solutions of differential equations, which was later published in 1896 in the *Quarterly Journal of Pure and Applied Mathematics* in 1896. Scott was sufficiently impressed by Maddison to write: "[She] has a powerful mind and excellent training." In acknowledgment of Scott's teaching skill, Maddison wrote an article on her for the *Bryn Mawr Alumnae Bulletin* in January 1932 entitled "Charlotte Angas Scott: An Appreciation."

In 1893, Maddison received her Bachelor of Science degree with Honours from the University of London. She was also given a resident mathematics fellowship at Bryn Mawr for the 1893–1894 school year. In 1895, she became the first person awarded the Mary E. Garrett Fellowship by Bryn Mawr to be used for study abroad. Maddison chose to attend the University of Göttingen, where she renewed her acquaintance with Grace Chisholm Young and met Annie Louis MacKinnon, who in 1894 had received her Ph.D. in mathematics from Cornell University. Both Maddison and MacKinnon were elected to the American Mathematical Society in 1897. Maddison concentrated on the lectures of **Felix Klein**, author of the *Erlanger Programm*, and **David Hilbert**, considered the greatest influence on geometry since **Euclid**. These lectures must have had an impact on Maddison because her field of specialization became algebraic geometry. In 1896, she published a translation of Felix Klein's work, "The Arithemitizing of Mathematics," in the *Bulletin of the American Mathematical Society*.

### Mathematician Turned Administrator

Maddison completed the work for her Ph.D. at Bryn Mawr and received the degree in 1896. Her dissertation was entitled "On Singular Solutions of Differential Equations of the First Order in Two Variables, and the Geometric Properties of Certain Invariants and Covariants of Their Complete Primitives." Concurrent with her studies, Maddison acted as assistant secretary to the president of Bryn Mawr. She then stepped into the dual role of reader in mathematics and secretary to the president, serving both positions for more than seven years. In 1904, she again accepted the tasks associated with two positions, becoming an associate professor and assistant to the president.

Although Maddison had never studied in Dublin, she was awarded a B.A. degree by the University of Dublin in 1905. That university was the first to award degrees to women in the British Isles. The degree was conferred on Maddison based on her work at Girton College.

Between 1910 and 1926, when she retired from teaching, Maddison remained the assistant to the president and jointly held the administrative position of Recording Dean. These duties left her little time for mathematics or research, which must have given her some cause for regret because in 1937 she wrote, "I confess to feeling ashamed of having deserted mathematics for a less rarified atmosphere of work among people and things." She always considered mathematics "the most perfect of the sciences" and felt that her loyalty remained steadfast with the discipline.

In 1897 Maddison became a member of the Daughters of the British Empire and in the same year she was elected to the American Mathematical Association. Maddison also joined the London Mathematical Society and remained a member throughout her lifetime. In spite of her considerable responsibilities as an administrator, Maddison found time to aid in the compilation of a study on women who had graduated from college. The study dealt with such issues as marriage, children, and occupations.

After her retirement from Bryn Mawr, Maddison returned to England. After a time she came back to Pennsylvania and it was there that she died in 1950. Upon her death, Bryn Mawr was bequeathed ten thousand dollars, with instructions that the gift is used for nonfaculty members as a pension fund in honor of the woman with whom she had such a long working relationship, President M. Carey Thomas.

## SELECTED WRITINGS BY MADDISON

"On Certain Factors of C– and P–Discriminants and Their Relation to Fixed Points on the Family of Curves," in *Quarterly Journal of Pure and Applied Mathematics* 26 (1893): 307–21.

"Note on the History of the Map Coloring Problem," in *Bulletin of the American Mathematical Society* 3 (1897): 257.

*Handbook of Courses Open to Women in British, Continental, and Canadian Universities* 1896; supplement 1897; Second edition 1899; supplement 1901

## FURTHER READING

### Books

Whitman, Betsy S. "Ada Isabel Maddison," in *Women of Mathematics*. Edited by Louise S. Grinstein and Paul J. Campbell. Westport, CT: Greenwood Press, 1987, pp. 144–46.

### Periodicals

Whitman, Betsy S. "Women in the American Mathematical Society before 1900." *Association for Women in Mathematics Newsletter* 13, no. 5 (September–October 1893): 7–9.

Williams, Mary. "Ada Isabel Maddison." (Handwritten manuscript, 5 pp. n.d.) The Mary Williams Collection, Schlesinger Library, Radcliffe College, Cambridge, Massachusetts.

### Other

"Ada Isabel Maddison." *Biographies of Women Mathematicians.* http://www.scottlan.edu/lriddle/women/chronol.htm (August 1997).

*—Sketch by Kelley Reynolds Jacquez*

# Vivienne Malone–Mayes
## 1932–1995
### African–American mathematics educator

Vivienne Lucille Malone–Mayes was a prominent mathematics educator who taught at Baylor

University in Waco, Texas, for nearly three decades. As a black woman who grew up in a segregated society, she was a pioneer in her field and an inspiration for younger students. Malone–Mayes was the fifth African–American woman to receive a doctorate in pure mathematics in the United States, and she was the first black full-time professor at Baylor.

Malone–Mayes was born on February 10, 1932, in Waco, Texas. Her father, Pizarro Ray Malone, was a visiting teacher in the Waco public schools for many years and also worked for the Urban Renewal Agency. Her mother, Vera Estelle Allen Malone, was a junior high school teacher. It is little wonder, then, that education was stressed in their home. Malone–Mayes later recalled that the only lie her parents ever encouraged her to tell was about her age when she started school. She was only five years old, but her parents told her to say she was six so that she would be admitted. Always an excellent student, Malone–Mayes graduated from her racially segregated high school in 1948 at age 16.

Leaving her home in Waco, Malone–Mayes traveled to Nashville, Tennessee, where she became a student at Fisk University. Her first ambition was to be a physician. However, she changed that goal after meeting her future husband, James Jeffries Mayes, a dental student. He convinced her that two doctors in the same family would never see each other, so Malone–Mayes switched her major from pre–med to mathematics. She received a B.A. degree in 1952. On September 1 of that year, she and Mayes were married.

Two years later, Malone–Mayes received an M.A. degree in mathematics, also from Fisk University. She promptly moved back to Waco, where she became chairperson of the math department at Paul Quinn College and her husband opened a dental practice. Malone–Mayes eventually spent seven years at Paul Quinn. She later chaired the math department at Bishop College in Dallas for a year. She and her husband had one daughter, Patsyanne. The couple divorced in 1985.

### A Career Filled with Firsts

By 1961, Malone–Mayes was eager to refresh her education and she applied to Baylor University. Her application was rejected on grounds of race. She turned instead to the University of Texas at Austin, which had already been required by federal law to desegregate. Malone–Mayes endured much emotional stress and social ostracism as a black female graduate student. She later recalled that many classmates liked to gather at a cafe that would not serve blacks; the closest she came to joining them was marching in picket lines protesting the cafe's policy. For her thesis topic, Malone–Mayes chose "A Structure Problem in Asymptotic Analysis." In 1966, she received a Ph.D. from the University of Texas, making her the second black, and the first black woman, to earn a doctorate in math from that institution.

Malone–Mayes was immediately hired as a professor by Baylor, the same university that had rejected her as a student just five years before. She remained in that position until 1994, when she was forced to retire due to ill health. During her years as a professor, Malone–Mayes continued to break racial barriers, becoming the first African–American elected to the executive committee of the Association for Women in Mathematics. She also served on the board of directors of the National Association of Mathematicians, a group oriented toward the black community. In 1988, she participated in a panel featuring prominent female mathematicians that was part of the American Mathematics Society's Centennial Celebration in Providence, Rhode Island.

Malone–Mayes had an active life outside of academia. For many years she served as youth choir director and organist at her local Baptist church. She also volunteered for a number of charitable organizations, and she served as advisor for a traditionally black sorority at Baylor. Her last years were marred by health problems, however. She was plagued by lupus, a chronic inflammatory disease that can damage multiple systems of the body. On June 9, 1995, Malone–Mayes died of a heart attack in Temple, Texas. In an obituary in the *Association for Women in Mathematics Newsletter*, **Etta Falconer** and Lee Lorch wrote of their friend and colleague: "With skill, integrity, steadfastness and love she fought racism and sexism her entire life, never yielding to the pressures or problems which beset her path. She leaves a lasting influence."

## SELECTED WRITINGS BY MALONE–MAYES

### Books

(With Howard Rolf) *Pre–calculus,* 1977

## FURTHER READING

### Periodicals

Falconer, Etta and Lee Lorch. "Vivienne Malone–Mayes: In Memoriam." *Association for Women in Mathematics Newsletter* (November–December 1995).

Simpson, Elizabeth. "'You Had to Make It All Alone': Black Baylor Teacher Recalls Road to Success." *Waco Tribune–Herald* (August 22, 1988): 1A, 3A.

**Other**

Falconer, Etta, Dr., and Dr. Lee Lorch. "Vivienne Malone–Mayes: In Memoriam." *Biographies of Women Mathematicians*. June 1997. http://www.scottlan.edu/lriddle/women/chronol.htm (July 22, 1997).

Miscellaneous newspaper clippings and press releases. The Texas Collection, Baylor University Library, Waco, TX.

*—Sketch by Linda Wasmer Smith*

# Benoit B. Mandelbrot
## 1924-
### Polish-born American geometer

Benoit B. Mandelbrot is a mathematician who conceived, developed, and named the field of fractal geometry. This field describes the everyday forms of nature—such as mountains, clouds, and the path traveled by lightning—that do not fit into the world of straight lines, circles, and smooth curves known as Euclidean geometry. Mandelbrot was also the first to recognize fractal geometry's value as a tool for analyzing a variety of physical, social, and biological phenomena.

Mandelbrot was born November 20, 1924, to a Lithuanian Jewish family in Warsaw, Poland. His father, the descendant of a long line of scholars, was a manufacturer and wholesaler of children's clothing. His mother, trained as a doctor and dentist, feared exposing her children to epidemics, so instead of sending her son to school, she arranged for him to be tutored at home by his Uncle Loterman. Mandelbrot and his uncle played chess and read maps; he learned to read, but he claims that he never did learn the whole alphabet. He first attended elementary school in Warsaw. When he was eleven years old, his family moved to France, first to Paris and then to Tulle, in south central France. When Mandelbrot entered secondary school, he was 13 years old instead of the usual 11, but he gradually caught up with his age group. His uncle Szolem Mandelbrojt, a mathematician, was a university professor, and Mandelbrot became acquainted with his uncle's mathematician colleagues. Mandelbrot's teenage years were disrupted by World War II, which rendered his school attendance irregular. From 1942 to 1944, he and his younger brother wandered from place to place. He found work as an apprentice toolmaker for the railroad, and for a time he took care of horses at a château near Lyon. He carried books with him and tried to study on his own.

After the war, at the age of 20, Mandelbrot took the month-long entrance exams for the leading science schools. Although he had not had the usual two years of preparation, he did very well. He had not had much formal training in algebra or complicated integrals, but he remembered the geometric shapes corresponding to different integrals. Faced with an analytic problem, he would make a drawing, and this would often lead him to the solution. He enrolled in Ecole Polytechnique. Graduating two years later, he was recommended for a scholarship to study at the California Institute of Technology. In 1948, after two years there, he returned to France with a master's degree in aeronautics and spent a year in the Air Force.

Mandelbrot next found himself in Paris, looking for a topic for his Ph.D. thesis. One day his uncle, rummaging through his wastebasket for something for Mandelbrot to read on the subway, pulled out a book review of *Human Behavior and the Principle of Least Effort,* by George Zipf. The author discussed examples of frequency distributions in the social sciences that did not follow the Gaussian "bell-shaped curve," the so-called normal distribution according to which statistical data tend to cluster around the average, scattering in a regular fashion. Mandelbrot wrote part of his 1952 University of Paris Ph.D. thesis on Zipf's claims about word frequencies; the second half was on statistical thermodynamics. Much later, Mandelbrot commented that the book review greatly influenced his early thinking; he saw in Zipf's work flashes of genius, projected in many directions yet nearly overwhelmed by wild notions and extravagance, and he cited Zipf's career as an example of the extraordinary difficulties of doing scientific work that is not limited to one field. At the time, Mandelbrot had read **Norbert Wiener** on cybernetics and **John von Neumann** on game theory, and he was inspired to follow their example in using mathematical approaches to solve long-standing problems in other fields.

Mandelbrot was invited to the Institute for Advanced Studies at Princeton University for the academic year 1953–54. On returning to Paris, he became an associate at the Institut Henri Poincaré. In 1955, he married Aliette Kagan, who later became a biologist; they had two children, Laurent and Didier. From 1955 to 1957 Mandelbrot taught at the University of Geneva. In 1957 and 1958 he was junior professor of applied mathematics at Lille University and taught mathematical analysis at Ecole Polytechnique. In 1958 he became a member of the research staff at the IBM Thomas J. Watson Research Center in Yorktown Heights, New York.

### Studies Statistical Irregularity

In the 1960s Mandelbrot studied stock market and commodity price variations and the mathematical models used to predict prices. A Harvard professor had studied the changes in the price of cotton over many years and had found that the changes in price did not follow the bell-shaped distribution. The variations appeared to be chaotic. Existing statistical models for stock-market prices assumed that the rise and fall was continuous, but Mandelbrot noted that prices may jump or drop suddenly. He showed that a model that assumes continuity in prices will turn out to be wrong. He used IBM computers to analyze the data, and he found that the pattern for daily price changes matched the pattern for monthly price changes. Statistically, the choice of time scale made no difference; the patterns were self-similar. Using this concept, he was able to account for a great part of the observed price variations, where earlier statistical techniques had not succeeded.

Shortly thereafter, IBM scientists asked for Mandelbrot's help on a practical problem. In using electric current to send computer signals along wires, they found occasional random mistakes, or "noise." They suspected that some of the noise was being caused by other technicians tinkering with the equipment. Mandelbrot studied the times when the noise occurred. He found long periods of error-free transmission separated by chunks of noise. When he looked at a noisy chunk in detail, he saw that it, in turn, consisted of smaller error-free periods interspersed with smaller noisy chunks. As he continued to examine chunks at smaller and smaller scales, he found that the chance of the noise occurring was the same, regardless of the level of detail he was looking at. He described the probability distribution of the noise pattern as self-similar, or scaling—that is, at every time scale the ratio of noisy to clean transmission remained the same. The noise was not due to technicians tinkering with screwdrivers; it was spontaneous. In understanding the noise phenomenon, Mandelbrot used as a model the Cantor set, an abstract geometric construction of **Georg Cantor** , a 19th-century German mathematician. The model changed the way engineers viewed and addressed the noise problem.

For centuries humankind has tried to predict the water level of rivers like Egypt's Nile in order to prevent floods and crop damage. Engineers have relied on such predictions in building dams and hydroelectric projects. In the 1960s, Mandelbrot studied the records of the Nile River level and found that existing statistical models did not fit the facts. He found long periods of drought along with smaller fluctuations, and he found that the longer a drought period, the more likely the drought was to continue. The resulting picture looked like random noise superimposed on a background of random noise. Mandelbrot made graphs of the river's actual fluctuations. He showed the unlabeled graphs to a noted hydrologist, along with graphs drawn from the existing statistical models and other graphs based on Mandelbrot's

statistical theories. The hydrologist dismissed the graphs from the old models as unrealistic, but he could not distinguish Mandelbrot's graphs from the real ones. For Mandelbrot, this experience illustrated the value of using visual representations to gain insight into natural and social phenomena. Other researchers found similar support for Mandelbrot's statistical model when they showed fake stock charts to a stockbroker; the stockbroker rejected some of the fakes as unrealistic, but not Mandelbrot's.

Early in this century, mathematicians and geometers created curves that were infinitely wrinkled and solids that were full of holes. Much later, Mandelbrot found their abstract mathematics useful in models for shapes and phenomena found in nature. He had read an article about the length of coastlines in which Lewis Fry Richardson reported that encyclopedias in Spain and Portugal differed on the length of the border between the two countries; Richardson found similar discrepancies—up to 20 percent—for the border between Belgium and the Netherlands. Mandelbrot took up the question in a paper he called "How Long Is the Coast of Britain?" The answer to the question, according to Mandelbrot, depended on the length of the ruler you used. Measuring a rocky shoreline with a foot ruler would produce a longer answer than measuring it with a yardstick. As the scale of measurement becomes smaller, the measured length becomes infinitely large. Mandelbrot also investigated ways of measuring the degree of wiggliness of a curve. He worked with programmers to develop computer programs to draw fake coastlines. By changing a number in the program, he could produce relatively smooth or rough coastlines that resembled New Zealand or those of the Aegean Sea. The number determined the degree of wiggliness and came to be known as the curve's fractal dimension.

Fascinated with this approach, Mandelbrot looked at patterns in nature, such as the shapes of clouds and mountains, the meanderings of rivers, the patterns of moon craters, the frequency of heartbeats, the structure of human lungs, and the patterns of blood vessels. He found that many shapes in nature—even those of ferns and broccoli and the holes in Swiss cheese—could be described and replicated on the computer screen using fractal formulas.

## Formulates Fractal Geometry

In Mandelbrot's reports and research papers during this period, he made clear that his methods were part of a more general approach to irregularity and chaos that was applicable to physics as well. Editors, however, usually preferred a more narrowly technical discussion. But then he was invited to give a talk at the Collège de France in 1973. Rather than selecting one of his many areas of research, he decided to explain how his many different interests fit together. Mandelbrot wanted a name for this new family of geometric shapes, which typically involved statistical irregularities and scaling. Looking through his son's Latin dictionary, he found the adjective *fractus,* meaning "fragmented, irregular," and the verb *frangere,* "to break," and he came up with *fractal.* His lecture aroused considerable interest and was published in expanded form in 1975 in French as *Les Objets Fractals: Forme, Hasard et Dimension.* Revised and expanded versions were published later in English in the United States as *The Fractal Geometry of Nature.* The publication of his book, which Mandelbrot called a manifesto and a casebook, attracted interest from researchers in fields from mathematics and engineering to economics and physiology. Mandelbrot remained at IBM as an IBM fellow but with various concurrent positions and visiting professorships at universities, including Harvard University, Massachusetts Institute of Technology, Yale University, Albert Einstein College of Medicine, and the University of California.

Using fractal formulas, computer programmers could produce artificial landscapes that were remarkably realistic. This technology could be used in movies and computer games. Among the first movies to use fractal landscapes were George Lucas's *Return of the Jedi,* for the surface of the Moons of Endor, *Star Trek II: The Wrath of Khan,* and *The Last Starfighter.* Some fractal formulas produced fantastic abstract designs and strange dragon-like shapes. Mathematicians of the early 20th century had done research in this area, but they did not have the advantage of seeing visual representations on a computer screen. The formulas were studied as abstract mathematical objects and, because of their strange properties, were called "pathological."

### Discovers "Mandelbrot Set"

In the 1970s, Mandelbrot became interested in investigations carried out during World War I by French mathematicians Pierre Fatou and Gaston Julia, the latter having been one of his teachers years before at Polytechnique. Julia had worked with mathematical expressions involving complex numbers (those which have as a component the square root of negative one). Instead of graphing the solutions of equations in the familiar method of **René Descartes**, Julia used a different approach; he fed a number into an equation, calculated the answer, and then fed the answer back into the equation, recycling again and again, noting what was happening to the answer. Mandelbrot used the computer to explore the patterns generated by this approach. For one set, he used a relatively simple calculation in which he took a complex number, squared it, added the original number, squared the result, continuing again and again; he plotted the original number on the graph only if its answers did not run away to infinity. The

figure generated by this procedure turned out to contain a strange cardioid shape with circles and filaments attached. As Mandelbrot made more detailed calculations, he discovered that the outline of the figure contained tiny copies of the larger elements, as well as strange new shapes resembling fantastic seahorses, flames, and spirals. The figure represented what came to be known as the Mandelbrot set. Representations of the Mandelbrot set and the related sets studied by Julia, some in psychedelic colors, soon appeared in books and magazines—some even in exhibits of computer art.

Through his work with fractals and computer projections of various equations, Mandelbrot had discovered tools that could be used by scientists and engineers for strengthening steel, creating polymers, locating underground oil deposits, building dams, and understanding protein structure, corrosion, acid rain, earthquakes, and hurricanes. Physicists studying dynamical systems and fractal basin boundaries could use Mandelbrot's model to better understand phenomena such as the breaking of materials or the making of decisions. If images could be reduced to fractal codes, the amount of data necessary to transmit or store images could be greatly reduced.

Fractal geometry showed that highly complex shapes could be generated by repeating rather simple instructions, and small changes in the instructions could produce very different shapes. For Mandelbrot, the striking resemblance of some fractal shapes to living organisms raised the possibility that only a limited inventory of genetic coding is needed to obtain the diversity and richness of shapes in plants and animals.

In 1982 Mandelbrot was elected a fellow of the American Academy of Arts and Sciences. In 1985 he received the Barnard Medal for Meritorious Service to Science, awarded every five years by the National Academy of Sciences for a notable discovery or novel application of science beneficial to the human race. In 1986 he received the Franklin Medal for his development of fractal geometry. In 1987 he became a foreign associate of the U. S. Academy of Sciences. In 1988 he received the Harvey Prize and in 1993 the Wolf Prize for physics for having changed our view of nature. He officially retired from IBM in 1993, but he continued to work at Yale and at IBM as a fellow emeritus, preparing a collection of his papers and doing further research in fractals.

### SELECTED WRITINGS BY MANDELBROT

*The Fractal Geometry of Nature,* 1982

## FURTHER READING

### Books

Albers, Donald J., and G. L. Anderson, editors. *Mathematical People: Profiles and Interviews.* New York: Birkhauser, 1985.

Briggs, John. *Fractals, the Patterns of Chaos: A New Aesthetic of Art, Science, and Nature.* New York: Simon & Schuster, 1992.

Gardner, Martin. *Penrose Tiles to Trapdoor Ciphers.* New York: W. H. Freeman, 1989.

Gleick, James. *Chaos: Making a New Science.* New York: Viking Penguin, 1987.

Peitgen, Heinz-Otto, and Dietmar Saupe, editors. *The Science of Fractal Images.* New York: Springer-Verlag, 1988.

Peitgen, Heinz-Otto, and P. H. Richter. *The Beauty of Fractals.* New York: Springer-Verlag, 1986.

### Periodicals

"Franklin Institute Honors Eight Physicists." *Physics Today* (April 1987): 101–102.

Gleick, James. "The Man Who Reshaped Geometry." *New York Times Magazine* (December 8, 1985): 64 ff.

"Interview: Benoit B. Mandelbrot" *Omni* (February 1984): 65–66, 102–107.

Jürgens, Hartmut, Heinz-Otto Peitgen, and Dietmar Saupe. "The Language of Fractals." *Scientific American* (August, 1990): 60–67.

"Tomorrow's Shapes: The Practical Fractal." *The Economist* (December 26, 1987): 99–103.

### Other

Mandelbrot, Benoit B., interview with C. D. Lord conducted August 17, 1993.

—*Sketch by C. D. Lord*

# Gregori Aleksandrovitch Margulis
## 1946-

### Russian group theorist

The study of Lie groups has proved to be extremely useful both within mathematics and in various other fields; by describing the relationships between algebraic, geometric and analytic structures, it has impacted on the fields of astrophysics, chemistry, unified field theory, and high-energy particle physics. Perhaps no one has done more to unlock

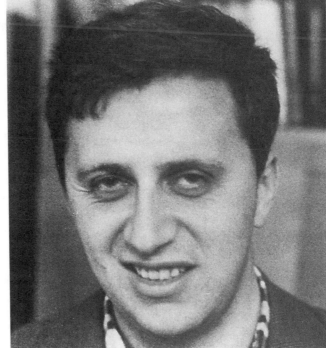

*Gregori Aleksandrovitch Margulis*

their secrets than the Russian mathematician Gregori Aleksandrovitch Margulis. Through his exploration of the Lie group's basic substructure known as a lattice, Margulis uncovered the various interrelationships between the lattice and the Lie group, allowing for a clearer definition of both. These results had such far-reaching effects that in 1978 the International Mathematics Union awarded him the highest honor in mathematics, the Fields Medal.

Margulis was born in Moscow in 1946, the son of Aleksander and Tsilya Osharenko Margulis. His father was a mathematician, and Margulis showed an early interest in both mathematics and chess. However, he largely lost interest in chess playing when he began his mathematical studies at Moscow University. He studied there until 1970, when he received his Ph.D. under the direction of Yakov G. Sinai. By the time he received his degree, Margulis had already made his first major contribution to mathematics, a partial proof of the Selberg conjecture.

**Atle Selberg**, a Norwegian, had conjectured that lattices, which are discrete subgroups of continuous topological groups known as Lie groups, were all arithmetic in nature. In other words, Selberg said that they behaved in predictable ways, not unlike integers. Proof of his conjecture would have phenomenal effects on the applications of Lie groups in general, because it would provide a link between the lattices and the larger Lie groups. Selberg, however, only succeeded in proving that lattices fall into two different cases, cocompact and noncocompact. Still, by doing this he paved the way for the work Margulis would undertake.

In 1968, while working with D. A. Kazhdan, Margulis managed to prove Selberg's conjecture for noncocompact lattices, which formed a major part of the Lie group. Margulis and Kazhdan proved that these lattices contained nontrivial unipotent elements—points which, through repeated multiplication, would result in a return to the original element. By using the orbits, or repeated multiplication paths, to describe the lattice of these noncocompact groups, Margulis proved their arithmeticity. This result, while important, was only the first step toward a complete proof of the Selberg conjecture.

### Proof of Conjecture Leads to Medal

Next, Margulis undertook the study of the more difficult cocompact lattices. Cocompact lattices have a smaller structure with fewer elements, and thus the idea of the nontrivial unipotent element does not apply. Margulis therefore had to develop a whole new approach for these lattices. Using results obtained earlier by G. D. Mostow, who showed that lattices are rigid or unchangeable, Margulis considered a certain lattice in an alternative setting. By combining elements of algebra, analysis and number theory, and then applying them to the lattice, Margulis was able to define the structure of not only this but all cocompact lattices. These included the most complex form, lattices based on the p-adic Lie groups, whose structure is related to prime numbers.

Margulis obtained these results while at Moscow's Institute for Transmission of Information, where he had worked since completing his education. Unlike many mathematicians, he did not have the opportunity to travel and lecture, due to restrictions imposed upon scientists by the Soviet government. This did not prevent Western mathematicians from recognizing the importance of what he was doing, though, and it came as no surprise to anyone familiar with Margulis's work when his name was announced as a recipient of the 1978 Fields Medal.

The International Congress of Mathematicians convened in Helsinki that year to award Fields Medals to Margulis and three others. Most of the mathematicians in attendance had heard of Margulis only through his work and were eagerly anticipating the opportunity to hear him speak. Margulis, however, was denied permission from the Soviet government to attend the conference. Jacques Tits, who spoke about Margulis's work at the conference, did not hide his dismay at the Soviet government's position and led the crowd in a standing ovation in honor of his achievements.

Back in the Soviet Union, Margulis continued his research into Lie groups. He explored various aspects of the groups and their subgroups and attempted to find applications for the work he had done. The many papers he wrote on this subject over the next ten years would form the basis for a 1989 textbook explaining group theory. He received a great deal of encouragement to write the book from Tits, who had praised him so highly in Helsinki.

With the breakup of the Soviet Union, Margulis was free to travel, and in 1990 he gave one of the plenary addresses at the International Congress of Mathematicians in Kyoto. He came to Harvard as a visiting professor that same year with his wife Raisa, a computer programmer, and their teenaged son. He decided to stay in the United States and gained permanent residency in 1991; after a brief appointment at Princeton's Institute for Advanced Study, he settled at Yale University.

## SELECTED WRITINGS BY MARGULIS

### Books

*Discrete Subgroups of Semisimple Lie Groups,* 1989

### Periodicals

"Discrete groups of motions of manifolds with nonpositive curvature." *Proceedings of the International Congress of Mathematics* 2 (1975).

## FURTHER READING

### Periodicals

"Four Mathematicians Receive Fields Medals." *Notices of the American Mathematical Society* (August 1978): 432.

Mostow, G. D. "The Fields Medals (I): Relating the Continuous and the Discrete." *Science* (October 20, 1978): 297–298.

Tits, Jacques. "The Work of Gregori Aleksandrovitch Margulis." *Proceedings of the International Congress of Mathematicians* (1979): 56–63.

### Other

Margulis, Gregori, interview with Paul Becker conducted March 12, 1994.

*—Sketch by Karen Sands*

*Andrei Andreevich Markov*

# Andrei Andreevich Markov
## 1856-1922
### Russian number theorist

Andrei Andreevich Markov's research covered a number of fields in mathematics, including number theory, differential equations, and quadrature formulas. He is best known, however, for his work in probability theory and his derivation of a powerful predictive tool now known as Markov chains. Although Markov himself saw few applications for this tool, Markov chains are now widely used in many fields of modern science.

Markov was born in Ryazan, Russia, on June 14, 1856. His mother was Nadezhda Petrovna, daughter of a government worker, and his father was Andrei Grigorievich Markov, an employee of the state forestry department who also managed a private estate. Markov suffered from poor health as a child, walking with crutches until the age of ten. He was not a particularly good student, although he demonstrated an interest and skill in mathematics at an early age. While still in high school he wrote a paper on the integration of linear differential equations that drew the attention of members of the mathematics faculty at the University of St. Petersburg. Markov entered that university in 1874, where he studied with the renowned **Pafnuty L. Chebyshev**. Markov formed a long-term working relationship with Chebyshev, with

whom he wrote a number of papers. In one of his earliest papers, Markov reworked one of Chebyshev's theorems, called the central limit theorem, and corrected certain errors his teacher had made.

Markov received his bachelor's degree in 1878 for a thesis on differential equations and continuing fractions, for which he was also awarded a gold medal. Markov then stayed on at St. Petersburg to work for his master's degree, which was granted in 1880, then for his doctorate, which he received in 1884. His doctoral thesis was also on continuing fractions, and this was a subject that would remain central to much of his career.

Markov began teaching at the University of St. Petersburg in 1880 while still a graduate student. By 1886 he was named extraordinary professor of mathematics and in 1893 full professor, a post he would hold until his retirement in 1905. Markov also held parallel appointments in the St. Petersburg Academy of Sciences during this period. He was elected an adjunct member of the academy in 1886, extraordinary academician in 1890, and an ordinary academician in 1896. During his years at St. Petersburg Markov pursued a rather wide variety of topics in mathematics, including the search for minima in indefinite quadratic functions; the evaluation of limits for functions, integrals, and derivatives; and the method of moments. It was not until 1907, however, after his retirement from St. Petersburg, that he completed the work in probability theory for which he is now best known.

### Markov Chains

His most important contribution to probability theory began when he was writing a textbook on probability calculus, originally published in 1900. During preparation of the book, Markov encountered a particular type of probabilistic event, in which it is sometimes possible to predict the future status of some collection of random events if certain information is available about the present status of the sequence of events. As an example, the movement of gas molecules in a container is random. It can never be predicted exactly what any one molecule will do at any given moment. But, Markov showed, there are certain circumstances under which a later state of molecules in the container can be predicted if certain conditions exist among the molecules now. The sequence of events under which this situation can occur is called a Markov chain.

Markov apparently saw few practical applications for his theorem, mostly because of the state of science at that time, when it was believed that natural laws determined most events. Within a decade, however, the nature of science had undergone a dramatic revolution. Phenomena were seen to be the result of the *probable* behavior of fundamental particles and waves, though these actions often could not be proved. Markovian analysis became useful for predicting this type of probable behavior, and today Markov's work finds applications in a countless number of ways in the biological and physical sciences and in technology.

Beyond his academic work, Markov was also involved with Russian politics in the years leading up to the Russian revolution. He vigorously supported the liberal movement that swept through the country in the early twentieth century. In 1902, for example, he protested the action of Czar Nicholas II in withholding membership in the St. Petersburg Academy from dissident writer Maxim Gorky. Markov refused government honors later offered to him and in 1907 resigned from the academy in protest of the czar's opposition to government reform. During the revolution of 1917 Markov volunteered to teach mathematics without pay at the remote village of Zaraisk. He became ill shortly after his return to St. Petersburg and died there in 1922.

## SELECTED WRITINGS BY MARKOV

### Books

*Differential Calculus* 2 volumes, 1889–1891
*Probability Calculus*, 1900

### Periodicals

"Functions Generated by Developing Power Series in Continuing Fractions." *Duke Mathematical Journal* 7 (1940): 85–96.

## FURTHER READING

### Books

Travers, Bridget E., editor. *World of Scientific Discovery.* Detroit: Gale Research, 1994, pp. 421–422.
Youschekevitch, Alexander A. "Andrei Andreevich Markov," in *Dictionary of Scientific Biography.* Volume IX. Edited by Charles Coulston Gillispie. New York: Charles Scribner's Sons, 1975, pp. 124–130.

—*Sketch by David E. Newton*

# Walter S. McAfee
## 1914-1995
### African-American astrophysicist

As the mathematician for Project Diana under the direction of the U.S. Army's Signal Corps, Walter

*Walter S. McAfee*

S. McAfee made the essential calculations that led to the first human-made contact with the moon, a radar signal sent on January 10, 1946. McAfee's role in the success of the project laid the groundwork for the space communications and exploration that was to come.

Walter Samuel McAfee was born in Ore City, Texas, on September 2, 1914. He was the second of nine children born to Luther F. McAfee, a carpenter, and Susie A. (Johnson) McAfee. His mother had studied to be an elementary schoolteacher, but never taught professionally. Instead, she made certain her children knew how to read and write before they entered school. McAfee received his bachelor's degree from Wiley College in 1934, and went on to Ohio State University for his master's degree in 1937. Unable to afford further graduate work, McAfee turned to teaching, and from 1939 to 1942 he taught physics at Champion Junior High School in Columbus. There, he met Viola Winston, a French teacher, whom he married in 1941. They have two daughters, Diane McAfee, a teacher, and Marsha (McAfee) Morris, an actuary. McAfee died of cancer on February 18, 1995.

In May 1942, McAfee joined the theoretical studies unit of the Electronics Research Command, part of the U.S. Army's Signal Corps at Fort Monmouth, New Jersey. As a civilian physicist, he was to remain with the group in various capacities for more than 40 years, studying and experimenting in theoretical nuclear physics and electromagnetic theory, quantum optics, and laser holography. He retired from the Signal Corps in 1985.

Project Diana was an effort to bounce a radar signal off the moon's surface. It was not known at the time if a high–frequency radio signal could penetrate the Earth's ionosphere or stratosphere. Early experiments with low– and medium–frequency radio waves had failed. Another obstacle, besides the Earth's atmosphere, confronted the Project Diana scientists: how to accurately account for the moon's speed, which varies from 750 miles (1,207 km) per hour slower than the Earth to 750 miles per hour faster. As mathematician for the project, McAfee made the calculations. On January 10, 1946, a radar pulse was sent through the project's 40 square foot (3.6 square meters) antenna toward the moon. Two and a half seconds later, a faint radar echo was heard. They had succeeded.

The experiment's significance was not fully appreciated when it was made public two weeks later by the Signal Corps. One expert interviewed by the *New York Times* stated that other astronomers were then working on project that promised "far more startling" results. It was impossible to know then that Project Diana's success gave an important head start to space exploration and current technology.

In 1946, McAfee enrolled in the doctoral program at Cornell University, studying with theoretical physicist Hans Bethe. His studies were supported by a Rosenwald Fellowship. McAfee received his doctorate in physics from Cornell University in 1949.

The official news report of the accomplishment did not include McAfee's name, nor gave any hint of the role he had played. Recognition did not come to McAfee until 25 years later at the anniversary of Project Diana in 1971. Since then, however, he has been honored with the Stevens Award from the Stevens Institute of Technology, which he received in 1985. Wiley College established a Science Hall of Fame in 1982, and McAfee was one of the first graduates inducted. In an 1982 interview he said, "If the program bears fruit, and if my presence helped in some small way, then that shall have been reward enough."

Always interested in education, McAfee established a mathematics and physical science fellowship at Wiley College to encourage minority students in math and science. He was a member of the American Association for the Advancement of Science, the American Astronomy Society, the American Physical Society, the American Association of Physics Teachers, and was a senior member of the Institute of Electrical and Electronics Engineers.

## SELECTED WRITINGS BY MCAFEE

(With G.J. Iafrates and A. Ballato) "Electron Backscattering from Solids and Double Layers." *Journal of Vacuum Science Technology* 13, no. 4 (1976): 843–847.
"Determination of Energy Spectra of Backscattered Electrons by Use of Everhart's Theory." *Journal of Applied Physics* 47, no. 3 (1976): 1179–1184.

## FURTHER READING

"Contact with Moon Achieved by Radar in Test by the Army." *New York Times,* January 25, 1946, p. 1
"McAfee Named to Wiley's Science Hall of Fame." *ERADCOM Currents,* May 1982.
"Original Participants Mark Diana's 25th Anniversary." *Army Research and Development Newsmagazine,* January/February 1971.

### Other

McAfee, Walter S., interview with F.C. Nicholson conducted February 9, 1994.
McAfee, Viola, interview with Fran Nicholson Hodgkins conducted May 1, 1997.

*—Sketch by Fran Hodgkins*

---

# Margaret Dusa McDuff
## 1945–
### English symplectic topologist

State University of New York at Stony Brook professor and Royal Society fellow Dusa McDuff won the first Ruth Lyttle Satter Prize in Mathematics in 1991. This prize recognizes outstanding contribution to mathematics research by a woman in the previous five years. Even as a graduate student McDuff was particularly creative. Her Ph.D. thesis used operator theory to solve a well–known problem about the factors of Von Neumann algebras. As interpreted by Pam Davis in her poster "Squeezing the Phase Space," McDuff's current work in geometry applies to the classical physics of orbiting bodies in space. Which properties are intrinsic to geometric shape, and which to the descriptive coordinates used for the entire system are still being determined.

McDuff was born Margaret Dusa Waddington on October 18, 1945, and her earliest aspiration was to be a farmer's wife. Although she was born in London,

*Margaret Dusa McDuff*

she grew up in Scotland, where both her parents worked. Her father was a professor of genetics and an author, and her mother, who came from a long line of intelligent, productive women, was an architect who worked for the civil service. Dusa bears the name of her maternal grandmother, a writer and political activist. McDuff's sister became an anthropologist at King's College, Cambridge. Despite the fact that so many women in her family were successful in their chosen occupations, McDuff herself felt isolated. At the girl's school she attended she discovered mathematics, a field that did not at the time seem to offer her any role models.

McDuff's earliest years of formal study in mathematics were marked by some passivity. Instead of accepting a scholarship to Cambridge, she took a bachelor of science degree at the University of Edinburgh in order to stay with her boyfriend, whom she later married and whose surname she adopted. However, he returned the favor when she became a graduate student, and followed her to Cambridge. McDuff's Ph.D. under G. A. Reid in functional analysis was published, and stood as her best work for many years.

While her husband was in Moscow for a time conducting archival research in Russian poetry, McDuff was able to study with the group theorist Israel M. Gel'fand. She was inspired to write two mathematical papers in Russian that were published in Soviet journals during 1970.

## Expanding Professional Horizons

Upon returning to Cambridge to formally receive her doctorate, McDuff found it difficult to channel her energies. The still predominantly–male institution did not seem welcoming to a young, married, and soon pregnant woman. To escape from this situation, McDuff accepted a job at York University. While running the household and caring for her infant, McDuff found the time to collaborate with Graeme Segal to produce what she considered tantamount to a "second Ph.D." on homology fibrations and the group–completion theorem. This paper was published in 1976, the year McDuff moved to Warwick University for another lectureship.

Meanwhile McDuff's professional horizons were expanding. In 1974 she was invited to visit MIT to fill a position reserved for a female. There she found the company of other women mathematicians invigorating. She was encouraged to apply to the Institute for Advanced Study and also won a spot there for a year. McDuff soon trusted her instincts enough to give up her tenure–track job at Warwick for an untenured position at SUNY–Stony Brook. During this period the McDuffs had divorced.

McDuff's "Autobiographical Notes" specifically mention that "there are only a few people who are interested in what I did," indicating that even today some of her work is obscure to many mathematicians. She had to struggle alone to solidify her early work into a foundation for her subspeciality in foliations. After some years of dealing with a commuter relationship, McDuff married mathematician **John Milnor** and gave birth to another child. Around this time she became involved in the field of symplectic topology. She gained a full professorship in 1984, and took a sabbatical in Paris the next year to investigate her newest speciality. In 1991 McDuff began a two–year stint as head of the mathematics department at SUNY–Stony Brook.

The forums at which McDuff presents her findings are socially as well as scientifically noteworthy. In 1994, she took part in "A Celebration of Women in Mathematics," a two–day conference at MIT featuring all female speakers. A lecture McDuff gave at Oregon State University in 1996 was sent out over the Internet in between rounds of The First Micro–Robot World Cup Soccer Tournament. Her talk was accompanied by interactive notes later posted on the World Wide Web. Another online project is part of a series funded by the National Science Foundation, called "Visualizing Women in Mathematics, the Physical Sciences and Technology." For it McDuff, artist Pam Davis, and fellow SUNY faculty member Tony Phillips collaborated on a Web installation. McDuff's set of activities are coordinated around the construction of a hypercube from plastic straws. These studies in projective geometry and trigonometry are not only physically accessible to anyone with Internet access. Their levels of difficulty range from elementary school to college–level topics.

McDuff is also active in the offline world. She helps oversee publication of the *Journal of the American Mathematical Society* as an associate editor. In 1994 she was inducted into the Royal Society of London, one of two female mathematicians in its ranks. She was elected a foreign honorary member of the American Academy of Arts and Sciences in 1995, one of only eight mathematicians chosen that year. Most recently, she was given an honorary doctorate by the University of Edinburgh. McDuff is a proponent of calculus reform and is an active mentor in the WISE (Women in Science and Engineering) program at SUNY–Stony Brook. WISE offers support to female undergraduates who plan careers in mathematics, science, or engineering.

## SELECTED WRITINGS BY MCDUFF

### Books

(With D. Salamon) *J–Holomorphic Curves and Quantum Cohomology*, 1994
(With D. Salamon) *Introduction to Symplectic Topology*, 1995

### Periodicals

"A Countable Infinity of II$_1$ Factors." *Annals of Mathematics* 90 (1969): 362–71.

### Other

"An Introduction to the Vocabulary of Dimension." *Squeeze the Phase Space.* Online Introductory Activities to Accompany the Posters of Dusa McDuff. http://math.math.sunysb.edu/~tony/visualization/dusa/activities.html
"Some Autobiographical Notes."
http://math.math.sunysb.edu/~tony/visualization/dusa/dusabio.html
"Symplectomorphisms and the Flux Homomorphism." *Pacific Northwest Geometry Seminar.* Online Lecture Notes for the 1996 Fall Meeting. http://www.math.washington.edu/~lee/PNGS/96–fall/

## FURTHER READING

### Books

*American Men & Women of Science.* Nineteenth edition. New Providence, NJ: Bowker, 1994.

### Other

Love, Lauren. "Dusa McDuff." *Biographies of Women Mathematicians.* June 1997. http://www.agnesscott.edu/lriddle/women/chronol.htm (July 1997).
Phillips, Tony. "Dusa McDuff: Biographical Sketch." http://math.math.sunysb.edu/~tony/visualization/dusa/bio.html

*—Sketch by Jennifer Kramer*

# Menaechmus
## c. 380–c. 320 B.C.
### Greek geometer

Menaechmus was a renowned Greek geometer whose efforts to solve the problem of doubling the cube led him to discover the conic sections. He is also known for writing about the foundations and mechanics of geometry, differentiating between two meanings of the word "elements," and arguing against the distinction between theorems and problems. In his history of Greek geometry, Proclus includes Menaechmus among those who have "made the whole of geometry still more perfect."

What little is known about Menaechmus has been gleaned from fragmented reference by ancient writers. Nothing is known of his birthplace or family, except that he had a brother, Dinostratus, who was also a mathematician and the first to use the trisectrix of Hippias for squaring the circle. Menaechmus is also believed to be Menaechmus of Suidas and Eudocia, a Platonic philosopher who wrote three books on Plato's *Republic.*

According to Proclus, Menaechmus was a tutor of Alexander the Great, which would indicate that he lived in middle of the fourth century B.C. Proclus also writes that Menaechmus was a pupil of Eudoxus in Athens and an associate of Plato. He probably belonged to the mathematical school of Cyzicus and, at some point, may have been the school's director. In a famous anecdote passed down by Stobaeus and the grammarian Serenus, Alexander the Great asked Menaechmus to teach him geometry but to do so in an easy way. Menaechmus replied, "O King, for traveling through the country there are private roads and royal roads, but in geometry there is one road for all." A similar story has been told about **Euclid** and **Claudius Ptolemy.**

### Discovers the Conic Sections

The ancient Greeks had long been fascinated by the problem of duplicating, or doubling, the cube, also called the Delian Problem. In a letter to Ptolemy III, the Greek mathematician **Eratosthenes** recounts how the Delians requested that the geometers who were part of Plato's Academia solve the problem. Hippocrates had already proposed that a cube could be doubled by finding the two mean proportionals (geometric means) between two given lines or between a number and its double. **Archytas** of Tarentum then discovered the proportional between two given lines.

It was Menaechmus who took the next step by focusing the geometric structure of the cone. By slicing a right circular cone with a plane perpendicular to one of its elements, he discovered the conic sections (parabolas, hyperbolas, and ellipses) in order to solve the problem. According to a commentary by Eutocius on **Archimedes**' treatise, *On the Sphere and Cylinder,* Menaechmus developed two solutions to the problem. In one solution, he showed that the two mean proportionals between the lines could be obtained by slicing cones with a plane and creating curves and then determining the intersection of the parabola and hyperbola. (It is highly unlikely that Menaechmus used the terms parabola and hyperbola, which are believed to have been introduced later by **Apollonius**.) His second solution was to intersect one parabola by another at right angles to it. Unfortunately, Menaechmus' own descriptions of his work are lost.

Menaechmus' approach to doubling the cube was not without its critics, chief among them being Plato himself. Plutarch reports that Plato criticized Eudoxus, Archytas, Menaechmus, and their school of mathematics for using instruments to draw curves. As a result, Plato accused these mathematicians of reducing the doubling of the cube to "mechanical contrivances," thus "perverting" geometry to "things of the sense" rather than keeping its focus on the "eternal and incorporeal" world.

Although there is no recorded proof that Menaechmus had developed a mechanical device for drawing curves, it is possible that a mechanical solution to the problem credited to Plato by Eutocius was, in fact, developed by Menaechmus. Regardless of whether or not Menaechmus used mechanical instruments, there is no evidence that the Greeks held any previous knowledge of conic sections, and Menaechmus has been credited with their discovery.

While his most famous accomplishment in mathematics focused on the practical application of using conics for doubling the cube, Menaechmus was also devoted to the philosophy of mathematics. He was the first to bring attention to the two senses of the word "element" and may have helped establish its proper use. Menaechmus pointed out that, in one sense, the word element refers to a proposition leading to another proposition. The word could also be used to mean a simple, fundamental component of a composite entity. As a result, Menaechmus clarified that propositions are elements only when they have a basic link to the conclusion.

According to Proclus, Menaechmus was also among those mathematicians who recognized that many so-called conversions of propositions are not really conversions. Menaechmus and his school also objected to the distinction between mathematical problems and theorems and stated they were all problems, the only difference being in the nature of the questions and the answers sought. Menaechmus may also have been interested in mathematical astronomy and is credited with Callippus for explaining the movements of the heavenly bodies based on a system of concentric spheres. In his commentary on Plato's *Republic,* Menaechmus discuses the distaff of the Fates and postulates a larger number of spheres than those contained within the planetary theories of Eudoxus and Callippus.

As a pioneer of geometrical theories and a philosopher of mathematics, Menaechmus proved himself an accomplished mathematician respected by both his contemporaries and those who followed. His genius is most strongly epitomized by his being the first to slice cones and establish curves as an integral part of mathematical theory. His discovery of conics led to centuries of abstract mathematical studies and eventually were applied by **Johannes Kepler, Issac Newton**, **Albert Einstein** and others to reveal how planets and comets move through the heavens.

## FURTHER READING

### Books

Allman, George Johnston. *Greek Geometry from Thales to Euclid.* London: Longmans, Green, & Co., 1889, pp. 153–79.

Bulmer–Thomas, Ivor. "Menaechmus," in *Dictionary of Scientific Biography.* Volume IX. Edited by Charles Coulston Gillispie. New York: Charles Scribner's Sons, 1975, pp. 268–77.

Heath, Sir Thomas. *A History of Greek Mathematics.* Volume I. Oxford: Clarendon Press, 1921, pp. 251–55.

Terry, Leon. *The Mathmen.* New York: McGraw–Hill Book Company, 1964, pp. 131–33.

*—Sketch by David A. Petechuk*

# Hugues Charles Robert Méray
## 1835–1911
### French number theorist and educator

Hugues Charles Robert Méray was not considered a leading mathematician by his contemporaries, but is renowned for one significant contribution to his field. In 1869 Méray was the first to publish a theory of irrational numbers that correctly anticipated the later, more popular theory developed by **Georg Cantor**, whose theory is one of the main steps in the arithmetization of analysis. In developing his theory of irrational numbers, Méray consciously built upon the earlier work of **Joseph–Louis LaGrange**, believing he could firmly establish what LaGrange had merely proposed.

Méray was born on November 12, 1835 in Chalon–sur–Saône, France, and received his formal mathematical training at the École Normale Supérieure there, commencing studies in 1854. From 1857 until 1859 he taught at the lycée of St. Quentin and then retired for seven years to a small village near his birthplace. In 1866 he came out of retirement to accept a position as a lecturer at the University of Lyons and, the following year, accepted a position as professor at the University of Dijon, where he remained for the rest of his teaching career.

Méray first presented his theory of irrational numbers in an article entitled "Remarques sur la nature des quantités définies par la condition de servir de limites à des variable données," published in *Revue des sociétés savantes des départements* in 1869. The article captured little attention, however, partially because the journal was relatively obscure. Today, although credit for the theory is commonly given to Cantor, Méray's work is considered of great historical significance. While **Karl Weierstrass** had introduced the theory in his lectures, Méray's article marked the first time the theory had been published. Méray's theory was again published in 1873 in his *Nouveau précis d'analyse infinitésimale*, but again his theory was largely ignored. The focus of the *Nouveau précis* was the presentation of a theory of functions of complex variables. The theory was based on the notion of a power series, another concept developed earlier by Weierstrass, although it is not believed that Méray was aware of Weierstrass' work either in this area or, specifically, with regard to the theory of irrational numbers. The *Nouveau précis* is also noted for its rigorous structure, an approach not yet common in Méray's time.

Méray died in Dijon in 1911.

## SELECTED WRITINGS BY MÉRAY

"Remarques sur la nature des quantités définies par la condition de servir de limites à des variable donnéss," 1869
*Nourveau précis d'analyse infinitésimale*, 1873

## FURTHER READING

### Books

Djugak, P. "The Limit Concept and Irrational Numbers: Ideas of Charles Méray and Karl Weierstrass," in *Studies in the History of Mathematics*. Moscow, 1973, 176–80.
Robinson, Abraham. "Hugues Charles Robert Méray," in *Dictionary of Scientific Biography*. Volume IX. Edited by Charles Coulston Gillispie. New York: Charles Scribner's Sons, 1970, pp. 307–08.

### Other

"Hugues Charles Robert Méray." *MacTutor History of Mathematics Archive.* http://www–groups.dcs.st–and.ac.uk/~history/Mathematicians/index.html. (August 1997).

—*Sketch by Kristin Palm*

# Nicolaus Mercator
## 1619(?)–1687
### Danish mathematical astronomer

Truly a man of many talents, Nicolaus Mercator is one of those historical figures who has not received the accolades he deserves for his scientific thought and discoveries. Much of his work in mathematics, planetary motion and education have gone unnoticed with other scientists of his time like **Isaac Newton** and **Christiaan Huygens** taking much of the historical focus. Nonetheless, Mercator's work has survived the test of time.

### Education and Early Work

Mercator was born Niklaus Kauffmann in Eutin, Schleswig–Holstein, Denmark, about 1619; his birthplace is now part of Germany. Mercator's father was Martin Kauffman, a schoolteacher at Oldenburg in Holstein, Denmark. Martin started teaching school there in 1623 and died when Niklaus was just a teenager in 1638. He was able to bestow an education to Niklaus at his school and with this foundation,

*Nicolaus Mercator*

Nicolaus graduated from the University of Rostock and won a position as a faculty member in 1642. During his time at Rostock, Mercator wrote several textbooks on astronomy and trigonometry. He also taught at the University of Copenhagen during this time, but it is not known exactly how his time was split between these two universities.

Mercator's early work is only documented by these textbooks, but they are evidence to his firm reign on the mathematic essentials like geometry, algebra, and trigonometry. His 1651 book, *Trigonometria sphaericorum logarithmica*, explained applications of logarithms to trigonometry. This volume also tabulated the logarithms of the standard trigonometric functions in one–minute intervals. Two more texts written in 1651, *Cosmographia* and *Astronomia*, explained the physical geography of the Earth and gave a mathematical introduction to astronomy. In these texts Mercator was one of many mathematicians who made the distinction between rational and irrational numbers. In his 1653 publication titled *Rationes mathematicae*, Mercator drew a parallel between rational and irrational numbers and harmony and dissonance in the study of music.

One of Mercator's last papers while in Denmark was a 1653 work titled *De emendatione annua*. In this paper, Mercator suggested reforming the 365–day year into 12 months of unequal days; he suggested 29, 29, 30, 30, 31, 31, 32, 31, 31, 31, 30, and 30 days each. This work was labeled "propagandist" and he took criticism for it. Of course, today it is obvious that his suggestion had merit, since today's calendar uses a similar system of months with unequal days. Despite the controversy, his paper attracted the attention of Oliver Cromwell in England, and although it is unclear whether Mercator actually received an invitation from Cromwell, he eventually left Denmark for London.

## Mercator in London

Mercator was unable to secure a university position in England, but he lived there for almost thirty years working as a private math tutor. During his long stay in England, he was known as Nicolaus Mercator, and eventually dropped his Germanic surname. Despite not being affiliated with a major university, he became acquainted with other British mathematicians. His mathematical contacts, along with his development of the marine chronometer, led to his election in November 1666 as one of the first members of the Royal Society of London, a group of the most notable English scientific thinkers.

One of Mercator's most important books was published ten years after his move to England in 1664. Titled *Hypothesis astronomica nova*, Mercator combined **Johannes Kepler**'s laws of planetary motion which states that planet travel in elliptical orbits around the sun, with his own belief relating the motion of the planets with the eccentricity of the planet's orbit. Some historians believe that given Newton's precarious grasp of the mechanics of planetary motion, Mercator's theories may have been the source for some of Newton's understanding.

## Mercator's Equation

Mercator was best known however, for his 1668 book *Logarithmotechnia* where he constructed logarithmic tables starting with the basic mathematical identity

$$\text{if } a^b = c \text{ then } b = \log_a c$$

and the application of the inequality

$$\left(\frac{a+px}{a-px}\right) < \left(\frac{a+x}{a-x}\right)^p, \quad p = \frac{1}{2}, \frac{1}{3}, \cdots$$

In an addition to the manuscript in 1667, Mercator described the equation that carries his name, a series expansion that computes the area under a hyperbola. Gregory of St. Vincent proved that the area under a hyperbola $y = 1/(1+x)$ from $x = 0$ to $x = b$ is equal to $\ln(1+b)$. Mercator was able to expand this equation into an infinite series using Gregory's method of long division followed by integration, showing that

$$\int \frac{dx}{1+x} = \frac{x}{1} - \frac{x^2}{2} + \frac{x^3}{3} - \frac{x^4}{4} + \cdots$$

Although this equation has been named for Mercator, it was also discovered independently by Huddle and Newton at about the same time.

### Other Talents

In addition to a sharp mathematical mind, Mercator was also bledded with mechanical prowess. His invention of the chronometer indicates his expertise in the theory of Gerard Mecator's maps. In fact, he was so sure of his mastery of these theories, he wagered his profits from the chronometer in a challenge published in the Royal Society journal that no one could challenge him in knowledge of Mercator's work. Mercator was also responsible for the design of what was later called the Huygenian pendulum watch. He also registered barometric measurements with the English Royal Society in the early part of 1667.

Mercator was described by his friends as a soft-tempered, introverted man of short stature with black hair and dark eyes. Perhaps that is why he is not as well known as many of the other mathematicians with more flair and bigger ego of the seventeenth century. In 1683 Mercator accepted a commission to plan the fountains of Versaille, France but after moving to France he had a falling out with his financial supporter. Nicolas Mercator died in 1687 in Paris at the age of 68.

## SELECTED WRITINGS BY MERCATOR

### Books

*Logarithmotechnia,* 1668

### Periodicals

"Certin Problems Touching Some Points of Navigation." *Philisophical Transactions of the Royla Society,* June 4, 1666, pp 215-218.

## FURTHER READING

### Books

Aubrey, John. *Letters . . . and Lives of Eminent Men, II*, London: 1813, pp 450–451.

### Periodicals

Hofmann, J. E. "Nicolaus Mercator (Kauffman), sein Leben und Wirken, vorzugsweise als Mathematiker." *Akademie der Wissenschaften und der Literatur in Mainsz* (1950).

—*Sketch by Roger Jaffe*

# Helen Abbot Merrill
## 1864–1949
### American analyst and educator

Helen Abbot Merrill took up mathematics as a vocation and avocation at a time when women were rarely visible in the field. She was most active as an instructor, co–writing textbooks as well as publishing articles on pedagogy. Merrill also wrote a "mathematical amusement" book, a populist work for young readers entitled *Mathematical Excursions.* Even in her spare time she was devoted to broader aspects of her profession, joining many mathematical, academic, and scientific organizations, and serving as vice–president of both the Mathematical Association of America and the American Mathematical Society.

Merrill was born March 30, 1864 in Orange, New Jersey, near Thomas Edison's facilities at Llewellyn Park. Her family traced its ancestry back to 1633, when Nathaniel Merrill settled in Massachusetts. Her father, George Dodge Merrill, had many business concerns, including being an inventor. Her siblings included a sister, Emily, and two brothers, Robert and William, who both grew up to become Presbyterian ministers.

### An Experimental Situation

Merrill began high school in 1876 at Newburyport, and entered Wellesley College six years later in one of the first graduating classes in the history of the college. Originally, her major was classical languages, an interest she would keep up throughout her life. However, as a freshman she committed to mathematics. At that time, Wellesley was still considered something of an experiment, but Merrill responded to the close–knit atmosphere and wrote a history of her graduating class as a commemorative booklet. Merrill graduated with a B.A. after four years and began her career as a teacher. Working at the Classical School for Girls in New York, Merrill was allotted courses in Latin and history as well as mathematics. She was assigned to a variety of students, including "mill girls" from New Brunswick, New Jersey, and immigrant children in the Germantown section of Philadelphia.

In 1893, Merrill was asked to return to Wellesley, this time as an instructor, in exchange for a stipend and housing. Helen Shafer, who had hired Merrill, allowed her time off intermittently for graduate studies at the universities of Chicago, Göttingen, and eventually Yale. Merrill earned a Ph.D. from Yale in 1903 and her thesis on "Sturmian" differential equations was published the same year. Merrill moved up from instructor to associate professor status at Wellesley. The college benefitted directly from Merrill's excursions, as she introduced courses in functions and descriptive geometry for her undergraduate students based on her graduate work.

## "Flowery," Not Thorny Paths

Merrill dedicated herself to providing a "flowery path" for her young charges to follow into the normally thorny subject of mathematics. She did not lower her standards for undergraduates; in fact, the courses Merrill taught were often in subjects generally offered only at the graduate level. However, she was quick to offer tailored assistance to any young woman she considered a diamond in the rough.

After being promoted to full professor in 1916, Merrill was appointed head of the mathematics department the next year. She was particularly active as associate editor of the Mathematical Association of American's monthly newsletter, member of the executive council, and later vice president in 1920. With a fellow MAA member, Clara E. Smith, who would also serve as vice president of the group, Merrill authored two textbooks. Merrill remained at Wellesley until her retirement, when she was named a Lewis Atterbury Stimson professor.

Merrill was also an amateur historian, fulfilling archival duties at Wellesley. She was elected as an executive committee member of the National Historical Society. Her interest in music and language led her to become as student again, taking summer courses at the University of California at Berkeley. She also traveled across Europe and the Americas. Merrill retired as professor emerita in 1932, and died at her home in Wellesley on May 1, 1949.

## SELECTED WRITINGS BY MERRILL

### Books

(With Clara E. Smith) *Selected Topics in College Algebra,* 1914
(With Clara E. Smith) *A First Course in Higher Algebra,* 1917
*Mathematical Excursions: Side Trips along Paths Not Generally Traveled in Elementary Courses in Mathematics,* 1933

### Periodicals

"On Solutions of Differential Equations Which Possess an Ooscillatoin [*sic*] Theorem," in *Transactions of the American Mathematical Society* 4 (1903): 423–33.
"Why Students Fail in Mathematics," in *Mathematics Teacher* 11 (1918): 45–56.
"Three Mathematical Songs," in *Mathematics Teacher* 25 (1932): 36–37.

## FURTHER READING

### Books

Green, Judy and Jeanne LaDuke. "Women in American Mathematics: A Century of Contributions." *A Century of Mathematics in America*, Volume 2. Edited by Peter Duren. Providence, RI: American Mathematical Society, 1989, pp. 384, 386.
Henrion, Claudia. "Helen Abbot Merrill." *Women of Mathematics.* Edited by Louise S. Grinstein and Paul J.Campbell. Westport, CT: Greenwood Press, 1987, pp. 147–151.
*The National Cyclopedia of American Biography.* Volume 42. Reprint. Ann Arbor, MI: University Microfilms, 1967–71, pp. 171–72.
Siegel and Finley. *Women in the Scientific Search: An American Bio–bibliography, 1724–1979.* Metuchen, NJ: The Scarecrow Press, Inc., 1985, pp. 214–15.

### Periodicals

"Helen A. Merrill of Wellesley, 85." *The New York Times* (May 3, 1949): p. 25.
"Helen Abbot Merrill." *Yale University Obituary Record* (July 1, 1949): 142.

### Other

"Helen Abbot Merrill." *Biographies of Women Mathematicians.* June 1997. http://www.scottlan.edu/lriddle/women/chronol.htm (July 22, 1997).

*—Sketch by Jennifer Kramer*

# Winifred Edgerton Merrill
## 1862–1951
### American mathematics educator

Winifred Edgerton Merrill forever secured her place in history by becoming the first American woman to be awarded a Ph.D. in mathematics. This confirmation is further conspicuous because she was the first woman to receive any sort of degree of any type from Columbia University. Merrill was born September 24, 1862, in Ripon, Wisconsin, to parents Emmet and Clara (Cooper) Edgerton. Merrill's American ancestry dates back to 1632, when Richard Edgerton left England and settled in Saybrook, Connecticut. Richard Edgerton is credited as one of the founders of Norwich, Connecticut.

Merrill was educated by private tutors before entering Wellesley College in Massachusetts. Wellesley was founded in 1871 as a seminary for women to be educated mainly as teachers. Because of the Civil War, the number of male students had decreased; coupled with the growing system of education initiated in the United States, and the fact that female teachers were paid about one–third that of male teachers, women were in high demand as teachers for public schools. Merrill graduated in 1883. She married Frederick J.H. Merrill, a mining geologist, in 1887 and they raised four children, Louise, Hamilton, Winifred, and Edgerton.

## Victory and Recognition

When Merrill graduated from Wellesley, she astonished the academic community by applying to Columbia University. Women were expected to, at most, secure a bachelor's degree then become instructors in the public school system. Merrill's admission to Columbia was granted only after the trustees met on several occasions to discuss the situation. As a condition of being the first woman to participate in the graduate mathematics program, she was obligated to keep the instruments used by the men as well as herself dust free and to pursue her studies in such a way that it would not discomfit her male peers. The academic trustees also made it clear that her admission was by no means meant "to set a precedent" and Merrill's acceptance was merely an "exceptional case." The male students in one of her classes were so disturbed by the idea of studying alongside a woman that they entered into a conspiracy with the professor, asking him to choose the most difficult textbook available for the course. The professor agreed to the choice. The plan gave Merrill no cause for concern, however, since she had already studied the book while at Wellesley College.

Merrill was awarded her Doctor of Philosophy degree *cum laude* in 1886 "in consideration of the extraordinary excellence of the scientific work" even though the university had not yet approved the dispensing of degrees to women. Merrill's doctorate dissertation was entitled "Multiple Integrals: a.) Their Geometrical Interpretation in Cartesian Geometry, in Trilinears and Triplanars, in Tangenials, Quaternions, and in Modern Geometry; b.) Their Analytic Interpretation in the Theory of Equations Using Determinants, Invariants, and Covariants as Instruments in the Investigation."

## An Educator and Pioneer for Women's Education

At about the same time that Merrill entered Columbia she began teaching at Sylvanus Reed's School in New York City. Three years later, in 1886, she was appointed vice principal and held the position until she married. In 1888, Merrill was one of five who helped in the founding of Barnard College, located on land adjacent to Columbia University. Barnard became part of Columbia in 1900 and remains the undergraduate liberal arts school for women. The music, classics, mathematics, physics, and religion departments are joint departments with Columbia. Merrill served on the Board of Trustees for a number of years but gave up her membership after her husband objected to frequent obligatory meetings with two male lawyers in private offices.

The demands of married life and child rearing kept Merrill away from teaching for a time. She taught for one year at the Emma Willard School in Troy, New York in 1894, then did not resume her career until 1906, when she accepted a position in Yonkers at Highcliff, then founded her own girls' school, Oakesmere, when she was in her forties and served as its principal for 13 years. The school was located in New Rochelle, New York, then later moved to Mamaroneck. Merrill directed the school until 1928, watching the enrollment increase over the years from eight to 148 students. The school was recognized for its high academic standards and in 1912 another school, Oakesmere Abroad, was founded in Paris, France. After 1928, Merrill made her home in New York City, assuming the role of director for the Three Arts Wing of the Barbizon Club. During the same time, she served as an alumna trustee and fundraiser for Wellesley College. She served as editor for the publications *Historical Vistas* and *Renaissance Vistas* in 1930. Her experience and interest in education also prompted her to write articles and give public speeches on various issues of education. Merrill was a member of many educational and political associations, including the Women's Graduate Club of Columbia University, the Woman's Organization for National Prohibition, and the American Association for the Advancement of Science.

In 1933, the college that had her donning a dust cloth 50 years earlier honored her as the inaugural woman graduate of Columbia, and recognized her efforts in furthering the education of women by installing her portrait in the Philosophy Hall. An inscription under the portrait reads, "She opened the door." The portrait was presented by Merrill's 1883 graduating class from Wellesley. Merrill died on September 6, 1951, in Stratford, Connecticut.

## FURTHER READING

### Books

Farnes, Patricia and G. Kass–Simon. *Women of Science: Righting the Record*. Bloomington: Indiana University Press.

*National Cyclopedia of American Biography*. Volume 41. New York: James T. White & Co., 1956.

*Marin Mersenne*

### Periodicals

Green, Judy and Jeanne LaDuke. "Women in the American Mathematical Community: The Pre–1940 Ph.D.'s." *The Mathematical Intelligencer* 9, no. 1 (1987): 11–13.

### Other

"Winifred Edgerton Merrill." *Biographies of Women Mathematics.* June 1997. http://www.scottlan.edu/lriddle/women/chronol.htm (July 21, 1997).

—*Sketch by Kelley Reynolds Jacquez*

# Marin Mersenne
## 1588–1648
### French number theorist and writer

Had it not been for the tireless efforts of French Minimite friar Marin Mersenne, communication describing the discoveries in science would have never been dispersed to the far corners of the mathematical and scientific worlds during the 17th century. During this critical period in the history of mathematics there were no scientific publications, bulletins, or newsletters. Instead, information was disseminated via scientific discussion circles such as Accademia dei Lineei (host to **Galileo**) and Accademia del Cimento in Italy, the Invisible College in England, and by written correspondence. Fortunately, Mersenne had a particular penchant for correspondence and a personal interest in the advancement of mathematical knowledge. It was said that to notify Mersenne of a discovery was the same as informing all of Europe.

### Correspondent Extraordinary

Mersenne was born on September 8, 1588, near Oize, Sarthe, France, and later entered into service for the Roman Catholic Church as a devoted teacher in 1611. He attended school with **Réne Descartes** and the two men remained friends throughout their lives. It was Mersenne who defended Descartes' philosophy against his critics in the Church and it was to him that Descartes wrote first whenever he developed a new theory in mathematics or philosophy. It was also through Mersenne that Descartes gained fame in the circles of European intellectuals. Mersenne was also a staunch defender of Galileo, assisting him in translations of some of his mechanical works. And it was Mersenne, Galileo's representative in France, who circulated Galileo's question of the path of falling objects on a rotating Earth, leading Descartes to put forth the equiangular or logarithmic spiral $r = e^{a\theta}$ as the possible path. This is just one of many instances in which Mersenne acted as correspondent and intermediary for the mathematical breakthroughs of the 17th century.

Very often Mersenne was give privileged previews of works that were later lauded as masterpieces; Descartes' *Le monde* is one example. The treatise was given to Mersenne as a New Year's gift in 1634 while all the rest of Paris, aware of its creation, had to wait until after Descartes' death to read it.

In addition to his responsibilities as a teacher of philosophy and theology at Nevers and Paris, Mersenne also conducted weekly scientific discussions, from which the French Académie Royale des Sciences was established in 1699 in Paris. While credit for starting two separate discussion groups at around the same time is usually given to Descartes and **Blaise Pascal**, it was actually Mersenne who originated the gathering of great minds to exchange information. One proof of this is that the discussions led by "Father Mersenne" were already in place at the time Pascal began to join them when he was only 14 years old. Another is that Descartes spent much of his time living in Holland and outside of Paris.

Mersenne's writings was not exclusive to acting as the broker for the mathematical world, however. He wrote extensively on physics, mechanics, navigation, geometry, and philosophy. A portion of his life's work was also devoted to editing the compositions of

*John Milnor*

**Euclid, Apollonius, Archimedes**, Theodosius, and Menelaus, as well as other ancient Greek mathematicians.

### Mersenne Remembered

Much is owed to Mersenne for his work as correspondent within the mathematical world, but the reason he is still known today can be attributed to his "Mersenne numbers." Although Mersenne has been called "the famous amateur of science and mathematics," his contribution did much to launch the discoveries of those who succeeded him. Mersenne published his *Cogitata Physico–Mathematica* in 1644 and it is for this paper that he is best known. In the paper, he asserted that a particular formula could be used for finding prime numbers, positive numbers which are only divisible by 1 and the number itself. In the paper Mersenne gave no reasons for why he believed his number theories were correct and later the formulas were proved to be incomplete. Many of the large numbers he alleged to be prime were not. Even though Mersenne was not successful in his attempt to create an ironclad formula, his conjectures were the stimulus and basis for later research into the theory of numbers and the search for large prime numbers (called Mersenne primes).

Mersenne spent his life as a staunch supporter of experimentation and was indirectly responsible for the invention of the pendulum clock. He suggested to **Christiaan Huygens** that he experiment with timing objects rolling down a slanted surface by using a pendulum. Galileo had previously noted the characteristic timekeeping property of the pendulum but through Mersenne's suggestion it was Huygens who developed the general application of the pendulum as a time controller in clocks in 1656.

*Cogitata Physico–Mathematica* also contained explanations of some of Mersenne's experiments in the field of physics. These experiments were, no doubt, stimulated by his correspondence with Galileo, Huygens, **Pierre de Fermat**, and other men of science, but it appears that Mersenne was better at chronicling and appending the work of others than he was at immortalizing his own. The last of Mersenne's papers were published in 1644, containing a condensation of mathematics. Mersenne died in Paris in 1668 at the age of 60.

### FURTHER READING

Asimov, Isaac. *Asimov's Biographical Encyclopedia of Science and Technology.* Garden City, NY: Doubleday & Company, Inc., 1972.

Ball, W.W. Rouse. *A Short Account of the History of Mathematics.* New York: Dover Publications, Inc., 1960.

Bell, E.T. *The Development of Mathematics.* New York: McGraw–Hill Book Company, Inc., 1945.

Boyer, Carl B. *A History of Mathematics.* Second edition. Revised by Uta C. Merzbach. New York: John Wiley & Sons, Inc., 1991.

Smith, David Eugene. *History of Mathematics.* Volume 1. New York: Dover Publications, Inc., 1958.

Struik, Dirk J. *A Concise History of Mathematics.* New York: Dover Publications, Inc., 1967.

*—Sketch by Kelley Reynolds Jacquez*

# John Milnor
## 1931-
### American topologist

John "Jack" Milnor, one of the leading topologists in the second half of the 20th century, has studied the generalized, multidimensional surfaces known as manifolds and has had a particular interest in working out the relationships among the various ways of viewing these spaces. In 1956, he astounded the mathematical world with the first example of manifolds that are different when considered in the context of calculus but are equivalent when viewed geometrically. For this discovery, Milnor received the Fields Medal in 1962, the highest honor in mathematics. He is regarded as a master of using current algebraic techniques to analyze complex geometrical objects.

John Willard Milnor was born in Orange, New Jersey, on February 20, 1931, to John Willard and Emily (Cox) Milnor. He published his first paper, "On the Total Curvature of Knots," in 1950 while he was an undergraduate at Princeton University. It has been said that Milnor mistook an unsolved conjecture written on the board for the homework assignment, and his simple yet ingenious solution was the catalyst for this paper. His proof that the total curvature of a non-trivial, three-dimensional knot is at least $4\pi$ was described by Lisa Goldberg and Anthony Phillips in *Topological Methods in Modern Mathematics* as a simple and elegant argument that foreshadowed his future work. He received his A.B. degree in 1951 and continued his doctoral work at Princeton.

In 1954, Milnor received his Ph.D. under the direction of Ralph Fox, a mathematician known as the dean of American knot theorists (knot theory is a branch of topology developed in the 20th century). The following year Milnor was named the Higgins Lecturer at Princeton. In 1962, he was promoted to full professor and designated Princeton's Henry Putnam University Professor of mathematics.

## Discovers "Exotic" Spheres

Milnor's primary work concerned the study of manifolds. These spaces arise in topology (a branch of mathematics related to geometry) and are the analogs of curves and surfaces. Near any point, a coordinate system can be introduced so that the immediate neighborhood of the point looks like ordinary Euclidean space (although it may be of any finite dimension). The coordinate systems for various points can be different but must fit together in a continuous fashion. In 1956, Milnor published the paper "On Manifolds Homeomorphic to the 7-Sphere," which presented the first example of a pair of manifolds that are equivalent from the standpoint of continuity (i.e., are homeomorphic) but are distinct from the standpoint of smoothness (i.e., are not diffeomorphic). The 7-sphere (a seven-dimensional sphere) is both continuous (without gaps) and smooth (without corners); however, the "exotic" 7-sphere Milnor described was continuous yet could not be deformed into a smooth surface. As **Michael Atiyah** said in his 1975 Bakerian Lecture, Milnor demonstrated that "there are two different schemes of classification, one (topology) using continuity and the other (differential topology)

requiring also continuity of tangents (derivatives). Moreover, Milnor showed that there are precisely 28 different seven-dimensional 'spheres'."

This revolutionary discovery of exotic spheres earned Milnor the Fields Medal in 1962 and election to the National Academy of Sciences the following year. Over the next decade, Milnor continued to study these exotic spheres and found a way to combine them. In fact, this has become such a fruitful field for research that his original paper was chosen for the 1982 Steele Prize as a publication of "fundamental or lasting importance in its field, or a model of important research." In 1989, Milnor was awarded the Wolf Price in mathematics in recognition of his "ingenious and highly original discoveries in geometry from the algebraic, combinatorial, and differentiable viewpoint."

## Explains Complex Ideas Clearly

In addition to his mathematical innovations, Milnor is known for the clarity of his lectures. Jim Stasheff, a former student, wrote in *Topological Methods in Modern Mathematics* that Milnor is "at once composer and performer, expositor and creator." Goldberg and Phillips described the "unrelenting clarity of his presentation [and] his uncanny ability to find an interesting mathematical kernel in the most obtuse question from the class." He has been invited to give numerous lectures and presentations, including a 30-minute address at the 1958 International Congress of Mathematicians (ICM) in Edinburgh, and an hour-long address in Stockholm at the 1962 ICM.

Milnor was a visiting professor at the University of California, Berkeley during the academic year 1959-60, and at the University of California at Los Angeles in 1967-68. He spent two years as professor of mathematics at the Massachusetts Institute of Technology, and then returned to Princeton in 1970 as a professor at the Institute for Advanced Study. Around that time, Milnor became interested in the use of computer graphics to experiment with the new field of dynamical systems. Although he did not begin to publish his work in this area until 1985, his ideas were influential through preprints of his work in progress and his collaboration with other mathematicians.

In 1989, Milnor accepted the position of director at the newly formed Institute for Mathematical Sciences at the State University of New York at Stony Brook. In June of 1991, the Institute and the Mathematics Department at Stony Brook organized a conference in honor of Milnor's 60th birthday. The participation by 220 mathematicians (including 100 graduate students) from around the world testifies to Milnor's lasting influence on mathematics.

### Brings Insight to Diverse Fields

Milnor's work in dynamical systems evolved into complex dynamics, in which "real" problems are analyzed using "complex" techniques. He seems to be too busy to completely develop all of his ideas. Milnor's popular books have come about at the urging of graduate students who volunteered to write up their notes from his lectures, and who spent a great deal of effort incorporating changes and clarifications that he suggested after reviewing what they had written. By means of the resulting mimeographed documents (and in some cases published versions), Milnor's ability to organize new and complicated fields of mathematics was influential far beyond his immediate students. Goldberg and Phillips wrote that his 1989 article, "Self-Similarity and Hairiness in the Mandlebrot Set," describes a set of conjectures with such clarity that the statement actually suggests a method of proof.

Still active in his research, Milnor has made notable contributions in the areas of algebraic and differential topology, differential geometry, game theory, algebraic K-theory, singularities of complex hypersurfaces, and dynamical systems. At the symposium celebrating Milnor's 60th birthday, his former student Jonathan Sondow said, "many of his papers contain startling discoveries: they revealed unsuspected distinctions between related structures." Another speaker, Hyman Bass, said, "The legacy of Milnor's work lies not only in its profound expansion of our knowledge, but also in its revelation of new mathematical landscapes. On several occasions his work has virtually defined new fields and launched their research agendas."

## SELECTED WRITINGS BY MILNOR

### Books

*Morse Theory: Based on Lecture Notes by M. Spivak and R. Wells,* 1963
*Topology from the Differential Viewpoint,* 1965
*Collected Papers,* 1994

### Periodicals

"On the Total Curvature of Knots." *Annals of Mathematics* 52 (1950): 248–57.
"On Manifolds Homeomorphic to the 7-Sphere." *Annals of Mathematics* 64 (1956): 399–405.

## FURTHER READING

### Books

*Topological Methods in Modern Mathematics: A Symposium in Honor of John Milnor's Sixtieth Birthday.* Edited by Lisa R. Goldberg and Anthony V. Phillips. Houston: Publish or Perish, Inc., 1993.

*Hermann Minkowski*

### Periodicals

Atiyah, M. F. "Bakerian Lecture, 1975: Global Geometry." *Proceedings of the Royal Society of London* 347 (1976): 291–299.
Ewing, J. H., et al. "American Mathematics from 1940 to the Day before Yesterday." *American Mathematical Monthly* 83 (1976): 503–516.
"John W. Milnor." *Notices of the American Mathematical Society* 29 (October 1982): 507.

### Other

John W. Milnor home page. http:// math.math.sunysb.edu/.jack/

   *—Sketch by Robert Messer and Loretta Hall*

# Hermann Minkowski
## 1864-1909
### Russian-born German analyst

In spite of a relatively short career, Hermann Minkowski played an important role in the development of modern mathematics. His work formed the basis for modern functional analysis, and he did much to expand the knowledge of quadratic forms. Minkowski also developed the mathematical theory known as the geometry of numbers and laid the mathematical foundation for **Albert Einstein**'s theory of relativity by pioneering the notion of a four-dimensional space-time continuum.

Minkowski was born in Alexotas, Russia, on June 22, 1864, of German parents. The family returned to their native Germany in 1872, to the city of Königsberg, where Minkowski spent the rest of his childhood and also attended university. His brother, Oskar Minkowski, became famous as the physiologist who discovered the link between diabetes and the pancreas.

### Achieves Renown at an Early Age

Even as a student at the University of Königsberg, Minkowski demonstrated a rare mathematical talent. In 1881, the Paris Académie Royale des Sciences offered a prize, the Grand Prix des Sciences Mathématiques, for a proof describing the number of representations of an integer as a sum of five squares of integers—a proof that, unbeknownst to the Académie, the British mathematician H. J. Smith had already outlined in 1867. Minkowski produced the proof independently, while Smith filled in the details of his outline and submitted it. In 1883, both Smith and Minkowski received the prize. At that time, the 19–year–old Minkowski was two years away from receiving his doctorate from the University of Königsberg. The work contained in his 140–page solution was, in fact, considered a better formulation than Smith's because Minkowski used more natural and more general definitions in arriving at his proof.

While he was a university student, Minkowski began a lifelong friendship with fellow student **David Hilbert,** who would eventually edit Minkowski's collected works. After receiving his doctorate from the University of Königsberg in 1885, Minkowski taught at the University of Bonn until 1894. Returning to teach at the University of Königsberg for two years, he then taught at the University of Zurich until 1902. One of his closest colleagues at Zurich was a former teacher, A. Hurwitz, who is best known for his theorem on the composition of quadratic forms.

### Develops Geometry of Numbers

Throughout his life, Minkowski worked on the arithmetic of quadratic forms, particularly in $n$ variables. According to mathematician Jean Dieudonné, who profiled him for the *Dictionary of Scientific Biography,* Minkowski made two important contributions to this field. One was a characterization of equivalence of quadratic forms with rational coefficients, under a linear transformation with rational coefficients. The other, published in 1905, completed **Charles Hermite**'s theory of reduction for positive definite $n$-ary quadratic forms with real coefficients by finding a unique reduction form for each equivalence class. Pursuing the results of his 1905 paper, Minkowski developed a more geometric style of work, which led to what Dieudonné called his "most original achievement'—the geometry of numbers.

In 1889, Minkowski had introduced the geometrical concept of volume into his work on ternary quadratic forms. This technique involved centering ellipses on lattice points in the plane and looking at the areas of the ellipses as the lattice points increased in number. The limiting case as the number of non-overlapping ellipses in the plane approaches infinity gave Minkowski an estimate of the minimal solution of a specified quadratic equation in two variables. Using this type of geometrical technique, he was able to prove various theorems about numbers without performing any numerical calculations, a feat that was praised by Hilbert as "a pearl of Minkowski's creative art," reported Harris Hancock in the introduction to his book *Development of the Minkowski Geometry of Numbers.*

Minkowski generalized the technique to ellipsoids and other convex shapes (such as cylinders and polyhedrons) in three dimensions. This work led to investigations in packing efficiency (how to fill up space most densely with given shapes), a topic that has applications in chemistry, biology, and other sciences. Generalizing further to various types of convex objects in $n$-dimensional space, he produced numerous results in number theory through geometry. As Dieudonné wrote, "Long before the modern conception of a metric space was invented, Minkowski realized that a symmetric convex body in an $n$-dimensional space defines a new notion of 'distance' on that space and, hence, a corresponding 'geometry'"—a development that laid the foundation for modern functional analysis.

Hancock wrote of Minkowski, "His grasp of geometrical concepts seemed almost superhuman." However, he also noted that Minkowski's publications and notes were often incomplete and poorly explained. In part this was simply Minkowski's style, although it was complicated by his early death that brought an abrupt end to his works in progress. He died at the age of 44 from a ruptured appendix in Göttingen, Germany, on January 12, 1909. Hancock wrote his book to "reconstruct and clarify much that Minkowski would have done, had he lived."

### Builds Foundation for the Theory of Relativity

At Hilbert's urging, the University of Göttingen created a new professorship for Minkowski in 1902. It was during his tenure at Göttingen that Minkowski turned his attention to relativity theory. Einstein, who published his initial work in relativity in 1905, had taken nine classes from Minkowski in Zurich, more than he had taken from anyone else (even the physics professor). The two men had no particular liking for one another. Lewis Pyenson wrote in *Archive for History of Exact Sciences* that "By the time he graduated in 1900 . . . Einstein had become indifferent to Minkowski's approach to mathematics and physics," and "Minkowski later thought that his own interpretation of the principle of relativity was superior to Einstein's because of Einstein's limited mathematical competence."

In 1905, Minkowski participated in an electron theory seminar that discussed the current theories of electrodynamics. With this background, he studied the competing theories of subatomic particles proposed by Einstein and Hendrik Lorentz. Minkowski was the first to realize that both theories led to the necessity of visualizing space as a four-dimensional, non-Euclidean, space-time continuum. Dieudonné wrote that Minkowski "gave a precise definition and initiated the mathematical study [of this four-space]; it became the frame of all later developments of the theory and led Einstein to his bolder conception of generalized relativity."

Rather than a mathematical adjustment of Einstein's theory, Minkowski developed his ideas as an alternate theory. As soon as Minkowski's first publication on relativity appeared, Pyenson wrote, "Einstein turned to the only part of Minkowski's paper that contained a physical prediction" and showed how it failed to account for a known phenomenon. The rivalry between the two men was carried on at a level that few recognized; the differences between their derivations were subtle. Ultimately, Einstein used Minkowski's ideas to develop his general theory of relativity, which was published seven years after Minkowski's death.

## SELECTED WRITINGS BY MINKOWSKI

*Raum und Zeit,* 1909
*Gesammelte Abhandlungen.* Edited by David Hilbert. 1911
*Geometrie der Zahlen,* 1968

## FURTHER READING

### Books

Dieudonné, J. "Hermann Minkowski," in *Dictionary of Scientific Biography.* Volume IX. Edited by Charles Coulston Gillispie. New York: Charles Scribner's Sons, 1974, pp. 411–14.
Hancock, Harris. *Development of the Minkowski Geometry of Numbers.* 2 volumes. New York: Dover, 1964.

"Hermann Minkowski." *The New Encyclopaedia Britannica.* Micropaedia Volume 8. Chicago: Encyclopaedia Britannica, Inc., 1997, p. 166.

### Periodicals

Pyenson, Lewis. "Hermann Minkowski and Einstein's Special Theory of Relativity." *Archive for History of Exact Sciences* 17 (1977): 71–95.

### Other

"Minkowski." http://www.vma.bme.hu/mathhist/ Mathematicians/Minkowski.html (July 8, 1997).

—*Sketch by Maureen L. Tan and Loretta Hall*

# Abraham de Moivre
## 1667–1754
### Anglo–French probabilist and analyst

De Moivre's name is attached to a basic result in trigonometry, although he never stated it in the form in which it is usually cited. On a larger scale, his contributions to probability and statistics helped to lay the foundations for them as mathematical disciplines. Despite the quality of his work, de Moivre was never able to secure regular employment as a mathematician. His productivity is remarkable in light of the many other occupations he had to undertake in order to make a living.

Abraham de Moivre was born into a Huguenot family on May 26, 1667 in Vitry–le–François, France. His father was a surgeon outside the nobility, so it is usually assumed that the "de" in de Moivre's name was assumed when he came to England. After studying at both Catholic and Protestant schools, he went to Paris in 1684 to study under Jacques Ozanam, one of the great teachers of mathematics at the time. Following the revocation of the Edict of Nantes, the document that had guaranteed religious freedom for Huguenots, de Moivre appears to have been imprisoned for some time in Paris, but within a few years he joined the large number of Huguenot refugees, preferring the linguistic unfamiliarity of England to the religious hostility of their native land.

De Moivre's foreign ancestry may have kept him from achieving academic advancement, since such individuals as poet Alexander Pope and **Isaac Newton** attested to de Moivre's achievements in mathematics. After Newton had left Cambridge for London to serve as Master of the Mint, he often referred prospective students to de Moivre for instruction.

After arriving in England, de Moivre took an active part in the discussions of the Royal Society, his first paper being communicated in 1695 by **Edmond Halley**. Within two years de Moivre had become a Fellow of the Royal Society, and he had 15 papers published in the *Philosophical Transactions of the Royal Society* over the next 50 years. He appears to have made a living by tutoring and by offering advice to gamblers and underwriters who patronized Slaughter's Coffee House on Fleet Street.

### Creates an Audience for Probability

Although de Moivre's first paper submitted to the Royal Society was on calculus, he maintained his interest in other branches of mathematics even after having required a reputation in probability. The theorem that bears his name refers to the nth power of an expression involving trigonometric functions and the imaginary unit. First stated in 1722, it stresses the close connection between the algebra and the geometry of complex numbers.

De Moivre's work on probability was originally published in Latin in the *Philosophical Transactions*, but it became better known following publication as the *Doctrine of Chances*. In 1718, it went through two subsequent editions, and the third edition has been called the first modern probability textbook. In *Doctrine of Chances,* de Moivre strengthens the result given by **Jakob Bernoulli** in his *Ars Conjectandi*, often called the law of large numbers. The law deals with how far removed from genuine frequencies the proportions of results of an experiment can be as the number of repetitions increases. From the mathematical point of view, de Moivre's proof of the law is an improvement over Bernoulli's (which was only published posthumously). De Moivre employed the law further as an argument for the existence of design in the universe. The problem is that his argument depended on the existence of stable frequencies in the first place, so that it could scarcely also be used to establish them. Presumably, de Moivre could not imagine a universe in which frequencies did not become stable in the long run, but that does not amount to a proof.

The tone of the *Doctrine of Chances* is relatively light, as de Moivre tailored his work for readers who were not attuned to mathematics. Among the problems to which he gave his attention was the one–dimensional random walk. Given the numbers from 0 to a + b lined up in order, if one begins on the number a and has a probability of going up one step of p, the question is how long it will take to run into one of the barriers (0 or a + b). De Moivre's solution is generally considered an improvement of **Christiaan Huygens'** work.

De Moivre's work on annuities was published in 1725 and also marked a sizable improvement over the previous material on the subject. In 1735, he was made a fellow of the Berlin Academy and in the year of his death a fellow of the Paris Académie as well. He died in London on November 27, 1754, after having seen probability transform into a discipline that had become mathematically respectable. Although Newton never turned his mathematical attention to probability, the work of de Moivre suggests the direction the subject would have taken in mathematically.

## SELECTED WRITINGS BY DE MOIVRE

*The Doctrine of Chances*, 1718
*Annuities upon Lives*, 1725
*Miscellanea Analytica de Seriebus et Quadraturis*, 1730

## FURTHER READING

Clerke, Agnes Mary. "Abraham De Moivre," in *Dictionary of National Biography*. Volume 13. Edited by Leslie Stephen and Sidney Lee. Oxford: Oxford University Press, 1968, pp. 563–564.

Hacking, Ian. "Abraham De Moivre," in *Dictionary of Scientific Biography*. Volume IX. Edited by Charles Coulston Gillispie. New York: Charles Scribner's Sons, 1973, pp. 452–455.

Hald, Anders. *A History of Probability and Statistics and Their Applications before 1750*. New York: John Wiley & Sons, 1990.

—*Sketch by Thomas Drucker*

# Gaspard Monge
## 1746–1818
### French geometer and educator

Considered the founder of descriptive and differential geometry, Gaspard Monge was one of the most famous mathematicians of his day and renowned for his application of geometry to problems of construction. A man of many interests, Monge also worked in chemistry and physics, as well as education, training a generation of mathematicians and paving the way for the extension of his theories to projective geometry. A fervent republican, Monge was a supporter of the French Revolution, and was for a time minister of the navy. He helped to establish the metric system and was a founder of the École Polytechnique, which he

*Gaspard Monge*

directed in its early years. Made a count in 1808, Monge is sometimes known as Comte de Peluse. His major work, *Geometrie descriptive*, leads directly to the methods employed in modern mechanical drawing known as orthographic projection.

Monge was born on May 9, 1746, in Beaune, France, to Jacques Monge and Jeanne Rousseaux. The elder Monge was a knife grinder and peddler with a belief in the power of education. Young Monge was considered the genius of the family and excelled at the college of the Orations in Beaune which he attended until 1762. From 1762 to 1764, he attended college in Lyons where, at age 16, he was made a physics instructor. On vacation in Beaune in 1762, Monge spent his free time by completing a map of the town, developing both the means of such large scale projection as well as the surveying equipment necessary for its completion. This work caught the eye of an officer of engineers who recommended Monge for a place at the military school at Mezieres, even though as the son of a commoner, he would never be eligible for an officer's commission.

### From Military Engineer to Geometer

Monge made an early name for himself by devising a plan for gun emplacements in a proposed fortress. He managed to substitute a geometrical process for such a calculation, avoiding the cumbersome arithmetic techniques then in use, and it was this breakthrough which in part led to his contribu-

tions in descriptive geometry. Initially, Monge's plan was ignored, as the officer in charge felt there was some trickery involved, though when finally his plan was examined, it was found to be of such value that he was pledged to secrecy. For the next fifteen years, Monge's method of descriptive geometry—representing three–dimensional objects in two dimensions— was kept under wraps as a military secret. Monge was made a professor at the École Royale du Genie in Mezieres, and from 1768 to 1783 taught both physics and mathematics, developing the field of geometry in service to the solution of construction and mechanical problems: everything from fortifications to scaffolding and general architecture.

Monge married Catherine Huart in 1777, and the couple eventually had three daughters. With his election to the French Académie Royale des Sciences in 1780, he began to spend more time in Paris, dividing his time between the capital city and Mezieres. He participated in research of sponsored by the Académie and presented several papers there. In 1783, Monge was named examiner of naval cadets, a position which made it impossible for him to continue his professorship in Mezieres. For the next nine years he was busy with tours of inspection of the naval schools and his academic duties in Paris. The most fruitful years of research were behind him.

### Monge and the Revolution

At the outbreak of revolution in 1789, Monge was one of the best known French scientists. His support of the Revolution, however, was kept relatively quiet until 1792 with his membership in several revolutionary clubs. With the fall of the monarchy and the takeover of the Legislative Assembly, Monge was appointed minister of the navy, a position he held for a scant eight months. His politics were considered too moderate for the men in charge, and thereafter Monge's political activity was at a minimum. He did, however, continue to work for the democratic goals of the Revolution, even after the suppression of the Académie in 1793. Working with the Temporary Commission on Weights and Measures, Monge helped to develop the metric system. Lecturing at the École Normale in Paris in 1794, he was first allowed to teach his methods of descriptive geometry in public. Heeding an appeal for scientific men to come to the aid of French industry in support the war effort, Monge even supervised foundries and wrote a factory handbook. Important for the future course of mathematics, Monge was influential in preparing the way for the École Polytechnique, which opened in 1795. His lectures in infinitesimal geometry held there were eventually printed first as *Feuilles d'analyse appliquee a la geometrie*, and later as *Application de l'analyse a la geometrie.* This text established the algebraic principles of three–dimensional geometry and helped to revolutionize engineering design.

In 1796, Monge left for what would be a protracted absence from France, first on a mission to Italy, and thereafter in service to Napoleon Bonaparte in Egypt, where he was assigned various technical and scientific tasks, including the establishment of the Institut d'Egypt in Cairo. With the defeat of Napoleon by the British fleet, Monge escaped back to France. During these years, he had continued his analytical researches, adding chapters to his *Application de l'analyse a la geometrie,* making it in effect a textbook of differential geometry with his introduction of the idea of lines of curvature in space.

Back in Paris, Monge resumed his duties at the École Polytechnique, and with the advent of the new century, honors and awards started coming his way. He was named a grand officer of the Legion of Honor in 1804, president of the Senate in 1806, and made a count in 1808. His duties as a senator accordingly took away from his work at the École Polytechnique, yet he was still able to see his monumental *Application de l'analyse a la geometrie* published in 1807. With the ultimate defeat of Napoleon and his abdication, Monge fled France for a time. Stripped of his honors and professional position by the restored Bourbon monarchy, Monge returned to Paris in 1816, and lived out the last two years of his life reviled by many for his part in the Bonaparte regime. At his death in 1818, and despite government censure, many in the scientific community, both his students and colleagues, paid Monge respect by defying a government ban and placing a wreath on his grave the day after his funeral.

### A Lasting Legacy

Monge is remembered primarily for his development of descriptive geometry, which he developed from 1766 to 1775, and its practical applications in the fields of construction and architecture. Though much of this was a codification of what others had done before, Monge made a system of such projective techniques. Such techniques paved the path ultimately for the development of projective geometry. Monge also pioneered techniques in analytic and infinitesimal geometry, a particular favorite of his, researching the properties of surfaces and of space curves. In particular, Monge is known for the application of the calculus to the examination of the curvature of surfaces. Additionally, Monge made important progress in the field of differential equations.

Less well known is Monge's work in mechanics, chemistry, and technology. In addition to books on the theory of machines, Monge also worked in caloric theory, acoustics, and optics. Perhaps his most famous work in chemistry dealt with the composition of water. In all, Monge's accomplishments in mathematical research and technology have won him a lasting position in the canon of mathematical greats. He was,

as E. T. Bell described him in *Men of Mathematics,* "a born geometer and engineer with an unsurpassed gift for visualizing complicated space–relations."

## SELECTED WRITINGS BY MONGE

*Traite elementaire du statique,* 1788
*Geometrie descriptive,* 1799
*Application de l'analyse a la geometrie,* 1807

## FURTHER READING

### Books

Aubry, Paul V. *Monge, le savant ami de Napoleon: 1746–1818.* Paris: 1954.
Bell, E. T. *Men of Mathematics.* New York: Simon and Schuster, 1937, pp. 183–205.
Coolidge, Julian Lowell. *A History of Mathematical Methods.* New York: Dover Publications, 1963.
*L'oeuvre scientifique de Gaspard Monge.* Paris: Paris Presses universitaires de France, 1951.
Taton, René. "Gaspard Monge," in *Dictionary of Scientific Biography.* Volume IX. Edited by Charles Coulston Gillispie. New York: Charles Scribner's Sons, 1974, pp. 469–78.

### Periodicals

Booker, J. "Gaspard Monge and His Effect on Engineering Drawing and Technical Education." *Transactions of the Newcomen Society* 34 (1961–1962): 15–36.

*—Sketch by J. Sydney Jones*

# Cathleen Synge Morawetz
## 1923-
### Canadian-born American applied mathematician

In 1984 Cathleen Synge Morawetz became the first woman in the United States to head a mathematical institute, the Courant Institute of Mathematical Sciences at New York University. Since receiving her Ph.D. from the university in 1951, her research on fluid dynamics and transonic flow has been influential in the fields of aerodynamics, acoustics, and optics. In 1993 she was elected president of the American Mathematical Society.

Born on May 5, 1923, to mathematician John Synge and Eleanor Mabel Allen Synge, Morawetz grew up in Toronto, Canada, and won a scholarship in mathematics to the University of Toronto. She served in a wartime post in 1943–44 before earning her B.A. in 1945. That same year she married chemist Herbert Morawetz (with whom she eventually had four children) and began graduate studies at the Massachusetts Institute of Technology (MIT), from which she received her master's degree in 1946. At a time when few scientific fields were open to women, Morawetz briefly considered working for Bell Laboratories in New Jersey but decided to pursue a Ph.D. instead.

While editing the book *Supersonic Flow and Shock Waves,* written by New York University mathematicians **Richard Courant** and Kurt Friedrichs, she became interested in the phenomenon of transonic flow, or the behavior of air at speeds approximating that of sound. Spurred by the urgency of this research (in the late years of World War II, the new jet engine enabled aircraft to approach supersonic flight) she decided to pursue a Ph.D. in the subject. Working under Friedrichs at New York University (NYU), she wrote her doctoral thesis on imploding shock waves and received her degree in 1951. After a brief stint as a research associate at MIT, she joined NYU's mathematics faculty as a research assistant in 1952, publishing scores of scientific articles during her upward climb to full professor in 1965. Morawetz also rose within NYU's Courant Institute of Mathematical Sciences, where she received the ultimate honor of being named the institute's director in 1984—the first woman ever to head a mathematical institution of that stature in the United States. She has also served as the editor of several scientific journals, including the *Journal of Mathematical Analysis and Applications* and *Communication in Pure and Applied Mathematics.*

Morawetz continued research in the field of fluid dynamics, the study of forces exerted upon liquids and gases and the effects of those forces on motion. She conclusively demonstrated that all airplane wings produce shock waves if they are moving fast enough. Her findings have contributed to important advances in aerodynamics, acoustics, and optics. Similarly, her work on transonic flow has been applied in two widely different areas: the use of seismic waves (movement caused by vibrations within the earth) in prospecting for oil and the development of medical imaging techniques.

A member of the American Academy of Arts and Sciences, Morawetz has received several major awards, including two Guggenheim Fellowships, the Gibbs Lectureship of the American Mathematical Society, and the Lester R. Ford Award from the Mathematical Association of America. She also holds honorary degrees from Smith College, Brown University, Princeton University, and Duke University. In 1994, in addition to being the only woman member of the Applied Mathematics Section of the National Academy of Sciences, Morawetz was elected the second woman president to head the American Mathematical Society. In a university press release, NYU President L. Jay Oliva summarized the esteem in which she is held by her peers: "Morawetz is an outstanding mathematician, and has long been one of the leading lights at our prestigious Courant Institute. I know that the American Mathematical Society will benefit greatly by her considerable acumen and compelling leadership."

## SELECTED WRITINGS BY MORAWETZ

(With I. Kolodner) "On the Non-existence of Limiting Lines in Transonic Flows." *Communication in Pure and Applied Mathematics* (February 1953): 97–102.
"Energy Flow: Wave Motion and Geometrical Optics." *Bulletin of the American Mathematical Society* (July 1970): 661–74.
"The Mathematical Approach to the Sonic Barrier." *Bulletin of the American Mathematical Society* (March 1982): 127–45.
"Giants." *American Mathematical Monthly* (November 1992): 819–28.

## FURTHER READING

### Books

Patterson, James D. "Catherine Synge Morawetz," in *Women of Mathematics.* Edited by Louise S.Grinstein and Paul J. Campbell. Westport, CT: Greenwood Press, 1987, pp. 152–55.

### Periodicals

"Cathleen Morawetz, First Woman to Head a Mathematics Institute." *Series in Applied Mathematics News* 17, number 4 (1984): 5.
Kolata, Gina Bari, "Cathleen Morawetz: The Mathematics of Waves." *Science* 206 (1979): 206–07.

### Other

"NYU Mathematician Cathleen Morawetz Elected Next President of 30,000-Member American Mathematical Society." New York University press release, December 20, 1993.

*—Sketch by Nicholas Pease*

*Louis Joel Mordell*

# Louis Joel Mordell
## 1888–1972
### American–born English geometer and educator

Louis Mordell, who is best known for his work in "pure mathematics" and the development of the "finite basis theorem," became fascinated with mathematics at an early age. Born in Philadelphia, Pennsylvania on January 28, 1888 to Hebrew scholar Phineas Mordell and Annie Feller Mordell, both Jewish immigrants from Lithuania, Mordell was introduced to complex mathematical concepts through used books he purchased at a bookstore. He already had a solid grasp of mathematics when he entered Central High School in Philadelphia at the age of 14, and he completed that school's four–year mathematics course in two years.

In 1907, Mordell placed first on Cambridge University's scholarship examination and accepted a scholarship to St. John's College at the British school. His interest in Cambridge dated back to his self–taught days, when the used books he studied presented numerous examples from Cambridge scholarship and tripos papers. Two years later, he took the first part of the mathematical tripos and placed third, behind P. J. Daniell and E. H. Neville. Following this success, he pursued research in the theory of numbers. There was little interest in this area of study in

England at that time, and Mordell was largely self–taught. His focus was the integral solutions of the Diophantine equation $y^2=x^3+k$, which dates back to the 1600s and the mathematician **Pierre de Fermat**. Mordell, however, greatly expanded the work in this area, determining the solubility for many new values of $k$, among other things. His work earned him a Smith's Prize but this high honor did not immediately lead to a college fellowship.

Mordell continued his research and acted as a tutor, and in 1913 joined the faculty of Birkbeck College in London. In 1916 he married Mabel Elizabeth Cambridge, with whom he later had a daughter and a son. Mordell remained at Birkbeck College until 1920, and much of his work there focused on modular forms, which resulted in two important advances. His first discovery involved the tau function—the set of coefficients of a specific modular form—which had previously been introduced by **Srinivasa Ramanujan**. Mordell proved Ramanujan's theory that the tau function has the property of multiplicativity. The theory was studied, proven, and popularized by Erich Hecke in 1937. The second discovery involved the representation of integers as the sum of a fixed number of squares of integers. Mordell's work in this area was also furthered by Hecke.

Mordell accepted a lecturer position at Manchester College of Technology in 1920, where he remained for two years. It was during this period that Mordell developed the finite basis theorem, based on earlier work by **Jules Henri Poincaré**. It is this work for which Mordell is best known, and it was later furthered by **André Weil**. In his theorem, Mordell determined that there exist a finite number of rational points on any curve of genus greater than unity. "Mordell's conjecture," as this determination was commonly called, was proven by Gerd Faltings in 1983.

In 1922, Mordell transferred to the University of Manchester, and in 1923 was named Fielden professor of pure mathematics and head of the mathematics department at that school. In 1924, while still a citizen of the United States, he was named to the Royal Society. Mordell became a British subject in 1929. Mordell remained at Manchester until 1945 and the school became known as a leading center for mathematics during his tenure. Mordell himself explored the theory of numbers and made significant advances in the area of the geometry of numbers while at Manchester. His work during this period was political as well as scientific, as he devoted much time to assisting European refugees. Mordell even secured temporary and permanent positions at the university for some of the immigrants.

Mordell left Manchester in 1945 to succeed **Godfrey Harold Hardy** as Sadleirian professor of pure mathematics at Cambridge. At this time he also became a fellow of St. John's College at Cambridge. Mordell advanced Cambridge's reputation as a highly reputable research school

before his retirement in 1953. He continued to work with other mathematicians, especially beginners, but also used the time to travel. Cambridge continued to be his home, however, and he died there after a brief illness in 1972.

## SELECTED WRITINGS BY MORDELL

### Books

*Diophantine Equations,* 1969

### Periodicals

"The Diophantine Equation $y^2 - k = x^3$," in *Proceedings of the London Mathematical Society,* second series, 13 (1913): 60–80.
"Reminiscences of an Octogenarian Mathematician," in *American Mathematical Monthly* 78 (November 1971): 952–961.

## FURTHER READING

### Periodicals

Cassels, John W.S., "Louis Joel Mordell." *Biographical Memoirs of Fellows of the Royal Society* 19 (1973): 493–520.
Davenport, Harold. "Louis Joel Mordell." *Acta Arithmetica* 9 (1964): 1–22.

—*Sketch by Kristin Palm*

# Shigefumi Mori
## 1951-
### Japanese mathematician and algebraic geometer

Shigefumi Mori has made important contributions to the field of algebraic geometry. His major work, the creation of a minimal model theory for certain algebraic varieties, has been dubbed Mori's Program. Writing in the *Proceedings of the ICM–90,* Heisuke Hironaka called Mori's Program "the most profound and exciting development in algebraic geometry during the last decade." For research on his program, Mori has received numerous awards, including the 1990 Fields Medal.

Mori was born in Nagoya, Japan, on February 23, 1951. His father, Fumio Mori, and mother, Fusa Mori, were business people who ran a trading company. Mori attended Kyoto University where he received his B.A. in 1973, his M.A. in 1975, and his Ph.D. in 1978. While working on his dissertation, Mori was appointed an assistant instructor at the university. He would teach there until 1980, also spending part of the time as a visiting professor at Harvard University in the United States.

In 1979, Mori published his first major results, a proof of the Hartshorne conjecture, which stated that a certain class of algebraic varieties are projective in nature. In other words, these varieties or sets of solutions to given polynomial equations could be described using projective geometry. Although Hartshorne, the mathematician who first put forth the theory, had felt certain that he was correct, he had not been able to prove his conjecture. Mori proved the conjecture true, and his approach was only his first step in classifying three-dimensional varieties. Previous to Mori's research, one- and two-dimensional algebraic varieties had been classified, or grouped, according to the way they behaved under certain conditions. However, it was generally believed that three-dimensional varieties, which involved four variables, were impossible to categorize because they could not be easily represented or described. With the success of his work on the Hartshorne conjecture, Mori had enough evidence to proceed in his attempts to completely classify all three-dimensional algebraic varieties, the project that would eventually become know as Mori's Program.

### Work in Third Dimension Leads to International Fame

In 1980, Mori returned to his hometown of Nagoya in order to accept a position as lecturer there. However, he soon found he was much in demand, especially after the publication of his paper entitled "Threefolds Whose Canonical Bundles Are Not Nef" that same year. "Nef" was Mori's abbreviation for "numerically effective." Essentially, this paper split the three-dimensional algebraic varieties into two groups, those that are nef and those that are "not nef." Using an extended version of his proof of Hartshorne's conjecture, Mori was able to classify all the not-nef cases into a small number of possibilities, which meant that all the not-nef cases had been reduced to their minimal model, making them easy to describe. Because of the recognition this paper brought him, over the course of the next few years, Mori lectured in Italy and Poland, and he did research at Princeton's Institute for Advanced Studies (IAS) as well as the Max Planck Institute in Germany.

During this period, Mori continued his efforts to finish his program. The nef cases turned out to be far more difficult than the not-nef cases, and he had to invent a entirely new terminology to describe these varieties. In some cases he had to "blow up', or expand, a piece of the surface of the variety; in other cases, he had to perform divisorial contraction, the opposite operation. He "flipped" varieties into simpler ones and "flopped" to avoid problems. In every case, he was trying to reach the ultimate goal of achieving the minimal model, or simplest description, of each variety.

Throughout the 1980s, Mori worked on classifying the nef cases of three-dimensional algebraic varieties. He received several awards, including the 1983 Japan Mathematical Society's Iyanaga Prize, the 1984 Chunichi Culture Prize, and the 1989 Inoue Science Prize. In the United States, C. Herbert Clemens and János Kollár, both of the University of Utah, arranged for Mori to spend one-quarter of each academic year there beginning in 1987.

Finally, in January of 1988, Mori published a paper which proved the existence of minimal models for all three-dimensional algebraic varieties. In just ten years, Mori had completed what many said could never be done. In recognition of this outstanding feat, Mori received both the American Mathematical Society's Cole Prize and the Japan Academy Prize, as well as the highest honor in mathematics, the Fields Medal. He accepted the Fields Medal at the 1990 International Congress of Mathematicians, held that year in his native Japan.

Taking the Fields Medal as a challenge to achieve even further success, Mori began work the following year on a project with Kollár and Y. Miyaoka to further define the varieties he had studied. While continuing to spend part of the academic year in Utah, Mori moved with his wife and four children to Kyoto University's Research Institute for Mathematics and Science. Here he would continue his study of algebraic varieties, a study which had already marked him as a leader in the field of algebraic geometry.

## SELECTED WRITINGS BY MORI

### Books

"Birational Classification of Algebraic Threefolds." *in Proceedings of ICM–90, Kyoto,*1991, pp. 235–248.

### Periodicals

"Threefolds Whose Canonical Bundles Are Not Numerically Effective." *Annals of Mathematics* 116 (1982): 133–176.
"Flip Theorem and the Existence of Minimal Models for 3-Folds." *Journal of the American Mathematical Society* 1 (1988): 117–253.

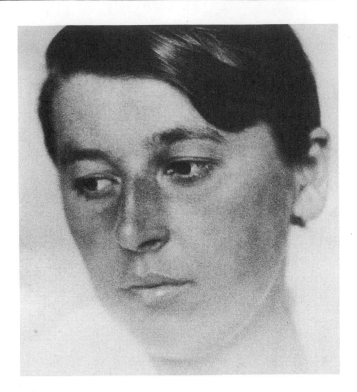

*Ruth Moufang*

## FURTHER READING

### Periodicals

Clemens, H. C., and J. Kollár. "Shigefumi Mori." *Notices of the American Mathematical Society* 37 (November 1990): 1213–14.
Hironaka, Heisuke. "On the Work of Shigefumi Mori." *Proceedings of ICM–90, Kyoto.* New York: Springer-Verlag, 1991, pp. 19–25.
"Shigefumi Mori Awarded 1990 Cole Prize in Algebra." *Notices of the American Mathematical Society* 37 (February 1990): 118–19.

*—Sketch by Karen Sands*

# Ruth Moufang
## 1905–1977
### German algebraic geometer

Ruth Moufang employed algebra to study forms of projective geometry, devising what are now called Moufang planes and Moufang loops. Because of the restrictive policies of the Nazi Party, Moufang was forced to work within the German industrial–military war machine during World War II. Later, when democracy was restored in her country, she became the first female to attain the rank of full professor in Germany at the University of Frankfurt.

Moufang was born on January 10, 1905, to Dr. Eduard and Else Fecht Moufang of Darmstadt, Germany. She had an older sister, named Erica, and her father worked as a scientific consultant for the chemical industry. At the Realgymnasium in Bad Kreuznach, Moufang was introduced to mathematics. The family was financially secure and progressive enough to encourage both girls to pursue to higher education. Erica became an artist. In 1925, Moufang began studying mathematics at the University of Frankfurt, eventually passing the teacher's examination in 1929. By 1930, she had settled on projective geometry as her specialty, studying under Max Dehn. Her doctoral degree was impressive enough to earn her a fellowship in Rome in 1931–32. Upon her return to Germany Moufang accepted a teaching post at the University of Königsberg for a year, and another back at Frankfurt where she remained for two years.

### Hitler's Male Order

Moufang's doctoral work had extended **David Hilbert**'s axioms of plane geometry into the field of projective geometry, especially in terms of a special case eventually called the Moufang plane. This was a major step forward, one that especially impressed one of her peers at Königsberg, the knot theorist Reidermeister. Moufang was the third female to earn a certificate to lecture at the university level (called a *habilitation*) yet was notified a month later that women could not be ratified. She would be allowed to continue work in the university system only if she could secure a research position with no contact with the student body.

Because Moufang could not find such a position in academia, she accepted a research position at the Krupps Institute in Essen in the fall of 1937. She was one of the first women in Germany with a Ph.D. degree to join an industrial firm. Moufang remained there for nearly 10 years, mainly writing on elasticity theory. After the war, she was invited back to Frankfurt and was granted the *venia legendi* in 1946, that part of habilitation conferring the right to teach. Finally, Moufang moved up the academic ladder from lecturer to associate professor, and in early 1957, became the first woman in Germany to be appointed as a full professor in mathematics. Her best known work occurred before she went to the Krupps Institute. She did not publish much during her 10 years in Frankfort but she did supervise many Ph.D. students. Altogether, she published over a dozen paper in pure and applied mathematics between 1932 and 1955. Moufang died on November 26, 1977.

Moufang forged a new mathematical specialty, the algebraic analysis of projective planes, that drew upon a mixture of geometry and algebra. David Hilbert had proved the existence of "non–Desarguian planes" at the turn of the century. These became the subject of Moufang's Ph.D. thesis. A projective plane that can be coordinatized by an alternative division ring is today called a Moufang plane. Chandler and Magnus have described Moufang's work in projective geometry as "based on a powerful geometric intuition and on development of difficult algebraic techniques." Her pioneering work stimulated further research in algebra and projective geometry.

Other papers concerned nonassociative algebraic structures, now called "Moufang loops," so christened by one of the mathematicians who extended her work on alternate division rings. These loops have been used to study properties of cubic hypersurfaces in projective space. Moufang also ventured into the mathematical physics of continuum mechanics. Although she only published one item solely devoted to group theory, in it she applied the study of a particular type of group to number theory, knot theory, and basic geometry as well. Because Moufang did not speak or write English, all of her writing only exists in the original German.

## SELECTED WRITINGS BY MOUFANG

"Zur struktur der projectiven Geometrie der Ebene," in *Mathematische Annalen* 105 (1931): 536–601.
"Zur struktur von Alternativkorpern," in *Mathematische Annalen* 110 (1934): 416–30.

## FURTHER READING

### Books

Chandler, Bruce and Wilhelm Magnus. *The History of Combinatorial Group Theory: A Case Study in the History of Ideas.* New York: Springer–Verlag, 1982.
Mehrtens, H. "Moufang, Ruth," in *Dictionary of Scientific Biography.* Volumes XVII–XVIII, Supplement II. Edited by Frederic L. Holmes. New York: Charles Scribner's Sons, 1990, p. 659.

## Periodicals

Srinivasan, Bhama. "Ruth Moufang, 1905–1977." *The Mathematical Intelligencer* 6, no. 2 (1984): 51–55.

## Other

"Ruth Moufang." *MacTutor History of Mathematics Archive.* http://www–groups.dcs.st–and.ac.uk/~history/Mathematicians/Moufang.html (July 1997).

"Ruth Moufang." *Biographies of Women Mathematicians.* http://www.scottlan.edu/lriddle/women/chronol.htm (July 22, 1997).

*—Sketch by Jennifer Kramer*

http://www.agnescott.edu/ lriddle/women/women.htm

*John Napier*

# John Napier
## 1550–1617
### Scottish inventor of logarithms

John Napier is best remembered as the inventor of the first system of logarithms. A logarithm is the power to which a number, called the base, must be raised to produce a given number. Napier's work with logarithms was described in two Latin treatises, *Mirifici logarithmorum canonis descriptio* (1614; "A Description of the Wonderful Canon of Logarithms") and *Mirifici logarithmorum canonis constructio* (1619; "The Construction of the Wonderful Canon of Logarithms"). This work was immediately recognized by other mathematicians as a great advance. Three hundred years later, in an essay marking the publication anniversary of the *Descriptio*, P. Hume Brown wrote that Napier's most notable achievement "has given him a high and permanent position in the history of European culture."

Napier, the eighth laird of Merchiston, was born in 1550 at Merchiston Castle near Edinburgh, Scotland. He came from a line of influential noblemen and statesmen. His father, Sir Archibald Napier, was a prominent public figure who was allied with the Protestant cause. Among other roles, Sir Archibald served for more than 30 years as Master of the Mint. Sir Archibald's first wife and Napier's mother, Janet Bothwell, was the daughter of an Edinburgh burgess.

Napier entered the University of St. Andrews at age 13, which was typical of wellborn boys of the time. However, he was a dropout, staying at the university for only a short while. It is likely that Napier then traveled to the European continent to continue his studies, although little is known of this period. By 1571, however, he had returned to Scotland and was living at Gartnes, where he built a castle. In 1572 Napier married Elizabeth Stirling. The couple had two children, Archibald and Joanne, before Elizabeth's death in 1579. Napier later married Agnes Chisholm, with whom he had ten children. These included a son named Robert, who eventually served as his father's literary executor. The family stayed in Gartnes until 1608, when Napier inherited Merchiston Castle upon Sir Archibald's death.

### The "Marvelous Merchiston"

Napier lived an active life as a Scottish landowner. He was an amateur scientist who never held a professional post. Nevertheless, his varied accomplishments earned him the nickname of "Marvelous Merchiston." He experimented with fertilizers to improve his land, and he invented a hydraulic screw and revolving axle that could be used to remove water from flooded coal pits. Not surprisingly, Napier was also intrigued by the religious and political controversies of his day. A staunch Protestant, he published *A Plaine Discovery of the Whole Revelation of St. John* in 1593. This book, a virulent attack on Catholicism that concluded the Pope was the Antichrist, was read widely and translated into several languages.

Napier's preoccupation with defending his faith and country prompted him to design various weapons. These included burning mirrors for setting enemy ships on fire, an underwater craft, and a tank–like vehicle. He invested much time and money in these projects, even building prototypes in some cases. In one instance, Napier was said to have built and tested an experimental rapid–fire gun that he claimed would have killed 30,000 Turks without the loss of a single Christian.

## The Invention of Logarithms

Astronomy was another of Napier's passions, and it was this pursuit that led to his greatest discovery. Napier's astronomy research required him to do a number of tedious calculations involving trigonometric functions. Over the course of more than two decades, he gradually developed and refined new ideas for speeding such calculations through the use of logarithms. The *Descriptio* briefly explained his invention and presented the first logarithmic tables. The *Constructio*, published after his death, described how the tables had been computed.

The *Descriptio* attracted the attention of English mathematician **Henry Briggs**, who traveled to Edinburgh to discuss the new tables. Napier and Briggs worked on such improvements as using the base 10. The result was Briggs's development of a standard form of logarithmic table that remained in common use until the advent of calculators and computers. Thanks to his efforts, logarithms were quickly adopted by mathematicians throughout Europe.

## Analogies, Rules, and Bones

Napier made other advances in spherical trigonometry. The so-called "Napier's analogies" were formulas for solving spherical triangles. "Napier's rules of circular parts" were ingenious rules for stating the interrelationships of the parts of a right spherical triangle. In addition, Napier pioneered the use of the decimal point to separate the whole number part of a number from its fractional part. As one of the first to use decimal fractions, he helped to popularize them.

Napier's contributions to mathematics did not end there, however. His concern with simplifying calculations led him to invent mechanical aids for doing arithmetic, described in *Rabdologia* (1617). Among these aids were rods with numbers marked off on them, often known as "Napier's bones" because they were usually made of bone or ivory. Using the rods, multiplication became a process of reading the appropriate figures and making minor adjustments. In addition, Napier invented other types of rods for extracting square roots and cube roots.

It is not clear when Napier first began to dabble in mathematics. However, some early writings, dealing mainly with arithmetic and algebra, seem to date from the period just after his first marriage. These writings were collected and transcribed after Napier's death by his son Robert. They were first published in 1839 by descendant Mark Napier under the title *De arte logistica*, revealing the keenness of Napier's mathematical ruminations. Among other subjects, it appears that he investigated imaginary roots of equations.

## Mathematician or Magician?

During his lifetime, Napier was reputed to be not only a mathematician, but a magician as well. It was rumored that he possessed supernatural powers, and that he owned a black rooster as a spiritual familiar. Given his wide-ranging interests, Napier was almost never idle. A combination of overwork and gout finally led to his death at Merchiston on April 4, 1617. His burial place is uncertain, but it is probably at the old church of St. Cuthbert's parish in Edinburgh.

In 1914, the 300th anniversary of the publication of the *Descriptio* was commemorated by the Royal Society of Edinburgh. In his inaugural address, Lord Moulton lauded Napier as one who "stands prominent among that small band of thinkers who by their discoveries have substantially increased the powers of the human mind as a practical agent." In 1964 Napier University, named for the mathematician, was founded in Edinburgh. Among its campuses is one at Merchiston, which houses courses in science, technology, and design.

## SELECTED WRITINGS BY NAPIER

### Books

*Rabdology.* Translated by William Frank Richardson, 1990. (Originally published in Latin in 1617.)

*The Construction of the Wonderful Canon of Logarithms.* Translated by William Rae Macdonald, 1966. (Originally published in Latin in 1619.)

## FURTHER READING

### Books

Ball, W.W. Rouse. *A Short Account of the History of Mathematics.* New York: Dover Publications, 1960, pp. 235–237.

Baron, Margaret E. "John Napier." *Biographical Dictionary of Mathematicians.* Volume III. New York: Charles Scribner's Sons, 1991, pp. 1774–1778.

Brown, P. Hume. "John Napier of Merchiston." *Napier Tercentenary Memorial Volume.* Edited by Gargill Gilston Knott. London: Longmans, Green and Company, 1915, pp. 33–51.

Moulton, Lord. "Inaugural Address. The Invention of Logarithms: Its Genesis and Growth." *Napier Tercentenary Memorial Volume.* Edited by Gargill Gilston Knott. London: Longmans, Green and Company, 1915, pp. 1–32.

*John Forbes Nash*

Porter, Roy, editor. *The Biographical Dictionary of Scientists*. Second edition. New York: Oxford University Press, 1994, pp. 505–506.

### Periodicals

Gridgeman, N.T. "John Napier and the History of Logarithms." *Scripta Mathematica* (1973): 49–65.

—*Sketch by Linda Wasmer Smith*

# John Forbes Nash
## 1928–
### American algebraist and game theorist

John Forbes Nash is considered one of America's most eminent mathematicians, having been awarded the 1994 Nobel Prize in Economic Science for his work in game theory, which he shared with economists John C. Harsanyi and Reinhard Selten.

### A Genius of the Postwar Era

Nash was born in Bluefield, West Virginia, and came of age during the Depression. His mother, Lillian, was a Latin teacher, and his father, John Sr.,

an electrical engineer. The Depression did not affect the Nash family as severely as other families in West Virginia; they lived in a upper–class home down the road from the local country club. Nash and his younger sister, Martha, were both well educated. He was an avid reader, skilled at chess, and enough of an accomplished whistler to whistle complete melodies of Bach. Curiosity and problem solving were his passions even as a child. One of Nash's mathematics teachers told his parents he was having trouble in class when in actuality the young prodigy was solving math problems in a different way than his teacher could formulate.

In 1945, Nash entered the Carnegie Institute of Technology (now Carnegie–Mellon University), in Pittsburgh, Pennsylvania. Nash completed the requirements for a bachelor of science degree in two years, and immediately began graduate study. In 1948, Nash began his doctoral work at Princeton University, at that time the home of such renowned scholars as **Albert Einstein** and **John von Neumann**. It was in his second year at Princeton that Nash wrote his thesis paper which laid the mathematical foundations of game theory and, 45 years later, earned him the Nobel Prize.

### What Is Game Theory?

Game theory was first invented by John von Neumann and Oskar Morgenstern, an economist at Princeton, who wrote of their discovery in *The Theory of Games and Economic Behavior* in 1944. Game theory became a very popular concept during the Cold War. As the nuclear arms race between the United States and the USSR escalated, a methodology which could predict the possible outcome of a worst or best case scenario was a priceless commodity. Von Neumann and Morgenstern provided a theory based on pure rivalries where the gain for one side exactly balances the loss for the other side (called a "zero–sum situation"). In his doctoral thesis, Nash focused on nonzero–sum noncooperative games involving two or more players when the players are in direct competition. He proved that under appropriate conditions, there always exists stable equilibrium strategies for the players. In such a collection of strategies, no player can increase his gain by changing his strategy while the other players' strategies remain fixed. This concept is now known as the Nash equilibrium.

In 1952 Nash became an instructor at the Massachusetts Institute of Technology and began working on a series of papers, the first of which involved real algebraic varieties. His work was important to a theorem of Artin and Mazur concerning how the number of periodic points for a smooth map can increase as a function of the number of points. Nash's work allowed Artin and Mazur to translate the

16666

problem into an algebraic one of counting solutions to polynomial equations.

Nash also tackled one of the fundamental problems in Riemannian geometry, the isometric embedding problem for Riemannian manifolds. What Nash did was to introduce an entirely new method into nonlinear analysis, which enabled him to prove the existence of isometric embeddings. Nash also developed deep and significant results about basic local existence, uniqueness, and continuity theorems for parabolic and elliptic differential equations.

Nash was diagnosed with paranoid schizophrenia at the age of 30 and spent the greater part of what could have been his most productive years in and out of mental institutions. In 1966, during a period of remission, Nash published his last paper, which was a continuation of his work on the isometric embedding theorem.

In 1994, Nash shared the Nobel Prize in Economic Science for his work in game theory. Of his achievements **John Milnor** writes: "It is notoriously difficult to apply precise mathematical methods in the social sciences, yet the ideas in Nash's thesis are simple and rigorous, and provide a firm background, not only for economic theory but also for research in evolutionary biology, and more generally for the study of any situation in which human or nonhuman beings face competition or conflict."

## SELECTED WRITINGS BY NASH

### Books

(With C. Kalisch, J. Milnor, and E. Nering) "Some Experimental *N*-Person Games," in *Decision Processes,* edited by R.M. Thrall, C.H. Coombs, and R.L. Davis, 1954.

### Periodicals

"Equilibrium Points in *N*-person Games." *Proceedings of the National Academy of Sciences USA* 36 (1950), 48–49.
"The Bargaining Problem." *Econometrica* 18 (1950): 155–162.
"Real Algebraic Manifolds." *Annals of Mathematics* 56 (1952): 405–421.
"Two–Person Cooperative Games." *Econometrica* 21 (1953): 128–140.
"Results on Continuation and Uniqueness of Fluid Flow." *Bullentin of the American Mathematical Society* 60 (1954): 165–166.
"A Path Space and the Stiefel–Whitney Classes." *Proceedings of the National Academy of Sciences USA* 411 (1955): 320–321.
"The Imbedding Problem for Riemannian Manifolds." *Annals of Mathematics* 63 (1956): 20–63.

"Continuity of Solutions of Parabolic and Elliptic Equations." *American Journal of Mathematics* 80 (1958): 931–954.
"Analyticity of the Solutions of Implicit Function Problems with Analytic Data." *Annals of Mathematics* 84 (1966): 345–355.

## FURTHER READING

### Books

Casti, John L. *Five Golden Rules, Great Theories of 20th–Century Mathematics—and Why They Matter.* New York: John Wiley & Sons, Inc., 1996.

### Periodicals

Milnor, John. "A Nobel Prize for John Nash." *The Mathematical Intelligencer* 17, no. 3 (Summer 1995): 11–17.
Nasar, Sylvia. "The Lost Years of a Nobel Laureate." *New York Times* (November 13, 1994).
Warsh, David. "Economist Share Nobel Trio Pioneered Use of Game Theory in the Field." *The Boston Globe* (October 12, 1994): 43.

*—Sketch by Tammy J. Bronson*

# Evelyn Nelson
## 1943–1987
### Canadian algebraist

Though Evelyn Nelson's mathematical career was cut short by cancer, she published about 40 papers on her main research topics, algebra, and theoretical computer science. She was specifically concerned with equational compactness, model theory, and formal language theory. A generous teacher and researcher, Nelson spent her career at McMaster University.

Nelson was born Evelyn Merle Roden in Hamilton, Ontario, Canada, the daughter of Russian immigrants. A gifted student, Nelson attended Westdale High School in Hamilton. With her parents' constant encouragement, she began her undergraduate education at the University of Toronto in the Honours course of Mathematics–Physics–Chemistry. Two years later, she transferred to the Honours Mathematics program at McMaster University in Hamilton. Soon after her transfer she married Mort Nelson, and they had two daughters together. They later divorced.

Nelson began publishing her research quite early, beginning with her master's thesis. It was published in 1967 as "A finiteness criterion for partially ordered semigroups and its application to Universal Algebra," in *Canadian Journal of Mathematics.* She completed her university education at McMaster, earning her Ph.D in 1970 with a thesis titled "The lattice of equational classes of commutative semigroups." It was also published, though in a modified form, in the *Canadian Journal of Mathematics.*

Nelson wrote over 40 papers, primarily in the general area of universal algebra. The topic of her thesis, the lattice of equational classes of semigroups, continued to be explored in her first five papers. Her other papers ranged from various aspects of equational compactness to partially ordered universal algebra subject to conditions originating in theoretical computer science.

Nelson also taught mathematics at her alma mater, McMaster, throughout her career. She first lectured as a postdoctoral fellow (1970–73), then as a research associate (1973–78). She was finally promoted to the tenure track level as an associate professor in 1978 and became a full professor in 1983. Nelson also served as chair of the Computer Science Unit from 1982 to 1984, but could not continue in the position when Computer Science became a full–fledged department because of her increasing illness.

Nelson served the mathematical community and McMaster University in other ways. She edited the prestigious journal *Algebra Universalis*, and refereed articles and reviewed other scholar's work for several other journals. She also held many memberships in academic societies, among them the American Mathematical Society and the Canadian Mathematical Society. She also served on some committees for the latter. Nelson was active in the McMaster community outside of the mathematics department as well.

In 1987, Nelson died of cancer after battling the disease for many years.

## SELECTED WRITINGS BY NELSON

"Finiteness of semigroups of operators in Universal Algebra." *Canadian Journal of Mathematics* 19 (1967): 764–68.
"The lattice of equational classes of commutative semigroups." *Canadian Journal of Mathematics* 19 (1971): 875–95.

## FURTHER READING

Adamek, Jiri. "A Farewell to Evelyn Nelson." *Cahiers de topologie et géométrie différentielle catégoriques* (1988): 171–74.

*Hanna Neumann*

Banaschewski, B. "Evelyn M. Nelson: An Appreciation." *Algebra Univeralis* 26 (1989): 259–64.
"Evelyn Nelson." *Biographies of Women Mathematicians.* June 1997. http://www.scottlan.edu/lriddle/women/chronol.htm (July 22, 1997).

—*Sketch by Annette Petruso*

# Hanna Neumann
## 1914–1971
### German algebraist and educator

Hanna Neumann was a well–traveled mathematician who instilled a humble sense of dignity to the mathematics programs that she headed. She fled Nazi Germany to England, where she obtained a doctorate in mathematics and went on to solve many problems, including the finite basis problem for varieties generated by a finite group.

Born in Berlin on February 12, 1914, Neumann was the youngest of three children of Hermann and Katharina von Caemmerer. Her father was a historian who was killed in World War I, leaving only a war pension to support the family. To earn extra money, Neumann coached younger children in academics and evolved into an organized and studious pupil as a result. She entered the University of Berlin in 1932 and became inspired by her mathematics professor, Ludwig Bieberbach. He influenced her studies toward geometry, but it would be professors Erhard Schmidt and Issai Schur that would introduce her to analysis and algebra. Her future husband, Bernhard H. Neumann, was also a mathematics student at the university. They became secretly engaged in 1934, but due to the rise of Nazism in Germany, Bernhard (who was Jewish) moved to England and their plans were put on hold. Neumann remained in Germany and lost her job at the Mathematical Institute as a result of her activities to protect other Jewish lecturers. Neumann was advised that because of her political stance, she should avoid the oral exam on "political knowledge," which was required for a doctorate, and switch to the *Staatsexamen* final. This exam was a written essay and she chose as her topic the epistemological basis of numbers in Plato's later dialogues. She graduated with distinctions in mathematics and physics.

### Teaches in England

To escape the Nazis, Neumann moved to England and began working in Bristol. Her natural talent for learning new languages benefitted her career by being able to read journals and books in a number of different languages. She experimented with finite plane geometries, wrote several papers, and presented them at lecture courses. Neumann also contributed an explanation of the two types of quadrangles found in finite planes: those whose diagonal points are collinear, and those that are not (the Fano configuration).

In 1938, Neumann moved to Cardiff, England, and married Bernhard in secret to protect his parents, who were still living in Germany, from any reprisals. In 1939, their first child, Irene, was born. The family was soon uprooted because of their classification as "least restricted" aliens, whom are barred from living along the coast. Moving to Oxford, Neumann resumed her doctoral studies while pregnant with her second child, Peter. The acute housing shortage grew as European refugees swamped England, forcing Neumann and her children to move into a rented caravan in 1942. It would be here, by candlelight, that she would write her doctoral thesis, submitting it in 1943. Her third child, Barbara, was born shortly after, and when the war ended the entire family were able to move back to Cardiff. Bernhard was decommissioned from the army and returned to his job at the University College in Hull, where Neumann began her teaching career as a temporary assistant lecturer of applied mathematics. Their fourth and fifth children, Walter and Daniel, were born in 1946 and 1951, respectively. She continued to work at Hull for the next 12 years, changing the curriculum toward the more pure mathematics that she herself had been trained in.

In 1955, Neumann received a D.Sc. from Oxford based on her research and papers, two of which were published in the *American Journal of Mathematics*. A highly motivated individual, she chose to work as the secretary of a local United Nations Association in her spare time. Her husband had moved to Manchester to lecture in 1948. As she continued to search for a job that would bring her and her husband closer together, the faculty of Technology of the University of Manchester finally added an honors program that would enable her wish to come true.

### Breaks New Ground

Neumann was hired as senior lecturer at the University of Manchester in 1958 and set about making very abstract ideas more accessible for her students. She worked with a model building group, lectured on prime numbers and the dissection of rectangles into incongruent squares, and advised several graduate students. Neumann went beyond the confines of a teacher/student relationship and regularly invited staff and students over to her home for coffee and discussions. She mentored many individuals who would themselves go on to become successful mathematicians, including John Bowers, Jim Wiegold, and Chris Houghton.

In the following summer, Neumann toured the universities in Hungary, lecturing on her research on groups and analysis. She also attended the 12th British Mathematical Colloquium, held in 1960, and was asked to speak about wreath products (a group construction). A group theory, such as the wreath products, consists of a set of elements subject to one binary operation and meeting these three requirements: (1) It is a closed system; (2) It obeys the associative law; (3) There is an identity element, e, such that $e \odot x = x$ for any element x in the group; ($\odot$ is the binary operation).

Upon her return to England, Neumann began preparing for a joint study leave with her husband to the Courant Institute of Mathematical Sciences in New York. Her three sons accompanied them and the oldest, Peter, began to study under Gilbert Baumslag, one of the professors at Courant. During the course of the year, Peter and his parents solved the problem of the structure of the semigroup of varieties of groups, demonstrating that it is free. Neumann also presented a number of lectures and taught a graduate course on varieties of groups. During this time, Bernhard Neumann was invited to organize a research department of mathematics at the Australian National University and Hanna was offered a post as a reader.

### The Next Frontier

Neumann began her new job in Australia in 1964. She overhauled the department and trained its teachers on subject matter covered in the new syllabuses. The material reflected the changes in mathematics worldwide, including more emphasis on pure mathematics and take–home assignments that promoted the use of ideas and theorems firsthand. Due to the difficulty of the subject matter, Bernhard's research students helped part–time with tutoring.

A conference on the theory of groups was organized by Neumann in 1965. The next year, she finished a monograph about group varieties. With the royalties, she was finally able to buy a good camera and revive her hobby of photography, collecting shots of flowers and trees. She also enjoyed extensive bike rides, later taking to four wheel drives with her husband in the outback.

In January of 1966, the Australian Association of Mathematics Teachers was formed and Neumann was elected to the position of vice–president. She took on the responsibility of bringing together math teachers representing all areas of Australia. In 1967, she became president of the Canberra Mathematical Association and helped prepare pamphlets for teachers; her most famous one was on probability. During a sabbatical leave in 1969, Neumann wrote letters to publicize the inadequacy of Australian mathematics programs compared to the rest of the world and the need for reorganization of the content in many courses. She also visited the United States on a National Science Foundation Senior Foreign Scientist Fellowship. During her stay at Vanderbilt University in Nashville, Tennessee, Neumann and Ian Dey solved the problem on the free product of finitely many finitely–generated Hopf groups.

Neumann's return to Australia was short lived, for she accepted a lecture tour of Canada in 1971, speaking at the universities of British Columbia, Calgary, Alberta, Saskatchewan, and Manitoba. All of this travel and intense lecturing exhausted her and she fell ill on November 12, checking into a hospital only to lapse immediately into a coma. Neumann did not regain consciousness and died two days later at age 57. After her death, a memorial fund was created from supporters all over the world to further Neumann's courageous and inspiring teaching.

### FURTHER READING

Newman, M. F. "Hanna Neumann," in *Women of Mathematics: A Biobibliographic Sourcebook.* Edited by Louise S. Grinstein and Paul J. Campbell. Westport, CT: Greenwood Press, 1987, pp. 156–160.

Newman, M.F. and G. E. Wall. "Hanna Neumann." *Journal of the Australian Mathematical Society* 17 (1974): 1–28.

*Records of the Australian Academy of Science* 3, no. 2. (1975).

*Sir Isaac Newton*

### Other

"Hanna Neumann." *Biographies of Women Mathematicians.* June 1997. http:www.scottlan.edu/lriddle/women/chronol.htm (July 22, 1997).

*—Sketch by Nicole Beatty*

# Isaac Newton
## 1643–1727
### English mathematician and physicist

Isaac Newton was one of the leading mathematical and scientific geniuses of the 17th and 18th centuries, best known for his far–reaching discoveries in mathematics, physics, and optics. Among Newton's many achievements was his invention of the calculus (the German philosopher and mathematician **Gottfried Wilhelm Leibniz** would independently invent it a few years later), the formulation of the three laws of motion and the universal theory of gravitation, and he proved that sunlight is the combination of many colors. Newton also served as master of the Royal Mint in London and president of the Royal Society of London for many years.

Newton was born on January 4, 1643. He was so small and sickly at birth that two women who were sent into town to get supplies for him delayed coming back because they thought he would not live more than a few hours. His father, who died a few months before his son was born, was also named Isaac. His mother, Hannah Ayscough, remarried three years after her son's birth to Reverend Barnabas Smith, and they had three children.

### A Solitary but Energetic Youth

After his mother left to live with Reverend Smith, Newton stayed behind on the family farm with his maternal grandmother. His was a solitary childhood, fostering a habit that persisted for the rest of his life. When Newton was 12, he went to the grammar school at Grantham. There he studied Latin, and learned it so well that later he could write it as fluently as English. Early in his stay at Grantham, Newton was involved in a prophetic incident. A boy kicked him in the stomach on the way to school one day. Newton challenged him to a fight after school, and even though he was much smaller physically, Newton won the fight, and he rubbed his opponent's nose on the church wall. Later in his life, Newton was known to be easily provoked to retaliation, and he was usually very determined, nasty, and successful against his opponents.

When not in school, Newton spent his time at Grantham making sundials, drawings, and wooden models. He constructed a model windmill which included a mouse on a treadmill to supply power. He created a four–wheeled cart for himself, which he powered by turning a crank he had installed. Sometimes Newton spent so much time on these inventions that he fell behind at school. When that happened, he applied himself to his studies and regained his high academic standing in his class.

When Newton was almost 17 years old, his mother called him home to learn to manage the family farm. He was a disaster as a farmer. Newton was more interested in building models and reading, and he had no patience for watching the livestock or anything to do with farming. His schoolmaster at Grantham and his uncle William Ayscough had noticed his brilliance and successfully persuaded his mother to let him finish grammar school.

### Made Mathematical History at Cambridge

In June 1661, Newton was admitted to Cambridge University. He found the traditional Aristotelian curriculum at Cambridge so unrewarding that he never finished any of the assigned books, but he did read prodigiously about philosophy, science, and mathematics. Newton had learned little about mathematics at Grantham, but at Cambridge he devoured books by **Euclid, René Descartes, Galileo,** and **Johannes Kepler** and took copious notes. When

Newton was working on an interesting problem, he completely forgot about eating or sleeping. He worked so hard on mathematics and physics that within a year he began to record original insights in his notebooks. In 1665, Newton received his bachelor's degree. In the summer of 1665, the plague came to Cambridge, and the university shut down until the spring of 1667. Newton went home and spent his next two years laying the foundations for the calculus, which he called the "fluxional method."

The calculus is a mathematical tool that allows people to solve problems about the limits and rates of changes in dynamic relationships among such factors as time, velocity, position, force, distance, and so on. If one factor is a function of the other, (i.e., dependent upon it—the faster you go, the more distance you cover), then information about an unknown factor can be determined if the other factor is known. Since many factors in human experience are related, the calculus is crucial in many specialities, for example, civil engineering (e.g., what is the maximum stress a building can tolerate if it is made of x materials and is subjected to hurricane winds of y miles per hour?), celestial navigation (e.g., where will a satellite be at a certain time if it is traveling at x speed at y latitude?), and economics (e.g., what will a corporation's profit be if its costs go up at rate x, its sales go up at rate y, and inflation goes down at rate z?).

No one knew that Newton had developed the calculus at home in 1665 and 1666. In 1669, however, he made some of his work known to a few people, including **Isaac Barrow**, the Lucasian professor of mathematics at Cambridge. Barrow was impressed. Newton received his master's degree in 1668, and when Barrow resigned his professorship in 1669, Newton took his place. The Lucasian professor gave a series of lectures every term, but Newton's lectures were so difficult that few or no students attended them. When no one attended, he would lecture for fifteen minutes to the empty room and leave.

Newton's work in 1669 and the early 1770s was on optics and a theory of colors. To aid in his studies, Newton built a reflecting telescope which magnified 40 times. The Royal Society was so impressed by its quality that they elected him a member in January 1672. In February 1672, the Royal Society published an article by Newton on his theory of colors. Robert Hooke, another scientist, challenged Newton in a very superior tone. Newton's response to Hooke was vicious. Newton had thought about colors for years, had conducted many experiments, and knew his subject well. He had no patience for Hooke's tone or ignorance. Newton's disputes with Hooke and others made him withdraw into silence, but while he was silent, Leibniz was publishing his own work on the calculus.

In 1684, Newton was stirred to more work on mathematics and physics by **Edmund Halley**. In 1687, Newton published his *Philosophiae Naturalis Principia Mathematica*, which formulated the three laws of motion and the universal principle of gravitation. It remains today one of the most famous works of western science.

### Quarrels with Leibniz

Newton's most famous quarrel was with Gottfried Wilhelm Leibniz over who first invented the calculus. In 1684, Leibniz published a version of the calculus which had a superior system of notation, making it easier to use. When Newton's *Principia* and other works by English mathematicians were published after Leibniz's calculus, mathematicians throughout Europe assumed that Newton had taken his method from Leibniz.

Newton, however, had invented the calculus in 1665 and 1666, but did not publish his work for years after Leibniz did. Leibniz received letters from Newton in 1671 and 1676 discussing mathematics, and he also saw one of Newton's unpublished manuscripts in London in 1676 and took many notes on it. The issue of who first invented the calculus came to a head in 1699 when Leibniz was indirectly accused of plagiarizing Newton. There followed years of accusations on both sides, which diminished after Leibniz died in 1716. Many scholars today believe that Leibniz did not plagiarize Newton and that both men discovered the calculus separately, though Newton discovered it earlier. Leibniz's version, however, was easier to use and consequently caught on faster than Newton's.

### Improving the Royal Mint

Tired of academia, Newton accepted an offer to become warden of the Royal Mint in 1696. He was so well organized that he improved the Mint's operations enormously. Newton was promoted to master of the Mint in 1700 and became wealthy because he earned a commission on the coins minted.

Newton was elected president of the Royal Society from 1703 until his death. In 1705, Newton was knighted by Queen Anne. In the years before his death, he suffered from a number of physical ailments. Newton died in London on March 20, 1727.

## SELECTED WRITINGS BY NEWTON

### Books

*Philosophiae Naturalis Principia Mathematica*, 1687
*Optiks*, 1704

## FURTHER READING

### Books

Berlinski, David. *A Tour of the Calculus.* New York: Pantheon Books, 1995.

Hall, A. Rupert. *Isaac Newton: Adventures in Thought.* Oxford: Blackwell, 1992.

Schrader, Dorothy V. "The Newton–Leibniz Controversy Concerning the Discovery of the Calculus," in *From Five Fingers to Infinity: A Journey through the History of Mathematics.* Edited by Frank J. Swetz. Chicago: Open Court, 1994, pp. 509–520.

Westfall, Richard S. *Never at Rest: A Biography of Isaac Newton.* New York: Cambridge University Press, 1980.

*—Sketch by Patrick Moore*

# Nicholas of Cusa
## 1401–1464
### German geometer, theologian, philosopher, and prelate

Nicholas of Cusa, also known as Cusanus and Nikolaus von Cusa, was born Nicholas Krebs, or Chrypffs (Kryfts). While Nicholas led a life of service to the Roman Catholic Church, rising to the position of cardinal, he wrote several important philosophical works, including mathematical treatises, that profoundly influenced Western thought. Although better known as a philosopher than a mathematician, his interest in mathematics, far from being purely academic, reflected his passionate desire to understand the universe and situate it in the context of his mystical conception of God.

### Works for Church Unity

Nicholas studied law and mathematics at the University of Padua, receiving his doctorate in Canon Law. Active in Cologne in the 1420s, he studied theology, pursued historical and legal research, and gained fame for discovering copies of missing Latin manuscripts, including 12 comedies by the Roman playwright Plautus and the *Natural History* by Pliny the Elder. Having developed an interest in various philosophical traditions, including Greek thought, Scholasticism, and mysticism, Nicholas focused on Plato's *Parmenides* (more precisely, Proclus's commentary of Plato's seminal cosmological dialogue) and the works of the Catalan mystic, poet, and philosopher Ramón Llull.

Ordained a priest, probably around 1430, Nicholas participated in the Council of Basel, convened in 1431, where he worked to prevent the political fragmentation of the Church. In Basel, he met the noted Italian humanist Enea Silvio Piccolomini, who later rose to the papacy and ruled as Pius II from 1458 to 1464. In 1437 Nicholas traveled to Constantinople as member of a delegation whose task was to effect a reunification of the Western and Byzantine Churches. The mission failed, but Nicholas continued with his diplomatic work, for which Pope Nicholas V rewarded him with a cardinal's hat in 1450, naming him bishop of Brixen. Nicholas's career as cardinal was turbulent, as he mediated many conflicts between Church and secular authorities. Around 1460, overwhelmed by his work as mediator in Germany, he went to Rome, where his friend was Pope. He spent his final years preaching, working as the Pope's representative, writing, and studying.

## Applies Mathematics as a Theological Tool

Nicholas's work as a mathematician is in many ways defined by his unique position in European intellectual history. On the one hand, as a theologian and churchman, he is a typical representative of the medieval belief that theology is the science to which all other branches of intellectual enquiry are subordinated. On the other hand, Nicholas anticipated, and exemplified, certain traits that are identified with the dawning Renaissance, traits that included receptiveness to new—and forgotten—ideas, interest in unorthodox spiritual traditions (such as mysticism), and scientific curiosity. Familiar with the writings of **Euclid** and **Archimedes**, and conversant with the Pythagorean mysticism of numbers, he used his knowledge of geometry and arithmetics to illustrate some of his key philosophical concepts. For example, in his principal work *De docta ignorantia* (*On Learned Ignorance*), completed in 1440, Nicholas builds his entire theory of knowledge on the basis of the principle of the *coincidentia oppositorum* (the harmony of opposites). For example, when Nicholas states that in geometry, curvature and straightness are effectively reconciled when one imagines that the circumference of an infinitely large circle coincides with its diameter, or that is no difference between greatness and smallness, he is, paradoxically, using mathematical ideas in a discourse that transcends mathematics.

It is true that Nicholas unsuccessfully tackled the problem of squaring the circle, but it seems unfair to call him the "misguided circle-squarer." Indeed, squaring the circle, equating maxima and minima, may seem unreasonable in the universe of traditional mathematics, but Cusanus was attempting to imagine a geometry of absolute infinity, where boundaries and differences cease to exist. In other words, according to Nicholas, infinity is not of this world. As Alexandre

Koyré explains, Nicholas abandons the traditional idea of a finite physical world, but does not define it as infinite. Neither infinite nor finite, Nicholas's world is *interminatum*, or interminate, which means that it is without boundaries, but also bereft of a necessary structure that would render it intelligible. According to Nicholas, only God, which he, following an ancient Hermetic formula, defined as the "sphere whose center is everywhere, and circumference nowhere," is infinite, and this infinity transcends anything the human mind can ascertain, including mathematics.

Although sometimes hailed as a man ahead of his time and a precursor of modern mathematics and astronomy, Nicholas, in accordance with the profoundly theocentric nature of 15th-century thought, considered mathematics and all human knowledge as significant manifestations of human ignorance. But humankind's fundamental ignorance, Nicholas asserted, does not nullify scientific knowledge, including mathematical knowledge: the existence of a limitless world poses a challenge for the seeker of knowledge. However, as Koyré, has asserted, Nicholas's mathematics does not anticipate the modern mathematics of infinity, and his absolute Reality, or God, cannot be mathematized, quantified, or rationally comprehended. In addition, Nicholas's non-geocentric world view certainly does not point to a Copernican, or heliocentric, view, since, for Nicholas, the only possible center of our world is God. In fact, according to Nicholas, science and mathematics may describe the world, but only God defines, and constitutes, its essence:

"The world has no circumference; for if it had a centre and a circumference there would be some space and some thing beyond the world, suppositions which are wholly lacking in truth. Since, therefore, it is impossible that the world should be enclosed within a corporeal centre and a corporeal boundary, it is not within our power to understand the world, whose centre and circumference are God. And though this world cannot be infinite, nevertheless it cannot be conceived as finite, since there are no limits within which it could be confined. The earth, therefore, which cannot be the centre, cannot be wholly without motion. . . . And just as the world has no centre, so neither the sphere of the fixed stars nor any other is its circumference."

## SELECTED WRITINGS BY NICHOLAS OF CUSA

*De docta ignorantia libri tres*, 1913
*Opera*, 3 volumes, 1514

## FURTHER READING

Boyer, Carl B. *A History of Mathematics.* Second edition. Revised by Uta C. Merzbach. New York: John Wiley & Sons, 1991.

Burtt, E. A. *The Metaphysical Foundations of Modern Science.* Revised edition. New York: Doubleday, 1954.

Coplestone, Frederick. *Ockham to Suárez.* Volume 3: *A History of Philosophy.* Garden City, NY: Image Books, 1985.

Koyré, Alexandre. *From the Closed World to the Infinite Universe.* Baltimore: The Johns Hopkins University Press, 1957.

Lovejoy, Arthur O. *The Great Chain of Being: A Study of the History of an Idea.* Cambridge, MA: Harvard University Press, 1942.

Sheldrake, Rupert. *The Presence of the Past: Morphic Resonance and the Habits of Nature.* New York: Vintage Books, 1989.

—*Sketch by Zoran Minderovic*

# Nicomachus of Gerasa
## c. 60–c. 100
### Greek arithmetician, musical theorist, and philosopher

Born in Gerasa, Palestine (now Jerash, Jordan), Nicomachus is mainly known for his work in arithmetic. He has been credited with founding Greek arithmetic as a separate mathematical discipline, and is also considered a notable representative of Neo–Pythagorean philosophy.

### Formulates Arithmetical Theorem

Before discussing Nicomachus' work in arthimetic, it is important to understand, as Carl B. Boyer explains, that the Greek term *arithmetike*, although derived from the word *arithmos*, which means number, did not refer to actual calculation, but rather denoted theoretical, speculative thinking about numbers. Nicomachus' ideas about numbers is pretty much defined by **Pythagoras,** or, more precisely, the Neo–Pythagorean tradition that he represented. In his *Introduction to Arithmetic*, Nicomachus discusses various kinds of numbers (odd, even, prime, composite, figurate, perfect), but also considers numbers as divine entities with apparently anthropomorphic qualities, such as, for instance, goodness. However, his discussion of odd and even numbers, which for Pythagoreans raised fundamental ontological questions, also contains practical aspects, including the

following theorem. By grouping the odd integers in the following manner: $1; 3 + 5; 7 + 9 + 11; 13 + 15 + 17 + 19$, Nicomachus discovered that the sums were equal to the cubed integers. For example, $1^3 = 1; 2^3 = 3 + 5; 3^3 = 7 + 9 + 11; 4^3 = 13 + 15 + 17 + 19$, etc.

Nicomachus' world view is a fascinating blend of mathematics and philosophy, incorporating the main currents of Greek scientific, idealistic, and mystical thought. "In his system," Frederick Copleston has written, "the Ideas existed before the formation of the world (Plato), and the Ideas are numbers (Plato again). But the Number–Ideas did not exist in a transcendental world of their own: rather, they were Ideas in the Divine Mind, and so patterns or archetypes according to which the things of this world were formed (cf. Philo the Jew, Middle Platonism and Neo–Platonism)."

### Proposes Eclectic Approach to Musical Theory

In his *Manual of Harmonics*, Nicomachus discusses musical scales, notes, and intervals. Following the Pythagorean tradition as well as his interest in numerical ratios, he defines notes and intervals on the basis of numerical ratios. However, he also seems to have accepted the idea, developed by the noted music theorist Aristoxenus, that in music, for example, in determining whether an interval is consonant, auditive perception plays a crucial role. It is interesting that Nicomachus attempts to reconcile two opposing traditions in music theory.

## SELECTED WRITINGS BY NICOMACHUS

*The Manual of Nicomachus the Pythagorean,* translated with a commentary by Flora R. Levin, 1994

## FURTHER READING

Armstrong, A. H., editor. *The Cambridge History of Later Greek and Early Medieval Philosophy.* Cambridge, UK: Cambridge University Press, 1967.

Boyer, Carl B. *A History of Mathematics.* Second edition. Revised by Uta C. Merzbach. New York: John Wiley & Sons, Inc., 1991.

Copleston, Frederick. *A History of Philosophy.* Volume 1: *Greece and Rome.* Garden City, NY: Image Books, 1963.

—*Sketch by Zoran Minderovic*

# Nicomedes
## fl. late third or early second century B.C.
### Greek geometer

While Nicomedes has retained a secure place in the history of mathematics, nothing is known about his life except that he may have lived in the Greek city–state of Pergamon. He is best known for his discovery of an important lemma, famous among mathematicians as the "lemma of Nicomedes." Traces of his book, *On Conchoid Lines*, which is now lost, survive in commentaries by mathematicians throughout history.

### Uses Lemma to Solve Delian Problem

A *lemma*, which in Greek literally means *thing taken* or *assumption*, is in fact a minor theorem that can be used in the proof of a more important theorem. Nicomedes discovered his lemma while attempting to solve geometrically the age–old Delian problem, or doubling the cube. In principle, Greek mathematicians knew that in order to duplicate a cube, one must find two mean proportionals between two lines. An algebraic representation of the problem, which surpassed the competence of Greek mathematicians, would be $x^3 = 2a^2$, where $x$ indicates the unknown side of the new, duplicated cube. Algebraically, the problem can be solved in the following manner:

if $a : x = x : y = y: 2a$, then

$x^2 = ay$

$x^4 = a^2y^2$

$y^2 = 2ax$, and, therefore:

$x^4 = 2a^3x$, which amounts to $x^3 = 2a^3$

Having understood that finding two mean proportionals was the first step in his task, Nicomedes proceeded to do so by using straightedge and compass. To complete the construction, he began with two lines, creating and bisecting a variety of resultant triangles. The lemma, which emerged during this process, identifies two particular lines found in two of the derived triangles as equal. Nicomedes later realized that the lemma enabled him to arrive at the necessary mean proportionals. His lemma was used by **François Viète** as a means of solving equations of the third and the fourth degrees. **Isaac Newton**, in his *Arithmetica universalis*, frequently mentions the lemma in his discussion of higher equations.

### Describes Conchoid Curve

In mathematical nomenclature, the Greek word *gramme* refers to all lines, either curved or straight, and mathematicians often use descriptive terms to identify particular types of curves. The curve usually connected to Nicomedes is the conchoid. Because of the curve's shape, its name is derived from the Greek word *koghos,* meaning shellfish. The conchoid is constructed by pivoting a ruler around a fixed point on a given line. If one end of the ruler is fixed at a point, the other end will mark points, which, when traced, will produce a conchoid. Nicomedes demonstrated geometrically that many conchoids can be drawn in relation to a line and also showed that this curve is asymptotic to the line. The conchoid, which in Cartesian notation is represented as $(x - a)^2 (x^2 + y^2) - b^2x^2 = 0$, greatly facilitates the problem of angle trisection, another seemingly unsolvable problem of Greek mathematics. In other to trisect an acute angle, the geometer must first create a rectangle, double its diagonal, and move it so it comes between the two other sides of the rectangle. However, this operation, which is performed by moving the ruler, is dramatically simplified if the correct conchoid relative to either of these sides is introduced.

## FURTHER READING

Heath, T. L. *A History of Greek Mathematics.* Oxford: Clarendon Press, 1921.

*—Sketch by Zoran Minderovic*

# Emmy Noether
## 1882-1935
### German-born American algebraist and educator

Emmy Noether's innovative approach to modern abstract algebra not only produced significant results, but it inspired highly productive work by students and colleagues who emulated her technique. Dismissed from her university position at the beginning of Nazi rule in Germany—for she was both Jewish and female—Noether emigrated to the United States, where she taught at several universities and colleges. In a letter published in the *New York Times,* **Albert Einstein** eulogized her as "the most significant creative mathematical genius thus far produced since the higher education of women began."

Noether was born on March 23, 1882, in Erlangen, Germany. Her first name was Amalie, but she was known by her middle name of Emmy. Her mother, Ida Amalia Kaufmann Noether, came from a wealthy family in Cologne. Her father, **Max Noether,**

*Emmy Noether*

was an eminent mathematics professor at the University of Erlangen who worked on the theory of algebraic functions. Two of her three brothers became scientists—Fritz was a mathematician and Alfred earned a doctorate in chemistry.

Noether was educated as a typical German girl of her era. Besides learning to cook and do household chores, she took piano lessons and enjoyed going to dances. Since girls were not eligible to enroll in the gymnasium (college preparatory school), Noether attended the Städtischen Höheren Töchterschule, where she studied arithmetic and languages. In 1900, Noether passed the Bavarian state examinations, becoming certified to teach French and English at female-only institutions.

### Enters a Man's World

Rather than seeking a language teaching position, Noether decided to undertake university studies. Since she had not graduated from a gymnasium, she first had to pass an entrance examination. At the University of Erlangen, Noether obtained her instructors' permission to audit courses during 1900 and 1902, and in 1903 she passed the matriculation exam. Noether attended the University of Göttingen for a semester and heard such notable mathematicians as **Hermann Minkowski, Felix Klein,** and **David Hilbert.** Noether enrolled at the University of Erlangen when women were accepted in 1904. At Erlangen, she studied with Paul Gordan, a mathematics professor

who was also a family friend. Noether completed her dissertation, entitled "On Complete Systems of Invariants for Ternary Biquadratic Forms," and was awarded her Ph.D., *summa cum laude,* in 1908.

Throughout her career, Noether faced severe discrimination because of her gender, but she studied and worked in whatever capacity she was allowed. An inheritance from her father apparently allowed her to work for little or no pay when necessary. From 1908 until 1915, Noether did research at the Mathematical Institute of Erlangen, where she served as dissertation adviser for two students. She occasionally delivered lectures for her father, who suffered lingering effects of a childhood case of polio. During this period, Noether began to work with Ernst Fischer, an algebraist who directed her toward the broad theoretical style characteristic of Hilbert.

### Formulates the Mathematics of Relativity

Klein and Hilbert invited Noether to join them at the Mathematical Institute in Göttingen. They were working on the mathematics of the newly announced general theory of relativity, and they believed Noether's expertise would be helpful. Albert Einstein later wrote an article for the 1955 *Grolier Encyclopedia,* in which he characterized the theory of relativity by the basic question: "How must the laws of nature be constituted so that they are valid in the same form relative to . . . an arbitrary transformation of space and time?" It was precisely this type of invariance under transformation on which Noether focused her mathematical research.

In 1918, Noether proved two theorems that formed a cornerstone for general relativity. They were very general statements, though rigorous and succinct, that mathematically validated certain relationships suspected by physicists of the time. One, now known as Noether's Theorem, established the equivalence between an invariance property and a conservation law. The other involved the relationship between an invariance and the existence of certain integrals of the equations of motion.

While Noether was proving these profound and useful results, she was working without pay at Göttingen University, where women were not admitted to the faculty. Hilbert, in particular, attempted to obtain a position for her but could not persuade the historians and philosophers on the faculty to vote in a woman's favor. He arranged for her to teach, however, by announcing classes in mathematical physics under his name and letting her lecture in his place. By 1919, regulations were eased, and she was designated a *Privatdozent* (a licensed lecturer who could receive fees from students but not from the university). In 1922, Noether was given the title of unofficial associate professor, an honorary position bearing no responsibilities or income. She was, however, hired as an adjunct teacher and paid a modest salary.

A keen mind and infectious enthusiasm for mathematical investigations made Noether an effective teacher, although her classroom technique was nontraditional. Rather than simply lecturing, she conducted discussion sessions in which she and her students would jointly explore some topic. She loved to spend free time with her students, especially taking long walks while discussing their personal interests or pursuing some mathematical topic. She would sometimes become so engrossed in the conversation that her students would have to remind her to watch for traffic.

### Lays the Foundations of Abstract Algebra

Brilliant mathematicians often make their greatest contributions early in their careers; Noether was one of the notable exceptions to that rule. She began producing her most powerful and creative work at about the age of 40. Her change in style started with a 1920 paper on noncommutative fields (systems in which an operation such as multiplication yields a different answer for a x b than for b x a). During the years that followed, Noether developed a very abstract and generalized approach to the axiomatic development of algebra. As Weyl attested, "she originated above all a new and epoch-making style of thinking in algebra."

Noether's 1921 paper on the theory of ideals in rings is considered to contain her most important results. It extended the work of **Julius Dedekind** on solutions of polynomials and laid the foundation for modern abstract algebra. Rather than working with specific operations on sets of numbers, this branch of mathematics looks at general properties of operations. Because of its generality, abstract algebra represents a unifying thread connecting such theoretical fields as logic and number theory with applied mathematics useful in chemistry and physics.

During the winter of 1928-1929, Noether lectured as a visiting professor at the University of Moscow and the Communist Academy. She held a similar position during the summer of 1930 at Frankfurt. Noether was the main speaker at one of the section meetings of a conference of the International Mathematical Congress in 1928 in Bologna. In 1932, she addressed the Congress' general session during its conference in Zurich. Noether enjoyed learning about other cultures when she traveled.

When **Hermann Weyl** joined the Göttingen faculty in 1930, he tried unsuccessfully to obtain a better position for Noether because (as he later said in a eulogy delivered at Bryn Mawr and published in *Scripta Mathematica*) "I was ashamed to occupy such a preferred position beside her whom I knew to be my superior as a mathematician in many respects." In 1932, Noether again was honored by those outside her own university when she was named co-winner of the Alfred Ackermann-Teubner Memorial Prize for the Advancement of Mathematical Knowledge.

### Enters Exile

During the early 1930s, many considered the flourishing Mathematical Institute in Göttingen the premier mathematical center in the world, and its most active element was Noether's group of algebraists. On Noether's fiftieth birthday, the algebraists held a celebration, and her colleague Helmut Hasse dedicated in her honor a paper validating one of her ideas on noncommutative algebra. This successful and congenial environment ended in 1933 when the Nazi Party seized absolute power in Germany. Within months, anti-Semitic policies became law throughout the country, and on April 7, 1933, Noether was formally notified that she could no longer teach at the university. She was a dedicated pacifist, and Weyl later recalled, "[Noether's] courage, her frankness, her unconcern about her own fate, her conciliatory spirit were, in the midst of all the hatred and meanness, despair and sorrow surrounding us, a moral solace."

For a time, Noether continued to meet informally with students and colleagues, inviting groups to her apartment. By summer, the Emergency Committee to Aid Displaced German Scholars was entering into an agreement with Bryn Mawr, a women's college in Pennsylvania, to offer her a professorship. The Rockefeller Foundation provided a grant matching the Committee's appropriation, thus funding Noether's first year's salary.

In the fall of 1933, Noether was supervising four graduate students at Bryn Mawr. Starting in February 1934, she also delivered weekly lectures at the Institute for Advanced Study at Princeton University. She bore no malice toward Germany, and maintained friendly ties with her former colleagues. With her characteristic curiosity and good nature, she settled into her new home in America. Her English was adequate for conversation and teaching, although she occasionally lapsed into German when concentrating on technical material.

During the summer of 1934, Noether visited Göttingen to arrange shipment of her possessions to the United States. She returned to Bryn Mawr in the early fall, having received a two-year renewal on her teaching grant. In the spring of 1935, Noether underwent surgery to remove a uterine tumor. The operation was a success, but four days later, she suddenly developed a very high fever and lost consciousness. She died on April 14, apparently from a postoperative infection. Her ashes were buried near the library on the Bryn Mawr campus.

Over the course of her career, Noether supervised a dozen graduate students, wrote 45 technical publications, and inspired countless other research results through her habit of suggesting topics of investigation to students and colleagues. After World War II, the University of Erlangen attempted to show Noether the honor she had deserved during her lifetime. In 1958, the institution presented a conference commemorating the 50th anniversary of Noether's doctorate, and in 1982—the 100th anniversary year of Noether's birth—the university dedicated a memorial plaque to her in its Mathematics Institute. That same year, the Emmy Noether Gymnasium, a coeducational school emphasizing mathematics, the natural sciences, and modern languages, opened in Erlangen.

## SELECTED WRITINGS BY NOETHER

*Collected Papers,* 1983

## FURTHER READING

### Books

Brewer, James W. and Martha K. Smith, editors. *Emmy Noether: A Tribute to Her Life and Work.* New York: Marcel Dekker, 1981.

Kramer, Edna E. *The Nature and Growth of Modern Mathematics.* Princeton, NJ: Princeton University Press, 1981, pp. 656–60.

Magill, Frank N., editor. *Great Events from History II.* Volume 2. Pasadena, CA: Salem Press, 1991, pp. 650–54, 716–19.

Noether, Gottfried E. "Emmy Noether (1882-1935)." *Women in Mathematics.* Edited by Lynn M. Osen. Cambridge, MA: MIT Press, 1979, pp. 141–52.

Perl, Teri. *Math Equals: Biographies of Women Mathematicians.* Menlo Park, CA: Addison-Wesley, 1978, pp. 172–78.

Srinivasan, Bhama and Judith D. Sally. *Emmy Noether in Bryn Mawr.* New York: Springer–Verlag, 1983.

### Periodicals

Kimberling, Clark H. "Emmy Noether." *The American Mathematical Monthly* (February 1972): 136–49.

### Other

Taylor, Mandie. "Emmy Noether." *Biographies of Women Mathematicians.* June 1977. http://www.scottlan.edu/lriddle/women/chronol.htm (July 21, 1997).

*—Sketch by Loretta Hall*

*Max Noether*

# Max Noether
## 1844–1921
### German algebraist

Today, Max Noether's career as a mathematician is perhaps overshadowed by his more famous daughter, **Emmy Noether**. Still, at the height of his career, Noether led the German school of algebraic and geometric mathematics from the University of Erlangen. Noether's primary interests were algebra and algebraic geometry, and he is responsible for the development of algebraic function theory. His work on curves and related theorems inspired many Italian geometrists.

Noether was born in Mannheim, Germany, on September 24, 1844, the son of Hermann (an iron wholesaler) and Amalia (nee Würzburger) Noether. With his four siblings, Noether began his education in Mannheim. He contracted polio at age 14, and could not use his legs for two years. Because of this illness, Noether was handicapped for life. He continued his education at home, following the Gymnasium curriculum.

In 1865, Noether began his university studies in astronomy at the Mannheim Observatory, although he did not stay there long. He entered the University of Heidelberg in 1865, studying mathematics. He also studied at the University of Gissen and the University of Göttingen, before earning his doctorate (without a dissertation) from Heidelberg in 1868. Noether was the first member of his family to earn this degree. Noether began teaching at Heidelberg as a (privatdozent) lecturer from 1870 to 1874. In 1874, Noether became an associate professor at the University of Erlangen.

The mathematical research Noether did in this early part of his career, from approximately 1869 to 1879, is arguably his most important. His 1869 paper, "Über Flächen, welche Scharen rationaler Kurven besitzen" ("Concerning Surfaces which have families of Rational Curves"), began this work, containing what came to be known as Noether's fundamental theorem. In 1871, he published a proof that improved on work done by one of his inspirations as a mathematician, Antonio Cremona. In it, Noether proved the result that a plane Cremona transformation can be built up from a sequence of quadratic and linear transformations. Two years later, Noether published another theorem, perhaps his most famous result, concerning two algebraic curves and their intersection under certain so-called Noether conditions. The conditions concern the complexity of contact of curves and their common multiple points. This result could also be extended to surfaces and hypersurfaces, though Noether himself did not prove it.

After 1880, Noether's private life became more domesticated, perhaps accounting for the changes in Noether's research. Noether married Ida Amalia Kaufmann in 1880. Together they had four children, three of whom became scientists. His eldest daughter, Amalie Emmy Noether, born in 1882 and familiarly called Emmy, became a mathematician whose work built on and surpassed her father's. She had three younger brothers: Alfred, born in 1883, who became a chemist; Fritz, born in 1884, also a mathematician; and Gustav, born in 1889.

Still, Noether continued to do his original research and teaching at Erlangen, where he served as an associate professor until 1888. That year he became a full professor. From about 1879 onward, his mathematical work focused on refining his earlier publications. This work also inspired and opened up fields for other mathematicians, especially in Italy.

Some of Noether's later publications that are widely known include the 1882 publication on algebraic curves, *Zur Grundlegung der Theorie der Algebraischen Ramkurven.* In 1894 he coauthored *Bericht über der Entwicklung der algeraischen Funktionen,* with Alexander von Brill. This treatise concerned algebraic functions.

Noether was also involved with the journal *Mathematische Annalen.* In addition to publishing in it almost continuously from 1870–1921, he served as its director for a time. Noether also wrote many of the biographic obituaries for this journal, and other publications. These pieces also gained recognition and were known for their thoroughness and insight.

Noether's important mathematical contributions did not go unnoticed by his peers. In 1903, he became a member of the French Académie Royale des Sciences. He was also a member or honorary member of many other academic societies, among them the London Mathematical Society, the Academy of Berlin, and the Royale Accademia dei Lincei.

The last decade of his life saw Noether lose his wife and retire from teaching. When Ida Noether died 1915, Noether's daughter Emmy started to fill in for her father's lectures as necessary. Four years later, in 1919, he retired from teaching, and was given an emeritus professorship.

Noether died in Erlangen on December 13, 1921. In his obituary published in the *Proceedings of the London Mathematical Society,* the author describes the three main ways Noether contributed to his field: "By the new and fruitful ideas contained in his original researches, by the patient investigation and encouragement he gave to other writers, and by his acutely critical and detailed historical work."

## SELECTED WRITINGS BY NOETHER

"Über Flächen, welche Scharen rationaler Kurven besitzen" ("Concerning Surfaces Which Have Families of Rational Curves"), 1869
*Zur Grundelgung der Theorie der Algebraischen Ramkurven,* 1882
(With Alexander von Brill) *Bericht über der Entwicklung der algeraischen Funkionen,* 1894

## FURTHER READING

Kimberling, Clark. "Emmy Noether and Her Influence," in *Emmy Noether: A Tribute to Her Life and Work.* Edited by James W. Brewer and Martha K. Smith. New York: Marcel Dekker, Inc., 1981, pp. 3–8, 20.
Kramer, Edna E. "Max Noether," in *Dictionary of Scientific Biography.* Volume X. Edited by Charles Coulston Gillispie. New York: Charles Scribner's Sons, 1970, pp. 137–39.
"Max Noether." *Proceedings of the London Mathematical Society.* Volume 21, 2nd Series (1921): xxxvii–xlii.

*—Sketch by Annette Petruso*

*Sergei Novikov*

# Sergei Novikov
## 1938-
### Russian algebraic topologist

Sergei Novikov, a mathematician whose interests range from topology to theoretical physics, has made important contributions to several fields. He has worked on finding links between high-level mathematics and theoretical physics, but he is best known for his research on Pontryagin classes, which led to the classification of certain types of manifolds. For this work, Novikov was awarded both his own country's Lenin Prize and the international honor of the Fields Medal.

Sergei Petrovich Novikov was born into a mathematical family on March 20, 1938, in Gorky, Russia. His father, Petr Sergeevich Novikov, founded the Soviet Union's school of mathematical logic and made major contributions to the fields of set theory, algebra, and function theory. His mother, Lyudmila Vsevolodovna Keldysh, was also a mathematician; her research was concentrated in set theory and topology, a field in which her son would later excel. (Topology is a branch of mathematics concerned with properties of geometric configurations that are not altered by changes in shape—sometimes described as the study of continuity.) Novikov entered Moscow University as an undergraduate in 1955, and very early on he showed signs of mathematical brilliance.

His first paper, a short work on a part of topological theory, was published when he was only 21 years old; his second, published a year later, held the beginnings of his later research on manifolds. Novikov graduated from the Faculty of Mathematics and Mechanics of Moscow University in 1960, and he entered the Steklov Institute of Mathematics as a doctoral student.

### Breakthrough Research Leads to Top Award

Under the supervision of M. M. Postnikov, Novikov began his research at Steklov, where he would stay for five years. In 1962, he married Eleonora Tsoi (with whom he had three children). In 1964, he received his Ph.D., and a year later was awarded his doctor of science degree, an achievement equivalent to promotion to the rank of full professor. At Steklov, Novikov uncovered the crucial ideas of his manifold research. A manifold can be broadly described as a topological space containing a set of items—for example, a plane can be understood as a two-dimensional manifold of points. At the time of Novikov's research, three different categories of manifolds were discussed: differentiable, piece-wise linear (also known as combinatorial), and topological. In differentiable manifolds, those on which calculus is performed, the items in the set are connected by curves or twists. Piece-wise linear manifolds, on the other hand, are connected by straight lines. This distinction was already recognized at the time of Novikov's work. However, very little was known about topological manifolds. Did they behave more like differentiable manifolds or piece-wise linear ones? This was the question Novikov set out to answer.

Novikov began by looking at the Pontryagin classes. On a differentiable manifold, a Pontryagin class is a structure related to the manifold that is unchanged, or invariant, when the manifold is manipulated. These classes can be used to describe the amount of twisting or curvature present in the manifold. However, in 1957, French mathematician **René Thom** and others had proved that, when using real or rational numbers, the Pontryagin classes are piece-wise linear invariants; in other words, they are a characteristic feature of a piece-wise linear manifold, just as they are of a differentiable manifold. Finally, in 1965, Novikov was able to show that they were also topologically invariant. This dramatic result demonstrated that topological manifolds are, in the last analysis, most similar to piece-wise linear ones, a result that had great impact of future work in topology. An impressive aspect of this research was that Novikov was relatively isolated from other mathematicians working in this field at the time, and yet he solved a problem that had puzzled the entire mathematical community.

Moscow University, where Novikov had been teaching since 1964, recognized the importance of his work by appointing him to a full professorship in 1966. Other recognitions soon followed. In 1967, he won the Soviet Union's Lenin Prize, and in 1970, the International Mathematical Union awarded him the Fields Medal, the most prestigious honor a mathematician can receive.

### Diversifies Work to Unify Related Fields

After receiving the Fields Medal, Novikov turned his attention to a new topic, theoretical physics. He told the Symposium on the Current State and Prospects of Mathematics in 1991 that while deriving new results in fields such as dynamical systems, partial differential equations, and topology, "no matter how far I was moving into mathematics, I was not able to answer my basic question, concerning the goal of what we were doing." In physics, he saw a purpose and decided to use the research being done in modern mathematics to foster the progress of theoretical physics. His extensive research after 1971 varied between articles strictly concerned with algebraic geometry and those on modern mathematical physics. Interestingly, Novikov's attention to theoretical physics had a reciprocal benefit. In his 1991 symposium address, he recalled "I was not able to find anything substantially new and interesting in topology for myself until 1980 . . . The new era began when topology started to apply discoveries of physicists inside topology itself." In particular, Novikov concentrated on equations involving "solitons." By way of explanation, D. V. Feldman wrote in *Choice,* 'Nonlinear differential equations generally do not admit exact solutions; that is what chaos theory is all about. Nevertheless, there are important exceptions, [such as] the nonlinear Schrödinger equation; the term 'soliton' refers to certain solutions of such equations."

In 1975, Novikov accepted a position as head of the mathematics department at the L. D. Landau Institute. In 1981 he received belated recognition for work he had done over ten years earlier. As part of his research on manifolds, he had studied foliations, which are decompositions of manifolds into smaller ones, called leaves. Leaves could be either open or closed, but the closed type was most mathematically interesting because at the time its existence had not been proved. Novikov, using geometric proofs, established the existence of closed leaves in the case of a sphere. The mathematical community began using this research immediately, but it was not until 1981 that Novikov was honored for the work with the Lobachevsky International Prize of the Academy of Sciences of the Soviet Union.

In 1983, in addition to his professorship at Moscow University and his directorship of the Landau Institute, Novikov accepted the position of head of the Department of Geometry and Topology at Moscow University. From 1985 to 1996, he served as president of the Moscow Mathematical Society, and from 1986 to 1990, he was also a vice-president of the International Association in Mathematical Physics. In addition, the mid-1980s saw the publication of two important mathematical texts by Novikov and his colleagues, one in mathematical physics called *Theory of Solitons: The Inverse Scattering Methods* and one in geometry called *Modern Geometry: Methods and Applications.* Throughout his career, Novikov searched for ways to link the three major areas of modern-day pure mathematics: calculus, geometry, and topology; his 1990 work, *Basic Elements of Differential Geometry and Topology,* demonstrated his success in establishing such connections.

In 1992, Novikov began spending time as a visiting professor at the University of Maryland in the United States. The U.S. National Academy of Sciences granted him the rank of foreign associate in 1994, in recognition of his stature as "Russia's most distinguished topologist and (together with V. I. Arnold) one of the two most active and influential mathematicians in the senior generation in Russia." In the fall of 1996, he emigrated to the United States, accepting a full-time faculty position in the Department of Mathematics and the Institute for Physical Science and Technology at the University of Maryland.

## SELECTED WRITINGS BY NOVIKOV

(With B. A. Dubrovin and A. T. Fomenko) *Modern Geometry: Methods and Applications,* 2 volumes. 1984
*Basic Elements of Differential Geometry and Topology,* 1991
"Rôle of Integrable Models in the Development of Mathematics," in *Mathematical Research Today and Tomorrow: Viewpoints of Seven Fields Medalists,* 1992, pp. 13–28.
*Solitons and Geometry (Lezioni Fermaine),* 1995

## FURTHER READING

### Periodicals

Feldman, D. V. Review of *Solitons and Geometry. Choice* 33 (November 1995): 501.
"Sergei Petrovich Novikov (On His Fiftieth Birthday)." *Russian Mathematical Surveys* 43 (1988): 1–10.

### Other

"Serguei Novikov." http://mech.math.msu.su/ . million/novikov.htm (July 15, 1997).

*—Sketch by Karen Sands and Loretta Hall*

Moscow State in 1947. She received her doctorate in 1954, and shortly thereafter began the professorship in mathematics at Moscow State University that she holds today.

In 1972 Oleinik was promoted to the head chair of differential equations in her department. Her specialty is partial differential equations. Although her classes can carry such intimidating titles as "Asymptotic Properties of Solutions of Nonlinear Parabolic and Elliptic Equations and Systems," she is generous with her ideas and more than willing to give her students the right start, according to one former student. Igor Oleinik remembers how much time his "PDE" professor was willing to give to her students despite her busy schedule. Because they were fellow Ukrainians, and Oleinik is as common a name there as Smith is in America, Igor had to put up with a little teasing from fellow class members who joked that he was really Olga Oleinik's grandson.

Oleinik's studies in mathematics cover broad areas of physics, such as the interactivity of liquids or gases in porous substances, the thermodynamics of bodies in different phases, as well as problems in elasticity and homogenization.

Oleinik is married and has one son, Dmitri. She holds an honorary doctorate from Rome University, granted in 1981, and is also an honorary member of the Royal Society of Edinburgh. She is a member of various societies throughout Europe. Her awards include the medal of the College de France and a "first degree" medal from Prague's Charles University, as well as various prizes from Russian institutions.

*Olga Oleinik*

# Olga Oleinik
## 1925–

**Russian algebraic geometer and mathematical physicist**

Olga Oleinik is a prolific writer and educator with eight books and nearly 60 graduate students to her credit. She teaches mainly in Russia at Moscow State University, but she also travels to colloquia in America, and has held classes as a visiting scholar at such institutions as the University of South Carolina. Oleinik has also contributed more than 300 papers to a variety of professional journals and is a member of the Russian Academy of Sciences. She has made important findings in the area of algebraic geometry in projective space.

Olga Arsenievna Oleinik was born in Kiev, Ukraine on July 2, 1925. Her parents, Arseniy Ivanovich and Anna Petrovna lived in an area known as Matusov. Olga's early years spanned times of great upheaval and difficulty in the Soviet Union, especially World War II. She did, however, earn a degree from

## SELECTED WRITINGS BY OLEINIK

**Books**

*Homogenization of Differential Operators and Integral Functionals,* 1994
*Mathematical Problems in Elasticity and Homogenization,* 1992
*Some Asymptotic Problems of the Theory of Partial Differential Equations,* 1995

## FURTHER READING

**Books**

*Who's Who in the World, 1993–1994.* Eleventh edition. New Providence, NJ: Marquis, 1992.

**Other**

Oleinik, Igor, in an electronic mail interview with Jennifer Kramer, conducted July 12, 1997.

"Olga Oleinik." *The Emmy Noether Lectures of the American Association for Women in Mathematics. Profiles of Women in Mathematics.* http://www.math.neu.edu/awm/noetherbrochure/ Oleinik96.htm

—*Sketch by Jennifer Kramer*

# Nicole d'Oresme
## c. 1320–1382
**French algebraist, geometer, physicist, philosopher, and theologian**

A mathematician and physicist of great originality and theoretical audacity, Nicole Oresme is known for several important theoretical and empirical discoveries. In algebra, working in proportion theory, he suggested the possibility of irrational proportion, an idea which, as Carl B. Boyer has remarked, "is the first hint in the history of mathematics of a higher transcendental function." In an ingenious application of geometrical coordinates to the study of motion, Oresme introduced a mathematical methodology which centuries later developed into analytical geometry.

Oresme was educated at the University of Paris, where his studies included the liberal arts and theology, qualifying him to teach theology. Among his teachers was the noted theologian, philosopher, and physicist Jean Buridan, who, owing to his interest in empirical research, played a crucial role in the development of 14th-century science. Oresme taught theology, but also found time for other activities. Having befriended the French *dauphin*, or crown prince, who ruled France as Charles V from 1364 to 1380, he undertook several royal diplomatic missions. In addition, Oresme held a variety of ecclesiastical posts, obtaining the bishopric of Lisieux in 1377, apparently as a reward from Charles V for translating a number of Aristotelian texts from Latin into French. In 1380, he established residency in Lisieux, where he spent his final years.

Strongly influenced by the rational and empirical spirit of 13th- and 14th-century philosophy and science, exemplified by such thinkers as William of Ockham who defined intuition and experience as the foundations of knowledge, Oresme approached the much-discussed scientific problems of his time from an empirical standpoint. In particular, he wished to explain the phenomenon of motion. The starting point of medieval kinematics was Aristotle, who, after observing movement in nature, concluded that the necessary conditions for motion were an object, force, and resistance. Encouraged by the sixth-century commentator John Philoponus, who questioned Aristotle's inclusion of resistance in his definition of motion, 14th-century scientists strove to create a more comprehensive theory. At Oxford University, a group of scholars affiliated with Merton College, formulated what is now known as the Merton rule, which stated that if object A moves at a uniform rate of acceleration for a fixed period of time, the distance it will cover will be the same as that of object B moving constantly, within the same time-frame, at the speed if A at midpoint between its initial and final points.

## Creates Graphical Representation of the Merton Rule

Oresme had the idea, though perhaps not an original one, to illustrate the Merton rule by a graph. He drew a horizonal line on which he marked points separating time intervals of equal lengths. At each of these points he drew a perpendicular line representing the velocity of object A at that time. Because of the uniform acceleration, the endpoints of each vertical line would lie along a straight line.

Supposing that object A starts at rest, then the final graphical representation of this process will be a right triangle whose area represents the distance covered by object A. If $T$ is the total length of time and $V$ is the final velocity of object A, then the total distance of object A would be given by the formula for the area of a triangle, namely $D=TV/2$. Now the speed of A at the midpoint of the time interval is given by $V/2$, so the total distance of object B moving at the constant speed $V/2$ is also $D=TV/2$. By such an argument Oresme gave a geometric verification of the Merton rule. Moreover, as Boyer had indicated, Oresme's graphical representation directly leads to **Galileo**'s observation that as time is divided into equal segments, the distances covered by uniformly accelerated objects will have ratios represented by odd numbers. For example, if there are four equal time segments, the ratio will be 1 : 3 : 5 : 7.

Coordinates had been used before Oresme (e.g., by **Apollonius**), but, as Boyer emphasizes, Oresme's originality lies in his ability to use coordinates to represent a variable quantity. He knew, as his diagram of the Merton rule indicates, that a function of one unknown can be drawn as a curve (in his example, the curve is a straight line, because in linear functions the curve is linear), but he apparently did not grasp, as Boyer remarks, that, in the context of a coordinate system, every plane curve can be represented as a function of variable. Therefore, while Oresme certainly anticipated the modern analytical geometry firmly established by the work of **René Descartes**, he did not venture beyond linear function. However, it is clear that his use of the coordinate system, although limited, constitutes the necessary epistemological foundation of analytical geometry.

### Expands Bradwardine's Theory of Proportion

Oresme's work on proportions was strongly influenced by his older contemporary Thomas Bradwardine, mathematician, philosopher, theologian, and Archbishop of Canterbury. Bradwardine, like many other Scholastic philosophers, found Aristotle's theory of motion too speculative. Aristotle believed that the speed of a moving object is directly proportional to its moving force, and inversely proportional to the resistance of the medium in which it was moving. In order to avoid the paradoxical situation, theoretically possible according to Aristotle's formula, Bradwardine proposed a different formula. Still defining velocity, in an Aristotelian fashion, as $V = F/R$, where $F$ = force, and $R$ = resistance, he postulated that in order to double a given velocity $V$, $F/R$ should be squared, and to obtain a triple velocity. $F/R$ should be cubed, etc. In other words, $2V = (F/R)^2$, or, $nV = (F/R)^n$. In Bradwardine's formula, the exponent was by definition a positive integer. Oresme, however, greatly expanded Bradwardine's proportion theory by introducing new categories of exponents, including rational fractions and negative integers. His work with different exponents indicates that he understood rules such as (expressed in modern notation) $x^m \times x^n = x^{m+n}$ and $(x^m)^n = x^{mn}$. To facilitate his work with fractional exponents, Oresme invented a graphical notation to represent, for example, numbers such as $x^{2/3}$. Systems of graphical notations were used, because Oresme, and medieval and early Renaissance mathematicians in general, lacked the symbolic apparatus which enables us to represent $x^{2/3}$ as the cube root $x^2$. Finally, Oresme's calculations including irrational exponents, such as, in modern notation, $x^{\sqrt 2}$ reflects his extraordinary intuitive grasp of the possibility of irrational proportions.

### Questions Geocentric System

Intrigued by the question of the Earth's movement, Oresme persuasively argued against the dogmatically accepted idea that the Earth was a stationary object. For example, he pointed out that the perceived movement of the heavens around a stationary Earth may be deceptive. He even questioned biblical evidence for the sun's rotation around the Earth, such as the statement (in Joshua 10: 13), that God made the sun stand still, so the Israelites could

kill more of their fleeing enemies. In fact, Oresme, who dismissed astrology and argued that even miracles can be explained scientifically, rejected the biblical argument, declaring, as Frederick Copleston has pointed out, that it as an anthropomorphic fallacy to ascribe such typically human behavior—e.g. demonstrations of physical power—to God. It is important to point out, however, that such criticism stayed within the confines of the accepted Aristotelian division between a celestial world beyond the planetary system ruled by spiritual, incorporeal forces, and the lower world, which is affected by physical forces. According to Maurice de Wulf, Oresme, perhaps influenced by the statement made by St. Thomas Aquinas that the geocentric system is merely a hypothesis, taught, around 1362, just as the Scotist philosopher François de Meyronnes had suggested in the early 1320s, that it would be a *melior dispositio*, a "better arrangement" if the earth moved and the heavens stood still. However, after weighing the evidence, Oresme rejected heliocentrism as a hypothesis lacking empirical evidence.

## SELECTED WRITINGS BY D'ORESME

*Livre du ciel et du monde d'Aristotle,* 1943
*Quaestiones super geometriam Euclidis,* 1961
*Nicole Oresme and the Kinematics of Circular Motion: Tractatus de commensurabilitate vel incommensurabilitate motuum celi, with Critical Edition,* 1971, (Translated by Edward Grant)
*Nicole Oresme and the Marvels of Nature: A Study of His De causis mirabilum with Critical Edition,* 1985, (Translated by Bert Hansen)

## FURTHER READING

Boyer, Carl B. *A History of Mathematics.* Second edition. Revised by Uta C. Merzbach. New York: John Wiley & Sons, 1991.

Copleston, Frederick. *Ockham to Suarez.* Volume 3: *A History of Philosophy.* Garden City, NY: Image Books, 1963.

Clagett, Marshall. *The Science of Mechanics in the Middle Ages.* Madison: University of Wisconsin Press, 1959.

Crombie, A. C. *Science in the Later Middle Ages and Early Modern Times: XIII–XVII Centuries.* Volume 2: *Medieval and Early Modern Science.* Second revised edition. Garden City, NY: Doubleday, 1959.

De Wulf, Maurice. *Philosophy and Civilization in the Middle Ages.* New York: Dover, 1953.

Eamon, William. *Science and the Secrets of Nature: Books of Secrets in Medieval and Early Modern Culture.* Princeton, NJ: Princeton University Press, 1994.

Eves, Howard. *Foundations and Fundamental Concepts of Mathematics.* Third edition. Mineola, NY: Dover, 1990.

Lindberg, David B. *The Beginnings of Western Science: The European Tradition in Philosophical, Religious, and Institutional Context, 600 B.C. to A.D. 1450.* Chicago: University of Chicago Press, 1992.

—*Sketch by Zoran Minderovic*

# William Oughtred
## 1574–1660
### English clergyman and arithmetician

William Oughtred was an Anglican clergyman who spend a great deal of time teaching mathematics to various students. He was the author of several books on mathematics and is generally credited with inventing both the linear and circular slide rules. Oughtred also used many unique mathematical shorthand notations, including the notation × for multiplication and :: for proportion.

William Oughtred was born in Eton, Buckinghamshire, England, on March 5, 1574. His father was Benjamin Oughtred, who taught writing at Eton School. His father is sometimes referred to as the Reverend Oughtred, although there is little evidence he actually practiced as a clergyman.

Oughtred was educated as a king's scholar at Eton and entered King's College of Cambridge University when he was 15 years old. He received his B.A. degree from King's College in 1596 and his M.A. degree in 1600. Although very little mathematics was taught at either Eton or Cambridge, Oughtred developed a strong interest in the subject and studied it on his own at night after completing his regular studies.

In 1603, Oughtred was ordained an Anglican priest and became the vicar of Shalford in 1604. In 1610, he became the rector of Albury at an annual salary of £100. Oughtred held this position until his death in 1660. Oughtred began instructing students in mathematics in addition to his church duties sometime in the 1620s. He often took in private pupils who lived in his house while they received instruction. Oughtred neither asked for nor received any money for his services, stating that he was adequately provided for by his salary as a clergyman.

In 1628, he was asked to tutor Lord William Howard, who was the son of the Earl of Arundel. The Earl became a patron of Oughtred's and encouraged him to publish his work on mathematics. While tutoring young Lord Howard, Oughtred wrote a 100–page book summarizing all that was known at the time about arithmetic and algebra. Known by its shortened title, *Clavis mathematicae*, the book was first published in 1631. Although it was probably intended to be a reference textbook for his young student, the book quickly received favorable attention from the mathematics and physical sciences community in Europe. In this book, Oughtred used a wide variety of mathematical shorthand notations to denote quantities, powers, ratios, and the concepts of "greater than" and "less than." Many of these notations were overly complicated and did not gain general acceptance. However, two of his notations did—the notation × for multiplication, and :: for proportion—and these symbols continue to be used today.

Encouraged by the positive reviews of *Clavis mathematicae*, Oughtred wrote several other books. In 1632, he published *The Circles of Proportion and the Horizontal Instrument* in which he described a circular slide rule as an instrument for use in navigation. Unfortunately, the same instrument had been described two years earlier in a book published by one of Oughtred's students, Richard Delamain. Both men quarreled over who had been the first to invent the instrument, without any apparent resolution, although history has generally awarded the honor to Oughtred. Oughtred is also given credit as the inventor of the first linear slide rule, and there is evidence that he had actually designed such a device as early as 1621. Over the next 25 years, Oughtred published six more books, including *Trigonometria*, that dealt with both plane and spherical triangles and contained tables listing the values of trigonometric and logarithmic functions to seven places.

Politically, Oughtred was a staunch Royalist. During the English Civil War he was almost sequestered by the Puritan government, but was spared at the insistence of influential friends. Oughtred died on June 30, 1660, in Albury, Surrey, England shortly after learning that King Charles II had been restored to the British throne.

## SELECTED WRITINGS BY OUGHTRED

*Arithmeticae in numeris et speciebus instituto . . . quasi clavis mathematicae est* (generally known as *Clavis mathematicae*), 1631; second edition 1648

*The Circles of Proportion and the Horizontal Instrument,* 1632
*A Description and Use of the Double Horizontal Dial,* 1636
*The Solution of All Spherical Triangles,* 1651
*Trigonometria,* 1657

**FURTHER READING**

American Council of Learned Societies. *Biographical Dictionary of Mathematicians.* New York: Charles Scribner's Sons, 1991, pp. 1894–1895.

Cohen, I.B., editor. *Album of Science: From Leonardo to Lavoisier, 1450–1800.* New York: Charles Scribner's Sons, 1980, p. 78.

Struik, D.J., editor. *A Source Book in Mathematics: 1200–1800.* Princeton: Princeton University Press, 1986, p. 93.

—*Sketch by Chris Cavette*

*Blaise Pascal*

# Blaise Pascal
## 1623–1662
### French geometer, probabilist, physicist, inventor, and philosopher

Although Blaise Pascal can be seen in retrospect as an important scientist in his time, he was a controversial figure to his contemporaries. There is no doubt that Pascal was of a superior intelligence, but he was a modest man, embarrassed by his own genius. The reason for his contemporaries' doubt was perhaps in part due to the fact that much of his work was not published in his lifetime, which limited how his accomplishments could be viewed. Still, Pascal gave the world, mathematical and otherwise, many things: he opened up new forms of calculus, projective geometry, probability theory, and he designed and manufactured the first calculating machine run by cogs and wheels.

Pascal was born in Clermont (now known as Clermont–Ferrand), Auvergne, France, on June 19, 1623. He was the son of a mathematician and civil servant Ètienne and Antoinette (nee Bégon) Pascal. His mother died when Pascal was three. With his two sisters, Gilberte (Madame Périer after marrying Florin Périer in 1641) and Jacqueline, Pascal was educated at home, primarily by his father. Pascal was part of a very tightly knit family, and he was especially close to his sisters. Pascal's sister Jacqueline was a child literary prodigy, and, despite being sickly all his life, Pascal was also a gifted child. In 1631, the family moved to Paris, where Pascal began his mathematical education when he was 10 or 11. His father insisted that his education start with the study of ancient languages, before learning geometry.

### Publishes First Paper at 16

At age 13, Pascal and his father began attending discussions in Paris with a group of scientists and mathematicians, such as **René Descartes,** called the Académie Parisienne. By the time he was 16 years old, Pascal had already done a significant amount of his mathematical ground work. He continued his studies in Rouen where the family moved in 1639, when his father was appointed to the tax office there. Pascal continued to go to Paris occasionally while living in Rouen, and it was on one of these trips that he presented one of his important mathematical discoveries.

Published in 1640 as a pamphlet, *Essai sur les coniques* was a vital step in the development of projective geometry and it contained what came to be known as Pascal's mystic hexagram. Pascal began writing this treatise on conic sections to clarify the 1639 publication of **Gérard Desargues'** *Brouillon project d'une atteinte aux evenements des rencontres du cone avec un plan* ("Experimental project aiming to describe what happens when the cone comes into contact with a plane"). Desargues had written his book in a manner that was very difficult to understand, even for other mathematicians.

But as Pascal began to work with the propositions that Desargues made, he went beyond what Desargues accomplished. Pascal developed his own theorem which he used to deduce some 400 propositions as corollaries. His theorem, describing a figure known as Pascal's mystic hexagram, states that the three points of intersection of the pairs of opposite sides of a hexagon inscribed in a conic are collinear. When he presented his findings to the Académie Parisienne, Descartes could not believe a 16–year–old boy had written this work. Only part of the manuscript was published in the 1546 essay, but a whole manuscript did exist at one time.

### Develops Calculator

Soon after this pamphlet was published, in 1641, Pascal's health began to decline. He suffered from headaches (perhaps caused by a deformed skull), insomnia, and indigestion but he continued his work. To help with his father's lengthy tax work in Rouen, Pascal worked on what became the first manufactured calculator from 1642 to 1644. This machine could automatically add and subtract, using cogged wheels to do the calculations. The invention was patented and Pascal received a monopoly by a royal decree dated May 22, 1649. He wanted to manufacture these machines as a full scale business enterprise but it proved too costly. The basic principle behind Pascal's calculator was still used in this century before the electronic age.

The year 1646 was key in Pascal's life. He became part of an anti-Jesuit Catholic sect called Jansenism, which believed in predestination and that divine grace was the only way to achieve salvation. He persuaded his family to join him, and the influence of Jansenism played a dominant role in the rest of the life.

Pascal also began doing work in physics, conducting experiments in atmospheric and barometric pressure, and vacuums. Pascal used the theories of Evangelista Torricelli as a starting point for his work. In Pascal's experiments, he had his brothers-in-law climb Puy de Dôme with tubes filled with different liquids to test his theories. The results were not just more information about atmospheric pressure, for Pascal invented the syringe and the hydraulic press based on them. More importantly, he delineated what came to be known as Pascal's principle, which says that pressure will be transmitted equally throughout a confined fluid at rest, regardless of where the pressure is applied.

Pascal published his some of his results in 1647 under the title *Experiences nouvelles touchant le vide*, and in 1648 as *Récit de la grande experience sur l'equilibre des liqueurs*. He completed the work already done on hydrostatics theory, bringing together the mechanics of both fluids and rigid bodies. His whole treatise on this subject was not published until a year after his death.

### Retreats from Secular World

The Pascal family returned to Paris in 1647. His father died there in 1650, and his sister Jacqueline entered the Jansenism Convent at Port-Royale. Pascal himself had a profound religious experience four years later on the night of November 23, 1654. That night, Pascal was nearly killed in a riding accident. A few months later, Pascal left the secular world to live in the Port-Royal Convent. He did a little more work in mathematics and science, but primarily published religious philosophy.

Before Pascal's religious experience, earlier in 1654, he and **Pierre de Fermat** began writing each other about problems on dice and other games of chance. This correspondence laid the foundation for the mathematical theory of probability.

During his work on probability, Pascal made a comprehensive study of the arithmetic triangle. Although this triangle of numbers was more than 600 years old, Pascal used it so ingeniously in his probability studies that it became known as Pascal's triangle. His work on the binomial coefficients that make up the triangle helped to lead **Isaac Newton** to his discovery of the general binomial theorem.

Pascal's last mathematical work was on the cycloid, the curve traced by the motion of a fixed point on the circumference of a circle rolling along a straight line. This curve was known as far back as the early 16th century, but in 1658, over the course of eight intensive days of effort, Pascal solved many of the remaining problems about the geometry of the cycloid. The "theory of indivisibles," a forerunner of integral calculus, allowed him to find the area and center of gravity of any segment of the cycloid. He also computed the volume and surface area of the solid of revolution formed by rotating the curve around a straight line. As was customary in those days, once having found these solutions, Pascal proposed a challenge to other mathematicians to solve a set of problems about the cycloid and offered two prizes. Neither prize was ever awarded and Pascal eventually published his own solutions.

The fruition of Pascal's correspondence with Fermat came in 1658, when he was trying to forget the pain of a toothache. Pascal came up with solutions to problems related to the curve cycloid, also known as roulette. He solved the problems using what became known as Pascal's arithmetic triangle (also known as the triangle of numbers) to calculate probability. His results were published in 1658 as *Lettre circulaire relative a la cycloïde*. This work played a major role in the development of calculus, both differential and integral. With this framework, areas and volumes could be calculated, and infinitesimal problems could be solved.

Though Pascal had been sickly all his life, his health became much worse later in 1658. His last project was developing a public transportation system of carriages in Paris in the first part of 1662. He did not live to see the system running. He died in his sister Gilberte's home on August 19, 1662, probably of a malignant stomach ulcer. Before his death, he may have parted company with his Jansenist friends.

Pascal was often underestimated in his time, and the bulk of his work was published posthumously. For example, *Traité de la pesanteur de la masse de l'air* was published in 1663, and *Traité du triangle arithmétique* in 1665, although these are only two of many.

## SELECTED WRITINGS BY PASCAL

*Essai sur les coniques,* 1640
*Lettre circulaire relative a la cycloïde,* 1658
*Traité du triangle arithmétique,* 1665

## FURTHER READING

Abbott, David, editor. *The Biographical Dictionary of Scientists: Mathematicians.* New York: Peter Bedrick Books, 1986, pp. 102–03.

Asimov, Issac. *Asimov's Biographical Encyclopedia of Science and Technology.* Second revised edition. New York: Doubleday & Company, Inc., 1982, pp. 130–32.

Bishop, Morris. *Pascal: The Life of Genius.* New York: Reynal & Hitchcock, 1936, 398 pp.

Daintith, John, Sarah Mitchell, Elizabeth Tootill, and Derek Gjersten, editors. *Biographical Encyclopedia of Scientists.* Volume 2. Second edition. Philadelphia: Institute of Physics Publishing, 1994, pp. 686–87.

Mesnard, Jean. *Pascal: His Life and Works.* New York: The Philosophical Library, Inc., 1952, 211 pp.

Taton, René. "Blaise Pascal." *Dictionary of Scientific Biography.* Volume X. Edited by Charles Coulston Gillispie. New York: Charles Scribner's Sons, 1970, pp. 330–42.

Williams, Trevor I., editor. *A Biographical Dictionary of Scientists.* Third edition. London: Adam & Charles Black, 1982, pp. 405–06.

*—Sketch by Annette Petruso*

---

# Moritz Pasch
## 1843–1930
### German geometer

Moritz Pasch's mathematical work provided one of the foundations for modern mathematics, especially geometry. In fact, Pasch was the first mathematician since **Euclid** who presented geometric elements in relationships defined by abstract, formal axioms. By doing this, he discovered that there were a certain amount of assumptions in Euclid's work that had been previously overlooked. In conjunction with this work, Pasch also developed Pasch's axiom, which deals with lines and triangles. He played a central role in the development of the axiomatic method that is a central feature of 20th century mathematics.

Pasch was born in Breslau, Germany (now Worclaw, Poland) on November 8, 1843. He attended the Elisabeth Gymnasium in Breslau, and completed his education in Berlin and Giessen. Pasch began studying chemistry then switched to mathematics at the behest of a mentor. He published his dissertation in 1865.

Pasch spent his teaching career at one institution, the University of Giessen. He began there in 1870 as a lecturer, then moved rapidly up the ranks to a full professorship in 1875. In 1888, Pasch became chair of the department. Pasch also served the university in administrative capacities, serving as dean in 1883, and, in the 1893–94 academic year, as rector (a position equivalent to that of president). Leading a relatively simple life, Pasch also married at one point and with his wife, had two daughters, though his wife and one daughter died young.

As a mathematician, Pasch spent the first 17 years of his career studying algebraic geometry, then moved onto foundations. He published his most important work, *Vorlesungen über Neuere Géométrie,* in 1882. In this treatise, Pasch addressed descriptive geometry from an axiomatic point of view. His work is distinguished because he eschewed physical interpretation of mathematical terms in favor of purely formal axioms. He also defined what is now known as Pasch's axiom, which states that in a triangle ABC, if a line enters through side AB and does not pass through the C, then it must leave the triangle either between B and C, or between C and A.

Pasch retired from the University of Giessen in 1911 to devote himself to mathematical research full time. In 1923 his work was recognized by his peers when Pasch received two honorary doctoral degrees from the University of Freiburg and the University of Frankfurt. He died seven years later in Bad Hamburg while on vacation on September 20, 1930.

## SELECTED WRITINGS BY PASCH

*Vorlesungen über Neuere Géométrie,* 1882

## FURTHER READING

Dehn, Max and Friedrich Engel. "Moritz Pasch." *Jahresberichte der Deutschen Mathematiker–Vereinigung* 44 (1934): 120–42.

*Giuseppe Peano*

Seidenberg, A. "Moritz Pasch." *Dictionary of Scientific Biography*. Volume X. Edited by Charles Coulston Gillispie. New York: Charles Scribner's Sons, 1970, pp. 343–45.

—*Sketch by Annette Petruso*

# Giuseppe Peano
## 1858-1932
### Italian logician

Giuseppe Peano served most of his adult life as professor of mathematics at the University of Turin. His name is probably best known today for the contributions he made to the development of symbolic logic. Indeed, many of the symbols that he introduced in his research on logic are still used in the science today. In Peano's own judgment, his most important work was in infinitesimal calculus, which he modestly described as "not . . . entirely useless." Some of Peano's most intriguing work involved the development of cases that ran counter to existing theorems, axioms, and concepts in mathematics.

Peano was born in Spinetta, near the city of Cuneo, Italy, on August 27, 1858. He was the second of five children born to Bartolomeo Peano and the former Rosa Cavallo. At the time of Peano's birth, his family lived on a farm about three miles from Cuneo, a distance that he and his brother Michele walked each day to and from school. Sometime later, the family moved to Cuneo to reduce the boys' travel time.

At the age of 12 or 13, Peano moved to Turin, some 50 miles (80 km) south of Cuneo, to study with his uncle, Michele Cavallo. Three years later he passed the entrance examination to the Cavour School in Turin, graduating in 1876. He then enrolled at the University of Turin and began an intensive study of mathematics. On July 16, 1880, he passed his final examinations with high honors and was offered a job as assistant to Enrico D'Ovidio, professor of mathematics at Turin. A year later he began an eight-year apprenticeship with another mathematics professor, Angelo Genocchi.

### Writes "Genocchi's" Textbook on Calculus

Peano's relationship with Genocchi involved one somewhat unusual feature. In 1883 the publishing firm of Bocca Brothers expressed an interest in having a new calculus text written by the famous Genocchi. They expressed this wish to Peano, who passed it on to his master in a letter of June 7, 1883. Peano noted to Genocchi that he would understand if the great man were not interested in writing the book himself and, should that be the case, Peano would complete the work for him using Genocchi's own lecture notes and listing Genocchi as author.

In fact, that was just Genocchi's wish. A little more than a year later, the book was published, written by Peano but carrying Genocchi's name as author. Until the full story was known, however, many of Genocchi's colleagues were convinced that Peano had used his master's name to advance his own reputation. As others became aware of Peano's contribution to the book, his own fame began to rise.

Peano's first original publications in 1881 and 1882 included an important work on the integrability of functions. He showed that any first-order differential equation of the form $y' = f(x, y)$ can be solved provided only that $f$ is continuous. Some of these early works also included examples of a type of problem of which Peano was to become particularly fond, examples that contradicted widely accepted and fundamental mathematical statements. The most famous of these, published in 1890, was his work on the space-filling curve.

### Derives the Space-Filling Curve

At the time, it was commonly believed that a curve defined by a parametric function would always be limited to an arbitrarily small region. Peano showed, however, that the two continuous parametric functions $x = x(t)$ and $y = y(t)$ could be written in such a way that as $t$ varies through a given interval, the graph of the curve covers every point within a given area. Peano's biographer Hubert Kennedy points out that Peano "was so proud of this discovery that he had one of the curves in the sequence put on the terrace of his home, in black tiles on white."

Peano's first paper on symbolic logic was an article published in 1888 in which he continued and extended the work of **George Boole** , the founder of the subject, and other pioneers such as Ernst Schröder, H. McColl, and **Charles S. Peirce** . His magnum opus on logic *Arithmetices principia, nova methodoexposita* (*The Principles of Arithmetic, Presented by a New Method*), was written about a year later. In it, Peano suggested a number of new notations including the familiar symbol $\in$ to represent the members of a set. He wrote in the preface to this work that progress in mathematics was hampered by the "ambiguity of ordinary language." It would be his goal, he said, to indicate "by signs all the ideas which occur in the fundamentals of arithmetic, so that every proposition is stated with just these signs." In succeeding pages, then, we find the introduction of now familiar symbols such as $\cap$ for "and," $\cup$ for "or," $\supset$ for "one deduces that," $\ni$ for "such that," and $\Pi$ for "is prime with." Also included in this book were Peano's postulates for the natural numbers, an accomplishment that Kennedy calls "perhaps the best known of all his creations."

In 1891 Peano founded the journal *Rivista di matematica* ( *Review of Mathematics* ) as an outlet for his own work and that of others; he edited the journal until its demise in 1906. He also announced in 1892 the publication of a journal called *Formulario* with the ambitious goal of bringing together all known theorems in all fields of mathematics. Five editions of *Formulario* listing a total of 4,200 theorems were published between 1895 and 1908.

By 1900 Peano had become interested in quite another topic, the development of an international language. He saw the need for the creation of an "interlingua" through which people of all nations—especially scientists—would be able to communicate. He conceived of the new language as being the successor of the classical Latin in which pre-Renaissance scholars had corresponded, a *latino sine flexione,* or "Latin without grammar." He wrote a number of books on the subject, including *Vocabulario commune ad latino-italiano-français-english-deutsch* in 1915, and served as president of the Akademi Internasional de Lingu Universal from 1908 until 1932.

While still working as Genocchi's assistant, Peano was appointed professor of mathematics at the Turin Military Academy in 1886. Four years later he was also chosen to be extraordinary professor of infinitesimal calculus at the University of Turin. In 1895 he was promoted to ordinary professor. In 1901 he resigned his post at the Military Academy, but continued to hold his chair at the university until his death of a heart attack on April 20, 1932.

Peano had been married to Carla Crosio on July 21, 1887. She was the daughter of the painter Luigi Crosio and was particularly fond of the opera. Kennedy points out that the Peanos were regular visitors to the Royal Theater of Turin where they saw the premier performances of Puccini's *Manon Lescaut* and *La Bohème*. The couple had no children. Included among Peano's honors were election to a number of scientific societies and selection as knight of the Crown of Italy and of the Orders of Saints Maurizio and Lazzaro.

## SELECTED WRITINGS BY PEANO

### Books

(Angelo Genocchi, Ghostwriter) *Calcolo differenziale e principii di calcolo integrale, pubblicato con aggiante dal Dr. Giuseppe Peano,* 1884
*Calcolo geometrico secondo l'Ausdehnungslehre di H. Grassmann, preceduto dalle operazioni della logica deduttiva,* 1888
*Arithmetices principia, nova methodo exposita,* 1889
*Gli elementi di calcolo geometrico,* 1891
*Lezioni di analisi infinitesimale,* 1893
*Notations de logique mathématique,* 1894
*Vocabulario commune ad linguas de Europa,* 1909
*Fundamento de Esperanto,* 1914
*Vocabulario commune ad latino-italiano-français-english-deutsch,* 1915

## FURTHER READING

### Books

Kennedy, Hubert C. "Giuseppe Peano," in *Dictionary of Scientific Biography*. Volume X. Edited by Charles Coulston Gillispie. New York: Charles Scribner's Sons, 1975, pp. 441–444.
*Selected Works of Giuseppe Peano*. Translated and edited with a biographical sketch and bibliography by Hubert C. Kennedy. Toronto: University of Toronto Press, 1973.

*—Sketch by David E. Newton*

*Karl Pearson*

# Karl Pearson
## 1857-1936
### English statistician

Karl Pearson is considered the founder of the science of statistics. He believed that a true understanding of human evolution and heredity required mathematical methods for analysis of the data. In developing ways to analyze and represent scientific observations, he laid the groundwork for the development of the field of statistics in the 20th century and its use in medicine, engineering, anthropology, and psychology.

Pearson was born in London, England, on March 27, 1857, to William Pearson, a lawyer, and Fanny Smith. At the age of nine, Karl attended the University College School, but was forced to withdraw at sixteen because of poor health. After a year of private tutoring, he went to Cambridge, where the distinguished King's College mathematician E. J. Routh met with him each day at 7 A.M. to study papers on advanced topics in applied mathematics. In 1875, he was awarded a scholarship to King's College, where he studied mathematics, philosophy, religion, and literature. At that time, students at King's College were required to attend divinity lectures. Pearson announced that he would not attend the lectures and threatened to leave the college; the requirement was dropped. Attendance at chapel services was also

required, but Pearson sought and was granted an exception to the requirement. He later attended chapel services, explaining that it was not the services themselves, but the compulsory attendance to which he objected. He graduated with honors in mathematics in 1879.

After graduation, Pearson traveled in Germany and became interested in German history, religion and folklore. A fellowship from King's College gave him financial independence for several years. He studied law in London, but returned to Germany several times during the 1880s. He lectured and published articles on Martin Luther, Baruch Spinoza, and the Reformation in Germany, and wrote essays and poetry on philosophy, art, science, and religion. Becoming interested in socialism, he lectured on Karl Marx on Sundays in the Soho district clubs of London, and wrote hymns for the Socialist Song Book. Pearson was given the name Carl at birth, but he began spelling it with a "K," possibly out of respect for Karl Marx.

During this period, Pearson maintained his interest in mathematics. He edited a book on elasticity as it applies to physical theories and taught mathematics, filling in for professors at Cambridge. In 1884, at age twenty-seven, Pearson became the Goldsmid Professor of Applied Mathematics and Mechanics at University College in London. In addition to his lectures in mathematics, he taught engineering students, and showed them how to solve mathematical problems using graphs.

In 1885, Pearson became interested in the role of women in society. He gave lectures on what was then called "the woman question," advocating the scientific study of questions such as whether males and females inherit equal intellectual capacity, and whether, in the future, the "best" women would choose not to bear children, leaving it to "coarser and less intellectual" women. He joined a small club which met to discuss questions of morality and sex. There he met Maria Sharpe, whom he married in 1890. They had three children, Egon, Sigrid, and Helga. Maria died in 1928, and Pearson married Margaret V. Child, a colleague at University College, the following year.

### Develops Statistical Methods to Study Heredity

Pearson was greatly influenced by Francis Galton and his 1889 work on heredity, *Natural Inheritance.* Pearson saw that there often may be a connection, or correlation, between two events or situations, but in only some of these cases is the correlation due not to chance but to some significant factor. By making use of the broader concept of correlation, Pearson believed that mathematicians could discover new knowledge in biology and heredity, and also in psychology, anthropology, medicine, and sociology.

An enthusiastic young professor of zoology, W. F. R. Weldon, came to University College in 1891, further influencing Pearson's direction. Weldon was interested in Darwin's theory of natural selection and, seeing the need for more sophisticated statistical methods in his research, asked Pearson for help. The two became lunch partners. From their association came many years of productive research devoted to the development and application of statistical methods for the study of problems of heredity and evolution. Pearson's goal during this period was not the development of statistical theory for its own sake. The result of his efforts, however, was the development of the new science of statistics.

Remaining at the University College, Pearson became the Gresham College Professor of Geometry in 1891. His lectures for two courses there became the basis for a book, *The Grammar of Science,* in which he presented his view of the nature, function, and methods of science. He dealt with the investigation and representation of statistical problems by means of graphs and diagrams, and illustrated the concepts with examples from nature and the social sciences. In later lectures, he discussed probability and chance, using games such as coin tossing, roulette, and lotteries as examples. He described frequency distributions such as the normal distribution (sometimes called the bell curve because its graph resembles the shape of a bell), skewed distributions (for which the graphed design is not symmetrical), and compound distributions (which might result from a mixture of the two). Such distributions represent the occurrence of variables such as traits, events, behaviors, or other incidents in a given population, or in a sample (subgroup) of a population. They can be graphed to illustrate where each subject falls within the continuum of the variable in question.

Pearson introduced the concept of the "standard deviation" as a measure of the variance within a population or sample. The standard deviation statistic refers to the average distance from the mean score for any score within the data set, and therefore suggests the average amount of variance to be found within the group for that variable. Pearson also formulated a method, known as the chi-square statistic, of measuring the likelihood that an observed relation is in fact due to chance, and used this method to determine the significance of the statistical difference between groups. He also developed the theory of correlation and the concept of regression analysis, used to predict the research results. His correlation coefficient, also known as the Pearson $r$, is a measure of the strength of the relationship between variables and is his best-known contribution to the field of statistics.

Between 1893 and 1901 Pearson published 35 papers in the *Proceedings* and the *Philosophical Transactions* of the Royal Society, developing new statistical methods to deal with data from a wide range of sources. This work formed the basis for much of the later development of the field of statistics. He was elected to the Royal Society in 1896, was awarded the Darwin Medal in 1898, and, in 1903, was elected an Honorary Fellow of King's College and received the Huxley Medal of the Royal Anthropological Institute.

**Establishes Journal and Compiles Statistical Tables**

In 1901, Pearson helped found the journal *Biometrika* for the publication of papers in statistical theory and practice. He edited the journal until his death. His research often required extensive mathematical calculation, which was carried out under his direction by students and staff mathematicians in his biometric laboratory. Since high-speed electronic computers had not yet been invented, performing the calculations by hand was tedious and time-consuming. The laboratory staff produced tables of calculations which Pearson made available to other statisticians through *Biometrika,* and later as separate volumes. Access to these tables made it possible for others to carry out statistical research without the support of a large staff, and, again, proved to be a valuable contribution to the early development of the field of statistics.

Pearson became the Galton Professor of Eugenics in 1911, and headed a new department of applied statistics as well as the biometric laboratory and a eugenics laboratory, established to study the genetic factors affecting the physical and mental improvement or impairment of future generations. During World War I, Pearson's staff served Britain's interest by preparing charts showing employment and shipping statistics, investigating stresses in airplane propellers, and calculating gun trajectories. From 1911 to 1930, Pearson produced a four-volume biography of Francis Galton. In 1925, he founded the journal *Annals of Eugenics,* which he edited until 1933. In 1932, Pearson was the first foreigner to be awarded the Rudolf Virchow Medal by the Anthropological Society of Berlin. He retired in 1933 at age 77, and received an honorary degree from the University of London in 1934. Pearson died on April 27, 1936, in Coldharbour, Surrey.

Pearson produced more than three hundred published works in his lifetime. His research focused on statistical methods in the study of heredity and evolution but dealt with a range of topics, including albinism in people and animals, alcoholism, mental deficiency, tuberculosis, mental illness, and anatomical comparisons in humans and other primates, as well as astronomy, meteorology, stresses in dam construction, inherited traits in poppies, and variance in sparrows' eggs. Pearson was described by G. U. Yule as a poet, essayist, historian, philosopher, and statistician, whose interests seemed limited only by the chance encounters of life. Colleagues remarked on his boundless energy and enthusiasm. Although some saw him as domineering and slow to admit errors, others praised him as an inspiring lecturer and noted his care in acknowledging the contributions of the members of his lab group. For Pearson, scientists were heroes. The walls of his laboratory contained quotations from Plato, **Blaise Pascal**, Huxley and others, including these words from Roger Bacon: "He who knows not Mathematics cannot know any other Science, and what is more cannot discover his own Ignorance or find its proper Remedies."

## SELECTED WRITINGS BY PEARSON

*The Ethic of Freethought*, 1888
*The Grammar of Science*, 1892
*The Chances of Death and Other Studies in Evolution*, 1897
*Tables for Statisticians and Biometricians*, 1914
*The Life, Letters, and Labours of Francis Galton* four volumes, 1914, 1924, 1930

## FURTHER READING

### Books

Froggatt, P. *Modern Epidemiology: The Pearsonian Legacy.* New Lecture Series No. 54. Belfast: The Queen's University, 1970.

Haldane, J. B. S. *Karl Pearson, 1857–1957, The Centenary Celebration at University College, London, 13 May 1957.* (Privately issued by Biometrika Trustees, 1958.)

Pearson, E. S. *Karl Pearson: An Appreciation of Some Aspects of His Life and Work.* Cambridge, England: Cambridge University Press, 1938.

### Periodicals

Camp, Burton H., "Karl Pearson and Mathematical Statistics." *Journal of the American Statistical Association* (December 1933): 395–401.

"Karl Pearson." *Obituary Notices of Fellows of the Royal Society* 2, number 5 (December 1936): 72–110.

Pearl, Raymond. "Karl Pearson, 1857–1936." *Journal of the American Statistical Association* (December 1936): 653–664.

—*Sketch by C. D. Lord*

*Charles Sanders Peirce*

# Charles Sanders Peirce
## 1839–1914
### American logician

Charles Sanders Peirce remains one of the enigmatic figures in the history of American science. He made substantial contributions to a number of fields, especially logic, but his use of unusual terminology makes it difficult to appraise much of his work. As the project of publishing his collected writings continues, it may become possible to do justice to this many–sided thinker.

Charles Sanders Peirce was born on September 10, 1839 in Cambridge, Massachusetts. His father, Benjamin Peirce, was not only a professor of mathematics at Harvard University but also perhaps the most accomplished American mathematician of his generation. Peirce's early education outside the home was at various private schools in Boston and Cambridge, and he showed an interest in puzzles and chess problems. By the age of 13, he had read Archbishop Whately's *Elements of Logic*, perhaps a hint of the interests to come. Peirce entered Harvard in 1855, and the results were not impressive. Although he succeeded in graduating four years later, it was with a class rank of 71 out of 91. Upon graduation Peirce obtained a temporary position with the United States Coast Survey, which was to be his employer for most of his working life. His contributions to geodesy were many, and his service to the Coast Survey have been recognized with a memorial.

In the early 1860s Peirce studied under Louis Agassiz at Harvard, but his work with the Coast Survey proved to have had its benefits. He had become a regular aide to the Survey in 1861, which resulted in his exemption from military service. He was an assistant to the Coast Survey from 1867 to 1891, but that did not prevent his continuing researches in other areas. In particular, not only did he observe a solar eclipse in the United States in 1869, a year later he led an expedition to Sicily to observe a solar eclipse from a position that he had selected.

### Broadens the Scope of Logic

Peirce had developed a technical competence in mathematics that came in handy when he turned to logic. As an example of a result in mathematics itself, he succeeded in showing that of linear associative algebras (a subject to which his father had devoted a book) the only three that had a uniquely defined operation of division were real numbers, complex numbers, and the quaternions of Sir **William Rowan Hamilton**. Perhaps the most significant innovation he made in logic was the extension of Boolean algebra to include the operation of inclusion. The most widely influential treatise on the algebra of logic was produced by the German mathematician Ernst Schröder beginning in 1890, and he displayed a detailed familiarity with Peirce's work. In fact, had Peirce made the effort to produce a systematic account of the subject before Schröder, it would be easier to measure the importance of Peirce's contributions.

One of the factors that played a role in Peirce's interest in logic and its algebraic expression was his having taken a position in 1879 at Johns Hopkins University in Baltimore. During the five years that he worked at the university, he stayed on at the Coast Survey. As Nathan Houser remarked in an article about Peirce, "during those years Peirce was a frequent commuter on the B & O Railroad between Baltimore and Washington." Peirce's first paper on the algebra of logic was published in the *American Journal of Mathematics* in 1880. The period 1880 to 1885 saw Peirce's introduction of two ideas to mathematical logic: truth–functional analysis and quantification theory. Truth–functional analysis is the ancestor of the technique used by the Austrian philosopher Ludwig Wittgenstein to serve as the basis for logic in his *Tractatus Logico–philosophicus*. Quantification theory was at the heart of the logical apparatus introduced by **Gottlob Frege** for the reduction of mathematics to logic. It is difficult to imagine two more crucial contributions at the time, although Peirce's share of the recognition suffers by virtue of the scattered nature of his contributions.

### Develops an Alternative Philosophy of Science

Peirce was never one to limit his scientific investigations to a single discipline. In 1879, he determined the length of the meter based on a wavelength of light. This provided a natural alternative to the standard meter bar on deposit in Paris. Three years later, he worked on a mathematical study of the relationship between variations in gravity at different points on the Earth's surface and the shape of the Earth. Better known is his role in serving as an advocate of a philosophy of science called pragmatism. Peirce's pragmatism was heavily dependent on the idea of inference to the best explanation. In other words, what existed was determined by what was needed for successful scientific practice at the time. While neither realists nor their opponents were happy with Peirce's position, it has continued to offer an alternative. In particular, philosophers of science with an inclination to take the history of science seriously find Peirce's approach one of the few that take change in one's scientific models to heart.

In light of Peirce's contributions in so many areas of science and mathematics the puzzle remains of why he was unable to secure an academic position commensurate to his abilities. One factor may have been domestic: he married Harriet Melusina Bay in 1862 and was divorced from her in 1883, the year of his second marriage (to Juliette Froissy of Nancy, France). Peirce and his first wife had been separated since 1876 and public sentiment was on her side. More generally, however, Peirce's personality tended to go between extremes, and it was difficult for others to adjust to his mood swings. He was quick to enter into disputes (and frequently with the wrong party) and was easily influenced by others.

Peirce spent his later years in Milford, Pennsylvania, removed from the centers of intellectual life. He had been asked to resign from the Coast Survey in 1891 and for the rest of his life his income was uncertain, despite prodigious periods of writing. Even his philosophy, to which he continued to devote his best efforts, was neglected, if only as a result of his remoteness from university settings. He died in Milford on April 19, 1914, having made contributions across the intellectual map but more to the benefit of the discipline than his own.

## SELECTED WRITINGS BY PEIRCE

*Collected Papers*, volumes 1–6. Edited by Charles Hartshorne and Paul Weiss, 1935
*Collected Papers*, volumes 7–8. Edited by Arthur W. Burks, 1958
*Writings of Charles S. Peirce*, 1982–.

## FURTHER READING

Brent, Joseph. *Charles Sanders Peirce: A Life.* Bloomington: Indiana University Press, 1993.

Eisele, Carolyn. "Charles Sanders Peirce," in *Dictionary of Scientific Biography.* Volume X. Edited by Charles Coulston Gillispie. New York: Charles Scribner's Sons, 1973, pp. 482–488.

Hookway, Christopher. *Peirce.* London: Routledge and Kegan Paul, 1985.

Houser, Nathan. "Peirce and the Law of Distribution," in *Perspectives on the History of Mathematical Logic.* Edited by Thomas Drucker. Boston: Birkhauser, 1991, pp. 10–32.

Mlsak, Cheryl I. *Truth and the End of Inquiry.* Oxford: Oxford University Press, 1985.

Weiss, Paul. "Charles Sanders Peirce," in *Dictionary of American Biography.* Volume 7. Edited by Dumas Malone. New York, Charles Scribner's Sons, 1962, pp. 398–403.

*—Sketch by Thomas Drucker*

# Rózsa Péter
## 1905-1977
### Hungarian logician

Rózsa Péter was one of the early investigators in the field of recursive functions, a branch of mathematical logic. Recursive functions are those mathematical functions whose values can be established at every point, for whole numbers one and above. These functions are used to study the structure of number classes or functions in terms of the complexity of the calculations required to determine them, and have useful applications to computers and other automatic systems. Péter wrote two books and numerous papers on recursive functions, which are related to **Alan Turing**'s theory of algorithms and machines and to **Kurt Gödel**'s undecidability theorem of self-referential equations. Péter also wrote a popular treatment of mathematics, *Playing with Infinity*, which was translated into 14 languages. A teacher and teacher-training instructor before her appointment to a university post, Péter won national awards for her contributions to mathematics education and to mathematics.

Péter was born in Budapest on February 17, 1905. She received her high school diploma from Mária Terézia Girls' School in 1922, then entered the university in Budapest to study chemistry. Although her father, an attorney, wanted her to stay in that field, Péter changed to the study of mathematics. One of her classmates was László Kalmár, her future teacher and colleague. Péter graduated from the university in 1927, and for two years after graduation she had no permanent job, but tutored privately and took temporary teaching assignments.

In 1932, Péter attended the International Mathematics Conference in Zurich, where she presented a lecture on mathematical logic. She published papers on recursive functions in the period 1934 to 1936, and received her Ph.D. summa cum laude in 1935. In 1937, Péter became a contributing editor of the *Journal of Symbolic Logic*. Péter lost her teaching position in 1939 due to the Fascist laws of that year; Hungary was an ally of Nazi Germany and held similar purges of academics. Nevertheless, she published papers in Hungarian journals in 1940 and 1941.

In 1943, Péter published her book, *Playing with Infinity*, which described ideas in number theory, geometry, calculus and logic, including Gödel's undecidability theory, for the layman. The book, many copies of which were destroyed by bombing during World War II, could not be distributed until 1945. The war claimed the life of Péter's brother, Dr. Nicholas Politzer, in 1945, as well as the lives of her friend and fellow mathematician, Pál Csillag, and her young pupil, Káto Fuchs, who had assisted Péter with *Playing with Infinity*.

In the late 1940s, Péter taught high school and then became Head of the Mathematics Department of the Pedological College in Budapest. She also wrote textbooks for high school mathematics. In the 1950s, Péter published further studies of recursive functions; her 1951 book, *Recursive Functions*, was the first treatment of the subject in book form and reinforced her status as "the leading contributor to the special theory of recursive functions," as S. C. Kleene observed in the *Bulletin of the American Mathematical Society*.

Péter was appointed Professor of Mathematics at Eötvös Loránd University in Budapest in 1955, where she taught mathematical logic and set theory. The official publication of *Playing with Infinity* occurred in 1957. In this version, she wrote in the preface, she tried "to present concepts with complete clarity and purity so that some new light may have been thrown on the subject even for mathematicians and certainly for teachers."

In the 1960s and 1970s, Péter studied the relationship between recursive functions and computer programming, in particular, the relationship of recursive functions to the programming languages Algol and Lisp. Péter retired in 1975. She continued her research, however, publishing *Recursive Functions in Computer Theory* in 1976.

Péter's awards included the Kossuth Prize in 1951 for her scientific and pedological work, and the State Award, Silver Degree in 1970 and Gold Degree in 1973. Péter was a member of the Hungarian

Academy of Sciences and was made honorary President of the János Bolyai Mathematical Association in 1975. Interested in literature, film and art as well as mathematics, Péter translated poetry from German and corresponded with the literary critic Marcel Benedek. She noted in *Playing with Infinity* that her mathematical studies were not so different from the arts: "I love mathematics not only for its technical applications, but principally because it is beautiful; because man has breathed his spirit of play into it, and because it has given him his greatest game—the encompassing of the infinite." Péter died on February 17, 1977.

## SELECTED WRITINGS BY PÉTER

*Playing with Infinity: Mathematics for Everyman.* Translated by Z. P. Dienes, 1962
*Recursive Functions.* Third revised edition, translated by István Földes, 1967
*Recursive Functions in Computer Theory.* Translated by I. Juhász, 1981

## FURTHER READING

### Books

"Algorithms and Recursive Functions." *Mathematics at a Glance.* Second edition. New York: Van Nostrand Reinhold, 1989, pp. 340–342.

### Periodicals

Császár, Akos. "Rózsa Péter: February 17, 1905-February 16, 1977." *Matematikai Lapok* (published in Hungarian), 25 (1974): 257–258.
Dömölki, Bálint, et al. "The Scientific Work of Rózsa Péter." *Matematikai Lapok* (published in Hungarian), 16 (1965): 171–184.
Kleene, S.C. "Rekursive Funktionen." *Bulletin of the American Mathematical Society* (March 1952): 270–272.
Nelson, D. "Rózsa Péter, Rekursive Funktionen." *Mathematical Reviews* (1952): 421–422.
"Playing with Infinity': Telegraphic Reviews." *American Mathematical Monthly* 84, (February 1977): 147.
Robinson, Raphael M. "Rózsa Péter. Rekursive Funktionen." *Journal of Symbolic Logic* 16 (1951): 280–282.
Ruzsa, Imre Z. and János Urbán, "In Memoriam Rózsa Péter." *Matematikai Lapok* (published in Hungarian), 26 (1975): 125–137.
Sudborough, I. Hal. "Rózsa Péter, Rekursive Funktionen in der Komputer-Theorie." *Mathematical Reviews* 55, no. 6926.

Turán, Pál. "To the Memory of Mathematician Victims of Fascism." *Matematikai Lapok* (published in Hungarian), 26 (1975): 259–263.

### Other

Kocsor, Klára, notes on articles in Hungarian, January, 1994.

—*Sketch by Sally M. Moite*

# Bartholomeo Pitiscus
## 1561–1613
### German trigonometricist and theologist

A theologist by trade and a strong influence in the Calvinist government of his time, Bartholomeo Pitiscus also essentially coined the term "trigonometry." The term comes from the title of his book *Trigonometria*, which consists of three parts, including five chapters devoted to plane and spherical geometry, now known as plane and spherical trigonometry. In addition to its contribution to mathematical nomenclature, the text is highly regarded and is especially noteworthy because in it Pitiscus used all six of the trigonometric functions.

Pitiscus was born August 24, 1561, in Grünberg, Silesia (now Zielona Góra, Poland), but the details of his upbringing and mathematical education are not known. He studied Calvinist theology at Zerbst and Heidelberg, and, in 1584, was appointed to tutor 10–year–old Friedrich der Aufichtige, known as Frederick IV, elector Palatine of the Rhine. Pitiscus was subsequently appointed court chaplain at Breslau and court preacher to Frederick. Frederick took control of the government of the Palatinate after his uncle, John Casimir, died in 1592 and Pitiscus remained influential in the government's defense of Calvinism under his former pupil.

*Trigonometria: sive de solutione triangulorum tractatus brevis et perspicuus* was published in 1595 as the final section of A. Scultetus' *Sphaericorum libri tres methodicé conscripti et utilibus scholiis expositi.* The work was published in revised form in 1600 under the title *Trigonometriae sive de dimensione triangulorum libri quinque.*

*Trigonometriae* is divided into three sections, the first of which contains five books on plane and spherical trigonometry. The second section provides tables for all six of the trigonometric functions, carrying them out to five or six decimal places. The third section of the text, "Problem varia" contains 10 books of problems in geodesy, measuring of heights, geography, geometry, and astronomy. An enlarged edition containing the original first and third sections of *Trigonometriae* was published in 1609 with expanded tables at the end of the text. A third edition was published in 1612 with an expanded "Problem varia" section including problems related to architecture. The original *Trigonometria* was translated into English by R. Handson in 1614.

Further details of Pitiscus' life are not known. He died on July 2, 1613, in Heidelberg, Germany.

## SELECTED WRITINGS BY PITISCUS

*Trigonometria*, 1595; revised in 1600 as *Trigonometriae*

## FURTHER READING

### Books

Zeller, M.C. *The Development of Trigonometrie from Regiomontanus to Pitiscus* (Ph.D. thesis, University of Michigan, 1944). Ann Arbor, MI: Lithoprinted by Edwards Brothers 1944, pp. 102–04.

*—Sketch by Kristin Palm*

# Vera Pless
## 1931–
### American combinatorial analyst

Vera Pless took her early interest in the pure algebra of ring theory and applied it to the combinatorics of error–correcting computer codes in the developing field of computer science. This line of inquiry cuts to the quick of a computer's weakness—faulty data transmission. As one of the first lay people to recognize this challenge, Pless naturally became a leader in what is now being called discrete mathematics. As projects with computers demand more fidelity (as in compact discs) and greater scale (as with the Sojourner Mars mission), coding theory becomes more central to all technologies.

Vera Pless was born Vera Stepen in Chicago on March 5, 1931. Her parents were Russian immigrants who settled in the predominantly–Jewish west side of the city. A friend of the family, who was a graduate student at the University of Chicago, taught Pless calculus when she was 12 years old. He saw a bright kid with a future in math, but may not have realized that she would not welcome a scholarship to the College at the University of Chicago. The fifteen–year–old Pless wanted to play the cello, but acceded to her father's wish that she study a practical subject.

Pless took some inspiration from the woman some call the founder of modern algebra, **Emmy Noether.** "I passed my master's exam two weeks after I got married," in 1952, she states plainly enough in her autobiographical essay. Although her husband was pursuing a Ph.D. in physics, Pless accepted a research associateship at Northwestern University to pursue her own Ph.D. They both relocated to Boston while she was completing her thesis on ring theory and defended her thesis two weeks before giving birth in 1957.

Unfortunately, at that time Boston academia did not welcome a female Ph.D. with two children and a working husband. "I heard people say outright, 'I would never hire a woman'," Pless remembers. However, the nearby Air Force Research Labs in Cambridge were very much in need of her algebraic skills. Error correcting codes were of most immediate use to the military's intelligence and security programs. Pless welcomed the chance to work as part of a group and found the workshops with Andrew Gleason of Harvard University beneficial as well. Later on, she was allowed maternity leave for her third child. Pless' work there included designing programs to compare tracks left behind in experimental bubble chambers.

After the U.S. Congress passed the Mansfield Amendment banning basic research within the defense sector of the government, Pless returned to the possibility of an academic career. It was a struggle to get back into the mode of papers and lecturing. Nonetheless, three years as a research associate at M.I.T., working on such subjects as encryption, led her in 1975 to a position at the University of Illinois, back in her hometown.

### Greedy Codes

Pless has worked as a visiting professor, researcher, and scientist at The Technion in Haifa, the Argonne, and a number of American universities including M.I.T. The National Science Foundation (NSF) funded her stay at Caltech during 1985–86 with its visiting professorship for women grant, only the most recent example of NSF aid to Pless' career. There, she organized bimonthly coding seminars over the academic year. Over the years Pless has sat on several thesis committees and personally directed 15 graduate and postdoctoral students, sometimes three at once.

Such a busy schedule has not prevented Pless from figuring out new ways her students can use technology and innovative means of studying computer programs, as Laura Monroe attests. As Monroe's thesis advisor in 1995, Pless encouraged her to apply her ideas about "greedy codes" in computer science to some of the most powerful computers available, including a CRAY–C90 (UNICOS). "As an advisor," Monroe concluded, "she was always interested in her students' new ideas and enthusiastic about their progress."

The dual responsibilities of teaching and research have called for innovation on Pless' part. She started the CAMAC computer system at M.I.T. and brought it out west with her when she relocated in 1975, where she continues to extend and develop it with associate Thom Grace. The system has gained a measure of prominence because of its use in countless projects and being mentioned in over 20 publications. CAMAC I proved so popular that CAMAC II was started by another colleague, Jeff Leon, with Pless' aid.

### The All–Professor Team

Pless writes, reviews, referees, and edits extensively, having held concurrent editorships in several publications during the 1980s and 1990s. Some of this is borne of necessity, as her field is still very new. With Joel Berman, she designed the first college course in discrete mathematics based on their lecture notes. Pless' *Introduction to The Theory of Error-Correcting Codes* carries the highest rating in the Mathematical Association of America's Library Recommendations for Undergraduate Mathematics. From 1993 to 1994, the U.S. Department of Education retained her for a paid position as part of their Algebra Initiative.

Pless' professional service includes advisory and elected positions with a variety of mathematics and science groups, helping to oversee projects of the National Science Foundation, the National Research Council, and to review the funding of Fulbright scholars and National Security Agency proposals. As a member of the American Mathematical Society, Pless has served on committees concerning the human rights of mathematicians, academic freedom, and employment security. She is often invited and fully funded to speak at various professional meetings, seminars, and colloquia. Not all of Pless' honors are so formal. In 1993, the *Chicago Tribune* included her in their "All-Professor Team of Academic Champions."

By the middle of 1997, Pless' list of publications on aspects of information theory and code design numbered over 100, written either singly or with such illustrious figures as John H. Conway and **John Thompson**.

## SELECTED WRITINGS BY PLESS

### Books

*An Introduction to The Theory of Error-Correcting Codes*, 1983; revised edition 1990

(Editor) *Handbook on Coding* (in publication)

### Periodicals

"Continuous Transformation Ring of Biorthogonal Bases Spaces." *Duke Mathematics Journal* 25, MR 20, No. 2630 (1958).

"Power Moment Identities on Weight Distributions in Error Correcting Codes." *Information and Control*, MR 41, No. 6614 (1963).

"Attitudes About and of Professional Women: Now and Then," in *Career Guidance for Women Entering Engineering: Proceedings of an Engineering Foundation Conference*. Edited by Nancy Fitzroy. August 1973.

### Other

"Short Autobiography of Vera Pless." *Pless Papers*. University of Illinois at Chicago, January 1997.

## FURTHER READING

### Periodicals

Birman, Haimo, Landau, Srinivasan, Pless, and Taylor. "In Her Own Words: Six Mathematicians Reflect on Their Lives and Careers." *Notices of the American Mathematical Society* 38 No. 7 (1991): pp. 702–706.

### Other

"F. Jessie MacWilliams." *A Survey of Coding Theory*. http://www.math.neu.edu/awm/noether-brochure/MacWilliams80.html

"MAA Prizes Presented in Orlando." *Notices of the AMS* (April, 1996): http://e-math.ams.org/notices/199604/comm–maa.html

Monroe, Laura, email interview with Jennifer Kramer conducted July 10–18, 1997.

*—Sketch by Jennifer Kramer*

# Julius Plücker
## 1801–1868
### German geometer and physicist

One of the outstanding figures of 19th–century science, Julius Plücker is remembered for his seminal work in analytical geometry, and for his discoveries in physics. He is regarded as the co–discoverer of cathode rays (with his student, J. W. Hittorf), and he is credited with laying the foundation of spectroscopy.

*Julius Plücker*

## Perfects Analytical Geometry

Plücker studied in Bonn, Heidelberg, and Paris, eventually securing a teaching position at the University of Bonn. A professor of mathematics from 1836 to 1847, he taught physics from 1847 until 1868. The early decades of the 19th century were an auspicious time for mathematicians, like Plücker, who directed their creative energy to geometry. Firstly, German mathematicians greatly benefitted from the foundation of *Journal fur die reine und angewandte Mathematik* ( *Journal for Pure and Applied Mathematics*) in 1826. Founded by August Leopold Crelle, an engineer, the journal, which strongly supported research in pure mathematics, provided Plücker with the means to make his ideas known to the European scientific community. In 1827, **Karl Gauss** established the field of differential geometry, which studies curved surfaces, or surfaces in three–dimensional space. Crelle's journal and Gauss's research in geometry provided a favorable intellectual atmosphere for Plücker, who focused on analytic geometry.

By choosing to work in analytic geometry Plücker entered a field of age–old controversy, for mathematicians, particularly in Germany, had strong opinions about geometry: the proponents of synthetic geometry believed that geometry should be practiced with straightedge and compass only, without recourse to algebra. It is not surprising, therefore, that these "pure" geometers questioned the algebraic methodology of analytic geometry. Unlike synthetic geometry, analytic geometry fully utilizes algebra. Thus, in

analytical geometry, the plane is defined by a coordinate system, and a point has two coordinates (an $x$ coordinate and a $y$ coordinate). Lines are defined by equations (a straight line is represented by the linear equation $ax + by = c$), while quadratic equations are used for circles (e.g., $x^2 + y^2 + ax + by + c = 0$, where $a$ and $b$ are the coordinates of the circle's center, and $c$ determines the radius), ellipses, parabolas, and hyperbolas. The general two–variable function used to represent a curve is $f(x, y) = 0$.

Plücker's first book, the first volume of *Analytisch–geometrische Entwicklungen* (*Developments in Analytical Geometry*) was published in 1828; the second volume appeared in 1831. In his book, Plücker essentially demonstrated that analytical geometry was capable of replicating the results of pure geometry. Furthermore, in the second volume, Plücker introduced a new system of coordinates that had been discovered, independently, by **August Möbius**, **Karl Wilhelm Feuerbach**, and E. Bobillier. What made this new system, also called homogenous coordinates, revolutionary was the fact that it added a third coordinate to the Cartesian two. Unlike his predecessors, Plücker defined the three coordinates, $x$, $y$, and $t$ as the distances of a point within a triangle to the sides of that triangle. However, by introducing a third coordinate, Plücker effectively transcended the traditional concept of space: one could not count on a point having a single set of coordinates. If, for example, one imagines the third coordinate as $t = time$, it is clear that a point $p$ has more than one set of homogenous coordinates, such as $x$, $y$, $t$, as well as $fx$, $fy$, $ft$, where $f$ represents any factor.

Furthermore, the introduction of the third coordinate $t$ permits the transformation of Cartesian equations into trilinear equations. For example, if a curve is represented as $f(x, y) = 0$ in the Cartesian system, in Plücker's system the equation for that curve becomes $f(x / t, y / t) = 0$. Because zero is not permissible as a divisor, one would assume, in the new equation, that $t \neq 0$. As long as this is the case, it is safe to assume that, in a plane, as in the Cartesian system, an infinity of infinite lines can be drawn through a point $p$. These lines radiate in all possible directions, forming a solar image. Consequently, points on an infinite line, or "points at infinity," can lie on any line going through $p$. Plücker, however, knowing that the coordinates $x$, $y$, 0 cannot be translated into any pair of Cartesian coordinates, nevertheless introduced the seemingly impossible idea of $t = 0$. $T$ thus becomes a line, sort of like $x$ and $y$, but with one essential distinction: $t$, in Plücker's system, is the *only* line containing points at infinity. The sun–like image of lines projecting from a center is replaced by the image of a single line; instead of being everywhere, points of infinity are on line $t$. However, the many lines from the Cartesian solar image have not disappeared: in Plücker's system, they are all represented by a single line.

In 1829, Plücker introduced another revolutionary idea into geometry: he replaced the point by the straight line as the fundamental element in geometry. As Howard Eves explains, "A point now instead of having coordinates, possesses a linear equation—namely, the equation satisfied by the coordinates of all the lines passing through that point. The double interpretation of a pair of coordinates as either point coordinates or line coordinates and of a linear equation as either the equation of a line or the equation of a point furnishes the basis of Plücker's analytical proof of the principle of duality of plane projective geometry." If we remember that $t$ can be interpreted as $t = time$, the idea of interchanging a straight line and a point appears less outlandish. Indeed, if $x$ is a point, $xt$ can easily be imagined as a line. Plücker's geometry, more so than pure geometry, provided a solid base for work in projective geometry. However, in Germany, analytic geometry, despite Plücker's brilliant work, failed to receive the recognition it deserved, probably owing to the tremendous authority of **Jakob Steiner**, the proponent of synthetic geometry, whose allies included, surprisingly, **Karl Jacobi**, who was not in the synthetic geometry camp.

### Turns to Physics

Unwilling to work in Steiner's shadow, Plücker ceased his geometrical research without fully developing his geometry into a system of more than three dimensions. Leaving his work unfinished, he took over a professorship in physics. Plücker's work in physics initially focused on magnetism, particularly the effect of magnetic fields on objects. This work was a prelude to one of the great discoveries of 19th-century physics—cathode rays. In 1854, the well-known maker of scientific instruments Johann Heinrich Wilhelm Geissler perfected his vacuum tube. Placing electrodes in the vacuum tube, one at each end, Plücker drove electric current between the two electrodes. As a result of this experiment, which Plücker performed in 1858, a green glow appeared in the tube. Although lacking empirical proof, Plücker correctly attributed the glow to cathode-rays.

In addition, Plücker conducted pioneering research in spectroscopy, finding that chemical substances had individual spectral lines, thus opening the field of spectroscopic chemical analysis. In 1862, he discovered that a chemical element's spectral pattern may be affected by temperature changes. Plücker's work in physics, particularly his discovery that cathode-rays are affected by an electromagnetic field, opened new vistas in this scientific discipline, inspiring research that resulted in J. J. Thomson's discovery of the electron.

Plücker returned to mathematical research in 1865. He completed his line–based geometry of four–dimensional space and also developed an appropriate notation system. Plücker's notation marks six homogenous coordinates as $l$, $m$, $n$, $\lambda$, $\mu$, $\nu$, where $l\lambda + m\mu + n\nu = 0$. According to René Taton, the "originality and elegance of the new notation drew the attention of many mathematicians to the importance of ruled geometry, the properties of congruences and complexes, and their possible applications to the solution of certain differential equations, and to problems in geometrical optics, statics, and vector analysis."

Underestimated by his German colleagues, Plücker's brilliant contributions to science earned him the admiration of his English colleagues, who awarded him the Copley Medal in 1868.

## SELECTED WRITINGS BY PLÜCKER

*Analytisch–geometrische Entwicklungen*, 2 vols., 1828–31
*Neue Geometrie des Raumes, gegründet auf der geraden Linien als Raumelement*, 1868–69
*System der Geometrie des Raumes in neuer analytischer Behandlungsweise*, 1846
*Theorie der algebraischen Kurven*, 1939

## FURTHER READING

Boyer, Carl B. *A History of Mathematics*. Second edition. Revised by Uta C. Merzbach. New York: John Wiley & Sons, 1991.
Eves, Howard. *Foundations and Fundamental Concepts of Mathematics*. Third edition. Mineola, NY: Dover, 1997.
Kline, Morris. *Mathematical Thought from Ancient to Modern Times*. New York: Oxford University Press, 1972.
Taton, René. *History of Science*. Vol. 3: *Science in the Nineteenth Century*. New York: Basic Books, 1965.

*—Sketch by Zoran Minderovic*

# Jules Henri Poincaré
## 1854-1912
**French number theorist, topologist, and mathematical physicist.**

Jules Henri Poincaré has been described as the last great universalist—"the last man," E. T. Bell

*Jules Henri Poincaré*

wrote in *Men of Mathematics,* "to take practically all mathematics, pure and applied, as his province." He made contributions to number theory, theory of functions, differential equations, topology, and the foundations of mathematics. In addition, Poincaré was very much interested in astronomy, and some of his best known research is his work on the three-body problem, which concerns the way planets act on each other in space. He worked in the area of mathematical physics and anticipated some fundamental ideas in the theory of relativity. He also participated in the debate about the nature of mathematical thought, and he wrote popular books on the general principles of his field.

Poincaré was born in Nancy, France, on April 29, 1854. The Poincaré family had made Nancy their home for many generations, and they included a number of illustrious scholars. His father was Léon Poincaré, a physician and professor of medicine at the University of Nancy. Poincaré's cousin Raymond Poincaré was later to serve as prime minister of France and as president of the republic during World War I. Poincaré was a frail child with poor coordination; his larynx was temporarily paralyzed when he was five, as a result of a bout of diphtheria. He was also very bright as a child, not necessarily an advantage in dealing with one's peers. All in all he was, according to James Newman in the *World of Mathematics,* "a suitable victim of the brutalities of children his own age."

Poincaré received his early education at home from his mother and then entered the lycée in Nancy. There he began to demonstrate his remarkable mathematical talent and earned a first prize in a national student competition. In 1873, he was admitted to the École Polytechnique, although he scored a zero on the drawing section of the entrance examination. His work was so clearly superior in every other respect that examiners were willing to forgive his perennial inability to produce legible diagrams. He continued to impress his teachers at the École, and he is reputed to have passed all his math courses without reading the textbooks or taking notes in his classes.

After completing his work at the École, Poincaré went on to the École des Mines with the intention of becoming an engineer. He continued his theoretical work in mathematics, however, and three years later he submitted a doctoral thesis. He was awarded his doctorate in 1879 and was then appointed to the faculty at the University of Caen. Two years later, he was offered a post as lecturer in mathematical analysis at the University of Paris, and in 1886 he was promoted to full professor, a post he would hold until his death in 1912.

### Discovers Automorphic Functions

One of Poincaré's earliest works dealt with a set of functions to which he gave the name Fuchsian functions, in honor of the German mathematician Lazarus Fuchs. The functions are more commonly known today as automorphic functions, or functions involving sets that correspond to themselves. In this work Poincaré demonstrated that the phenomenon of periodicity, or recurrence, is only a special case of a more general property; in this property, a particular function is restored when its variable is replaced by a number of transformations of itself. As a result of this work in automorphic functions, Poincaré was elected to the French Académie des Sciences in 1887 at the age of 32.

For all his natural brilliance and formal training, Poincaré was apparently ignorant of much of the literature on mathematics. One consequence of this fact was that each new subject Poincaré heard about drove his interests in yet another new direction. When he learned about the work of **Georg Bernhard Riemann** and **Karl Weierstrass** on Abelian functions, for example, he threw himself into that work and stayed with the subject until his death.

Two other fields of mathematics to which Poincaré contributed were topology and probability. With topology, or the geometry of functions, he was working with a subject which had only been treated in bare outlines, and from this he constructed the foundations of modern algebraic topology. In the case of probability, Poincaré not only contributed to the mathematical development of the subject, but he also wrote popular essays about probability which were widely read by the general public. Indeed, he was elected to membership in the literary section of the French Institut in 1908 for the literary quality of these essays.

### Contributes to Three-Body Problem

In the field of celestial mechanics, Poincaré was especially concerned with two problems, the shape of rotating bodies (such as stars) and the three-body problem. In the first of these, Poincaré was able to show that a rotating fluid goes through a series of stages, first taking a spheroidal and then an ellipsoidal shape, before assuming a pear-like form that eventually develops a bulge in it and finally breaks apart into two pieces. The three-body problem involves an analysis of the way in which three bodies, such as three planets, act on each other. The problem is very difficult, but Poincaré made some useful inroads into its solution and also developed methods for the later resolution of the problem. In 1889, his work on this problem won a competition sponsored by King Oscar II of Sweden.

The work Poincaré did in celestial mechanics was part of his interest in the application of mathematics to physical phenomena; his title at the University of Paris was actually professor of mathematical physics. Of the roughly 500 papers Poincaré wrote, about 70 dealt with topics such as light, electricity, capillarity, thermodynamics, heat, elasticity, and telegraphy. He also made contributions to the development of relativity theory. As early as 1899 he suggested that absolute motion did not exist. A year later he also proposed the concept that nothing could travel faster than the speed of light. These two propositions are, of course, important parts of **Albert Einstein**'s theory of special relativity, first announced in 1905.

As he grew older, Poincaré devoted more attention to fundamental questions about the nature of mathematics. He wrote a number of papers criticizing the logical and rational philosophies of **Bertrand Russell**, **David Hilbert**, and **Giuseppe Peano**, and to some extent his work presaged some of the intuitionist arguments of **L. E. J. Brouwer**. As E. T. Bell wrote: "Poincaré was a vigorous opponent of the theory that all mathematics can be rewritten in terms of the most elementary notions of classical logic; something more than logic, he believed, makes mathematics what it is."

Poincaré died on July 17, 1912, of complications arising from prostate surgery. He was 58 years old. During his lifetime he had received many of the honors then available to a scientist, including election to the Royal Society as a foreign member in 1894. Poincaré was married to Jeanne Louise Marie Poulain D'Andecy, with whom he had four children, one son and three daughters.

## SELECTED WRITINGS BY POINCARÉ

### Books

*Electricité et Optique,* second edition, 1901
*La science et l'hypothèse,* 1906
*The Value of Science,* 1907
*Science et méthode,* 1908

## FURTHER READING

### Books

Bell, E. T. *Men of Mathematics.* New York: Simon & Schuster, 1937, pp. 526–554.
Dieudonné, Jean. "Jules Henri Poincaré," in *Dictionary of Scientific Biography. Volume I.* Edited by Charles Coulston Gillispie. New York: Charles Scribner's Sons, 1975, pp. 51–61.
Jones, Bessie Zaban, editor. *The Golden Age of Science.* New York: Simon & Schuster, 1966, pp. 615–637.
Newman, James R. *The World of Mathematics.* Volume 2. New York: Simon & Schuster, 1956, pp. 1374–1379.
Williams, Trevor, editor. *A Biographical Dictionary of Scientists.* New York: Wiley, 1974, pp. 422–423.

*—Sketch by David E. Newton*

# George Pólya
## 1887-1985
### Hungarian-born American number theorist

The career of George Pólya was distinguished by the discovery of mathematical solutions to a number of problems originating in the physical sciences. He made contributions to probability theory, number theory, the theory of functions, and the calculus of variations. He also cared about the art of teaching mathematics; he worked with educators, advocating the importance of problem solving, for which the United States gave him a distinguished service award. Pólya continued to do innovative research well into his nineties, but he is probably best known for his book on methods of problem solving, called *How to Solve It,* which has been translated into many languages and has sold more than one million copies.

Pólya was born in Budapest, Austria-Hungary (now Hungary), on December 13, 1887, the son of Jakob and Anna Deutsch Pólya. As a boy, he preferred geography, Latin, and Hungarian to mathematics. He liked the verse of German poet Heinrich

*George Pólya*

Heine and translated some into Hungarian. His mother urged him to become a lawyer like his father, and he began to study law at the University of Budapest, but soon turned to languages and literature. He earned teaching certificates in Latin, Hungarian, mathematics, physics, and philosophy, and for a year he was a practice teacher in a high school. Though physics and philosophy interested Pólya, he decided to study mathematics on the advice of a philosophy professor. In an interview published in *Mathematical People: Profiles and Interviews,* Pólya explained how he chose a career in mathematics: "I came to mathematics indirectly.... It is a little shortened but not quite wrong to say: I thought I am not good enough for physics and I am too good for philosophy. Mathematics is in between."

Pólya received his doctorate in mathematics from the University of Budapest in 1912 at the age of 24 with a dissertation on the calculus of probability. Traveling to Germany and France, he was influenced by the work of eminent mathematicians at the University of Göttingen and the University of Paris. In 1914, he took his first teaching position, at the Eidgenössische Technische Hochschule in Zurich, Switzerland; he taught there for twenty-six years, becoming a full professor in 1928.

During World War I, Pólya was initially rejected by the Hungarian Army because of an old soccer injury. When the need for soldiers increased later in the war, however, he was asked to report for military service, but by this time he had been influenced by the pacifist views of British mathematician and philosopher **Bertrand Russell** and refused to serve. As a result, Pólya was unable to return to Hungary for several years, and he became a Swiss citizen. He married Stella Vera Weber, the daughter of a physics professor, in 1918.

Pólya proved an important theorem in probability theory in a paper published in 1921, using the term "random walk" for the first time. Many years later, a display demonstrating the concept of a random walk was featured in the IBM pavilion at the 1964 World's Fair in New York; it recognized the work of Pólya and other distinguished scholars. Pólya's work with **Gabor Szegö**, also a Hungarian, resulted in *Problems and Theorems in Analysis,* published in 1925. The problems in the book were grouped not according to the topic but according to the methods that could be used to solve them. Pólya and Szegö continued to work together and they published another book, *Isoperimetric Inequalities in Mathematical Physics,* in 1951.

Pólya was awarded the first international Rockefeller Grant in 1924 and spent a year in England at Oxford and Cambridge, where he worked with English mathematician **Godfrey Harold Hardy** on *Inequalities, Inequalities* which was published in 1934. During the 1930s, Pólya frequently visited Paris to collaborate on papers with Gaston Julia. He received another Rockefeller Grant in 1933, this time to visit the United States, where he worked at both Princeton and Stanford universities. In 1940, after World War II had begun in Europe, Pólya left Switzerland with his wife and emigrated to the United States. He taught for two years at Brown University and for a short time at Smith College before joining Szegö at Stanford University in 1942, where he would remain until his retirement in 1953 at age 66. Pólya became a U.S. citizen in 1947.

Before leaving Europe, Pólya had begun writing a book on problem solving. Observing that Americans liked "how-to" books, Hardy suggested the title *How to Solve It.* The book, Pólya's most popular, was published in 1945; it examined discovery and invention and discussed the processes of creation and analysis. Although he officially retired in 1953, Pólya continued to write and teach. Another book on heuristic principles for problem solving, at a more advanced level of mathematics, was published in 1954, entitled *Mathematics and Plausible Reasoning.* A third book on problem solving, *Mathematical Discovery,* was published in 1962. In 1963, Pólya was the recipient of the distinguished service award from the Mathematical Association of America. The citation, as quoted in the *Los Angeles Times,* read: "He has given a new dimension to problem-solving by emphasizing the organic building up of elementary steps into a complex proof, and conversely, the decomposition of mathematical invention into smaller steps. Problem solving *a la Pólya* serves not only to develop mathematical skill but also teaches constructive reasoning in general."

Pólya also became interested in the teaching of mathematics teachers, and he taught in a series of teacher institutes at Stanford University supported by the National Science Foundation, General Electric, and Shell. His film, "Let Us Teach Guessing," won the Blue Ribbon from the Educational Film Library Association in 1968. In 1978, the National Council of Teachers of Mathematics held problem-solving competitions in *The Mathematics Student;* they named the awards the Pólya Prizes.

Pólya made significant contributions in many areas, including probability, geometry, real and complex analysis, combinatorics, number theory, and mathematical physics. Perhaps one indication of the breadth of his accomplishments is the range of fields that now contain concepts bearing his name. For example, the "Pólya criterion" and the "Pólya distribution" in probability theory; and "Pólya peaks," the "Pólya representation," and the "Pólya gap theorem" in complex function theory. Pólya's writings have been praised for their clarity and elegance; his papers were called a joy to read. The Mathematical Association of America established the Pólya Prize for Expository Writing in the *College Mathematics Journal.* Pólya's papers were collected and published by MIT Press in 1984.

Pólya received honorary degrees from the University of Wisconsin at Milwaukee, the University of Alberta, the University of Waterloo, and the Swiss Federal Institute of Technology. He was a member of the American Academy of Arts and Sciences, the National Academy of Sciences of the United States, the Hungarian Academy of Sciences, the Academie Internationale de Philosophie des Sciences in Brussels, and a corresponding member of the Académie Royale des Sciences in Paris. The Society for Industrial and Applied Mathematics named an honorary award after him, the Pólya Prize in Combinatorial Theory and Its Applications.

Among his colleagues, Pólya was known as a kind and gentle man, full of curiosity and enthusiasm. In honoring him, Frank Harary praised his depth, his versatility, and his speed and power as a mathematician. The mathematician N. G. de Bruijn wrote of Pólya in *The Pólya Picture Album: Encounters of a Mathematician:* "All his work radiates the cheerfulness of his personality. Wonderful taste, crystal clear methodology, simple means, powerful results. If I would be asked whether I could name just one mathematician who I would have liked to be myself, I have my answer ready at once: Pólya." Pólya suffered a stroke at age ninety-seven and died in Palo Alto, California, on September 7, 1985.

## SELECTED WRITINGS BY PÓLYA

### Books

(With Gabor Szegö) *Problems and Theorems in Analysis,* 1925
(With Godfrey Harold Hardy and J. E. Littlewood) *Inequalities,* 1934
*How to Solve It,* 1945
(With Szcgö) *Isoperimetric Inequalities in Mathematical Physics,* 1951

## FURTHER READING

### Books

Albers, Donald J., and G. L. Alexanderson, editors. *Mathematical People: Profiles and Interviews.* Boston: Birkhauser Boston, 1985.
Alexanderson, G. L., editor. *The Pólya Picture Album: Encounters of a Mathematician.* Boston: Birkhauser Boston, 1987.
Szegö, Gabor, editor. *Studies in Mathematical Analysis and Related Topics: Essays in Honor of George Pólya.* Stanford University Press, 1962.

### Periodicals

Dembart, Lee. "George Pólya, 97, Dean of Mathematicians, Dies." *Los Angeles Times* (September 8, 1985): 3, 29.

*—Sketch by C. D. Lord*

# Jean–Victor Poncelet
## 1788–1867
### French geometer

Regarded as one of the fathers of modern projective geometry, Jean–Victor Poncelet's development of the concept of poles and polar lines, in the context of conics, was the forerunner of the principle of duality. His development of distinctions between projective and metric properties served as a foundation for modern structural concepts.

Poncelet was born in July 1, 1788, in Metz, Lorraine, France, to Claude Poncelet, a wealthy landowner and advocate at the Parliament of Metz, and Anne–Marie Perrein. While he was Claude Poncelet's natural son, he was not legitimized as such until later in life. Poncelet's earliest studies took place in the small town of Saint–Avold, where he lived with a family to which he had been entrusted. Poncelet

*Jean-Victor Poncelet*

returned to Metz in 1804 and continued his studies at the city's lycée (the equivalent of elementary school). A highly successful student, he entered Metz's École Polytechnique in the fall of 1807, where he spent three years, having fallen behind a year due to poor health. His instructors at the Polytechnique included **Gaspard Monge**, S. F. Lacroix, and A. M. Ampère.

In 1810 Poncelet entered the French corps of military engineers, and graduated from the École d'Application of Metz in early 1812. He then conducted fortification work on Walcheren, a Dutch island, and served in the Russian campaign as lieutenant, assigned to the engineering general staff. In November 1812 Poncelet was taken prisoner by the Russians and held hostage at a camp on the Volga River at Saratov until June 1814. During his captivity, Poncelet studied mathematics. Because he had no access to mathematics books, he reconstructed elements of pure and analytic geometry, and then began research on the projective properties of conics and systems of conics. This preliminary research, published in 1862 in his *Applications d'analyse et de géométrie*, served as the basis for Poncelet's most important work.

Poncelet returned to France upon his release and was appointed captain of the engineering corps at Metz. For the next 10 years, he contributed to several topography and fortification projects, while concentrating on building and organizing an engineering arsenal. During this time, Poncelet developed a new

model for a variable counterweight drawbridge. He also devoted ample time to the continuation of his research on conics. In addition, he began to publish articles in the *Annales de mathématiques pures et appliquées*. His *Traité des propriétés projectives des figures*, written in consultation with Servois and published in 1822, set forth several fundamental concepts of projective geometry, including the cross–ratio, perspective, involution, and circular points at infinity.

During this time Poncelet also presented his theories on conics to the Académie Royale des Sciences, hoping to advance the idea that geometric concepts could be generalized using elements at infinity and imaginary elements. His theories were based on the concept of an "ideal chord" and a "principle of continuity." The referee of the Académie rebuffed Poncelet's theories, primarily the "principle of continuity," and Poncelet set out to prove his critic wrong. He was sidetracked, however, when he was offered a position as professor of mechanics applied to machines at the École d'Application de l'Attilerie et du Génie at Metz, which commenced in January 1825. He continued to publish in journals, but by and large abandoned his work in projective geometry.

As a professor and practitioner of mechanics, Poncelet focused on improving the efficiency of machines through a combination of mathematical theory and direct application. His work led to significant improvements in the operation of turbines and waterwheels, with his innovations more than doubling the efficiency of the waterwheel. Poncelet also became more involved in the civic life of Mertz, becoming a member of that city's municipal council and secretary of the Conseil Général of the Moselle in 1830. He soon moved from his native home, however, when in 1834 he was named scientific rapporteur for the Committee of Fortifications, editor of the *Mémorial de officier du génie,* and a member of the mechanics section of the Académie Royale des Sciences (an elected position), necessitating a move to Paris. He accepted a teaching post at the Faculty of Sciences in Paris as well in 1837, and he began to publish his courses and papers in book form at this time.

In 1842, Poncelet married Louise Palmyre Gaudin, and he planned to retire from the military soon after, in 1848. His plans were disrupted by the French Revolution of 1848, however, during which he was appointed brigadier general and assigned to a post of commandant at the École Polytechnique. There, he focused mainly on curriculum matters and management of riot situations which erupted among students in May and June of 1848. He remained at the institution until the fall of 1850. He also continued his work with the Conseil Général, although he resigned his position with the Académie. Poncelet finally retired in October 1850 and then focused on publishing his works. He died in Paris on December 22, 1867.

## SELECTED WRITINGS BY PONCELET

"Essai sur les propriétés projectives des sections coniques," 1820

*Applications d'analyse et de géométrie*, 2 volumes, 1862–1864

*Traité des propriétés projectives des figures*, 1822; second edition 1865–66

## FURTHER READING

"Jean–Victor Poncelet." *MacTutor History of Mathematics Archive.* http://www–groups.dcs.st–and.ac.uk/~history/index.html (July 1997).

Taton, René. "Jean–Victor Poncelet," in *Dictionary of Scientific Biography*. Volume XI. Edited by Charles Coulston Gillispie. New York: Charles Scribner's Sons, 1970, pp. 76–82.

—*Sketch by Kristin Palm*

*Claudius Ptolemy*

# Claudius Ptolemy
## c.70–c.130
### Greek geometer, astronomer, and geographer

Claudius Ptolemy was a famed Greek scholar whose work in astronomy and geometry helped form the basis of trigonometry. Ptolemy's earliest and most famous work was the *Almagest*, which focused on astronomy. In it, he developed geometrical proofs, which included the first notable value for $\pi$ since **Archimedes**' time, to explain the motion of the sun, the moon, and the planets. Although his geocentric theory of astronomy mistakenly identified the Earth instead of the sun as the center of the solar system, this work served as the authoritative text on astronomy for over 1,400 years. Ptolemy also wrote works on music, optics, and astrology. His treatise, *Geography*, survived for centuries as a primary reference source and included the first practical use of parallels and longitudes.

Ptolemy and his life remain a mystery outside of his scientific written works. Estimates of the era in which he lived are based on his recorded astronomical observations. The name Ptolemy was common in Egypt, leading to the belief that he was a native of that country. His first name, Claudius, suggests that he was recognized as a Roman citizen, probably due to a legacy handed down through his Greek ancestors. The exact date and cause of Ptolemy's dead are unknown.

Ptolemy probably lived in Alexandria, Egypt, where he made most of his astronomical observations. The city's famous library provided him with access to the largest collection of accumulated knowledge during his time. From his introduction to the *Almagest*, Ptolemy appears to adhere to the Aristotelian school of philosophy, but he was also influenced by other philosophies, such as stoicism. To his credit, Ptolemy acknowledged that he based much of his work on that of previous scholars, especially the astronomical theories of **Hipparchus**.

### Advances the Use of Trigonometry through Astronomy

Like Hipparchus before him, Ptolemy's interest in astronomy and measurements involving the "heavenly" bodies led to his important contributions in geometry and trigonometry. Although Hipparchus is credited as the founder of trigonometry, Ptolemy's earliest known work, the *Almagest*, comprehensively systemized the sum of Greek knowledge in both trigonometry and astronomy.

Ptolemy's Greek title for the *Almagest* was *Syntaxis*, translated as "mathematical compilation." It was quickly recognized as a ground breaking work and came to be called "the great compilation" by the Greeks, probably to distinguish it from earlier and more elementary works on astronomy. It eventually became known as the *Almagest* through early Arabic translations and subsequent Latin translations as the "almagesti" or "almagestum."

Consisting of 13 books, the *Almagest* includes original theories and proofs, including a method of calculating the chords of a circle. These chords were used by the Greeks in spherical geometry and required trigonometry to make various calculations for solar, lunar, and planetary positions and eclipses of the sun and moon. Ptolemy's model of astronomy used combinations of circles, known as epicycles, or small circular orbits centered on a larger circle's circumference. Although his geocentric theory that the heavenly bodies rotated around a central, stationary Earth was eventually discredited, the theorems and system that he advanced in this work provided an admirable and ready means to construct tables representing the movements of the sun, moon, and planets. Its geometrical representations of these movements represent a testimonial to both Ptolemy and the great mathematical thinkers he borrowed from. The *Almagest* also included a catalogue of stars based largely on the catalogue developed by Hipparchus.

Ptolemy also wrote several other works on astronomy. In his *Handy Tables*, he improved upon and systematically organized the Table of Chords from the *Almagest* into one volume for practical use. Ptolemy also wrote two treatises that demonstrated the Greeks had mastered more than basic "classical" geometry. In the *Analema* and *Planisphaerium*, Ptolemy uses both trigonometry and other mathematical techniques to discuss the projection of points on the celestial sphere and stereographic projection. These works include methods for constructing sundials and the basis for the astrolabe, an astronomical instrument used during the Middle Ages. In the *Planetary Hypotheses*, he carries on his astronomical work from the *Almagest* and includes new theories of planetary latitudes and for determining the size of planets and their distances from the Earth. Ptolemy also wrote other astronomical works, which are either lost or only partially survive.

### Establishes Scientific Foundation for Geography

Ptolemy's second great scientific work was *Geography*, an early attempt to map the world as it was known in his time. Although much of it is inaccurate, the book placed the accumulated Greek and Roman knowledge of the Earth's geography on a solid scientific foundation. Ptolemy's inaccuracies were inevitable, considering that modern surveying techniques were unknown and Ptolemy himself knew little of the world outside of the Roman empire. As a result, Ptolemy had to rely on information from travelers, including soldiers in the Roman army.

Much of *Geography* is based on the work of Marinus of Tyre, as well as Hipparchus and Strabo. Ptolemy's important contribution in the work was his introduction of the first systematic use of latitudes and longitudes. Not only did the book contain a number of maps, Ptolemy included directions followed by geographers until the Renaissance for creating maps. Unfortunately, Ptolemy miscalculated the Earth's circumference and many of his estimates of latitude were incorrect. Nevertheless, *Geography* became a profoundly influential work whose popularity led to the creation of opulent manuscripts for the wealthy.

Like many of the learned men of his day, Ptolemy displayed a wide ranging interest in the sciences. His treatise, *Optics*, uses mathematical equations to develop theories of light and refraction (especially as it pertains to the planets and stars) and also discusses vision and mirrors. His work *Harmonica* is only partially intact and represents a treatise on musical theory, including the mathematical intervals of notes.

In the time of Ptolemy, astrology was also considered a science, and Ptolemy wrote a four-book work called *Tetrabibilios*. Always the educator, Ptolemy viewed this work as the logical follow up to the *Almagest* and created it as a textbook for casting horoscopes. To provide this work with a scientific basis, he attempted to relate his efforts in astronomy to astrology.

While modern day science has proven many of Ptolemy's works to be inaccurate, he deserves his place in history as a great intellect. On the basis of the longtime worldwide interest in his works, Ptolemy had a profound impact on the thoughts and philosophies of many generations. A compiler of knowledge and teacher, Ptolemy exhibited a clarity of style and ingenuity, which made both the *Almagest* and *Geography* standard textbooks well into the Renaissance period and beyond. As a result, Ptolemy, who was the last great astronomer of the Alexandrian school, was long recognized as the authority on "all things in heaven and on Earth."

## FURTHER READING

### Books

Gow, James. *A Short History of Greek Mathematics*. New York: Chelsea Publishing Company, 1968, pp. 202–301.

Heath, Thomas L. *The Manual of Greek Mathematics*. New York: Dover Publications, 1963. pp. 402–14.

Magill, Frank N., editor. *The Great Scientists*. Danbury, CT: Grolier Educational Corporation, 1989, pp. 32–37.

Toomer, G. J. "Claudius Ptolemy," in *Dictionary of Scientific Biography*. Volume XI. Edited by Charles Coulston Gillispie. New York: Charles Scribner's Sons, 1975, pp. 186–206.

*—Sketch by David A. Petechuk*

# Pythagoras of Samos
## c.580–500 B.C.
### Greek geometer and philosopher

Pythagoras' wide-ranging interests in mathematics, music, and astronomy mark him as a seminal figure in early western civilization's intellectual development. In the realm of mathematics, he developed the Pythagorean theorem and discovered irrational numbers. Pythagoras also founded a philosophical and religious school steeped in mysticism and secrecy which left its mark on poets, artists, scientists, and philosophers down to the 20th century.

Pythagoras was born about 580 B.C., probably in Samos, Greece, where he grew up. His father, Mnesarchus, may have been an engraver of seals or a merchant. Living in a prosperous seaport that was the center of learning and art during the Golden Age of Greece stimulated Pythagoras' thirst for knowledge. However, conflicting reports and the lack of written records leave much of his early life a matter of conjecture. Some reports indicate that he had at least two elder brothers, was a champion athlete, and a child prodigy.

Pythagoras is said to have studied in Greece under Creophilus and Pherecydes of Syros, and **Thales** of Miletus. In his early twenties, he traveled to Egypt, where he probably learned geometry and was initiated in Egyptian philosophy and science. Pythagoras then may have traveled to Babylon, or have been taken their as a prisoner of the Persians following their invasion of Egypt around 525 B.C. If Pythagoras did live in Babylon, he probably gained a more in-depth understanding of mathematics and geometry. For example, Babylonians had developed reciprocals and square roots for solving equations involved in mathematical astronomy.

While it is uncertain whether Pythagoras traveled further east, many see his philosophical foundations as being closely aligned to Eastern mysticism and philosophy. Pythagoras eventually returned to his home, only to find it ruled by the tyrant Polycrates. As a result, Pythagoras eventually moved to Croton in southern Italy, where he founded his famous school.

### Establishes School of Philosophy and Science

Pythagoras established his school in Croton around 529 B.C. Focusing primarily on religion, philosophy, and mathematics, the school was known as a "homakoeion," meaning a gathering place for people to learn. The school's success can be attributed to the charismatic personality of Pythagoras. In a relatively short period of time, he established a large following, broken up into two groups—the "akousmatikoi," who primarily studied the philosophical teachings of Pythagoras, and the "mathematikoi," who focused on theoretical mathematics.

Students of Pythagoras' school, known as Pythagoreans, followed a number of strict rules. For instance, they were conscientious vegetarians, took a vow of silence for the first five years of their membership, and kept no written records. Liberal in nature, Pythagoras' school was open to everyone, including women, who it allowed to share in the instructing. Following Pythagoras' own belief in simplicity and disdain for worldly honors, Pythagoreans took no credit for their own work or discoveries, attributing all findings either to their master or the group.

Numbers were the very foundation of Pythagorean philosophy, which maintained that numbers were both mystical in nature and had a reality of their own outside of the human mind. Impressed by the ratios that existed in musical harmonies, astronomy, and geometrical shapes, the Pythagoreans developed a theory that all things, in essence, were numbers and related through numbers. Out of this belief, they developed certain representations for numbers: 1 was a point, 2 a line, 3 a surface, and 4 a solid. Moral qualities were also numbers, with 4 representing justice and 10 (known as the tetractys) representing the sum of all nature due to its being the sum of 1+2+3+4 (the point, line, plane, and solid).

While the Pythagoreans' mystical belief in numbers is largely ignored today, their belief that one could penetrate the secrets of the universe through numbers led them to conduct a careful study of mathematical theory, focusing on the principals of geometry.

### Scientific Contributions in Math, Music, and Astronomy

The nonexistence of written records attributable to Pythagoras or his followers has made it difficult for historians to determine what contributions were made directly by Pythagoras himself. However, most historians agree that Pythagoras was likely responsible for several basic tenets in math and science today.

Although the Babylonians knew about the relationship between the legs of a right triangle and the hypotenuse centuries before the birth of Pythagoras, he was the first to provide a general geometric proof of this relationship. The theorem states that the square of the measure of the hypotenuse of a right triangle is equal to the sum of the squares of the two legs. Considering that deductive geometry was in its infancy, this theorem was remarkable for its time. **Euclid**, who Pythagoras strongly influenced, stated the theorem, which is called the Pythagorean theorem, in book one of his *Elements*.

The Pythagorean theorem also led to the discovery of irrational numbers, which is considered to be one of the greatest discoveries of mathematical antiquity. Pythagoras discovered that the hypotenuse of the isosceles right triangle with legs of unit length is equal to the square root of two, which cannot be accurately represented by a rational number (a number that can be expressed either as a whole number, integer, or fraction). The irrational number cannot be represented by any ratio of integers, and in its decimal form does not terminate and does not repeat a certain pattern. Interestingly, the discovery of irrational numbers led to the contradiction of many existing mathematical proofs and theorems. According to legend, the Pythagoreans viewed irrational numbers as symbolic of the unspeakable.

Pythagoras was also the first to discover the mathematical basis of music. For Pythagoras and his followers, music was more than entertainment; it was an integral part of their philosophy and religion. Music was also seen as having an almost mystical affect on humans in terms of soothing the psyche and the soul. Pythagoras found that strings of the lyre produced harmonious tones when the ratios of the lengths of the strings are whole numbers. As a result, Pythagoras discovered the numerical ratios which determine the concordant scale in music. This discovery strengthened the Pythagoreans' belief that all things are numbers and ordered by numbers. Perhaps the most famous outgrowth of this belief is Pythagoras' theory that the heavenly bodies in the cosmos are separated in regular intervals, like the law of harmony. As a result, Pythagoras believed in a vast universal or cosmic harmony (ratio), which led to his doctrine of the "Music of the Spheres."

In the realm of astronomy, Pythagoras was the first to declare the Earth spherical in nature. It is reported that he made this discovery when he noticed the round shadow thrown by the Earth on the moon during a lunar eclipse. Pythagoras also deduced that the sun, moon, and planets have movements of their own, that is, that they rotate on their own axes and orbit a central point. However, Pythagoras mistakenly identified this central point as Earth. Eventually, the Pythagoreans deduced that Earth orbited around another central point, but never identified this point as the sun. The Pythagoreans also correctly deter-

mined that the Earth's rotation around this "central point" took 24 hours.

## The Later Years

In his own time and the first few centuries following his death, Pythagoras was revered by many, with some accounts of his deeds reaching mythic proportions (such as being able to walk on water). His influence can be found in such famous Greek scientists and philosophers as Euclid and Plato. However, Pythagoras was also viewed with disdain by some, primarily because of his philosophies (such as transmigration of the soul, or reincarnation), which resembled the philosophy of the Orient rather than the Greek philosophy of his day.

Although Pythagoras and his school prospered, their controversial philosophies and strong political influence created enemies. With the rise of the democratic party in southern Italy, Pythagoras and his students became the objects of persecution. The democrats considered the Pythagoreans as elitists who placed themselves above others and were also suspicious of their secret rites. This mistrust came to a head sometime between 500 B.C. and 510 B.C., resulting in the destruction of the Pythagorean school and campus. The exact fate of Pythagoras himself remains uncertain. According to some accounts, Pythagoras was killed during this attack. Other accounts indicate that he escaped and lived to be nearly 100 years old. In keeping with Pythagoras' mythic stature, he is also said to have ascended bodily into heaven.

Regardless of the fate of Pythagoras, his followers continued his teachings in other lands, eventually returning to Italy to reestablish the Pythagorean school in Tarentum. In terms of influence, the Pythagoreans outlasted all other philosophies of ancient Greece. They established a strong presence and/or influence both in Egypt and ancient Rome, where the Senate erected a statue in honor of Pythagoras as "the wisest and bravest of Greeks."

## FURTHER READING

Gorman, Peter. *Pythagoras: A Life*. London: Routledge & Kegan Paul, 1979.

Heath, Thomas L. *The Manual of Greek Mathematics*. New York: Dover Publications, 1963, pp. 36–72.

Smith, David Eugene. *History of Mathematics*. Volume 1. New York: Dover Publications, 1958, pp. 69–77.

Terry, Leon. *The Mathmen*. New York: McGraw–Hill Book Company, 1964, pp 42–73.

*—Sketch by David A. Petechuk*

Q

*Lambert Quetelet*

# Adolphe Quetelet
## 1796–1874
### Belgian statistician

Lambert–Adolphe–Jacques Quetelet was one of the individuals most responsible in the 19th century for the quantification of arguments in the physical and social sciences. He was born in Ghent on February 22, 1796, and was educated at the Lycee of Ghent. At age 19 he was appointed an instructor in mathematics at the Royal College in Ghent. In 1819 he was the first person to receive a doctorate from the University of Ghent for a dissertation on conic sections, and in the same year he moved to Brussels to take the chair of mathematics at the Athenaeum. Quetelet was elected to the Belgian Royal Academy in 1820. He married a Mlle. Curtet and they had two children.

In the early stages of his career, however, there was nothing to suggest that Quetelet's fame would spread throughout Europe. In 1824, he spent three months in Paris and learned two things: probability and how to run an observatory. As soon as he returned to Belgium, he began to campaign for the construction of a Royal Observatory. Successful in his efforts, Quetelet was appointed director of the Observatory in 1828. At the observatory, he could pursue celestial mechanics in the same manner as **Pierre Simon Laplace**, whose work he greatly admired.

### "The Average Man"

In 1835, Quetelet published *Sur l'homme et le developement de ses facultés* ("On Man and the Development of his Faculties"). He explores two topics: the importance of gathering quantitative data in order to answer questions about human problems, and the usefulness of the idea of "l'homme moyen" (the average man). Quetelet reveals a general understanding that the reliability of an average increases with the size of the population. What appears to have given the book some of its brilliance was Quetelet's vision of the average as something at which nature is aiming. Deviations from the average were seen as errors, although the notion of deviation as an object of study was familiar to Quetelet. This point also rendered Quetelet's perspective an easy subject for attack by the next generation of statisticians, who recognized the importance of variability in connection with the theory of evolution and other areas. The English statistician Francis Galton often criticized Quetelet's focus on the average rather than the deviations.

To some extent, Quetelet's portrayal of social phenomena as expressed in his book reflected the sources of his approach. On the one hand, he had a law of large numbers that had come down to him from more mathematically–minded statisticians, which indicated that when an experiment was repeated, the more reliable is the outcome. On the other hand, he also had a Laplacean view of some deterministic mechanics underlying the phenomena. As a result, in every area Quetelet found the accumulation of data essential for the purpose of recognizing the normal distribution underneath. The law of large numbers became the central principle of all science for Quetelet, and at the very least that encouraged him to promote data–gathering in every field.

In 1846, Quetelet published a volume of letters on the theory of probability. He was active in the reform of scientific teaching in Belgium and had become permanent secretary of the Brussels Academy in 1834. He was instrumental in the founding of the Statistical Society of London and was the first foreign member of the American Statistical Association. At a time when statistics began to play a role in settings outside the calculation of annuities and games of chance, Quetelet spoke for the statistical community. In addition to his scientific work, he wrote an opera and published poetry and popular essays.

Quetelet died on February 17, 1874 in Belgium. His funeral was distinguished by the presence of many scientists from abroad and his memory was honored by the erection of a monument in Brussels. For someone who seldom used mathematics, Quetelet had acquired quite a reputation as a mathematician. The "average man" as Quetelet envisioned may have been a figment of the imagination, but the recognition of the importance of data–gathering was a timely lesson for a scientific culture about to undergo a probabilistic revolution.

## SELECTED WRITINGS BY QUETELET

*Sur l'homme et le développement de ses facultés,* 1835

*Lettres á S.A.R. le Duc Régnant de Saxe–Coburg et Gotha, sur la theorie des probabilites,* 1846

## FURTHER READING

Daston, Lorraine J. *Classical Probability in the Enlightenment.* Princeton, NJ: Princeton University Press, 1988.

Freudenthal, Hans. "Adolphe Quetelet," in *Dictionary of Scientific Biography.* Volume XI. Edited by Charles Coulston Gillispie. New York: Charles Scribner's Sons, 1973, pp. 236–238.

Stigler, Stephen M. *The History of Statistics.* Cambridge, MA: Harvard University Press, 1986.

*—Sketch by Thomas Drucker*

*S. I. Ramanujan*

# S. I. Ramanujan
## 1887-1920
### Indian number theorist

S. I. Ramanujan was a self-taught prodigy from India. His introduction to the world of formal mathematics and subsequent fame arose from his correspondence and collaboration with the renowned British mathematician **Godfrey Harold Hardy** . In his short but prolific career Ramanujan made several important contributions to the field of number theory , an area of pure mathematics that deals with the properties of and patterns among ordinary numbers. Three quarters of a century after his death, mathematicians still work on his papers, attempting to provide logical proofs for results he apparently arrived at intuitively. Many of his theorems are now finding practical applications in areas as diverse as polymer chemistry and computer science, subjects virtually unknown during his own times.

Srinivasa Iyengar Ramanujan, born on December 22, 1887, was the eldest son of K. Srinivasa Iyengar and Komalatammal. He was born in his mother's parental home of Erode and raised in the city of Kumbakonam in southern India, where his father worked as a clerk in a clothing store. They were a poor family, and his mother often sang devotional songs with a group at a local temple to supplement the family income. Ramanujan received all his early education in Kumbakonam, where he studied English while still in primary school and then attended the town's English-language school. His mathematical talents became evident early on; at eleven he was already challenging his mathematics teachers with questions they could not always answer. Seeing his interest in the subject, some college students lent him books from their library. By the time he was thirteen, Ramanujan had mastered S. L. Loney's *Trigonometry,* a popular textbook used by students much older than him who were studying in Indian colleges and British preparatory schools. In 1904, at the age of 17, Ramanujan graduated from high school, winning a special prize in mathematics and a scholarship to attend college.

### Pursues Mathematics Independently

Shortly before he completed high school, Ramanujan came across a book called *A Synopsis of Elementary Results in Pure and Applied Mathematics.* This book, written by British tutor G. S. Carr in the 1880s, was a compilation of approximately five thousand mathematical results, formulae, and equations. The *Synopsis* did not explain these equations or provide proofs for all the results; it merely laid down various mathematical generalizations as fact. In Ramanujan, the book unleashed a passion for mathematics so great that he studied it to the exclusion of all other subjects. Because of this, although Ramanujan enrolled in the Fine Arts (F.A.) course at the local Government College, he never completed the course. He began to spend all his time on mathematics, manipulating the formulae and equations in Carr's book, and neglected all the other subjects that were part of his course work at the college. His scholarship was revoked when he failed his English composition examination. In all, he attempted the F.A. examinations four times from 1904 to 1907. Each time he failed, doing poorly in all subjects except mathematics.

During these four years, and for several more, Ramanujan pursued his passion with single-minded devotion. He continued to work independently of his teachers, filling up sheets of paper with his ideas and results which were later compiled in his famous

*Notebooks.* Carr's book had merely been a springboard to launch Ramanujan's journey into mathematics. While it gave him a direction, the book did not provide him with the methods and tools to pursue his course. These he fashioned for himself, and using them he quickly meandered from established theorems into the realms of originality. He experimented with numbers to see how they behaved, and he drew generalizations and theorems based on these observations. Some of these results and conclusions had already been proved and published in the Western world, though Ramanujan, sequestered in India, could not know that. But most of his work was original.

Meanwhile, his circumstances changed. Without a degree, it was very difficult to find a job, and for many of these years Ramanujan was desperately poor, often relying on the good graces of friends and family for support. Occasionally he would tutor students in mathematics, but most of these attempts were unsuccessful because he did not stick to the rules or syllabus. He habitually compressed multiple steps of a solution, leaving his students baffled by his leaps of logic. In July 1909 he married Janaki, a girl some ten years his junior. Keeping with local customs and traditions, the marriage had been arranged by Ramanujan and Janaki's parents. Soon afterwards, he traveled to Madras, the largest city in South India, in search of a job. Because he did not have a degree, Ramanujan presented his notebooks as evidence of his work and the research he had been conducting in past years. Most people were bewildered after reading a few pages of his books, and the few who recognized them as the work of a genius did not know what to do with them. Finally, Ramachandra Rao, a professor of mathematics at the prestigious Presidency College in Madras, reviewed the books and supported him for a while. In 1912 Ramanujan secured a position as an accounts clerk at the Madras Port Trust, giving him a meager though independent salary.

During this time, Ramanujan's work caught the attention of other scholars who recognized his abilities and encouraged him to continue his research. His first contribution to mathematical literature was a paper titled "Some Properties of Bernoulli's Numbers," and it was published in the *Journal of the Indian Mathematical Society* c. 1910. However, Ramanujan realized that the caliber of his work was far beyond any research being conducted in India at the time, and he began writing to leading mathematicians in England asking for their help.

### Sends Letter to Hardy

The first two mathematicians he approached were eminent professors at Cambridge University, and they turned him down. On January 16, 1913, Ramanujan wrote to Godfrey Harold Hardy, who agreed to help him. Hardy was a fellow of Trinity College, Cambridge, and he specialized in number theory and analysis. Although he was initially inclined to dismiss Ramanujan's letter, which seemed full of wild claims and strange theorems with no supporting proofs, the very bizarreness of the theorems nagged at Hardy, and he decided to take a closer look. Along with J. E. Littlewood, he examined the theorems more thoroughly, and three hours after they began reading, they both decided the work was that of a genius. "They must be true because, if they were not true, no one would have had the imagination to invent them," Hardy is quoted as saying in Robert Kanigel's book *The Man Who Knew Infinity.*

Hardy now set about the task of bringing Ramanujan to England. In the beginning, Ramanujan resisted the idea due to religious restrictions on traveling abroad, but he was eventually persuaded to go. Ramanujan spent five years in England, from 1914 to 1919, during which time he enjoyed a productive collaboration with Hardy, who personally trained him in modern analysis. Hardy described this as the most singular experience of his life, says Kanigel in *The Man Who Knew Infinity:* "What did modern mathematics look like to one who had the deepest insight, but who had literally never heard of most of it?" Ramanujan was to receive several laurels during this period, including a B.A. degree from Cambridge, and appointments as Fellow of the Royal Society (at 30 he was one of the youngest ever to be honored thus) and Trinity College. But the English weather affected Ramanujan's health, and he contracted tuberculosis. In 1919 he returned to India, where he succumbed to the disease, dying on April 26, 1920.

Until the very end, Ramanujan remained passionately involved in mathematics, and he produced some original work even after his return to India. His great love for the subject and his genius are perhaps best exemplified in an incident described by Hardy in his book *A Mathematician's Apology.* He related that while visiting Ramanujan at a hospital outside London, where he lay ill with tuberculosis, Hardy mentioned the number of his taxicab, 1729. Hardy thought it a rather dull number. "No, Hardy! No, Hardy!" Ramanujan replied. "It is a very interesting number. It is the smallest number expressible as the sum of two cubes in two different ways." Kanigel reported another comment Hardy made later. He said that had Ramanujan been better educated, "he would have been less of a Ramanujan and more of a European professor and the loss might have been greater." Ramanujan himself attributed his mathematical gifts to his family deity, the goddess Namagiri. A deeply religious man, he combined his passion with his faith, and he once told a friend that "an equation for me has no meaning unless it expresses a thought of God."

## SELECTED WRITINGS BY RAMANUJAN

### Books

*Notebooks,* 2 volumes, 1957
*The Lost Notebook and Other Unpublished Papers,*
1988

### FURTHER READING

### Books

Hardy, G. H. *A Mathematician's Apology.* Cambridge, England: Cambridge University Press, 1940.
Hardy, G. H. *Ramanujan: Twelve Lectures on Subjects Suggested by His Life and Work.* Cambridge, England: Cambridge University Press, 1940.
Kanigel, Robert. *The Man Who Knew Infinity: A Life of the Genius Ramanujan.* New York: Macmillan, 1991.

### Periodicals

Hardy, G. H. "Obituary, S. Ramanujan," *Nature* (June 17, 1920): 494–95.
Seshu Iyer, P. V., and Ramachandra Rao, R. "The Late Mr. S. Ramanujan, B.A., F.R.S.," *Journal of the Indian Mathematical Society* (June 1920): 81–86.

### Other

Sykes, Christopher. *Letters from an Indian Clerk* (documentary), BBC.

*—Sketch by Neeraja Sankaran*

# Robert Recorde
## 1510–1558
### Welsh–English mathematician, physician, and author

Recorde is considered the founder of the English school of mathematical writers, and was the first English writer on arithmetic, geometry and astronomy. Recorde was the first mathematician to introduce algebra in England. Most of his works were prevalent until the end of the 16th century, and some until the end of the 17th century. Many of his writings were in poetic form as an aid to students in remembering the rules of operation.

Recorde was the second son of Thomas Recorde, a second–generation Welshman, and Rose Johns of Montgomeryshire. He received a B.A. from Oxford University in 1531 and in the same year became a member of All Souls College—a chantry and graduate foundation for theology, civil and canon law, and medicine. He later moved to Cambridge, where he received his M.D. degree in 1545, and became physician to Edward VI and Queen Mary. Records indicate that Recorde had been licensed in medicine twelve years previously at Oxford. His university career and any other degrees are lost to history.

### Physician/Author

Three physicians during this period contributed greatly to mathematics: **Nicolas Chuquet**, **Girolamo Cardano** and Recorde. Recorde was the most influential within his own country, as he established the English mathematical school. He wrote in the vernacular, which may have limited his effect on the Continent. Recorde's first existent mathematical work was the *Grounde of Artes* (1541), which was made popular for its use of the abacus and algorithm for commercial use. Recorde's first edition dealt with only whole numbers with fundamental operations, progression, and counter reckoning. In 1552, the text was expanded to include fractions and alligation. He wrote *The Pathway to Knowledge* (1551), a translation of **Euclid's** first four volumes of *Elements*. The text separated construction from theorems, on the premise students need to understand at the onset, both what is being taught and why. His *The Gate of Knowledge* was a text on measurement and use of the quadrant, but has been lost, and perhaps never published. *Castle of Knowledge* (1556) dealt with construction and use of the sphere. The text was a study of standard authorities, and offered corrections of errors he felt were caused by others' lack of knowledge of Greek.

Recorde is best known for his *The Whetstone of Witte* (1557), which contained the "second part of arithmetic" promised in *The Ground of Artes*, and contained elementary algebra through quadratic equations. The text contains his introduction of the sign $=$ for equals, which he selected because nothing could be more equal than two parallel straight lines. In this text he also explained how the square root in algebra could be extracted.

Aside from his mathematical textbooks, he wrote *The Urinal of Physick* (1547), which was a short but systematic work which included figures and descriptions. The text was a complete assessment of urines, and contained nursing practices intended for the medical profession.

### An Outstanding Scholar

Recorde was a skillful teacher, as well as an outstanding scholar of mid–16th century English. He was as familiar with Greek and medieval texts as he was with contemporary developments. Recorde emphasized learning Greek in order to understand other sources, and developed English equivalents for Latin and Greek terms to simplify his teachings. Recorde was not internationally known, mainly because his works were in English and on an elementary level, but his texts were the standard in Elizabethan England. While the majority of his time was spent in the mathematical sciences, Recorde was also highly skilled in rhetoric, philosophy, literature, history, cosmography (the science that studies the whole order of nature), and all areas of natural history.

### Government Duties Lead to Trouble

Earlier in his career, Recorde served his government in other capacities. In 1549, he was appointed comptroller of the Bristol Mint. Trouble ensued when he refused to divert monies intended for King Edward to the armies under Lord John Russell and Sir William Herbert (later earl of Pembroke). Herbert accused him of treason and he was confined for 60 days, and the mints were forced to stop production. The incident lead to later career problems.

From 1551 to 1553, Recorde was surveyor of the mines in Ireland, in charge of the Wexford silver mines, and technical supervisor of the Dublin Mint. The mines were unprofitable and Recorde was recalled in 1553. In 1556, he tried to regain reinstatement at Court, but earlier he had charged malfeasance as commissioner of mints against Pembroke. Unfortunately, for him, Queen Mary and King Philip, sided with Pembroke, and Pembroke sued for libel with damages against Recorde. Recorde apparently was unable to pay the sum and was imprisoned at King's Bench prison, where he penned his will, leaving a little money to his relatives, and died in prison. Not until 1570 was his estate compensated for monies due him when he was recalled as surveyor of mines in 1553.

Recorde had five daughters and four sons. He was also an active supporter of the Protestant Reformation.

### FURTHER READING

Ball, W. W. R. *A Short Account of the History of Mathematics*. New York: Dover Publications, Inc., 1960, p. 214.

Boyer, Carl B. *A History of Mathematics*. New York. John Wiley & Sons, Inc., 1968, p. 297.

Easton, Joy B. "Robert Recorde," in *Dictionary of Scientific Biography*. Volume XI. Edited by Charles Coulston Gillispie. New York: Charles Scribner's Sons, 1981, pp. 338–40.

*Mina S. Rees*

Katz, V.J. *A History of Mathematics*. New York: HarperCollins, 1990, pp.325–327.

Smith, D.E. *History of Mathematics*. New York: Dover Publications, Inc., 1958, pp. 411–412.

*—Sketch by Corinne Johnson*

# Mina S. Rees
## 1902-1997
### American applied mathematician

Mina S. Rees was the founding president of the Graduate Center of the City University of New York, and was the first woman elected to the presidency of the American Association for the Advancement of Science. She has been recognized by both the United States and Great Britain for organizing mathematicians to work on problems of interest to the military during World War II. After the war she headed the mathematics branch of the Office of Naval Research, where she built a program of government support for mathematical research and for the development of computers.

Rees was born in Cleveland, Ohio, on August 2, 1902, to Moses and Alice Louise (Stackhouse) Rees. Educated in New York public schools, Rees received her A.B. summa cum laude from Hunter College in New York City in 1923, and taught at Hunter College High School from 1923 to 1926. She completed an M.A. at the Teacher's College of Columbia University in 1925, and became an instructor at the Mathematics Department of Hunter College the following year. She continued her training in mathematics at the University of Chicago, where she received a fellowship for 1931 to 1932, and earned a Ph.D. in mathematics in 1931 with a dissertation on abstract algebra. Returning to Hunter, Rees was promoted to assistant professor in 1932 and associate professor in 1940.

In 1943, in the midst of World War II, Rees joined the government as a civil servant, working as executive assistant and a technical aide to Warren Weaver, the chief of the Applied Mathematics Panel (AMP) of the National Research Committee in the Office of Scientific Research and Development. The AMP, located in New York City, established contracts with mathematics departments at New York University, Brown, Harvard, Columbia and other universities. Under these contracts, mathematicians and statisticians studied military applications such as shock waves, jet engine design, underwater ballistics, air-to-air gunnery, the probability of damage under anti-aircraft fire, supply and munitions inspection methods, and computers. In 1948, Rees was awarded the U.S. President's Certificate of Merit and the British King's Medal for Service in the Cause of Freedom for her work during the war.

### Joins Office of Naval Research

From 1946 to 1953, Rees worked for the Office of Naval Research (ONR), first as head of the mathematics branch and then, from 1950, as director of the mathematics division. Under Rees, the ONR supported programs for research on hydrofoils, logistics, computers, and numerical methods. Rees emphasized the study and development of mathematical algorithms for computing. The ONR supported the development of linear programming and the establishment in 1947 of an Institute for Numerical Analysis at the University of California at Los Angeles, and also worked with other military and civilian government agencies on the acquisition of early computers. In addition, the ONR funded university research programs to build computers, such as Project Whirlwind at MIT, lead by Jay Forrester, and the Institute for Advanced Study project under **John von Neumann** . The ONR also awarded grants to support applied and basic mathematical research.

In 1953, Rees returned to Hunter College as Dean of Faculty and Professor of Mathematics. She was married in 1955, to Dr. Leopold Brahdy, a physician. In 1961, she was appointed dean of graduate studies for the City University of New York (CUNY), which established graduate programs by pooling distinguished faculty from the City Colleges, including Hunter. The following year, Rees became the first recipient of the Award for Distinguished Service to Mathematics established by the Mathematical Association of America. Rees was appointed provost of the Graduate Division in 1968 and the first president of the Graduate School and University Center in 1969. By the time Rees retired as emeritus president in 1972, CUNY's graduate school had created 26 doctoral programs and enrolled over two thousand students. During her post-war years at Hunter and CUNY, Rees served on government, scientific, and educational advisory boards and held offices in mathematical, scientific, and educational organizations. She became the first female president of the American Association for the Advancement of Science in 1971. In 1983 Rees received the Public Welfare Medal of the National Academy of Sciences, an award that confers honorary membership in that organization. Rees died in 1997.

## SELECTED WRITINGS BY REES

"The Nature of Mathematics," in *Science* (October 5, 1962): 9–12

"The Mathematical Sciences and World War II," in *American Mathematical Monthly* (October 1980): 607–621

"The Computing Program of the Office of Naval Research, 1946–1953," in *Annals of the History of Computing* 4, number 2 (April 1982): 102–120

## FURTHER READING

### Books

Dana, Rosamond, and Peter J. H. Hilton. "Mina Rees," in *Mathematical People*. Edited by Donald J. Albers and G. L. Alexanderson. Boston: Birkhauser, 1985, pp. 256–265.

### Periodicals

"Award for Distinguished Service to Mathematics." *American Mathematical Monthly* (February 1962): 185–187.

*—Sketch by Sally M. Moite*

# Regiomontanus
## 1436–1476
### German trigonometrist and astronomer

Regiomontanus was the pen name of Johann Müller, responsible for what some call the "rebirth of

*Regiomontanus (Johann Müller)*

mathematics and protocol of astronomy, then still indistinguishable from astrology.

### A Deathbed Pledge

Muller's teacher, Georg von Peurbach, was a reformist who decried existing Latin translations of Greek texts from which he was forced to teach. Peurbach was particularly unsatisfied with a reference called *The Alfonsine Tables*. Regiomontanus became part of the faculty during 1457 and joined forces with his former teacher to improve these source materials. A visiting Church official, Cardinal Bessarion, came to their aid in 1460 by contracting Peurbach to produce a new Latin translation of **Claudius Ptolemy**'s *Amalgest* from the Greek instead of from its intermediate Arabic translation.

By the spring of 1461 Peurbach died at the age of 38, before the work was completed, and his last wish was that his student carry on his work. Bessarion became Regiomontanus' benefactor, teaching him Greek and taking him to Rome in search of ancient manuscripts. The cardinal's deeper motive was to forge common ground for the Greek and Latin factions of the Catholic Church. At various intellectual centers of Italy, they consulted with leading astronomers, particularly members of the Averroists. Regiomontanus was appointed professor of astronomy to take Peurbach's place, and finished the *Epitome* as promised. He even added commentary and revised calculations in some places, restoring omitted mathematical passages in others. Ironically, it would not see publication for decades after Regiomontanus' own death.

### Eclipses and Comets

Regiomontanus' 1464 effort, *De triangulis omnimodis libri quinque*, presented what was likely the first trigonometrical formula for the area of a triangle, spearheading the application of algebraic and algorithmic solutions to geometric problems. This publication sported the still innovative notations for sine and cosine. Simultaneously, Regiomontanus was drawing up charts for trigonometric functions, particularly sine tables for use in the days prior to the invention of logarithms. His longitudinal studies bore the *Tabulae Directionum*, or "tables of directions" for fixed celestial objects in relation to the apparent movement of the heavens. This work introduced the tangent function, with which Regiomontanus would later compute a table of tangents. After moving back to Nuremberg to assume the post of Royal Astronomer to King Matthias Corvinus of Hungary, he found yet another sponsor. Bernard Walther was an amateur astronomer who funded an observatory and workshop, and commissioned various scientific tools of Regiomontanus' design for their use. Upon securing a printing press, Regiomontanus gave his old teacher's

trigonometry" during the century following his death. Although he was a practicing astrologer, he also was one of the first Europeans to closely observe certain celestial objects and events previously tied to superstitions. Regiomontanus published his own and others' translations of classical and Arabic texts with his then novel printing press. He predicted that sailors would one day be able to navigate according to the position of the moon, a process that eventually did arise during the Age of Exploration. (Some speculate that Regiomontanus' early death prevented him from anticipating Copernicus.) Once dismissed as a comparatively undisciplined technician, Regiomontanus is now recognized as the West's answer to Nasir al–Din or Nasir Eddin, having brought plane and spherical trigonometry within the purview of pure mathematics and out of its apprenticeship to astronomy.

Regiomontanus was born in an area once known variously as Franconia or East Prussia, now part of Russia, on June 6, 1436. His father was a miller, but nothing else is known of the family. Regiomontanus studied dialectics for two or three years at the University of Leipzig before entering the University of Vienna at age 14 under the name Johannes Molitoris de Königsberg, but quickly came to be known by a Latin variant of his hometown. Königsberg meant King's Mountain, so the translation *regio monte* yielded Regiomontanus. He was awarded a bachelor's degree in one year. After two years at Vienna Regiomontanus elected to study the work precedence, making Peurbach's *Theoriae plane-tarum novae* one of the earliest mass–produced science texts. In 1474 Regiomontanus printed his own *Ephemerides*, an almanac of daily planetary positions and eclipses for the years 1475 to 1506. This was the reference Christopher Columbus used as leverage against the Jamaican aborigines, who found his prediction of a lunar eclipse awe–inspiring. More notably, one of the objects Regiomontanus had recorded during his research was the comet of 1472, renamed Halley's Comet upon its third return 210 years later.

As recorded in the *Nuremberg Chronicle*, Regiomontanus was summoned to Rome with hopes of reforming the Julian calendar. This was an ongoing problem many scientists, including Regiomontanus, had already pondered. Pope Sixtus IV awarded him the bishopric of Ratisbon, but he fell victim to a plague in the wake of the Tiber river flood of 1476. Regiomontanus' papers were inherited by Walther, who unfortunately did not disseminate them. Upon his death they were acquired by the Nuremberg Senate, where they languished for nearly 30 more years. Regiomontanus' projects in mathematics and science were extended by intellectual successors like Germany's Johann Werner and Belgium's Philip van Lansberge.

Regiomontanus was also a university professor, lecturing on Virgil and Cicero . Major classical works translated by Regiomontanus included the *Conics* of **Apollonius** and some writings of **Archimedes**. He also salvaged the only known extant portion of an uncopied manuscript by Diophantus. Regiomontanus' mathematical writing is generally classified as rhetorical, even careless, because of his habit of referring to proofs without actually presenting them. Of one in *De triangulis* he signs off, "...I leave it to you for homework."

Regiomontanus rejected the possibility that the Earth rotated around the sun; however, being an early Renaissance man, not all of his beliefs were so fixed. It is suggested that Novara, a teacher of Copernicus, came into possession of a letter from Regiomontanus that implied a primitive Copernicanism, but this has not been proven. Regardless, Copernicus was inspired by Regiomontanus' critical notes to Ptolemy's *Amalgest* to revise the milennium–old view of the planet Earth and the solar system it shares.

## SELECTED WRITINGS BY REGIOMONTANUS

*De triangulis omnimodis libri quinque*, 1464 (published in 1533)
*Tabulae Directionum*, 1475 (published in 1490)

## FURTHER READING

### Books

Asimov, Isaac. *Asimov's Biographical Encyclopedia of Science and Technology.* Second Revised Edition. Garden City, NY: Doubleday, 1982, pp. 69–70.
Boyer, Carl B. *A History of Mathematics.* NY: John Wiley & Sons, Inc., 1968, pp. 299–304.
Eaves, Howard. *An Introduction to the History of Mathematics.* 4th Edition. Holt, Rinehart & Winston, 1976, p. 214.
————. *Great Moments in Mathematics (After 1650).* Mathematical Association of America, Inc., 1980, p. 223.
Porter, Roy, editor. *The Biographical Dictionary of Scientists.* Second Edition. New York: Oxford University Press, 1994, p. 502.
Williams, Trevor I., editor. *A Biographical Dictionary of Scientists.* Third Edition. New York: John Wiley & Sons, 1982, pp. 437–38.
Zinner, E. *Regiomontanus: His Life and Work.* Amsterdam: 1990.

### Periodicals

Kren, C. "Planetary latitudes, the Theorica, and Regiomontanus." *Isis* 68 (1977), pp. 194–205.

### Other

"Georg Peurbach." *MacTutor History of Mathematics Archive.*
http://www–groups.dcs.st–and.ac.uk/~history/Mathematicians/Peurbach.html (July 1997).
"Johann Muller Regiomontanus." *MacTutor History of Mathematics Archive.*
http://www–groups.dcs.st–and.ac.uk/~history/Mathematicians/Regiomontanus.html (July 1997).

*—Sketch by Jennifer Kramer*

# Vincenzo Riccati
## 1707–1775
### Italian geometer and engineer

Born into a family of mathematicians and a member of the Jesuit order, Vincenzo Riccati studied integration and differential equations as his father had before him. Among Riccati's contributions to mathematics were some of the earliest and most extensive studies of hyperbolic functions and infinitesimal calculus. Riccati also was a talented hydraulic engineer who did vital work in flood prevention in northern Italy.

Riccati was born January 11, 1707, in Castelfranco Veneto, Italy. He was the second son of Count Jacopo Francesco Riccati, a noted mathematician in his own right. Two of Riccati's younger brothers, Francesco and Giordano, were also important mathematicians. Riccati received most of his early education at home and with the Jesuits. He himself became a Jesuit on December 20, 1726.

For the next ten years of his life, Riccati moved within Italy a number of times. In 1728, Riccati was teaching literature at the Piacenza–based Jesuit Colegea. Then, in succession, he moved to Padua in 1729, then Parma in 1734, then on to Rome (where he did some theological study), and finally settled in Bologna in 1739, which became his home for the next three decades.

While in Bologna, Riccati taught mathematics at the College of San Francesco Saviero. It was here that Riccati researched and published most of his important mathematical contributions.

Between 1757 and 1762 Riccati published two volumes of his seminal *Opuscula ad res physicas et mathematicas pertinentia*. In the first volume, he made a significant contribution to trigonometry by introducing the hyperbolic function, and a number of characteristics including their standard addition formulas, how they related to the exponential function and their derivatives. Riccati also used hyperbolic functions to solve cubics.

Riccati continued the work began on hyperbolic functions in *Opuscula* in *Institutiones analyticae*, which was written in collaboration with Girolamo Saladini, and published in three volumes from 1765 to 1767. Riccati and Saladini published their work on hyperbolic functions a number of years before **Johann Heinrich Lambert**, who is usually given the credit for this work. Also, in these volumes, Riccati and Saladini did some of the first work on infinitesimal calculus. They also researched Grandi's "rose curves" and delineated their simpler, improved answer to al'Haitam's problem concerning the reflection of light rays at certain points on a circular mirror.

Though Riccati made significant contributions to mathematics on his own, he did not forget the influence of his father. In 1758, he and his brother Giordano collected and published four volumes of their father's work in *Opere del Conte Jacopo Riccati*. Riccati also shared his father's interest in hydraulic engineering. Government commission in hand, Riccati designed flood control measures for areas near a number of Italian rivers: the Reno, Po, Adige, and Brenta. This project prevented serious flooding in both the Venice and Bologna areas. Riccati won many accolades for his engineering feats. One of the most

prestigious was being named a member—one of the first—of the Società dei Quaranta.

Towards the end of Riccati's life the Jesuit order was suppressed by Pope Clement XIV in 1773, whereupon Riccati returned to his family in Treviso. He died there two years later on January 17, 1775.

## SELECTED WRITINGS BY RICCATI

*Opuscula ad res physicas et mathematicas pertinentia*, 1757–1762
(With Giordano Riccati) *Opere del Conte Jacopo Riccati*, 1758
(With Girolamo Saladini) *Institutiones analyticae*, 1765–1767

## FURTHER READING

Natucci, A. "Vincenzo Riccati," in *Dictionary of Scientific Biography*. Volume XI. Edited by Charles Coulston Gillispie. New York: Charles Scribner's Sons, 1970, pp. 401–02.

—*Sketch by Annette Petruso*

# Georg Friedrich Bernhard Riemann
## 1826–1866
### German topologist and mathematical physicist

Intuitive and original, Georg Friedrich Bernhard Riemann had a gift for understanding connections among apparently heterogeneous phenomena. Almost all his works proved to be the beginning of new, productive research. He anticipated the work of James Clerk Maxwell in electrodynamics, and introduced the novel concept of curvature in a two–dimensional direction, providing the non–Euclidean geometry of space necessary for **Albert Einstein**'s theory of general relativity.

Riemann worked for many years in failing health in a university system where abject poverty was the lot of those without independent means. When he died of tuberculosis at age 40, however, he had known, worked with, and influenced the great mathematicians and physicists of his day. Riemann was shy and self–effacing and recognition for his work came slowly during his lifetime; awareness of his truly striking achievements was to come later as his work

*Georg Friedrich Bernhard Riemann*

was validated and as it stimulated the work of others. Today, Riemann's work, which often incorporates methods of physics, is accepted as some of the most remarkable in the history of all science, not only in mathematics. A number of theorems, methods, concepts, and conjectures carry his name.

Riemann was born September 17, 1826, in Breselenz in the kingdom of Hanover, now Germany. His parents were Friedrich Bernhard Riemann, a Protestant minister who had served in the Napoleonic wars, and Charlotte Ebell, the daughter of a court councillor. He was the second child of six, two boys and four girls. Riemann's family suffered from tuberculosis. His mother and three sisters, and eventually Riemann himself, died of the disease.

Riemann was taught primarily by his father and began making mathematical calculations as early as age six. By age ten he had surpassed his father and also his schoolmaster Shulz, who was employed to tutor him in mathematics. In 1840, at age 14, Riemann went to live with his grandmother in Hanover, where he studied a classical curriculum at the gymnasium, or secondary school. In 1842, his grandmother died and he transferred to the gymnasium at Löneburg, closer to his father's parish at Quickborn. Schmalfuss, the director at Löneburg, noted Riemann's interest in mathematics and lent him several classic books, including **Leonhard Euler**'s works and **Adrien–Marie Legendre**'s *Theory of Numbers*, which Riemann studied, mastered and returned

in a few days. He was to use some of this information in his later work.

## Studies Mathematics at Göttingen and Berlin

In 1846, Riemann entered Göttingen University as student of theology and philology, but his talent and superior knowledge in mathematics was obvious, and he soon persuaded his father to let him devote himself to mathematics. The faculty at Göttingen included the "king of mathematicians" **Karl Friedrich Gauss**, who lectured on the method of least squares. It also included Moritz A. Stern, who lectured on the numerical solution of equations and definite integrals, and the astronomer Carl Wolfgang Benjamin Goldschmidt, who lectured on terrestrial magnetism. Stern was later to recall that Riemann "already sang like a canary." The quality of teaching at Göttingen was low and elementary, however, and students were not included in the creative thinking of the teachers.

After only a year at Göttingen, Riemann spent two years in Berlin, where democratic ideas were afoot and the faculty at the University of Berlin shared ideas in open discussions with students. The mathematics faculty included **Peter Dirichlet, Karl Jacobi, Jakob Steiner**, and Ferdinand Eisenstein. Riemann attended Dirichlet's lectures on the theory of numbers, definite integrals, and partial differential equations, and Jacobi's on analytical mechanics and higher algebra. With Eisenstein, Riemann discussed the possibility of introducing complex variables into investigations of elliptic functions. Riemann was attracted to Dirichlet's methods of logical analysis of fundamental questions and avoidance of long computations. Riemann's stay in Berlin was coincident with the March 1848 insurrection in Prussia, and at the height of the revolution Riemann joined other students in guarding the palace of King Friedrich IV for 24 hours. When the king was deemed safe, he returned to the university.

In 1849, Riemann returned to Göttingen to work on his doctoral dissertation under the supervision of Gauss. He also attended the lectures of physicist Wilhelm Weber and assisted in his laboratory, and studied with the natural philosopher Johann Friedrich Herbart, absorbing concepts that would appear in his later mathematical works. The university system in which he found himself had no provision for talented scientists without independent means, and Riemann spent the next ten years in a period of "honored starvation" and declining health, obtaining very little money for his lectures. During this time, he developed his ideas about the nature of complex functions and Abelian integrals, manifolds, and the foundations of geometry. He also worked on a group of papers in mathematical physics.

## Presents Work on Complex Variables

In 1851, Riemann completed his doctoral dissertation on a general theory of functions of a complex variable, introducing topological methods into the theory. In this paper, which introduced what are known as Riemann surfaces, he gave a magnificently simple and panoramic view of a subject to which Gauss and **Augustin–Louis Cauchy** had contributed, but which had been previously tortuous and difficult to access. Gauss, who was usually aloof and critical, was very pleased with his student's work, describing Riemann has having a "gloriously fertile originality." In his theory, Riemann employed a modification of a procedure attributed to Dirichlet. A criticism of this procedure by **Karl Weierstrass** delayed full acceptance of Riemann's approach. Riemann's explanations had been clear to physicists but because of the nature of the proofs had presented difficulties for mathematicians. Only in 1900 was Riemann's work fully validated by mathematician **David Hilbert**.

In 1853, to qualify as a "Privatdozent" or university lecturer, Riemann was required to present a postdoctoral thesis to the Council of Göttingen University. He chose to present a thesis on Fourier series, a problem Dirichlet had addressed, and in it introduced the Riemann integral. Masterfully written, this work established Riemann as Dirichlet's equal. Riemann was also required to give a lecture, for which he had to submit three topics to the university faculty. Two were on electricity, something Riemann had thought about for a long time. To his surprise and dismay, at Gauss' recommendation the faculty chose his third topic, on the foundations of geometry. Riemann's lecture, given in 1854, was brilliant, however. In it he introduced the main ideas of modern differential geometry.

As a new lecturer, Riemann was delighted when a relatively large class of eight students appeared to hear him speak about partial differential equations and their application to physical problems. Riemann was attracted by many problems at once. In the ensuing few years he developed material which did for algebraic geometry what he had done for differential geometry in his postdoctoral lecture. He also wrote on electrodynamics and the propagation of waves.

In 1855, Gauss died and Dirichlet was brought in to take his place. Dirichlet, Riemann's friend and collaborator managed to obtain a small government stipend for Riemann with some difficulty. At 200 thalers, it amounted to about one–tenth of a full professor's salary. Two years later, Riemann, whose health was rapidly declining, was made extraordinary (or assistant) professor and his salary was raised to 300 thalers. Shortly thereafter, his father died and the family home at Quickborn was broken up.

When Dirichlet died in 1859, Riemann succeeded him as full professor. For the first time he had a salary sufficient to support himself, and he began to receive well–deserved honors. He was elected a corresponding member of the Berlin Academy of Sciences in "physical–mathematics" based on his doctoral dissertation and his theory of Abelian functions. In gratitude he submitted to the Academy an article "On the Number of Primes Not Exceeding a Given Bound." He also became a corresponding member of the Bavarian Academy of Sciences and a full member of the Göttingen Academy of Sciences. In 1860 he visited Paris where he met with the eminent French mathematicians **Charles Hermite**, Joseph Bertrand, Victor Puiseux, Charles Briot, and Jean Claude Bouquet.

## Spends Final Years in Italy

In 1862, Riemann married Elise Koch, a friend of his sister's. A month thereafter he had an attack of pleurisy and his tuberculosis achieved the upper hand. His friends secured a leave and money for travel, and Riemann continued his studies in the less–harsh climate of Italy, with intermittent returns to Göttingen. In Pisa, he was welcomed by the mathematicians Enrico Betti and **Eugenio Beltrami**, who were inspired by his ideas in their own work. In 1863, Riemann's daughter, Ida, was born in Pisa. In November of 1865 he returned again to Göttingen and was able to work a few hours each day through the cold, wet winter. In March of 1866 he was elected a foreign member of the Paris Académie Royale des Sciences, and in June he was elected a member of the Royal Society of London. It was clear that Riemann was dying, however, and to conserve his remaining days, he returned to Italy.

On July 20, 1866, Riemann died in Selasca, Italy, on Lake Maggione, with his wife beside him. He was buried in nearby Biganzola. At the time of his death he was attempting a unified explanation of gravity and light (a pursuit that is still elusive) and a theory of mechanics of the human ear, an investigation inspired by Hermann von Helmholtz.

In 1876, Riemann's works were collected and published by Heinrich Weber and Riemann's friend, Richard Dedekind, who was one of three who attended Riemann's lectures of 1855–1856. Although his published work consists of only one volume, his ideas continue to inspire mathematical inquiry.

## SELECTED WRITINGS BY RIEMANN

*The Collected Works of Bernhard Riemann.* Edited by Heinrich Weber, with the assistance of Richard Dedekind. With a supplement edited by M. Noether and W. Wirtinger, and a new introduction by Hans Lewy, 1953.

*Julia Bowman Robinson*

## FURTHER READING

### Books

Monastyrsky, Michael. *Riemann, Topology and Physics*. Translated by James King and Victoria King. Edited by R. O. Wells, Jr. Boston, Birkhauser, 1986.

Stillwell, John. *Mathematics and Its History*. New York: Springer–Verlag, 1989, pp. 215–219.

*—Sketch by Jill Carpenter*

# Julia Bowman Robinson
## 1919-1985
### American logician and number theorist

Julia Robinson was instrumental in the solution of **David Hilbert**'s tenth problem; over a period of two decades, she developed the framework on which the solution was constructed. In recognition of her accomplishments, she became the first woman mathematician elected to the National Academy of Sciences, the first female president of the American Mathematical Society, and the first female recipient of the MacArthur Foundation Prize for contributions to mathematics. Under pressure to provide information about her life, she finally asked her sister, Constance Reid, to write something for her. Just before she died, she and Reid collaborated on her "autobiography," which was printed in *The College Mathematics Journal*. Robinson concluded the article with this remark: "Rather than being remembered as the first woman this or that, I would prefer to be remembered, as a mathematician should, simply for the theorems I have proved and the problems I have solved."

Robinson was born December 8, 1919, in St. Louis, Missouri. Her mother, Helen Hall Bowman, died two years later; Robinson and her older sister went to live with their grandmother near Phoenix. The following year their father, Ralph Bowman, retired and joined them in Arizona after losing interest in his machine tool and equipment business. He expected to support his children and his new wife, Edenia Kridelbaugh Bowman, with his savings. In 1925, the family moved to San Diego; three years later a third daughter was born.

### Health Problems Did Not Deter Interest in Math

At the age of nine, Robinson contracted scarlet fever, and the family was quarantined for a month. They celebrated the end of isolation by viewing their first talking motion picture. The celebration was premature, however, as Robinson soon developed rheumatic fever and was bedridden for a year. When she was well enough, Robinson worked with a tutor for a year, covering the required curriculum for the fifth through eighth grades. She was fascinated by the tutor's claim that it had been proven that the square root of 2 could not be calculated to a point where the decimal began to repeat. Her interest in mathematics continued at San Diego High School; when she graduated with honors in mathematics and science, her parents gave her a slide rule that she treasured and named "Slippy."

Robinson entered San Diego State College at the age of 16. She majored in mathematics and prepared for a teaching career, being aware of no other mathematics career choices. At the beginning of her sophomore year, her father found his savings depleted by the Great Depression and committed suicide. With help from her older sister and an aunt, Robinson remained in school. She transferred to the University of California in Berkeley for her senior year and graduated in 1940. Robinson later recalled that at Berkeley she felt like the proverbial ugly duckling who suddenly realized she was really a swan, as she found teachers and fellow students who shared her excitement about mathematics.

In December 1941, she married Raphael Robinson, a faculty member in the mathematics department at Berkeley. At that time, she was a teaching assistant, having completed her master's degree earlier that year. When the new year began, however, the school's nepotism rule prevented her from teaching in the mathematics department. Instead, she worked in the Berkeley Statistical Laboratory on military projects. Robinson became pregnant but lost the baby; because of heart damage caused by the rheumatic fever, her doctor warned against future pregnancies. Having her hopes of motherhood crushed, Robinson endured a period of depression that lasted until her husband rekindled her interest in mathematics.

### Mathematics Anchors Her Life

In 1947, Robinson embarked on a doctoral program under the direction of **Alfred Tarski**; throughout her career, she regarded him as a logician of the caliber of **Kurt Gödel**. In her dissertation, Robinson proved the algorithmic unsolvability of the theory of the rational number field, and her Ph.D. was conferred in 1948.

While working at the RAND Corporation in 1949-1950, Robinson developed an iterative solution for the value of a finite two-person zero-sum game. The problem involves two individuals playing an infinite sequence of games, with the assumption that each player would keep track of how his opponent was playing and would respond with a move that would bring him the largest possible payoff. Robinson proved that the optimal payoffs for the two players will converge to the value of the game as the number of plays increases. Her only contribution to game theory is still considered a fundamental theorem in the field.

In late 1948, Tarski discussed an idea about diophantine equations (polynomial equations of several variables, with integer coefficients, whose solutions are to be integers) with Raphael Robinson, who shared it with his wife. By the time she realized it was directly related to the tenth problem on David Hilbert's famous list (to find an effective method for determining whether a given diophantine equation is solvable with integers), she was too involved in the topic to be intimidated by its stature. For the next 22 years, Robinson attacked various aspects of the problem, building a foundation for a 1970 proof by Russian mathematician Yuri Matijasevich, who showed that the desired general method for determining solvability does not exist.

### Humbly Accepts Honors

Robinson's fame from the Hilbert problem solution resulted in her appointment as a full professor at Berkeley in 1976, although she was expected to carry only one-fourth of the normal teaching load. The heart damage left from her childhood bout with rheumatic fever was surgically corrected in 1961, but Robinson's health remained impaired.

Throughout her career, Robinson had been fascinated by Hilbert's tenth problem. In her autobiography, she recalled that on her birthday each year, "when it came time for me to blow out the candles on my cake, I always wished, year after year, that the Tenth Problem would be solved—not that I would solve it, but just that it would be solved. I felt that I couldn't bear to die without knowing the answer." When the solution was finally published, Robinson downplayed her role in making it possible. However, according to an article in *The Mathematical Intelligencer,* a 1969 paper of Robinson's triggered Yuri Matijasevich's solution. Matijasevich wrote, "Somewhere in the Mathematical Heavens there must have been a God or Goddess of Mathematics who would not let me fail to read Julia Robinson's new paper. . . . I saw at once that [she] had a fresh and wonderful new idea." In fact, the solution followed so naturally from Robinson's work that some scholars wondered why she had not completed the solution herself. Aware of such questions, she responded in her autobiography that "no one else saw it either. There are lots of things, just lying on the beach as it were, that we don't see until someone else picks one of them up. Then we all see that one."

In recognition of her valuable contributions to the solution of this famous problem, Robinson received several significant awards. One of these, her election to the presidency of the American Mathematical Society, prompted some soul searching on her part. She was a quiet person who tended to avoid the limelight, but she accepted this position because of her dedication to gender equality. She put aside her shyness and poured her energy into the post. It was during the society's annual conference while she was president that she learned she was suffering from leukemia. After battling the disease for nearly a year, she died in Berkeley on July 30, 1985.

## SELECTED WRITINGS BY ROBINSON

(With Martin Davis and Yuri Matijasevich) "Hilbert's Tenth Problem: Diophantine Equations: Positive Aspects of a Negative Solution." *Mathematical Developments Arising From Hilbert's Problems.* Edited by Felix F. Browder. 1976, pp. 323–78, plus loose erratum.

## FURTHER READING

### Books

Reid, Constance, with Raphael M. Robinson. "Julia Bowman Robinson," in *Women of Mathematics.* Edited by Louise S. Grinstein and Paul J. Campbell. Westport, CT: Greenwood Press, 1987, pp. 182–89.

**Periodicals**

"Julia Bowman Robinson, 1919-1985." *Notices of the American Mathematical Society* (November 1985): 738-42.

Matijasevich, Yuri. "My Collaboration with Julia Robinson." *The Mathematical Intelligencer* (Fall 1992): 38-45.

Reid, Constance. "The Autobiography of Julia Robinson." *The College Mathematics Journal* (January 1986): 2-21.

Smorynski, C. "Julia Robinson, In Memoriam." *The Mathematical Intelligencer* (Spring 1986): 77-79.

**Other**

Bricker, Julie. "Julia Robinson." *Biographies of Women Mathematicians.* June 1997. http://www.scottlan.edu/lriddlc/women/chronol.htm (July 21, 1997).

—*Sketch by Loretta Hall*

# Michel Rolle
## 1652-1719
### French algebraist

Michel Rolle was a largely self-educated mathematician who taught himself both algebra and Diophantine analysis, a method for solving equations with no unique solution. He won early repute when he solved a problem set by the mathematician Jacques Ozanam, but his real passion lay in the field of the algebra of equations. His most famous work was the *Traite d'algebre* of 1690, in which he not only invented the modern notation for the $n$th root of x, but also expounded on his "cascade" method to separate the roots of an algebraic equation. He 's best remembered for the theorem which bears his name, Rolle's Theorem, which determines the position of roots in an equation.

Little is known about Rolle's youth. He was born in Ambert, Bass-Auvergne in France on April 21, 1652. His father was a shopkeeper and the young Rolle had only a basic elementary education before he began to earn a living as a scribe for a notary, and then for various attorneys near Ambert. The allure of the big city brought Rolle to Paris by age 24, and an early marriage and family kept him in secretarial and accountancy work. But his curious mind led him to a study of algebra and of Diophantine analysis. The work of the third-century Alexandrian mathematician, Diophantus, had been translated into Latin by Bachet de Meziriac, and Rolle made himself familiar with the solutions to both determinate equations, for which there exists a unique solution, and indeterminate equations. Diophantus had provided a method of analysis for indeterminate equations, and Rolle became an expert in this field.

### Early Success Wins Patronage

Rolle was also familiar with the work of another Frenchman, a contemporary of his and also a self-taught mathematician, Jacques Ozanam whose interests were in recreational mathematics—puzzles and tricks—and analysis. In 1682, Rolle published "an elegant solution" to a problem set by Ozanam, according to Jean Itard in *Dictionary of Scientific Biography*. The problem was as complex as the solution. Ozanam proposed finding four numbers which corresponded in the following ways: the difference of any two of these numbers not only make a perfect square, but also equal the sum of the first three. Rolle's solution as published in *Journal des scavans*, brought him notoriety and some patronage. The controller general of finance, Jean-Baptiste Colbert, took notice of him and arranged an honorary pension for Rolle so that he could continue with his mathematical work. Rolle also became the tutor to the son of a powerful minister, Louvois, and for a short time held an administrative post at the ministry of war.

### Develops Rolle's Theorem

Rolle became a member of the Academie Royal des Sciences in 1685. He published in 1690 his most famous work, *Traite d'algebre,* a book that has remained well known to this day. Algebra was a departure for Rolle, away from Diophantine analysis in which he had won early recognition, and into the algebra of equations. Rolle broke new ground in many aspects with this work. His new notation for the $n$th root of a number was first published in *Traite d'algebre*, and thereafter became the standard notation. Rolle also made important advances in systems of affine equations, which assign finite values to finite quantities, building on the work of Bachet de Meziriac, another contemporary whose work on mathematical tricks and puzzles led to the development of the field of recreational mathematics. But primarily the *Traite d'algebre* has retained its fame for the exposition of Rolle's method of "cascades." Utilizing a technique developed by Dutch mathematician Johann van Waveren Hudde to find multiple roots of an equation, or the highest common factor of a polynomial, Rolle was able to separate the roots of an algebraic equation. His "cascades" correspond to the derivatives of the highest common factor of a polynomial. Though never clearly defined, it is implied that such "cascades" are the result of the following: if in an equation $f(x) = 0$, $f(x)$ is multiplied by a progression, then the simplified result equated to zero would be such a "cascade." However, mathematicians of the day complained that Rolle's theory of cascades was given with too little proof.

To rectify that, Rolle published *Demonstration d'une Methode pour resoudre les Egalitez de tous les degrez* in 1691, essentially a little known work which further demonstrated the method of "cascades." However, in the process, Rolle developed the calculus theorem which bears his name. According to this theorem, if the function $y = f(x)$ is differentiable on the open interval from a to b and continuous on the closed interval from a to b, and if $f(a) = f(b) = 0$, then $f'(x) = 0$ has at least one real root between a and b. It was not for another century and a half that Rolle's name was given to the theorem he developed, a theorem that is a special case of the mean–value theorem, one of the most important theorems in calculus.

### The Pensionary Geometer

Rolle's work encompassed also a certain flair for Cartesian geometry, though he did break with Cartesian techniques in other respects. In 1691, he was one of a vanguard to go against the Cartesian grain in the order relation for the set of real numbers, noting, for example, that –2a was a larger quantity than –5a. He also described the calculus as a collection of ingenious fallacies. Rolle published another important work on solutions of indeterminate equations in 1699, *Methode pour resoudre les equations indeterminees de l'algebre,* the year that he became a *pensionnaire geometre* at the Academie. This was a post that carried with it a regular salary which further enabled him to devote more time to mathematics. At this time, members of the Academie were drawing up sides on the value of infinitesimal analysis, in which a variable has zero as a limit. Rolle became one of its most outspoken critics. Though by 1706 he finally and formally recognized the value of the new techniques, his very opposition and criticism had served to help develop the new discipline of infinitesimal analysis.

Two years later, in 1708, Rolle suffered a stroke, which diminished his mental powers. Though he recovered, a second stroke in 1719 killed him. He is remembered not only for Rolle's Theorem and the method of "cascades" for separating and determining the position of roots in an algebraic equation, but also for his break with Cartesian techniques.

### SELECTED WRITINGS BY ROLLE

#### Books

*Traite d'algebre, ou principes generaux pour resoudre les questions de mathematique,* 1690

### FURTHER READING

Disckson, Leonard E. *History of the Theory of Numbers.* Volume 2, New York: Chelsea, 1952, pp. 45, 447.

Itard, Jean. "Rolle, Michel." *Dictionary of Scientific Biography.* Volume XI. Edited by Charles Coulston Gillispie. New York: Charles Scribner's Sons, 1975, pp. 511–513.

Loria, G. *Storia delle matematiche.* Milan: 1950, pp. 670–3.

Smith, David Eugene. *A Source Book in Mathematics.* New York: McGraw–Hill, 1929, pp. 253–60.

*—Sketch by J. Sydney Jones*

# Mary Ellen Rudin
## 1924–
### American topologist

Mary Ellen Rudin's mathematical specialty is set theoretic topology, which comprises a modern, abstract geometry that deals with the construction, classification, and description of the properties of mathematical spaces. Rudin's approach is often to construct examples to disprove a conjecture. As an incoming freshman at the University of Texas, Rudin was chosen by the topologist R. L. Moore and trained almost exclusively by the unorthodox "Moore method" of active and competitive mathematical problem–solving. Rudin credits Moore with building her confidence that given the axioms, she should be able to solve any problem, even if it involves building a complicated structure.

Rudin was born December 7, 1924, in Hillsboro, Texas, to Irene Shook and Joe Jefferson Estill. Her father was a civil engineer and her mother, before she married, was a teacher. Rudin's parents were from Winchester, Tennessee, and both of her grandmothers were graduates of Mary Sharp College in Winchester. Advanced education was valued in both families, and her parents expected that she would go to the university to "do something interesting."

Rudin grew up in Leakey, Texas, a small isolated town in the hills of southwest Texas, where her father worked on road building projects. Her childhood surroundings were simple and primitive, and as a child she had lots of time to think and to play elaborate, complicated and imaginative games, something she says contributed to her later success as a mathematician. Rudin's performance was generally in the middle of her class of five students, and she expected she would make C's at the university; she made A's.

### Is Trained by the R. L. Moore "Method"

Rudin had no special course of study in mind when she went to the University of Texas at Austin. On her first day, Moore helped her register for classes, asked about her mathematical background, and enrolled her in his mathematics class. Although she took courses in other fields and was good at them, she continued to study with Moore through her B.S. degree in 1944 and her Ph.D. degree in 1949, and had a class from him every single semester. Rudin has said, "I am a mathematician because Moore caught me and demanded that I become a mathematician."

Moore, known for his unorthodox Socratic teaching style, preferred his students to be naive; he required that they be unspoiled by mathematical terminology, notation, methods, results, or ideas of others. He also required his students to actively think, rather than passively read. Moore forbade them to read the work of others and sometimes removed books from the library so that his students would not see them. He never referred to the work of others; rather, he gave definitions and required his students to prove theorems, some that had been solved, some that had not. Students were required to think about problems just as research mathematicians do. In the classroom, Moore called on the weakest students first, then proceeded through the class to the top students. Rudin was generally at the top of the class.

Rudin solved one of the unsolved problems as her thesis research, finding a counterexample to a well-known conjecture. At the time she wrote her thesis, she had never seen a mathematics paper. While the Moore method produced students who were independent, confident and creative, there were lacunae in their knowledge of mathematics and deficient in their mathematical language. Rudin has used a more traditional approach in her own teaching, requiring her students to learn as much as possible about what has been done by others, but she acknowledges that her students are not always as confident as she was.

### Research at Duke, Rochester, and Wisconsin

After Rudin received her Ph.D. in 1949, Moore told her she would be going to Duke University. At Duke, she worked on a problem related to Souslin's conjecture and began to be known for her work. She also met mathematician Walter Rudin at the university and married him in 1953. Together, they went to the University of Rochester where Walter had a position, and Rudin taught part-time and researched mostly as she pleased until 1958, when they moved to Madison, Wisconsin. Rudin held a similar position at the University of Wisconsin; she was a lecturer from 1959 to 1971, when she was appointed full professor.

Rudin has likened facility in research in mathematics to a career in music. "It must done every day," she says. "If you don't play for three years, you're not likely to be of concert pianist quality when you start playing again, if ever again." She also notes that mathematical research requires a high tolerance for failure; successes are much less frequent than failures; she may have three exciting breakthroughs in a year. Rudin prefers to work on topological problems while lying on the couch in her Frank Lloyd Wright–designed home, surrounded by the activities of her family. Rudin's productivity has been strong and consistent; she has almost 90 scientific papers and book chapters to her credit.

Rudin's counterexamples, she says, are "very messy" topological spaces that show that some ideas you thought were true, are not. She compares her many–dimensional examples, which are difficult for some people to imagine, to a business problem that has 20 aspects. The aspects are the dimensions of the problem, the number of factors taken into account.

Rudin has been at the University of Wisconsin since 1959, where she assembled a strong research group in topology. She is the first to hold the **Grace Chisholm Young** Professorship, which she assumed in 1981. Her visiting professorships include stints in New Zealand, Mexico, and China. Moore imbued Rudin with a sense of responsibility for publication and responsibility to the mathematical community, and she has held offices and worked on numerous committees of the American Mathematical Society, the Mathematical Association of America, the Association for Women in Mathematics, and the Association for Symbolic Logic. Over the years she has received several research grants from the National Science Foundation, and has served on mathematical advisory boards for the National Academy of Sciences, the National Science Foundation, and the United Nations.

Rudin and her husband have four children, and she notes that they are all skilled in pattern recognition. Now professor emerita, Rudin continues to lecture widely, produce vital papers in her field, and to promote and speak about women in mathematics. She is a Fellow of the American Academy of Arts and Sciences, has received the Prize of Niewe Archief voor Wiskunk from the Mathematical Society of the Netherlands, and has been awarded four honorary doctorates. In 1995 Rudin was elected to the Hungarian Academy of Sciences.

## FURTHER READING

### Books

Albers, Donald J., Gerald L. Alexanderson and Constance Reid. *More Mathematical People: Contemporary Conversations.* Boston: Harcourt Brace Jovanovich, 1990, pp. 282–303.

### Periodicals

Ford, Jeff. "Geometry with a Twist." *Research Sampler* (Spring 1987): 20–23.

### Other

Carr, Shannon. "Mary Ellen Rudin." *Biographies of Women Mathematicians.* June 1997. http://www.scottlan.edu/lriddle/women/chronol.htm (July 21, 1997).

—*Sketch by Jill Carpenter*

# Christoff Rudolff (also transliterated as Rudolf)
## 1499(?)–1545(?)
### Polish–born Austrian algebraist

Although few details are available regarding the life of Christoff Rudolff, his impact on the field of algebra is widely known. Rudolff is the author of *Coss*, the first comprehensive German algebra book. Algebra was referred to as the cossic art during Rudolff's lifetime, and algebraists were known as cossists. The word "cosa" means "thing" and it was used to stand for the unknown.

Rudolff is believed to have been born around 1499 in Jauer, Silesia (now Jawor, Poland). His references in *Coss* suggest that he received his early mathematical training through the study of printed books, and solutions presented in *Coss* indicate that he had studied the work of Johannes Wildman, Robert of Chester, a treatise compiled by Johann Vögelin, and Regensburg algebra. He attended the University of Vienna in Austria between 1517 and 1521, and studied algebra. Rudolff remained in Vienna after completing his studies and taught as a private mathematics instructor. Although not affiliated with the university, he was permitted to use its library and conducted much of the research for *Coss* there. When the book was published, his critics accused him of stealing his examples from works in the University of Vienna library. **Michael Stifel** defended Rudolff's work in his preface to a new

edition of *Coss.* Rudolff planned to publish a version of *Coss* in Latin, but never did so. He did publish two additional books on mathematics in German, *Künstliche Rechnung mit der Ziffer und mit de Zahlpfennigen,* in 1526, and *Exempelbüchlin* in 1530.

But it is *Coss* that stands as Rudolff's seminal work and contribution to his field. *Coss* is written in two parts, with the first 12 chapters covering topics that readers must first familiarize themselves before studying the solution of equations, or algebra. Topics covered in these chapters include the rule of three, square and cube roots, summation formulas, algebraic polynomials, and roots, binomials, and residues.

The latter half of *Coss,* which is divided into three sections, covers first– and second–degree equations and their higher degree variations. Unlike algebraists before him, Rudolff bases his theories on the assumption of the existence of only eight distinct rules, rather than 24. Rudolff later recanted his explanation of the sixth rule ($ax^2 + c = bx$), which he had contended lent itself to only one solution.

The second section of the latter half of *Coss* listed rules for solving equations, and the third section offered more than 400 examples of algebraic problems involving both abstract numbers and equations applicable to daily life. These are presented in a manner resembling that used in recreational mathematics and some introduce a second unknown. Problems in which there are more unknowns than equations are deemed "indeterminate." Nonetheless, Rudolff provided all the possible solutions to these problems.

Rudolff concludes *Coss* with three cubic problems for which he does not offer solutions, stating that he wished the problems to provoke further research. An addendum to *Coss* contains a "verbal computation," $3 + \sqrt{2}$, which Stifel suggested hinted at the then unknown solution to the cubic equation.

In *Künstliche Rechnung mit der Ziffer und mit den Zahlpfennigen,* Rudolff addresses the concept of computing with whole numbers and fractions for beginners, as well as the rule of three and 300 example problems related to the industrial culture which was developing around him. The *Exempelbüchlin* contains 293 problems, tables of measurements, a list of gauging symbols, and hints for solving problems.

With his published works, most notably *Coss,* Rudolff made several contributions to modern mathematics, including incorporating the use of decimal fractions and introducing symbols for the second, third, and fourth roots. The works are also considered precursors to exponential arithmetic and the concept of logarithms.

Despite its significance today, the popularity of *Coss* was short lived, although *Künstliche Rechnung* was well known and was reprinted in 1588. Rudolff is believed to have died in Vienna, Austria in 1545.

## SELECTED WRITINGS BY RUDOLFF

*Coss*, 1525
*Exempelbüchlin*, 1530
*Künstliche Rechnung mit der Ziffer und mit de Zahlpfennigen*, 1526

## FURTHER READING

### Books

Vogel, Kurt. "Christoff Rudolff," in *Dictionary of Scientific Biography*. Volume XI. Edited by Charles Coulston Gillispie. New York: Charles Scribner's Sons, 1970, pp. 598–600.

### Other

"Christoff Rudolff." *MacTutor History of Mathematics Archive.* http://www–groups.dcs.st–and.ac.uk/~history/Mathematicians/index.html (July 1997).

—*Sketch by Kristin Palm*

# Paolo Ruffini
## 1765–1822
### Italian algebraist and educator

Paolo Ruffini made significant contributions in the areas of medicine and philosophy, as well as mathematics, where he developed the theory that a quintic equation cannot be solved by radicals. This theory later came to be known as the Abel–Ruffini theorem, named after Ruffini and the Norwegian mathematician **Niels Abel**, who published the first accepted version of this theorem. Ruffini also played a crucial role in the development of group theory.

Ruffini was born September 22, 1765, in Valentano, Papal States (also known as Valentano, Viterbo and now part of Italy), to Basilio Ruffini, a physician, and his wife, Maria Francesca Ippoliti. He moved with his parents to Modena, Duchy of Modena (now also part of Italy) when he was a teenager and attended the University of Modena, where he studied medicine, philosophy, literature, and mathematics. He practiced mathematics with many well–known instructors and mathematicians, including Luigi Fantini, who taught geometry, and Paolo Cassiani, who

taught infinitesimal calculus. The well–rounded Ruffini excelled in all his classes, so much so that when Cassiani took a leave from his foundations of analysis course in the fall of 1787 to accept a position as councillor of the Este domains, Ruffini, at that time still a student, instructed the class himself.

Ruffini graduated from the university with a degree in philosophy and medicine the following summer, and soon after that obtained a degree in mathematics as well. That fall he was appointed professor of the foundations of analysis. He remained in that position until 1791, when he replaced Fantini as professor of the elements of mathematics. That same year Ruffini obtained his license to practice medicine by the Collegiate Medical Court of Modena. In addition to work in the academic field, he also began to practice medicine.

Ruffini's professional situation changed drastically in 1796, however, when Napoleon's troops occupied Modena and he was appointed representative from the department of Panaro to the Junior Council of the Cisalpine Republic, against his wishes. Ruffini was relieved of his post in 1798 and returned to scientific research, but was banned from teaching or holding a public office after he refused to swear an oath of allegiance to the republic, citing religious reasons. Ruffini continued to practice medicine and conduct mathematical research, and during this time published what later became known as the Abel–Ruffini theorem, which stated that a general algebraic equation of greater than the fourth degree could not be solved using only radicals, such as square roots, cube roots, etc.

The theorem was first published in *Teoria generale delle equazioni* in 1799, and later in revised form in *Riflessioni intorno alla soluzione delle equazioni algebriche generalie* in 1813. Upon publication, the theorem was met with skepticism among other mathematicians, with the exception of **Augustin–Louis Cauchy**, who told Ruffini in an 1821 letter that his work deserved the attention of the mathematics community and that he wholeheartedly agreed with Ruffini's findings. The theorem gained general acceptance when it was proven by Abel in 1824.

Ruffini also developed the theory of substitutions, considered a significant contribution to the theory of equations. This theory laid the foundation for **Évariste Galois'** general theory of the solubility of algebraic equations and served as an important forerunner of group theory.

Ruffini later developed the basic rule for determining the quotient and the remainder resulting from the division of a polynomial in the variable x by a binomial of the form x–a, using approximation by means of infinite algorithms. Despite a growing trend toward acknowledging the uncertainty of the foundations of infinitesimal analysis, Ruffini, with the support of Cauchy and Abel, demanded rigor of himself and others. During his tenure as president of the Societa Italiana dei Quaranta, he refused to accept to papers by Giuliano Frullani because they relied on series for which the convergence had not been demonstrated.

Ruffini moved on to apply his manner of thinking to philosophical and biological matters, determining that the faculties of the soul could not be measured because they do not correspond to magnitudes. Ruffini returned to academia with the fall of Napoleon in Modena in 1814. At this time, Ruffini was appointed rector of the university, as well as chair of the mathematics and practical medicine departments.

Ruffini continued as a practicing physician as well, treating impoverished and royal citizens of Modena alike. He contracted typhus while aiding victims of the 1817–18 typhus epidemic, but recovered enough to write about the experience from a medical and personal perspective. In 1822, however, he contracted chronic pericarditis, accompanied by a serious fever, and he died on May 10 of that year.

## SELECTED WRITINGS BY RUFFINI

*Sopra la determinazione delle radici delle equazioni numeriche di qualunque grado*, 1804
*Teoria generale delle equazioni in cui si dimostra impossibile la soluzione algebrica delle equazioni generali di grado superiore al quarto*, 2 volumes, 1799

## FURTHER READING

### Books

Carruccio, Ettore. "Paolo Ruffini." *Dictionary of Scientific Biography*. Volume XI. Edited by Charles Coulston Gillispie. New York: Charles Scribner's Sons, 1970, pp.598–600.

### Periodicals

Ayoub, R.G. "Paolo Ruffini's Contributions to the Qunitic." *Archive for History of Exact Science* 23 (1980).

### Other

"Paolo Ruffini." *MacTutor History of Mathematics Archive.* http//www–groups.dcs.st–and.ac.uk/~history/Mathematicians/index.html (August 26, 1997).

—*Sketch by Kristin Palm*

*Bertrand Russell*

# Bertrand Russell
## 1872-1970
### English logician and philosopher

A seminal figure in 20th-century mathematical logic and philosophy, Bertrand Russell produced more than seventy books and pamphlets on topics ranging from mathematics and logic to philosophy, religion, and politics. He valued reason, clarity, fearlessness, and independence of judgment, and held the conviction that it was the duty of the educated and privileged classes to lead. Having protested against Britain's participation in World War I and against the development of nuclear weapons, Russell was imprisoned twice for his convictions. In writing his own obituary, he described himself as a man of unusual principles, who was always prepared to live up to them.

Bertrand Arthur William Russell was born on May 18, 1872, in Ravenscroft, Trelleck, Monmouthshire, England. His family tree can be traced back to John Russell, a favorite courtier of Henry VIII. His grandfather, Lord John Russell, served as Prime Minister under Queen Victoria. Bertrand Russell's father, Lord Amberley, served in Parliament briefly, but was defeated because of his rejection of Christianity and his advocacy of voting rights for women and birth control. His mother was Kate Stanley, the daughter of a Liberal politician. Orphaned before the age of five, Russell was raised by his paternal grandmother, a Scotch Presbyterian with strong moral standards who gave duty and virtue greater priority than love and affection. His grandmother did not have confidence in the moral and religious environment at boarding schools, so, after briefly attending a local kindergarten, he was taught by a series of governesses and tutors. Later he wrote in his *Autobiography* that he had been influenced by his grandmother's fearlessness, her public spirit, her contempt for convention, and her indifference to the opinion of the majority. On his twelfth birthday she gave him a Bible in which she had written her favorite texts, including "Thou shalt not follow a multitude to do evil." It was due to her influence, he felt, that in later life he was not afraid to champion unpopular causes.

### Intrigued by Euclid

When Russell was eleven, his brother Frank began to teach him Euclidean geometry. As he recounts in his *Autobiography,* "This was one of the great events of my life, as dazzling as first love. I had not imagined that there was anything so delicious in the world." Russell was greatly disappointed, however, to find out that **Euclid** started with axioms which were not proved but were simply accepted. Russell refused to accept the axioms unless his brother could give a good reason for them. When his brother told him that they couldn't continue unless he accepted the axioms, Russell relented, but with reluctance and doubt about the foundations of mathematics. That doubt remained with him and determined the course of much of his later work in mathematics.

At age 15, Bertrand left home to take a training course to prepare for the scholarship examination at Trinity College, Cambridge. Lonely and miserable, he considered suicide but rejected the idea because, as he later said, he wanted to learn more mathematics. He kept a secret diary, written using letters from the Greek alphabet, in which he questioned the religious ideas he had been taught. After a year and a half, he took the Trinity College examination and won a scholarship. One of the scholarship examiners was **Alfred North Whitehead.** Apparently impressed with Russell's ability, Whitehead arranged for him to meet several students who soon became his close friends. He was invited to join a group, "The Apostles," which met weekly for intense discussions of philosophy and history. His chief interest and chief source of happiness, he said, was mathematics; he received a first class degree in mathematics in 1893. He found, however, that his work in preparing for the exams had led him to think of mathematics as a set of tricks and ingenious devices, too much like a crossword puzzle; this disillusioned him. Hoping to find some reason for supposing mathematics to be true, he turned to philosophy. He studied the philosophy of idealism, which was popular in Cambridge at the time, concluding that time, space and matter are all illusions and the world resides in the mind of the beholder. He took the Moral Science examination in 1894 and received an honors degree.

At 17, Russell had fallen in love with Alys Pearsall Smith, an American from a wealthy Philadelphia Quaker family, and after graduation they became engaged. His grandmother did not approve. Hoping he would become interested in politics and lose interest in Alys, she arranged for him to become an attaché at the British Embassy in Paris. He found his work at the Embassy boring, and upon his return three months later, he and Alys were married.

In 1895, Russell wrote a dissertation on the foundations of geometry, which won him a fellowship at Trinity College and enabled him to travel and study for six years. He and Alys went to Berlin to study the German Socialist movement and the writings of Karl Marx. During his travels, he formulated a plan to write a series of books on the philosophy of the sciences, starting with mathematics and ending with biology, growing gradually more concrete, and a second series of books on social and political questions, working to more abstract issues. During his long life, he managed to fulfill much of this plan. As a result, according to biographer Ronald Clark, no other Englishman of the twentieth century was to gain such high regard in both academic and non-academic worlds.

Russell returned to London in 1896 and lectured on his experiences to students of the London School of Economics and the Fabian Society, publishing his studies as *German Social Democracy,* the first of his numerous books and pamphlets. In this publication he demonstrated his ability to investigate a subject quickly and then present it in clear and convincing language. He and Alys traveled to the United States, where they visited Alys's friend, the poet Walt Whitman. Russell lectured on non-Euclidean geometry at Bryn Mawr College, where Alys had been a student, and at Johns Hopkins University. In 1900, he was asked to lecture on German mathematician **Gottfried Wilhelm von Leibniz** at Cambridge. He wanted to show that mathematical truths did not depend on the mathematician's point of view. He re-examined the philosophy of idealism and abandoned it, concluding that matter, space and time really did exist. He published his views in *A Critical Exposition of the Philosophy of Leibniz.*

### Seeks Basic Principles of Mathematics and Logic

Russell then started work on an ambitious task—devising a structure which would allow both the simplest laws of logic and the most complex mathematical theorems to be developed from a small number of basic ideas. If this could be done, then the axioms which mathematicians accepted would no longer be needed, and both logic and mathematics would be part of a single system. At a conference in Paris in 1900, he met **Giuseppe Peano**, whose book on symbolic logic held that mathematics was merely a highly developed form of logic. Russell became interested in the possibility of analyzing the fundamental notions of mathematics, such as order and cardinal numbers, using Peano's approach. He published in 1903; its fundamental thesis was that mathematics and logic are identical.

Applying logic to basic mathematical concepts and working with **Georg Cantor**'s proof that a class of all classes cannot exist, Russell formulated his famous paradox concerning classes that are members of themselves—such as the class of classes. According to Russell's paradox, it is impossible to answer the question of whether the class containing all classes that are not members of themselves is a member of itself—for if it is a member of itself, then it does not meet the terms of the class; and if it is not, then it does. (This is the same kind of paradox as is found in the statement "Everything I say is a lie.") In developing this Theory of Types, however, Russell realized the absurdity of asking whether a class can be a member of itself. He concluded that if classes belong to a particular type, and if they consist of homogenous members, then a class cannot be a member of itself. In planning *The Principles of Mathematics,* an exposition of his ideas on mathematics and logic, Russell decided on two volumes, the first containing explanations of his claims and the second containing mathematical proofs. His former teacher, Alfred Whitehead, had been working on similar problems. Consequently, the two scholars decided to collaborate on the second part of the task. For nearly a decade they worked together, often sharing the same house, sending each other drafts and revising each other's work. The result, *Principia Mathematica,* was a separate work published in three volumes, the last in 1913. Before it was completed, Russell was elected to the Royal Society, of which Whitehead was already a member.

Russell was persuaded to run for Parliament in 1907 as a supporter of voting rights for women; he lost the election. Appointed a lecturer in logic and the principles of mathematics at Trinity College in 1910, he published a short introduction to philosophy, *The Problems of Philosophy,* in 1912, and *Our Knowledge of the External World* in 1914. Invited to Cambridge, Massachusetts, to give the Lowell Lectures and a course at Harvard University in 1914, he also lectured in New York, Chicago, and Ann Arbor, Michigan.

### Imprisoned for Opposing World War I

Russell was an outspoken critic of England's participation in World War I. He worked with the No-Conscription Fellowship to protest the drafting of young men into the army. In 1916 he gave a series of lectures in London which were published as *Principles of Social Reconstruction.* They were also published in the United States as *Why Men Fight: A Method of Abolishing the International Duel.* He was invited to give a lecture series at Harvard based on the book, but he was denied a passport because he had been convicted of writing a leaflet criticizing the imprisonment of a young conscientious objector. As a result of his conviction, he was dismissed from his lectureship at Trinity College. He wrote an open letter to Woodrow Wilson, the President of the United States, urging him to seek a negotiated peace rather than go to war. Since Russell's mail was being intercepted by the British government, he sent the letter with a young American woman. The story made the headlines in the *New York Times.*

Russell wrote articles for *The Tribunal,* published by the No-Conscription Fellowship. In one article, he predicted that the consequences of the war would include widespread famine and the presence of American soldiers capable of intimidating striking workers. He was charged with making statements likely to prejudice Britain's relations with the United States and was sentenced to six months in prison. Before his imprisonment, he wrote *Roads to Freedom: Socialism, Anarchism and Syndicalism.* While in prison he wrote *Introduction to Mathematical Philosophy,* which explained the ideas of *The Principles of Mathematics* and *Principia Mathematica* in relatively simple terms.

In 1920, Russell visited Russia, where he was disappointed with the results of the 1917 revolution, and China, where he lectured at the National University of Peking. The following year, Russell divorced his wife Alys and married Dora Black. His son John Conrad was born, and in 1922 his daughter Katharine Jane was born. In 1927 he and Dora established Beacon Hill School for their children and others. He traveled to the United States for lecture tours in 1924, 1927, 1929 and 1931, speaking on political and social issues. He divorced Dora in 1935 and married Patricia Spence the following year. A son, Conrad Sebastian Robert, was born in 1937.

Russell lived in the United States during World War II, from 1938 to 1944, lecturing at the University of Chicago, the University of California at Los Angeles, Bryn Mawr, Princeton, and the Barnes Foundation at Merion, Pennsylvania. He was invited to teach at New York City College, but the invitation was revoked due to objections to his atheism and unconventional personal morality. He continued to publish works on philosophy, logic, politics, economics, religion, morality and education. In 1944, he returned to Trinity College, where he had been offered a fellowship. He published *A History of Western Philosophy,* participated in radio broadcasts in England, and lectured in Norway and Australia. He was awarded the Order of Merit by the King in 1949 and the Nobel Prize for Literature in 1950. At the age of 80, he divorced Patricia Spence and married Edith Finch.

Acutely aware of the dangers of nuclear war, Russell served as the first president of the Campaign for Nuclear Disarmament in 1958, and as president of the Committee of 100 in 1960. As a member of the Committee of 100, he encouraged demonstrations against the British government's nuclear arms policies; for this he was sentenced to two months in prison, which was reduced to one week for health reasons. In his ninth decade, Russell established the Bertrand Russell Peace Foundation, published his *Autobiography,* appealed on behalf of political prisoners in several countries, protested nuclear weapons testing, criticized the war in Vietnam, and established the War Crimes Tribunal. He died on February 2, 1970, at the age of 98.

## SELECTED WRITINGS BY RUSSELL

*German Social Democracy,* 1896
*An Essay on the Foundations of Geometry,* 1897
*A Critical Exposition of the Philosophy of Leibniz,* 1900
*The Principles of Mathematics,* 1903

(With Alfred North Whitehead) *Principia Mathematica,* three volumes, 1910-13
*Our Knowledge of the External World as a Field for Scientific Method in Philosophy,* 1914
*Principles of Social Reconstruction,* 1916
*Roads to Freedom: Socialism, Anarchism and Syndicalism,* 1918
*Introduction to Mathematical Philosophy,* 1919
*A History of Western Philosophy,* 1945
*The Autobiography of Bertrand Russell,* three volumes, 1967-69

## FURTHER READING

Ayer, A. J. *Bertrand Russell.* New York: Viking, 1972.
Clark, Ronald. *Bertrand Russell and His World.* New York: Thames & Hudson, 1981.
Gottschalk, Herbert. *Bertrand Russell.* Translated by Edward Fitzgerald. Roy Publishers, 1965.
Schilpp, Paul Arthur, editor. *The Philosophy of Bertrand Russell.* The Library of Living Philosophers, 1946.

*—Sketch by C. D. Lord*

# Alice T. Schafer
## 1915–
### American algebraist and educator

Alice T. Schafer was born in Richmond, Virginia, on June 18, 1915. Her mother died during childbirth and Schafer was sent to the countryside area of Scottsburg, Virginia, to live with family friends. She was raised by Pearl Dickerson, a woman Schafer considered her mother and who was supportive of Schafer's ambitions throughout her life.

As a child, Schafer wanted to write novels, but by the third grade she became intrigued with mathematics after a teacher expressed concern that she would not able to master long division. Schafer not only mastered long division, but took math courses throughout her years of primary education. In her senior year of high school Schafer asked the principal to write a letter of recommendation for a scholarship to study mathematics in college. He was only willing to write the letter if she would promise to major in history instead.

After graduating from high school Schafer enrolled at the University of Richmond, where, at the time, the classes for men and women were held on separate sides of the campus. Mathematics classes were not offered on the female side of the campus, however, and as the only woman majoring in math, Schafer had to walk to the men's area of the campus to receive instruction. Women were also not allowed in the main library at the university. Books has to be sent over to the women's section of the campus upon request. In Schafer's junior year she questioned this policy and was finally allowed to sit in the library to read *Cyrano de Bergerac*, a book that was not available for circulation. After reading the book, Schafer was asked not to make any more requests to visit the library. The following summer she was offered a job in the library alphabetizing books. Schafer later quipped that she took the job because she needed the money.

Also in her junior year at Richmond, Schafer won the mathematical prize competition in Real Analysis but received no congratulatory praise from the chairman of the prize committee. Schafer graduated Phi Beta Kappa in 1936 when she was 21 years old with a degree in mathematics.

### A Long and Varied Career

Schafer was offered a position at a high school in Glen Allen, Virginia, where she taught for three years.

In 1939 she was awarded a fellowship at the University of Chicago and attained her Ph.D. in 1942. (Her dissertation was titled "Projective Differential Geometry.") While at the university, Schafer studied with Ernest P. Lane and Adrian Alberts.

Schafer's first teaching job as a Ph.D. was at Connecticut College, where she taught such classes as linear algebra, calculus, and abstract algebra for two years before accepting a position with Johns Hopkins University in the Applied Physics Laboratory doing research for the war effort. She was the only woman on the five member team of scientists.

Schafer married Richard Schafer, a professor of abstract algebra at MIT, in 1942. They have two children, Richard Stone Schafer, born in 1945, and John Dickerson Schafer, born in 1946. Between the years 1945 and 1961 Schafer taught at such institutions as the University of Michigan, the Drexel Institute of Technology, and Swarthmore College, among others. In 1962, she joined the staff at Wellesley College and stayed there until her forced retirement at the age of 65 in 1980. The retirement proved to be short–lived, however, when in that same year Schafer went to Harvard University as a consultant and teacher of mathematics in the management program. She again became a professor of mathematics, this time at Marymount University, in 1989 and retired in 1996.

### A Trailblazer for Women

Schafer has been a visiting professor and lecturer at various colleges, including Brown University and Simmons College. Schafer was the first woman to receive a Honorary Doctor of Science degree from the University of Richmond in 1964 and was presented with the Distinguished Alumna Award, Westhampton College, at the University of Richmond in 1977. During her tenure at Wellesley College, Schafer and two other women professors succeeded at implementing the black studies department there. She is also a cofounder of the Association for Women in Mathematics (AWM), along with Mary W. Gray, and others. The organization was established in 1971 to encourage women to study and seek careers in the mathematical sciences; currently, it boasts a membership of 4,500 from the United States and abroad. More than 300 academic institutions are supporting members of AWM in the United States. The AWM is open to both women and men. Schafer served as its president from 1973 to 1975 and remains active on various committees. A prize was established in her name in 1989 and is given annually to an undergraduate woman for excellence in mathematics.

Schafer is the author of eight articles concerning the progress of women in the field of mathematics and affirmative action. Her other articles include research on space curves and theorems on finite groups. A book published by the American Mathematical Society includes talks given by Schafer at their "100 Years of Annual Meetings Celebration." Her fields of specialization are abstract algebra (group theory). Schafer has three times been the leader of the delegation of women mathematicians to China, the last of which was the U.S.–China Joint Conference on Women's Issues held in Beijing between August 24 and September 2, 1995.

Schafer currently lives in Arlington, Virginia, with her husband.

## SELECTED WRITINGS BY SCHAFER

### Books

"Women and Mathematics," in *Mathematics Tomorrow*. Edited by Lynn Arthur Steen. New York: Springer–Verlag, 1981.

### Periodicals

"The Neighborhood of an Undulation Point of a Space Curve." *American Journal of Mathematics* 70 (1948): 351–363.

"Mathematics and Women: Perspectives and Progress." *American Mathematical Notices* (September 1991).

(With M.W. Gray) "Guidelines for Equality: A Proposal." *Academe* (December 1981).

"Two Singularities of Space Curves." *Duke Mathematical Journal* (November 1994): 655–670.

### Other

"Alice T. Schafer." *Biographies of Women Mathematicians.* June 1997. http://www.scottlan.edu/lriddle/women/chronol.htm (July 22, 1997).

Schafer, Alice T., interview with Kelley Reynolds Jacquez conducted May 8, 1997.

*—Sketch by Kelley Reynolds Jacquez*

# Charlotte Angas Scott
## 1858–1931
### English analytical geometer and educator

Charlotte Angas Scott was the first mathematics department head at Bryn Mawr College and a mem-

*Charlotte Angas Scott*

ber of its founding faculty. She developed the curriculum for graduate and undergraduate math majors and upgraded the minimum mathematics requirements for entry and retention at Bryn Mawr. Scott also initiated the formulation of the College Entrance Examination Board in order to standardize such requirements nationwide. At one time, she was the only woman featured in the first printing of *American Men of Science*, and the only mathematician in another venerable reference book, *Notable American Women 1607–1950*. Scott was one of the main organizers of the American Mathematical Society and the only female to leave such an extensive mark on its first 50 years of existence.

### Breaking "The Iron Mould"

Scott was born in Lincoln, England, on June 8, 1958, to Caleb and Eliza Exley Scott. As the daughter of the president of Lancashire College, she was provided with mathematics tutors as early as age seven. Her father and grandfather, Walter, were both social reformers as well as educators, and encouraged her to "break the iron mould" and seek a university education. At the age of 18 Scott won a scholarship to Hitchin College, now known as Girton College, the women's division of Cambridge University.

As the 19th century drew to a close, Girton was still an anomaly. Scott and her classmates numbered 11. The young women had to walk three miles to Cambridge to attend classes with those lecturers who allowed them in their classroom, but they had to sit behind a screen where they could not see the blackboard. At that time, a woman caught unescorted on Cambridge campus grounds could be sent to The Spinning House, a prison for prostitutes both active and suspected. A hint of the future could be seen in the changing attitudes of male undergraduates and graduate students, however. Tutors offered to prepare female undergraduates for the "Tripos," a grueling oral examination that lasted over a week's time. The first female student took the mathematics Tripos in 1872, and thereafter more women applied for permission to take the tests along with their male counterparts. Scott took the examination in January 1880, placing eighth. Although university policy kept her accomplishment a secret, the news spread throughtout the campus. The awards ceremony was disrupted by a crowd of young men shouting "Scott of Girton!" over the name of the man honored in her place. Scott was later "crowned with laurels" in a private ceremony. In February 1881 Cambridge reversed its policy and women were allowed to take examinations with the male students.

### Double Duty

Scott remained at Girton as a lecturer for four years while finishing her graduate studies at the University of London. The algebraist **Arthur Cayley**, a leader in coeducational reform, became Scott's mentor and recommended her for jobs as well as guiding her graduate research. Her doctorate was the first of its kind to be earned by a British female. In the nascent specialty of analysis within algebraic geometry, Scott focused on analyzing singularities in algebraic curves. Both of Scott's degrees at London were of the highest rank.

In 1885 Scott emigrated to the United States, where she joined the faculty of Bryn Mawr College in Pennsylvania. Founded that same year by the Society of Friends, Bryn Mawr was the first women's college to offer graduate degrees. Between 1885 and 1901 Scott successfully lobbied for a series of reforms to the admissions policies and entrance procedures at Bryn Mawr. Once the College Entrance Examination Board was instituted with her help, she served as Chief Examiner from 1902 to 1903. Scott's dedication was finally rewarded in 1909 with Bryn Mawr's first endowed chair and a formal citation.

### Helps Organize the AMS

In 1891 Scott was one of the first women to join the New York Mathematical Society, which later evolved into the American Mathematical Society (AMS) in 1895. Scott served on the council that oversaw this transition and received an "acclaimed review" from the group. She would serve on the AMS council again (between 1899 and 1901), and as vice–president in 1905.

Also, in 1899 Scott became the coeditor of the *American Journal of Mathematics*. She would continue to edit and peer review for this publication until two years after her official retirement from Bryn Mawr. Her influence spread internationally with her proof of **Emmy Noether**'s "fundamental theorem," an accomplishment which helped place Bryn Mawr and American mathematics on the world map.

### "Auntie Charley"

During her long sojourn in America, Scott was visited regularly by her father and younger brother, Walter, while her sisters remained in England. "Auntie Charley," as Scott was called, would travel to Europe during spring and summer breaks to visit with her expanding circle of relatives and with mathematicians in major European cities. Scott's own personal life was more circumspect, clouded further by the fact that all her personal correspondence was apparently disposed of or lost. She traveled often to Baltimore to visit her close friend Frank Morley, but she never married.

As Scott aged, the deafness that had plagued her since her student days became a stumbling block. However, even rheumatoid arthritis could not dampen her ambitions although it disrupted her publications' schedule for many years. On the advice of a physician Scott took up gardening, only to breed a new species of a chrysanthemum.

Scott retired at age 67, after a 40–year career as one of the many European women who could only find work as scientists and mathematicians in the United States. She stayed on voluntarily at Bryn Mawr until the following year, however, when her last doctoral student graduated. Scott continued mentoring younger mathematicians and inspired another generation of women to follow in her footsteps. She died in November 1931 in Cambridge, England, and was buried next to her cousin Eliza Nevins in St. Giles's Churchyard. Her textbook on analytical geometry, having gone through a second edition in 1924, would be reissued in a third edition 30 years after her death.

## SELECTED WRITINGS BY SCOTT

### Books

*An Introductory Account of Certain Modern Ideas and Methods in Plane Analytical Geometry.* First Edition, 1894 (republished as *Projective Methods in Plane Analytical Geometry* in 1961)

*Cartesian Plane Geometry, Part I: Analytical Conics*, 1907

**Periodicals**

"A Proof of Noether's Fundamental Theorem." *Mathematische Annalen* 52 (1899): 592–97.

## FURTHER READING

### Books

Eves, Howard. *An Introduction to the History of Mathematics*. Sixth Edition. Philadelphia, PA: Saunders College Publishing, 1990.

Green, Judy and Jeanne LaDuke. "Contributors to American Mathematics: An Overview and Selection," in *Women of Science: Righting the Record*. Edited by G. Kass–Simon and Patricia Farnes. Bloomington and Indianapolis: Indiana University Press, 1990.

————. "Women in American Mathematics: A Century of Contributions," in *A Century of Mathematics in America*. Volume 2. Providence, RI: American Mathematical Society, 1989, pp. 379–389.

Kenschaft, Patricia Clark. "Charlotte Angas Scott," *Women of Mathematics*. Edited by Louise S. Grinstein and Paul J. Campbell. New York: Greenwood Press, 1987, pp. 193–203.

Lehr, Marguerite. "Charlotte Angas Scott." *Notable American Women, 1607–1950*. Volume 3. Cambridge, MA.: Belknap Press of Harvard University, 1971, pp. 249–250.

Ogilvie, Marilyn Bailey. "Charlotte Angas Scott," *Women in Science*. Cambridge, MA.: MIT Press, 1986, pp. 158–59.

Rossiter, Margaret W. *Women Scientists in America: Struggles and Strategies to 1940*. Baltimore, MD: Johns Hopkins University Press, 1982.

**Periodicals**

Katz, Kaila and Patricia Kenschaft. "Sylvester and Scott." *The Mathematics Teacher* 75 (1982): 490–494.

Kenschaft, Patricia C. "The Students of Charlotte Angas Scott." *Mathematics in College* (Fall 1982): 16–20.

————. "Why Did Charlotte Angas Scott Succeed?" *Association for Women in Mathematics Newsletter* 17, no. 2 (1988): 9–11.

————. "Charlotte Angas Scott 1858–1931." *College Mathematics Journal* 18 (March 1987): 98–110.

Maddison, Isabel and Marguerite Lehr. "Charlotte Angas Scott: An Appreciation." *Bryn Mawr Alumni Bulletin* 12 (1932): 9–12.

**Other**

Chaplin, Stephanie. "Charlotte Angas Scott." http://www.agnesscott.edu/lriddle/women/chronol.htm (July 1997).

"Charlotte Angas Scott." http://www–groups.dcs.st–and.ac.uk/~history /Mathematicians/Scott.html (July 1997).

—*Sketch by Jennifer Kramer*

# (Takakazu) Seki Kōwa
## 1642–1708
**Founder of Japanese mathematics**

Though time and the secrecy which once surrounded Japanese schools have obscured his life and work, Seki Kōwa is widely regarded as the father of modern mathematics in Japan. Some biographers see him as the **Isaac Newton** of Japan, for he tackled many of the same problems as the English mathematician, inventing his own system of notation and an early form of calculus in the process. He and Newton were the same age, in fact, but the isolation of Japan in the 17th century kept Seki Kōwa's work from becoming more widely known.

### A Prodigy, Born to the Samurai Class

Seki Kōwa was born in March 1642 in Fujioka in the Kozuke province of Japan. His father, Nagaakira Utiyama, was a Samurai. Details about his mother are unknown. For reasons that are not clear, the infant boy was soon adopted by another noble family, that of accountant Seki Gorozayemon, who gave him the name by which he is now known. By the time Seki Kōwa was five years old, the boy's natural abilities with mathematics were so apparent that members of the household called him "divine child." He grew up in an age when arithmetic was a leisure–time diversion in Japan and was apparently able to listen in to the discussions of elders and correct their math. A servant in the household who was studying mathematics is believed to have begun Seki Kōwa's formal education at the age of nine. But accounts of this vary widely, and many scholars believe that Seki Kōwa was largely self–taught. His library included the great works of Japanese and Chinese mathematicians and he clearly built upon their methods.

### The "Arithmetical Sage" Draws Students

"Seki [Kōwa] wrote very much but printed little," biographer Yoshio Mikami writes in *The Development of Mathematics in China and Japan*. But his one major work, *Hatubi sanpō*, brought him great notoriety upon its publication in 1674. In it, he solved 15 supposedly unsolvable problems that had been published only four years before. In the style of the times, however, Seki Kōwa published only the final equations, not the methods he used to attain them. Even the students who studied with him were kept in the dark about many methods of the man they called "The Arithmetical Sage," but that was typical of Japanese education at that time. In 1685, one of Seki Kōwa's students published a work that further explained how the answers had been arrived at in *Hatubi sanpō*.

What is known about Seki Kōwa's work is that he was refining Chinese mathematics and bringing Japanese problem–solving into the modern age, conquering many of the same problems that Newton was working on in England at the same time. The Chinese techniques of Seki Kōwa's day relied on counting rods or "suan–ch'ou." With these, one could solve an equation with a single variable. Seki Kōwa developed a written system that allowed him to express known and unknown variables, much as letters are used in algebra today. He approximated values rather than solving for them directly, however, as there was no algebra at the time. He came up with a method of approximating the root of a numerical equation that turned out to be quite similar to one developed by Newton. And he calculated the value of $\pi$ accurately to perhaps 20 digits, using an approximation based on measuring a polygon with an infinite number of sides drawn within a circle. His work on circles is perhaps his most significant, according to *Dictionary of Scientific Biography* essayist Akira Kobori, and it came to be known as Enri, or principle of the circle. Seki Kōwa developed a formula for measuring ellipses, oblong circles formed by cutting across a cylinder at an angle. He is also credited with developing determinants, a table of numbers in columns and rows that is used to solve simultaneous equations, at about the same time as the method appeared in Europe.

Being a descendant of the Samurai class, Seki Kōwa became a public official, serving as examiner of accounts to the Lord of Koshu. Again, there is a parallel to the life of Isaac Newton, who served as master of the mint under Queen Anne. When the Lord of Koshu became Shōgun, Seki Kōwa became a Shōgunate samurai. In the twilight of his life, he was named master of ceremonies in the shōgun's household. But it was largely a figurehead post, as he had become too old and infirm to carry out many duties. Seki Kōwa died at age 66 on October 24, 1708, and was buried in the Buddhist cemetery at Ushigome in Tokyo. He left no children. Eighty years after his death, his tomb was rebuilt by mathematicians in his school and inscribed with the title he earned quite early in life: The Arithmetical Sage.

## SELECTED WRITINGS BY SEKI KŌWA

### Books

*Hatubi sanpō, 1674
*Katuyo sanpō, 1709
[*Both are held at the Mathematical Institute, Faculty of Science, Kyoto University]
*Sekiryu sanpōsitibusyo* ("Seven Books on the Mathematics of Seki Kōwa's School"), 1709

## FURTHER READING

### Books

Kobori, Akira. "Seki Kōwa." in *Dictionary of Scientific Biography* . Volume XII. Edited by Charles Coulston Gillispie. New York: Charles Scribner's Sons, 1975, pp. 290–92.
Mikami, Yoshio. *The Development of Mathematics in China and Japan.* 1913. Reprint, New York: Chelsea Publishing Company, 1961.
Smith, David E. and Mikami, Yoshio. *A History of Japanese Mathematics.* Chicago: Open Court Publishing Co., 1914.

### Other

"Takakazu Seki Kōwa," in *The MacTutor History of Mathematics Archive.* http://www–groups.dcs.st–and.ac.uk/~history/Mathematicians/Seki.html (June 1997).

—*Sketch by Karl Leif Bates*

# Atle Selberg
## 1917–
### Norwegian number theorist

Atle Selberg's most newsworthy achievement came in 1950, when he won the Fields Medal for an elementary proof of the 18th century conjecture known as the Prime Number Theorem. Since this achievement, Selberg has been affliated with the Institute for Advanced Studies (IAS) at Princeton, New Jersey. However, he continues to venture out to visit with children in their early teens, and to take part in special math–oriented programs held around the country.

Selberg was born in Langeslund, Norway, on June 14, 1917, to Ole Michael and Anna Kristina (nee Skeie) Selberg. He received all his early schooling there and earned a Ph.D. from the University of Oslo in 1943. Selberg was resident fellow at the university for a total of five years. However, he would soon become part of that wave of European emigres who enriched American mathematics after World War II.

Selberg married Hedvig Liebermann on August 13, 1947, the year he relocated to the United States. They would eventually have two children, Ingrid Maria and Lars Atle. Aside from a year–long stint at Syracuse University in New York, Selberg did not teach full–time in America. He was granted membership to the IAS and by 1951 was a member of the faculty there. Selberg was one of the three top European number theorists working in America, including Hungarian mathematician **Paul Erdös**, with whom he discovered the proof for the Prime Number Theorem.

Originally, Selberg and Erdös agreed to publish back to back papers in the same journal, but Selberg jumped ahead at the last minute. The fact that he published first led to his being awarded the Fields Medal by the International Mathematics Union. Selberg received more awards, including the Wolf Prize in 1986 and an honorary commission of Knight Commander with Star from the Royal Norwegian Order of St. Olav in 1987. He also holds an honorary doctorate from the University of Trondheim in his home country.

### Math Camp

As Selberg approached age 80, he traveled widely, taking part in commemorative and experimental events. For example, Selberg was invited to give the keynote address at the 1996 Seattle Mathfest celebrating the 100th anniversary of the Prime Number Theorem. That same month, he visited the "Math Camp" at the University of Washington. The Math Camp is an annual summer workshop designed to bring together and inspire bright youngsters from the Americas, Europe, and Asia. Selberg was one of many distinguished speakers invited, who had made significant strides in their particular areas of mathematics.

Selberg's specialty is in generalizing the works of others, an important step towards solidifying the foundations of number theory. He arrived at generalizations of Viggo Brun's sieve methods, and also summarized the Prime Number Theorem to include all prime numbers in an arbitrary arithmetic progression. His investigations also hinge on group theory and analysis. Not all of Selberg's work follows others, though; his conjecture that Lie groups are arithmetical was eventually proven in 1968 by Russian mathematicians Gregori Aleksandrovitch Margulis and D.A. Kazhdan.

## SELECTED WRITINGS BY SELBERG

### Books

"The General Sieve–Method and Its Place in Prime–Number Theory," in *Proceedings of the International Congress of Mathematicians, Cambridge, MA, 1950, Volume 1*, 1952, pp. 286–292.

"Recent Developments in the Theory of Discontinuous Groups of Motions of Symmetric Spaces," in *Proceedings of the 15th Scandinavian Congress, Oslo, 1968.* Lecture Notes in Mathematics, Volume 118, 1970, pp. 99–120.

### Periodicals

"On the Zeros of Riemann's Zeta–Function." *Skrifter utgitt av Det Norske Videnskaps–Akademi i Oslo I. Mat.–Natruv. Klasse* no. 10 (1942): 1–59.

"An Elementary Proof of the Prime Number Theorem." *Annals of Mathematics* 50, no. 2 (1949): 305–15.

## FURTHER READING

### Books

Albers, Alexanderson, and Reid. *International Mathematical Congresses: An Illustrated History 1893–1986.* Revised edition. New York: Springer–Verlag, 1986.

Sands, Karen. "Gregori Aleksandrovitch Margulis," in *Notable Twentieth Century Scientists.* Volume 3. Edited by Emily J. McMurray. Detroit, MI: Gale Research, Inc., 1995, p. 1318.

Tarwater, Dalton, editor. *The Bicentennial Tribute to American Mathematics: 1776–1976.* Mathematical Association of America, Inc. 1977.

*World Who's Who in Science.* First Edition. Edited by Allen G. Debus. Chicago: Marquis, 1968, p. 1519.

### Other

Peterson, Ivars. "Math Camp." *Ivars Peterson's MathLand* (August 26, 1996). http://www.maa.org/mathland/mathland_8_26.html

"Atle Selberg." *MacTutor History of Mathematics Archive.* http://www–groups.dcs.st–and.ac.uk/~history/Mathematicians/Erdös.html (July 1997).

*—Sketch by Jennifer Kramer*

*Jean-Pierre Serre*

# Jean-Pierre Serre
## 1926-
### French topologist

Jean-Pierre Serre received a Fields Medal for his work in topology, the study of geometric figures whose properties are unaffected by physical manipulation. He has received international acclaim for both his theoretical contributions to mathematics and the clarity of his writings. Serre has authored a dozen books and numerous technical articles in the areas of topology, analytic geometry, algebraic geometry, group theory, and number theory.

Serre was born in Bages, France, on September 15, 1926, to Jean and Adèle (Diet) Serre. Both pharmacists, his parents instilled in him an early interest in chemistry. That interest eventually gave way to mathematics, however, when Serre began reading his mother's calculus books. By the age of 15, he was teaching himself the fundamentals of such topics as derivatives, integrals, and series. During high school at the Lycée de Nîmes, Serre found a practical use for his mathematical talents. He told C. T. Chong and Y. K. Leong (for an article published in *The Mathematical Intelligencer*) that some of the older students at the boarding house where he lived had a tendency to bully him. Even though they were taking more advanced classes than he was, he pacified them by doing their math homework for them. "It

was as good a training as any," he recalled. In 1944, Serre won first prize in the Concours Général, a national mathematics competition.

In 1945, having passed the competitive entrance examination, Serre entered the prestigious École Normale Supérieure in Paris. Soon after enrolling at the institution, he decided to abandon his plans to become a high-school mathematics teacher and to concentrate instead on research. It was not until then, he later told Chong and Leong, that he realized he could earn a living as a research mathematician. Serre married Josiane Heulot, a chemist, in 1948, and they have one daughter. Between 1948 and 1954, he held various positions at the Centre National de la Recherche Scientifique in Paris. His earliest research was in the field of topology; he was awarded his doctorate from the Sorbonne in this subject in 1951 after writing a dissertation on homotopy groups. After two years on the faculty of the University of Nancy, he became professor and chair of the department of algebra and geometry at the Collège de France, which remained his home throughout his career. Serre retired in 1994, assuming the position of honorary professor.

### Produces Award-Winning Research

The Fields Medal, given once every four years by the International Congress of Mathematics, is intended to recognize outstanding contributions to mathematics by a researcher under age 40. It was awarded to Serre in 1954, when he was 28 years old. The citation read that Serre "achieved major results on the homotopy groups of spheres, especially in his use of the method of spectral sequences. Reformulated and extended some of the main results of complex variable theory in terms of sheaves" (a sheaf is a group of planes that pass through a common point). In fact, Serre's work on homotopy groups (classes of functions that are equivalent under a continuous deformation) was the origin of the loop space method in algebraic topology, which led quickly to numerous results. In 1952, he lectured on homotopy groups at Princeton University, discussing the topic's extension, called C-theory.

Since receiving the Fields Medal, Serre has explored other topics in mathematics, including complex variables, cohomology, algebraic geometry, and number fields. He explained to Chong and Leong that he finds it easy and natural to move gradually from one topic to another as he perceives their relationships to each other. Serre hesitates to categorize some of his work as being in one field or another, saying "such questions are not group theory, nor topology, nor number theory: they are just mathematics."

Serre is a member of the French, Dutch, Swedish, and U. S. academies of science and has been made an honorary fellow of the Royal Society of London and the London Mathematical Society. In addition to the Fields Medal, he has been awarded the Balzan Prize in 1985 and the Medaille d'Or of the Centre National de la Recherche Scientifique.

### Wins Praise for Exposition

In 1995, Serre was awarded the Leroy P. Steele Prize for Mathematical Exposition by the American Mathematical Society. Specifically, the prize was given for his book *A Course in Arithmetic,* which was originally published in 1970. However, the citation noted: "Any one of Serre's numerous other books might have served as the basis of this award. Each of his books is beautifully written, with a great deal of original material by the author, and everything smoothly polished." It went on to mention that many of Serre's books have become standard references in their respective topic areas. In general, the citation continued, "they are alive with the breath of real mathematics, and are an example to all of how to write for effect, clarity, and impact." *A Course in Arithmetic* was praised for both its presentation of basic topics in algebra and its explanation of the link between function theory and the combinatorial aspects of elementary number theory, which formed a basis for modern developments in number theory and the geometry of numbers.

In his acceptance of the Steele Prize, Serre explained that the book had been a compilation of notes for various lectures he had given during the early 1960s. "Strangely enough," he said, "the different pieces fitted well together; I was especially pleased with the way algebraic and analytic arguments complemented each other." With characteristic humor, he recalled some problems, typographical as well as technical, that had plagued both the original French and English versions of *A Course in Arithmetic,* leading him to refer to the earlier editions as "A Curse in Arithmetic."

Another of Serre's books, *Abelian L-Adic Representations and Elliptic Curves,* was reissued in 1989 (it was originally published in 1968). In reviewing the book for the *Bulletin of the American Mathematical Society,* Kenneth Ribet noted that, although numerous advances had arisen in the field since it was first published, the original text was certainly not outmoded. "For one thing, . . . it's the only book on the subject. More importantly, it can be viewed as a toolbox which contains clear and concise explanations of fundamental facts about a series of related topics . . . . The tools introduced in this book have been, and will continue to be, extremely useful in other contexts."

In a more recent book, published in 1992, Serre presents a historically based explanation of the inverse Galois problem and explores its applications in algebraic number theory, arithmetic geometry, and

coding theory. According to Michael Fried, who reviewed *Topics in Galois Theory* for the *Bulletin of the American Mathematical Society,* Serre was heard to remark at a friend's birthday celebration in 1990, "The inverse Galois problem gives us excuses for learning a lot of new mathematics."

Not only have Serre's books proven to be popular. Serre is often invited to lecture at universities around the world, including Bonn, Göttingen, Mexico, Moscow, and Singapore. He has made numerous visits to Harvard and the Institute for Advanced Study at Princeton.

When Chong and Leong asked Serre about the style of writing he prefers, he replied, "precision combined with informality!" Although noting that such writing is rare, he mentioned **Michael Atiyah** and **John Milnor** as examples of mathematicians whose work is accessible to nonmathematicians. Elaborating on his views about good exposition, he suggested that mathematical papers should contain more side remarks and open questions, which are often more interesting than the theorems themselves. "Alas," he said, "most people are afraid to admit that they don't know the answer to some question, and as a consequence they refrain from mentioning the question, even if it is a very natural one. What a pity! As for myself, I enjoy saying 'I do not know.'" Indeed, it is that admission that leads to investigation and discovery.

## SELECTED WRITINGS BY SERRE

*A Course in Arithmetic,* 1973
*Oeuvres: Collected Papers.* 3 volumes, 1986
*Abelian L-Adic Representations and Elliptic Curves,* 1989
*Topics in Galois Theory,* 1992

## FURTHER READING

Chong, C. T., and Y. K. Leong. "An Interview with Jean-Pierre Serre." *The Mathematical Intelligencer* 8 (1986): 8–12.

Fried, Michael. Review of *Topics in Galois Theory,* in *Bulletin of the American Mathematical Society,* n.s. 30 (January 1994): 124–35.

"Jean-Pierre Serre: 1995 Steele Prize for Mathematical Exposition." *Notices of the American Mathematical Society* 42 (November 1995): 1290–91.

Ribet, Kenneth A. Review of Abelian L-Adic Representations and Elliptic Curves, in *Bulletin of the American Mathematical Society,* n.s. 22 (January 1990): 214–18.

*—Sketch by Loretta Hall and David E. Newton*

# Wacław Sierpiński
## 1882–1969
### Polish topologist and logician

Wacław Sierpiński and his colleagues are credited with revolutionizing Polish mathematics during the first half of the 20th century. They took a couple of relatively new fields of mathematics and devoted whole journals to them. Although detractors had opined that such an experiment could not succeed, the mathematical heritage of the Polish community between the world wars has left a legacy of results, problems, and personalities, chief among them being Sierpiński.

Sierpiński was born in Warsaw on March 14, 1882. His father was Constantine Sierpiński, a successful physician, and his mother was Louise Łapinska. Sierpiński received his secondary education at the Fifth Grammar School in Warsaw, where he studied under an influential teacher named Włodarski. From there, he entered the University of Warsaw and began studying number theory under the guidance of G. Voronoi. In 1903, Sierpiński's work in mathematics was recognized by a gold medal from the university, from which he graduated the next year. After graduation, he taught in secondary schools, a standard career path due to the shortage of positions available to Poles under Russian rule. In that capacity, he was involved in the school strike that occurred during the revolution of 1905. Even though the strike was not wholly unsuccessful, Sierpiński resigned his teaching position and moved to Kraków.

In 1906 Sierpiński received his doctorate from the Jagiellonian University in Kraków. Two years later he passed the qualifying examination to earn the right to teach at Jan Kazimierz University in Lwów, to which he had gone at the invitation of one of the faculty. There, Sierpiński offered perhaps the first systematic course in set theory, the subject of his investigations for the next 50 years. In 1912 he gathered his lecture notes and published them as *Zarys Teorii Mnogości* ("Outline of Set Theory"). Sierpiński's texts were recognized by prizes from the Academy of Learning in Kraków.

With the outbreak of World War I in 1914, Sierpiński was interned by the Russians, first at Vyatka, then in Moscow. This internment was not particularly severe, for while he was in Moscow, Sierpiński was accorded a cordial reception by the leading Russian mathematicians of the era. In fact, he was able to conduct some joint research with N. Lusin during this period in the field of set-theoretic topology.

### Blends Set Theory with Topology

The area of set-theoretic topology in which Sierpiński worked depended on a few basic notions.

One of these is that of a closed set, or a set that includes its boundary. A simple example is the interval of real numbers between 0 and 1, including both endpoints. Related to the notion of a closed set is that of an open set, one which does not include its boundary. An example of an open set is the interval between 0 and 1, not including either 0 or 1. If one takes that interval and includes 0 but not 1, then the set is neither closed nor open. Much time was spent investigating the results of combining open and closed sets in various infinite combinations. The entrance of infinity is what required the use of methods and ideas from set theory.

When the war ended Sierpiński returned to Lwów, but in the fall of that year he was appointed to a position at Warsaw. He devoted a number of papers to the set-theoretic topics of the continuum hypothesis and the axiom of choice. The continuum hypothesis, which had been known to **Georg Cantor**, claimed that there were no infinite numbers between the number of integers and the number of real numbers. If the continuum hypothesis is present, it reduces the complexity of the hierarchy of infinite numbers. The axiom of choice had been a matter for much discussion at the turn of the 19th century, allowing for the possibility of making an infinite number of choices simultaneously. Some distinguished French mathematicians like **Émile Borel** questioned the meaningfulness of such a choice, and one of the early consequences of investigation into the axioms of set theory was the discovery that the axiom of choice was equivalent to a number of other propositions of set theory. Sierpiński took an agnostic position with respect to the axiom, using it in proofs and also trying to eliminate it wherever possible.

It was during the period between the world wars that Sierpiński, in conjunction with several Polish colleagues, created what has since become known as the Polish school of mathematics. The subjects that dominated the Polish school were logic and set theory, topology, and the application of these subjects to questions in analysis. To make certain that there would be an audience for the work in these areas, the journal *Fundamenta Mathematicae* was founded in 1919. Although by the end of the 20th century a profusion of specialized journals within mathematics had sprung up, *Fundamenta* was the first of its kind and was greeted with some suspicion about its likelihood for survival. The quality of its papers was high, the contributors were international, and the problems proposed and solved were substantial. As a permanent record of the Polish mathematical school, *Fundamenta Mathematicae* supplements the reminiscences of those who took part in the work.

### Presides over the Polish Mathematical Community

Sierpiński often led the Polish delegations to international congresses and conferences of mathematicians. One of the most ambitious projects in which he was involved was a Congress of Mathematicians of Slavic Countries, whose very existence attests to a political consciousness side–by–side with the mathematical one. The event took place in Warsaw in 1929 and was chaired by Sierpiński.

During World War II Sierpiński was in Warsaw, holding classes in whatever secret settings were available. A good deal of the discussion went on in Sierpiński's home, where his wife did her best to make guests feel as comfortable as possible in such troubled times. In 1944, the Nazis took control of Warsaw and Sierpiński was taken by the Germans to a site near Kraków. After the latter city was liberated by the Allies, he held lectures at the Jagiellonian University there before returning to Warsaw.

The period after the war was marked by further honors to Sierpiński as his students dominated the mathematical landscape. He served as vice president of the Polish Academy of Sciences from its inception and was awarded the Scientific Prize (First Class) in 1949, as well as the Grand Cross of the Order of Polonia Restituta in 1958. There were not many mathematicians in any country who approached his publication records of more than 600 papers in set theory and approximately 100 articles about number theory.

Sierpiński died in Warsaw on October 21, 1969. His mathematical textbooks educated an entire generation, and he helped to lay the foundations of the discipline of set–theoretic topology. Sierpiński's legacy was in establishing a Polish mathematical community, a contribution at the same time to mathematics and to national identity.

## SELECTED WRITINGS BY SIERPIŃSKI

*Hypothèse du continu*, 1934
*Cardinal and Ordinal Numbers*, 1958
*Elementary Theory of Numbers*, 1964

## FURTHER READING

Kuratowski, Kazimierz. *A Half Century of Polish Mathematics*. Warsaw: Polish Scientific Publishers, 1980, pp. 167–173.
Kuzawa, Sister Mary Grace. *Modern Mathematics: The Genesis of a School in Poland*. New Haven, CT: College and University Publishers, 1968.

*—Sketch by Thomas Drucker*

*Stephen Smale*

# Stephen Smale
## 1930-
### American topologist

Stephen Smale has worked in a number of mathematical fields, including topology and dynamical systems. In the 1970s he became especially interested in the application of mathematical theory to economics. More recently, his research has focused on computer science and its relevance to fundamental principles of mathematics. In recognition of his work in these fields, Smale has been honored with the Veblen Prize, the Fields Medal, and the Chauvenet Prize.

Smale was born in Flint, Michigan, on July 15, 1930. His parents were Lawrence and Helen Smale. Smale has described his father as an "armchair revolutionary" who was once expelled from Albion College in Michigan for publishing a newsletter whose political and social views were offensive to college administrators. The elder Smale never returned to college and eventually became an assistant in the ceramics laboratory at the AC Sparkplug factory in Flint.

### Becomes Politically Active at an Early Age

In 1935 when Smale was five years old, his family moved to a small farm outside of Flint and he began school in a one-room schoolhouse. In high school Smale's primary interest was chemistry, and for a while he planned on a career in that field. He also became politically active during his high-school years, organizing a protest against the omission of evolution from his biology class curriculum. The protest was not very successful, as he told Donald J. Albers and Constance Reid in *More Mathematical People,* "I am not sure I succeeded in getting even one other person to sign the petition."

In 1948 Smale enrolled at the University of Michigan planning to major in physics rather than chemistry. He did poorly in physics, however, and gradually drifted into mathematics because it was a subject in which he had always excelled. Still, until his second year in graduate school he did not make a firm decision to pursue a career in mathematics. In 1953 he added an M.S. to the B.S. he had received the year before and, three years later, he received his Ph.D., also from the University of Michigan.

During his college career Smale was involved in number of activities beyond his academic studies. As he told Albers and Reid, he "[played] a lot of chess and . . . a lot of Go [a Japanese game similar to chess or checkers]." Smale was also a member of the Communist party and was very active in opposition to the Korean War. His primary motivation for staying in college, he later admitted, was to avoid being drafted into fighting a war to which he was strongly opposed.

### Begins Work in Topology

The mathematical topic that first attracted Smale's attention was topology, the mathematical study of the properties of figures that are not affected by changes in shape or size. For a period of about seven years beginning in graduate school, Smale worked on topological questions, a field that he later described as "very fashionable" at the time. He chose to pursue research in topology, he told Albers and Reid, because it was "just there." Despite this attitude, his work proved valuable, and he was awarded the prestigious Fields Medal of the International Mathematical Union in 1966 for his research in the field.

Smale accepted an appointment as professor of mathematics at Columbia University in 1961, the same year he made a surprising decision: He chose to abandon his work in topology and begin research in a new field, that of dynamical systems. Dynamical systems refers to mathematical methods for dealing with changes that take place in some real or abstract system over time. For the next five years Smale vacillated among a number of mathematical fields, including the calculus of variations and infinite dimensional manifolds. He also decided to leave Columbia University and accepted a position as professor of mathematics at the University of Califor-

nia in Berkeley in 1964 where he began to concentrate once again on dynamical systems. In 1970 Smale experienced yet another career transformation that was prompted by a series of conversations he had with Gerard Debreu, a Nobel laureate in economics. From these conversations, Smale began to work on the applications of mathematical theory to economic systems.

Smale remained politically active during his years at Columbia and Berkeley. He was outspoken in his opposition to the war in Vietnam and helped organize, along with Jerry Rubin and others, the "days of protest" held against the war. However, his political attitudes have undergone an evolution over the past two decades. After a series of trips to Russia in the 1960s, Smale became disillusioned with communism and eventually found himself in agreement with at least some principles of conservative political philosophy. He recently described himself to Albers and Reid as being "consistently . . . against the military" and having beliefs that are "more radical than those of most liberals," but "conservative part of the time."

In recent years, Smale has become interested in computer sciences. He has expressed an interest in bringing the mathematics used in this field into closer relationship with mainstream mathematics. Although his objective is not yet defined, Smale feels that establishing such a relationship may result in some revolutionary changes in the nature of mathematics itself.

Smale married Clara Davis, a classmate at the University of Michigan, in 1955. The couple has two children, Laura, a biological psychologist, and Nat, a mathematician. Smale has a number of leisure interests including a large mineral collection and a 43-foot, ocean-going ketch. In addition to his Fields Medal, Smale has been given the 1966 Veblen Prize of the American Mathematical Society and the 1988 Chauvenet Prize of the Mathematical Association of America.

## SELECTED WRITINGS BY SMALE

(With M. W. Hirsch) *Differential Equations, Dynamical Systems and Linear Algebra,* 1974
*The Mathematics of Time: Essays on Dynamical Systems, Economic Processes, and Related Topics,* 1980

## FURTHER READING

### Books

Albers, Donald J., Gerald L. Alexanderson, and Constance Reid, editors. *More Mathematical*

*Willebrord Snell*

People. New York: Harcourt, 1991, pp. 305–323.

—*Sketch by David E. Newton*

# Willebrord Snell
## 1580?–1626
### Dutch mathematical physicist

Willebrord Snell is best known for his discovery regarding the refraction of light rays. This discovery, known as Snell's law, demonstrates that when a ray of light passes from a thinner element such as air, into a denser element, such as water or glass, the angle of the ray bends to the vertical. Snell's law—a key revelation in the science of optics—was formulated after much experimentation in 1621. This is expressed as $\sin i = \mu \sin r$ ($i$=angle of incidence, $r$=angle of refraction and $\mu$ = a constant). However, he did not publish his findings, and the law did not appear in print until **René Descartes** discussed it (without giving Snell credit) in his *Dioptrique* in 1637. Snell also determined a formula to measure distances using trigonometric triangulation. His method, developed in 1615, used his home and the spires of Leiden churches as reference points. (In 1960, a plaque recognizing his work was placed on his home.) In an age of world

exploration, this was very important work, because it contributed to improved accuracy in the art of mapmaking. Using the triangulation method, Snell measured the Earth's meridian for the first time, and also attempted to measure the size of the Earth using this method. He set down the principles of spherical trigonometry that determine the length of a meridian arc when measuring any base line. Snell's writings on his triangulation method were presented in *Eratosthenes batavus* (1617). His observations of comets sighted in 1585 and 1618 are described in his *Cyclometricus de circuli dimensione* (1621). His last works, *Canon triangulorum* (1626) and *Doctrina triangulorum*, published after his death in 1627, also addressed the measuring of distances through plane and spherical trigonometry.

Snell was born in Leiden, Holland, in 1591. He was the son of Rudolph Snellius (Snel van Royen), a professor of mathematics at the University of Leiden, and Machteld Cornelisdochter. He became interested in mathematics at an early age, but initially entered the University of Leiden to study law. Snell married Maria De Lange, daughter of a Schoonhoven burgher master, in 1608. They had eighteen children; only three survived. In 1613, after his father's death, he became mathematics professor at the University. He also taught astronomy and optics.

Snell occupied himself in his early career with travel throughout Europe and the translation of numerous mathematical treatises. In his travels, he consulted with such eminent scientists as Tycho Brahe, **Johannes Kepler**, and Michael Maestlin, Kepler's astronomy teacher. Snell's Latin translation of the German *Wisconstighe Ghedachtenissen* was published as *Hypomnemata mathematica* in 1608; books of **Apollonius** based on plane loci were also translated during this time. Other translations of books on mathematics published by Snell were Ramus' *Arithmetica* (1613), and *Geometria* (1622).

### Invents Triangulation Method to Measure the Earth

The book translations were done at a time when Snell was very involved in applying mathematics to determine the shape and size of the Earth and the exact position of points on its surface. This work was to occupy him from 1615 onward. The mathematics he used is called plane trigonometry, which deals with calculations in two dimensions for measuring distances that cannot be measured directly, such as across lakes or vast expanses of land. Plane trigonometry began to be used in the mid–15th century—not surprising that a method so useful in geography, surveying, and mapmaking would be needed at a time when Europeans were undertaking their systematic maritime journeys of discovery. Snell's method, triangulation, had been used by Tycho Brahe and also by

Gemma Frisius in 1533, but Snell advanced it to such a higher level he became known as the father of triangulation. Using his house and the spires of churches as starting points, he determined angles with a quadrant more than 7 feet long (213 cm). Building on a network of triangles, he calculated the distance between two towns—Alkmaar and Bergen–op–Zoom — located on the same meridian approximately 80 miles (130 km) from each other. Taking this calculation further, he was able to determine the radius of the Earth. The calculations for this difficult work were tedious, given that he computed his measurements without the help of logarithms. In spite of that, they remain surprisingly accurate. In 1624, he published *Tiphys batavus*, a work on navigation which parallels his work on determining land measurements. In it, his geometric theories hint of the  differential triangle developed by later mathematicians, especially **Blaise Pascal**.

### Light Refraction Studies Become Snell's Law

Snell's work in astronomy resulted in two books: *Cyclometricus de circuli dimensione*, and *Concerning the Comet*, the latter published in 1618. His observations demonstrate his shorter method for determining comet distance and movements. Coinciding with this work was his study of light refraction. In his experimentation, he referred to Kepler's writings, as well as those of Risner. The phenomenon of refraction was known as far back as **Claudius Ptolemy**, but Snell's explanation—which states that the ratio of the sines of the angles of the incident and refracted rays to the normal is a constant—became known as Snell's law of optical refraction. The constant determines how much a ray of light will be bent as it travels through one medium to another. Some controversy surrounds the fact that, although Snell never published his findings, they appear uncredited in René Descartes *Dioptrique* in 1637, more than ten years after Snell's death. Some have accused Descartes of plagiarism, but it is unknown whether Descartes knew of Snell's discovery and passed it off as his own, or if his knowledge was gained independent of Snell's work.

Snell died in Leiden on October 30, 1626, and he was buried in the Pieterskerk. A monument was erected there in his honor.

### SELECTED WRITINGS BY SNELL

*Eratosthenes batavus*, 1617
*Concerning the Comet*, 1618
*Cyclometricus de circuli dimensione*, 1621
*Tiphys batavus*, 1624
*Canon triangulorum*, 1626
*Doctrina triangulorum*, 1627

## FURTHER READING

### Books

Asimov, Isaac. *Asimov's Biographical Encyclopedia of Science and Technology*. Garden City, New York: Doubleday & Company, Inc., 1972.

Derbus, Allen G., editor. *World Who's Who in Science*. Chicago: Marquis Who's Who, Inc., 1968, p. 1573.

Hollingdale, Stuart. *Makers of Mathematics*. London: Penguin Books, 1989.

Porter, Roy, editor. *The Biographical Dictionary of Scientists*. New York: Oxford University Press, 1994, pp. 630–631.

Struik, Dirk J. "Willebrord Snell," in *Dictionary of Scientific Biography*. Volume XII. Edited by, Charles Coulston Gillespie. New York: Charles Scribners' Sons, 1975, pp. 499–502.

Travers, Bridget, editor. *World of Scientific Discovery*. Detroit: Gale Research, Inc., 1994, pp. 403–404; 607–608; 654.

*—Sketch by Jane Stewart Cook*

# Mary Fairfax Somerville
## 1780–1872
### Scottish–born English mathematics writer

Mary Fairfax Somerville was called the "Queen of nineteenth–century Science" by the London *Morning Post* on her death. She is best known for her work explaining the mathematical and scientific works of others. Somerville was one of the first two women admitted to the Royal Astronomical Society, and a portrait bust of her is on display in the great hall of the Royal Society in London. Somerville College of Oxford University is named after her.

### Rebels Against the Norm in Pursuing Education

Somerville was born on December 26, 1780, in Jedburgh, Scotland. Her father was an Englishman, Vice–Admiral Sir William George Fairfax, and her mother was Margaret Charters Fairfax, a Scotswoman. Because girls in Somerville's era were often discouraged from pursuing a formal education, Somerville's earliest schooling consisted of reading the Bible. She disliked playing with dolls, but she loved to roam the Scottish seacoast and countryside. Her father worried about her carefree behavior, and sent her away to boarding school when she was ten. Somerville's experience at the school was a horror. For one assignment, she had to memorize an entire page of a dictionary and recite back the words in their

*Mary Fairfax Somerville*

correct order, spelling each properly, citing its meaning, and naming its part of speech. She lasted twelve months. When Somerville returned home, she had learned a little arithmetic, grammar, and French, and also developed an interest in reading.

Most important, however, was Somerville's insatiable curiosity. One day she had noticed some Xs and Ys printed in a fashion magazine and realized the symbols were from a problem in algebra. No one in her immediate circle knew what algebra was, but Somerville found out herself. She asked a male friend to purchase an algebra text and **Euclid** 's *Elements of Geometry* and began to study them, which upset her parents. Somerville's father forbade her to study mathematics, fearing that she would have to be straitjacketed, because a woman he knew had gone insane trying to understand longitude. Her mother confiscated all of Somerville's candles so she could not read at night. Somerville, however, had a partial victory. She had already read and memorized six of Euclid's books.

In 1804, Somerville married her cousin, Captain Samuel Greig, a Russian naval officer assigned as consul in England, who had little use for science and learning opportunities for women. They had two sons, Woronzow, who lived into middle age, and William George, who died in his youth in 1814. Greig died in 1807, leaving Somerville financially secure.

Somerville returned to Scotland and began to study astronomy and mathematics. She made friends

with some scholars at the University of Edinburgh, who recommended her books on mathematics and offered encouragement. Somerville began submitting solutions to problems in a mathematics journal, and in 1811 won a silver medal for solving a mathematical problem.

In 1812, Somerville married another cousin, Dr. William Somerville, who encouraged her in her studies and helped her get published. They had three daughters, Margaret, Martha, and Mary, and one son, Thomas, who died in infancy. Dr. Somerville, a surgeon, was a ranking member of the English medical establishment, and Somerville met many important scientific people in London and Europe. Her friendships with practicing scientists and mathematicians would prove valuable.

### Begins Successful Writing Career

Somerville began her publishing career in 1826 with a scientific paper on the magnetic power of sunlight in the *Philosophical Transactions of the Royal Society*. Other scientific articles followed. Her greatest fame, however, came from her explanations of the technical works of other mathematicians and scientists. Her first book, published in 1831, was *The Mechanism of the Heavens*, which explained **Pierre Simon Laplace**'s book on celestial mechanics, and added many of Somerville's own ideas. Celestial mechanics is best understood by people who have a working knowledge of calculus, but Somerville made certain that intelligent laypeople with no training in advanced mathematics could understand her book. She had approached the project of writing the book with some anxiety because she did not have a college education. Somerville was so fearful about the book's success that she insisted that her work be kept secret, and if her editors did not like the manuscript, she wanted it destroyed. The book was a smashing success. It was the first accurate synthesis of celestial mechanics that an educated layperson could understand.

Somerville's next book, *On the Connection of the Physical Sciences*, was published in 1834. It too was a great success and was published in ten editions from 1834 to 1877. The astronomer John Couch Adams, who co–discovered the planet Neptune, said that he got the idea to search for Neptune from an observation he read about the odd orbital movements of Uranus in Somerville's text.

Somerville's husband began to experience health problems in the late 1830s, and the family moved to Italy. She published her most successful book, *Physical Geography*, in 1848. It was the first major work in England to focus on the physical surface of the Earth, including its landforms, soils, climates, and vegetation.

Somerville's husband died in 1860, and her son passed away in 1865. At her husband's death, Somerville was in her mid–eighties, but remained productive. In 1869, she published her last major book, *On Molecular and Microscopic Science*, a two–volume summary of recent advances in physics and chemistry. In that same year, Somerville received a gold medal from the Royal Geographical Society, and was elected to the American Philosophical Society.

Somerville died peacefully in Naples, Italy, on November 29, 1872 at the age of 92. Her father's fears that mathematics would drive her insane proved unfounded. Somerville continued to study algebra for four to five hours a day, and she worked on a mathematical article the day before she died.

## SELECTED WRITINGS BY SOMERVILLE

*The Mechanism of the Heavens*, 1831
*On the Connection of the Physical Sciences*, 1834
*Physical Geography*, 1848
*On Molecular and Microscopic Science*, 1869

## FURTHER READING

### Books

Osen, Lynn M. "Mary Fairfax Somerville." *Women in Mathematics*. Cambridge, MA: MIT Press, 1974, pp. 95–116.
Patterson, Elizabeth C. *Mary Somerville and the Cultivation of Science, 1815–1840*. Boston: Martinus Nijhoff Publishers, 1983.
"Mary Fairfax Somerville," in *Women of Mathematics: A Biobiographic Sourcebook*. Edited by Louise S. Grinstein and Paul J. Campbell. Westport, CT: 1987, pp. 208–216.
Perl, Teri. *Math Equals: Biographies of Women Mathematicians*. New York: Addison–Wesley, 1978, pp. 83–99.

### Periodicals
Patterson, Elizabeth C. "The Case of Mary Somerville: An Aspect of Nineteenth–Century Science." *Proceedings of the American Philosophical Society* (June 1974): 269–275.
Wood, Shane. "Mary Fairfax Somerville." *Biographies of Women Mathematicians*. June 1997. http://www.scottlan.edu/lriddle/women/chronol.htm (July 21, 1997).

—*Sketch by Patrick Moore*

*Richard Stanley*

# Richard Stanley
## 1944–
### American algebraic topologist and educator

Richard Stanley is a pioneer in the field of combinatorics, a specialty not highly regarded when Stanley was emerging as a mathematician, but one that has gained greater acceptance throughout his career. His *Enumerative Combinatorics,* which is based on concepts presented in commutative algebra, homological algebra and algebraic topology, is one of the seminal texts in this area.

Stanley was born in New York City on June 23, 1944, to Alan Stanley, a chemical engineer, and Shirley Stanley, a homemaker. Stanley did not remain in New York City long, spending his early childhood in Tahawus, New York, a city that was later moved to another location building by building. During his time in Tahawus, he explains in a document on his early years available on his World Wide Web site on the Internet, he dreamed of being a ventriloquist. As an adolescent, then living in Lynchburg, Virginia, he set his sights on astronomy, which triggered his interest in mathematics as well. He credits an elderly resident of Lynchburg—who taught him the standard synthetic algorithm for finding the square root of a positive real number—with getting him started. Around the same time, he also began to read the mathematics section of his father's *Handbook of Chemistry and*

*Physics*, which detailed the cube root algorithm. Just before his family's move to Savannah, Georgia, when the younger Stanley was 13, he shifted his interest from astronomy to nuclear physics until he met a classmate who was working on determining synthetic algorithms for finding roots higher than the cubic. His interest was piqued and he began to read widely on mathematical topics in addition to his high school mathematics courses. Stanley often worked independently of his classmates since only basic courses were available at his school.

Stanley earned his Bachelor of Science degree in mathematics in 1966 from the California Institute of Technology and went on to receive a Ph.D. in mathematics from Harvard University in 1971, with G.C. Rota, a professor at the Massachusetts Institute of Technology (MIT), as his thesis advisor. He also worked as the C.L.E. Moore Instructor of Mathematics at MIT in the final year of his Ph.D. program. While at Harvard, Stanley met his wife Doris, who became a French teacher. The Stanleys have two children.

After receiving his Ph.D., Stanley conducted postdoctoral studies through a Miller Research Fellowship at the Berkeley campus of the University of California. In 1973, he returned to MIT as an assistant professor in mathematics and, in 1979 was named a full professor of applied mathematics, eventually serving as chair of the Applied Mathematics Committee. He relinquished that post when he accepted a visiting professor position at Berkeley, where he spent a year as the Chern Visiting Professor in the Mathematical Sciences Research Institute, over which Stanley served as chair from 1993 to 1997. Stanley also spent a year as a visiting professor at the University of California at San Diego for the 1978–79 academic year.

Stanley has held other visiting professor positions across the world, including short stints at the University of Stockholm, University of Augsburg, and Tokai University in Japan. In addition, Stanley spent a semester as the Göran Gustafsson Professor at the Royal Institute of Technology and the Institute Mittag–Leffler in Sweden in 1992. He was recruited for this position by Anders Björner, one of Sweden's leading mathematicians, for whom Stanley had served as an advisor when Björner spent time at MIT as a visiting student.

Among his awards and honors, Stanley received a Guggenheim Fellowship in 1983, which he used to fund research on reduced combinatorics. He has published widely on the topic of combinatorics, including authoring the definitive text on the topic, *Enumerative Combinatorics*. As of 1997, he was at work on a second volume of *Enumerative Combinatorics*, which he had spent 11 years researching at that time. In addition to teaching, Stanley works as a consultant for Bell Telephone Laboratories in Murray Hill, New Jersey, and serves on editorial boards for several prestigious mathematical journals, including the *Journal on Algebraic and Discrete Methods*, *Studies in Applied Mathematics*, and *PanAmerican Mathematics*.

In an interview with Kristin Palm, Stanley discussed the qualities that drew him to his profession. "To me it's just a very elegant, beautiful structure. It's been developed for thousands of years. You can go exploring on your own with it and add to it," he explained.

## SELECTED WRITINGS BY STANLEY

*Combinatorics and Commutative Algebra*, second edition, 1996
*Enumerative Combinatorics*, 1986

## FURTHER READING

Richard Stanley. http://www–math.mit.edu/~rstan (August 26, 1997).

### Other

Stanley, Richard, in an interview with Kristin Palm, conducted July 11, 1997.

—*Sketch by Kristin Palm*

# Karl Georg Christian von Staudt
## 1798–1867
### German geometer

Karl von Staudt is considered an innovator in mathematics. Although he published only a few well–considered publications in his lifetime, he has earned this reputation in part because of his work in projective geometry. He developed the first complete theory of points, lines and planes in projective geometry. Von Staudt also did some work with Bernoulli numbers, and is also well known for the construction of a 17–sided regular inscribed polygon using compasses for his only tools.

Born Karl Georg Christian von Staudt on January 24, 1798, he was the son of a municipal councilman, Johann Christian von Staudt, and his wife, Maria Albrecht. Von Staudt was born in what was then the Imperial Free City of Rothenberg. (The city

*Karl Georg Christian von Staudt*

became part of Bavaria and was renamed Rothenburg ob der Tauber while von Staudt was a child. It later became part of Germany).

Von Staudt's education was comprehensive. He attended Ansbach's Gymnasium from 1814 to 1817. He received his university education at Göttingen from 1818 until 1822, where he primarily focused on astronomy, both theoretical and practical. He determined the orbit of a comet discovered in 1821. Von Staudt also became acquainted with **Karl Friedrich Gauss** and his work in number theory while at Göttingen. He earned his doctorate in 1822 from the University of Erlangen, based on the merits of his astronomical work done at Göttingen.

In the same year, 1822, von Staudt began his mathematical teaching career after being qualified to teach in Munich. From approximately 1822–27, he taught in Würzberg, at both a secondary school and the University there. In 1827 he transferred to Nürnberg, where he again taught at two schools, another secondary school and the local Polytechnic, until 1835. While in Nürnberg, in 1832, he married Jeanette Drechsler, with whom he had two children, a son, Eduard, and a daughter, Mathilde. Von Staudt's wife died in 1848, and he never remarried.

In 1835 Von Staudt became a full professor at the University of Erlangen, fulfilling a professional goal. He led the school's mathematical department until his death. He published on Bernoulli numbers as early as 1840 and 1845, and delineated what came to be known as the Staudt–Clausen theorem. This theorem defined a formation law for Bernoulli numbers.

Von Staudt's most important research to the history of mathematics focused on projective geometry (called the geometry of position). Projective geometry was created to supplant the older, dominant Euclidean geometry. In his most famous work on the subject, *Geometrie der Lage* (1847), von Staudt was the first to do pure projective geometry without reference to measurements or lengths. *Beiträge zur Geometrie der Lage* (published three times between 1856 and 1860), supplements von Staudt's 1847 book. Here, he applied his discoveries in projective geometry to synthetic geometry, and also introduced the formulas for one, two, and three dimensional complex projective spaces.

Von Staudt suffered from asthma in the final years of his life. He died on June 1, 1867, in Erlangen, Bavaria.

## SELECTED WRITINGS BY VON STAUDT

*Die Geometrie der Lage*, 1847
*Beiträge zur Geometrie der Lage*, 1856–1860

## FURTHER READING

### Books

Burau, Wener. "Karl Georg Christian von Staudt." *Dictionary of Scientific Biography.* Volume XII. Edited by Charles Coulston Gillispie. New York: Charles Scribner's Sons, 1970, pp. 4–6.

Coolidge, Julian Lowell. *A History of Geometrical Methods.* New York: Dover Publications, 1963, pp. 97–101.

Klein, Felix. *Development of Mathematics in the 19th Century.* Translated by Robert Hermann. Brookline, MA: Math Sci Press, 1979, pp. 121–29.

### Periodicals

Noether, Max. "Zur Erinnerung an Karl Georg Christian von Staudt." *Jahresberichte der Deutschen Mathematiker–Vereinigung* 32 (1923): 97–119.

*—Sketch by Annette Petruso*

*Jakob Steiner*

# Jakob Steiner
## 1796–1863
### Swiss geometer

Jakob Steiner, unschooled until age 18, was one of the founders of and greatest contributors to the field of projective or modern geometry. Building on the work of such greats as **Gerard Desargues** and **Gaspard Monge**, the Swiss geometer discovered both the Steiner surface and the Steiner theorem, and extrapolated the work of the French geometer, **Jean Poncelet**, to develop the Poncelet–Steiner theorem, all of which helped to build the developing field of synthetic or projective geometry. A firm believer in the power of geometry, he pronounced that the calculations involved in algebra and analysis simply replaced real thinking, while geometry stimulated it. Made wealthy by his lectures on geometry, at his death Steiner bequeathed a third of his fortune to the Berlin Academy to establish the Steiner Prize.

Steiner was born on March 18, 1796, in the village of Utzensdorf, near Bern, Switzerland. The last of eight children of Niklaus Steiner and Anna Barbara Weber, Steiner grew up without formal schooling, working instead on the family farm and business where his natural skills with numbers was greatly needed. He did not learn to write until age 14, and against his parents' will, he finally left home four years later to attend a school in Yverdon run by Johann Heinrich Pestalozzi, the Swiss educational reformer whose theories laid the foundation for modern elementary education. The Pestalozzi method reacts to the highly individual needs of each learner and scorns rote memorization in favor of concrete experience and critical thinking. It was a method Steiner took to eagerly, so eagerly that within a year–and–a–half of studying with Pestalozzi, Steiner himself was teaching mathematics. His studies and teaching experience at Yverdon left a lasting legacy with Steiner: the desire, as a researcher, to discover a scientific unity in his chosen field of mathematics, and the ability, as a teacher, to encourage independent thinking on the part of his students.

### Establishes a Career in Berlin

Steiner left Yverdon, Switzerland, in 1818 for Heidelberg, Germany, where he studied at the university. To pay his way as a student, Steiner gave private lessons in mathematics. At the university, he attended lectures by Ferdinand Schweins in combinatorial analysis, and also studied algebra as well as differential and integral calculus. Early papers from 1821, 1824, and 1825 demonstrate the influence of his studies in Heidelberg. At the suggestion of a friend, however, Steiner decided to leave Heidelberg for the Prussian capital of Berlin in 1821.

His lack of education and stubborn will played against Steiner initially in Berlin. To receive a teaching license, he was forced to take examinations in various subjects. Steiner was not totally successful, receiving a partial license in mathematics and employment for a short time at a gymnasium, or high school. He gave private instruction to make a full living, and attended the University of Berlin from 1822 to 1824. Thereafter, he taught at a technical school in Berlin until 1834, when he was appointed a professor at the University of Berlin. He held this post until his death 29 years later.

A blunt, somewhat corrosive personality, Steiner was known for both the startling nature of his lectures and the originality of his research. The important phase of the latter began with an 1826 publication of "Einige geometrische Betrachtungen" in the new *Journal fur die reine änd angewandte Mathematik,* founded by his friend, August Crelle. In total, Steiner contributed 62 articles to that journal. Those articles, in addition to his longer works, *Systematische Entwicklung der abhangigkeit geometrischer Gestalten voneinander* of 1832, and the posthumously published *Vorlesungen uber synthetische Geometrie* and *Allgemeine Theorie uber das Beruhren und Schneiden der Kreise und der Kugeln,* together dealt with Steiner's passion for reforming geometry and also made public the basics of what has come to be called projective geometry. Steiner's articles and books have been gathered in his *Gesammelte Werke.*

## Steiner Unifies Geometrical Theory

It was Steiner's great dream to discover the commonality between seemingly unrelated theorems, providing a simple and logical way of deducing such theorems. As he noted in his *Systematische Entwicklung*, "Here the main thing is neither the synthetic nor the analytic method, but the discovery of the mutual dependence of the figures and of the way in which their properties are carried over from the simplest to the more complex ones." All of Steiner's work draws from one basic principle, or so contended F. Gonseth in the foreword to Steiner's posthumously published *Allgemeine Theorie*. That principle is the stereographic projection of the plane onto the sphere.

Steiner noted in *Systematische Entwicklung* the principle of duality in projective geometry is a principle normally considered a property of algebra. But Steiner showed that in projective geometry that if two operations are interchangeable or dual, then whatever results are true for one are also true for the other. He contributed a plethora of original concepts that created the scaffolding of projective geometry. Among others, he discovered the Steiner surface, containing an infinity of conic sections; the Steiner theorem, proving that the intersection points of corresponding lines of two projective pencils (sets of geometric objects) of lines form a conic section. He also built on the work of Poncelet, in his Poncelet–Steiner theorem which shows that just one given circle with its center and straight edge are needed for any Euclidian construction. His investigation of conic sections and surfaces is at the very heart of modern geometry.

## Steiner the Man

Steiner never married. Over the course of a long career, many honors came his way. He received an honorary doctorate from the University of Königsberg in 1833 and was elected a member of the Prussian Academy of Science in 1834. He also became a corresponding member of the French Académie Royale des Sciences after a winter of lecturing in Paris in 1844 and 1845, and held membership in the Accademia dei Lincei as well. In the final decade of his life, a kidney ailment limited his lecturing to the winter months only, and he died in 1863, revered as both a mathematician and educator.

Over the years, Steiner was able to accumulate a relatively large estate. His lecturing proved popular and most of his savings came from this source. In addition to the bequest he left to the Berlin Academy for the establishment of a mathematics prize in his name, and to the large chunk of money left to his relatives, Steiner also left money to the school in his native village of Utzensdorf to establish prizes for students adept at mathematics. To his death, he never forgot the value of education, nor the bitter memory of the lack of it in his own youth.

## SELECTED WRITINGS BY STEINER

*Jakob Steiner's Gesammelte Werke.* Edited by K. Weierstrass. Berlin: 1881–1882.

## FURTHER READING

### Books

Burckhardt, Johann Jakob. "Jakob Steiner." *Dictionary of Scientific Biography*. Vol. XIII. Edited by Charles Coulston Gillispie. New York: Charles Scribner's Sons, 1976, pp. 12–22.

Geiser, Carl Friedrich. *Zur Erinnerung an Jakob Steiner.* Schiffhausen: 1874.

Graf, J. H. *Der Mathematiker Steiner von Utzensdorf.* Bern: 1897.

Kollros, Louis. *Jakob Steiner.* Basel: Birkhauser, 1947.

### Periodicals

Bierman, Kurt R. "Jakob Steiner." *Nova acta Leopoldina* 27, no. 167 (1963): 31–47.

Butzberger, F. "Jakob Steiner bei Pestalozzi in Yverdon." *Schweizerische padagogische Zeitschrift* 6 (1896): 19–30.

Emch, A. "Unpublished Steiner Manuscripts." *American Mathematical Monthly* 36 (1929): 273–275.

Lampe, Emil. "Jakob Steiner." *Biblioteca mathematica* 1 (1900): 129–141.

*—Sketch by J. Sydney Jones*

# Simon Stevin
## 1548–1620
### Dutch applied mathematician, physicist, and engineer

An encyclopedic mind and a prolific writer on many topics, Stevin, known as "the Dutch **Archimedes**," contributed to many areas of knowledge, particularly arithmetic, algebra, geometry, physics, optics, mechanics, astronomy, hydrostatics, musical theory, and military engineering. A dedicated empiricist, he experimentally refuted, years before **Galileo** (who received credit), the Aristotelian doctrine that heavy objects fall faster than light–weight objects. He also improved mathematical notation and cham-

*Simon Stevin*

pioned the decimal system, suggesting that decimal mensuration could be used in all spheres of life. In addition, Stevin was an exceptional prose stylist. He pioneered the use of his vernacular, Dutch, at a time when Latin was dominant as the language of science. Stevin significantly enriched the Dutch scientific vocabulary, coining words and phrases which have been become part of the Dutch lexicon.

Born in Bruges (now Belgium), in the southern Netherlands, Stevin was employed by the city administration as a bookkeeper. He made several trips abroad in the 1570s, settling in Leiden, in the north, in 1581. Already in his thirties, he proceeded to obtain a formal education at the University of Leiden, adding to independently acquired knowledge of science and engineering. Having left the Spanish–occupied southern provinces, Stevin prospered in the northern Netherlands, participating in the national struggle for liberation from Spanish rule. His engineering skills were recognized by the leaders of the Dutch liberation war and he became quartermaster–general of the army in the early 1600s. He was also engaged by Prince Maurice of Nassau, the stadholder (head of state), as his personal science and mathematics teacher. Always willing to put his scientific expertise in the service of the war effort, Stevin was very highly regarded by his compatriots. He married Catharine Cray in 1610; the couple had four children, one of whom, Hendrick, also became a scientist.

## Explains Decimal Fractions

Although decimal fractions had been known centuries before Stevin, he is credited with explaining the concept of decimal fractions, which, although known to proficient mathematicians, seemed unfamiliar to most mathematicians. He did not view decimal fractions as fractions, and proceeded to write their numerators without the denominators. For example, Stevin represented the value of $\pi$ (in modern notation: 3.1416...) as 3 0 1 1 4 2 1 3 6 4. The second digit—around which he drew a circle—of each pair was the power of ten of the implied divisor. Thus, 1 1 is 1/10; 4 2 is 4/100, etc. Stevin's notation may have been cumbersome, but his idea made sense, for, to cite Carl Boyer's example, it seems more natural to imagine 3 seconds as an integer (3) than as 3/60 of one hour.

Unlike his predecessors, who used words to represent operations, Stevin insisted whenever this was possible, on notational symbols. For example, he used circled numbers to represent exponents, circled 2 replacing $Q$ (for square), and circled 3 instead of $C$ (for cube). While still imperfect, Stevin's notation, as scholars suggest, points to the modern algebraic notation which was firmly established not long after his death.

## Corrects Misconceptions and Anticipates New Theories in Physics

In 1586, Stevin described his experiment with two leaden spheres, one ten times heavier than the other. Having dropped the spheres simultaneously from a height of 30 ft (914 cm) onto a wooden surface, Stevin observed that they both hit the surface at the same time. Historians, who agree that Stevin and not Galileo, should be credited for this revolutionary experiment, nevertheless accept Galileo as the founder of modern mechanics, because it was the Italian scientist who formulated the final theoretical consequences of Stevin's discovery. However, according to Stephen F. Mason, Stevin, who described his falling spheres, as the "experiment against Aristotle," effectively demolished Aristotelian mechanics, thereby creating the foundations of modern mechanics.

Stevin's book, *De Beginselen der Weegconst* (*Principles of the Art of Weighing*), continues the work begun by Archimedes (for example, refining the Greek scientist's explanation of buoyancy) and tackles such Archimedean subjects as levers and inclined planes, including his famous wreath of spheres. Stevin placed a wreath of spheres on two inclined planes *AB* and *BC*, *AB* being twice the length of *BC*. If we imagine *ABC* as a triangle, we will notice that, the spheres being spaced evenly and weighing the same, and if there are two spheres on *BC*, there will naturally be four on *AB*. Despite the different numbers of spheres on each side, the wreath does not move, because there is an inverse proportion between the effect of gravity and the length of an inclined plane. A different triangle, with *AB* three times the length of *BC*, would, according to Stevin's principle, would balance two spheres on one side *BC*, and six on the other. "Stevin," Mason has written, ". . . obtained an intuitive understanding of the parallelogram of forces, a method of finding the resultant action of a combination of two forces that are not in the same straight line nor parallel. The method . . . consists in the representation of the two forces, in magnitude and direction, by two straight lines originating from a common point: the resultant is then given by the diagonal of the parallelogram formed by drawing two other lines parallel to the first two." Once again, Stevin made a great discovery, leaving the precise theoretical formulation to later scientists. The parallelogram of forces was clearly explained by **Isaac Newton** and by the French mathematician Pierre Varignon in 1687.

### Suggests Equally–Tempered Musical Scale

In Stevin's time, the musical scale was not evenly tempered, which posed a serious problem for players of stringed instruments who had to contend with the paradox exemplified by the fact that a sequence of four pure fifths and a sequence of two octaves should, but do not result in, a perfect third: the interval created when the final tones of the two sequences are played together is not a perfect minor third. The interval of a pure octave is represented by the ratio 2 : 1, which means that if a frequency *f* is doubled, the new frequency is an octave higher than *f*. The pure fifth corresponds to the ratio 3 : 2, and the pure third to the ratios 5 : 4 (major) and 6 : 5 (minor). To resolve this paradoxical situation, which obliged musicians to constantly change tuning strategies, mathematicians and musicians proposed a variety of solutions, including the division of the octave into 12 mathematically, but perhaps not acoustically, equal semitones. Scholars believe that the modern, equally–tempered, scale originated with Stevin's idea that all semitones should be absolutely equal—their frequencies determined by the mathematical formula $2^{n/12}$, even if the consequences included less–than–perfect intervals.

Stevin may have been familiar with *Dialogo della musica antica e della moderna* (*Dialogue Concerning Ancient and Modern Music* [1581]) by Vincenzo Galilei, father of Galileo, which, among many other subjects, discussed the equally–tempered scale. However, scholars have suggested that the scale proposed by Stevin is the scale that was gradually accepted in the 17th and 18th centuries, particularly by players of keyboard instruments, and by composers such as Johann Sebastian Bach. His *Well–Tempered Clavier*, which contains preludes and fugues in all the 24 keys, constitutes the foundation of the classical, mathematical, conception of tonality, the embodiment of which

is the modern piano. In science, Stevin's idea was further refined by the 19th–century English mathematician and phonetician A.J. Ellis, who divided the octave (2 : 1) into 1200 cents, 100 cents representing an equally–tempered semitone.

### SELECTED WRITINGS BY STEVIN

*De Thiende*, 1585
*De Beghinselen der Weegconst*, 1586
*Les oeuvres mathematiques de Simon Stevin*, 1634
*The Principal Works of Simon Stevin*, 5 volumes, 1955–68

### FURTHER READING

Boyer, Carl B. *A History of Mathematics.* Second edition. Revised by Uta C. Merzbach. New York: Wiley, 1991.

Crombie, A. C. *Science in the Later Middle Ages and Early Modern Times.* Volume 2: *Medieval and Early Modern Science.* Second revised edition. Garden City, NY: Doubleday, 1959.

Grout, Donald J., and Claude V. Palisca. *A History of Western Music.* Fifth edition. New York: W. W. Norton, 1996.

Dijksterhuis, E. J. *Simon Stevin: Science in the Netherlands around 1600.* The Hague: Martinus Nijhoff, 1970.

Mason, Stephen F. *A History of the Sciences.* Revised edition. New York: Macmillan, 1962.

*—Sketch by Zoran Minderovic*

---

# Michael Stifel (also transliterated as Styfel, Styffel, Stieffell, Stifelius, and Stiefel)
## 1487?–1567
### German algebraist

Michael Stifel was a monk turned mathematician. His most famous work, *Arithmetica integra*, was published in 1544. The book contains all that was known at the time about arithmetic and algebra, supplemented by original contributions. His work contains binomial coefficients and the notation $+, -, \sqrt{\,}$. He invented logarithms using a different approach than **John Napier**, known for his inventions of logarithms. In 1545, Stifel published his second work, *Deutsche arithmetica* as a way to make algebra more accessible to German readers by not using foreign words.

Stifel was born in Esslingen in southwestern Germany, in 1487. His father was Conrad Stifel. Although Stifel's father's occupation is not known, he was respected, but not affluent. Nothing is known of his mother.

### Joins a Monastery

Stifel entered the Augustinian monastery at Esslingen and was ordained a priest in 1511. He did not conform to the Roman Catholic faith, however, as he did not like taking money from the poor. He became disillusioned with what he considered the clergy's declining morality and abuses relating to the administration of indulgences.

By 1520, Stifel was an early follower of Martin Luther. In his studies of the Bible, Stifel found numbers in Revelation and the Book of Daniel. He became so captivated by their mystical translation that he was able to manipulate the numbers to show that Pope Leo X was the beast written about in Revelation.

Stifel's superiors became suspicious of him after he granted absolution without receiving indulgence money and for composing a song honoring Martin Luther as well. He was forced out of the Monastery in 1522. Fleeing Esslingen, he sought refuge with the Lutherans, staying in the castle of the knight Hartmut von Kronberg, in the Taunus mountains. He then journeyed to Wittenberg to stay with Luther, where the two became friends. In 1523, Luther attained the position of Pastor for Stifel at the court of the Count of Mansfield. Due to anti–Lutheran pressure, Stifel was forced out of this position as well as several others. Stifel returned to Wittenberg about a year later.

In 1525, Stifel traveled to Castle Tollet in Upper Austria. He served as councillor, pastor and tutor to Dorothea Jorger, widow of nobleman Wolfgang Jorger, and their son, Christoph. Again, politics forced him to rejoin Luther. Luther secured a parish for Stifel at Lochau (now Annaberg), in 1528. Luther accompanied him there, where he married Stifel to the widow of a previous incumbent. In 1531–32 Esslingen had gone over to Protestantism, seizing the assets of the monastery. Since Stifel's parish was not well–endowed, Stifel tried to persuade the city council there to release the holdings he had relinquished when he joined the order. There is no indication if this was successful.

### Religion and Mathematics

While at Annaberg, Stifel began playing with numbers again. This activity showed his skill in recognizing number–theory relationships. Through further study of the Bible, Stifel believed he calculated the date of the end of the world as being October 18, 1533. On September 28, 1533, Stifel warned his congregation of the impending apocalypse and was subsequently arrested and dismissed from the parish.

In 1535, Stifel procured another parish in Holzdorf with help from Luther and Melanchthon. Holzdorf proved to be a productive place for Stifel. He concentrated on mathematics, earning his master's degree from the University of Wittenberg. Stifel also gave private lessons in mathematics to students including Melanchthon's son–in–law, Kasper Peucer. While at Holzdorf, Stifel wrote *Arithmetica integra* and *Deutsche arithmetica*. *Welsche Practick* (1546) was the last book Stifel published during the 12 years he was there.

In all of his books, Stifel made original contributions, not just explanations of existing knowledge of arithmetic and algebra, and his work prepared the way for further progress in these fields. Stifel was the first person to present a general method for solving equations. The method would replace the 24 rules traditionally given by the Cossists.

During the religious Schmalkaldic War of 1547, Lutheran duke Maurice of Saxony and Holy Roman Emperor Charles V attempted to take a region of Saxony out of Protestant control. The war forced the townspeople, including the clergy and Stifel, to flee. Stifel then journeyed to Prussia in 1549. In 1551, he found work as a pastor at Haberstroh, near Königsberg. During his time there, he lectured on mathematics and theology at the University of Königsberg. However, Stifel soon returned to Saxony after arguments with colleagues.

By 1559, Stifel had given up his pastorate. He obtained a position at the University of Jena, he lectured on arithmetic and geometry. Stifel died in Jena, Germany on April 19, 1567.

## SELECTED WRITINGS BY STIFEL

### Books

*Arithmetica integra*, 1544

*Deutsche aristhmetica, inhaltend Haussrechnung, Deutsche Coss, Kirchrechnung*, 1545

*Die schonen Exempeln der Coss. Durch Michale Stifel gebessert und sehr gemehrt*, 1552–1553

*Ein Rechen Buchlin vom End Christ, Apocalypsis in Apocalypsin*, 1532

## FURTHER READING

### Books

Cantor, M. *Vorlesungen uber Geschichte der Mathematik*. Volume II, Leipzig, 1913, pp.430–449.

Vogel, Kurt. "Michael Stifel," in *Dictionary of Scientific Biography*. Volume XIII. Edited by Charles Coulston Gillispie. New York: Charles Scribner's Sons, 1976, pp. 58–62.

**Periodicals**

Arnold, W. "Michael Stifel," in H. Wussing and W. Arnold, *Biographien bedeutender Mathematiker*. Berlin, 1983.

Hofmann, J.E. "Michael Stifel." *Jahbuch fur Geschichte der Oberdeutschen Reichsstadte*. 14 (1968): 30–60.

Jentsch, W. "Michael Stifel: Mathematicker und Mitstreiter Luthers." *Schriftenreihe fur Geschichte der Naturwissenschaften Technik und Medizin* 23 (1986): 11–33.

Reich, K. "Michael Stifel." *Mitt. Math. Ges. Hamburg* 14 (1995): 23–33.

**Other**

Bowen, Jonathan. "Early English Algebra." http://www.comlab.ox.ac.uk/oucl/users/jonathan.bowen/algebra/section3_2.html. (May 6, 1997).

"John Napier." *MacTutor History of Mathematics Archive*. http://www–groups.dcs.st–and.ac.uk/~history/Mathematicians/Napier.html. (May 30, 1997).

"Michael Stifel." MacTutor History of Mathematics Archive. http://www–groups.dcs.st–and.ac.uk/~history/Mathematicians/Stifel.html. (May 6, 1997).

Westfall, Richard S. "Stifel [Styfel], Michael." http://es.rice.edu/ES/humsoc/Galileo/Catalog/Files/stifel.html. (May 6, 1997).

—*Sketch by Monica L. Stevens*

# James Stirling
## 1692–1770
### Scottish number theorist

James Stirling was a mathematician best known for the formula on series expansion which bears his name, Stirling's formula. Stirling's work contributed greatly to the development of the theory of infinite series and to the branch of calculus known as infinitesimal calculus. Although his formal education was interrupted over political quarrels, Stirling worked both in Venice and London, doing early work on the extension of **Isaac Newton**'s theory of plane curves. His most influential work, also inspired by theories of Newton, *Methodus differentialis*, was published in 1730. In 1735 Stirling returned to his native Scotland, where he took a position at the Leadhills mining company, married, and concerned himself almost totally with mining affairs for the rest of his life.

Stirling was born in Garden, Stirlingshire, Scotland, in 1692. The third son of Archibald Stirling and Anna Hamilton, Stirling came from a staunchly Jacobite family, supporters of the restoration of the Stuarts after their overthrow in 1688. His father had once been imprisoned on charges of treason for his beliefs, but was later released. Such a background did not serve the young Stirling well when he left college in Glasgow to be admitted to Balliol College at Oxford University in 1711, the recipient of a scholarship donated by John Snell of Ayrshire. With the onset of the First Jacobite Rebellion of 1715, and Stirling's refusal to take an oath of allegiance to the king, he was deprived of his scholarship, and left Oxford, without graduating, in 1716.

### Early Work Wins Recognition

Stirling's time at Oxford had not been wasted, however. Connections with such notables as the mathematician and surgeon, John Arbuthnot, helped to get his first book published in 1717, with Newton himself among the names on the list of subscribers. *Lineae tertii ordinis Neutonianae*, to give the book its short title, not only recounted but also extended Newton's theory of plane curves of the third order. In this work, Stirling proved the 72 types of cubic curves that Newton had enumerated, and added four more of his own. He also solved a problem of **Gottfried Wilhelm von Leibniz** on curves. This book was dedicated to Nicholas Tron, another connection made during Stirling's Oxford years. Tron, the Venetian ambassador to London and a member of the Royal Society of London, secured a teaching position for Stirling in the Venetian Republic. Stirling left for Venice in 1718, but the position fell through. It is unclear just how long Stirling stayed in Venice or what his activities were while there. He did submit to the Royal Society an important early paper, "Methodus differentialis Newtoniana illustrata," from Venice in 1718, a paper which helped inspire his later, major work. There is also evidence that he attended the University of Padua in 1721. Additionally, it is believed that Stirling's departure from Venice may have been caused by him learning some of the trade secrets of the famous glass blowers. Such secrets were as closely guarded as computer innovations today, and Stirling's life may well have been in danger if he had access to such knowledge.

### The Prodigal Returns

Whatever the reason, it seems Stirling was back in Glasgow by 1722, but there is a gap in the biographical data until he reappears in London in 1724 or 1725. Part of the difficulties of accurately tracing the events of Stirling's life is that the name James Stirling is a rather common one in Scotland, and at least two other James Stirlings have been confused with the mathematician in some historical records. It is clear that Stirling remained in London for ten years, first teaching at and later becoming a partner in Watt's Academy in Covent Garden, also known as the Little Tower Street Academy. The academy was a respected school and affiliation with it helped to buoy Stirling's usual poor financial situation. Stirling also won, with Arbuthnot's support, a fellowship in the Royal Society in 1726.

While in London, Stirling continued his mathematical researches, moving from the study of cubic curves to the field of differences, also inspired by earlier work by Newton, particularly his *Analysis* of 1711. This work saw earlier publication in Stirling's 1718 paper, but by 1730 he had a full book on the subject, *Methodus differentialis: sive tractatus de summatione et interpolatione serierum infinitarum*. Published in Latin initially, the book was translated into English as well.

### Develops Stirling's Formula

Stirling's *Methodus differentialis* is divided in two parts, which follow a short introductory section. One part deals with summation, or the method of associating a sum with a divergent series, and the other with interpolation, or the estimation of an intermediate term in a sequence. The book also deals with infinite series and quadrature, the process of finding a square equal in area to any given surface, especially that bounded by a curve. It was in Proposition 28, Example 2, of *Methodus differentialis* that Stirling proposed an asymptotic, or increasingly exact formula, for $n!$ which bears his name: $n! = n^n e^{-n} \sqrt{(2\pi n)}$. This series expansion formula was also derived by **Abraham de Moivre**, although credit usually goes to Stirling who also supplied tables of coefficients for recurring series, creating what are known as Stirling Numbers of the first or second kind.

### Mathematician Turns Miner

Stirling found a new interest in the early 1730s with the study of the Earth and its gravitational forces, culminating with the 1735 publication of *On the Figure of the Earth and On the Variation of the Force of Gravity on Its Surface*. He also spent the summers of 1734–1736 in Scotland, in the employ of the Scottish Mining Company in their lead mines at Leadhills, Lanarkshire. In 1737, Stirling accepted a permanent appointment with the mines, at a salary of 210 pounds per year, a position he held for the rest of his life. Though involved primarily in administration, at which he was by all accounts notably successful,

Stirling also developed a system of shaft ventilation and of air blasts for smelters, which he published in 1745 in a paper entitled "Description of a Machine to Blow Fire by the Fall of Water." This paper may have been influenced by his earlier work with the Venetian glass blowers.

Stirling married Barbara Watson of Thirty Acres, a village near Stirlingshire. The exact date of the marriage is not known, though it is assumed it could not have taken place before 1745. The couple had one child, a daughter, who later married her cousin, Archibald Stirling, the succeeding manager at Leadhills. Stirling continued to communicate with other mathematicians, though his Jacobite principles prevented him from returning to the world of professional mathematics when a chair opened in Edinburgh in 1746 upon the death of his friend **Colin Maclaurin**. That same year, Stirling was honored by election to the Berlin Academy of Sciences. His last work of a scientific nature took place in 1752 when he conducted the first survey of the River Clyde for the Corporation of Glasgow, one of a series of surveys which were intended to improve the navigability of the river. Stirling was presented with a silver teapot by the Corporation in recognition of his services. The following year, Stirling's wife died, and in 1754 he resigned from the Royal Society, an event which appears to have been prompted by financial concerns. Stirling was heavily in arrears to the Society for his annual subscriptions.

Later in life, Stirling fell into ill health. Visiting Edinburgh for medical treatment in 1770, he died and was buried at Greyfriars Churchyard. He is remembered by a small plaque in the cemetery wall, and by his work on infinite series theory. By the time of Stirling's death, his most significant work was far behind him, relegated to the 1720s and 1730s, before taking his position at the lead mines.

## SELECTED WRITINGS BY STIRLING

### Books

*Lineae tertii ordinis Neutonianae, sive illustratio tractatus D. Neutoni de enumeratione linearum tertii ordinis*, 1717

*Methodus differentialis: sive tractatus de summatione et interpolatione serierum infinitarum*, 1730. Translated by Francis Holliday as *The Differential Method*, 1749

## FURTHER READING

### Books

Tweddle, Ian. *James Stirling*. Edinburgh: Scottish Academy Press, 1988.

*George Gabriel Stokes*

Tweedie, C. *James Stirling: A Sketch of His Life and Works along with His Scientific Correspondence.* Oxford: 1922.

Wallis, P. J. "Stirling, James." *Dictionary of Scientific Biography.* Volume XIII. Edited by Charles Coulston Gillispie. New York: Charles Scribner's Sons, 1975, pp. 67–70.

**Periodicals**

Tweddle, Ian. "James Stirling's Early Work on Acceleration of Convergence." *Archive for History of Exact Sciences* 45 (1992): 105–25.

—*Sketch by J. Sydney Jones*

# George Gabriel Stokes
## 1819–1903
### British mathematical physicist

Sir George Gabriel Stokes made important advances in the fields of hydrodynamics and optics. He also did significant work in wave theory, as well as the elasticity and light diffraction of solids. With his work on viscous fluids, he helped develop the theoretical foundation for the science of hydrodynamics. These equations, known as the Navier–Stokes equations (he shared credit with Claude Navier) describe the motion of viscous fluids. The word "fluorescence" entered the English language when Stokes first used it to explain his conclusions about the blue light emitting from the surface of colorless, transparent solutions. He then applied the phenomena of fluorescence to study light spectra. An important practical use for fluorescence was in the pharmacy, where British chemists used it—instead of relying on the availability of sunlight—to tell the difference among chemicals. Stokes is also considered a pioneer in scientific geodesy (publishing a major work on the variation of gravity at the Earth's surface in 1849), and spectrum analysis. In 1849, he assumed the Lucasian chair at Cambridge University, which he held until his death. Stokes was also very active in various scientific and academic societies. He served as president of the Cambridge Philosophical Society from 1859 to 1861, and was president of the Royal Society of London from 1885 to 1890. In 1887, he became a member of Parliament, representing the University of Cambridge, serving until 1891. The Royal Society awarded Stokes the Rumford Medal in 1852 for his work with fluorescence; he was also awarded the Copley Medal in 1893. His scientific contributions were recognized by a knighthood in 1889.

Stokes was born in Skreen, County Sligo, Ireland, on August 13, 1819. His family was of Anglo–Irish extraction, whose forebears included many clergy in the Church of Ireland. His father, Gabriel, was rector in Skreen, and his mother, Elizabeth Haughton Stokes, was the daughter of a rector. Stokes was the youngest of six children. He was educated by his father as a young boy, later attending a Dublin school. He went to Bristol College in England to prepare for university.

### Begins Long Tenure at Cambridge

In 1837, he entered Pembroke College, Cambridge, to study mathematics. He graduated in 1841, winning senior Wrangler and 1st Smith's prizeman academic honors. He was given a fellowship at Pembroke and remained at Cambridge the rest of his life. His appointment as Lucasian professor coincided with teaching at the Government School of Mines in London, which he did to increase his income—his Cambridge position being poorly endowed at that time. He married Mary Susanna Robinson in 1857, giving up his fellowship to marry, as was the custom at that time. They became the parents of three children, two sons and a daughter. He was able to regain a fellowship in 1859 when a change in University rules regarding married professors was made.

### Introduces Advances in Hydrodynamics

Pure mathematics was never the sole focus of Stokes' inquiry—his experimental work was equally important. For him, theory was placed in the service of answering specific physical questions, and his mathematical papers reflect his interest in developing problem–solving methods or proving out existing formulas. His investigations into hydrodynamics led him to test out and publish his theories on the internal friction of fluids. Although French physicists ( Claude Navier in particular) had already done work in this area, Stokes introduced new applications for their mathematical equations. In 1846, he presented a paper on hydrodynamic advances to the British Association for the Advancement of Science, which greatly enhanced his reputation. Later hydrodynamic experiments involved a study of the propagation of oscillatory waves in water, and led to a method of calculating their shape.

In 1850, Stokes published a major paper on hydrodynamics that included Stokes' Law governing the fall of an object through a liquid, using the action of pendulums to test out his fluid theories. Based on his observations, and also using the experiments conducted by others in the field, he was able to explain how clouds formed in the atmosphere. These conclusions, combined with his studies on surface gravity, caused Stokes to be regarded in England as an authority on the emerging science of geodesy. In 1854 he published Stokes' Theorem, a three–dimensional generalization of Green's Theorem in vector calculus.

To compilment his study of hydrodynamics, Stokes also delved into the principles of sound. He analyzed how sound was produced and the effect of wind on its intensity. Using this information, he explained why stringed instruments required sounding boards in order to transmit sound. Stokes also explained how sound is produced through telegraph wires.

### Examines Concept of Light Propagation

The wave theory of light also captured Stokes' attention. His early mathematical studies revolved around the nature of ether—then a concept for explaining how light is propagated. Although he attempted to work around conceptual flaws by connecting his work on elastic solids to the properties of ether, its existence was eventually disproved in later experiments by other scientists. However, Stokes' work in light diffraction bore fruit in many other areas. For example, in 1849, he developed instruments which allowed him to measure the amount of astigmatism in the human eye, and in 1851 he invented a device to analyze polarized light. Stokes subsequently published, in 1852, a mathematical formula for the characteristics and behavior of polarized light, known as Stokes' Parameters.

### Important Light Experiments Involve Fluorescence

Stokes used the results of his many experiments with fluorescence for further investigation into the properties of light and other elements of nature. One example is his use of fluorescence to study ultraviolet spectra. He eventually was able to determine that the dark spectral lines discovered by Joseph von Fraunhofer were lines of elements which absorbed light from the sun's outer crust. Stokes' last major work was the mathematical study on the dynamical theory of double refraction, which were presented to the British Association for the Advancement of Science in 1862.

Stokes is regarded as having great influence on later Cambridge physicists, not only through his own research, but also through his knowledge of the work physicists outside Britain were doing. His promulgation of French mathematical physics—at a time when Cambridge had little knowledge of it—is one example.

Because of the time–consuming administrative duties Stokes was responsible for in the several scientific societies of which he served as an officer, his scholarly output was curtailed toward the end of his career. Stokes was, however, quick to respond to the many queries he received from other physicists regarding the progress and problems of their work. His voluminous correspondence with his contemporaries, notably **George Green**, James Challis, and William Thomson, details his influence and guidance. From 1883 to 1885, he delivered a series of lectures on light at the University of Aberdeen, and a series of lectures on theology at Edinburgh from 1891 to 1983. These lectures were later published. Stokes died at Cambridge on February 1, 1903.

## SELECTED WRITINGS BY STOKES

### Books

*Mathematical and Physical Papers,* 5 volumes, 1880–1905
*Burnett Lectures. On Light. In Three Courses Delivered at Aberdeen in November, 1883, December, 1884, and November, 1885.*
*Natural Theology. The Gifford Lectures Delivered Before the University of Edinburgh in 1891.*

## FURTHER READING

### Books

Asimov, Isaac. *Asimov's Biographical Encyclopedia of Science and Technology.* Garden City, NY: Doubleday & Company, Inc., 1972.
Daintith, John, editor. *Biographical Encyclopedia of Scientists.* Volume 2. New York: Facts on File, Inc., 1981, pp. 759–760.

Debus, Allen G., editor. *Who's Who In Science.* Chicago: Marquis Who's Who, Inc., 1968, p. 1616.

Parkinson, E.M. "George Gabriel Stokes," in *Dictionary of Scientific Biography.* Edited by Charles Coulston Gillespie. Volume XIII. New York: Charles Scribner's Sons, 1976, pp. 74–79.

Porter, Roy, editor. *The Biographical Dictionary of Scientists.* New York: Oxford University Press, 1994, pp. 647–648.

Wilson, D.B. *Kelvin and Stokes: A Comparative Study in Victorian Physics*, 1987.

—*Sketch by Jane Stewart Cook*

# Alicia Boole Stott
## 1860–1940
### Irish–English geometer

Alicia Boole Stott is considered noteworthy for her famous relatives as much for her own discoveries that translate Platonic and Archimedean solids into higher dimensions. She is more likely described when mentioned in other contexts as the daughter of mathematician and Royal Medal recipient **George Boole** and **Mary Everest Boole.**

Stott was born on June 8, 1860, in Cork, Ireland, where her father held a professorship at Queen's College. When George Boole died of a fever in 1864, Stott's sisters were dispersed to live with relatives while their mother struggled to support herself in London. Stott was shuttled between her grandmother in England and a great–uncle in Ireland, and was not reunited with her sisters until she was more than 10 years old.

### Cardboard Models

Stott was well into her teens by the time she became seriously interested in mathematics. A family friend named Howard Hinton, soon to become her brother–in–law, introduced her to the tesseract, or four–dimensional hypercube. He not only offered Stott intellectual stimulation, he got her a job as secretary to an associate, John Falk. At that time, Hinton was working on a book that would eventually see publication in 1904.

In 1900, Stott (with the encouragement of Walter Stott, an actuary whom she later married) published an article on three–dimensional sections of hypersolids. They led an ordinary middle–class existence following their marriage and had two children, Mary

and Leonard. Walter took note of his wife's interests and introduced her to the work of Pieter Hendrik Schoute of the University of Groningen. The Stotts took a chance and wrote to him describing Stott's work.

Upon viewing photographs of Stott's cardboard models, Schoute elected to relocate to England from the Netherlands in order to collaborate with her. Over their 20–year relationship, Schoute arranged for the publication of Stott's own papers and cowrote others. Stott refined her approach towards deriving the Archimedean solids from the Platonic solids to improve upon **Johannes Kepler**'s. She also coined the term "polytope" as a name for a four–dimensional convex solid form.

Schoute's university colleagues were impressed enough to invite Stott to their tercentenary celebration in 1914, to bestow upon her an honorary doctorate. Unfortunately, the 71–year–old Schoute died before the event. At a loss, Stott resumed her role as homemaker for nearly 20 years.

Stott found another collaborator, H.S.M. Coxeter, a writer who specialized in the geometry of kaleidoscopes, in 1930. She was quite taken with these "magic mirrors" and the challenge they would present to more old–fashioned mathematicians. Stott was inspired to devise a four–dimensional analogue to two of the Archimedean solids, which she called the "snub 24–cell." This construction was not original to her, having been discovered earlier by Thorold Gosset. However, her cardboard models of it in its "golden ratio" relationship with the regular 24–cell are still stored at Cambridge University.

Stott was an animal lover who enjoyed bird watching. She became ill around the time England entered World War II and died on December 17, 1940.

### SELECTED WRITINGS BY STOTT

"On Certain Sections of the Regular Four–Dimensional Hypersolids," in *Verhandelingen der Koninklijke Akademie van Wetenschappen* (1.sectie) 7 (3) (1900): 1–21.

"Geometrical Deduction of Semiregular from Regular Polytopes and Space Fillings," in *Verhandelingen der Koninklijke Akademie van Wetenschappen* (1.sectie) 11 (1) (1910): 1–24.

### FURTHER READING

#### Books

Coxeter, H.S.M. "Alicia Boole Stott," in *Women of Mathematics.* Edited by Louise S. Grinstein and Paul J. Campbell. Westport, CT: Greenwood Press, 1987, pp. 220–24.

*James Joseph Sylvester*

**Other**

Frost, Michelle. "Mary Everest Boole." *Biographies of Women Mathematicians.* http://www.scottlan.edu/lriddle/women/chronol.htm (July 1997).

"Alicia Boole Stott." *MacTutor History of Mathematics Archive.* http://www-groups.dcs.st-and.ac.uk/~history/Mathematicians/Stott.html (July 1997).

—*Sketch by Jennifer Kramer*

# James Joseph Sylvester
## 1814–1897
### English algebraist and combinatoricist

James Joseph Sylvester was one of the most colorful mathematicians of the Victorian age. His role in giving an impetus to the American mathematical community in the late 19th century provided a sense of national independence from European scholarship.

Sylvester was born on September 3, 1814 in London, the son of Abraham Joseph. He was the youngest son out of a total of nine children and received his early education at Jewish schools in Highgate and Islington. Sylvester entered St. John's

College at Cambridge in 1831. His undergraduate career was interrupted by illness that kept him at home for several years. Finally, in 1837 he took the Tripos (the final mathematical examinations) and placed second. Ordinarily, a student in that situation would have proceeded to take his degree and to compete for the Smith's Prize. Since, however, Sylvester was Jewish, he could not subscribe to the Thirty-Nine Articles of the Church of England, a requirement at that time for receiving a degree from Oxford or Cambridge. Sylvester later earned both bachelor and master's degrees from Trinity College in Dublin, which did not have the same requirements. Some 35 years after passing the Tripos, Sylvester received his B.A. from Cambridge when the doctrinal requirements were dropped.

Sylvester began teaching at University College, London, as the colleague of **Augustus De Morgan**, but did not find the chair of natural philosophy a congenial one. By 1839, he was a Fellow of the Royal Society and two years later made his first venture at transatlantic teaching, accepting a position at the University of Virginia. This proved to be a complete fiasco, as Sylvester encountered intense anti-Semitism. In the absence of offers from other American universities, Sylvester returned to England and took a position as actuary and secretary to the Equity and Law Life Assurance Company ("assurance" is the English term for "insurance"). He took private pupils in mathematics as well, the most notable of them being Florence Nightingale. In 1846, three years after returning from the United States, Sylvester entered the Inner Temple, one of the Inns of Court designed to prepare students for the bar. He was called to the bar in 1850, but never made a success at it as did his colleague **Arthur Cayley**.

### Finds a Position in the Educational Establishment

In 1855, Sylvester finally managed to secure a position teaching mathematics in England. He was appointed to the chair of mathematics at the Royal Military Academy at Woolwich. That same year he became the editor of the *Quarterly Journal of Pure and Applied Mathematics*, which he transformed into one of the leading periodicals about mathematical research in England. He remained its editor until 1877, during which period he also served as president of the London Mathematical Society from 1866 to 1868.

As a mathematician Sylvester occasionally indulged in hasty generalizations, but his imagination at its best could create whole new fields of study. He did not take the trouble to study the contributions of contemporary mathematicians outside his circle, with the result that some of his work amounts to an independent rediscovery. He did a good deal of work in the area of invariant theory, the study of algebraic expressions that remain fixed under various classes of transformations. He certainly introduced an enormous amount of terminology—including "discriminant"—an activity of which he was especially proud.

In addition to his work on invariant theory, Sylvester deserves much credit for the early study of the subject of partitions. A partition is a way of breaking up a positive whole number as a sum of other positive whole numbers, so that $3 + 3$, $2 + 2 + 2$, and $5 + 1$ are all partitions of the number 6. Counting the number of partitions of various kinds required some inventiveness and Sylvester helped to get the subject well-launched. The subject of partitions turned out to have a good deal of importance in analysis, although Sylvester did not push this aspect of his studies very far.

### Returns to America

Sylvester had resigned from Woolwich in 1870 and was busily engaged in his mathematical work, but he could not resist an invitation from the eminent American physicist Joseph Henry to come to the newly-founded Johns Hopkins University in Baltimore. Henry assured Sylvester that he would be involved with serious students and have ample opportunities for research. He arrived in Baltimore in 1876 to find his expectations fulfilled. In addition to his work with some of the best students in the country, Sylvester also founded the *American Journal of Mathematics*. At this point in history there was no serious American publication devoted to mathematics and Sylvester's journal filled the void. In particular, to make certain that it would be taken seriously, he contributed 30 of his own papers to the journal. Despite his successes at Johns Hopkins, Sylvester could not pass up the opportunity to succeed H.J.S. Smith as Savilian Professor of Geometry at Oxford University.

Among the many honors Sylvester had received was the Royal Medal in 1861 and the Copley Medal in 1880, both from the Royal Society, and the De Morgan Medal from the London Mathematical Society in 1887. He was a corresponding member of numerous foreign societies and received four honorary degrees.

Sylvester died on March 15, 1897 and was buried at the Jewish Cemetery at Ball's Pond, London. In 1997, the London Mathematical Society held a memorial service on the centenary of his death. There was also a day's conference devoted during that centenary year to his life and the areas of mathematics to which he contributed. Without question, the mathematical communities on both sides of the Atlantic are beneficiaries of Sylvester's ideas and energy.

## SELECTED WRITINGS BY SYLVESTER

*Collected Mathematical Papers*. Edited by H.T. Baker, 1904–1912

## FURTHER READING

Andrews, George E. "J.J. Sylvester, Johns Hopkins and Partitions," in *A Century of Mathematics in America*. Volume 1. Edited by Peter Duren et al. Providence, R.I.: American Mathematical Society, 1988, pp. 21–40.
Bell, Eric. *Men of Mathematics*. New York: Simon and Schuster, 1937, pp. 378–405.
Matheson, Percy Ewing and Edwin Bailey Elliott. "James Joseph Sylvester," in *Dictionary of National Biography*. Volume 19. Edited by Leslie Stephen and Sidney Lee. Oxford: Oxford University Press, 1968, pp. 258–260.
North, J.R. "James Joseph Sylvester," in *Dictionary of Scientific Biography*. Volume XIII. Edited by Charles Coulston Gillispie. New York: Charles Scribner's Sons, 1973, pp. 216–222.

—*Sketch by Thomas Drucker*

# Gabor Szegö
## 1895–1985
### Hungarian-born American analyst and educator

Gabor Szegö was a product of the German–Hungarian mathematical school, something of a child prodigy who, by age 20, had published a seminal paper in an internationally recognized mathematical journal. By the age of 30, he had published some 30 noteworthy papers, and also coauthored with **George Polyá**, a fellow Hungarian, a famous mathematical problem book, *Problems and Theorems in Analysis*. Szegö most important theoretical work was in orthogonal polynomials and Toeplitz matrices, and he was, according to **Richard Askey** and Paul Nevai writing in *The Mathematical Intelligencer*, "one of the most prominent classical analysts of the twentieth century." Forced to immigrate to the United States to avoid Nazi persecution, Szegö helped build the mathematics department at Stanford University, where he taught from 1938 until his retirement.

Szegö was born on January 20, 1895 in Kunhegyes, then part of Austro–Hungary, to Adolf Szegö and Hermina Neuman. Szegö completed elementary school in the town of Szolnok, some 60 miles

*Gabor Szegö*

southeast of Budapest, and in 1912 enrolled in the Pázmány Péter University in the capital, now known as Eotvos Lorand University. It was in that same year that he won his first mathematic prize, the coveted Eotvos Competition. As Askey and Nevai noted in their article, it was fortunate for Szegö that he won such a prize at the outset of his studies, for as a Jew, he might not otherwise have secured a university post. He followed this impressive beginning with a further prize the next year, this time for a paper on polynomial approximations of continuous functions.

### World War I Interrupts Studies

Szegö's first publication in a notable journal came in 1913 when he published the solution of a problem given by George Polyá in *Archiv der Mathematik und Physik*. His first research paper, on the limit of the determinant of a Toeplitz matrix formed from a positive function, was published in 1915 and again inspired by a conjecture by Polyá. As Askey and Nevai stated, "Szegö spent another 45 years working on sharpening, extending, and finding applications of the results published in this article." It was during this time also that Szegö became a tutor to another mathematical prodigy, **John von Neumann.**

With the onset of World War I, however, this life of research came to a temporary halt. Szegö joined the cavalry, though he was a self–admitted poor horseman. He remained in the army from 1915 to 1919, even after the collapse of the Austro–Hungarian

Empire. He was able, while posted in Vienna, to finish his doctorate, and after the war he married Erzébet Anna Neményi, herself the holder of a Ph.D. in chemistry. The couple had two children, Peter and Veronica. But the effects of the war did not end with the Armistice. The early 1920s were a turbulent time in much of Central Europe, and Hungary was no exception. Szegö worked for a time as an assistant at the Technical University of Budapest, but by 1921 he and his family were forced to move to Germany in search of a more secure living. In Berlin, Szegö secured a post with the university for his work on orthogonal polynomial series, and became friends and worked with Issai Schur and **Richard von Mises**, among others. He continued his research and won the Julius Konig Prize in 1924, yet was only an associate professor without tenure at the University of Berlin.

### Writes a Classic Problem Book

Together with his old acquaintance Polyá, Szegö collaborated on "the best–written and most useful problem book in the history of mathematics," according to Askey and Nevai. In 1925, the two published *Aufgaben und Lehrsatze aus der Analysis*, a two–volume work later translated into English as *Problems and Theorems in Analysis*. Students are introduced to mathematical research via a series of problems in analysis, number theory, combinatorics, and geometry, such that the solution to a group of problems prepares the reader for independent research in that area. Publication of this book rightly placed Szegö as not only a noted researcher, but also as an educator with innovative ideas.

The following year, Szegö was invited to assume to position of full professor at the University of Könisgsberg, succeeding Knopp in that position. He remained at Könisgsberg for eight years, but increasingly the situation for Jews in Germany was becoming untenable. It slowly became apparent to Szegö and his wife that they would need to move once again, and this time not simply to a neighboring country.

### Emigrates to America

In 1934, Szegö secured, through the intercession of friends, a teaching position at Washington University in St. Louis. The money to pay for his salary was raised by a Rockefeller Grant, and by grants from local Jewish merchants as well as the Emergency Committee in Aid of Displaced Scholars. Szegö remained in St. Louis until 1938, during which time he advised five Ph.D. students, and worked on his seminal book, published in 1939, *Orthogonal Polynomials.* This book has become a standard reference work for both pure and applied mathematicians. Orthogonal polynomials are polynomials that first appeared in connection with numerical analysis and approximation theory. It was Szegö's accomplishment to reduce many of these problems to the asymptotic behavior of certain Toeplitz determinants, which come into consideration as a variable reaches a limit, most usually infinity. Applications of Szegö's theory have proved of significance in the fields of numerical methods, differential equations, prediction theory, systems theory, and statistical physics. Also, Szegö's work on Toeplitz matrices led to the concept of the Szegö reproducing kernel and the Szegö limit theorem.

### Builds Stanford Mathematics

In 1938, Szegö accepted an offer to lead the department of mathematics at Stanford University, in Palo Alto, California. He remained Head until 1953, and during those 15 years he succeeded in building the department to one of the most renowned in the world, bringing people such as Polyá, Loewner, Bergman, and Schiffer to the school, and helping to train an entire generation of new mathematicians. It was while he was at Stanford also, in 1940, that Szegö became a naturalized American citizen.

Szegö retired from Stanford in 1960 as Professor Emeritus. His wife died in 1968, and Szegö remarried in 1972 to Irén Vajda in Budapest. Ten years later, his second wife also died. In 1985, after several years of declining health, Szegö passed away, leaving behind a body of work of some 140 articles and six books authored or coauthored, as well as a generation of trained mathematicians to carry on his research in orthogonal polynomials, a field which he pioneered.

In 1995, the centenary of his birth, Szegö was posthumously honored by the dedication of a statue in his hometown of Kunhegyes, replications of which were also installed at Washington University and at Stanford.

## SELECTED WRITINGS BY SZEGÖ

### BOOKS

*Orthogonal Polynomials*, 1939
(With George Polya) *Problems and Theorems in Analysis*. Volumes 1 and 2. Translated by Dorothee Aeppli and C. E. Billingheimer, 1972–76.
*Gabor Szegö: Collected Papers*. Volumes 1–3. Edited by Richard Askey, 1982.

## FURTHER READING

### Books

Askey, Richard, "Gabor Szegö: A Short Biography." *Gabor Szegö: Collected Papers*. Volume 1. Edited by Richard Askey. Boston, Basel: Birkhauser, 1982, pp. 1–7.

### Periodicals

Askey, Richard, and Paul Nevai. "Gabor Szegö: 1895–1985," in *The Mathematical Intelligencer* 18, no. 3 (1996): 10–22.

*—Sketch by J. Sydney Jones*

# Richard A. Tapia
## 1939-
**Hispanic American computational mathematician and educator**

Richard A. Tapia is a nationally recognized educator who was named one of the 20 most influential leaders in minority mathematics education in 1990. His four-day workshops on computational science have also received national attention by bringing secondary and middle school teachers from schools with high minority enrollments to Rice University to learn about opportunities in the computational sciences. As associate director for minority affairs in the Office of Graduate Studies at Rice, Tapia has been extremely successful in producing minority Ph.D.s. He is one of only 27 Hispanic Americans currently holding a Ph.D. in mathematics. His technical area is computational science, particularly optimization theory and iterative methods for nonlinear problems.

Richard Alfred Tapia was born on March 25, 1939, in Santa Monica, California. His parents, Amado and Magda Tapia, immigrated from Mexico and became American citizens; neither of them finished high school because they needed to help support their families. As a young man, Tapia became an avid drag racer, attaining world record-holder status. He married Jean Rodriguez in 1959, and they have a son and a daughter. Mrs. Tapia, a former professional dancer who was diagnosed with multiple sclerosis in 1978, developed a new career teaching exercise techniques to people with physical disabilities.

Tapia entered the University of California at Los Angeles (UCLA) after spending two preparatory years in junior college. He earned his B.A., M.A., and Ph.D. (1967) degrees at UCLA. After completing his doctorate, he taught at the University of Wisconsin's Mathematics Research Center for two years. In 1970, Tapia took the position of assistant professor of mathematical science at Rice University in Houston, Texas, where he has remained. He was promoted to a full professor in 1976, and between 1978 and 1983, he served as chair of the department. In 1989 Tapia took on the additional responsibilities of associate director for minority affairs in Rice's Office of Graduate Studies and director of education and minority programs for the Center for Research on Parallel Computation (CRPC), a National Science Founda-

tion-supported Science and Technology Center headquartered at Rice. He has been a visiting professor at Baylor College and Stanford University. Tapia continues to hold the latter two positions, as well as his faculty position as Noah Harding Professor at Rice.

### Establishes Technical Reputation

Numerical optimization, a technique Tapia helped develop, is used in industry to predict and optimize their output. For example, using the power of parallel processing with computers, he helped a major airline solve a scheduling problem with 12 million variables, taking into account such factors as fuel consumption, consumer needs, personnel availability, and equipment maintenance. Tapia's biography on the Rice University web site notes that "he has contributed significantly to the algorithmic development and theory of quasi-Newton methods for general nonlinear programming problems." His recent work with a group of fellow Rice researchers on interior-point methods for linear programming has attracted international attention. The method involves creating a geometric model of the resources and constraints for a given problem, and then calculating potential solutions at certain key points. The Rice biography explains that "interior-point methods are effective algorithms for solving very large problems in science, industry, and business. They are currently being used to solve larger problems than have ever been solved previously or that anyone had ever expected would be solved."

Tapia has written more than 60 technical papers as well as a textbook, *Nonparametric Function Estimation, Modeling, and Simulation,* first published in 1978. In a review of the revised edition of the book, Randall Eubank wrote in the *Journal of the American Statistical Association* that the earlier version had played "an important role in stimulating research into nonparametric density and function estimation." In recognition of his technical work and his educational accomplishments, Tapia became the first Mexican-American to be elected to the National Academy of Engineering in 1992. He was one of 20 minority scientists featured on a 1996 PBS documentary series called *Breakthrough: The Changing Face of Science in America.*

Tapia has been concerned with the future role of mathematicians and mathematics in a world increasingly dominated by computers. When he served on the Strategic Planning Committee of the American Mathematics Society, he had been able to lead the debate on this important issue. It is imperative, he believes, that mathematics show itself to be relevant to the everyday world and be able to contribute to solving problems in engineering, medicine, and other applied disciplines.

### Promotes Minority Education

Cheryl Laird wrote in the *Houston Chronicle* that Tapia is "a higher-education version of Jaime Escalante, the inspiring Los Angeles teacher made famous in [the film] *Stand and Deliver*" and that he "can make a graduate-level lecture on computational mathematics sound fun even to a listener who only understands every fourth word." He excites audiences as young as third-grade students, and has won acclaim from his university students for his teaching prowess. *Hispanic Engineer Magazine* has selected Tapia as both Engineer of the Year (1996) and College Level Educator of the Year (1990). In 1996, he received a Presidential Award for Excellence in Science, Mathematics, and Engineering Mentoring in recognition of his efforts to attract minority students to technical studies.

Through the CRPC, Tapia developed Mathematical and Computational Sciences Awareness Workshops that inform elementary and secondary teachers about mathematics and science opportunities for minorities. He also directs the "Spend a Summer with a Scientist" program that allows minority students to help Rice faculty members conduct their research.

In 1990, the National Research Council of the National Science Foundation named Tapia one of the 20 most influential leaders in minority mathematics education. He is an active member of the Society for Industrial and Applied Mathematics, the Mathematical Programming Society, the American Mathematical Society, and the Society for the Advancement of Chicanos and Native Americans in Science (SACNAS). In addition, he serves on the Mathematical Association of America's Committee on Minority Participation in Mathematics and the National Academy of Sciences' Mathematical Sciences Education Board. He has been nominated by President Bill Clinton to serve on the National Science Board, which governs the National Science Foundation.

Tapia's curriculum vitae lists more than 120 activities (such as speeches, conferences, and panel discussions) he has done in support of minority and outreach education. In 1994, he became the first recipient of the A. Nico Habermann Award given by the Computer Research Association in recognition of his outstanding contribution to helping members of under represented groups within the computing research community.

### SELECTED WRITINGS BY TAPIA

(With James R. Thompson) *Nonparametric Function Estimation, Modeling, and Simulation,* 1978; revised edition 1990

### FURTHER READING

De Vries, Dan, and Melanie Cole. "Latinos in the Lab." *Hispanic* 9 (September 1996): 36–42.
Laird, Cheryl. "By the Numbers: It All Adds Up for Math Prof Richard Tapia." *Houston Chronicle* (April 25, 1996).
Mellado, Carmela C. "Math Education in the Computer Age." *Hispanic Engineer* (Spring 1991): 26–27.
"Presidential Mentoring Awards Announced." *Notices of the American Mathematical Society* (December 1996): 1539.
"1990 Award Winners." *Hispanic Engineer* (Conference Issue, 1990): 66, 68.

#### Other

"Professor Richard Tapia." *Rice University.* May 12, 1997. http://www.caam.rice.edu/.rat/ (June 10, 1997).

—*Sketch by Leonard C. Bruno and Loretta Hall*

# Alfred Tarski
## 1901-1983
### Polish American logician and algebraist

Alfred Tarski made considerable contributions to several areas of mathematics, including set theory and algebra, and his work as a logician led to important breakthroughs in semantics—the study of symbols and meaning in written and verbal communication. Tarski's research in this area yielded a mathematical definition of truth in language, and also made him a pioneer in studying models of linguistic communication, a subject that became known as model theory. Tarski's research also proved useful in the development of computer science, and he became an influential mentor to later mathematicians as a professor at the University of California at Berkeley.

Born Alfred Tajtelbaum in Warsaw, Poland (then part of Russian Poland), on January 14, 1901, Tarski was the elder of two sons born to Ignacy Tajtelbaum, a shopkeeper of modest means, and Rose (Iuussak) Tajtelbaum, who was known to have an exceptional memory. During his teens Tarski helped supplement

*Alfred Tarski*

the family income by tutoring. He attended an excellent secondary school, and although he was an outstanding student, he, surprisingly, did not get his best marks in logic. Biology was his favorite subject in high school, and he intended to major in this discipline when he first attended the University of Warsaw. However, as Steven R. Givant pointed out in *Mathematical Intelligence,* "what derailed him was success." In an early mathematics course at the university, Tarski was able to solve a challenging problem in set theory posed by the professor. The solution led to his first published paper, and Tarski, at the professor's urging, decided to switch his emphasis to mathematics.

Tarski received a Ph.D. from the University of Warsaw in 1924, the same year he met his future wife, Maria Witkowski. They got married on June 23, 1929, and later had two children, Jan and Ina. It is believed that the young mathematician was in his early-twenties when he changed his name from Tajtelbaum to Tarski. His son, Jan, told interviewer Jeanne Spriter James that this step was taken because Tarski believed that his new Polish-sounding name would be held in higher regard at the university than his original Jewish moniker. When Tarski was married, he was baptized a Catholic, his wife's religion.

### Early Struggles in Academia

Tarski served in the Polish army for short periods of time in 1918 and 1920. While working on

his Ph.D. he was employed as an instructor in logic at the Polish Pedagogical Institute in Warsaw beginning in 1922. After graduating he became a docent and then adjunct professor of mathematics and logic at the University of Warsaw beginning in 1925. That same year he also took a full-time teaching position at Zeromski's Lycee, a high school in Warsaw, since his income from the university was inadequate to support his family. Tarski remained at both jobs until 1939, despite repeated attempts to secure a permanent university professorship. Some have attributed Tarski's employment difficulties to anti-semitism, but whatever the reason, his lack of academic prominence created problems for the young mathematician. Burdened by his teaching load at the high school and college, Tarski was unable to devote as much time to his research as he would have liked. He later said that his creative output was greatly reduced during these years because of his employment situation. The papers he did publish in this period, however, quickly marked Tarski as one of the premiere logicians of the century. His early work was often concentrated in the area of set theory. He also worked in conjunction with Polish mathematician **Stefan Banach** to produce the Banach-Tarski paradox, which illustrated the limitations of mathematical theories that break a space down into a number of pieces. Other research in the 1920s and 1930s addressed the axiom of choice, large cardinal numbers, the decidability of Euclidean geometry, and Boolean algebra.

Tarski's initial research on semantics took place in the early–1930s. He was concerned here with problems of language and meaning, and his work resulted in a mathematical definition of truth as it is expressed in symbolic languages. He also provided a proof that demonstrated that any such definition of truth in a language results in contradictions. A London *Times* obituary on Tarski noted the groundbreaking nature of his work in this area, proclaiming that the mathematician's findings "set the direction for all modern philosophical discussions of truth." Tarski expanded this early work in semantics over the ensuing years, eventually developing a new field of study—model theory—which would become a major research subject for logicians. This area of study examines the mathematic properties of grammatical sentences and compares them with various models of linguistic communication.

Additionally, Tarski pursued research in many other areas of math and logic during his career, including closure algebras, binary relations and the algebra of relations, cylindrical algebra, and undecidable theories. He also made a lasting contribution to the field of computer science. As early as 1930 he produced an algorithm that was capable of deciding whether any sentence in basic Euclidian geometry is either true or false. This pointed the way toward later machine calculations, and has also had relevance in determining more recent computer applications.

### Outbreak of World War II leads to New Opportunities

In 1939 Tarski left Poland for a conference and speaking tour in the United States, intending to be gone for only a short time. Shortly after his departure, however, the German Army invaded and conquered Poland, beginning World War II. Unable to return to his homeland, Tarski found himself stranded in the United States without money, without a job, and without his wife and children who had remained in Warsaw. The family would not be reunited until after the war, and in the meantime, Tarski set about finding work in America. He first served as a research associate in mathematics at Harvard University from 1939 to 1941. In 1940 he also taught as a visiting professor at the City College of New York. He had a temporary position at the Institute for Advanced Study at Princeton beginning in 1941, and in 1942 he obtained his first permanent position in the United States when he was hired as a lecturer at the University of California at Berkeley. The university would remain his professional home for the rest of his career.

Tarski became an associate professor at the university in 1945, was appointed to the position of full professor the following year, and was named professor emeritus in 1968. Tarski's contributions to mathematics and science were enhanced by his role as an educator. He established the renown Group in Logic and the Methodology of Science at Berkeley, and over his long tenure he taught some of the most-influential mathematicians and logicians to emerge after World War II, including **Julia Bowman Robinson** and Robert Montague. His stature was further enhanced through his service as a visiting professor and lecturer at numerous U.S. and international universities. In 1973 Tarski ended his formal teaching duties at Berkeley, but he continued to supervise doctoral students and conduct research during the final decade of his life. He died in 1983 from a lung condition caused by smoking.

Tarski received many awards and honors throughout his career. He was elected to the National Academy of Sciences and the Royal Netherlands Academy of Sciences and Letters, and was also made a corresponding fellow in the British Academy. In 1966 he received the Alfred Jurzykowski Foundation Award, and in 1981 he was presented with the Berkeley Citation, the university's highest faculty honor. He also was awarded numerous fellowships and honorary degrees, and was a member in many professional organizations, including the Polish Logic Society, the American Mathematical Society, and the International Union for the History and Philosophy of Science.

## SELECTED WRITINGS BY TARSKI

### Books

*Introduction to Logic and to the Methodology of Deductive Sciences*, 1941
*Logic, Semantics, Metamathematics: Papers from 1923 to 1938*, 1956, revised edition. Edited by J. Corcoran, 1983
*Alfred Tarski: Collected Papers* 4 volumes. Edited by Steven R. Givant and Ralph N. McKenzie, 1986
(With Steven R. Givant) *A Formalization of Set Theory without Variables*, 1987

### Periodicals

"The Semantic Conception of Truth and the Foundations of Semantics," *Philosophy and Phenomenological Research* 4 (1944): 341–375

## FURTHER READING

### Books

*Dictionary of Scientific Biography.* Volume 18, supplement II, Scribner, 1990, pp. 893–896.
*Proceedings of the Tarski Symposium, An International Symposium to Honor Alfred Tarski on the Occasion of His Seventieth Birthday.* Volume 25. Washington, D.C.: American Mathematical Society, 1974.
Ulam, Stanislaw M. *Adventures of a Mathematician.* New York: Charles Scribner & Sons, 1976, pp. 29, 40, 114, 119, 122.

### Periodicals

Addison, John W. *California Monthly* (December 1983).
*Chicago Tribune* (October 30, 1983).
Givant, Steven R. "A Portrait of Alfred Tarski," *Mathematical Intelligence* 13, no. 3 (1991): 16–32.
*Journal of Symbolic Logic* 51 (1986); 53, 1988.
*Times* (London) (December 6, 1983): 16-G.
*Washington Post* (October 29, 1983).

### Other

Tarski, Jan, interviews with Jeanne Spriter James conducted November 1-4, and 21, 1993.

*—Sketch by Jeanne Spriter James*

*Tartaglia (Nicolò Fontana)*

# Tartaglia (Nicolò Fontana)
## 1499(?)–1557
### Italian algebraist

Tartaglia was a self–taught mathematician whose work in algebra led to the first generalized solution of cubic equations. He was neither prominent nor well connected and spent most of his life earning a meager salary as a school teacher. In addition to his work in algebra, Tartaglia also made contributions to the fields of fundamental arithmetic, ballistics, military engineering, cartography, and the development of certain instruments used in surveying. His Italian translations of some of the works of **Euclid** and **Archimedes** represented the first time this information had been available in a modern language.

Tartaglia's given name was Nicolò (or Niccolò) Fontana. He was born in Brescia, Italy, in 1499 or possibly 1500. His father, Michele, was a postal courier who worked for the government of Brescia. When his father died in about 1506, the family was left in poverty. In 1512, the French army attacked Brescia. During the attack a French soldier set upon young Nicolò with a saber and left him with five head wounds. Although Nicolò eventually recovered, his mouth was so disfigured that he could no longer speak clearly. As a result of this speech impediment, he was given the nickname Tartaglia, from the Italian word *tartagliare*, which means "to stammer." Rather than

being offended, Nicolò kept the nickname as a surname and used it for the rest of his life.

When he was about 14 years old, Tartaglia went to a tutor to learn the alphabet. By the time he reached the letter "k," he was no longer able to pay for lessons and subsequently taught himself from books. Tartaglia learned quickly and acquired enough skill at mathematics that he was able to obtain a position as a teacher of practical mathematics in Verona sometime between 1516 and 1518. He stayed in Verona until 1534 and rose to the position of headmaster of a school there. Some evidence indicates that Tartaglia had a family in Verona, although his income was barely enough to support them. In 1534, he moved to Venice to become a professor of mathematics. Except for a period of 18 months in 1548–1549, Tartaglia remained in Venice until his death.

### Discovers a Solution to Cubic Equations

Tartaglia's most notable contribution to mathematics came shortly after he moved to Venice. As was the custom of the times, persons in academic positions were often subject to challenges of their knowledge and skills. In 1535, Tartaglia was challenged to a contest by Antonio Fior. Fior had been an aspiring but below average student of the mathematician **Scipione dal Ferro**, before dal Ferro's death in 1526. Just before his death, dal Ferro had given Fior a method for solving the so–called "depressed cubic" equation, in which there was no second power term. In turn, Fior had kept this method secret until he decided to use it to boost his floundering career by challenging Tartaglia. For his part, Tartaglia already knew how to solve cubic equations that lacked the first power term, and so he readily accepted Fior's challenge. Each person was to submit 30 problems for the other to solve. Rather than give Tartaglia problems covering a wide range of mathematical topics, Fior gambled and gave Tartaglia nothing but depressed cubics. After many days, with the deadline drawing near, Tartaglia discovered the solution and quickly solved all 30 problems. Fior, on the other hand, solved only a few of the problems posed by Tartaglia and suffered a humiliating defeat.

Tartaglia now knew the solutions to two types of cubic equations, and this knowledge put him in an enviable position among his fellow mathematicians. One in particular, **Girolamo Cardano**, repeatedly beseeched Tartaglia to share his secrets. Finally, in 1539, Tartaglia gave Cardano the solutions on the condition that Cardano not reveal them until Tartaglia had published them himself. Armed with this new knowledge, Cardano set about to discover the solution to a generalized cubic equation, in which all of the terms were present. Working with his new student, Ludovico Ferrari, Cardano discovered a method for reducing any generalized cubic equation to a depressed cubic. Then, using Tartaglia's methods, they could solve the equation. Unfortunately, Cardano was still bound by his promise not to reveal Tartaglia's methods. Then, in 1543, Cardano discovered that it was dal Ferro who had originally solved the depressed cubic equation, not Tartaglia. With this information Cardano went ahead and published his solution to a generalized cubic equation in 1545, giving full credit to dal Ferro and Tartaglia for their work. Tartaglia was outraged at this perceived breech of trust and wrote hotly and offensively about Cardano. Ferrari rose to Cardano's defense and began a long series of acrimonious correspondence with Tartaglia, ending in a public debate in Ferrari's home town of Milan in 1548. Backed by a sympathetic and unruly audience, Ferrari easily defeated Tartaglia, who was forced to withdraw in fear of his life.

### Makes Contributions in Other Fields

In addition to his work in mathematics, Tartaglia's work drew interest in other areas. His work in ballistics included the determination that the maximum theoretical range of a projectile is obtained when the firing angle is 45 degrees. He also proposed the concept of firing tables in which range is correlated to firing angle, powder charge, and other factors. In surveying, he showed how to apply the compass and proposed the design of two instruments for determining the heights and distances to inaccessible points.

Tartaglia published his Italian translation of Euclid's *Elements* in 1543, and produced his translation of a Latin version of some of Archimedes' works the same year. He made additional translations of Archimedes' works in 1551, adding his own commentary. Tartaglia's translations contributed to the dissemination of knowledge by making these ancient works available in a modern language.

Tartaglia died impoverished in his home in Venice on December 13, 1557.

*Olga Taussky-Todd*

Cohen, I.B., editor. *Album of Science: From Leonardo to Lavoisier, 1450–1800.* New York: Charles Scribner's Sons, 1980, p. 77.

Dunham, W. *Journey through Genius: The Great Theorems of Mathematics.* New York: Wiley, 1990, pp. 135–136, 140–142.

Struik, D.J., editor. *A Source Book in Mathematics, 1200–1800.* Princeton: Princeton University Press, 1986, pp. 62–64.

*—Sketch by Chris Cavette*

### SELECTED WRITINGS BY TARTAGLIA

#### Books

*Nova scientia,* 1537
*Euclide Megarense,* 1543
*Opera Achimedis,* 1543
*Questi et inventioni diverse,* 1546
*Travagliata inventione,* 1551
*Generale trattato di numeri et misure,* 1556–1560

### FURTHER READING

#### Books

American Council of Learned Societies. *Biographical Dictionary of Mathematicians.* New York: Charles Scribner's Sons, 1991, pp. 2418–2421.

# Olga Taussky–Todd
## 1906–1995
### Austro–Hungarian–born American number theorist

Olga Taussky–Todd is best remembered for her research on matrix theory and algebraic number theory. Matrix theory is the study of sets of elements in a rectangular array that are subject to operations such as multiplication or addition according to specified rules. Number theory is the study of integers and their relationships. During a long, productive career, Taussky–Todd published over 200 research papers and other writings on a wide range of mathematical topics. In 1964 she was named "Woman of the Year" by the *Los Angeles Times*, and in 1970 she received the Ford Prize for an article on sums of squares. Taussky-Todd was also well known for her lectures. In 1981, she gave the **Emmy Noether** Lecture at the annual meeting of the Association for Women in Mathematics, taking as the subject of her talk the many aspects of Pythagorean triangles.

Taussky-Todd was born on August 30, 1906, in Olmütz, Austria-Hungary (now Olomouc, Czechoslovakia). She was the second of three daughters born to Julius David Taussky and Ida Pollach Taussky. Her father was an industrial chemist who also worked as a newspaper journalist. He encouraged his daughters to take their education seriously. Her mother was an intelligent person as well, but had little formal education. In an autobiographical essay published in *Mathematical People: Profiles and Interviews*, Taussky-Todd recalled of her mother: "She was rather bewildered about our studies and compared herself to a mother hen who had been made to hatch duck eggs and then felt terrified on seeing her offspring swimming in a pool." However, Taussky-Todd also noted that her mother was more willing than her father to accept the notion of girls actually using their educations later in life to earn a living.

Shortly before Taussky-Todd turned three, her family moved to Vienna. Midway through World War I, the family moved again, this time to Linz in upper Austria. Her father was manager of a vinegar factory there, and he often asked Taussky-Todd to help with such chores as calculating how much water to add to mixtures of various vinegars to achieve the right acidity. Taussky-Todd's best subjects in school were grammar and expository writing. As a girl in Linz, she began a lifelong hobby of writing poems.

### First Forays into Number Theory

During Taussky-Todd's last year of high school, her father died, leaving the family without an income. Taussky-Todd took jobs tutoring and working at the vinegar factory. The next year she entered the University of Vienna, determined to prove that her plan to study mathematics was a practical one. Among her professors was noted number theorist Philip Furtwängler. When the time came for Taussky-Todd to decide upon a thesis topic, Furtwängler suggested class field theory. In mathematics, a field is a set that has two operations, addition and multiplication. For each operation, the set is closed, associative, and commutative, and it has an identity element and inverses. As Taussky-Todd wrote in *Mathematical People*, "This decision had an enormous influence on my whole future . . . It helped my career, for there were only a very few people working in this still not fully understood subject. It was definitely a prestige subject."

In 1930, Taussky-Todd received her doctoral degree in mathematics. Based on her thesis, she was promptly offered a temporary post at the University of Göttingen in Germany, where she helped edit **David Hilbert**'s writings on number theory. She also edited Emil Artin's lectures on class field theory. By 1932, the growing political tensions in Germany made it unwise for Jews such as Taussky-Todd to stay there. She returned to Vienna, where she worked as a mathematics assistant. Among those she assisted was Hans Hahn, one of her former professors. Hahn had first introduced Taussky-Todd to functional analysis, the study of a particular type of function.

### Travels to United States and England

Taussky-Todd applied for and received a three-year science fellowship from Girton College at Cambridge University in England. It was agreed that she could spend the first year of the fellowship at Bryn Mawr College in Pennsylvania. Taussky-Todd took a few English lessons and embarked for the United States in 1934. At Bryn Mawr, she had the chance to work with Emmy Noether, whom she had earlier met at Göttingen. Noether, who was 24 years older than Taussky-Todd, was already an established figure in modern abstract algebra. Taussky-Todd enjoyed accompanying the older woman on her weekly trips to Princeton University whenever possible. However, she also found that Noether had a critical side. As Taussky-Todd recalled in *Mathematical People*, "She disliked my Austrian accent, my less abstract thinking, and she was almost frightened that I would obtain a [permanent] position before she would."

In 1935, Taussky-Todd traveled to Girton College at Cambridge, where she spent the last two years of her fellowship. The mathematical interests of her colleagues there did not quite match her own. However, she did get some much-needed practice at teaching in English. In 1937 she took a junior-level teaching position at one of the women's colleges at the University of London. The hours were arduous, but she still found time to attend professional seminars. It was at one such seminar that she met fellow mathematician John (Jack) Todd. The two were married in 1938.

### Growing Interest in Matrix Theory

Soon thereafter World War II broke out, bringing not only political but also personal upheaval. The newlyweds moved 18 times during the war. For a while, the couple lived in Belfast, Ireland, where Taussky-Todd first began to focus on matrix theory while teaching at Queen's University. A year later Taussky-Todd returned to work at her London college, which had since been relocated to Oxford for safety reasons. In 1943, she took a research job in aerodynamics with the Ministry of Aircraft Production. There she joined a group that was studying flutter problems in combat aircraft. Flutter refers to the self–excited oscillations of part of an airplane such as the wings. A corresponding problem in mathematics involves the stability of certain matrices. As a result, this job just strengthened Taussky–Todd's growing fascination with matrix theory.

In 1947 Taussky–Todd's husband accepted an invitation to come to the United States for a year and work for the National Bureau of Standards. Taussky–Todd also joined the staff of the bureau field station at the University of California at Los Angeles. After this first year, the couple briefly went back to London. They soon returned to work again for the National Bureau of Standards, however, this time in Washington, D.C. Taussky–Todd's title at the Bureau was mathematical consultant, and as she noted in *Mathematical People*, "this I truly was, because everybody dumped on me all sorts of impossible jobs, from refereeing every paper that was written by a member or visitor to the group, to answering letters from people . . ., to helping people on their research."

### Ends Personal Odyssey at CalTech

Taussky–Todd and her husband made one last major career move in 1957, accepting positions at the California Institute of Technology. In an autobiographical essay in *Number Theory and Algebra*, Taussky–Todd wrote: "It seemed to me as if an odyssey of 20 years (I left Cambridge, England, in 1937) had ended. I could at last work again with academic freedom and have Ph.D. students." Some of her students went on to play starring roles in the burgeoning of matrix theory that has occurred since the 1960s. In 1977, Taussky–Todd was made a professor emeritus at Caltech.

Taussky–Todd received a number of honors in the course of her prolific career. Upon her retirement, two journals, the *Journal of Linear Algebra* and the *Journal of Linear and Multilinear Algebra*, published issues dedicated to her. Going a step further, the *Journal of Number Theory* published an entire book, *Algebra and Number Theory*, dedicated to Taussky–Todd and two others. Taussky–Todd received the Gold Cross of Honor from the Austrian government in 1978. A decade later she was awarded an honorary D.Sc. degree by the University of Southern California.

Taussky–Todd died on October 7, 1995, at her home in Pasadena, California. An obituary by Myrna Oliver in the *Los Angeles Times* referred to her as "one of the most prominent women mathematicians in the United States." Indeed, a lifetime of contributions to both pure and applied mathematics across several specialty areas had earned her the respect of mathematicians of both genders and many nationalities.

## SELECTED WRITINGS BY TAUSSKY–TODD

"Olga Taussky–Todd: An Autobiographical Essay." *Mathematical People: Profiles and Interviews.* Edited by Donald J. Albers and G.L. Alexanderson, 1985, pp. 310–336.

## FURTHER READING

### Books

Luchins, Edith H. "Olga Taussky–Todd." *Women of Mathematics: A Biobibliographic Sourcebook.* Edited by Louise S. Grinstein and Paul J. Campbell. Westport, CT: Greenwood, 1987, pp. 225–235.

"Olga Taussky–Todd." *Number Theory and Algebra: Collected Papers Dedicated to Henry B. Mann, Arnold E. Ross, and Olga Taussky–Todd.* Edited by Hans Zassenhaus. New York: Academic Press, 1977, pp. xxxiv–xlvi.

### Periodicals

Oliver, Myrna. "Olga Taussky–Todd: Noted Mathematician." *Los Angeles Times* (December 3, 1995): A44.

### Other

Davis, Chandler. "Remembering Olga Taussky Todd." *Biographies of Women Mathematicians.* June 1997. http://www.scottlan.edu/lriddle/women/chronol.htm (July 22, 1997).

*—Sketch by Linda Wasmer Smith*

# Brook Taylor
## 1685–1731
### English mathematician

Brook Taylor was born in Edmonton, Middlesex, England, on August 18, 1865, the oldest child of a family considered part of the minor nobility. His mother, Olivia, was the daughter of Sir Nicholas Tempest. His father, John Taylor of Bifrons House, Kent, was a strict parent who alienated Taylor from the family when Taylor married a woman without his father's approval in 1721. The estrangement ended after two years when Taylor's wife died in childbirth and he returned to the family home. Taylor remarried in 1725, this time to a woman who enjoyed the family's approval. Unfortunately, Taylor's second wife also died in childbirth in 1730, leaving him to

*Brook Taylor*

raise their daughter, Elizabeth, alone. Following his father's death in 1729, Taylor inherited the family estate in Kent.

After being tutored at home in mathematics and the classics, Taylor was admitted to St. John's College, Cambridge, in 1709, where he completed a Bachelor of Laws degree. He was elected to Great Britain's oldest scientific community, the Royal Society, in 1712 and earned a Doctorate of Laws in 1714. He served as secretary to the Royal Society for four years then resigned, citing poor health as the impetus for his decision. It was rumored that in addition to suffering some health problems, Taylor also found the duties as a chronicler too restrictive and less than satisfying for his active mind.

### The Productive Years

While serving as secretary to the Royal Society, Taylor wrote many articles on such diverse subject as magnetism, capillarity (the study of the surface tensions in liquids which come in contact with solids), the barometer (using logarithmic function to explain atmospheric pressure at different altitudes), and the thermometer. Thirteen articles in all appeared in *Philosophical Transactions of the Royal Society* between 1712 and 1724 addressing theories on dynamics, hydrodynamics, and heat, among others. His correspondence to fellow mathematicians during this time contained a solution to astronomer **Johannes Kepler**'s second law of planetary motion. Taylor's first

significant paper also appeared in *Philosophical Transactions* in 1714 and produced a solution to the problem of the center of oscillation. Taylor's claim to priority of the solution was unjustly challenged by Swiss mathematician **Johann Bernoulli**. Although the dispute was later resolved, Taylor would again face accusations by Bernoulli concerning further mathematical work.

In 1715, Taylor published *Methodus Incrementorum Directa et Inversa* ("Direct and Indirect Methods of Incrementation"), which contained the first discourse dealing with the calculus of finite differences. The dissertation also contained the theorem for which Taylor is most famous—the "Taylor series." The Taylor series was the first general expression for the expansions of functions of a single variable in finite series. Both **Isaac Newton** and **Nicolaus Mercator** had devoted time to developing the series, but their work had been confined to particular cases. Again, Johann Bernoulli stepped in to dispute the priority of Taylor's work, claiming that he had made the discovery. Bernoulli's claim was again proven to be unfounded since his work had never reached the fruition of Taylor's. The importance of the Taylor series remained unrecognized until 1772. It was then that **Joseph–Louis La Grange** realized its importance and declared it the basic principle of differential calculus. The series has also been called the "Maclaurin series" and this series holds validity when $a$ is substituted for $0$ in the Taylor theorem. Still, Taylor deserves the credit since his treatise preceded **Colin Maclaurin**'s by 27 years, and Maclaurin's theorem was another case of addressing specifics instead of developing a basis just as Newton and others had done. Taylor did have one predecessor to his discovery. **James Gregory** had worked on the theorem some 40 years before Taylor, but Gregory's work was unknown to Taylor, hence eliminating any allegations of misappropriation on the part of Taylor.

In *Methodus*, Taylor also applied the calculus for the solution of several problems which had baffled previous investigators. He obtained a formula showing that the rapidity of vibration of a string varies directly as the weight stretching it and inversely concerning its own weight and length. Also, in *Methodus*, Taylor supplied a determination of the differential equation of the path of a ray of light passing through a heterogeneous medium. No further advance was made in the theory of refraction for more than 80 years.

### Artist, Musician, Philosopher

Also published in 1715 was Taylor's *Linear Perspective*, later followed by an updated version (1719) entitled *New Linear Perspective*, both of which set forth the basic principles of perspective, and contained the first general treatment of the principle of vanishing points. As with *Methodus*, much of this work to went unnoticed in Taylor's lifetime. Taylor's father had a keen interest in art and music so Taylor was introduced to and instructed in these arts throughout his childhood. As it turned out, he was himself a gifted artist and musician, leaving small wonder that he would eventually turn his attention to the study of artistic theory. St. John's College, Cambridge, is said to house an unpublished manuscript entitled *On Musick* among its Taylor archives.

The fact that so much of Taylor's work was recognized following his death can be directly attributed, at least in part, to Taylor himself. The brevity and obscurity of his writing made it difficult to comprehend. Concerning this too, Taylor's tireless rival Johann Bernoulli had something to say: "[The writing is] abstruse to all and unintelligible to artists for whom it was more especially written." As offensive as Bernoulli was to Taylor throughout their respective careers, it is interesting to note that Bernoulli cited some of Taylor's work dealing with string vibration in letters to his son Daniel. Still, Taylor often did leave out steps in his theorems, believing that others would recognize the end result as easily as he had. Regrettably, this was not the case, and it took the close examination of his successors to fully illuminate his contributions to the field of mathematics as well as the other sciences.

Taylor spent the last part of his writing career producing religious and philosophical texts. Several unfinished treatises were found among his papers concerning various religions. Taylor wrote nothing during the last ten years of his life, but his third and final completed book, entitled *Comtemplatio Philosophica,* was published and circulated after his death in 1731. His daughter, Elizabeth, had married Sir William Young, and it was their son who made certain that his grandfather's manuscript was published. Young authored a biography of his grandfather's life and works and attached it to *Comtemplatio* upon publication.

**FURTHER READING**

Ball, W. W. Rouse. *A Short Account of the History of Mathematics.* New York: Dover Publications, Inc., 1960.

Bell, E.T. *The Development of Mathematics.* Second Edition. New York: McGraw–Hill., 1945.

Boyer, Carl B. *A History of Mathematics.* Revised by Uta C. Merzbach. New York: John Wiley & Sons, 1989.

Jones, Phillip S. "Brook Taylor," in *Dictionary of Scientific Biography.* Volume XIII. Edited by Charles Coulston Gillispie. New York: Charles Scribner's Sons., 1972, pp. 265–68.

*Thales of Miletus*

Smith, David Eugene. *History of Mathematics.* Vol. I. New York: Dover Publications, Inc., 1958.

Struik, Dirk J. *A Concise History of Mathematics.* Third Revised Edition. New York: Dover Publications, Inc., 1967.

                              —*Sketch by Kelley Reynolds Jacquez*

# Thales of Miletus
## c.640–c.546 B.C.
### Greek geometer, astronomer, and philosopher

Thales of Miletus was the first known Greek philosopher and mathematician who introduced the basic concepts of geometry to the Greeks. A man of many talents, Thales was a successful merchant and founded his own school of philosophy in Miletus, which included teachings in mathematics and astronomy. Notably, Thales emphasis on "natural" rather than "supernatural" explanations for phenomena marks the beginning of scientific methodology and theoretical reasoning. Honored and revered as a man of knowledge and wisdom in his own time, Thales was named one of the Seven Wise Men of Greece.

Thales was born circa 640 B.C. to 620 B.C. in Miletus, a Greek island along the coast of Asia Minor. Since Thales left behind no writings, little is known about his life. As with many ancient thinkers, fact, fiction, and myth have become intermixed. His father was Examius and his mother Cleobuline. Growing up in a strategic caravan route from the East, Thales first made his mark as a highly successful merchant. According to one account, Thales made a fortune when he foresaw a good season for the olive crop and bought all the local olive presses, monopolizing olive oil market.

With his financial success, Thales traveled to Egypt, either to pursue business opportunities or to study Egyptian science and philosophy. Certainly, Thales learned the fundamentals of geometry while in Egypt. His aptitude for deductive reasoning is illustrated through the story of Thales' ability to calculate the height of the pyramids. According to legend, Thales was observing the Great Pyramid one day when the Pharaoh and his entourage came by. Intrigued by the foreigner's reputation as a man of wisdom, the Pharaoh asked Thales if he could calculate the pyramid's height. Thales may have accomplished the feat by using the "magic" of the Egyptian shadow stick, which, when placed in the ground, one could use it to calculate the height of an object by establishing a relationship between the height of the stick's shadow and the shadow of the object to be measured. A more plausible explanation is that Thales measured the pyramid's shadow at the exact time when his own shadow appeared to be the same length as himself, thus ensuring that the Great Pyramid's shadow accurately represented its true height.

Thales may also have traveled to Babylon and studied geometry and astronomy. He eventually returned to Miletus (possibly around 590 B.C.) and established a school where he taught science, astronomy, mathematics, and other subjects. Eventually, the philosophy of Thales became known as the Ionian School of Philosophy.

### Introduces Geometry to Greece

Thales' reputation as one of the founders of mathematics and applied geometry stems not only from his introduction of geometry to Greece, but in his establishing the need for deductive reasoning and proofs for demonstrating a statement or theorem's validity. The first written account of Thales' mathematical acumen comes from Proclus, who wrote a famous commentary on **Euclid**'s *Elements*. Prior to Thales, geometry was concerned primarily with measuring surfaces and solids. Thales developed geometry in both a practical and theoretical way by focusing on lines, circles, and triangles. Proclus credits Thales with five basic theorems of elementary geometry: (1)

The circle is bisected by its diameter; (2) the base angles of an isosceles triangle are equal; (3) pairs of vertical angles formed by two intersecting straight lines are equal; (4) an angle inscribed in a semicircle is a right angle; (5) two triangles are congruent if they have two angles and one side that are equal.

Whether or not Thales actually proved geometrical theorems is unknown. Of his first three theorems, historians debate whether Thales developed them through intuition or whether he discovered and demonstrated them scientifically. However, Proclus indicates that Thales did prove theorem number two even though by the time Euclid wrote his *Elements* this theorem was accepted as obvious and needed no proof. As for theorem number four, it is not known whether Thales developed a proof or demonstrated the theorem empirically.

Many scholars believe that Thales made practical use of his fifth, or congruence, theorem in his renowned ability to measure the distance of a ship from shore. One theory of how Thales accomplished this remarkable measurement for his day says that Thales used a tower of known height on shore, a plum line, and a carpenter's square (known as a gnomon). Driving a nail in the tower that corresponded to a line of sight to the ship, Thales determined the distance from the right–angled base of the gnomon to where it intersected the line of sight, thus creating an easily measured right triangle. Then, by dropping a plum to the ground and determining the height of the tower, he was able to calculate the size of a larger right triangle that was proportional to the smaller triangle. The distance to the ship would equal the calculated length of the larger right triangle's base. To accomplish this feat, it is assumed that Thales knew that the sides of equiangular triangles are proportional.

### Predicts Solar Eclipse

In addition to his interest in geometry and mathematics, Thales was fascinated by astronomy. A famous story concerning Thales' love of astronomy has him falling into a well while gazing at the stars. He is soon discovered by a slave girl who wryly comments that Thales was "so interested in the heavens he could not see what was in front of his own feet."

Thales may have developed his interest in astronomy during his reported travels to Babylon and studies under the Chaldean magi. He is reported to have written works focusing on the equinox and the solstice and is sometimes attributed with explaining solar eclipses. He is also reported to have written on nautical astronomy and to have advised navigators to depend on Ursa Minor (Little Bear) rather than Ursa Major (Great Bear) for accurately navigating their destinations.

In his own time, Thales' most noted feat of astronomical observation was his prediction of a solar eclipse in 585 B.C. as reported by Herodotus. Thales possibly learned the secret of predicting such events through the Babylonians, who had kept detailed records of astronomical events for many centuries. With these records, Thales may have been able to calculate that an eclipse might occur in that year. No matter how Thales made his famous prediction, he had a tremendous amount of luck, not the least of which was the fact that the eclipse happened to be visible in Miletus. Regardless, this prediction bolstered Thales' fame to new heights.

### Becomes One of the Seven Greek Sages

While he is remembered today primarily for his efforts in geometry and astronomy, Thales was also the first great Greek philosopher. In his Ionian school, philosophy, or the love of wisdom, took precedence over all other studies. Thales was renowned for his intimate knowledge of human nature and his humor. For example, the maxim "know thyself" is purported to originate with Thales. Other sage advice credited to Thales include his counsel that people could lead righteous lives by "refraining from doing what we blame in others" and his pronouncement that the strangest thing he had ever seen was "an aged tyrant."

Like most of his contemporaries, Thales mistakenly believed that the Earth was a flat disc on an infinite ocean. His belief that water was the origin of all life and matter was probably based on the knowledge that water is the essential element needed for the growth and nurturing of all organisms. Still, his belief that supposedly supernatural occurrences could be explained by natural events profoundly influenced the thinking and scientific explorations of those scholars who came after him. His most famous pupil was Anaximander, who was the first natural philosopher in the Milesian school of philosophy.

As a wealthy and learned man, Thales also played a vital role as a political counselor and statesman. Thales is reported to have convinced the Greek colonies (or Ionian city–states) to form an alliance to fight off Persian invaders. As a military advisor, he used his mathematical and engineering skills to help a Greek army cross a river by constructing a channel and diverting the river into it. Since the "Seven Wise Men of Greece" were predominantly statesmen, Thales was probably named one of the seven for these efforts. According to some accounts, Thales, who died circa 546 B.C., was the only one of the seven to be declared a Wise Man twice by the Oracle at Delphi, the second time for his prediction of the solar eclipse.

## FURTHER READING

### Books

Heath, Sir Thomas. *A Manual of Greek Mathematics.* New York, Dover Publications, 1931, pp. 80–91.

Smith, David Eugene. *History of Mathematics.* Volume I. New York: Dover Publications, 1958, pp. 65–68.

Terry, Leon. *The Mathmen.* New York: McGraw–Hill Book Company, 1964, pp. 9–41.

Van Der Waerden, L. Van. *Science Awakening.* New York, John Wiley & Sons, Inc., 1963, pp. 85–90.

### Periodicals

Dicks, D.R. "Thales." *Classical Quarterly* (1959): 294–309.

*—Sketch by David A. Petechuk*

# Theon of Alexandria
## c. 335–c. 400
**Greek astronomer, mathematician, commentator, editor, and poet**

Although an intellectual figure of certain stature and much praised by his contemporaries, Theon is primarily known as the father of **Hypatia**, the noted Neo–Platonist philosopher and mathematician who was murdered by a fanatical Christian mob in 415. As a mathematician, Theon exemplified the declining Alexandrian culture of the final period of antiquity. He concentrated on eclectic work, producing commentaries, consolidating acquired knowledge, and sometimes added minor elements to already established mathematical theorems.

Scholars agree that the 360s may have been the zenith of Theon's scientific career. His stature was certainly enhanced in 364, when he correctly predicted eclipses of the sun and the moon. In addition, as a respected scientist, Theon was a member of the Alexandrian Museum, or the *Musaion,* an independent institution dedicated exclusively to scientific and literary research. According to *Souda,* the Byzantine lexicon/biographical encyclopedia compiled in the 10th century, Theon was also a philosopher, despite the fact that, unlike his daughter, he did not teach philosophy. In addition, according to Maria Dzielska, astrological sources acknowledge his interest in magic and astrology.

### Provides Commentary of Ptolemy

Although Theon provided an edition of **Euclid**'s *Elements*, he is mainly remembered for his multi–volume commentary on **Claudius Ptolemy**'s *Almagest*. As a historical synthesis of Greek astronomy, the *Almagest* is regarded as the cornerstone of the geocentric theory, which dominated the scientific world until the late Middle Ages. Theon's commentary, while of minimal scientific importance, provides important information about lost astronomical works. A tireless compiler, commentator, and educator, Theon employed many young scholars as assistants. Among his associates, as Dzielska has stated, was also his daughter, whom he called *thygater* (daughter), instead of the common term for students, *teknon* (child).

### Contributes to Mathematics

Although Theon was not a mathematician of any considerable originality, his mathematical work was highly regarded by his contemporaries. Because he was a teacher, Theon generally focused on basic mathematical operations and known geometrical proofs. According to D. E. Smith, he described a method whereby sexagesimal fractions (within the numerical system which has 60 as base) are used to find to square roots.

### SELECTED WRITINGS BY THEON

*Almagest, Claudii Ptolemaei magnae constructionis . . . lib. xiii. Theonis Alexandrini in eosdem commentariorum lib. xi.,* 1538

### FURTHER READING

Boyer, Carl B. *A History of Mathematics.* Second edition. Revised by Uta C. Merzbach. New York: John Wiley & Sons, 1991.

Dzielska, Maria. *Hypatia of Alexandria.* Translated by F. Lyra. Cambridge: Harvard University Press, 1995.

Smith, D. E. *History of Mathematics.* Volume 1. New York: Dover, 1958.

—*Sketch by Zoran Minderovic*

*René Frédéric Thom*

# René Frédéric Thom
## 1923-
**French topologist and mathematical philosopher**

René Frédéric Thom is best known as the founder of catastrophe theory, a field with numerous applications in the exact and social sciences. Catastrophe theory provides models for the description of continuous processes that experience abrupt change. Thom is the recipient of the Fields Medal and the Grand Prix Scientifique de la Ville de Paris. He is a member of the French Académie Royale des Sciences and has been named a Knight of the Legion of Honor in France.

Thom was born on September 2, 1923, at Montbéliard, France, to Gustav Thom, a pharmacist, and Louise Ramel. He married Suzanne Heimlinger on April 9, 1949, and they have three children, Françoise, Elizabeth, and Christian. Thom's formal education took place at the Collége Cuvier in Montbeliard and later in Paris at the Lycée Saint-Louis. After earning a master's degree in mathematics and a doctorate in science from the École Normale Supérieure in Paris, Thom went to Princeton University in 1951. He taught in Grenoble from 1953 to 1954 and in Strasbourg for the next decade.

## Achieves Early Acclaim

In 1958, at the age of 35, Thom received the Fields Medal, the most prestigious prize in mathematics. When awarding the prize, the International Mathematical Union (IMU) cited Thom's 1954 invention of the theory of cobordism in algebraic topology. Cobordism is a classification scheme for manifolds (multidimensional topological surfaces) based on homotopy (a continuous deformation of one function into another). The IMU recognized Thom's use of this technique as a "prime example of a general cohomology theory." In a 1991 address to the Symposium on the Current State and Prospects of Mathematics, Thom recalled that after he received the Fields Medal, he experienced doubts about his ability to continue developing meaningful mathematical results. He decided to turn his attention from the more algebraic types of work and concentrate on singularities of differentiable maps, a topic he considered "more flexible and more concrete."

While he was teaching at Strasbourg University, Thom collaborated with a physicist on an investigation of caustics in optical geometry (i.e., rays reflected or refracted by curved solid surfaces), studying both their singularities and their deformations. He told the 1991 symposium that "to my surprise, I found that in caustics, organized by some very simple optical instruments like spherical mirrors and rectilinear diopters, one may find a singularity that should not theoretically exist." He eventually connected this phenomenon to **Pierre de Fermat**'s principle of optimality.

Since 1964, Thom has been a professor at the Institut des Hautes Etudes Scientifiques (IHES) at Bures-sur-Yvette, where he concentrates on research, with no teaching or administrative duties. In this environment, he wrote a carefully reasoned article opposing the popular movement to discard geometry from general mathematics education. His argument concerning the importance of geometry centered on the potential value of studying singularities of function maps. As a result of this exercise, Thom developed the theory of singularities.

## Formulates Catastrophe Theory

During the 1960s, Thom turned his attention from optical problems in physics to the biological topic of embryology. His objective was to apply transversality theory to natural science. Out of his mathematical analysis of current theories of cellular differentiation, he developed what he later called "my famous list of seven elementary catastrophes in space-time," a result that fit naturally into singularity theory. Applying a very abstract theorem about universal deformations of analytic sets proved by **Alexander Grothendieck**, a colleague at the IHES, Thom wrote a book attempting to explain the origin of natural forms. Delayed by the bankruptcy of the original publishing house, *Stabilité Structurelle et Morphogénése (Structural Stability and Morphogenesis)* finally appeared in print in 1972, although a few copies had already circulated in the mathematical community. E. Christopher Zeeman of the University of Warwick was fascinated with the contents of the book. Thom later told the 1991 symposium that "[Zeeman's] reflections on this subject led to a grandiose extension of the theory," extending it from four-dimensional space-time to any locally Euclidean space. Zeeman gave a couple of lectures on what he called "catastrophe theory," and wrote a fascinating account about it for *Scientific American* magazine.

Generally classified as a branch of geometry because variables and results are shown as curves or surfaces, catastrophe theory attempts to explain predictable discontinuities in output for systems characterized by continuous inputs—a task that cannot be done using differential calculus. Contrary to the implications of the theory's name, the "catastrophes" studied are not necessarily negative in nature; the term simply refers to sudden change. For example, when a balloon is steadily filled with air, it expands and changes its shape in a relatively uniform manner until the pressure in the balloon's interior reaches a critical value. Then, when the balloon undergoes more abrupt, but predictable, change, it pops. More complex phenomena, such as the refraction of light through moving water, the amount of stress that can be placed on a bridge, and the synergistic effects of the ingredients in drugs can also be effectively studied using catastrophe theory. The theory provides a universal method for the study of all jump transitions, discontinuities, and sudden qualitative changes.

Thom recalled, "as a result [of Zeeman's publicity], catastrophe theory . . . took off like a rocket, propelled by the principal media all over the world. This glory was short-lived, however, and the brief success of [catastrophe theory] soon fell to the slings and arrows of trans-Atlantic criticism." In fact, some scientists have hailed Thom's catastrophe theory as a tool more valuable to humanity than Newtonian theory, which considers only smooth, continuous processes. However, the theory became something of a fad in the 1970s and 1980s and was used in applications that the theory does not support. As a result of such indiscriminate application, the theory has at times been unjustly criticized as a cultural phenomenon or a metaphysical view rather than a legitimate branch of mathematics. Although it has been presented in a metaphysical vein as proof of the deterministic nature of the universe, catastrophe theory does not purport to abolish the indeterminacy that is central to nuclear physics. Nevertheless, catastrophe theory has been used to study abrupt systems changes in such diverse fields as hydrodynamics, geology, particle physics, industrial relations, embryology, economics, linguistics, civil engineering, and medicine.

### Turns to Philosophy

By his own account, Thom has been primarily a mathematical philosopher since the early 1970s, largely due to the controversy surrounding catastrophe theory. A major criticism of the theory was that it allows only qualitative rather than quantitative predictions. That is, given a specific application, the type of abrupt change could be predicted, but it could not be quantified in numerical terms. While admitting that this is a legitimate concern, Thom still sees value in the modeling techniques catastrophe theory makes available to fields such as psychology. A classical example described by Zeeman, for instance, involved predicting the action of a dog experiencing a combination of the emotions of rage and fear—emotions that (if acting alone) would cause the dog to attack or flee, respectively. What catastrophe theory provides is a mathematical framework for clearly describing the situation being modeled, much as algebra provides a shorthand for expressing relationships between numbers and variables in a way that facilitates manipulating them to obtain results.

Speaking from the perspective of 20 years of philosophical reflection, Thom said in his 1991 symposium address: "What is important in science is not the distinction between true and false. This might seem strange to mathematicians, but I will say that if I had the choice between an error which has an organizing power of reality (this could exist) and a truth which is isolated and meaningless in itself, I would choose the error and not the truth. There are many examples of errors which are scientifically important, and there are many, many examples of meaningless truths in science."

## SELECTED WRITINGS BY THOM

### Books

*Structural Stability and Morphogenesis,* 1976
"Leaving Mathematics for Philosophy," in *Mathematical Research Today and Tomorrow: Viewpoints of Seven Fields Medalists,* 1992. (Addresses given at the 1991 Symposium on the Current State and Prospects of Mathematics.)

### Periodicals

"Topological Models in Biology." *Topology* 8 (1969): 313–35.

## FURTHER READING

### Books

Arnold, V. I. *Catastrophe Theory.* Third edition. New York: Springer-Verlag, 1992.

*Alan Turing*

### Periodicals

Murphy, John W. "Catastrophe Theory: Implications for Probability." *American Journal of Economics and Sociology* 50 (April 1991): 143–48.
Zeeman, Christopher. "Catastrophe Theory." *Scientific American* (April 1976): 65–83.

### Other

"René Thom." *MacTutor History of Mathematics Archives.* http://www-groups.dcs.st-and.ac.uk/~history/Mathematicians/Thom.html (July 10, 1997).

—*Sketch by Loretta Hall and Maureen L. Tan*

# Alan Turing
## 1912-1954
### English algebraist and logician

Alan Turing is recognized as a pioneer in computer theory. His classic 1936 paper, "On Computable Numbers, with an Application to the Entscheidungs Problem," detailed a machine that served as a model for the first working computers. During World

War II, Turing took part in the top-secret ULTRA project and helped decipher German military codes. During this same time, Turing conducted ground-breaking work that led to the first operational digital electronic computers. Another notable paper was published in 1950 and offered what became known as the "Turing Test" to determine if a machine possessed intelligence.

Alan Mathison Turing was born on June 23, 1912, in Paddington, England, to Julius Mathison Turing and Ethel Sara Stoney. Turing's father served in the British civil service in India, and his wife generally accompanied him. Thus, for the majority of their childhoods, Alan and his older brother, John, saw very little of their parents. While in elementary school, the young Turing boys were raised by a retired military couple, the Wards. At the age of 13, Turing entered Sherbourne school, a boys' boarding school in Dorset. His record at Sherbourne was not generally outstanding; he was later remembered as untidy and disinterested in scholastic learning. He did, however, distinguish himself in mathematics and science, showing a particular facility for calculus. Turing also developed an interest in competitive running while at Sherbourne.

### Produces Prophetic Paper

Turing twice failed to gain entry to Trinity College in Cambridge, but was accepted on scholarship at King's College (also in Cambridge). He graduated in 1934 with a master's degree in mathematics. In 1936 Turing produced his first, and perhaps greatest, work. His paper "On Computable Numbers, with an Application to the Entscheidungs Problem," answered a logical problem staged by German mathematician **David Hilbert**. The question involved the completeness of logic—whether all mathematical problems could, in principle, be solved. Turing's paper, presented in 1937 to the London Mathematical Society, proved that some could not be solved. Turing's paper also contained a footnote describing a theoretical automatic machine, which came to be known as the Turing Machine, that could solve any mathematical problem—provided it was give the proper algorithms, or problem-solving equations or instructions. Although it may not have been Turing's intent at the time, his Turing Machine defined the modern computer.

After graduating from Cambridge, Turing was invited to spend a year in the United States studying at Princeton University. He returned to Princeton for a second year—on a Proctor Fellowship—to finish his doctorate. While there, he worked on the subject of computability with **Alonso Church** and other mathematicians. Turing and his associates worked with binary numbers (1 and 0) and Boolean Algebra, developed by **George Boole**, to develop a system of

equations called logic gates. These logic gates were useful for producing problem-solving algorithms such as would be needed by an automatic computing machine. From the initial paper exercise, it was a simple matter to develop logic gates into electrical hardware, using relays and switches, which could—theoretically, and in huge quantities—actually perform the work of a computing machine. As a sideline, Turing put together the first three or four stages of an electric multiplier, using relays he constructed himself. After receiving his doctorate, Turing had an opportunity to remain at Princeton, but decided to accept a Cambridge fellowship instead. He returned to England in 1938.

### Helps Crack German War Codes

Cryptology, the making and breaking of coded messages, was greatly advanced in England after World War I. The German high command, however, had modified a device called the Enigma machine that mechanically enciphered messages. The English found little success in defeating this method. The original Enigma machine was not new, or even secret; a basic Enigma machine had been in operation for several years, mostly used to produce commercial codes. The Germans' alterations, though, greatly increased the number of possible letter combinations in a message. The Allies were able to duplicate the modifications, but it was a continual cat-and-mouse game; each time Allied analysts figured out a message, the Germans' changes made all of their work useless.

In the fall of 1939, Turing found himself in a top-secret installation in Bletchley, where he played a critical role in the development of a machine that deciphered the Enigma's messages by testing key codes until it found the correct combinations. This substitution method was uncomplicated, but impractical to apply because possible combinations could range into the tens of millions. Here Turing was able to put his experience at Princeton to good use; no one else had bridged the gap between abstract logic theory and electric hardware as he had with his electric multiplier. Turing helped construct relay-driven decoders (which were called Bombes, after the ticking noise of the relays) that shortened the code-breaking time from weeks to hours. The Bombes helped uncover German movements, particularly the U-boat war in the Atlantic, for almost two years. Eventually, however, the Germans changed their codes and the new level of complexity was too high to be solved practically by electrical decoders. British scientists agreed that although a Bombe of sufficient size could be made for further deciphering work, the machine would be slow and impractical.

Yet other advances would prove advantageous for the decoding machine. Vacuum tubes used as switches (the British called them thermiotic valves) used no moving parts and were a thousand times faster than electrical relays. A decoder made with tubes could do in minutes what it took a Bombe several hours to accomplish. Thus work began on a device which was later named Collosus. Based on the same theoretical principles as earlier Bombes, Collosus was the first operational digital electronic computer. It used 1800 vacuum tubes, proving the practicality of this approach. Much information concerning Collosus remained classified by the British government in the early 1990s. Some claimed that Turing supervised the construction of the first Collosus.

Many stories were circulated about Turing during the war; mostly surrounding his eccentricity. Andrew Hodges noted in his book *Alan Turing, The Enigma,* "With holes in his sports jacket, shiny grey flannel trousers held up with an ancient tie, and hair sticking out at the back, he became the cartoonist's 'boffin —an impression accentuated by his manner of practical work, in which he would grunt and swear as solder failed to stick, scratch his head and make a strange squelching noise as he thought to himself." Unconvinced of England's chances to win the war, Turing converted all of his funds to two silver bars, buried them, and was later unable to locate them. He was horrified at the sight of blood, was an outspoken atheist, and was a homosexual. Still, for his unquestionably vital role in the British war effort, he was later awarded the Order of the British Empire, a high honor for someone not in the combat military.

In the waning months of the war, Turing turned his thoughts back to computing machines. He conceived of a device, built with vacuum tubes, that would be able to perform any function described in mathematical terms and would carry instructions in electronic symbols in its memory. This universal machine, clearly an embodiment of the Turing machine described in his 1936 paper, would not require separate hardware for different functions, only a change of instructions. Turing was not alone in his ambition to construct a computing machine. A group at the University of Pennsylvania had built a computer called ENIAC (Electronic Numerical Integrator and Computer) that was similar to, but more complex, than Colossus. In the process, they had concluded that a better machine was possible. Turing's design was possibly more remarkable because he was working alone out of his home while they were a large university research group with the full backing of the American military. The American group published well before Turing did, but the British government subsequently took a greater interest in Turing's work.

### Postwar Work

In June of 1945 Turing joined the newly formed Mathematics Division of the National Physical Laboratory (NPL). Here he finalized plans for his Automatic Computing Engine (ACE). The rather archaic term "engine" was chosen by NPL management as a tribute to **Charles Babbage**'s Analytical Engine (and also because it made a pleasing acronym). Turing, however, was unprepared for the inertia and politics of a bureaucratic government foundation. All of his previous engineering projects had been conducted during wartime, when time was of the essence and no budget constraints existed. More than a year after the ACE project was approved, though, no engineering work had been completed and there was little cooperation between participants. A scaled-down version of the ACE was finally completed in 1950. But Turing had already left NPL in 1948, frustrated at the slow pace of the computer's development.

In 1950 Turing produced a widely read paper titled "Computing Machinery and Intelligence." This classic paper expanded on one of Turing's interests— if computers could possess intelligence. He proposed a test called the "Imitation Game", still used today under the name "Turing Test." In the test, an interrogator was connected by teletype (later, by computer keyboard) to either a human or a computer at a remote location. The interrogator is allowed to pose any questions and, based on the replies, the interrogator must decide whether a human or a computer is at the other end of the line. If the interrogator cannot distinguish between the two in a statistically significant number of cases, then artificial intelligence has been achieved. Turing predicted that within fifty years, computers could be programmed to play the game so effectively that after a five-minute question period the interrogator would have no more than a seventy-percent chance of making the proper identification.

### Personal Troubles Mount

Turing's personal life deteriorated in the early 1950s. After leaving NPL he took a position with Manchester College as deputy director of the newly formed Royal Society Computing Laboratory. But he was not involved in designing or building the computer on which they were working. By this time, Turing was no longer a world-class mathematician, having for too long been sidetracked by electronic engineering, nor was he engineer: The scientific world seemed to be passing him by.

While at Manchester, Turing had an affair with a young street person named Arnold Murray, which led to a burglary at his house by one of Murray's associates. The investigating police learned of the relationship between Turing and Murray; in fact Turing did nothing to hide it. Homosexuality was a felony in England at the time, and Turing was tried and convicted of "gross indecency" in 1952. Because of his social class and relative prominence, he was sentenced to a year's probation and given treatments of the female hormone estrogen in lieu of serving a year in jail.

Turing committed suicide by eating a cyanide-laced apple on June 7, 1954. His death puzzled his associates; he had been free of the hormone treatments for a year, and, although a stigma remained, he had weathered the incident with his career intact. He left no note, nor had he given any hint that he had contemplated this act. His mother tried for years to have his death declared accidental, but the official cause of death was never seriously questioned.

## SELECTED WRITINGS BY TURING

### Periodicals

"On Computable Numbers, with an Application to the Entscheidungs Problem." *Proceedings of the London Mathematical Society* 42 (1937): 230–265.

"Computing Machinery and Intelligence." *Mind* 59 (1950): 433–460.

## FURTHER READING

### Books

Carpenter, B. S., and R. W. Doran, editors. *A. M. Turing's ACE Report of 1946 and Other Papers.* Cambridge, MA: MIT Press, 1986.

Hodges, Andrew. *Alan Turing, The Enigma.* New York: Simon and Schuster, 1983.

Shurkin, Joel. *Engines of the Mind.* New York: W.W. Norton, 1984.

Slater, Robert. *Portraits in Silicon.* Cambridge, MA: MIT Press, 1987.

Turing, Sara. *Alan M. Turing.* Heffers, 1959.

—*Sketch by Joel Simon*

broad and intellectually stimulating subject. After earning her B.S. degree in 1964, she became a National Science Foundation Graduate Fellow, pursuing graduate study in mathematics at Brandeis University. In 1965, she married Olke Cornelis Uhlenbeck, a biophysicist; they later divorced.

Uhlenbeck received her Ph.D. in mathematics from Brandeis in 1968 with a thesis on the calculus of variations. Her first teaching position was at the Massachusetts Institute of Technology in 1968. The following year she moved to Berkeley, California, where she was a lecturer in mathematics at the University of California. There she studied general relativity and the geometry of space-time, and worked on elliptic regularity in systems of partial differential equations.

In 1971, Uhlenbeck became an assistant professor at the University of Illinois at Urbana-Champaign. In 1974, she was awarded a fellowship from the Sloan Foundation that lasted until 1976, and she then went to Northwestern University as a visiting associate professor. She taught at the University of Illinois in Chicago from 1977 to 1983, first as associate professor and then professor, and in 1979 she was the Chancellor's Distinguished Visiting Professor at the University of California, Berkeley. An Albert Einstein Fellowship enabled her to pursue her research as a member of the Institute for Advanced Studies at Princeton University from 1979 to 1980. She published more than a dozen articles in mathematics journals during the 1970s and was named to the editorial board of the *Journal of Differential Geometry* in 1979 and the *Illinois Journal of Mathematics* in 1980.

In 1983, Uhlenbeck was selected by the John D. and Catherine T. MacArthur Foundation of Chicago to receive one of its five-year fellowship grants. Given annually, the MacArthur fellowships enable scientists, scholars, and artists to pursue research or creative activity. For Uhlenbeck, winning the fellowship inspired her to begin serious studies in physics. She believes that the mathematician's task is to abstract ideas from fields such as physics and streamline them so they can be used in other fields. For instance, physicists studying quantum mechanics had predicted the existence of particle-like elements called instantons. Uhlenbeck and other researchers viewed instantons as somewhat analogous to soap films. Seeking a better understanding of these particles, they studied soap films to learn about the properties of surfaces. As soap films provide a model for the behavior of

*Karen Uhlenbeck*

# Karen Uhlenbeck
## 1942-
### American geometer

Karen Uhlenbeck is engaged in mathematical research that has applications in theoretical physics and has contributed to the study of instantons, models for the behavior of surfaces in four dimensions. In recognition of her work in geometry and partial differential equations, she was awarded a prestigious MacArthur Fellowship in 1983.

Karen Keskulla Uhlenbeck was born in Cleveland, Ohio, on August 24, 1942, to Arnold Edward Keskulla, an engineer, and Carolyn Windeler Keskulla, an artist. When Uhlenbeck was in third grade, the family moved to New Jersey. Everything interested her as a child, but she felt that girls were discouraged from exploring many activities. In high school, she read American physicist George Gamow's books on physics and English astronomer Fred Hoyle's books on cosmology, which her father brought home from the public library. When Uhlenbeck entered the University of Michigan, she found mathematics a surfaces in three-dimensions, instantons provide analogous models for the behavior of surfaces in four-dimensional space-time. Uhlenbeck cowrote a book on this subject, *Instantons and 4-Manifold Topology,* which was published in 1984.

After a year spent as a visiting professor at Harvard, Uhlenbeck became a professor at the University of Chicago in 1983. Her mathematical interests at this time included nonlinear partial differential equations, differential geometry, gauge theory, topological quantum field theory, and integrable systems. She gave guest lectures at several universities and served as the vice president of the American Mathematical Society. The Alumni Association of the University of Michigan named her Alumna of the Year in 1984. She was elected to the American Academy of Arts and Sciences in 1985 and to the National Academy of Sciences in 1986. In 1988, she received the Alumni Achievement award from Brandeis University, an honorary doctor of science degree from Knox College, and was named one of America's 100 most important women by *Ladies' Home Journal.*

In 1987, Uhlenbeck went to the University of Texas at Austin, where she broadened her understanding of physics in studies with American physicist Steven Weinberg. In 1988, she accepted the Sid W. Richardson Foundation Regents' Chair in mathematics at the University of Texas. She also gave the plenary address at the International Congress of Mathematics in Japan in 1990.

Concerned that potential scientists were being discouraged unnecessarily because of their sex or race, Uhlenbeck joined a National Research Council planning group to investigate the representation of women in science and engineering. She believes that mathematics is always challenging and never boring, and she has expressed the hope that one of her accomplishments as a teacher has been communicating this to her students. "I sometimes feel the need to apologize for being a mathematician, but no apology is needed," she told *The Alcalde Magazine.* "Whenever I get a free week and start doing mathematics, I can't believe how much fun it is. I'm like a 12-year-old boy with a new train set."

## SELECTED WRITINGS BY UHLENBECK

(With D. Freed) *Instantons and 4-Manifold Topology* 1984

## FURTHER READING

### Periodicals

Benningfield, Damond. "Prominent Players." *The Alcalde Magazine* (September/October 1988): 26–30.

### Other

"Karen Uhlenbeck." *Biographies of Women Mathematicians.* http://www.scottlan.edu/lriddle/women/chronol.htm (July 22, 1997).

Uhlenbeck, Karen. "Some Personal Remarks on My Partly Finished Life." unpublished manuscript.

—*Sketch by C. D. Lord*

# Charles Jean Gustave Nicolas de la Vallée–Poussin 1866–1962

**Belgian number theorist**

Charles Jean Gustave Nicolas de la Vallée–Poussin was responsible for proving the prime number theorem. A prime number is a number that can be divided by only one and itself without producing a remainder, and de la Vallée–Poussin—like many others—set out to prove the relationship between prime numbers. In an article for *MAA Online* dated December 23, 1996, Ivars Peterson asserts: "In effect, [the prime number theorem] states that the average gap between two consecutive primes near the number $x$ is close to the natural logarithm of $x$. Thus, when $x$ is close to 100, the natural logarithm of $x$ is approximately 4.6, which means that in this range, roughly every fifth number should be a prime." De la Vallée–Poussin was additionally known for his writings about the zeta function, Lebesgue and Stieltjes integrals, conformal representation, algebraic and trigonometric polynomial approximation, trigonometric series, analytic and quasi–analytic functions, and complex variables. His writings and research, which were—and are—considered clear, stylish, and precise, were highly respected by his peers in academia and other well–placed individuals in Western society.

Despite the historical confusion posed by de la Vallée–Poussin's name (it is often rendered as Charles–Jean–Gustave–Nicolas, Charles–Jean Gustave Nicolas, Charles–Joseph, Vallée Poussin, etc.), the facts surrounding his origins are well known. De la Vallée–Poussin was born on August 14, 1866, in Louvain, Belgium. A distant relative of the French painter Nicolas Poussin, de la Vallée–Poussin's father was, like himself, an esteemed teacher at the University of Louvain. (The elder de la Vallée–Poussin, however, specialized in geology and mineralogy.) De la Vallée–Poussin's family was well–off, and as a child, he found encouragement and inspiration in fellow mathematician Louis–Philippe Gilbert (some sources identify him as Louis Claude Gilbert), with whom he would eventually work.

De la Vallée–Poussin enrolled at the Jesuit College at Mons in southwestern Belgium, where it is said he originally intended to pursue a career in the clergy. He ultimately, however, obtained a *diplôme d'ingenieur* and began to pursue a career in mathematics. In 1891, like his father and Gilbert, he became employed at the University of Louvain, where he initially worked as Gilbert's assistant. Gilbert's death the following year created an academic opening to which de la Vallée–Poussin was appointed in 1893, thereby earning him the title of professor of mathematics.

Although de la Vallée–Poussin was gaining recognition as early as 1892, when he won a prize for an essay on differential equations, he earned his first widespread fame four years later. In 1896, de la Vallée–Poussin capitalized on the ideas set forth by earlier mathematicians, notably **Karl Friedrich Gauss, Adrien Marie Legendre, Leonhard Euler, Peter Gustav Lejeune Dirichlet, Pafnuty Lvovich Chebyshev,** and **Georg Friedrich Bernhard Riemann**, and proved what is now known as the prime number theorem. (De la Vallée–Poussin shares this honor with **Jacques Hadamard**, who revealed his finding in the same year. Historians note, however, that de la Vallée–Poussin's and Hadamard's achievements were performed independently and that although both mathematicians used the Riemann zeta function in their work, they came to their conclusions in different ways.)

De la Vallée–Poussin revealed much of his groundbreaking work in a series of celebrated books and papers. His two–volume *Cours d'analyse infinitésimale* went through several printings, and the work was consistently edited between printings to offer updated information. Initially, the book was directed toward both mathematicians and students, and de la Vallée–Poussin used different fonts and sizes of types to differentiate between the audiences to whom a particular passage was directed. In the 1910s, de la Vallée–Poussin was preparing the third edition of this work, but this was destroyed by German forces, who invaded Louvain during World War I. De la Vallée–Poussin subsequently dedicated his 1916 *Intégrales de Lebesgue fonctions d'ensemble, classes de baire* to the Lebesgue integral to compensate for the material destroyed by the Germans. De la Vallée–Poussin continued to publish well into his eighties, and, like *Cours d'analyse infinitésimale,* many of his writings went through various reprintings and revisions. Almost all of de la Vallée–Poussin's writing have been praised for their originality and the clarity of his writing style.

De la Vallée–Poussin, who married a Belgian woman whom he met while vacationing in Norway in the late 1890s, died on March 2, 1962, in the city of his birth. During his lifetime, he was accorded many honors. In addition to the celebrations commemorating his 35th and 50th anniversaries as chair of mathematics at the University of Louvain, he was elected to various prestigious institutions including the French Académie Royales des Sciences, the International Mathematical Union, the London Mathematical Society, the Belgian Royal Academy, and the Legion of Honor. In 1928, the king of Belgium also awarded him the title of baron in recognition of his years of academic tenure and his professional achievements.

## SELECTED WRITINGS BY DE LA VALLÉE POUSSIN

### Books

*Cours d'analyse infinitésimale.* 2 volumes, 1903–1906

*Intégrales de Lebesgue fonctions d'ensemble, classes de baire,* 1916

*Leçons sur l'approximation des fonctions d'une variable réelle,* 1919

*Leçons de mécanique analytique,* 1924

*Les nouvelles méthodes de la théorie du potentiel et le problème généralisé de Dirichlet: Actualités scientifiques et industrielles,* 1937

*Le potentiel logarithmique, balayage et représentation conforme,* 1949

*(Much of de la Vallée–Poussin's works was revised and reprinted. This work, in particular, was heavily revised, with each new edition offering substantive changes and new information.)

## FURTHER READING

### Books

Burkhill, J. C. "Charles–Jean–Gustave–Nicolas de laVallée–Poussin," in *Dictionary of Scientific Biography.* Volume XIII. Edited by Charles Coulston Gillispie. New York: Charles Scribner's Sons, 1976, pp. 561–62.

Young, Laurence. *Mathematicians and Their Times: History of Mathematics and Mathematics of History.* Amsterdam: North–Holland Publishing Company, 1981, p. 306.

### Periodicals

Bateman, Paul T. and Harold G. Diamond. "A Hundred Years of Prime Numbers." *The American Mathematical Monthly* 103, no. 9 (November 1996): 729–41.

*Argelia Velez-Rodriguez*

Burkill, J. C. "Charles–Joseph de la Vallée Poussin." *Journal of the London Mathematical Society* 39 (1964): 165–75.

### Other

Peterson, Ivars. "Ivars Peterson's MathLand: Prime Theorem of the Century." *MAA Online* (December 23, 1996). http://www.maa.org.

—*Sketch by C. J. Giroux*

# Argelia Velez–Rodriguez
## 1936–
### Cuban–born American mathematics educator

Since leaving her native Cuba shortly after completing her Ph.D., Argelia Velez–Rodriguez has devoted her career to mathematics and physics education. She has been involved with math education programs of the National Science Foundation (NSF) since 1970 and became director of the Minority Science Improvement Program at the U.S. Department of Education in 1980.

After showing promise in mathematics as a girl, Velez–Rodriguez earned a bachelor's degree in 1955 from the Marianao Institute of Cuba and a Ph.D. in

1960 from the University of Havana. Her doctoral dissertation concerned the use of differential equations in figuring astronomical orbits. Her father, Pedro Velez, had worked in the Cuban Congress under Fulgencio Batista, the leader ousted by Fidel Castro in 1959.

Velez–Rodriguez's first teaching position in the United States was at Texas College, where she began teaching mathematics and physics in 1962. In 1972, she became a professor of math and served as the department's chair at Bishop College in Dallas. Velez–Rodriguez's research at the time focused on differential equations and classical analysis, and it was at Bishop that she first became involved with the NSF programs for improving science education. Velez–Rodriguez has also studied teaching strategies, with a particular focus on helping minorities and disadvantaged students learn mathematics. She directed and coordinated several NSF programs for high school and junior high school mathematics teachers.

Velez–Rodriquez was married to Raul Rodriguez in 1954 in Cuba and they had two young children when the family fled the country in 1962. "I had just finished my Ph.D.," she told contributor Karl Bates in an interview. Her son is now a surgeon, and her daughter is an engineer with a Harvard MBA. She and Rodriguez are divorced and she is a naturalized citizen of the United States.

## FURTHER READING

"Argelia Velez–Rodriguez." *Biographies of Women Mathematicians.* June 1997. http://www.scottlan.edu/lriddle/women/chronol.htm (July 22, 1997).

## Other

Velez–Rodriguez, Argelia, interview with Karl Leif Bates, conducted June 17, 1997.

—*Sketch by Karl Leif Bates*

# François Viète (also known as Franciscus Vieta)
## 1540–1603
### French algebraist

Many scholars consider François Viète the seminal algebraist of the 16th century. Although he worked in mathematics in his spare time and was

*François Viète*

regarded as an amateur mathematician, he has contributed numerous improvements, solutions, and other vital developments to mathematics. For example, Viète had a hand in the development of modern algebra by devising algebraic notation, using letters to denote quantities, both unknown and known. He also contributed to the theory of equations, and introduced algebraic terms, such as "coefficient," that are still used today. Among Viète's other significant mathematical accomplishments are: using algebra to solve longstanding geometrical problems; calculating $\pi$ to 10 places, an important accomplishment of his time; playing a role in the improvements brought by the development of Julian calendars; and having a public hand in the mathematical controversies of his era.

Viète was born in Fontenay–le–Comte, Poitou, France, to Étienne (a lawyer and notary) and his wife, Marguerite Dupont. After receiving his education in Fontenay—partly from monks at a local Franciscan cloister—he studied law at the university in Poiters beginning in 1556. Upon graduation with a bachelor's degree, he worked as a lawyer in Fontenay for four years. Viète was apparently successful in this career; he counted among his clients Queen Eleanor of Austria and Mary Stuart. During this period, Viète also began conducting mathematical research on the side, as well as cosmological and astronomical studies.

In 1564, Viète became closely involved with the aristocratic Soubise family, including Antoinette d'Aubeterre, who consulted him on legal matters. Viète stopped practicing law to serve as private secretary to d'Aubeterre. He moved with the family to La Rochelle where he tutored d'Aubeterre's daughter, Catherine de Parthenay, until her marriage in 1570. During this time, Viète decided to become a Huguenot (French Protestant). These years also mark the first period where Viète had a significant amount of time to devote to his mathematical research.

After his charge's marriage, Viète's primary occupation became service in the royal courts. He was appointed by King Charles IX to serve as a councilor to Brittany's parliament at Rennes in 1573, a post he held until 1580. He then served Charles IX and his successor, Henry III, in Paris as member of the privy counsel and the maître de requêtes (Master of Requests). In 1584, soon after Henry III came to the throne, Viète was forced from the royal court because of his Huguenot sympathies. He spent the next five years in cities such as Garnache, Fontenay, and Bigotière, where he devoted his full attention to mathematics and related research. This marks Viète's second fruitful period of mathematical work.

Viète's ties to the royal court allowed his work to be printed by the royal printer, Jean Mettayer, who published most of Viète's mathematical treatises. Viète's first important book, *Canon mathematicus seu ad triangula cum appendibus* ("Mathematical Laws Applied to Triangles") was published in 1579. One of Viète's goals as a mathematician was to give more credence to trigonometry in the mathematical world and this text contributed significantly to this cause. Using all six trigonomic functions, he solved problems on plane and spherical triangles.

In 1589, Viète was invited to the court of Henry III (then in Tours) as a counselor to the French parliament. In the same year, he was appointed a royal Privy Councilor, and broke the  code used by Philip II of Spain for internal messages when France was at war with Spain in 1589–90. Henry IV ascended to the French throne in 1589 upon Henry III's death, and, like his royal patron, Viète declared himself Catholic. He stayed in service at the court—following it to Paris in 1594—and remained there until his retirement in 1602.

These years also saw Viète publishing a series of important works. The most influential of these was his 1591 *In artem analyticam Isagoge* ("Introduction to the Analytical Arts"), an algebra text. In this work he introduced the first systematic use of symbolic algebraic notation. Viète demonstrated, for example, the value of symbols by using the plus and minus signs for operations, vowels for unknown quantities (variables), and consonants for known quantities (parameters). He also delineated new ways of solving cubic equations, using, among other approaches, trigonometry. This book helped algebra develop fur-

ther, and is perhaps one of the earliest identifiable "modern" algebra textbooks.

Two years later, in 1593, *Supplementum geometriae* was published. Although published in 1593, Viète had probably written this treatise in 1591. Among the many geometric topics in this and a second volume published the same year were solutions to the problem of the trisection of an angle and the corresponding cubic equation, the doubling of the cube, the construction of the regular heptagon inscribed in a circle, and the earliest known explicit expression for $\pi$ as an infinite product. The last significant work of Viète's published in his lifetime was *De numerosa potestatum puarum atque ad fectarum ad exegesin resolution e tractatcus* (1600). In it, he delineates a way of approximating roots of numerical equations. He also solved the fourth Apollonian problem, which concerns three circles and the construction of a fourth tangent to the other three.

Twelve years after his death, *De aquationem recognitione et emedatione libri duo* ("Concerning the Recognition and Emendation of Equations") was published. It was Viète's primary text dealing with the theory of equations; he developed methodology for solutions of second, third, and fourth degree equations. Viète also introduced the term "coefficient" in this treatise.

Viète died December 13, 1603 (or February 23, 1603, depending on the source) in Paris, France. It is known that he was twice married, first to Barbe Cotherau and later, after becoming a widower, to Juliette Leclerc. He had at least one child, but details concerning his marriages and children are sketchy.

## SELECTED WRITINGS BY VIÈTE

*The Early Theory of Equations: On Their Nature and Constitution. Translations of Three Treatises, by Viète, Girard, and De Beaume.* Translated by Robert Schmidt, 1986
*The Analytic Art.* Translated by T. Richard Witmer, 1993

## FURTHER READING

Abbott, David, editor. *The Biographical Dictionary of Scientists: Mathematicians.* New York: Peter Bedrick Books, 1986, p. 129.
Asimov, Issac. *Asimov's Biographical Encyclopedia of Science and Technology.* Second revised edition. New York: Doubleday & Company, Inc., 1982, pp. 89–90.
Busard, H.L.L. "François Viète," in *Dictionary of Scientific Biography.* Volume XIV. Edited by Charles Coulston Gillispie. New York: Charles Scribner's Sons, 1970, pp. 18–25.

*Richard von Mises*

Daintith, John, Sarah Mitchell, Elizabeth Tootill, and Derek Gjertsen, editors. *Biographical Encyclopedia of Scientists*. Volume 2. Philadelphia: Institute of Physics Publishing, 1994, p. 910.

Klein, Jacob. *Greek Mathematical Thought and the Origin of Algebra*. Translated by Eva Brann. Cambridge, MA: The MIT Press, 1968, pp. 150–85.

Stillwell, John. *Mathematics and Its History*. New York: Springer–Verlag, 1989, pp. 63–64.

Witmer, T. Richard. Translator's Introduction to *The Analytic Art*, by François Viète. Kent, OH: The Kent State University Press, 1993, pp. 1–10.

—*Sketch by Annette Petruso*

# Richard von Mises
## 1883-1953

### Austrian-born American statistician and aerodynamicist

Richard Martin Edler von Mises was an aerodynamicist and applied mathematician who made sig-

nificant contributions to the theory of probability and statistics, and also helped pioneer airplane design. A student of logical postitivism, von Mises believed in solving problems through rational reasoning. Although he has made valuable contributions to probability theory and statistics, his ideas have not been universally accepted.

Von Mises was born April 19, 1883, in Lemberg, Austria (now Lvov, Ukraine). His father, Arthur Edler von Mises, a railway engineer, held a doctorate of technical sciences. His mother, Adele Landau von Mises, came from a family of literary scholars. A younger brother died in infancy, and an older brother, Ludwig, became an economist. Although his family was Jewish, von Mises converted to Catholicism as a young man. The family lived in Vienna, except for occasional temporary relocations for the father's work assignments. During his teenage years, von Mises studied a classical and humanistic curriculum at the Akademische Gymnasium; he graduated in 1901 with distinction in Latin and mathematics. A serious student who loved reading poetry and philosophy, he also enjoyed skating, dancing, and tennis.

Early in his professional life, von Mises joined the "Vienna Circle," which promoted a logical positivist philosophy. Throughout his career, he believed in solving problems through logical reasoning based on experience. Von Mises studied German literature and published eight articles and books on Rainer Maria Rilke's life and poetry; his extensive collection of Rilke's works is now housed at Harvard University. Besides German, von Mises was fluent in several other languages, including Italian, French, Turkish, and English.

Von Mises studied mechanical engineering at the Vienna Technical University, publishing his first mathematical paper as an undergraduate. He received his doctorate from the University of Vienna in 1907 and lectured the following year at the German Technical University at Brünn. In 1909, he became Professor of Applied Mathematics at the University of Strassburg, Germany (now Strasbourg, France).

### Shaped the Development of Flight Mechanics

Von Mises began his career by investigating fluid mechanics. Dramatic advances in heavier-than-air flying machines drew his attention to aerodynamics and aeronautics. After learning to fly, von Mises taught a summer class in 1913 that was apparently the first university course in the mechanics of powered flight. His own contributions to the field included improvements in boundary layer flow theory and refinements in airfoil design.

When World War I began in 1914, von Mises returned to his homeland to volunteer for military service, becoming an officer in the new Flying Corps of the Austro-Hungarian Army. Eventually, he was assigned to teach at the Fliegerarsenal in Aspern, where he led a team that devised, built, and tested a 600-horsepower military airplane featuring a new wing profile of his own design.

At the end of the war, with Strassburg becoming part of France, von Mises lost his university position as well as much of his personal property. He moved to Germany and taught mathematics in Frankfurt and Dresden. In 1920, von Mises became professor of applied mathematics and director of the Institute for Applied Mathematics at the University of Berlin. He published nearly 150 technical works throughout his career and was particularly proud of having founded the journal *Zeitschrift für angewandte Mathematik und Mechanik* in 1921 and serving as editor for 12 years. While in Berlin, von Mises met **Hilda Geiringer**, a student from Vienna. She became his assistant, then his collaborator, and eventually his wife. Years later, after his death, Geiringer edited and completed some of his unfinished manuscripts.

### Formalized the Frequency Theory of Probability

Von Mises viewed applied mathematics as the crucial link between theory and scientific observation. He saw the field of statistics as fundamentally important: repetitions of an experiment yield a set of different measurements, which must be analyzed to obtain a single result for comparison with theoretical predictions. Statistics, in turn, is intimately related to probability, especially as von Mises developed the concept.

Recognizing that the existing theory of probability was vague and non-rigorous, von Mises developed a formal, axiomatic treatment of the subject. **Pierre S. Laplace**'s original definition of the probability of an event, formulated in 1820, was the ratio of the number of favorable outcomes to the number of possible outcomes, assuming each outcome to be equally likely. This worked adequately for artificial applications such as games of chance, but failed when applied to naturally occurring events (such as expressing the probability of rain on a given day). Half a century later, John Venn and others proposed defining the probability of an event as the relative frequency with which the event would actually occur "in the long run" (i.e., after an unlimited number of trials). This approach seemed promising, but it was not developed with sufficient precision to validate it.

In 1919, von Mises proposed two axioms that probability must satisfy. His axiom of convergence states that as a sequence of trials is extended, the proportion of favorable outcomes tends toward a definite mathematical limit. Such a limit must exist in order to define the probability of an event as the long-term relative frequency of its occurrence. The axiom of randomness, von Mises' most important new

concept, states that the limiting value of the relative frequency must be the same for all possible infinite subsequences of trials chosen solely by a rule of place selection within the sequence (i.e., the outcomes must be randomly distributed among the trials).

With probability clearly defined as a relative frequency, statistics comprises the system of strategies and tools for designing and evaluating experiments to determine the probability of a specific event. Von Mises' 1928 book, *Probability, Statistics and Truth,* offers an interesting explanation of the subjects for the lay reader. An inconsistency inherent in von Mises' treatment of probability implies that the observation of any finite number of trials is insufficient to determine the probability of an event. Nonetheless, his contributions significantly affected the development of modern probability theory and its relationship to statistics.

When Adolph Hitler became chancellor of Germany in 1933, von Mises moved to Turkey and taught at the University of Istanbul. In 1939, he left for the United States and joined the faculty at Harvard University. In 1944, von Mises was named Gordon McKay Professor of Aerodynamics and Applied Mathematics at Harvard, a position he held until his death from cancer on July 14, 1953, in Boston.

### SELECTED WRITINGS BY VON MISES

*Selected Papers of Richard von Mises.* 2 volumes. 1963
*Positivism: A Study in Human Understanding,* 1968
*Probability, Statistics, and Truth,* 1981

### FURTHER READING

Abbott, David, editor. *The Biographical Dictionary of Scientists: Mathematicians.* New York: P. Bedrick Books, 1985, pp. 130–31.
Gridgeman, Norman T. "Richard von Mises," in *Dictionary of Scientific Biography.* Volume IX. Edited by Charles Coulston Gillispie. New York: Charles Scribners' Sons, 1974, pp. 419–20.
Magill, Frank N., editor. *Great Events from History II,* Volume 2. Pasadena, CA: Salem Press, 1991, pp. 664–68.

*—Sketch by Loretta Hall*

*John von Neumann*

# John von Neumann
## 1903-1957
### Hungarian American mathematician

John von Neumann, considered one of the most creative mathematicians of the 20th century, made important contributions to quantum physics, game theory, economics, meteorology, the development of the atomic bomb, and computer design. He was known for his problem-solving ability, his encyclopedic memory, and his ability to reduce complex problems to a mathematically tractable form. Von Neumann served as a consultant to the United States government on scientific and military matters, and was a member of the Atomic Energy Commission. According to mathematician Peter D. Lax, von Neumann combined extreme quickness, very broad interests, and a fearsome technical prowess; the popular saying was, "Most mathematicians prove what they can; von Neumann proves what he wants." The Nobel Laureate physicist Hans Albrecht Bethe said, "I have sometimes wondered whether a brain like von Neumann's does not indicate a species superior to that of man."

Max and Margaret von Neumann's son Janos was born in Budapest, Hungary, on December 28, 1903. As a child he was called Jancsi, which later became Johnny in the United States. His father was a prosperous banker. Von Neumann was tutored at home until age ten, when he was enrolled in the Lutheran Gymnasium for boys. His early interests included literature, music, science and psychology. His teachers recognized his talent in mathematics and arranged for him to be tutored by a young mathematician at the University of Budapest, Michael Fekete. Von Neumann and Fekete wrote a mathematical paper which was published in 1921.

Von Neumann entered the University of Budapest in 1921 to study mathematics; he also studied chemical engineering at the Eidgenössische Technische Hochschule in Zurich, receiving a diploma in 1925. In those same years, he spent much of his time in Berlin, where he was influenced by eminent scientists and mathematicians. In 1926 he received a Ph.D. in mathematics from the University of Budapest, with a doctoral thesis in set theory. He was named *Privatdozent* at the University of Berlin (a position comparable to that of assistant professor in an American university), reportedly the youngest person to hold the position in the history of the university. In 1926 he also received a Rockefeller grant for postdoctoral work under mathematician **David Hilbert** at the University of Göttingen. In 1929 he transferred to the University of Hamburg. By this time, he had become known to mathematicians through his publications in set theory, algebra, and quantum theory, and was regarded as a young genius.

### Extends Theory in Pure Mathematics and Quantum Physics

In his early career, von Neumann focused on two research areas: first, set theory and the logical foundations of mathematics; and second, Hilbert space theory, operator theory, and the mathematical foundations of quantum mechanics. During the 1920s, von Neumann published seven papers on mathematical logic. He formulated a rigorous definition of ordinal numbers and presented a new system of axioms for set theory. With Hilbert, he worked on a formalist approach to the foundations of mathematics, attempting to prove the consistency of arithmetic. In about 300 B.C., **Euclid**'s *Elements of Geometry* had proved mathematical theorems using a limited number of axioms. Between 1910 and 1913, **Bertrand Russell** and **Alfred North Whitehead** had published *Principia Mathematica,* which showed that much of the newer math could similarly be derived from a few axioms. With Hilbert, von Neumann worked to carry this approach further, although in 1931 **Kurt Gödel** proved that no formal system could be both complete and consistent.

Hilbert was interested in the axiomatic foundations of modern physics, and he gave a seminar on the subject at Göttingen. The two approaches to quantum mechanics—the wave theory of Erwin Schrödinger and the particle theory of Werner Karl Heisenberg —

had not been successfully reconciled. Working with Hilbert, von Neumann developed a finite set of axioms that satisfied both the Heisenberg and Schrödinger approaches. Von Neumann's axiomatization represented an abstract unification of the wave and particle theories.

During this period, some physicists believed that the probabilistic character of measurements in quantum theory was due to parameters that were not yet clearly understood and that further investigation could result in a deterministic quantum theory. However, von Neumann successfully argued that the indeterminism was inherent and arose from the interaction between the observer and the observed.

In 1929 von Neumann was invited to teach at Princeton University in New Jersey. He accepted the offer and taught mathematics classes from 1930 until 1933, when he joined the elite research group at the newly established Princeton Institute for Advanced Study. The atmosphere at Princeton was informal yet intense. According to mathematician Stanislaw Ulam, writing in the *Bulletin of the American Mathematical Society,* the group "quite possibly constituted one of the greatest concentrations of brains in mathematics and physics at any time and place." During the 1930s von Neumann developed algebraic theories derived from his research into quantum mechanics. These theories were later known as von Neumann algebras. He also conducted research into Hilbert space, ergodic theory, Haar measure, and noncommutative algebras. In 1932 he published a book on quantum physics, *The Mathematical Foundations of Quantum Mechanics,* which remains a standard text on the subject. After becoming a naturalized citizen of the United States, von Neumann became a consultant to the Ballistics Research Laboratory of the Army Ordnance Department in 1937. After the attack on Pearl Harbor in 1941, he became more involved in defense research, serving as a consultant to the National Defense Research Council on the theory of detonation of explosives, and with the Navy Bureau of Ordnance on mine warfare and countermeasures to it. In 1943 he became a consultant on the development of the atomic bomb at the Los Alamos Scientific Laboratory in New Mexico.

### Develops Design for Stored-Program Computer

At Los Alamos, von Neumann persuaded J. Robert Oppenheimer to pursue the possibility of using an implosion technique to detonate the atomic bomb. This technique was later used to detonate the bomb dropped on Nagasaki. Simulation of the technique at the Los Alamos lab required extensive numerical calculations which were performed by a staff of twenty people using desk calculators. Hoping to speed up the work, von Neumann investigated using computers for the calculations and studied the

design and programming of IBM punch-card machines. In 1943 the Army sponsored work at the Moore School of Engineering at the University of Pennsylvania, under the direction of John William Mauchly and J. Presper Eckert, on a giant calculator for computing firing tables for guns. The machine, called ENIAC (Electronic Numerical Integrator and Computer), was brought to von Neumann's attention in 1944. He joined Mauchly and Eckert in planning an improved machine, EDVAC (Electronic Discrete Variable Automatic Computer). Von Neumann's 1945 report on the EDVAC presented the first written description of the stored-program concept, which makes it possible to load a computer program into computer memory from disk so that the computer can run the program without requiring manual reprogramming. All modern computers are based on this design.

Von Neumann's design for a computer for scientific research, built at the Princeton Institute for Advanced Study between 1946 and 1951, served as the model for virtually all subsequent computer applications. Those built at Los Alamos, the RAND Corporation, the University of Illinois and the IBM Corporation all incorporated, besides the stored program, the separate components of arithmetic function, central control (now commonly referred to as the central processing unit or CPU), random-access memory (or RAM) as represented by the hard drive, and the input and output devices operating in serial or parallel mode. These elements, present in virtually all personal and mainframe computers, were all pioneered under von Neumann's auspices.

In addition, von Neumann investigated the field of neurology, looking for ways for computers to imitate the operations of the human brain. In 1946 he became interested in the challenges of weather forecasting by computer; his Meteorology Project at Princeton succeeded in predicting the development of new storms. Because of his role in early computer design and programming techniques, von Neumann is considered one of the founders of the computer age.

### Formulates Game Theory and Its Application to Economics

While in Germany, von Neumann had analyzed strategies in the game of poker and wrote a paper presenting a mathematical model for games of strategy. He continued his work in this area while he was at Princeton, particularly considering applications of game theory to economics. When the Austrian economist Oskar Morgenstern came to Princeton, he and von Neumann started collaborating on applications of game theory to economic problems, such as the exchange of goods between parties, monopolies and oligopolies, and free trade. Their ambitious 641-page book, *Theory of Games and Economic Behavior,* was published in 1944. Von Neumann's work opened new channels of communication between mathematics and the social sciences.

Von Neumann and Morgenstern argued that the mathematics as developed for the physical sciences was inadequate for economics, since economics seeks to describe systems based not on immutable natural laws but on human action involving choice. Von Neumann proposed a different mathematical model to analyze strategies, taking into account the interdependent choices of "players." Game theory is based on an analogy between games and any complex decision-making process, and assumes that all participants act rationally to maximize the outcome of the "game" for themselves. It also assumes that participants are able to rank-order possible outcomes without error. Von Neumann's analysis enables players to calculate the consequences or probable outcomes of any given choice. It then becomes possible to opt for those strategies that have the highest probability of leading to a positive outcome. Game theory can be applied not only to economics and other social sciences but to politics, business organization and military strategy, to mention only a few areas of its usefulness.

### Serves as Advisor to Government and Military

After the war, von Neumann served as a scientific consultant for government policy committees and agencies such as the CIA and National Security Agency. He advised the RAND Corporation on its research on game theory and its military applications, and provided technical advice to companies such as IBM and Standard Oil. Following the detonation of an atomic bomb by the Soviets in 1949, von Neumann contributed to the development of the hydrogen bomb. He believed that a strong military capacity was more effective than a disarmament agreement. As chairman of the nuclear weapons panel of the Air Force scientific advisory board (known as the von Neumann committee), his recommendations led to the development of intercontinental missiles and submarine-launched missiles. Herbert York, the director of the Livermore Laboratory, said, "He was very powerful and productive in pure science and mathematics and at the same time had a remarkably strong streak of practicality [which] gave him a credibility with military officers, engineers, industrialists and scientists that nobody else could match."

In 1954 President Eisenhower appointed von Neumann to the Atomic Energy Commission. Von Neumann was hopeful that nuclear fusion technologies would provide cheap and plentiful energy. According to the chairman of the Commission, Admiral Lewis Strauss, "He had the invaluable faculty of being able to take the most difficult problem, separate it into its components, whereupon everything looked brilliantly simple, and all of us wondered why we had not been able to see through to the answer as clearly as it was possible for him to do." He received the Enrico Fermi Science Award in 1956, and in that same year the Medal of Freedom from President Eisenhower.

Von Neumann has been described as a genius, a practical joker, and a raconteur. Laura Fermi, wife of the associate director of the Los Alamos Laboratory Enrico Fermi, wrote that he was "one of the very few men about whom I have not heard a single critical remark. It is astonishing that so much equanimity and so much intelligence could be concentrated in a man of not extraordinary appearance."

Von Neumann married Mariette Kovesi, daughter of a Budapest physician, in 1929. Their daughter, Marina, was born in 1935. Mariette obtained a divorce in 1937. The following year, von Neumann married Klara Dan, from an affluent Budapest family. In 1955, von Neumann was diagnosed with bone cancer. Confined to a wheelchair, he continued to attend Atomic Energy Commission meetings and to work on his many projects. He died in 1957 at the age of 53.

## SELECTED WRITINGS BY VON NEUMANN

*Mathematische Grundlagen der Quanton-mechanik* (title means "Mathematical Foundations of Quantum Mechanics."), 1932
(With Oskar Morgenstern) *The Theory of Games and Economic Behavior*, 1944
*The Collected Works of John von Neumann.* Edited by A. H. Traub, 1963

## FURTHER READING

### Books

Aspray, William. *John von Neumann and the Origins of Modern Computing.* Cambridge, MA: MIT Press, 1990.
Glimm, James, John Impagliazzo, and Isadore Singer, editors. *The Legacy of John von Neumann: Proceedings of Symposia in Pure Mathematics,* Volume 50. American Mathematical Society, 1990.
Heims, Steve J. *John von Neumann and Norbert Wiener: From Mathematics to the Technologies of Life and Death.* Cambridge, MA: MIT Press, 1980.
Macrae, Norman. *John von Neumann.* New York: Pantheon, 1992.
Poundstone, William. *Prisoner's Dilemma.* New York: Doubleday, 1992.

**Periodicals**

Ulam, Stanislaw. "John von Neumann," *Bulletin of the American Mathematical Society* (May 1958): 1–49.

—*Sketch by C. D. Lord*

*John Wallis*

# John Wallis
## 1616–1703
### English algebraist

John Wallis was a founding member of the Royal Society, one of the oldest scientific organizations still in existence, and is considered by many the most influential British mathematician preceding **Isaac Newton**. He contributed the earliest forms, terms, and notations to nascent fields such as calculus and analysis. Wallis was the first to attempt to write a comprehensive history of British mathematics, striving to bring continuity to mathematical study and research. Among the many classical tracts Wallis translated and edited are two major works of **Archimedes**. He was also involved in government; as the Parliamentarians' cryptographer, Wallis was instrumental in deciphering enemy codes during the English Civil War. Not all of Wallis' discoveries were valid, as shown by the rare example of an attempt to analyze **Euclid**'s fifth postulate in 1663 that turned out to be a trivial proof.

Wallis was born on November 23, 1616, to John and Joanna (nee Chapman) Wallis, residents of Ashford in Kent. In 1622, Wallis' father, a rector, died, and three years later young Wallis left Ashford in order to escape an outbreak of the plague. In 1625, he attended boarding school in Ley Green, near Tenterden, and after five years there he transferred to Martin Holbeach in Essex county. During one Christmas holiday Wallis asked his brother to introduce him to arithmetic and he mastered it in two weeks. It turned out Wallis was a "calculating prodigy," who could solve problems like the square root of a 53–digit number to 17 places without notation.

Wallis' education continued at Emmanuel College in Cambridge, where he studied medicine and wrote one of the first papers on the circulatory system. He also took courses in physics and moral philosophy, for which he was awarded a B.A. in 1637 and a fellowship to Queen's College. During this time, civil unrest was building in England. Wallis became a minister of the Church of England in 1640, serving as private chaplain in Yorkshire and Essex. He would also eventually become a royal chaplain to the court of Charles II and bishop of Winchester as well. Upon his marriage to Susanna Glyde in 1645 Wallis was forced out of his fellowship and moved to London. For his service in deciphering Royalist letters during the Civil War, Oliver Cromwell appointed Wallis Savilian Professor of Geometry at Oxford.

In 1655 Wallis produced his first major work, *Arithmetica Infinitorum*, which systematized the analytical methods of **René Descartes** and **Bonaventura Cavalieri's** method of indivisibles. It became a standard reference, influencing many mathematicians who thereafter wrote on the subject, and is still considered a monumental text in British mathematics. In 1658 Wallis was appointed an official archivist at Oxford, and in the following year his treatise on conics, *Tractatus de sectionibis conicis*, was published. In it, Wallis attempted to simplify Cartesian geometry—a notoriously recondite work—for consumption and use by his peers.

Among the vast range of mathematical subjects Wallis tackled included the quadrature of many curves, an infinite approximation for $4/\pi$, negative and fractional exponents, and calculating the center of gravity in cycloids by using indivisibles. He was the first to use the symbol $\infty$ to indicate infinity and the capital S for sine, and introduced the mantissa in logarithms. In his treatise on cycloids Wallis was the first to recast conic sections as curves of the second degree. His approximation for $4/\pi$, a formula now famous in mathematics, used "interpolation," a word famous in mathematics, used "interpolation," a word peculiar to Wallis that became standardized.

*Algebra: History and Practice*, published in English in 1685, was significant not just as a historical document. Here, Wallis made the first recorded effort to graphically display the complex roots of a real quadratic equation. This intimated the geometric interpretations of complex numbers and trigonometry in use today in the Gaussian plane. He also prepared and published a second edition in 1693 as volume two of his *Opera Mathematica*. This enlarged version included the first systematic use of algebraic formulae—using numerical ratio to represent a given magnitude. That systematization had great impact on subsequent mathematical arguments about important unsolved problems in physics.

Wallis survived another political crisis, the Glorious Revolution (1688–1689), which led to the deposition of King James II. James' successor, William III, retained Wallis to decipher enemy communiques.

### "Dueling Pamphlets"

Wallis wrote on subjects other than mathematics, including physics, theology, etymology and linguistics, general philosophy, and pedagogy. He is recognized as the first hearing person to systematize a means of teaching deaf mutes. Wallis was well known as a partisan; he successfully lobbied against the adoption of the Gregorian Calendar and was opposed to crediting **Gottfried Wilhelm Leibnitz** as a co-founder of the calculus. Wallis launched into an argument with the philosopher Thomas Hobbes, ostensibly regarding the subject of geometry. The two men traded insults in pamphlets with such florid titles as "Due Correction for Mr. Hobbes, or School Discipline for not saying his Lessons." Wallis married and had one son and two daughters. His wife died in 1687.

Wallis kept his post as Savilian Professor until his death at Oxford on October 28, 1703. Christopher Wren and **Christiaan Huygens**, inspired by Wallis' use of analogy in his writings, extended his investigations to the next generation of European mathematicians.

### SELECTED WRITINGS BY WALLIS

*Arithmetica Infinitorum*, 1655
*Mathesis Universalis*, 1657
*Tractatus de sectionibis conicis*, 1659
*Mechanica*, 1669
*Algebra: History and Practice*, 1685
*Opera Mathematica*, 1693

## FURTHER READING

### Books

Abbott, David, editor. *The Biographical Dictionary of Scientists*. New York: Peter Bedrick Books, 1986, pp. 132–3.

Asimov, Isaac. *Asimov's Biographical Encyclopedia of Science and Technology*. Second Revised Edition. Garden City, NY: Doubleday, 1982, pp. 126–7.

*Biographical Encyclopedia of Scientists*. Edited by Urdang and Associates. New York: Facts on File, 1981, p. 823.

Eaves, Howard. *An Introduction to the History of Mathematics*. Fourth Edition. New York: Holt, Rinehart & Winston, 1976, pp. 317–18.

————. *Great Moments in Mathematics (After 1650)*. Mathematical Association of America, Inc., 1980, p. 66.

Porter, Roy, consultant editor. *The Biographical Dictionary of Scientists*. Second Edition. New York: Oxford University Press, 1994.

Ronan, Colin. *Astronomers Royal*. Garden City, NY: Doubleday & Co., Inc., 1969, p. 33.

Scott, J.F. *Mathematical Works of John Wallis*. Second Edition. New York: Chelsea Publishing Co., 1981.

Williams, Trevor I., editor. *A Biographical Dictionary of Scientists*. Third Edition. New York: John Wiley & Sons, 1982, pp. 541–42.

*World Who's Who in Science*. First Edition. Edited by Allen G. Debus. Chicago: Marquis, 1968, p. 1753.

### Periodicals

Pycior, H. "Mathematics and Philosophy: Wallis, Hobbes, Barrow and Berkeley." *Journal of the History of Ideas* 48 (1987): 265–287.

Yule, G.U. "John Wallis 1616–1703." *Notes and Records*, Royal Society of London 2 (1939).

### Other

"John Wallis." *MacTutor History of Mathematics Archive*.

http://www-groups.dcs.st-and.ac.uk/~history/Mathematicians/Wallis.html (July 1997).

—*Sketch by Jennifer Kramer*

# Edward Waring
## 1734(?)–1798
### English algebraist and physician

Edward Waring was an established 18th-century mathematician and theorist who did groundbreaking work in the areas of imaginary numbers and their roots. He is best known for the Cauchy ratio test and Waring's theorem, which is also known as Waring's Problem. In the former, Waring focused on the convergence and divergence of numerical series. In the theorem named after him, Waring postulated that any positive integer is the sum of not more than nine cubes, and is the sum of not more than 19 fourth powers. The result about cubes was not proved until 1910, and the result about fourth powers was not proved until 1986. Waring also posited that integers are composed of a specific combination of prime numbers. For example, an even integer, he asserted, was the sum of two prime numbers, whereas an odd number was either, in and of itself, a prime number or the sum of three primes. Waring was also known for his work with quartic curves and the approximation of imaginary roots.

Waring was born near Shrewsbury in Shropshire, a borough in western England, to John Waring, a wealthy farmer, and his wife, Elizabeth. Exact details about their oldest son's origins are sketchy: some sources say the mathematician was born in 1734, whereas others say 1736; Plealey has likewise been given as the more exact place of his birth, as has Old Heath.

Waring initially attended school in Shrewsbury before going on to Cambridge University, where he was enrolled as a sizar at Magdalene College in 1753. (A sizar, according to *Webster's Dictionary*, is a student who receives an academic stipend, often in return for performing work for other students.) In 1757, having established himself and excelled in the field of mathematics, Waring graduated with a bachelor of arts and the title of senior wrangler. Around this time, he was also granted a fellowship, and this enabled him to continue with his studies. He was likewise honored with membership in Cambridge's recently founded Hyson Club.

### Fights for Academic Recognition

The death of faculty member John Colson at Cambridge created a job opening for which Waring was nominated. Because of the complaints of some faculty members who faulted him for his age, work, and lack of teaching experience, Waring countered their opposition by showing some of his writings; he was ultimately offered the position, although he did not yet possess his master's degree, a requirement for this position. He eventually received the degree by royal mandate in 1760 and was awarded the title of "Lucasian professor of mathematics," (a position first held by **Isaac Newton**). Waring held this title for the remainder of his life and it would ultimately enable him to pursue other interests. Additional honors included Waring's induction into the Royal Society of

London a few years after his M.A. was conferred upon him, as well as that organization's Copley Medal.

During his tenure at Cambridge, Waring published several nonfiction works and treatises about mathematics, the earliest of which were published in Latin. Portions of his first work, *Miscellanea analytica de aequationibus algebraicis, curvarum proprietatibus, fluxionibus et serierum summatione* (1759; "Miscellany of Analysis"), were used by Waring when he was attempting to strengthen his position against those critics who deemed him too young to be the Lucasian professor of mathematics and were protesting his nomination. In this work and its 1762 revised version, *Miscellanea analytica de aequationibus algebraicis et curvarum proprietatibus,* Waring focused in part on his theory of numbers and their composition (as a cube, a fourth power, or some combination of either cubes or fourth powers). Subsequent works published in Latin include the 1770 *Meditationes algebraicae* ("Thoughts on Algebra") and *Proprietates algebraicarum curvarum* ("The Properties of Algebraic Curves"), which was printed in 1772. His later works—notably *On the Principle of Translating Algebraic Quantities into Probable Relations and Annuities* (1792) and *An Essay on the Principles of Human Knowledge* (1794)—were first printed in his native language. In addition to his books, Waring wrote numerous other small articles and pieces. Despite the erudition of the ideas presented in his writings, Waring's books were often faulted by critics for their brevity, lack of explanation, and apparent lack, at times, of clarity and organization. For example, *Miscellanea analytica* was once described as "one of the most abstruse books written in the abstrusest parts of Algebra," a quotation that has been attributed to both Glieg and Charles Hutton, the latter being the English mathematician who had determined the Earth's mean density.

### A Second Career and Personal Sacrifice

Perhaps because of the sometimes negative reception of his works upon their publication, it has since been noted that Waring's contributions stem from the originality of his work and research, and not necessarily as a writer or teacher of his ideas. Indeed, during his academic career, Waring spent little time teaching students (reports about his personality indicate that he would not have particularly suited to the job—he has been described as being somewhat arrogant, and it is even rumored that he once claimed that no one in his homeland was competent enough to read, let alone understand, his work) and actually devoted a portion of his life to medicine, eventually earning a medical degree from Cambridge in either the late 1760s or the early 1770s. (Although some sources disagree, it is often accepted that Waring participated in the dissection of corpses and was frequently found at work in the hospitals around

London and Cambridge. For a time, it is believed that he actually practiced medicine at various English hospitals before retiring from the profession in the 1770s.)

Whatever the reception of Waring's work in their published format, his ideas about mathematics and science were groundbreaking but frequently misunderstood. Although some of his work still remains unproved and much of it was ignored or disparaged by his contemporaries, his ideas remain highly respected for their innovativeness. Waring died on August 15, 1798, in Pontesbury, Shropshire.

## SELECTED WRITINGS BY WARING

### Books

*Miscellanea analytica de aequationibus algebraicis, curvarum proprietatibus, fluxionibus et serierum summatione,* 1759; revised edition 1762
*Meditationes algebraicae,* 1770
*Proprietates algebraicarum curvarum,* 1772
*Meditationes analyticae,* 1776
*On the Principle of Translating Algebraic Quantities into Probable Relations and Annuities,* 1792
*An Essay on the Principles of Human Knowledge,* 1794

## FURTHER READING

### Books

Cajori, Florian. *A History of Mathematics.* Second edition. New York: The Macmillan Company, 1919, pp. 248–49.
Scott, J. F. "Edward Waring," in *Dictionary of Scientific Biography.* Volume XIV. Edited by Charles Coulston Gillispie. New York: Charles Scribner's Sons, 1976, pp. 179–81.
W.P.C. "Edward Waring," in *The Dictionary of National Biography.* Volume XX. Edited by Sir Leslie Stephen and Sir Sydney Lee. London: Oxford University Press, 1937–38, pp. 840–42.

### Periodicals

"Queries with Answers." *Notes and Queries,* Second Series, XI (February 2, 1861): 89–90.
Stewart, Ian. "The Waring Experience." *Nature* 323, no. 23 (October 1986): 674.

*—Sketch by C. J. Giroux*

*Karl Theodor Wilhelm Weierstrass*

# Karl Theodor Wilhelm Weierstrass
## 1815–1887
### German analyst and number theorist

Karl Wilhelm Theodor Weierstrass was considered one of the greatest mathematical analysts of 19th century Europe. He is well known as a cofounder of the theory of analytic functions and their representation as power series. Weierstrass made crucial contributions to the arithematization of analysis and to the theory of real numbers. He showed the importance of uniform convergence, furthered the understanding of elliptic functions, and made contributions to the field of differential equations. Weierstrass' reputation for high standards of proof and definition is reflected in the modern development of calculus and analysis.

The eldest child of Wilhelm Weierstrass, a customs officer, and Theodora (nee Forst), Weierstrass was born in Ostenfeld, Germany, on October 31, 1815. The family soon moved to Westernkotten in Westphalia, where Weierstrass' three siblings, Peter, Klara, and Elise, were born. Shortly after Elise's birth in 1826, Weierstrass' mother died, and a year later his father remarried. His father, who had once been a teacher and had a reputation for righteousness and uncompromising authority, wished for his son to receive a solid education and pursue a bureaucratic career that might afford him a comfortable lifestyle.

Since there was no school in the village of Westernk-otten, Weierstrass was sent to study in nearby Muenster. At the age of 14 he entered the Gymnasium in Paderborn, from which he graduated in 1834. Earning awards in almost all of his subjects and finishing among the top in his class, Weierstrass appeared bound for the success his father had envisioned. It would not be long, however, before ambitions of a comfortable bureaucratic career were derailed. Weierstrass was directed by his father to attend the University of Bonn, where it was expected he would acquire an education in business and law. He enrolled at the university, but rather than taking his studies seriously, he embraced the sport of fencing and the habit of socializing in German pubs. After four years Weierstrass returned home without a degree. His father and siblings, reportedly shamed by his failure, decided he might salvage the family name by obtaining a teaching certificate. In 1839 he began studying toward this end at the Academy of Muenster.

Although Weierstrass had read *Celestial Mechanics* by **Pierre Simon Laplace** during his time in Bonn, he had, till 1839, shown no extraordinary interest in mathematics. At the Academy of Muenster, however, he came under the tutelage of Christof Gudermann, who taught a course in elliptic functions. Gudermann's special interest was in attempting to represent elliptic functions with power series, an approach to analysis that was to have a profound influence on Weierstrass' career. Weierstrass proved, in Muenster, to be an able and dedicated student. In 1841, when he took his examination for the teaching certificate, he asked Gudermann to provide a mathematical problem that might normally be presented to a doctoral candidate. Gudermann obliged, and proposed that Weierstrass find the power series developments of the elliptic functions. The examination results were so exemplary that Gudermann recommended Weierstrass be granted a university post, but the Academy was only able to grant him a teaching certificate.

## Conducts Research in Isolation

Weierstrass began his teaching career at the Gymnasium in Muenster in 1841. In 1842 he took a position as a teacher of mathematics and physics in Deutsche–Krone, West Prussia. That same year his first mathematical work, "Remarks on Analytical Factorials," was printed in a school publication. The narrow circulation of the journal prevented Weierstrass' significant original contribution to the field from being immediately recognized. It would be more than a decade before the paper came to the attention of a wider, more scholarly audience. During this period Weierstrass was forced to confine his mathematical research to his spare hours, devoting most of his energies to the daily demands of his secondary school teaching. Occasionally, he would work until dawn on a mathematical problem of interest, unaware

that the night had passed. Although he worked in isolation and had not yet established personal contact with the prominent European mathematicians of his day, he did manage to remain somewhat current in his readings. In 1848, Weierstrass moved to Braunsberg to teach at the Royal Catholic Gymnasium. By now he had taken an interest in the work of **Niels Henrik Abel**. Shortly after assuming his new teaching post, he published a paper in the gymnasium's journal entitled "Contributions to the Theory of Abelian Integrals." Once again, limited circulation meant his important discoveries would receive no immediate notice.

## Achieves Sudden Fame with Treatise on Abelian Functions

During the summer of 1853, Weierstrass spent his vacation in his father's home in Westernkotten, writing another paper on Abelian functions. This time, he chose to submit the work to the *Journal fur die reine and angewandte Mathematik*, a prestigious mathematical research publication begun in 1826 by August Leopold Crelle. The paper was accepted for publication and appeared in print in 1854, bringing Weierstrass instant acclaim. He had successfully solved a problem of hyper–elliptic integrals, and established himself as one of the truly great mathematical analysts. The University of Königsberg bestowed on Weierstrass an honorary doctorate, and the gymnasium in Braunsberg granted him a leave so that he could pursue mathematics full time. In 1856, he was appointed professor at the Royal Polytechnic School in Berlin. He received a joint appointment as an assistant professor at the University of Berlin and was awarded membership in the Berlin Academy. That same year his first paper, "Remarks on Analytical Factorials," was reprinted in Crelle's journal, finally receiving an appropriate audience.

In Berlin, Weierstrass rapidly took on a heavy research and teaching schedule, and his lectures on Abelian functions and transcendents were widely attended. Although he wrote very few manuscripts, his ideas on analytic functions became widely known through the dissertations and other writings of his students. Eventually, in 1886, G.H. Halphen, a student of Weierstrass, published a comprehensive discussion of Weierstrass' theory of elliptic functions, entitled *Theorie Des Fonctions Elliptiques et des leurs Applications*.

At the core of Weierstrass' mathematical research was his work on the theory of analytic functions based on power series and the process of analytic continuation. Weierstrass demonstrated that the integral of an infinite series is equal to the sum of the integrals of the separate terms when the series converges uniformly within a given region. In 1861, Weierstrass demonstrated a function that is continuous over an interval but does not possess a derivative at any point on this interval. Before this, it had been assumed that a continuous function must have a derivative at most points. In 1863 he provided a proof of a Gaussian theorem that complex numbers are the only commutative algebraic extensions of the real numbers. In the field of the calculus of variations, he brought clarity and rigor to necessary and sufficient conditions for elliptic functions. Toward the end of his career Weierstrass became interested in the astronomical problem of the stability of the solar system.

In addition to his numerous contributions to the field of mathematics, Weierstrass also earned a reputation as an excellent teacher and lecturer. He gathered about him a band of young students, many of whom themselves became successful mathematicians and propagators of Weierstrass' ideas. His students and the auditors in his seminars read like a who's who of 19th century mathematics. But undoubtedly, his most favorite pupil was the Russian mathematician **Sonya Kovalevskaya**. The two first met in 1870 when he was 55 and she was 20. Because the University of Berlin would not allow a woman to officially attend Weierstrass' lectures, he taught her privately for four years. Through his efforts, Kovalevskaya finally received her doctorate in absentia from Göttingen in 1874. They continued to correspond after her return to Russia and eventual appointment as professor of mathematics at the University of Stockholm until her untimely death in 1891.

Weierstrass remained at the University of Berlin for 30 years. His activities as a mathematician and lecturer in Berlin were interrupted from time to time because of chronic illness. He developed vertigo in the 1860s and later suffered from chronic bronchitis and phlebitis. In 1894 he became confined to a wheelchair. Weierstrass died of influenza, following a long illness, on February 19, 1897, in Berlin. A lifelong bachelor, he was buried in a Catholic cemetery alongside his two sisters, with whom he had lived for much of his adult life.

## SELECTED WRITINGS BY WEIERSTRASS

### Books

*Mathematische Werke*, 7 volumes, 1894–1927

### Periodicals

"Remarks on Analytical Factorials," 1842; reprinted 1856
"Contributions to the Theory of Abelian Integrals," 1848

## FURTHER READING

Bell, E.T. "Master and Pupil." *Men of Mathematics*. New York: Simon and Schuster, 1986, pp. 406–432.
Cajori, Florian. *A History of Mathematics*. New York: Chelsea Publishing Co., 1991.

—*Sketch by Leslie Reinherz*

# André Weil
## 1906-
### French algebraist, number theorist

André Weil is responsible for important advances in algebraic geometry, group theory, and number theory and belonged to the group of French mathematicians who published many important works under the collective pseudonym of Nicolas Bourbaki. Many of his peers in the 1950s considered him the finest living mathematician in the world. In 1980, he was presented with the Barnard Medal by Columbia University; prior recipients of the medal, which is awarded every five years, include **Albert Einstein** , Ernest Rutherford, and Neils Bohr. The prize recognizes outstanding accomplishment in physical or astronomical science or a scientific application of great benefit to humanity.

Weil was born May 6, 1906, in Paris, France, to free-thinking Jewish parents. His father, Bernard, was a physician, and his mother, Selma Reinherz Weil, came from a cultured Russian family. His sister was the famous writer, social critic, and World War II French Resistance activist, Simone Weil. When he was eight years old, Weil happened upon a geometry book and began to read it for recreation. By the time he was nine, he was absorbed in mathematics and was solving difficult problems. In her biography of Weil's sister, Simone Pétrement quotes Weil's mother as saying that at nine years of age André "is so happy that he has given up all play and spends hours immersed in his calculations." Weil's father was drafted into the military in 1914, and the family accompanied him to various medical assignments around France during World War I. At age 16, Weil was accepted at the elite École Normale Supérieure in Paris, where he received his doctorate in 1928. He also studied at the Sorbonne, the University of Göttingen, and the University of Rome. From 1930 to 1932, he taught at the Aligarh Muslim University in India. From 1933 to 1940, he was a professor of mathematics at the University of Strasbourg in France. In 1937, he married.

Weil was in Finland with his wife when France entered World War II. He believed that he could do France more good as a mathematician and refused to return to his home country and join the army. He was walking near an anti-aircraft gun emplacement when the Russians invaded Finland, and the Finns arrested him, thinking that he was a spy. The letters to Russian mathematicians in his room did not help his case, and for a while it appeared that he would be executed. The Finns, however, released Weil to the Swedes, who sent him to England, from where he returned to France to be imprisoned and tried for not reporting for military service. He was tried on May 3, 1940 and convicted, and he asked to be sent to the front. The court obliged, and he was to be sent to an infantry unit along the English Channel at Cherbourg. Weil's boat, however, wound up in a British port, and he made his way back to France later in 1940. He soon rejoined his wife, Eveline, and they escaped the war to the United States. Their daughter, Sylvie, was born on September 12, 1942. Weil taught at Haverford and Swarthmore colleges in the United States in 1941 and 1942, and at the University of São Paulo in Brazil from 1945 to 1947.

In 1947, Weil was recruited to the mathematics department at the University of Chicago, where he taught until 1958. One of his colleagues at Chicago was Irving Kaplansky, who gives a sense of Weil's personality in *More Mathematical People*: "There we were at Chicago, lucky enough to have André Weil, one of the greatest mathematicians in the world. There were several times in my life that I've, one way or another, got that feeling, my gosh, here is a tremendous mathematician.... He was very impatient with what he regarded as incompetence." Kaplansky added, "Then there is his extraordinary quickness.... You can take an area of mathematics that he presumably never heard of before and just like that he'll have something to say about it." From 1958 until his retirement, Weil taught at the Institute for Advanced Study at Princeton.

### Reveals Discovery of "Uniform Space"

Weil's mathematical innovations are highly technical and involve complex formulas. One of his discoveries was the concept of "uniform space," a kind of mathematical space that cannot be readily visualized like the three-dimensional space that we occupy in our daily lives. The *Science News Letter* pronounced Weil's discovery of uniform space one of the most important mathematical discoveries of 1939. In 1947, Weil developed some formulas in the field of algebraic geometry, which are known as the "Weil conjectures." Weil's conjectures, as Ian Stewart explains in *Scientific American*, "give formulas for the number of solutions to an algebraic equation in a finite field. In particular they allow one to deduce that a given equation does or does not have solutions; this

information can be transferred to analogous equations involving integers or algebraic numbers.... [They] are of fundamental importance in algebraic geometry."

Weil's algebraic and geometrical innovations of the first half of the 20th century were especially important for the technological innovations of the second half of the 20th century. Complex computer software that models black holes for astronomers, scientific graphics for research physicists, and special effects visualizations for Hollywood filmmakers all rely in part on mathematical innovations in algebra and geometry. As the *Science News Letter* said in 1939, in the decades to come, mathematical innovations like Weil's may lead to "some concept that will illumine the universe as glimpsed by the 200-inch telescope or the atom as created or smashed by the powerful cyclotron."

### Becomes Involved with Influential Group

In the mid–1930s, Weil and other important young French mathematicians—among them Jean Dieudonné, Claude Chevallier, and **Henri Cartan**—began to write a series of mathematical works under the pseudonym of Nicolas Bourbaki. As Paul Halmos said in *Scientific American*, one writer called Bourbaki a "polycephalic mathematician." The group has varied in number from ten to twenty and has been composed, predominantly, of those of French nationality. Their purpose was quite serious: to write a series of books about such fundamental mathematical areas as set theory, algebra, and topology. The resulting series of books, which to date number over thirty, was called the *Elements of Mathematics*. As Halmos said, "The main features of the Bourbaki approach are a radical attitude about the right order for doing things, a dogmatic insistence on a privately invented terminology, a clean and economical organization of ideas, and a style of presentation which is so bent on saying everything that it leaves nothing to the imagination." Their work has been very thorough (for example, it took them two hundred pages to define the number "1") and influential. Among other things, they inspired the "new math" that was introduced into American schools in the 1960s.

While their purpose is serious, the Bourbakians cultivate an atmosphere of mystery about their identities: they attempt to keep their names secret, they like to make up stories about themselves, and they love pranks. One story about their origin, which could well be a hoax, is that they got the idea for their name from the annual visit of a character named Nicolas Bourbaki to the École Normale Supérieure, where many of them were educated. This character was an actor who gave a mock-serious lecture on mathematics in double-talk. Some of their own double-talk consists of saying that the home institution of Nicolas Bourbaki is the "University of Nancago," a fusion of the Universities of Nancy and Chicago, where several members of the group teach. Another story reported is that the name was inspired by General Charles Denis Sauter Bourbaki, a colorful figure in the Franco-Prussian war. One of the group's pranks was to apply for a membership to the American Mathematical Society under the name of N. Bourbaki. They played another prank on Ralph P. Boas, the executive editor of *Mathematical Reviews*. Boas had said in one of the *Encyclopedia Britannica*'s annual *Book of the Year* volumes that Nicolas Bourbaki did not exist. The Bourbakians sent a letter to the editors of the *Britannica* complaining about Boas's charge. Later, as Paul Halmos said in *Scientific American*, the Bourbakians "circulated a rumor that Boas did not exist. Boas, said Bourbaki, is the collective pseudonym of a group of young American mathematicians who act jointly as the editors of *Mathematical Reviews*."

## SELECTED WRITINGS BY WEIL

*Foundations of Algebraic Geometry*, 1946
*Basic Number Theory,* third edition, 1974
*Number Theory*, 1984

## FURTHER READING

### Books

Albers, Donald J., Gerald Alexanderson, and Constance Reid. *More Mathematical People.* New York: Harcourt, 1990.
Pétrement, Simone. *Simone Weil: A Life.* Translated by Raymond Rosenthal. New York: Pantheon, 1976.

### Periodicals

Halmos, Paul R. "Nicolas Bourbaki." *Scientific American* (May 1957): 88–99.
"New Kind of Space, Year's Discovery in Mathematics." *Science News Letter* (January 21, 1939): 45–46.
Stewart, Ian. "Gauss." *Scientific American* (July 1977): 122–131.

—*Sketch by Patrick Moore*

# Mary Catherine Bishop Weiss
# 1930–1966
## American trigonometrist

Mary Weiss helped to create methods of harmonic analysis that apply to higher–dimensional geometry and one problem in lacunary series that defied solution for 20 years. Her work was recognized and supported by the National Science Foundation.

Weiss was born on December 11, 1930, to Albert Bishop, a mathematics professor, and his wife, Helen, in Wichita, Kansas. Both Weiss and her brother, Errett, would follow in their father's footsteps. Mr. Bishop had become an Army officer after graduating from the United States Military Academy at West Point, but began teaching at the university level following his retirement from the military. He died while Weiss and her brother were still very young, and the family moved to another region in Kansas to be closer to relatives. Weiss and her mother moved to the Chicago area when Errett entered the University of Chicago, where she also enrolled in the institution's experimental Laboratory School. During her undergraduate years she met and married a fellow student, Guido Weiss. They both earned their doctorates at the University of Chicago, both working in the same general area of inquiry under Antoni Zygmund. Zygmund would later become Weiss' colleague, co–writer, and editor, a collaboration which lasted until her death.

Weiss' Ph.D. thesis, completed in 1957, laid the foundation for three years of work on lacunary series, the subject of her first five published papers. During this time she provided a proof for a theorem of Raymond Paley's and solved a problem first posed by one of the founders of the Hardy–Littlewood series, John E. Littlewood. Her investigations held a lot in common with some tenets of probability theory involving random variables.

Weiss had a mathematician's temperament, in the sense that she could summon great concentration for days at a time when absorbed by some mathematical enigma. Yet she was highly sociable. During her career as a lecturer, she worked individually with students at De Paul, Washington, and Stanford universities as well as her alma mater. She read widely and held an appreciation for the fine arts, and she often participated in campus protests against the Vietnam War.

### Overseas Experiences

Between 1960 and 1961, Weiss and her husband took a sabbatical to Buenos Aires and Paris, yet it was a working vacation for both of them. Guido had been wrestling with Hardy spaces in the framework of classical mathematics. Once both of them attacked the topic they extended its domain onto the complex or Gaussian plane, and also into higher dimensions. Later during this period they invited J.P. Kahane to aid their continued efforts with general lacunary power series. Upon returning from sabbatical Weiss contacted Zygmund and **Alberto P. Calderón** about their discoveries in harmonic analysis, and the three wrote a paper on Calderón–Zygmund singular integral operators in higher dimensions. Weiss' central contribution was in proving the early assumptions of Zygmund and Calderón. She also continued work with her husband on the theory of Hardy spaces in higher dimensions with E.M. Stein. Here, she applied mainly geometric methods of analysis which have yet to be fully exploited by her successors.

During the academic year of 1965–66, the National Science Foundation funded Weiss' senior postdoctoral fellowship at Cambridge University, after which she returned to America. She began work that fall at the University of Illinois, but died on October 8, a few weeks after the semester had begun. In 1967, a symposium on harmonic analysis was held at the Edwardsville campus of Southern Illinois University in Weiss' honor. At this gathering, her mentor Zygmund presented a technical summary of her published work.

## SELECTED WRITINGS BY WEISS

"The Law of the Iterated Logarithm for Lacunary Trigonometric Series." *Transactions of the American Mathematical Society* 91 (1959): 444–469.

(With Guido Weiss). "A Derivation of the Main Results of the Theory of Hp Spaces." *Revista de la Union Matematica Argentina* 22 (1960): 63–71.

(With Alberto P. Calderón and Antoni Zygmund) "On the Existence of Singular Integrals," in *Singular Intervals: Proceedings of Symposia in Pure Mathematics. Volume 10.* Edited by Alberto P. Calderón, 1967.

"A Theorem on Lacunary Trigonometric Series," in *Orthogonal Expansions and Their Continuous Analogues.* Edited by D.T. Haimo, 1968, pp. 227–230.

## FURTHER READING

### Books

Pless, V. and Srinivansan, B. "Mary Catherine Bishop Weiss (1930–1966)." *Historical Encyclopedia of Chicago Women* (in production).

Weiss, Guido. "Mary Catherine Bishop Weiss," in *Women of Mathematics.* Edited by Louise S. Grinstein and Paul J. Campbell. Westport, CT: Greenwood Press, 1987, pp. 236–40.

Zygmund, A. "Mary Weiss: December 11, 1930–October 8, 1966," in *Orthogonal Expansions and Their Continuous Analogues.* Edited by D.T. Haimo. Carbondale, IL: Southern Illinois University Press, 1968, pp. xi–xviii.

### Other

"Mary Catherine Bishop Weiss." *Biographies of Women Mathematicians.*
http://www.scottlan.edu/lriddle/women/chronol.htm (July 1997).

*—Sketch by Jennifer Kramer*

# Caspar Wessel
## 1745–1818
### Danish–born Norwegian number theorist and surveyor

Casper Wessel made a significant contribution to mathematics, but his legacy is to be a footnote in histories of other great mathematicians. **Karl Friedrich Gauss** and **Jean Robert Argand** are most often given credit for expressing complex numbers as geometric shapes, but it was an idea that Wessel, a surveyor and map–maker, was first to develop. Because he was not a professional mathematician and made no great effort to publish his one breakthrough paper, it went largely unknown for a century.

Wessel was born on June 8, 1745, in Jonsrud in Akershus county, Norway (then part of Denmark), to Jonas Wessel, a church vicar, and Maria Schumacher, his wife. He attended the Christiania Cathedral School in Oslo from 1757 to 1763 and then spent a year at the University of Copenhagen. Wessel's career in cartography, or map–making, began in 1764, when he became an assistant to the Danish Survey Commission soon after leaving the University of Copenhagen. He earned a law degree in 1778 and rose to survey superintendent in 1798.

In 1797, Wessel published a little–noticed paper on the geometric representation of complex numbers, which he presented to the Royal Danish Academy of Sciences. Gauss reportedly had been working on the problem as early as 1799, but did not publish the idea until 1831. Argand, a bookkeeper, published the concept in 1806, and that work became very influential, hence many refer to a plane of complex numbers as the Argand plane or an Argand diagram. "And yet [Wessel's] exposition was, in some respects, superior to and more modern in spirit than Argand's," writes Phillip S. Jones in the *Dictionary of Scientific Biography*. Wessel's work lay forgotten for 98 years and was republished in French on the 100th anniversary of the original publication. "It is regrettable that it was not appreciated for nearly a century and hence did not have the influence it merited," writes Jones. Biographical information about Wessel is also lacking, as he had been deceased 79 years before his achievement was recognized.

Graphical representation of complex numbers is a key mathematical component of contemporary computer graphics and was a major breakthrough for mathematicians grappling with understanding higher concepts. "The simple idea of considering the real and imaginary parts of a complex number $a + bi$ as the rectangular coordinates of a point in a plane made mathematicians feel much more at ease with imaginary numbers," Howard Eves wrote in *An Introduction to the History of Mathematics*. "Seeing is believing, and former ideas about the nonexistence or fictitiousness of imaginary numbers were generally abandoned."

Some scholars also credit Wessel with advancing the idea of adding vectors in a three–dimensional space, a basic concept in modern physics. The mapmaker realized that nonparallel vectors could be added together by laying the terminal end of one line at the beginning of a second, and then summing them by drawing a line from the beginning of the first vector to the end of the second.

Although he is now best known for the work on complex numbers, Wessel was actually quite famous in his day as a geographer. His work as a surveyor was so highly regarded that he was awarded a medal by the Royal Danish Academy of Sciences and was knighted in Danebrog in 1815. He had retired in 1805, but continued working until 1812, when his rheumatism became too severe. Wessel died on March 25, 1818, in Copenhagen, Denmark.

## FURTHER READING

Crowe, Michael J. *A History of Vector Analysis.* South Bend: University of Notre Dame Press, 1967.

Eves, Howard. *An Introduction to the History of Mathematics with Cultural Connections.* Sixth edtion. New York: Saunders College Publishing, 1990.

Jones, Phillip S. "Caspar Wessel," in *Dictionary of Scientific Biography*. Volume XIV. Edited by Charles Coulston Gillispie. New York: Charles Scribner's Sons, 1976, pp. 279–81.

—*Sketch by Karl Leif Bates*

*Hermann Weyl*

# Hermann Weyl
## 1885-1955
### German-born American mathematical physicist

Hermann Weyl was one of the most wide-ranging mathematicians of his generation, following in the footsteps of his teacher **David Hilbert.** Weyl's interests in mathematics ran the gamut from foundations to physics, two areas in which he made profound contributions. He combined great technical virtuosity with imagination, and devoted attention to the explanation of mathematics to the general public. He managed to take a segment of mathematics developed in an abstract setting and apply it to certain branches of physics, such as relativity theory—a theory that holds that the velocity of light is the same for all observers, no matter how they are moving, that the laws of physics are the same in all inertial frames, and that all such frames are equivalent—and quantum mechanics—a theory that allows mathematical interpretation of elementary particles through wave properties. His distinctive ability was integrating nature and theory.

Claus Hugo Hermann Weyl was born on November 9, 1885, at Elmshorn, near Hamburg, Germany. The financial standing of his parents (his father, Ludwig, was a clerk in a bank and his mother, Anna Dieck, came from a wealthy family) enabled him to receive a quality education. From 1895 to 1904 he attended the Gymnasium at Altona, where his performance attracted the attention of his headmaster, a relative of an eminent mathematician of that time, David Hilbert. Weyl soon found himself at the University of Göttingen where Hilbert was an instructor. He remained there for the rest of his student days, with the exception of a semester at the University of Munich. He received his degree under Hilbert in 1908 and advanced to the ranks of privatdocent (an unpaid but licensed instructor) in 1910.

Weyl married Helene Joseph (known as Hella to the family) in 1913 and in the same year took a position as professor at the National Technical University (ETH) in Zurich, Switzerland. He declined the offer to be Felix Klein's successor at Göttingen, despite the university's central role in the mathematical world. It has been suggested that he wanted to free himself, somewhat, of the influence of Hilbert, especially in light of the fact that he had accepted an invitation to take a chair at Göttingen when Hilbert retired. In any case, he brought a great deal of mathematical distinction to the ETH in Zurich, where his sons Fritz Joachim and Michael grew up.

It is not surprising that Weyl's early work dealt with topics that which Hilbert held an interest. His *Habilitationsschrift* was devoted to boundary conditions of second-order linear differential equations. (The way the German educational system worked, it was necessary to do a substantial piece of original research beyond the doctoral dissertation in order to qualify to teach in the university. This "entitling document" was frequently the launching point of the mathematical career of its author.) In other words, he was looking into the way functions behaved on a given region when the behavior at the boundary was specified. His results were sufficient for the purpose of enabling him to earn a living, but he rapidly moved on to areas where his contributions were more innovative and have had a more lasting effect.

One of the principal areas of Weyl's research was the topic of Hilbert spaces. The problem was to understand something about the functions that operated on the points of Hilbert space in a way useful for analyzing the result of applying the functions. In particular, Weyl wanted to know where the functions behaved more simply than on the space as a whole, since the behavior of the function on the rest of the space could be represented in terms of its behavior on the simpler regions. Different kinds of functions behaved in radically different ways on a Hilbert space, so Weyl had to restrict his attention to a subclass of functions small enough to be tractable (for example, the functions could not "blow up") but large enough to be useful. His choice of self-adjoint, compact operators was justified by their subsequent importance in the field of functional analysis.

Among the areas he brought together were geometry and analysis from the 19th century and topology, which was largely a creation of the 20th century. Topology sought to understand the behavior of space in ways that require a less-detailed understanding of how the elements of a structure fit together than geometry demanded. One of the basic ideas of topology is that of a "manifold," first introduced by **G. F. B. Riemann** in his *Habilitationsschrift* as a student of **Karl Gauss**. Riemann had little material with which to work, while Weyl was able to take advantage of the work of Hilbert and the Dutch mathematician **Luitzen Egbertus Jan Brouwer**. This effort culminated in his 1913 book on Riemann surfaces, an excellent exposition on how complex analysis and topology could be used together to analyze the behavior of complex functions.

Weyl served briefly in the German army at the outbreak of World War I, but before this military interlude, he did research that led to one of his most important papers. He looked at the way irrational numbers (those that cannot be expressed as a ratio of two whole numbers) were distributed. What he noticed was that the *fractional* parts of an irrational number and its integral multiples seemed to be evenly distributed in the interval between O and 1. He succeeded in proving this result, and it is known as the Kronecker-Weyl theorem, owing half of its name to an influential number theorist who had had an effect on Hilbert. Although the result may seem rather narrow, Weyl was able to generalize it to sequences of much broader application.

During his time in Zurich, Weyl spent a year in collaboration with **Albert Einstein** and picked up a dose of enthusiasm for the relativity theory. Among the other results of this collaboration was Weyl's popular account of relativity theory, *Space, Time, Matter* (the original German edition appeared in 1918). In those early days of general relativity, which describes gravity in terms of how mass distorts space-time, the correct mathematical formulation of some of Einstein's ideas was not clear. He had been able to use ideas developed by differential geometers at the end of the nineteenth century that involved the notion of a tensor. A tensor can be thought of as a function on a number of vectors that takes a number as its value. Weyl used the tensor calculus that had been developed by the geometers to come up with neater formulations of general relativity than the original version proposed by Einstein. In later years, he took the evolution of tensors one step further while maintaining a strict mathematical level of rigor.

## Constructs New Foundations

One of the most visible areas in which Weyl worked after World War I was in the foundations of mathematics. He had used some of the topological results of the Dutch mathematician Brouwer in working on Riemann surfaces. In addition, he had looked at some of Brouwer's ideas about the philosophy of mathematics and was convinced that they had to be taken seriously. Although it was not always easy to understand what Brouwer was trying to say, it was clear that he was criticizing "classical" mathematics, that is, the mathematics that had prevailed at least since **Euclid**. One of the standard methods of proof in classical mathematics was the reductio ad absurdum, or proof by contradiction. If one wished to prove that P was true, one could assume that not-P was true and see if that led to a contradiction. If it did, then not-P must not be true, and P must be true instead. This method of proof depended on the principle that either P was true or not-P was true, which had seemed convincing to generations of mathematicians.

Brouwer, however, found this style of argument unacceptable. For reasons having to do with his understanding of mathematics as the creation of the human mind, he wanted to introduce a third category besides truth and falsity, a category we could call "unproven." In other words, there was more to truth than just the negation of falsity—to claim P or not-P, something had to be proven. This argument of Brouwer was especially directed against so-called nonconstructive existence proofs. These were proofs in which something was shown to exist, not by being constructed, but by arguing that if it didn't exist, a contradiction arose. For ordinary, finite mathematics it was usually easy to come up with a constructive proof, but for claims about infinite sets nonconstructive arguments were popular. If Brouwer's objections were to be sustained, a good part of mathematics even at the level of elementary calculus would have to be rewritten and some perhaps have to be abandoned.

This attitude aroused the ire of David Hilbert. He valued the progress that had been made in mathematics too highly to sacrifice it lightly for philosophical reasons. Although Hilbert had earlier expressed admiration for Brouwer's work, he felt obliged to negate Brouwer's philosophy of mathematics known as intuitionism. What especially disturbed Hilbert was Weyl's support of Brouwer's concepts, since Hilbert knew the mathematical strength of his former student. In the 1920s, while the argument was being considered, Hilbert was discouraged about the future of mathematics in the hands of the intuitionists.

Although Weyl never entirely abandoned his allegiance to Brouwer, he also recognized that Hilbert's program in the philosophy of mathematics was bound to appeal to the practicing mathematician, more than Brouwer's speculations. In 1927, responding to one of Hilbert's lectures concerned with the foundations of mathematics, Weyl commented on the extent to which Hilbert had been led to a reinterpretation of mathematics by the need to fight off Brouwer's

criticisms. The tone of Weyl's remarks suggested that he would not have been unhappy if Hilbert's point of view was to prevail. This flexibility with regard to the foundations of mathematics indicates that Weyl was sensitive to the changes in attitude that others ignored, but also may explain why Weyl never founded a philosophical school: he was too ready to recognize the justice of others' points of view.

In general, Weyl took questions of literature and style seriously, which goes far to explain the success of his expository writings. His son recalls that when Weyl would read poetry aloud to the family, the intensity and volume of his voice would make the walls shake. He kept in touch with modern literature as well as the classics of his childhood. While he continued to enjoy the poetry of Friedrich Hölderlin and Johann Wolfgang von Goethe, he also read Friedrich Nietzsche's *Also Sprach Zarathustra* and Thomas Mann's *The Magic Mountain*. He could cite quotations from German poetry whenever he needed them. For those of a psychologizing bent, it has even been argued that his fondness for poetry may be in line with his preference for intuitionism as a philosophy of mathematics. The kind of poetry he preferred spoke to the heart, and he used quotations to add a human dimension to otherwise cold mathematical writing.

### Leaves Germany for the United States

After he accepted a chair at Göttingen in 1930, Weyl did not have long to enjoy his return to familiar surroundings. In 1933 he decided that he could no longer remain in Nazi Germany, and he took up a permanent position at the Institute for Advanced Study, newly founded in Princeton, New Jersey. Although Weyl himself was of irreproachably Aryan ancestry, his wife was partly Jewish, and that would have been enough to attract the attention of the authorities. There may have been the additional attraction of the wealth of intellectual company available at the institute, between its visitors from all over the globe and permanent residents such as Einstein. Weyl took his official duties as a faculty member seriously, although his reputation could be terrifying to younger mathematicians unaware of the poet within.

Weyl's work continued to bridge the gap between physics and mathematics. As long ago as 1929, he developed a mathematical theory for the subatomic particle the neutrino. The theory was internally consistent but failed to preserve left-right symmetry and so was abandoned. Subsequent experimentation revealed that symmetry need not be conserved, with the result that Weyl's theory reentered the mathematical physics mainstream all the more forcefully. Another area for the interaction of mathematics and physics was the study of spinors, a kind of tensor that has proven to be of immense use in quantum mechanics. Although spinors had been known before Weyl, he was the first to give a full treatment of them. Perhaps it was this work that led Roger Penrose, one of the most insightful mathematical physicists of the second half of the twentieth century, to label Weyl "the greatest mathematician of this century."

One of the challenges of physical theories is to find quantities that do not change (are conserved) during other changes. **Felix Klein** in the 19th century had stressed the importance of group theory, then a new branch of mathematics, in describing what changed and what remained the same during processes. Weyl adapted Klein's ideas to the physics of the twentieth century by characterizing invariant quantities for relativity theory and for quantum mechanics. In a 1923 paper Weyl had come up with a suitable definition for congruence in relativistic space-time. Even more influential was his 1928 book on group theory and quantum mechanics, which imposed a model that would have been welcome to Klein due to the previously rather disjointed results assembled by quantum physicists. Weyl was an artist in the use of group theory and could accomplish wonders with modest mathematical structure.

After the end of World War II, Weyl divided his time between Zurich and Princeton. His first wife died in 1948, and two years later he married Ellen Bär. He took a serious view of the history of mathematics and arranged with Princeton to give a course on the subject. One of his magisterial works was a survey of the previous half-century of mathematics that appeared in the *American Mathematical Monthly* in 1951. Although he never became as fluent in English as he had in German, he retained a strong commitment to the public's right to be informed about scientific developments.

John Archibald Wheeler, an American physicist, called attention to Weyl's anticipation of the anthropic principle in cosmology. In a 1919 paper Weyl had speculated on the coincidence of the agreement of two enormous numbers of very different origin. In the 1930s this speculation had been given with the title "Weyl's hypothesis," although later authors referred more to its presence elsewhere. What cannot be denied is that the recent discussion of the anthropic principle concerning features necessary for human existence in the universe has, as Wheeler noted, taken up Weyl's point once again.

Weyl was unaware of the rules governing the length of time that a naturalized citizen could spend abroad at one time without losing citizenship. By inadvertence he exceeded the time limit and lost his American citizenship in the mid 1950s. To remedy the situation required an act of Congress, but there was no lack of help in securing it. In the meantime, Weyl celebrated his seventieth birthday in Zurich amid a flurry of congratulations. On December 8, 1955, as he was mailing some letters of thanks to well-wishers, he died of a heart attack. With his death passed one of the links with the great era of Göttingen as a mathematical center and one of the founders of contemporary mathematical physics, but even more, a mathematician who could convey the poetry in his discipline.

## SELECTED WRITINGS BY WEYL

### Books

*Die Idee der Riemannschen Fläche*, 1913
*Raum, Zeit, Materie*, 1918
*Philosophie der Mathematik und Naturwissenschaft*, 1926
*Grüppentheorie und Quantenmechanik*, 1928
*Symmetry*, 1952
*Selecta Hermann Weyl*, 1956
*Gesammelte Abhandlungen*, four volumes, 1968

## FURTHER READING

Chandrasekharan, K., editor. *Hermann Weyl: Centenary Lectures.* New York: Springer-Verlag, 1986.
Deppert, Wolfgang, and others, editors. *Exact Sciences and Their Philosophical Foundations.* New York: Peter Lang, 1988.
Dieudonné, J. "Hermann Weyl," in *Dictionary of Scientific Biography.* Volume XIV. Edited by Charles Coulston Gillispie. New York: Charles Scribner's Sons, pp. 281–285.

*—Sketch by Thomas Drucker*

# Anna Johnson Pell Wheeler
## 1883–1966
### American algebraist

A distinguished mathematics researcher and educator, Anna Johnson Pell Wheeler is best remembered for her interest and research in biorthogonal systems of functions and integral equations. Much of her work was in the area of linear algebra of infinitely many variables—an area which she studied her entire career. Wheeler struggled to gain equality with men in the field of mathematics. In 1910, she was only the second woman at the University of Chicago to receive a doctorate in mathematics. Finding a full-time teaching position was difficult, even though she was often more qualified than the male applicants. Her break came when she substituted for her incapacitated first husband, Alexander Pell, at the Armour Institute in Chicago. There, although she did not obtain a permanent position, she convinced her superiors of her competency. She was then hired as an instructor in mathematics at Mount Holyoke College in Hadley, Massachusetts, in 1911, leaving there in 1918 to take a position as associate professor at Bryn Mawr College in Pennsylvania. She remained at Bryn Mawr until her retirement in 1948, becoming head of the mathematics department in 1924. A champion of women in the field of mathematics, she urged her students to persevere toward terminal degrees despite the gender prejudices exhibited by authorities at colleges and universities at that time. It is significant, in her tenure at Bryn Mawr College, that seven of her graduate students received doctorates in mathematics.

During her career, Wheeler was active in many professional associations. She served on the council and the board of the American Mathematical Society, and was also a member of the Mathematical Association of America and the American Association for the Advancement of Science. (In 1927, Wheeler was invited by the American Mathematical Society to deliver their annual Colloquium Lectures—the only woman so honored until 1970.) In 1940, she received recognition from the Women's Centennial Congress as one of the 100 women honored who had succeeded in non-traditional careers. Continuing her support of women mathematicians, Wheeler helped **Emmy Noether**, the eminent German algebraist, to relocate to Bryn Mawr when she sought political asylum from Nazi Germany in 1933.

Wheeler, of Swedish heritage, was born in Hawarden, Iowa, on May 5, 1883. Her parents were Amelia (Frieberg) and Andrew Johnson. Her father was an undertaker and furniture dealer in the small town of Akron, Iowa. Anna attended the local high school there and received her undergraduate degree from the University of South Dakota in 1903. There, her exceptional ability in mathematics was observed by her professor, Alexander Pell (later to become her first husband). She furthered her education at the University of Iowa and at Radcliffe College, earning master's degrees from both institutions.

### Fellowship Allows Study in Europe

In 1906, Wheeler received an Alice Freeman Palmer Fellowship, allowing her to continue her studies at Göttingen University in Germany. There, she was guided in her work in integral equations by such eminent mathematicians as **Hermann Minkowski**, Felix Klein, and **David Hilbert**. Her thesis was completed under Hilbert's instruction, but for some reason—speculated to have been a dispute with Hilbert—she did not receive a degree from Göttingen.

Although Wheeler accepted the fellowship with the understanding she could not marry during its term, Pell joined her in Germany at the end of the year and they were married there in 1907. Her new husband had an interesting past. He was actually a Russian revolutionist named Sergei Degaev, who had fled his country after being implicated in the murder of an officer of the Russian secret police. After emigrating to the United States, Dagaev changed his name to Alexander Pell, and began his new life as a mathematics professor.

The Pells left Germany and returned to the University of South Dakota. Shortly after their return, Pell accepted a position to teach at the Armour Institute of Technology in Chicago. Wheeler completed the work for her Ph. D. at the University of Chicago, and when Pell suffered a stroke in 1911, she assumed his teaching duties. She had hoped for a permanent position with a Midwestern university, but was unsuccessful. Other than taking over her husband's classes, the closest Wheeler came to being employed while in Chicago was when she taught a course at the University the fall semester of 1910. When she did find permanent work, it was out east at Mount Holyoke, and later, at Bryn Mawr. Pell, who was a semi-invalid after his stroke, died in 1921. In 1925, she married Arthur Leslie Wheeler, a classics scholar who had just become professor of Latin at Princeton University in New Jersey. Wheeler continued to teach at Bryn Mawr, even though they lived in Princeton where her husband was teaching. They also enjoyed a summer home in the Adirondack Mountains, to which she often invited her students. When Arthur died suddenly in 1932, she moved back to Bryn Mawr, where she lived and taught for the rest of her life.

### Strengthens School's Reputation in Mathematics

Wheeler's work at Bryn Mawr took her beyond the classroom. She was well aware of the need to strengthen the reputation of the school's mathematics department, and set about doing so. She advised reducing teaching loads so that more research could be carried out by the faculty and encouraged professional collaboration and theoretical exchanges with other schools in the Philadelphia area. During this time of increasing administrative responsibilities, Wheeler remained active in publishing the results of her research into integral equations and functional analysis. Her Colloquium Lectures, however, were never published.

Although suffering from arthritis, Wheeler continued to participate in mathematics association meetings after her retirement. She died at age 82 at

Bryn Mawr on March 26, 1996, after suffering a stroke.

## SELECTED WRITINGS BY WHEELER

*Biorthogonal Systems of Functions,* 1911

## FURTHER READING

### Books

Bailey, Martha J. *American Women in Science: A Biographical Dictionary.* Santa Barbara, CA: ABC–CLIO, Inc., 1994, pp. 414–415.

Ogilvie, Marilyn Bailey. *Women in Science: Antiquity through the Nineteenth Century.* Cambridge, MA: MIT Press, 1986, pp. 173–174.

Sicherman, Barbara and Carol Hurd Green, editors. *Notable American Women: The Modern Period, A Biographical Dictionary.* Cambridge, MA: The Belknap Press, 1980, pp. 725–726.

### Periodicals

"Dr. Anna Pell Wheeler" (obituary). *The New York Times* (April 1, 1966): 35: 1.

### Other

"Anna Johnson Pell Wheeler." *Biographies of Women Mathematicians.* June 1977. http://www.scottlan.edu/lriddle/women/chronol.htm (July 22, 1997).

—*Sketch by Jane Stewart Cook*

*Alfred North Whitehead*

culminated in the publication of the three-volume *Principia Mathematica,* widely regarded as one of the most important books in mathematics ever written. In 1924 Whitehead became professor of philosophy at Harvard University, where he devoted his time to the development of a comprehensive and complex system of philosophy.

Whitehead was born on February 15, 1861, at Ramsgate in the Isle of Thanet, Kent, England. Both his grandfather, Thomas Whitehead, and his father, Alfred Whitehead, had been headmasters of a private school in Ramsgate. Alfred Whitehead had later joined the clergy and become vicar of St. Peter's Parish, about two miles from Ramsgate. In his autobiography, Whitehead said of his father that he "was not intellectual, but he possessed personality." Whitehead's mother was the former Maria Sarah Buckmaster, daughter of a successful London businessman. As a young man, Whitehead often traveled to London to visit his maternal grandmother.

### Educated at Sherborne and Trinity College

For the first fourteen years of his life, Whitehead was educated at home primarily by his father. Then, in 1875, he was sent to the public school at Sherborne in Dorsetshire. Whitehead described his education as traditional, with a strong emphasis on Latin and Greek. But, he continued, "we were not overworked," so that he had plenty of time for sports such as cricket and football, private reading, and a study of history.

# Alfred North Whitehead
## 1861-1947
### English American algebraist and logician

Albert North Whitehead began his career as a mathematician, but eventually became at least as famous as a philosopher. His first three books, *A Treatise on Universal Algebra, The Axioms of Projective Geometry,* and *The Axioms of Descriptive Geometry,* all dealt with traditional mathematical topics. In 1900, Whitehead first heard about the new system for expressing logical concepts in discrete symbols developed by the Italian mathematician **Giuseppe Peano** . Along with **Bertrand Russell** , his colleague and former student, Whitehead saw in Peano's symbolism a method for developing a rigorous, nonnumerical approach to logic. The work of these two men

At Sherborne he also had his introduction to science and mathematics, at which he excelled. In fact, he was apparently excused from some Latin requirements in order to have more time for his mathematical studies.

In 1880 Whitehead received a scholarship to continue his studies at Trinity College, Cambridge. While at Trinity, all of Whitehead's formal education was in the field of mathematics, the British system not having yet accepted the concept of a broad liberal education for all students. Still, he later wrote, his mathematics courses "were only one side of the education" he experienced at Trinity. Another side consisted of regular evening meetings with other undergraduates at which virtually all subjects were discussed. Whitehead later referred to these meetings as "a daily Platonic dialogue." Through these dialogues, Whitehead rapidly expanded his knowledge of history, literature, philosophy, and politics.

Whitehead was awarded his bachelor of arts degree in 1884 for a thesis on James Clerk Maxwell's theory of electromagnetism. A few months later he was elected a fellow of Trinity College and appointed assistant lecturer in mathematics. Whitehead was awarded his M.A. in 1887, and in 1903, was named senior lecturer. Two years later he was granted his doctor of science degree.

### Meets and Marries Evelyn Wade

While still at Trinity, Whitehead met Evelyn Willoughby Wade, described in the *Dictionary of American Biography* as the "daughter of impoverished Irish landed gentry" who was "witty, with passionate likes and dislikes, a great sense of drama, and . . . a keen aesthetic sense." Whitehead himself credits his wife with teaching him "that beauty, moral and aesthetic, is the aim of existence; and that kindness, and love, and artistic satisfaction are among its modes of attainment." The two were married on December 16, 1890. They later had four children, Thomas North in 1891, Jesse Marie in 1893, Eric Alfred in 1898, and an unnamed boy who died at birth in 1892. Eric later became a pilot with the Royal Flying Corps and was killed in March, 1918, during World War I.

Whitehead's first book, *A Treatise on Universal Algebra,* was begun in January, 1891, and published seven years later. The book was an attempt to expand on the works of three predecessors, **Hermann Grassmann**, **William Rowan Hamilton**, and **George Boole**, the founder of symbolic logic. Whitehead later wrote that Grassmann, in particular, had been "an original genius, never sufficiently recognized." All of Whitehead's future work on mathematical logic, he said, was derived from the contributions of these three men. Whitehead's book was a tour de force that earned him election to the Royal Society five years after its publication. In the book, Whitehead argues

that algebraic concepts have an existence of their own, independent of any connection with real objects.

### Begins a Long Working Relationship with Russell

The year of Whitehead's marriage, 1890, also marked the beginning of another long and fruitful relationship, with Russell. The two had met when Russell was still a freshman at Cambridge; Whitehead was one of his teachers. In later years their student-and-teacher relationship blossomed into a full-blown working relationship as professional colleagues. They were eventually to collaborate on a number of important mathematical works.

An important event in their association occurred in July, 1900, when they attended together the First International Congress of Philosophy in Paris. It was there that Whitehead and Russell were introduced to the techniques of symbolic logic developed by the Italian mathematician Giuseppe Peano. They immediately saw that Peano's symbolism could be used to clarify fundamental concepts of mathematics. When Russell returned to England, he began to incorporate Peano's approach into the book on which he was then working, *Principles of Mathematics.*

Before long, however, it occurred to Russell that he and his former teacher were both working on very similar topics, he on his *Principles of Mathematics* and Whitehead on a second volume of his *Universal Algebra.* The two agreed to start working together, with the result that the second volume of neither work ever appeared. Instead, they developed the three-volume masterpiece, *Principia Mathematica.* The fundamental concept behind the book was that the basic principles of mathematics can be derived in a strict way through the precise rules of symbolic logic. The *Principia Mathematica* has since been described by one of Whitehead's biographers, Victor Lowe, in the *Dictionary of American Biography* as "one of the great intellectual monuments of all time."

### Moves from Cambridge to London

In 1910, the year in which the first volume of *Principia Mathematica* appeared, Whitehead ended a 25-year teaching career at Trinity College and moved to London. He remained without an academic appointment for one year, during which time he wrote his *Introduction to Mathematics,* which James R. Newman has called "a classic of popularization" of mathematics. Whitehead then accepted an appointment as lecturer in applied mathematics and mechanics at University College, London, and two years later, was made reader in geometry there. In 1914 he was named professor of applied mathematics at the Imperial College of Science and Technology in Kensington.

The London period was for Whitehead a particularly busy time in the political and administrative arenas. He served on a number of faculty and governmental committees and was outspoken in his concern about educational reform. Perhaps his best-known remarks on this subject came in a 1916 address to the Mathematical Association, "The Aims of Education: A Plan for Reform." In this address, Whitehead pointed out that the narrow view of education in which the classics are taught to a select number of upper-class men had become outmoded in a world of a "seething mass of artisans seeking intellectual enlightenment, of young people from every social grade craving for adequate knowledge."

During the latter years of his London period, Whitehead's interests shifted from mathematics to the philosophy of science. The fourth volume of *Principia Mathematica,* dealing with the foundations of geometry, was never completed. Instead, Whitehead began to write on the philosophical foundations of science in books such as *An Enquiry Concerning the Principles of Natural Knowledge* in 1919, *The Concept of Nature* in 1920, and *The Principle of Relativity, with Applications to Physical Science* in 1922.

The main theme of these books was that there exists a reality in the physical world that is distinct from the descriptions that scientists have invented for that reality. Scientific explanations certainly have their functions, according to Whitehead, but they should not be construed as being the reality of nature itself.

### Invited to Join Harvard's Philosophy Department

As early as 1920, Harvard University had been interested in offering Whitehead a position in its philosophy department. For financial reasons, a firm offer was not made until 1924, when Whitehead was sixty-three years old. He accepted the offer partly because he was nearing mandatory retirement age at Imperial College and partly because he looked forward to the opportunity of expanding his intellectual horizons. On September 1, 1924, Whitehead's appointment at Harvard became official.

Until the Harvard post became available, Whitehead had remained rather strictly within the areas of mathematics and natural science. After 1924, however, he extended the range of his writings to include far broader topics. His first work published in the United States, *Science and the Modern World,* discussed the significance of the scientific enterprise for other aspects of human culture. The book was an instant professional and commercial success and earned Whitehead an immediate reputation as a profound thinker and a writer of great clarity and persuasiveness.

His next book, *Religion in the Making,* was the first of a number that carried Whitehead far beyond the fields of mathematics and science. Its publication in 1926 was followed by *Symbolism, Its Meaning and Effect* in 1927, *The Aims of Education and Other Essays* in 1929, *The Function of Reason,* also in 1929, and a half dozen more books over the next two decades. In recognition of his work, Whitehead was elected a fellow of the British Academy in 1931, and he was awarded the Order of Merit, the highest honor that Great Britain can bestow on a man of letters, in 1945. Whitehead retired from active teaching at Harvard in 1937, at which time he was named emeritus professor of philosophy. He died at his home in Cambridge, Massachusetts, on December 30, 1947.

## SELECTED WRITINGS BY WHITEHEAD

*A Treatise on Universal Algebra,* 1898
*The Axioms of Projective Geometry,* 1906
*The Axioms of Descriptive Geometry,* 1907
(With Bertrand Arthur William Russell) *Principia Mathematica,* three volumes, 1910–13
*An Introduction to Mathematics,* 1911
*An Enquiry Concerning the Principles of Natural Knowledge,* 1919
*The Concept of Nature,* 1920
*The Principle of Relativity, with Applications to Physical Science,* 1922
*Science and the Modern World,* 1925
*Religion in the Making,* 1926
*Symbolism, Its Meaning and Effect,* 1927
*The Aims of Education and Other Essays,* 1929
*The Function of Reason,* 1929
*Process and Reality: An Essay in Cosmology,* 1929
*Adventures of Ideas,* 1933
*Nature and Life,* 1934
*Modes of Thought,* 1938
*Essays in Science and Philosophy,* 1947
*Dialogues of Alfred North Whitehead,* edited by L. Price, 1954
*The Interpretation of Science: Selected Essays,* edited by A. H. Johnson, 1961

## FURTHER READING

Johnson, R. C. "Alfred North Whitehead," in *Dictionary of Literary Biography.* Volume C. Edited by Robert Baum. Charles Scribner's Sons, 1990, pp.306–315.
Newman, James R. *The World of Mathematics* Volume 1. New York: Simon & Schuster, pp. 395–401.
Price, Lucien. *Dialogues of Alfred North Whitehead.* New York: Little, Brown, 1954, pp. 3–20.
Schilpp, Paul Arthur. *The Philosophy of Alfred North Whitehead.* Tudor, 1941.

*—Sketch by David E. Newton*

*Norbert Wiener*

# Norbert Wiener
## 1894-1964
**American logician**

Norbert Wiener was one of the most original mathematicians of his time. The field concerning the study of automatic control systems, called cybernetics, owes a great deal not only to his researches, but to his continuing efforts at publicity. He wrote for a variety of popular journals as well as for technical publications and was not reluctant to express political views even when they might be unpopular. Perhaps the most distinctive feature of Wiener's life as a student and a mathematician is how well documented it is, thanks to two volumes of autobiography published during his lifetime. They reveal some of the complexity of a man whose aspirations went well beyond the domain of mathematics.

Wiener was born in Columbia, Missouri, on November 26, 1894. His father, Leo Wiener, had been born in Bialystok, Poland (then Russia), and was an accomplished linguist. He arrived in New Orleans in 1880 with very little money but a great deal of determination, some of it visible in his relations with his son. He met his wife, Bertha Kahn, at a meeting of a Browning Club. As a result, when his son was born, he was given the name Norbert, from one of Browning's verse dramas. In light of the absence of Judaism from the Wiener home (Norbert was fifteen before he

learned that he was Jewish), it is surprising that one of Leo Wiener's best-known works was a history of Yiddish literature.

As the title of the first volume of his autobiography *Ex-Prodigy* suggests, Wiener was a child prodigy. Whatever his natural talents, this was partly due to the efforts of his father. Leo Wiener was proud of his educational theories and pointed to the academic success of his son as evidence. Norbert was less enthusiastic and in his memoirs describes his recollections of his father's harsh disciplinary methods. He entered high school at the age of nine and graduated two years later. In 1906 he entered Tufts University, as the family had moved to the Boston area, and he graduated four years later.

Up until that point Wiener's education had clearly outrun that of most of his contemporaries, but he was now faced with the challenge of deciding what to do with his education. He enrolled at Harvard to study zoology, but the subject did not suit him. He tried studying philosophy at Cornell, but that was equally unavailing. Finally, Wiener came back to Harvard to work on philosophy and mathematics. The subject of his dissertation was a comparison of the system of logic developed by **Bertrand Russell** and **Alfred North Whitehead** in their *Principia Mathematica* with the earlier algebraic system created by Ernst Schröder. The relatively recent advances in mathematical research in the United States had partly occurred in the area of algebraic logic, so the topic was a reasonable one for a student hoping to bridge the still-existent gap between the European and American mathematical communities.

Although Wiener earned a Harvard travelling fellowship to enable him to study in Europe after taking his degree, his father still supervised his career by writing to Bertrand Russell on Norbert's behalf. Wiener was in England from June 1913 to April 1914 and attended two courses given by Russell, including a reading course on *Principia Mathematica*. Perhaps more influential in the long run for Wiener's mathematical development was a course he took from the British analyst **Godfrey H. Hardy**, whose lectures he greatly admired. In the same way, Wiener studied with some of the most eminent names in Göttingen, Germany, then the center of the international mathematical community.

Wiener returned to the United States in 1915, still unsure, despite his foreign travels, of the mathematical direction he wanted to pursue. He wrote articles for the *Encyclopedia Americana* and took a variety of teaching jobs until the entry of the United States into World War I. Wiener was a fervent patriot, and his enthusiasm led him to join the group of scientists and engineers at the Aberdeen Proving Ground in Maryland, where he encountered Oswald Veblen, already one of the leading mathematicians in the country. Although Wiener did not pursue Veblen's lines of research, Veblen's success in producing results useful to the military impressed Wiener more than mere academic success.

### Takes the Mathematical Turn

After the war two events decisively shaped Wiener's mathematical future. He obtained a position as instructor at the Massachusetts Institute of Technology (MIT) in mathematics, where he was to remain until his retirement. At that time mathematics was not particularly strong at MIT, but his position there assured him of continued contact with engineers and physicists. As a result, he displayed an ongoing concern for the applications of mathematics to problems that could be stated in physical terms. The question of which tools he would bring to bear on those problems was answered by the death of his sister's fiancé. That promising young mathematician left his collection of books to Wiener, who began to read avidly the standard texts in a way that he had not in his earlier studies.

The first problem Wiener addressed had to do with Brownian motion, the apparently random motion of particles in substances at rest. The phenomenon had earlier excited **Albert Einstein**'s interest, and he had dealt with it in one of his 1905 papers. Wiener took the existence of Brownian motion as a sign of randomness at the heart of nature. By idealizing the physical phenomenon, Wiener was able to produce a mathematical theory of Brownian motion that had wide influence among students of probability. It is possible to see in his work on Brownian motion, steps in the direction of the study of fractals (shapes whose detail repeats itself on any scale), although Wiener did not go far along that path.

The next subject Wiener addressed was the Dirichlet problem, which had been reintroduced into the mathematical mainstream by German **David Hilbert**. Much of the earliest work on the Dirichlet problem had been discredited as not being sufficiently rigorous for the standards of the late 19th century. Wiener's work on the Dirichlet problem produced interesting results, some of which he delayed publishing for the sake of a couple of students finishing their theses at Harvard. Wiener felt subsequently that his forbearance was not recognized adequately. In particular, although Wiener progressed through the academic ranks at MIT from assistant professor in 1924 to associate professor in 1929 to full professor in 1932, he believed that more support from Harvard would have enabled him to advance more quickly.

Wiener had a high opinion of his own abilities, something of a change from colleagues whose public expressions of modesty were at odds with a deep-seated conviction of their own merits. Whatever his talents as a mathematician, Wiener's expository standards were at odds with those of most mathematicians of his time. While he was always exuberant, this was often at the cost of accuracy of detail. One of his main theorems depended on a series of lemmas, or auxiliary propositions, one of which was proven by assuming the truth of the main theorem. Students trying to learn from Wiener's papers and finding their efforts unrewarding discovered that this reaction was almost universal. As Hans Freudenthal remarked in the *Dictionary of Scientific Biography,* "After proving at length a fact that would be too easy if set as an exercise for an intelligent sophomore, he would assume without proof a profound theorem that was seemingly unrelated to the preceding text, then continue with a proof containing puzzling but irrelevant terms, next interrupt it with a totally unrelated historical exposition, meanwhile quote something from the 'last chapter' of the book that had actually been in the first, and so on."

In 1926 Wiener was married to Margaret Engemann, an assistant professor of modern languages at Juniata College. They had two daughters, Barbara (born 1928) and Peggy (born 1929). Wiener enjoyed his family's company and found there a relaxation from a mathematical community that did not always share his opinion of the merits of his work.

During the decade after his marriage, Wiener worked in a number of fields and wrote some of the papers with which he is most associated. In the field of harmonic analysis, he did a great deal with the decomposition of functions into series. Just as a polynomial is made up of terms like $x$, $x^2$, $x^3$, and so forth, so functions in general could be broken up in various ways, depending on the questions to be answered. Somewhat surprisingly, Wiener also undertook putting the operational calculus, earlier developed by Oliver Heaviside, on a rigorous basis. There is even a hint in Wiener's work of the notion of a distribution, a kind of generalized function. It is not surprising that Wiener might start to move away from the kind of functions that had been most studied in mathematics toward those that could be useful in physics and engineering.

In 1926 Wiener returned to Europe, this time on a Guggenheim fellowship. He spent little time at Göttingen, due to disagreements with **Richard Courant**, perhaps the most active student of David Hilbert in mathematical organization. Courant's disparaging comments about Wiener cannot have helped the latter's standing in the mathematical community, but Wiener's brief visit introduced him to Tauberian theory, a fashionable area of analysis. Wiener came up with an imaginative new approach to Tauberian theorems and, perhaps more fortunately, with a coauthor for his longest paper on the subject. The quality of the exposition in the paper, combined with the originality of the results, make it Wiener's best exercise in communicating technical mathematics, although he did not pursue the subject as energetically as he did some of his other works.

In 1931 and 1932 Wiener gave lectures on analysis in Cambridge as a deputy for G. H. Hardy. While there, he made the acquaintance of a young British mathematician, R. E. A. C. Paley, with whom a collaboration soon flourished. He brought Paley to MIT the next academic year and their work progressed rapidly. Paley's death at the age of 26 in a skiing accident early in 1933 was a blow to Wiener, who received the Bôcher prize of the American Mathematical Society the same year and was named a fellow of the National Academy of Sciences the next. Among the other areas in which Wiener worked at MIT or Harvard were quantum mechanics, differential geometry, and statistical physics. His investigations in the last of these were wide ranging, but amounted more to the creation of a research program than a body of results.

### Creates Cybernetics

The arrival of World War II occupied Wiener's attention in a number of ways. He was active on the Emergency Committee in Aid of Displaced German Scholars, which began operations well before the outbreak of fighting. He made proposals concerning the development of computers, although these were largely ignored. One of the problems to which he devoted time was antiaircraft fire, and his results were of great importance for engineering applications regarding filtering. Unfortunately, they were not of much use in the field because of the amount of time required for the calculations.

Weiner devoted the last decades of his life to the study of statistics, engineering, and biology. He had already worked on the general idea of information theory, which arose out of statistical mechanics. The idea of entropy had been around since the nineteenth century and enters into the second law of thermodynamics. It could be defined as an integral, but it was less clear what sort of quantity it was. Work of Ludwig Boltzmann suggested that entropy could be understood as a measure of the disorder of a system. Wiener pursued this notion and used it to get a physical definition of information related to entropy. Although information theory has not always followed the path laid down by Wiener, his work gave the subject a mathematical legitimacy.

An interdisciplinary seminar at the Harvard Medical School provided a push for Wiener in the direction of the interplay between biology and physics. He learned about the complexity of feedback in animals and studied current ideas about neurophysiology from a mathematical point of view. (Wiener left out the names of those who had most influenced him in this area in his autobiography as a result of an argument.) One area of particular interest was prosthetic limbs, perhaps as a result of breaking his arm in a fall. Wiener soon had the picture of a computer as a prosthesis for the brain. In 1947 he agreed to write a book on communication and control and was looking for a term for the theory of messages. The Greek word for messenger, *angelos,* had too many connections with angels to be useful, so he took the word for helmsman, *kubernes,* instead and came up with *cybernetics.* It turned out that the word had been used in the previous century, but Wiener gave it a new range of meaning and currency.

Cybernetics was treated by Wiener as a branch of mathematics with its own terms, like signal, noise, and information. One of his collaborators in this area was **John von Neumann** , whose work on computers had been followed up much more enthusiastically than Wiener's. The difference in reception could be explained by the difference in mathematical styles: von Neumann was meticulous, while Wiener tended to be less so. The new field of cybernetics prospered with two such distinct talents working in it. Von Neumann's major contribution to the field was only realized after his death. Wiener devoted most of his later years to the area. Among his more popular books were *The Human Use of Human Beings* in 1950 and *God and Golem, Inc.* in 1964.

In general, Wiener was happy writing for a wide variety of journals and audiences. He contributed to the *Atlantic, Nation,* the *New Republic,* and *Collier's,* among others. His two volumes of autobiography, *Ex-Prodigy* and *I Am a Mathematician,* came out in 1953 and 1956, respectively. Reviews pointed out the extent to which Wiener's memory operated selectively, but also admitted that he did bring the mathematical community to life in a way seldom seen. Although Wiener remarked that mathematics was a young man's game, he also indicated that he felt himself lucky in having selected subjects for investigation that he could pursue later in life. He received an honorary degree from Tufts in 1946 and in 1949 was Gibbs lecturer to the American Mathematical Society.

In 1964 Wiener received the National Medal of Science. On March 18, while travelling through Stockholm, he collapsed and died. A memorial service was held at MIT on the June 2, led by Swami Sarvagatananda of the Vedanta Society of Boston, along with Christian and Jewish clergy. This mixture of faiths was expressive of Wiener's lifelong unwillingness to be fit into a stereotype. He was a mathematician who talked about the theology of the Fall. He did not discover that he was Jewish until he was in graduate school but found great support in the poems of Heinrich Heine. Nevertheless, his intellectual originality led him down paths subsequent generations have come to follow.

## SELECTED WRITINGS BY WIENER

### Books

*Cybernetics,* 1948
*The Human Use of Human Beings,* 1950
*Ex-Prodigy,* 1953
*I Am a Mathematician,* 1956
*God and Golem, Inc.,* 1964
*Selected Papers,* 1964

## FURTHER READING

### Books

Freudenthal, Hans. "Norbert Wiener," in *Dictionary of Scientific Biography.* Volume XIV. Edited by Charles Coulston Gillispie. New York: Charles Scribner's Sons, 1970–1978, pp. 344–347.

Heims, Steve J. *John von Neumann and Norbert Wiener.* MIT Press, 1980.

Masani, P. R. *Norbert Wiener.* Birkhäuser, 1990.

—*Sketch by Thomas Drucker*

*Andrew J. Wiles*

# Andrew J. Wiles
## 1953-
### English-born American number theorist

Andrew J. Wiles conquered the most famous unsolved problem in mathematics—Fermat's Last Theorem, a conjecture that has frustrated professional and amateur mathematicians for more than 300 years. While the proof was a personal and professional victory of monumental proportions, the lasting value of Wiles' solution is the application of his innovative techniques to a wide range of other problems. Wiles was respected as a talented mathematician well before he announced his triumph on June 23, 1993, at the age of 40. John H. Coates, Wiles' former graduate studies advisor, wrote in *Notices of the AMS* that by 1986, Wiles already stood "amongst the select few over the last 150 years who have made profound contributions to algebraic number theory." Since proving Fermat's Last Theorem, Wiles has received many prestigious awards, including the Wolf, Schock, and Cole Prizes, and the National Academy of Sciences Award in Mathematics.

Andrew John Wiles was born in Cambridge, England, on April 11, 1953. His father, a professor of theology, taught at Oxford University. After earning a bachelor's degree from Oxford, Wiles pursued graduate studies at Cambridge University, earning his master's degree in 1977 and his Ph.D. in 1980. He came to the United States in 1982 to become a professor of mathematics at Princeton University. In 1988, Wiles received the Whitehead Prize from the London Mathematics Society, and he returned to England for two years as a research professor at Oxford. Wiles returned to Princeton in 1990, where he continues to work.

### Fermat's Last Theorem

Fermat's Last Theorem (FLT) is compelling because it can be stated so simply that a child can understand it, yet it has stumped professional and amateur mathematicians for centuries. Wiles became intrigued with it when he was ten years old and continued to work on it throughout his teen years. During his university years, however, he realized it was more complicated than he had initially thought and pragmatically turned his attention to other interesting topics.

FLT concerns an equation similar to the well-known Pythagorean theorem: $a^2 + b^2 = c^2$. The integers 3, 4, and 5 satisfy this equation ($3^2+4^2=5^2$), as do many other sets of integers. In 1637, **Pierre de Fermat** jotted a note in the margin of his copy of Diophantus' *Arithmetic,* asserting that the analogous equation $a^n+b^n=c^n$ (where a, b, and c represent integers) can never be true if $n>2$. In his note, Fermat asserted that he had found a marvelous proof, which the margin was too small to hold. In fact, Fermat later proved the theorem for n=4, as well as an extension of the form $a^4+b^4=c^2$. However, although he lived until 1665, he never again referred to a proof of the general theorem, leading to modern speculation that he had discovered a flaw in his original, supposed proof. Incidentally, FLT was not literally the last theorem proposed by Fermat, but its title reflects the fact that it was his last conjecture to remain unsolved.

More than 100 years after Fermat stated his famous theorem, **Leonhard Euler** proved it for n=3, demonstrating that there is no combination of integers that satisfy the equation $a^3+b^3=c^3$. Another century later, **P. G. Lejeune Dirichlet** proved the equation for n=5 and n=14. Other mathematicians whittled away at the problem, with Ernst E. Kummer proving in 1847, for example, that the theorem holds for all but three values of n less than 100. Even though a general proof was elusive, John Horgan wrote in *Scientific American,* "Techniques employed in these proofs have become standard tools in number theory, which has itself become vital to cryptography, error-protection codes and other applications." Modern computer-assisted calculations confirmed that the equation cannot be satisfied for any value of n less than 4 million, but this result was not established in the traditional sense of mathematical proof.

Fermat had originally proved his theorem for n=4 by inventing his technique of infinite descent. The theory of elliptic curves, however, played a key role in the final solution of this problem. In 1955, a Japanese mathematician named Yutaka Taniyama proposed a conjecture linking elliptical curves to modular forms of equations; the conjecture was refined by Goro Shimura of Princeton University and became known as the Taniyama-Shimura conjecture. During the 1980s, Gerhard Frey, a German mathematician, suggested that an elliptical curve might be used to represent all of the solutions to Fermat's equation; if so, proving the Taniyama-Shimura conjecture might indirectly prove FLT. In 1986, Kenneth Ribet, an American mathematician, accomplished the required formulation. This rekindled Wiles' hope that FLT was vulnerable, and he immediately dedicated himself to finding the solution. For the next seven years, other than fulfilling his teaching duties at Princeton, Wiles worked in solitude in an office in the attic of his home, pursuing his childhood quest.

### Elliptical Route to Solution

An elliptic curve (whose graph is *not* an ellipse) is defined by an equation of the form $y^2=x(x-A)(x+B)$, where A and B are non-zero integers such that $C=A+B\neq0$. What Taniyama had done was treat elliptic equations, not as equalities, but as congruences modulo some prime number. (Two numbers are congruent modulo $p$ if they each have the same remainder when divided by $p$.) Reducing an elliptic curve modulo $p$ (i.e., dividing all values of x and y by $p$ and retaining only the remainder) is useful because it results in a curve that has only a finite number of rational points. Furthermore, counting the number of solutions modulo $p$ for many prime values of $p$ can reveal other information about the elliptic curve. Some of this information is contained in a related mathematical expression called an L-series. A seemingly unrelated concept is that of modular forms, which are analytic functions defined on the upper half of the complex plane; they have nice properties such as being essentially invariant under certain transformations. What the Taniyama-Shimura conjecture asserted was that any elliptic curve can be associated with a modular form that has the same L-series. This provided a new way of analyzing elliptic curves.

Wiles believed he could prove that the elliptical curve representing the solutions to Fermat's equation could not exist and thereby prove FLT to be true. Wanting to develop the proof completely on his own, he told only his wife and one trusted colleague what he was working on. Although it took seven years to complete the solution, he did not get discouraged because he felt that he was making continual progress. By May of 1993 he had a proof that was complete except for one single but critical special case. While reading a paper by Harvard mathematician Barry Mazur, Wiles found a technique that would help him over this last hurdle.

Wiles announced his proof in an unusual manner. His withdrawal from most professional activities within the mathematics community had made his presentations at conferences rare. He asked to give three lectures at a small mathematics meeting in Cambridge, England, and rumors of an exciting announcement abounded, though the title of his lectures gave no hint about FLT. In his first lecture, Wiles remained secretive about the final outcome of his series of talks, but twice as many people attended his second lecture. He concluded the third lecture with a proof of a major portion of the Taniyama-Shimura conjecture. Almost as an afterthought, he added that this proved Fermat's Last Theorem. Word of Wiles' proof spread throughout the mathematical world at the speed of electronic mail, and both technical journals and the general press hounded him for interviews.

Only a few mathematicians in the world are qualified to analyze and confirm the entire 200-page proof. Reviewers did find some gaps, but Wiles corrected them quickly—except for one that took him 18 months and the help of a former graduate student, Richard Taylor. The proof is now considered complete.

### Personal Aspects

During the seven years of his all-out assault on FLT, Wiles virtually withdrew from his professional activities to concentrate on his quest. He is quoted in the *MacTutor History of Mathematics* web site as saying, "my wife has only known me while I have been working on Fermat. I told her a few days after I got married. I decided that I really only had time for my problem and my family . . . I found with young children [he has two daughters] that it was the best possible way to relax. When you're talking to young children, they're simply not interested in Fermat."

When asked how he felt about completing his proof, Wiles told *People Weekly,* 'there is a sense of loss, actually." There are still many challenging avenues for Wiles (and other mathematicians) to pursue, however. In a series of talks Wiles gave in early 1996 to the joint meetings of the American Mathematical Society and the Mathematical Association of America, he proposed several significant questions generated by his famous proof that should be studied. As for his own future work, he told Horgan that "I have a preference for working on things that nobody else wants to or that nobody thinks they can solve. I prefer to compete with nature rather than be part of something fashionable. . . . I'm afraid I've made [elliptic curves] so fashionable that I may have to move onto something else."

## SELECTED WRITINGS BY WILES

"Modular Elliptic Curves and Fermat's Last Theorem." *Annals of Mathematics* 141 (May 1995): 443ff.

(With Richard Taylor) "Ring–Theoretic Properties of Certain Hecke Algebras." *Annals of Mathematics* 141 (May 1995): 553ff.

## FURTHER READING

### Books

Ford, David N. "Andrew J. Wiles," in *Notable Twentieth-Century Scientists.* Volume 4. Edited by Emily J. McMurray. Detroit: Gale Research Inc., 1994, pp. 2202–03.

### Periodicals

"Andrew J. Wiles." *Current Biography Yearbook 1996* (March 1996): 624–27.

Cipra, Barry. "Fermat Prover Points to Next Challenges." *Science* 271 (March 22, 1996): 1668–69.

"Fermat's Last Theorem Finally Yields." *Science* 261 (July 2, 1993): 32–33.

"Princeton Mathematician Looks Back on Fermat Proof." *Science* 268 (May 26, 1995): 1133–34.

Coates, John. "Wiles Receives NAS Award in Mathematics." *Notices of the AMS* 43 (July 1996): 760–63.

Horgan, John. "Fermat's MacGuffin." *Scientific American* 269 (September 1993): 30, 34.

Ribet, Kenneth A., and Brian Hayes. "Fermat's Last Theorem and Modern Arithmetic." *American Scientist* 82 (March-April 1994): 144–56.

"The 25 Most Intriguing People of 1993–Andrew Wiles." *People Weekly* (December 27, 1993): 104.

"1997 Cole Prize." *Notices of the AMS* 44 (March 1997): 346–47.

### Other

"Andrew John Wiles." *MacTutor History of Mathematics Archive.* http://www-groups.dcs.st-and.ac.uk/~history/Mathematicians/Wiles.html (July 12, 1997).

—*Sketch by Loretta Hall*

# J. Ernest Wilkins, Jr.
## 1923-
### African-American mathematical physicist

A distinguished applied mathematician and nuclear engineer, J. Ernest Wilkins, Jr., has enjoyed a diverse career spanning governmental, industrial, and academic positions. He was involved in the Manhattan Project— the top-secret quest to construct a nuclear bomb during the 1940s—and was a pioneer in nuclear reactor design. He served as President of the American Nuclear Society, and contributed to the mathematical theory of Bessel functions, differential and integral equations, the calculus of variations, and to optical instruments for space.

Jesse Ernest Wilkins, Jr. was born in Chicago on November 27, 1923, the son of J. Ernest Wilkins, Sr., and Lucile Beatrice Robinson Wilkins. The senior Wilkins was a prominent lawyer who was president of the Cook County Bar Association in 1941–42, and an Assistant Secretary of Labor in the Eisenhower administration. Wilkins' mother was a schoolteacher with a master's degree. Both parents remained active in the Methodist church. Wilkins' two brothers became lawyers, but Wilkins preferred mathematics, entering the University of Chicago at the age of thirteen, and becoming the youngest student ever admitted to that institution. He completed his baccalaureate in 1940, his master's in 1941, and, by the age of nineteen, had earned his doctoral degree.

Wilkins went to the Institute for Advanced Study on a Rosenwald scholarship in 1942, then taught at Tuskegee Institute in 1943–44. He returned to the University of Chicago, where he worked on the Manhattan Project at the Metallurgical Laboratory from 1944 to 1946. Wilkins spent the bulk of his career in industry, however, starting with a position as Mathematician at the American Optical Company in Buffalo, New York, in 1946. He left to become a Senior Mathematician at the Nuclear Development Corporation of America (NDA), later United Nuclear Corporation, in White Plains, New York, in 1950. There he became Manager of the Physics and Mathematics Department in 1955 and later Manager of Research and Development. While he was at NDA, Wilkins earned a B.M.E. degree in 1957, and an M.M.E. in 1960, from New York University.

Beginning in the sixties, Wilkins held various offices in the American Nuclear Society, becoming President of the organization during 1974–75. In the early sixties, Wilkins moved to the General Atomic Division of General Dynamics Corporation in San Diego, where he remained until 1970. Wilkins next took an academic position as Distinguished Professor of Applied Mathematical Physics at Howard University in Washington, DC., remaining in that position for the next seven years.

In 1977 Wilkins went to EG&G Idaho in Idaho Falls, where he was Associate General Manager and then Deputy General Manager. Leaving in 1984, Wilkins was an Argonne Fellow at Argonne National Laboratory in 1984 and 1985, and although retired beginning in 1985, he remained active as a consultant. In 1990, Wilkins joined Clark Atlanta University as Distinguished Professor of Applied Mathematics and Mathematical Physics.

### Completes Gamma Ray Research

During his career Wilkins published roughly a hundred papers and reports on pure and applied mathematics, nuclear engineering, and optics. He is best-known for studies with Herbert Goldstein on gamma-ray penetration, the results of which are used for the design of nuclear reactor and radiation shielding, and of neutron absorption, which produced the Wigner-Wilkins approach to estimating the distribution of neutron energies in nuclear reactors. Wilkins also wrote papers on reactor operation and design and heat transfer. In addition, Wilkins continued to write on optical optimization problems and did interesting work on the estimation of the number of real roots of polynomials with random coefficients.

Wilkins served on advisory committees on scientific and engineering education for the National Academy of Engineering, the National Research Council, and other organizations and universities. Wilkins married Gloria Stewart in 1947. They had two children, Sharon and J. Ernest III. Wilkins remarried in 1984.

## SELECTED WRITINGS BY WILKINS

### Books

(With Robert L. Hellens and Paul E. Zweifel) "Status of Experimental and Theoretical Information on Neutron Slowing-Down Distributions in Hydrogenous Media," in *Proceedings of the International Conference on the Peaceful Uses of Atomic Energy.* Volume 5 1956, pp. 62–76.

"The Landau Constants," in *Progress in Approximation Theory,* 1991, pp. 829–842.

"Mean Number of Real Zeroes of a Random Trigonometric Polynomial, II," in *Topics in Polynomials and Their Applications,* 1993, pp. 581–594.

### Periodicals

(With Herbert Goldstein and L. Volume Spencer) "Systematic Calculations of Gamma-Ray Penetration." *Physical Review* 89 (1953): 1150.

(With G.B. Melese-d'Hospital) "Steady-state Heat Conduction in Slabs, Cylindrical and Spherical Shell with Non-uniform Heat Generation." *Nuclear Engineering and Design* 24 (1973): 62–77.

(With K.N. Srivastava) "Minimum Critical Mass Nuclear Reactors, Part I and Part II." *Nuclear Science and Engineering* 82 (1982):307–315, 316–324.

(With J. N. Kibe) "Apodization for Maximum Central Irradiance and Specified Large Rayleigh Limit of Resolution, II." *Journal of the Optical Society of America A, Optics and Image Science* 1 (1984): 337–343.

## FURTHER READING

### Books

Glasstone, Samuel and Alexander Sesonske. *Nuclear Engineering.* New York: Van Nostrand, 1955.

*In Black and White: A Guide to Magazine Articles, Newspaper Articles and Books Concerning more than 15,000 Black Individuals and Groups.* Volume 2, third edition. Detroit: Gale, 1980, p. 1040.

### Periodicals

*Ebony* (February 1958):60–67.

—*Sketch by Sally M. Moite*

*Shing-Tung Yau*

# Shing-Tung Yau
## 1949-
### Chinese-born American differential geometer

Shing-Tung Yau has made fundamental contributions to differential geometry that have influenced a wide range of scientific disciplines, including astronomy and theoretical physics. With Richard Schoen, Yau solved a longstanding question in **Albert Einstein**'s theory of relativity by proving that the sum of the energy in the universe is positive; their proof has provided an important tool for understanding how black holes form. Yau won the Fields Medal in 1982, the highest award in mathematics, for his proofs of the positive mass conjecture and the Calabi conjecture in algebraic geometry, and for other accomplishments in the realm of differential equations. He received the 1994 Crafoord Prize from the Royal Swedish Society in recognition of his "development of nonlinear techniques in differential geometry leading to the solution of several outstanding problems."

Yau was born April 4, 1949, in Swatow, in southern China, the fifth of eight children of Chen Ying Chiou and Yeuk-Lam Leung Chiou. Within the year, the Communists had overthrown the government and the family fled to Hong Kong, where Yau's father, a respected economist and philosopher, obtained a position at a college that would later be part of Hong Kong University. Yau's mother sold handcrafted goods to help support the family, for professors were poorly paid. During high school in Hong Kong, Yau told F. C. Nicholson in a 1994 interview that much emphasis was placed on mathematics, partly because the laboratories for the sciences were ill equipped. He credits his father, who died when Yau was 14 years old, with encouraging him to study mathematics, and he has retained a passion for it. He explained to a reporter for the *Harvard Gazette,* "It's clean, clear-cut, beautiful, and has a lot of applications."

Yau entered the Chinese University of Hong Kong, earning his undergraduate degree in 1968. One of his professors had attended the University of California at Berkeley and suggested he study there. Assisted by an IBM fellowship, Yau attended Berkeley, studying under Shiing-Shen Chern, the legendary geometer (Yau would later edit a collection of papers honoring his teacher). Yau completed his Ph.D. in mathematics in 1971 at the age of 22. He spent the following year at the Institute for Advanced Study (IAS) at Princeton, and then he accepted a position as assistant professor of mathematics at the State University of New York in Stony Brook. In 1974, he became a full professor of mathematics at Stanford University. Yau married Yu Yun Kuo, a fellow Berkeley student, in 1976; they eventually had two children. After leaving Stanford in 1979, Yau returned to the IAS, where he stayed until 1984, when he accepted the positions of professor and chairman of the mathematics department at the University of California at San Diego.

### Focuses on Differential Geometry

Differential geometry, Yau's primary field of interest, was developed during the 1800s. It uses derivatives and integrals to describe geometric objects such as surfaces and curves. Differential geometry is particularly concerned with geometrical calculations in spaces of many dimensions. The simplest kind of geometry would be one and two dimensional, analyzing figures such as squares or circles; the geometry of a three-dimensional figure, such as a cube or a

**519**

cylinder, is more complicated. Differential geometry is primarily concerned with geometrical figures in four or more dimensions. It is possible to visualize a four-dimensional figure by thinking of a three-dimensional object that changes over time, such as a raindrop falling and splashing on the ground. It is very difficult to visualize images of higher-dimensional figures (called manifolds), although they are important in physical systems characterized by five or more variables.

In 1991, Yau was invited to join six other Fields Medal laureates in presenting their assessment of the present state and the future prospects in mathematics. Yau, whose mother had just died, was unable to attend the symposium in Barcelona; however, he sent his thoughts in a paper on "The Current State and Prospects of Geometry and Nonlinear Differential Equations." In it, he noted that these fields have become "fundamental tools" useful in many subjects, including particle physics, general relativity, inverse problems in quantum chemistry, biological modeling, weather prediction, information theory, computer graphics, and robotics. Concerning the application he was most familiar with, Yau wrote, "Modern theoretical physics has provided a very effective leadership for many activities in geometry. It is mysterious that such a theory can provide deep intuition in mathematics and yet such intuitions rarely lead to a direct rigorous proof."

When he was elected to the National Academy of Sciences in 1993, Yau was cited for "revolutionizing the subject of differential geometry with his use and mastery of the partial differential equations basic to the subject. His influence extends widely into subjects as varied as mathematical physics, algebraic geometry, differential topology and the more technical aspects of partial differential equations."

## Solves Famous Physics Conjectures

Yau's first major contribution to differential geometry was his proof of the Calabi conjecture, which concerns how volume and distance can be measured, not in four dimensions, but in five or more. As is often the case when a famous problem is solved, the solution suggests other intriguing questions. In a 1984 address commemorating the 25th anniversary of the Institut des Haute Études Scientifiques (IHES), **Michael Atiyah** described Yau's work as being among the most notable in current geometry and analysis. Concerning the Calabi conjecture, he noted that a special case of it establishes the existence of a particular metric on certain types of four-dimensional surfaces, adding that "As yet very little is known about the Yau metric beyond its existence. Further investigation seems important."

One of the most important applications of differential geometry is Einstein's theory of relativity:

Einstein used differential geometry in his original calculations, and it was central to his theory of gravity. The general theory of relativity includes the conjecture (i.e., an unproven assertion) that in any isolated physical system the total energy, including gravity and matter, would be positive. Called the positive mass conjecture, this was fundamental to the theory of relativity although no one had succeeded in proving it. In 1979, Yau and Richard Schoen proved Einstein's positive mass conjecture by applying methods Yau devised. The proof was based on their work with minimal surfaces. A minimal surface is one in which a small deformation creates a surface with some larger area—soap films are often used as an example of minimal surfaces. Curves and surfaces in differential geometry are generally described by differential equations, but minimal surfaces must be described by partial, nonlinear differential equations, which are far more difficult to work with. Schoen and Yau's proof analyzed how such surfaces behave in space-time and showed that Einstein had correctly defined mass. Their methods allowed for the development of a new theory of minimal surfaces in higher dimensions, and they have had an impact on topology, algebraic geometry, and general relativity.

In 1987, Yau joined the faculty of Harvard University as a professor of mathematics. During 1991 and 1992 (concurrent with his position at Harvard), he served as the Wilson T. S. Wang Distinguished Visiting Professor at the Chinese University of Hong Kong and Special Chair of National Tsing Hua University in Hsinchu, Taiwan. He has received numerous prestigious awards, including the Crafoord and Veblen Prizes, and was granted an honorary doctorate from Harvard University. In addition, Yau holds an honorary professorship at the City University of Hong Kong. Among his many professional affiliations, Yau is a member of the American Physical Society, the American Academy of Arts and Sciences, and the Society for Industrial and Applied Mathematics. He has published more than 150 scientific papers and has served as editor-in-chief of the *Journal of Differential Geometry* and the *Asian Journal of Mathematics* (an English language quarterly journal launched in March 1997 to stimulate research throughout Asia). He has also served as editor of both *Communications in Mathematical Physics* and *Letters in Mathematical Physics.*

In 1997, President Bill Clinton presented Yau with the National Medal of Science. The award, authorized by the U.S. Congress and administered by the National Science Foundation, is based on the total impact of an individual's work in an area of science or mathematics. As reported in the *Harvard Gazette,* Yau's citation recognizes "profound contributions to mathematics that have had a great impact on fields as diverse as topology, algebraic geometry, general relativity, and string theory. His work insightfully combines two different mathematical approaches and has resulted in the solution of several longstanding and important problems in mathematics."

## SELECTED WRITINGS BY YAU

### Books

"The Current State and Prospects of Geometry and Nonlinear Differential Equations," in *Mathematical Research Today and Tomorrow: Viewpoints of Seven Fields Medalists.* Addresses to 1991 Symposium on the Current State and Prospects of Mathematics, 1992.

(Editor) *Chern: A Great Geometer of the Twentieth Century,* 1992

### Periodicals

(With R. Schoen) "On the Proof of the Positive Mass Conjecture in General Relativity." *Communications in Mathematical Physics* 65 (1979): 45–76.

(With R. Schoen) "The Existence of a Black Hole Due to the Condensation of Matter." *Communications in Mathematical Physics* 90 (1983): 575–79.

## FURTHER READING

### Periodicals

Atiyah, Michael. "Geometry and Analysis in the 1980s." *Michael Atiyah: Collected Works.* Volume 1. Oxford: Clarendon Press, 1988, pp. 319–21.

Cromie, William J. "Three Awarded Nation's Highest Science Honor." *Harvard Gazette* (May 1, 1997).

"Shing-Tung Yau: Breaking through Geometric Barriers." *Harvard Gazette* (November 5, 1987).

### Other

"S. T. Yau's Home Page." *Chinese University of Hong Kong.* http://www.math.cuhk.hk/.yau/ (June 10, 1997).

—*Sketch by F. C. Nicholson and Loretta Hall*

# Grace Chisholm Young
## 1868-1944
### English applied mathematician

A distinguished mathematician, Grace Chisholm Young is recognized as being the first woman to

*Grace Chisholm Young*

officially receive a Ph.D. in any field from a German university. Working closely with her husband, mathematician **William Henry Young**, she produced a large body of published work that made contributions to both pure and applied mathematics.

Grace Emily Chisholm Young was born on March 15, 1868, in Haslemere, Surrey, England, to Anna Louisa Bell and Henry William Chisholm. Her father was a British career civil servant who (following his own father) rose through the ranks to become the chief of Britain's weights and measures. Grace Emily Chisholm was the youngest of three surviving children. Her brother, Hugh Chisholm, enjoyed a distinguished career as editor of the eleventh edition of the *Encyclopaedia Britannica.*

As befitted a girl of her social class, Young received an education at home. Forbidden by her mother to study medicine—which the youngster wanted to do—she entered Girton College, Cambridge (one of two women's colleges there) in 1889. She was 21 years of age, and the institution's Sir Francis Goldschmid Scholar of mathematics. In 1892 she graduated with first-class honors, then sat informally for the final mathematics examinations at Oxford; there, she placed first. In 1893 she transferred to Göttingen University in Germany, where she attended lectures and produced a dissertation entitled "The Algebraic Groups of Spherical Trigonometry" under noted mathematician **Felix Klein**. In 1895 she became the first woman to receive a Göttingen doctorate in any subject. The degree bore the distinction magna cum' laude.

She returned to London and married her former Girton tutor, William Henry Young, who had devoted years to coaching Cambridge students. After the birth of their first child, the Youngs moved to Göttingen. There, William Young began a distinguished research career in mathematics, which would be supported in large part by the work of his wife. Grace Chisholm Young studied anatomy at the university and raised their six children, while collaborating with her husband on mathematics in both co-authored papers and those published under his name alone. In 1905 the pair authored a widely regarded textbook on set theory. Grace Chisholm Young's most important work was achieved between 1914 and 1916, during which time she published several papers on derivates of real functions; in this work she contributed to what is known as the Denjoy-Saks-Young theorem.

The Young family lived modestly, and William Young traveled frequently to earn money by teaching. In 1908, with the birth of their sixth child, the Youngs moved from Göttingen to Geneva. William Young continually sought a well-paying professorship in England, but he failed to obtain such a position; in 1913 he obtained a lucrative professorship in Calcutta, which required his residence for only a few months per year, and after World War I he became professor at the University of Wales in Aberystwyth for several years. Switzerland, however, remained the family's permanent home.

With advancing years, Grace Chisholm Young's mathematical productivity waned; in 1929 she began an ambitious historical novel, which was never published. Writing fiction was but one of her many varied interests, which included music, languages, and medicine. She also wrote children's books, in which she introduced notions of science. Her children followed the path she had pioneered, becoming accomplished scholars of mathematics, chemistry, and medicine. Her son Frank, a British aviator, died during World War I.

Grace Chisholm Young had lived with her husband's extended absences for her entire married life, and the spring of 1940 found them separated again: she in England, and he in Switzerland. From that time onward, neither spouse was able to see the other again—both were prevented from doing so by the downfall of France during the war. William Young died in 1942, and Grace Chisholm Young died of a heart attack in 1944.

## SELECTED WRITINGS BY YOUNG

(With William Henry Young) *The First Book of Geometry*, 1905, reprinted, 1969

(With William Henry Young) *The Theory of Sets of Points*, 1906, reprinted, 1972

**Periodicals**

"On the Form of a Certain Jordan Curve." *Quarterly Journal of Pure and Applied Mathematics* 37 (1905): 87–91.
(With William Henry Young) "An Additional Note on Derivates and the Theorem of the Mean." *Quarterly Journal of Pure and Applied Mathematics* 40 (1909): 144–145.
"A Note on Derivatives and Differential Coefficients." *Acta Mathematica* 37 (1914): 141–154.
(With William Henry Young) "On the Reduction of Sets of Intervals." *Proceedings of the London Mathematical Society* 14 (1914): 111–130.
"On the Solution of a Pair of Simultaneous Diophantine Equations Connected with the Nuptial Number of Plato." *Proceedings of the London Mathematical Society* 23 (1925): 27–44.

## FURTHER READING

**Books**

Grinstein, Louise S., and Paul J. Campbell, editors. *Women of Mathematics: A Biobibliographic Sourcebook.* Westport, CT: Greenwood Press, 1987, pp. 247–254.

**Periodicals**

Cartwright, M. L. "Grace Chisholm Young." *Journal of the London Mathematical Society* 19 (1944): 185–192.
Grattan-Guinness, Ivor. "A Mathematical Union: William Henry and Grace Chisholm Young." *Annals of Science* 29 (1972): 105–186.

—*Sketch by by Lewis Pyenson*

# Lai–Sang Young
## 1952–
### Chinese dynamical systems analyst

Lai–Sang Young keeps a low profile outside the field of mathematics, but her investigations into the statistical parameters of dynamical systems broke through a mathematical bottleneck that had existed for several years. For this achievement, Young received the 1993 Ruth Lyttle Satter Prize in mathematics. She also won a National Science Foundation Faculty Award for Women Scientists and Engineers, which funds her teaching and research over a number of years.

Lai–Sang Young was born in Hong Kong in 1952. Before emigrating to the United States at an unknown date, her mainly Cantonese education included English classes from a variety of schoolteachers. Her higher education was received in America, however. Young earned a bachelor's degree from the University of Wisconsin at Madison in 1973, and both her M.A. and Ph.D. from the University of California at Berkeley over the ensuing five years.

Young began a teaching career that took her from Northwestern in 1978, to Michigan State in 1980, then on to concurrent posts at the University of Arizona and the University of California at Los Angeles (UCLA). As a recipient of a Sloan Fellowship during 1985–86, she taught at the Universitat Bielefeld in Germany. Young also held visiting positions at institutions at Warwick in England and Berkeley's Mathematical Science Research Institute, and at the Institute for Advanced Study in Princeton, New Jersey, during the 1980s.

### Specializes in Ergodic Theory

Young specializes in a field that is fairly new in applied mathematics. In the realm of systems analysis, a rigorous computational method to measure the complexity of non–uniform systems is under current investigation. The subspecialty devoted to the statistical properties of dynamical systems is called "ergodic theory." The probability of the recurrence of any state in a model and the chance that any sample is equally representative of the whole make up the key questions in ergodic theory. The goal is to be able to model deterministic systems in other disciplines, such as anthropology, with a model that is random yet statistically regular.

With this goal in mind, Young has focused on the dynamics of strange attractors. A Henon attractor is a fractal with two variables, an *a* parameter and a *b* parameter, that looks something like a fuzzy sketch of a boomerang. It is considered one of the simplest dynamic systems that does not possess a stable periodic cycle. In a joint paper with a colleague, Young showed that a subset of the fractal turns out to have "a common distribution to the limit." This statistical norm is what allows the interdimensional figure to show up on a computer screen as a boomerang shape. Young produced similar work with quadratic maps, another set of simple yet nonuniform systems, that clarified previously contradictory evidence. Finding predictability in chaos was an achievement termed "both unexpected and deep" in Young's Satter Prize citation.

### "Embarrassing" Lack of Examples Leads to Breakthrough

Since the 1970s, mathematicians have been able to measure "uniformly hyperbolic" systems, and by the 1980s a number of researchers had come up with an ergodic theory for non–uniform counterparts. This work built a platform for generalizing about both types in terms of being able to predict an outcome for an attractor regardless of any initial condition. That means that no matter where researchers start, they wind up with the same general results. However, a natural invariant measure—the necessary statistical pattern—could not be found for a specific non–uniform case. As Young had stated during her response to the Satter Prize citation, the "lack of examples was starting to get a little embarrassing" for everyone. Investigations into the Henon attractor maps, considered the most likely suspects at that juncture, were already underway when Young was invited to join. She and a teammate constructed invariant measures for the parameters of a Henon, considered the first examples of their kind.

Young has participated in a number of events that bring mathematicians together in formal and informal settings. She delivered the invited address at the 1985 meeting of the American Mathematical Society (AMS). In 1994, she presented at a workshop on lattice dynamics and ergodic theory at the Mathematics Research Centre of Warwick. As the second recipient of the Satter Prize, Young kept the tradition, begun with **Margaret Dusa McDuff**, of helping to select the next winner, **Sun–Yung Alice Chang**, in 1995. Young was most recently part of the AMS Committee on Summer Institutes and Special Symposia, to organize the 1997 Summer Research Institute on differential geometry and control at the University of Colorado.

Gary Froyland, an Australian Ph.D. who shares Young's interest in topological entropy, was at first surprised at how approachable she was. Froyland's supervisor at the University of Western Australia asked Young, considered a world leader in her field, to spare some time at a conference. Young not only met and offered some pertinent explanations to the student, she invited him to UCLA. She also helped him find accommodations there and cover his expenses. During his stay, Froyland was impressed with Young's humor, clarity of mind, and teaching style. In her lectures and consultations Young strives to keep attention to precisely accurate detail without sacrificing the accessibility of everyday examples. Froyland reports that Young is a bit of a mimic, at least when it comes to imitating the accents of all her former English teachers, including a range of British natives and an American priest. The two mathematicians also discovered a shared interest in table tennis. A hint of Young's high personal standards came to light when she admitted she was once ranked among the top 20 female ping–pong players in the United States, but retired at the age of 25 when she felt her reflexes were beginning to slow.

## SELECTED WRITINGS BY YOUNG

(With Huyi Hu) "Nonexistence of SBR Measures for Some Diffeomorphisms that Are 'Almost Anosov'." (31K, TeX) *The Mathematical Physics Preprint Archive.* http://www.ma.utexas.edu/mp_arc/a/97–30

## FURTHER READING

### Periodicals

"1993 Ruth Lyttle Satter Prize." *Notices of the American Mathematical Society* 40, no. 3 (March 1993): 229–30.

### Other

Froyland, Gary, in an electronic mail interview with Jennifer Kramer conducted July 12, 1997.
Gary Froyland Web Page. http://maths.uwa.edu.au/~gary/mycv.html (August 15, 1997).
"Lai–Sang Young." *Biographies of Women Mathematicians.* http://www.scottlan.edu/lriddle/women/chronol.htm (August 15, 1997).

—*Sketch by Jennifer Kramer*

# William Henry Young
## 1863–1942
### English applied mathematician and educator

William Henry Young made advances in several areas of mathematics, but his most significant contribution was the development of a calculus approach that has been adopted by nearly all authors of advanced calculus textbooks since 1910. Together with his wife, **Grace Chisolm Young**, he published widely on a variety of mathematical topics.

The eldest child of Henry Young and Hephzibah Jeal, Young was born in London, England, on October 20, 1863, into a family that had been prominent in banking for several generations. He received his early education at the City of London School. The headmaster of the school, Edwin A. Abbott, author of a mathematical fantasy novel entitled *Flatland*, recognized Young as a skilled mathematician. Young chose to focus on mathematics in his college career at the Peterhouse College of Cambridge University, where he commenced study in 1881. He was expected to place as the senior wrangler, or first place, on the 1884 mathematical tripos, but he only placed fourth. Young later contended that he did not place as expected because he refused to limit his scope to the intensive study of mathematics required to excel on the tripos, and, in addition to mathematics, chose to pursue other intellectual avenues as well as athletics. He studied the works of Molière and competed for a Smith prize in the field of theology rather than mathematics. Young won the prize.

From 1886 to 1892, Young was a fellow of Peterhouse, although he held no official post at the school. Rather than immersing himself in research, Young chose the more lucrative path of private mathematics instruction of undergraduates. During this time, he also laid the foundation for a professional and romantic partnership with one of his students, Grace Chisholm, a student at the Girton College of Cambridge who obtained her first class degree in mathematics in 1892 and obtained senior wrangler status on the mathematical tripos. They were married in 1896.

The Youngs lived primarily in Göttingen, Germany, where Grace earned her doctorate, until 1908, when they moved to Switzerland. They lived first in Geneva and later moved to Lausanne. Young did not seriously pursue research until after his marriage, however, beginning around 1900. Between then and 1924, however, he wrote more than 200 papers with his wife. He devoted significant time to the study of real functions, independently discovering what came to be known as Lebesgue integration. Unfortunately, he did not make the discovery until two years after **Henri Lebesgue**, and although Young's definition of integration was different in form from Lebesque's, it was basically equivalent to it. Young made significant contributions to the further development of this area of study, however, most notably by establishing a method of monotone sequences as used in the Stieltjes integral. Young also made advances in Fourier analysis and measure theory.

Young's career was also noteworthy because of his professional collaboration with his wife, whose doctorial thesis focused on algebraic groups of spherical trigonometry. The Youngs embarked on projects together prior to their marriage, with Grace conducting most of the research while William provided financial support through his private tutoring. Due to the educational climate at the time, most papers written by the couple were attributed solely to William. This arrangement, with Grace undertaking the bulk of the research, may have hindered William later in life, as his lack of research credits may have been a factor in his inability to obtain positions of the high quality he desired.

One of the most important works to come out of the Youngs' collaboration was their second textbook, *The Theory of Sets of Points*. While set study is now a foundation for studies in all areas of mathematics, it was not yet commonly regarded at the time. On the whole, the mathematical community of the time overlooked this contribution, although **Georg Cantor**, one of the foremost experts in modern set theory, hailed the work.

Young returned to instruction in 1913, when he accepted part-time chair positions at the universities of Calcutta and Liverpool, and a position as professor at Aberystwyth. He was named an honorary doctor of the universities of Calcutta, Geneva, and Strasbourg. Among his many honors was the Sylvester Medal of the Royal Society, bestowed upon him in 1928. He also served as president of the International Union of Mathematicians from 1929 to 1936.

With the travels he undertook to earn a living, Young frequently found himself away from home for extended periods of time. The spring of 1940 found Young in Switzerland and his wife in England. Although they anticipated no problems reuniting, the escalation of World War II and the downfall of France intervened. For the last two years of his life, Young was separated from his family, which in addition to Grace included two sons and three daughters (a third son was killed in 1917). He died on July 7, 1942, in Lausanne, Switzerland.

In addition to their large body of research, his and Grace's legacy included two children who contin-ued study in the field of mathematics, Professor Laurence Chisholm Young and Dr. Rosalind Cecily Tanner, who both practiced pure mathematics.

## SELECTED WRITINGS BY YOUNG

### Books

(With Grace Chisholm Young) *The First Book of Geometry,* reprinted 1969

(With Grace Chisholm Young) *The Theory of Sets of Points,* 1906; second edition 1972

*The Fundamental Theorems of the Differential Calculus,* 1910

## FURTHER READING

### Books

Pycior, Helena M., Nancy G. Slack, and Pnina G. Abir–Am, editors. *Creative Couples in the Sciences.* New Brunswick: Rutgers University Press, 1996, pp. 126–40.

### Periodicals

"William Henry Young." *Journal of the London Mathematical Society* 17 (1942): 218–37.

"William Henry Young." *Obituary Notices of Fellows of the Royal Society of London* 3 (1943): 307–23.

—*Sketch by Kristin Palm*

# Zeno of Elea
## c.490–c.430 B.C.
### Greek logician and philosopher

Zeno of Elea was a Greek philosopher and logician whose development of paradoxical philosophical arguments about motion greatly influenced mathematical thought. One of the last major proponents of the Eleatic school of philosophy, he created his paradoxes to defend and confound this school of thought's detractors. His importance in early Greek philosophy and mathematics was noted by both Plato and Aristotle, who credited Zeno with the creation of dialectics.

Zeno was born in Elea, a southern Italian city, in the fifth century B.C. Although the exact date is unknown, historians place his birth at around 490 B.C., based largely on references by Plato in his book *Parmenides*. Plato describes Zeno as "tall and fair to look upon" and a favorite disciple of Parmenides, a Greek philosopher. He also reports that Zeno was about 40 years old when he accompanied his teacher to Athens in 449 B.C. In Athens, Zeno met and greatly impressed a young Socrates. Legend indicates that Zeno stayed for a number of years in Athens, where he made a good living by teaching philosophy. Athenian statesmen Pericles and Callias possibly studied under him.

According to Diogenes Laëritus, Zeno also created a unique cosmology in which he proposed the existence of several worlds. However, he was predominantly known for his paradoxes, which instigated centuries of philosophical debate. Little else is known about Zeno's life. He returned to live and work in Elea, where he eventually was caught up in a political intrigue that led to his death.

### Develops Philosophical Paradoxes

Zeno's teacher, Parmenides, was without question the greatest influence on his life. The founder of the Eleatic school of philosophy, Parmenides believed that there is only the "One" unchangeable being who constitutes the universe and argued that people's senses were mistaken in interpreting a world of motion and change. Parmenides was attacked and ridiculed by some, including the Pythagoreans. Zeno set out to defend his mentor's view of the world by creating a series of philosophical paradoxes.

At a very young age, Zeno wrote his famous and only known work, *Epicheiremata*. Only a few small fragments, amounting to about 200 words, survive. Fortunately, through Plato, Aristotle, Proclus, and other ancient authors, historians have been able to piece together the fundamentals of Zeno's work, which reportedly was published without his consent.

Zeno's unique approach to defending his master does not take the usual course of trying to prove Parmenides' position. Instead, he set out to create proofs that would reveal the absurd and self–contradictory nature of his opponents' views. The result was a series of paradoxes that profoundly influenced both philosophical and mathematical thought.

Zeno's philosophical puzzles focus on the goal of proving that our sense of motion, time, and change are based on illusion. According to Plato and Proclus, Zeno developed 40 different paradoxes (of which only eight survive) designed to disprove, confound, and agitate Parmenides' detractors. Among the most famous is the paradox of "Achilles," who was a mythological Greek hero who was the "swiftest of mortals." This paradox states that Achilles could not catch a crawling tortoise with a head start because Achilles must always reach a point at which the tortoise has started from and already passed. The "Dichotomy" and "Arrow" paradoxes similarly focus on the illusion of motion as represented by points in space or time. The paradox of the arrow, for example, states that an arrow can never reach it target because an object can only reach one point at a time and to be someplace (or at some point) an object must be at rest. If the arrow as at rest at each point, then motion forward is impossible.

Simplistically, these paradoxes may appear to be mere philosophical "brain teasers," but they represent the foundation of a type of philosophical argument called dialectic. This method of deduction allows one to work out contradictory conclusions from a single postulate or proposition that is assumed to be true. The dialectic approach to philosophy influenced the work of Plato, Aristotle, Immanuel Kant, Georg Hegel, **Bertrand Russell**, and even **Albert Einstein**, who, like Zeno, was interested in the concepts of space and time.

### Influences 20th Century Mathematics

Although developed primarily as a means of defending his teacher's philosophical views, Zeno's paradoxes also created difficulties for mathematicians. Some historians believe Zeno's paradoxes may have been directed toward the Pythagoreans (followers of the great Greek mathematician Pythagoras) and their handling of the infinitesimal in geometry. Eudoxius of Cnidus, who was a member of Plato's academy, may also have been influenced by Zeno in his development of a theory of proportions to accurately deal with the infinitesimal. However, since Greek mathematicians had not developed a strong concept of convergence or infinity, they mostly chose to ignore the mathematical dilemmas presented by the paradoxes. This early neglect was reinforced by Aristotle's declaration (without proof) that Zeno's paradoxes were merely fallacies.

Although Zeno's influence on early Greek mathematics probably was minimal, many of his paradoxes, like the "Achilles" paradox, relate to the concepts of infinity and the continuum, which would become major areas of interest in calculus. But it would take 19th– and 20th–century mathematicians to clearly understand the mathematics of Zeno's paradox, which is based on the distinction between measures of intervals and the number of points they contain. With the development of this concept, Zeno's paradoxes established an important influence on the work 19th–century German mathematician **Karl Weierstrass**, who helped start the "mathematical renaissance" and developed the notion of uniform convergence. The dilemmas presented by the paradoxes continued to exert a strong hold on mathematical thought through the work of Bertrand Russell, German mathematician Lewis Carroll, who introduced the concept of infinite sets, and many others.

Despite being dubbed the first "great doubter" and "man of destiny" in mathematics, Zeno was primarily a philosopher and not a mathematician. It is a testament to Zeno's keen intellect that later mathematicians went on to form numerous mathematical theories and problems from his work. While Cantor's theories on infinite sets led to explaining some of Zeno's paradoxes, the philosophical problems he presented have remained of interest and debate.

As reported by Plato, Zeno admitted that his book developed from the "pugnacity" of a younger man defending his mentor. Zeno, who was reported to be an active participant in politics, exhibited a similar contentiousness later in life when he "defended" good government by joining a plot against the tyrant Nearchus. The plot failed, and Zeno was arrested. According to legend, Zeno was tortured to reveal the names of his co-conspirators. Instead, the unyielding Zeno named the tyrant's own friends. Another legend says that Zeno bit off his tongue and spit it at his interrogators rather than betray his friends. He was eventually executed around 440 B.C.

## FURTHER READING

### Books

Burnet, John. *Early Greek Philosophy*. New York: Meridian Books, 1964, pp. 310–320.

Lee, H.D.P. *Zeno of Elea: A Text, with Translation and Notes*. Amsterdam: Adolf M. Hakkert, 1967.

Magill, Frank N., editor. *Great Lives from History: Ancient and Medieval Series*. Pasadena, CA: Salem Press, 1988, pp. 2403–2406.

Von Fritz, Kurt. "Zeno of Elea," in *Dictionary of Scientific Biography*. Volume XIV. Edited by Charles Coulston Gillispie. New York: Charles Scribners' Sons, 1975, pp. 607–613.

*—Sketch by David A. Petechuk*

# Milestones in the History of Mathematics

**50,000 B.C.**
Primitive humans leave behind evidence of their ability to count. Paleolithic people in central Europe make notches on animal bones to tally.

**c.25,000 B.C.**
Geometric designs are made by early humans.

**c.15,000 B.C.**
Cave dwellers in the Middle East make notches on bones to keep count and possibly to track the lunar cycle.

**c.8000 B.C.**
Clay tokens are used in Mesopotamia to record numbers of animals. This eventually develops into the first system of numeration. (See c.3500 B.C.)

**4700 B.C.**
The calendar may have been introduced in Babylonia at this time.

**c. 3500 B.C.**
Sumerians introduce a new type of token to represent numbers of animals or quantities of grain.

**3500 B.C.**
The Egyptian number system reaches the point where they now can record numbers as large as necessary by introducing new symbols.

**3100 B.C.**
An Egyptian royal mace is found dating to this period which contains hieroglyphics indicating amounts in the hundreds of thousands and millions.

**3000 B.C.**
Hieroglyphic numerals are in widespead use in Egypt.

**3000 B.C.**
Sumerian clay tokens are standardized and begin to represent different numerals.

**2773 B.C.**
The calendar is thought to have been introduced in Egypt by this time.

**c.2600 B.C.**
The Great Pyramid at Giza (near present-day Cairo, Egypt), is erected. Covering 13 acres and containing over 2 million stone blocks, it is a testament to the impressive mathematical and engineering skills of the Egyptians.

**c.2400 B.C.**
Mathematical tablets dated to this period are found at Ur, a city of ancient Sumer (present-day Iraq).

**c.2400 B.C.**
Positional notation comes into use in Mesopotamia by theSumerians. It has a base of 60 instead of 10 and uses cuneiform symbols.

**c.2200 B.C.**
Many mathematical tablets found at Nippur, a city of ancient Sumer in Babylonia, are dated to this time.

**c.2000 B.C.**
Mathematicians in Mesopotamia learn to solve quadratic equations, or equations in which the highest power is two.

**c.2000 B.C.**
Babylonians regularly use the rules of mensuration or measuring.

**c.1850 B.C.**
The Moscow or Golenishev papyrus dated to this period shows that the Egyptians know cipherization and possess a considerable knowledge of geometry. Along with the Rhind papyrus (see c.1650 B.C.), this becomes a major source of information about Egyptian mathematics.

**c.1650 B.C.**
The Rhind papyrus (also known as the Ahmes papyrus) is prepared by the Egyptian scribe Ahmes, or Ahmose, which contains solutions to simple equations. Discovered in the 19th century, it becomes a primary source of knowledge about early Egyptian mathematics, describing their methods of multiplying and dividing.

**c.1350 B.C.**
The Chinese use decimal numerals.

**c.1350 B.C.**
The Rollins papyrus, containing elaborate bread accounts, is dated to this time. It indicates how large numbers are used in everyday,

practical ways.

**1167 B.C.**
The Harris papyrus is dated to this year. It is prepared by Ramses IV when he assumes the Egyptian throne, and offers a practical accounting of the temple wealth accumulated by his father, Ramses III.

**c.1150 B.C.**
Won-wang (1182-1135 B.C.) writes the *I-king* which contains some sections on mathematics.

**c.1105 B.C.**
The *Chou-pei* is written and is the oldest extant Chinese mathematical work. It indicates that the Chinese had knowledge of the Pythagorean theorem of the right triangle.

**876 B.C.**
The first known reference to the usage of the symbol for zero is made in India.

**c.585 B.C.**
Thales of Miletus (c.640-c.546 B.C.), Greek geometer and philosopher, converts Egyptian geometry into an abstract study. He removes mathematics from a sole consideration of practical or concrete problems and goes about proving mathematical statements by a regular series of logical arguments. Doing this, Thales invents deductive mathematics. He is also credited with discovering several geometrical theorems.

**c.500 B.C.**
Chinese rod numerals first appear.

**c.500 B.C.**
Pythagoras of Samos (c.580-c.500 B.C.), Greek geometer and philosopher, formulates the idea that the entire universe rests on numbers and their relationships. Several mathematical principles and discoveries, such as being the first to deduce that the square of the length of the hypotenuse of a right triangle is equal to the sum of the squares of the lengths of its sides, has been attributed to him. Pythagoras also studies sound and finds that the relationship of pitch is correlated with the string length of a musical instrument.

**c.500 B.C.**
The *Sulvasutras,* an East Indian or Hindu religious text, is dated to this period. Its title means "Rules of the Cord" for it contains thegeometric rules for the construction of square and rectangular altars by rope stretching.

**c.500 B.C.**
Hippasus of Metapontum (fl. c. 400 B.C.), Greek geometer, constructs the circumscribed sphere called a dodecahedron, a regular solid with twelve pentagon faces.

**c.450 B.C.**
Zeno of Elea (c.490-c.425 B.C.), Greek philosopher, offers his four paradoxes of motion. All are fallacies based on the assumption that space and time are infinitely divisible and that sensory information should be ignored, but they are extremely important as they stimulate mathematical thought.

**c.440 B.C.**
Hippocrates of Chios (c.470-c.410 B.C.), Greek geometer, writes his *Elements,* the first known textbook on geometry. He arranges the known propositions of geometry in a scientific fashion and his work may have served as a model for Euclid's *Elements.*

**c.440 B.C.**
Anaxagoras of Clazomenae (c.499-c.428 B.C.), Greek geometer and philosopher, is the first Greek known to attempt the quadrature of the circle. Although his contribution to this problem is not known, no other problem exercises a longer fascination than that of constructing a square equal in area to a given circle.

**c.430 B.C.**
Antiphon, Greek philosopher, offers what becomes called his method of exhaustion as a way of squaring the circle.

**c.425 B.C.**
Hippias of Elis (fl. c. 5th cent. B.C.), Greek geometer, discovers the quadratrix, a curve that will not only trisect a given angle but multisect it into any number of equal parts.

**c.425 B.C.**
Theodorus of Cyrene (c.460-369 B.C.), Greek mathematician, shows that square roots of 3 and other non-square numbers to 17 are irrational.

**c.400 B.C.**
Archytas of Tarentum (c.428-c.350 B.C.), Greek mathematician and philosopher, is the first Greek mathematician to apply mathematics to mechanics. He distinguishes harmonic from geometrical and arithmetical progressions and contributes to a theory of acoustics.

**c.380 B.C.**
Plato (c. 427-347 B.C.), Greek philosopher, presents his philosophy of mathematics. He holds that nature is pervaded by mathematical harmony and mathematical structure, and that a

knowledge of mathematics is essential if the mind is to be trained properly. Platonism in science has come to mean an emphasis on a priori, abstract mathematical thinking.

**c.375 B.C.**
Theaetetus (c.417-c.369 B.C.), Greek mathematician, systematizes the study of irrational numbers and shows that they exist in large numbers.

**c.370 B.C.**
Eudoxus (c.400-c.347 B.C.), Greek astronomer and mathematician, introduces mathematics into astronomy and creates the general theory of proportion later used by Euclid. Eudoxus also solves several difficult geometrical problems.

**c.360 B.C.**
Autolycos of Pitane, Aeolian mathematician and astronomer, writes on the geometric properties of spheres.

**c.350 B.C.**
Menaechmus (c.380-c.320 B.C.), Greek geometer, is the first to consider systematically the geometry of the cone. He shows that ellipses, parabolas, and hyperbolas are all curves produced by theintersection of a cone and a plane.

**c.350 B.C.**
Dinostratus (c.390-c.320 B.C.), Greek geometer and brother of Menaechmus, contributes to calculating the area of a circle and attempts to square the circle.

**c.335 B.C.**
Eudemus of Rhodes, Greek mathematician, writes an early history of geometry.

**c.320 B.C.**
Aristaeus the Elder (c.360-c.300 B.C.), Greek philosopher and geometer, writes texts on conics and regular solids that are later used by Euclid.

**c.300 B.C.**
Euclid (c.300 B.C.), Greek mathematician, writes a textbook on geometry called the *Elements* that becomes the standard work on its subject for over 2,000 years. Although not an original work, it codifies into a single book all the known mathematical work of the epoch. It also uses a logically-arranged method of axioms and postulates that are both elegant and simple. Until the 19th century, Euclid's axioms are taken as absolute truths.

**c.260 B.C.**
Conon of Samos (c.280-c.220 B.C.), Greek mathematician, succeeds Euclid at Alexandria and becomes the teacher of Archimedes. Conon may be the first to study the mathematical curve that later becomes known as the "spiral of Archimedes."

**c.250 B.C.**
Asoka (272-232 B.C.), king of the Maurya empire in India, erects stone pillars in several major cities that contain the earliest preserved representation of the number symbols used today.

**c.250 B.C.**
The Mayan people of Central America and southern Mexico develop a number system based on 20 and which also possesses a zero. Their numbers are written vertically.

**c.240 B.C.**
Nicomedes (c.300-c.240 B.C.), Greek geometer, discovers the conchoid curve which he uses to solve the problems of doubling the cube and bisecting an angle.

**c.230 B.C.**
Dositheus of Pelusium, Greek geometer, succeeds Conon at themathematical school in Alexandria.

**c.230 B.C.**
Eratosthenes (c.285-c.205 B.C.), Greek astronomer, develops a system for determining prime numbers that becomes known as the "sieve of Eratosthenes."

**c.225 B.C.**
Archimedes of Syracuse (c.287-c.212 B.C.), Greek geometer and engineer, calculates the most accurate arithmetical value for $\pi$ to date. He also uses a system for expressing large numbers that uses an exponential-like method. Archimedes finds areas and volumes of special curved surfaces and solids and is able to determine the ratio of a cylinder to the sphere it circumscribes. He also devises the first mathematical exposition of the principle of composite movements, and is able to calculate square roots by approximation. Archimedes is often considered the greatest mathematical genius of antiquity.

**c.225 B.C.**
Apollonius of Perga (c.262-c.190 B.C.), Greek geometer, writes a treatise on conic sections which offers 400 theorems, many of them original. He also introduces the names parabola, ellipse, and hyperbola.

**c.180 B.C.**
Diocles (c.240-c.180 B.C.), Greek geometer, invents the cissoid, an ingenious geometrical construction he uses to solve the problem of doubling the cube.

**c.180 B.C.**
Hypsicles of Alexandria (fl. c. 2nd cent. B.C.), mathematician and astronomer, writes on the inscription of regular solids in a sphere. He is also credited with introducing the 360° circle into Greek mathematics.

**152 B.C.**
Chinese mathematician Chang Ts'ang dies. Among his many writings is the *Chiu-chang Suan-shu* or "Arithmetic in Nine Sections," the greatest arithmetical classic of China. Its original date is unknown, and Chinese tradition ascribes it to the 27th century B.C. (See 263)

**c.150 B.C.**
Perseus, Greek mathematician, discovers spiric curves or plane sections of surfaces that are generated by the revolution of a circle around the axis in its plane without passing through the center.

**c.140 B.C.**
Hipparchus of Rhodes (c.180-c.125 B.C.), Greek astronomer, creates trigonometry. In his calculations to measure the size and distance of the sun and moon, he works out an accurate table of chords or trigonometric sections.

**c.100 B.C.**
Negative numbers are used in China.

**c.100 B.C.**
Theodosius of Bythinia, Greek geometer, writes on the geometry of the sphere.

**c.60 B.C.**
Geminus of Rhodes (c.130-c.70 B.C.), Greek astronomer and mathematician, writes a comprehensive work on mathematics in which he classifies the different mathematical sciences and gives a history of Greek mathematics.

**6 A.D.**
Liu Hsin, Chinese scholar and astronomer, writes his treatise on the calendar and is the first person known to use decimal fractions.

**c.75 A.D.**
Hero (or Heron) of Alexandria (c.65-c.125), Greek mathematician and engineer, gives algebraic solutions to first and second degree equations. He also approximates square and cube roots and gives a formula for the area of a triangle as the function of its sides. In optics, Hero demonstrates that the angle of incidence equals the angle of reflection.

**c.100 A.D.**
Nicomachus of Gerasa (c.60-c.90 B.C.), Greek mathematician, writes his *Introductio Arithmetica*, the first exhaustive work in which arithmetic is treated independently of geometry. It also contains the Pythagorean theory of numbers.

**c.100**
Menelaus of Alexandria (fl. c. 1st cent. A.D.), Greek geometer, writes his *Sphaerica* in which he gives the first definition of a spherical triangle and proves the theorems on thecongruence of spherical triangles. He is also the first to separate trigonometry from stereometry and astronomy.

**c.125**
Theon of Smyrna, mathematician, writes a treatise on the mathematical rules necessary for the study of Plato.

**c.150**
Claudius Ptolemy (c.70-c.130), Greek geometer and astronomer, writes his *Megale Mathematike Syntaxis* ("Great Mathematical Composition") that comes to be known as the *Almagest*. Besides the astronomical system it contains, it also extends the work of Hipparchus on trigonometry, especially spherical trigonometry.

**c.150**
Claudius Ptolemy gives the first notable value for π after Archimedes.

**c.250**
Diophantus of Alexandria (c.210-c.290), Greek algebraist, is the first Greek to write a significant work on algebra, *Arithmetica*, which is the earliest treatise on algebra partially extant. He becomes best known for Diophantine equations, or equations (indeterminate) that have no single set of solutions. Diophantus is also the first Greek mathematician to treat fractions as numbers and is credited with introducing the symbol for minus.

**263**
Liu Hui, Chinese mathematician, writes a commentary on the ancient Chinese mathematical work, *Chiu-chang Suan-shu.* He also is known for his use of red computing rods for positive numbers and black for negative numbers.

**c.320**
Pappus of Alexandria (fl. c.320), Greek geometer, writes what comes to be known as the *Mathematical Collection* in which he summarizes all acquired knowledge of Greek mathematics. This is the last of the large compendiums as well as the best source for Greek mathematics.

**c.320**
Iamblichus (c.250-c.330), Syrian mathematician and

philosopher, continues Pythagorean number theory and emphasizes its mystical aspects. He also ascribes to Pythagoras the discovery of amicable or friendly numbers. Two numbers are amicable if each is the sum of the proper divisors of the other. These numbers will play a major role in the history of magic, sorcery, and astrology.

**c.325**
Serenus, Egyptian geometer, studies the plane sections of cylinders (recognizing them as ellipses) and demonstrates various propositions concerning the areas and volumes of various types of cones.

**c.390**
Theon of Alexandria, the last recorded member of the Museum at Alexandria, publishes Euclid's *Elements* with his own notes and commentary. Theon's copy was the oldest version of Euclid's work until 1808, when a 10th century manuscript of Euclid's is discovered, predating Theon's. This copy is the oldest version of the *Elements*. Theon is also known as the father of Hypatia. (See c.400)

**c.400**
Hypatia of Alexandria (c.370-415), Greek geometer and philosopher, writes commentaries on Diophantus and Apollonius. As the recognized head of the Neoplatonist school of philosophy at Alexandria, she is noted as the only woman scholar of ancient times and the first woman mentioned in the history of mathematics. Hypatia's death by a fanatical Christian mob leads to the departure of many scholars from Alexandria and marks the beginning of its decline as a major center of ancient learning.

**c.460**
Proclus (410?-485), Greek mathematician, writes commentaries on Euclid and Ptolemy and gives an historical outline of the evolution of Greek geometry. As the Neoplatonist head of the Academy at Athens, Proclus is eventually driven into exile by Christians because of his pagan beliefs.

**480**
Tsu Ch'ung-chih (430-501), Chinese mathematician, determines the value of π accurate to 6 decimal places.

**499**
Aryabhata the Elder (476-c.550), Hindu mathematician and astronomer, writes his *Aryabhatiya*, which contains his significant description of the Indian numerical system. Aryabhata also works on series, permutations, and arithmetic progressions, gives an accurate value for π, and is one of the earliest to use algebra.

**c.500**
Ammonios, Alexandrian philosopher, is the first to divide

mathematics into arithmetic, geometry, astronomy, and music. This division or classification will form the famous "quadrivium" of Western medieval education.

**c.500**
The Akhmim Papyrus is written in Egypt. Composed in Greek, it is a manual on practical arithmetic and is the oldest extant treatise on this subject. It is of Byzantine origin and was authored by a Christian.

**510**
Anicius Manlius Severinus Boethius (c.480-524), Roman philosopher, writes summaries of various scientific subjects, including arithmetic, music, and geometry. He also introduces an improved abacus and uses numerals of Indian origin. Boethius may have been the last Roman scholar who was literate in Greek.

**c.530**
Simplicius (fl. c. 530), Greek Neoplatonic philosopher and the last head of the Platonic Academy, writes commentaries on Euclid and Hippocrates of Chios and explains Antiphon's quadrature of the circle.

**c.560**
Eutocius of Ascalon (born c.480), Byzantine mathematician, writes commentaries of Apollonius and Archimedes and offers insight into the calculating methods used in the school of Alexandria.

**625**
Wang Hs'iao-t'ung, Chinese mathematician, writes his *Ch'i-ku Suan-ching* in which numerical cubic equations appear for the first time in Chinese mathematics.

**628**
Brahmagupta (c.598-c.665), Hindu algebraist and astronomer, writes his *Brahma-sphuta-siddhanta* in which he applies algebraic methods to astronomical problems.

**c.650**
Li Ch'un-feng, Chinese mathematician, comments on ancient Chinese arithmetic and works on problems in algebra and geometry.

**662**
Severus Sebokt, a Syrian bishop, makes the earliest known mention of Hindu numerals outside of India. He writes of Hindu computations "which excel the spoken word. . .and are done with nine symbols."

**710**

Venerable Bede (673-735), English scholar, writes on numerous mathematical subjects and uses mathematics in his work on the calendar in which he determines the date for Easter. Bede also writes on finger reckoning which is a system of expressing numbers by various positions of the fingers and hands.

**775**
Alcuin (732-804), English scholar, joins the court of the Frankish emperor Charlemagne (742-814), and establishes an educational system for the empire. He writes on a number of mathematical topics and is credited with a collection of mathematical puzzle problems that are often reproduced.

**820**
Al-Khwarizmi (c.780-c.850), Arab algebraist and astronomer, writes a mathematical treatise introducing the Arabic word "al-jabr" which becomes transliterated into "algebra." His own name is distorted by translation into "algorism," which comes to mean the art of calculating or arithmetic. Al-Khwarizmi also uses Hindu numerals, including zero, and when his work is translated into Latin and published in the West, those numerals are called "Arabic numerals."

**c.825**
Al-Hajjai Ibn Yusuf (c.786-833), Arab mathematician, produces the first Arabic translation of Euclid's *Elements*.

**c.825**
First information on the Hindu numerical system is found in a treatise titled *Algoritmi de Numero Indorum* which is translated from the Arabic. (See 976)

**850**
Mahavira, Indian mathematician, writes his *Ganita-Sara-Sangraha* which details Hindu geometry and arithmetic and is the only Hindu work dealing with ellipses.

**c.850**
Al-Kindi (c.801-866), Arab mathematician and astronomer, writes a work on Hindu numerals as well as a treatise on the geometry of optics.

**860**
Al-Mahani, Persian mathematician and astronomer, is the first to state the Archimedean problem of dividing a sphere into two segments according to a given ratio in the form of a cubic equation.

**870**
Thabit Ibn Qurra (836-901), Arab mathematician,

translates most of the major Greek mathematical works. His treatise on amicable or friendly numbers is the first example of original mathematical work in the Middle East.

**900**
Abu Kamil Shudja (850-930), Arab geometer, improves upon Al-Khwarizmi's algebra as applied to geometry and his work is later drawn upon by Western mathematicians.

**c.900**
Qusta ben Luga al-Ba'labakki, Greek translator, translates the works of Diophantus, Aristarchus, and Hero and develops the use of tangents and cotangents in trigonometry.

**c.900**
Abu Kamil Shudja (850-930), Egyptian mathematician, writes a treatise on algebraic methods applied to geometry in a very advanced manner.

**c.900**
Al-Battānī (also known as Albatgenius, 858?-929), Arab astronomer, uses spherical trigonometry to a great degree in his astronomy and establishes several new theorems.

**c.912**
Ahmad Ibn Yusif, Egyptian mathematician known in the West as Hametus, writes his *Liber de Proportione et Proportionalitate* which greatly influences medieval mathematical thinking.

**914**
Abu Utman, Arab mathematician and physician, translates the tenth book of Euclid's *Elements*.

**c.940**
Ibrahim Ibn Sinan (908-946), Arab mathematician, astronomer, and physician, develops the most elegant method for the quadrature of the parabola that is accomplished prior to the invention of integral calculus.

**c.950**
Al-Imrani, Arab mathematician and astronomer, comments on the work of Abu Kamil and makes a Latin translation of his work that becomes very influential in the West.

**c.950**
Al-Kazin, Arab mathematician and astronomer also called Alchazin, becomes the first Arab to solve the cubic equation by means of conic sections.

**976**
Traces of the Hindu numerical system are found in the

*Codex Virgilianus* written this year. It only mentions numerals however, and does not develop them.

**c.980**
Gerbert (c.945-1003), French scholar, helps initiate the rebirth of learning in Europe, especially in mathematics. As one of the first Christians to study in the Moslem schools of Spain, he reintroduces the abacus to the West and uses Arabic numerals with the exception of zero. Gerbert is not only a profound scholar but in 999 he is elected pope, taking the name Sylvester II.

**c.980**
Abu'l-Wefa (940-998), Persian mathematician and astronomer, publishes commentaries on the Greek mathematicians and invents a method for computing sine tables which give the sine of half a degree correct to nine decimal places. He also introduces the new trigonometric functions of the secant and cosecant.

**c.988**
Al-Kuhi, Arab mathematician and astronomer, solves and discusses the conditions of solvability of equations higher than the second degree.

**c.1000**
Alhazen (c.965-1039), Arab physicist and mathematician, gives an original solution to the cubic equation by means of conic intersections. He uses geometry and algebra in his work on optics.

**c.1000**
Heriger, abbot of Lobbes in Hainaut (present-day Belgium), writes about the abacus.

**c.1000**
Abbo of Fleury (c.947-1004), writes *De Numero et Pondere Super Calculum Victorii*, a commentary on ancient computation tables.

**c.1020**
Alkarki, Arab algebraist, writes a treatise on algebra considered the best produced by a Middle Eastern scholar. He is also the first to work with higher roots.

**1030**
Al-Biruni (973-1048), Persian mathematician, astronomer, and philosopher, writes a major summary of mathematics and astronomy as well as the best medieval account of Hindu numerals. He also investigates geometric progressions and studies what becomes known as albirunic problems (the trisection of an angle and other problems that are solvable with only a ruler and compass). Al-Biruni is regarded as one of the greatest scientists of the Islamic world.

**c.1100**
Omar Khayyam (1048?-1131?), Persian astronomer, mathematician, and poet, writes a book on algebra and is the one person who does the most to raise to a method the solution of algebraic equations by intersecting conics.

**c.1116**
Abraham bar Hiyya Savasorda (c.1065-c.1136), Jewish philosopher, mathematician, and astronomer living in Spain, leads the revival of Hebrew as a written language of western Mediterranean Jews. A Latin translation of his geometry textbook is used throughout Europe.

**1120**
Adelard of Bath (c.1090-c.1150), English scholar, produces the first Latin translation of Euclid's *Elements* taken from an Arabic version. This serves as the major geometry textbook in the West. He also introduces Arabian trigonometry to the West and uses Arabic numbers.

**c.1134**
Plato of Tivoli, Italian translator, translates several Arabic mathematical works and is the first to offer in Latin the complete solution of a quadratic equation.

**c.1140**
Johannes Hispalensis, Spanish translator, translates Arabic arithmetics into Castilian and they are then translated into Latin by others.

**c.1145**
Robert of Chester, English translator, translates Arab mathematician Al-Khwarizmi's book on algebra into Latin, making it accessible to Western readers.

**c.1150**
Atscharja Bhaskara (1114-1185), Indian mathematician, writes his *Siddhanta S'iromani* in which he gives several approximations for π.

**1175**
Gerard of Cremona (c.1114-1187), Italian scholar, translates Claudius Ptolemy's *Alamgest* and Euclid's *Elements* into Latin for Western scholars.

**1202**
Leonardo Fibonacci (1170?-1250?), Italian number theorist also known as Leonardo da Pisa, writes his book about the abacus called *Liber Abaci* in which he explains the use of Arabic (Hindu) numerals, the importance of positional notations, and explains the use of zero. This work is instrumental in ending the ancient Roman system of numerical notation.

**1220**
Leonardo Fibonacci writes *Practica Geometriae* which offers much of Euclid's geometry along with Greek trigonometry.

**1225**
Leonardo Fibonacci writes *Liber Quadratorum* in which he uses algebra based on the Arabic system.

**c.1225**
Jordanus Nemorarius, German algebraist and physicist, writes an algebraic work containing rules and problems leading to linear and quadratic equations. He also develops new propositions in the theory of numbers.

**c.1230**
Johannes de Sacrobosco (d. 1244 or 1256), English mathematician and astronomer also known as John of Holywood or of Halifax, writes *De Arte Numerandi seu Algorismus Vulgaris*. This simple and practical book of arithmetic greatly influences the acceptance of Hindu-Arabic numerals in Europe.

**c.1240**
Johannes de Sacrobosco writes *Algorismus Vulgaris* which becomes a standard university text.

**1247**
Ch'in Kiu-shao, Chinese mathematician, writes *Su-shu Chiu-chang*, which discusses indeterminate equations. He also uses the symbol 0 for zero.

**1248**
Li Yeh, Chinese mathematician, writes his *T'se-yuan Hai-ching* in which he designates negative numbers by drawing a cancellation mark across the symbol.

**c.1250**
Nasir al-Din (1201-1274), Persian mathematician and astronomer also known as Nasir Eddin, is the first Moslem to write a text on trigonometry as a study separate from astronomy. He also publishes a proof of the parallel postulate.

**1260**
Johannes Campanus (d.1296) Italian mathematician, produces a new Latin translation of Euclid's *Elements* from the Arabic, which becomes the authoritive version of that work for the next 200 years. (See 1482)

**1268**
Roger Bacon (c.1220-c.1292), English scholar, publishes *Opus Majus*, in which he demonstrates little mathematical ability but fully appreciates its importance and value to science.

**c.1270**
William of Moerbeke (c.1215-c.1286), Flemish scholar, translates the mathematical works of Archimedes. Although his translations are extremely literal, they are well received and serve as the basis for the first printed versions in the 16th century. (See 1543)

**1299**
Yang Hui, Chinese mathematician, writes *Suan-hsiao Chi-meng* in which Chinese algebra reaches its highest point. In the 17th century, this work stimulates Japanese mathematics.

**1299**
A law is passed in Florence, Italy, forbidding the use of Hindu-Arabic numbers by bankers. Authorities believed such numbers are more easily forged than Roman numerals. Similar laws were passed throughout Europe.

**c.1300**
Maximos Planudes (c.1260-c.1310), Byzantine mathematician and translator, writes a treatise on Hindu arithmetic which introduces Hindu-Arabic numerals into the Byzantine Empire.

**1321**
Levi ben Gerson (1288-1344), Jewish-French mathematician, writes *Sefer ha Mispar*, an arithmetic in Hebrew.

**1325**
Thomas Bradwardine (c.1290-1349), English mathematician, writes the first of his many works on arithmetic and geometry. He also devises mathematical formulae for physical laws which led to the development of quantitative measurement of physical processes.

**c.1330**
Richard of Wallingford (c.1292-1336), English mathematician, writes about arithmetic and trigonometry.

**1340**
Johannes de Lineriis of France writes *Algorismus de Minutiis* in which he uses fractions and includes the practice of placing the numerator over the denominator, using a horizontal line dividing the two.

**1344**
Gregory of Rimini (c.1323-1358), Italian philosopher and theologian also known as Gregorio Novelli da Rimini, discusses the concepts of continuity and infinity in mathematics.

**c.1360**

Nicole Oresme (c.1323-1382), French mathematician, writes a mathematical work that contains the first known use of fractional exponents or fractional powers. In another text, he locates points by coordinates, which anticipates modern coordinate geometry.

**c.1400**
Al-Kashi, Persian mathematician and astronomer, writes a short work on arithmetic and geometry and is the first to use decimal fractions.

**1429**
Al-Kashi computes $\pi$ to 16 decimal places.

**1436**
Leon Battista Alberti (1404-1472), Italian mathematician, writes *Trattato della Pittura*, which lays the groundwork for the new science of perspective founded on rigorous geometrical principles.

**1445**
Leon Battista Alberti writes *De Re Aedificatoria*, which introduces the use of multiple mathematical considerations into architecture. It is not published until 1485.

**c.1450**
Nicholas of Cusa (1401-1464), German scholar, writes treatises on the quadrature of the circle and the theory of numbers as well as calendar reform.

**c.1453**
Georg von Peurbach (1423-1461), Austrian mathematician and astronomer, lectures in Vienna on mathematics and writes an arithmetic containing a table of sines.

**1464**
Johann Müller (1436-1476), German astronomer and mathematician also known as Regiomontanus, writes *De Triangulis Omnimodes*, which is the first systematic European exposition of plane and spherical trigonometry as considered independently of astronomy.

**1478**
The first dated and printed arithmetic appears. Entitled *Arte dell'Abbaco*, it was published anonymously in Treviso, Italy.

**1478**
Piero Della Francesca (c.1420-1492), Italian painter, writes his *De Prospectiva Pingendi* which is the first true treatise on perspective. His actual paintings reflect his intimate knowledge and study of geometry and mathematics.

**1482**

The first printed edition of Euclid's *Elements, Elementa Geometriae*, is published in Venice, Italy. Johannes Campanus translated Euclid's work into Latin about 200 years earlier.

**1484**
Nicholas Chuquet (1445-1500), French mathematician, writes an arithmetic known as *Triparty en la Science des Nombres*. In this very advanced work, he discusses rational and irrational numbers, the theory of equations, and the role of the symbol for zero. Chuquet is also the first to use the radical sign with an index to indicate roots.

**1484**
Pietro Borghi, Italian mathematician, publishes *Arithmetica*, which focuses on the commercial aspects of mathematics. It becomes a highly popular and influential work.

**1489**
Johann Widmann (c.1460-c.1500), German mathematician, publishes *Behende und Hubsche Rechenung*, which is the first printed book to contain plus and minus symbols. They are not used as symbols of operation but merely to indicate excess and deficiency.

**1491**
Filippo Calandri of Italy publishes *Pictagoras Arithmetrice Introductor*, which contains the first printed example of the modern process of long division. It also contains the first illustrated mathematical problems published in Italy.

**1494**
Luca Pacioli (c.1445-1517), Italian mathematician, publishes *Summa de Arithmetica, Geometrica, Proportioni et Proportionali*, which contains the first printed example of a detailed description of double-entry bookkeeping.

**1500**
The use of Hindu-Arabic numerals in Europe has overtaken Roman numerals for most computational purposes.

**c.1505**
Scipione Dal Ferro (1465-1526), Italian mathematician, solves algebraically the cubic equation. He does not publish the solution, which is in keeping with the secretive practices of the time.

**1511**
Charles de Bouelles (c.1470-c.1553), French geometer and number theorist also known as Charles Bouvelles or Carolus Bovillus, publishes *Le Livre et de l'Art et Science de Géometrie*, which is the first geometry text to be published in French.

**1512**
Juan de Ortega (c.1480-c.1568), Spanish mathematician, publishes his *Tractado Subtilismo d'Arithmetica y de Geometria*, which is one of the early Spanish arithmetics.

**1514**
Jakob Köbel (1470-1533), German mathematician, publishes his *Rechenbiechlin*, a very popular arithmetic that was reprinted in 22 editions. It is influential in disseminating knowledge of Hindu-Arabic numerals in the West.

**1515**
Gaspar Lax (1487-1560), Spanish-French mathematician, publishes *Arithmetica Speculativa* and *Proportiones* in Paris.

**1522**
Cuthbert Tunstall (1474-1559), English scholar, publishes*De Arte Supputandi*, which is the first separate arithmetic printed in England.

**1522**
Adam Riese (c.1492-1559), German mathematician, publishes *Rechnung auff der Linihen und Federn*, an arithmetic that becomes so popular and useful that it gives rise to the well-known German phrase "nach Adam Riese," which is used to indicate a correct calculation. He advocates replacing the old counters with the more modern written computation.

**1522**
Johannes Werner (1468-1528), German mathematician, publishes *Elementis Conicis*, which is the first work on conic sections to appear in Christian Europe.

**1525**
Albrecht Dürer (1471-1528), German artist, publishes *Underweysung der Messung*. In this practical treatise on geometry, Dürer also teaches the theory and method of perspective, which he regarded as an important branch of mathematics. It is known today as projective geometry.

**1525**
Christoff Rudolff (c.1500-1545), German algebraist, publishes *Die Coss*, which is the first German textbook on algebra. He also invents the radical sign for square roots.

**1534**
Peter Apian (1495-1552), German astronomer also known as Petrus Apianus, publishes *Instrumentum Sinuum sive Primi Mobilis*, which contains a table of sines for every minute of arc. These are the first such tables to be printed.

**1535**
Tartaglia (1499-1557), Italian mathematician also known as Niccolò Fontana, demonstrates in a public forum his correct solution to the cubic equation. He later discloses in confidence his secret methods to another Italian mathematician, Girolamo Cardano, who then proceeds to publish the solution. (See 1545)

**1543**
Tartaglia edits William of Moerbeke's translation of Archimedes' writings. This is the first printed version of these works, and it is through Tartaglia that they become fully known to Western scholars.

**1543**
Robert Recorde (c.1510-1558), Welsh-English mathematician, publishes *The Ground of Artes*, which is one of the more significant early arithmetics written in English.

**1544**
Michael Stifel (c.1486-1567), German algebraist, publishes *Arithmetica Integra*, which contains the beginnings of a theory of exponents and of logarithms.

**1545**
Girolamo Cardano (1501-1576), Italian algebraist, publishes *Ars Magna*. In this treatise on algebra, he discloses the discovery of the rule which allows the cubic equation to be solved, despite having pledged his secrecy to Tartaglia (see 1535). Although Cardano does give credit to Tartaglia, the method becomes known as "Cardano's rule." This work also contains the first major Western treatment of negative numbers and the first usage of a complex number in a computation.

**c.1545**
Girolamo Cardano writes *De Ludo Aleae* in which he begins the study of probability theory. (See 1654)

**1546**
Tartaglia publishes *Quesiti ed Invenzioni Diverse* in which he applies mathematics to mechanics.

**1549**
Jacques Peletier (1517-1582), French mathematician and philosopher also known as Peletarius, publishes *L'Arithmétique*, which combines theoretical and commercial arithmetic. He also observes that the root of an equation is the divisor of the last term.

**c.1550**

**538**

Rheticus (1514-1576), German mathematician whose given name is Georg Joachim von Lauchen, completes his trigonometric tables with sines calculated out to 10 places. He is also the first to define the trigonometric functions as ratios of the sides of a right triangle.

**1551**
Robert Recorde publishes *The Pathwaye to Knowledge*, an adaptation of the first four books of Euclid's *Elements*.

**1556**
The first work on mathematics printed in the New World is a small commercial compendium by Juan Diez. It is produced in Mexico City.

**1556**
Tartaglia publishes the first of his six-part *Il General Trattato di Numeri et Misure*, which contains a complete account of commercial arithmetic and is considered the best Italian work on arithmetic produced in the 16th century.

**1557**
Robert Recorde publishes *The Whetstone of Witte* in which the modern symbol for equality (=) is used for the first time.

**1558**
Frederigo Commandino (1509-1575), Italian mathematician, translates the mathematical works of Archimedes and brings the ancient method of integration within the reach of contemporary mathematicians.

**1559**
Jean Buteo (1492-c.1565), French geometer, publishes *Logistica*, which is one of the earliest works on geometry in French.

**1567**
Pedro Nunes (1502-1578), Portuguese geometer, publishes his *Livro de Algebra en Arithmetica y Geometria*. He is considered the foremost geometer in Portugal at this time.

**1570**
John Dee (1527-1608), English mathematician, astrologer, and alchemist, writes his *Mathematicall Praeface* to Billingley's translation of Euclid and offers an influential survey of the mathematics of his era and its practical applications.

**1570**
Henry Billingsley, English mathematician, makes the first complete English translation of Euclid's *Elements*.

**1571**
Thomas Digges (c.1543-1595), English mathematician, publishes posthumously his father's work entitled *A Geometrical Practise Named Pantometria*, a work of practical mathematics geared especially to surveying. His father, Leonard Digges (c.1520-c.1571?), was one of England's most significant early mathematicians.

**1572**
Frederigo Commandino (1509-1575), Italian mathematician, publishes an important Latin translation of Euclid from the Greek. This version serves as the basis of many subsequent translations of Euclid's work.

**1572**
Rafaello Bombelli (1526-1573), Italian algebraist, publishes a book on algebra in which he points out the reality of the term imaginary roots and lays the foundation for a better knowledge of imaginary quantities. This is the first consistent theory of imaginary numbers to be used.

**1575**
Francesco Maurolico (1494-1575), Italian geometer and optician, publishes his *Opuscula Mathematica* in which he works on conic sections and introduces secants into trigonometrical calculations.

**1579**
François Viète (1540-1603), French mathematician, publishes *Canon Mathematicus seu Ad Triangula*, which is the first book in Western Europe to develop systematically methods for solving plane and spherical triangles with the aid of all six trigonometric functions. He also calculates $\pi$ correctly to 9 decimal places.

**1583**
Christopher Clavius (1537-1612), German mathematician and astronomer, publishes his *Epitome Arithmeticae Practicae*. This popular book is instrumental in promoting mathematical knowledge. Clavius also contributes to calendar reform.

**1583**
Thomas Finck (1561-1656), Danish mathematician and physician, publishes his *Geometriae Rotundi* in which the words "tangent" and "secant" appears for the first time. Finck is credited with their invention.

**1585**
Simon Stevin (c.1548-c.1620), Dutch mathematician also known as Simon Stevinus, publishes his *De Thiende*, which contains the first comprehensive system of decimal fractions and their practical applications.

**1591**

François Viète publishes his *Isagoge in Artem Analyticum*, which is the foundation of modern algebra. Viète contributes to the development of symbolic algebra as he is the first to introduce the practice of using vowels to represent unknown quantities and consonants to represent known ones. This is then considered the first work on symbolic algebra.

**1593**
Adriaen van Roomen (1561-1615), Dutch mathematician, correctly calculates the value of π to fifteen decimal places.

**1594**
John Napier (1550-1617), Scottish mathematician, first conceives of the notion of obtaining exponential expressions for various numbers, and begins work on the complicated formulas for what he eventually calls "logarithms." (See 1614)

**1600**
Bartholomeo Pitiscus (1561-1613), German mathematician, publishes his *Trigonometriae; sive, De Dimensione Triangulorum* in which he coined the term "trigonometry."

**1600**
Guidobaldo Dal Monte (1545-1607), Italian geometer and astronomer, publishes his *Perspectivae* in which he helps establish the new method of perspective on a firm geometric basis.

**1600**
Thomas Harriot (1560-1621), English algebraist and astronomer, introduces the symbols for equality and inequality, using " >" to mean greater than and "<" to denote less than. He is also the founder of the English school of algebraists and is the first to factor equations.

**1600**
Joost Bürgi (1552-1632), Swiss algebraist, conceives of the idea of logarithms independently of John Napier but publishes his discovery in 1620, six years after Napier. Most agree that Napier had conceived the idea first, and while his approach to logarithms is geometrical, Bürgi's is algebraic. (See 1614)

**1600**
Adriaen Metius (1571-1635), Dutch mathematician also known as Adriaanszoon, obtains information from Chinese texts to approximate the value of π.

**1603**
Marino Ghetaldi (1566-1626) publishes his *Archimedes Promotus* in which he uses algebra to solve geometric problems. Ghetaldi's other publications anticipates analytical

geometry.

**1604**
Johannes Kepler (1571-1630), German astronomer, suggests the possibility that parallel lines may meet at infinity. He is best known for employing his mathematical knowledge in devising his three planetary laws of motion.

**1609**
Johannes Kepler advances the development of the geometry of the ellipse as he attempts to prove that planets move in elliptical orbits.

**1610**
Ludolph van Ceulen (1540-1610), Dutch mathematician, determines the value of π to 20 places.

**1613**
Pietro Antonio Cataldi (1552-1626), Italian mathematician, publishes his *Trattato del Mondo Brevissimo di Trovare la Radica Quadra delli Numeri* in which he discusses his invention of the common form of continued fractions. He also develops the first symbolism for continued fractions and shows how to use them to determine square roots.

**1613**
Willebrord van Roijen Snell (1591-1626), Dutch mathematician, succeeds his father as professor of mathematics at the University of Leiden and applies his prodigious mathematical talents to astronomy. He is the first to use triangulation to measure the meridian and also discovers the law of refraction.

**1613**
François d'Aguilon (c.1566-1617), Dutch mathematician, publishes his *Opticorum* in which he gives the earliest description of stereographic principles and first uses the term "stereographic projection."

**1614**
John Napier publishes his *Mirifici Logarithmorum Canonis Descriptio* in which he originates the concept of logarithms as a mathematical device to aid in calculations. His tables of logarithms simplify routine calculations to a great degree and especially revolutionize astronomical calculations.

**1615**
Johannes Kepler publishes his *Stereometria Doliorum* in which he studies the volumes of solids of revolution, especially volumes generated by revolutions of conics.

**1615**

François Viète has his *De Aequationum Recognitione et Emendatione* posthumously published. This is his main work on the theory of equations, and gives methods for solving third, fourth, and fifth degree equations.

**1620**

Edmund Gunter (1581-1626), English mathematician, coins the terms "cosine" and "cotangent." Cosine is an abbreviation of "complemental sine." He also invents Gunter's Scale, a forerunner of the slide rule.

**c.1620**

Marin Mersenne (1588-1648), French mathematician, begins his correspondence with the great mathematicians of his day and serves as a link between them, disseminating their ideas and accomplishments throughout Europe. This is before any scientific journals are published.

**1621**

Willebrord van Roijen Snell devises a trigonometric improvement of the classical method for computing π.

**1621**

Claude-Gaspar Bachet de Meziriac (1581-1638), French mathematician, publishes his translation of the *Arithmetica* by Diophantus. This work contributes to the advancement of number theory during the 17th century.

**1622**

William Oughtred (1575-1660), English arthemetician, invents the straight logarithmic slide rule. It consists of two rulers alongside which logarithmic scales are laid. The rulers can then slide along each other. He describes his invention in print 10 years later.

**1624**

Henry Briggs publishes his *Arithmetica Logarithmica* in which heinvents the common or Briggsian logarithms whose base of 10 make them easier to use. John Napier, the inventor of logarithms, agrees with this. Briggs also invents the modern method of long division.

**1627**

Ezechiel de Decker, Dutch mathematician, publishes a table of logarithms that is complete for all natural numbers up to 100,000. His work fills much of the gaps left by Henry Briggs.

**1629**

Albert Girard (1595-1632), French mathematician, publishes a work on trigonometry that contains the first use of the abbreviations "sin," " tan," "sec" for sine, tangent, and secant. Healso is the first to employ brackets in algebra.

**1629**

Pierre de Fermat (1601-1665) becomes a pioneer in calculus, having applied the idea of infinitesimals to questions of quadrature and maxima and minina, as well as to the construction of tangents. He publishes little in his life however, and regards mathematics as an avocation. Fermat's accomplishments are such that he is called the "prince of amateurs" in mathematics. (See c.1629)

**c.1629**

Pierre de Fermat reconstructs the theory of conic sections of Apollonius and works out equations for geometric figures that will lead to the invention of analytic geometry. Although René Descartes is credited with the invention and full development of analytic geometry, Fermat develops it earlier but does not publish his findings. (See 1637)

**1630**

Christoph Grienberger (1561-1636), Austrian-born mathematician and astronomer, computes the value of π to 39 places.

**1630**

Johann Faulhaber (1580-1635), German mathematician, publishes his *Canonem Logarithmicum* in which he disseminates in Germany a knowledge of logarithms.

**1631**

Richard Delamain, English mathematician also known as the Elder, publishes his *Grammelogia; or, the Mathematicall Ring* which describes a circular slide rule that is different from William Oughtred's rectilinear rule.

**1631**

William Oughtred publishes his *Clavis Mathematicae* which contains a large amount of mathematical symbolism, including the notation "x" for multiplication and "::" for proportion.

**1631**

Thomas Harriot has his *Artis Analyticae Praxis* published posthumously. This is an exhaustive study on the theory of equations.

**1634**

Pierre Herigone, French mathematician, publishes his *Cursus Mathematicus* in which he develops his own system of mathematical notation and demonstrates a complete understanding of the possibilities of mathematical symbolism.

**1635**

Paul Guldin (1577-1643), Swiss mathematician, publishes his *Centrobaryca.*. He introduces the centrobaric method which gives the relation between volume and area of a revolving figure. It is later known as Guldin's theorem, although it was actually first stated by the 4th century Greek mathematician Pappus in his *Mathematical Collections*.

**1635**

Bonaventura Cavalieri (1598-1647), Italian mathematician, publishes *Geometria Indivisibilibus*. He expounds on his method of indivisibles, a forerunner of integral calculus. Cavalieri is also the one of the first to recognize and popularize the value of logarithms.

**1635**

Bonaventura Cavalieri publishes his *Trigonometria Plana et Sphaerica* in which he proves that for any triangle on a sphere, the sum of the three angles exceeds 180° but is no greater than 450°.

**1636**

Pierre de Fermat introduces the modern theory of numbers. He offers theories on everything from perfect and amicable numbers, figurate numbers, magic squares, and above all, prime numbers.

**1637**

René Descartes (1596-1650), French philosopher and mathematician, publishes his *Discours de la Méthode*, which contains a 106-page essay titled *La Géometrie*. In this essay, Descartes introduces analytic geometry by demonstrating how geometric forms may be systematically studied by analytic or algebraical means. He is the first to use the letters near the beginning of the alphabet for constants and those near the end for variables and also introduces the use of exponents and the square root sign. (See 1659)

**c.1637**

Pierre de Fermat writes in the margin of a book a reference to what comes to be known as "Fermat's Last Theorem." Also called his "Great Theorem," this remains the most famous unsolved problem in mathematics (until possibly solved in 1994). Fermat says he has a proof for the particular problem posed, but that the margin is too small to include it there.

**1638**

Galileo Galilei (1564-1642), Italian astronomer and physicist, publishes *Discorsi e Dimonstrazioni Matematiche Intorno a Due Nuove Scienze* in which he points out a fundamental difference between infinite and finite classes of numbers.

**1639**

Gérard Desargues (1593-1662), French geometer, publishes

*Brouillon Projet* which becomes a classic in the early development of synthetic projective geometry. Although this book is totally ignored for over two centuries, its theory of involution and transversals laid the basic rules of modern synthetic geometry.

**1639**

René Descartes studies simple polyhedrons and essentially begins the study of topology, the branch of mathematics concerned with properties of geometric configurations which are unaltered by elastic deformations. (See 1679)

**1640**

Blaise Pascal (1623-1662), French mathematician and physicist, publishes *Essai sur les Coniques*, a one-page treatise in which he makes important contributions to the study of conic sections. In this paper, Pascal stated his theorem that theopposite sides of a hexagon inscribed in a conic intersect in three collinear points.

**1640**

Evangelista Torricelli (1608-1647), Italian physicist, discovers the isogonic center of the triangle. He solves the problem of determining a point in the plane of a triangle such that the sum of its distances from the three vertices be a minimum. This is the first notable point of the triangle discovered since the period of ancient Greek mathematics.

**c.1640**

Gilles Personne de Roberval (1602-1675), French geometer, develops Cavalieri's indivisibles and applies them to a method of drawing tangents and for determining the area of a cycloid and the volume generated by it. He becomes well known for his discoveries in the field of higher plane curves and the geometry of infinitesimals.

**1644**

Marin Mersenne publishes his *Cogita-Physico-Mathematica* in which he suggests a formula that will yield prime numbers. These "Mersenne numbers" are not always correct, but they stimulate research into the theory of numbers.

**1647**

Bonaventura Cavalieri publishes *Exercitationes Geometricae Sex* in which he calculates areas by dividing regions into increasingly smaller parts. This leads eventually to the development of the calculus.

**1647**

Grégoire de Saint-Vincente (1584-1667), Dutch geometer, publishes *Opus Geometricum Quadraturae Circuli et Sectiones Coni* in which he makes several basic discoveries on conic sections. He is also the first to use a geometric series on the paradoxes of Zeno.

**1648**

Abraham Bosse (1602-1676), French engraver, publishes *Leçons de Géometrie et de Perspective Practique* in which he states a fundamental principle of projective geometry that is now known as Desargues' Theorem.

**1650**

Pietro Mengoli (1625-1686), Italian mathematician, publishes *Novae Quadraturae Arithmeticae* in which he does work on the infinite series. He also later demonstrates the convergence of reciprocals of triangular numbers.

**1651**

André Tacquet (1612-1660), Dutch mathematician, publishes *Cylindricorum et Annualarium* in which he uses and further develops Archimedes' method of exhaustion for finding volumes and areas.

**1653**

Blaise Pascal writes his *Traité du Triangle Arithmétique* which is not published until 1665. Pascal is the first European to discover the arithmetic triangle (which the Chinese knew of centuries before), and it is in this treatise that he introduces mathematical induction into mathematics.

**1654**

Pierre de Fermat exchanges letters with Blaise Pascal in which they discuss the basic laws of probability and essentially found the theory of probability. (See 1657)

**1655**

John Wallis (1616-1703), English mathematician, publishes *Arithmetica Infinitorum*, in which hesystematizes and extends the work of Descartes and Cavalieri. He is also the first to extend the notion of exponents to include negative numbers and fractions.

**1656**

John Wallis is the first to use the sideways figure eight symbol ∞ for infinity. He is also the first to interpret imaginary numbers geometrically.

**1657**

Christiaan Huygens (1629-1695), Dutch physicist and astronomer, writes what is the first formal treatise on probability, basing his work on the ideas brought forth during the Pascal-Fermat correspondence. Huygens also introduces the important concept of "mathematical expectation." (See 1713)

**1657**

Bernhard Frenicle De Bessy (1602-1675), French number theorist, publishes *Solutio Duorum Problematum Circa Numeros Cubos et Quadratos* in which he works on Pythagorean numbers. De Bessy discovers that the number of magic squares increases with the order.

**1657**

Frans van Schooten, Jr. (1615-1661), Dutch mathematician, publishes his *Exercitationes Mathematicae*. He is best known as the teacher of Christiann Huygens, the editor of François Viète's works, and the Latin translator of René Descartes' *La Geometrie*.

**1658**

Blaise Pascal solves the problem of the quadrature of the cycloid. Having given up mathematics after a religious experience, he takes up this problem to distract himself from a painful toothache.

**1658**

Jan (or Johann) Hudde (1628-1704), Dutch geometer, is the first to use three variables in analytic geometry as well as the first to use a letter as a symbol for a negative quantity.

**1659**

Florimond de Beaune (1601-1652), French mathematician, publishes *Cartesii Geometrium* in which he translates Descartes' intentionally obscure rendering of *La Géometrie* into understandable Latin with explanatory notes.

**1659**

Blaise Pascal uses the pseudonym A. Dettonville to write about methods of using infinitesimals. He applies them to practical mathematical problems and takes a major step toward the development of the calculus.

**1659**

Pierre de Fermat first introduces the proof by infinite descent.

**1659**

René François Walter de Sluse (1622-1685), Dutch geometer, publishes *Mesolabum* in which he writes on the cycloid and solutions to third and fourth degree equations. He also discovers a method for drawing tangents to all geometrical curves.

**1659**

Pietro Mengoli (1625-1686), Italian geometer, publishes *Geometria Speciosa* in which he gives the first convergent series of logarithms.

**1661**

Stefana degli Angeli (1623-1697), Italian mathematician, publishes *De Infinitorum Cochlearum Mensuris* in which he develops and applies the method of indivisibles.

**1662**
William Brouncker (1620-1684), English mathematician, becomes the first president of the newly-created Royal Society of London. He also is known for having given the first infinite series for the area of an equilateral hyperbola.

**1662**
John Graunt (1620-1674), English statistician, publishes *Natural and Political Observations. . .Upon the Bills of Mortality* in which he is the first to apply mathematics to the integration of vital statistics. As the first to establish life expectancy and to publish a table of demographic data, Graunt is considered the founder of vital statistics.

**1664**
Isaac Newton (1642-1727), English scientist and mathematician, discovers the binomial theorem either this year or the next, but does not communicate his discovery until 1676 in two letters to German natural philosopher Henry Oldenburg (c.1615-1677).

**1666**
Gottfried Wilhelm von Leibniz (1646-1716), German logician and philosopher, writes his essay, "De Arte Combinatoria," which begins the study of symbolic logic. Since Leibniz calls for a "calculus of reasoning" in mathematics, this is traditionally viewed as the inauguration of symbolic and mathematical logic.

**1667**
James Gregory (1638-1675), Scottish mathematician and astronomer, is the first to study systematically the convergent series and distinguishes between convergent and divergent series.

**1668**
James Gregory publishes two books on geometry that display his work on integration and infinite series. This is another step toward the calculus of Newton and Leibniz.

**1668**
Nicolaus Mercator (c.1620-1687), German mathematician and astronomer, publishes *Logarithmotechnia* in which he is the first to calculate the area under a curve using the newly-developed analytical geometry.

**1669**
Isaac Newton writes "De Analysi per Aequationes Numero Terminorum Infinitas," a paper which he circulates privately and which contains the first systematic account of discovery of fluxional (differential and integral) calculus. (See 1671)

**1669**
Isaac Barrow (1630-1677), English geometer and theologian,

publishes *Lectiones Opticae et Geometricae* in which he introduces the differential triangle and develops a method of tangents that differs from differential calculus in its notation. Barrow is also the first to note the reciprocal relationship between differentiation and integration.

**1669**
Christopher Wren (1632-1723), English architect and scientist, writes on the geometry and surfaces of curves and discovers that on the hyperboloid of revolution of one sheet there are two families of generating lines.

**1671**
Isaac Newton writes "Methodus Fluxionum et Serierum Infinitarum," which amplifies further his invention of the calculus but which he does not publish. (See 1736)

**1673**
Christiaan Huygens publishes *Horologium Oscillatorium* in which he describes his invention of the pendulum clock. This work contains several fundamental contributions to mathematics (involving centripetal force and thependulum) in addition to its major technical accomplishments.

**1673**
Gottfried Wilhelm von Leibniz begins his development of differential and integral calculus independently of Isaac Newton.

**1674**
Seki Kōwa (1642-1708), Japanese mathematician, publishes his only book, the *Hatsubi Sampo*. Considered Japan's greatest mathematician, he is believed to have originated the theory of determinants around 1683.

**1674**
Vincenzo Viviani (1622-1703), Italian geometer, publishes *Elementi d'Euclide* which is one of his several commentaries on the works of the ancient Greek mathematicians. Viviani is the leading geometer of his time.

**1678**
Giovanni Ceva (1647-1734), Italian geometer, publishes his *De Lineis Rectis se Invicem Secantibus Constructio Statica* in which he gives static and geometric proofs for concurrency of straight lines through vertices of triangles (Ceva's theorem). Cevi also discovers theorems on the theory of transversals.

**1679**
Gottfried Wilhelm von Leibniz publishes *Characteristica Geometrica*, which marks the beginnings of the study of "analysis situs," or topology. (See 1847)

**1680**

Jan Hudde writes on the theory of equations and gives an ingenious rule for finding multiple roots of a polynomial that is in use today, where the roots of the highest factor of the polynomial and its derivative can be found.

**1683**

Ehrenfried Walter Tschirnhausen (1651-1708), German algebraist, advances the solution of equations of any degree by removing all the terms except the first and last.

**1684**

Gottfried Wilhelm von Leibniz publishes an account of his discovery of the calculus in the new scientific journal *Acta Eruditorum*. He discovers it independently of Isaac Newton, although later than him. Newton, however, publishes his discovery after Leibniz in 1687. Of the two systems, Leibniz's proves more convenient and flexible because of its well-devised system of notation or symbolism.

**1685**

John Wallis publishes *De Algebra Tractatus: Historicus & Practicus*, which is considered the first serious history of mathematics.

**1685**

Philippe de la Hire (1640-1718), French mathematician, writes three works on conic sections, the best of which is his *Sectiones Conicae* published this year.

**1687**

Isaac Newton publishes *Philosophiae Naturalis Principia Mathematica* which, besides its laws of motion and universal gravitation, contains the first published account of Newton's invention of the calculus. This landmark work is generally considered the beginning of applied mathematics. (See 1704)

**1690**

Jakob Bernoulli (1654-1705), Swiss mathematician, discovers the so-called isochrone or isochronous curve along which a body will fall with uniform vertical velocity.

**1691**

Michel Rolle (1652-1719), French algebraist, publishes *Méthode pour Resoudre les Egalitez* in which "Rolle's theorem" is introduced. It is from this theorem that calculus texts usually derive the highly useful "theorem of mean value."

**1692**

Vincenzo Viviani solves the trisection of an angle problem by using an equilateral hyperbola.

**1693**

Edmund Halley (1656-1742), English astronomer, compiles the first set of detailed mortality tables of the sort presently used in the insurance industry, making possible the statistical study of life and death.

**1696**

Guillaume François Antoine de L'Hospital (1661-1704), French mathematician, publishes *Analyse des Infiniment Petits*, which is the first textbook on the calculus invented by Leibniz to appear. This work does much to make it calculus better known and it also contains Johann Bernoulli's method of finding the limiting value of a fraction whose numerator and denominator tend simultaneously to zero.

**1697**

Johann Bernoulli (1667-1748), Swiss mathematician, solves the problem of the brachistochrone—the determination of the curve of quickest descent of a weighted particle moving between two given points in a gravitational field. This is a major step toward a theory of the calculus of variations.

**1699**

Abraham Sharp (1651-1742), English mathematician, correctly calculates $\pi$ to 71 decimal places.

**1703**

Pierre Raymond de Montmort (1678-1719), French mathematician, publishes *Essai d'Analyse sur Les Jeux de Hazard* in which he applies the calculus of finite differences to probability theory.

**1704**

Isaac Newton returns to his calculus and publishes his paper on the quadrature of curves entitled "Tractatus de Quadratura Curvarum." Newton wrote it between 1692 and 1693 but typically waits several years before publishing.

**1706**

Edmund Halley restores the lost Book VIII of Apollonius' *Conic Sections* by inference.

**1706**

William Jones (1675-1749) English geometer, publishes *Synopsis Palmariorum Matheseos* in which he first uses the 16th letter of the Greek alphabet ($\pi$) as the symbol for the ratio of the circumference to the diameter of a circle. (See 1737)

**1706**

John Machin (1680-1751), English mathematician, calculates the value of $\pi$ to 100 decimal places.

**1707**

Gabriel Manfredi (1681-1761), Italian mathematician, publishes *De Constructione Aequationum Differentialium Primi Gradus*, which is the first Italian work dedicated to pure integral calculus.

**1713**

The first full-length treatment of the theory of probability appears in the posthumously-published *Ars Conjectandi* by the Swiss mathematician, Jakob Bernoulli (1654-1705). In this work, he reproduces Huygens' earlier treatise and also includes a general theory of permutations and combinations, all accompanied by mathematical theorems.

**1715**

Brook Taylor (1685-1731), English mathematician, publishes *Methodus Incrementorum Directa et Inversa,* which is the first treatise on the calculus of finite differences. Taylor's invention of finite differences adds a new branch to mathematics.

**1716**

Giulio Carlo Fagnano (1682-1766), Italian geometer, contributes to the study of the elliptic as he proves that two arcs of any ellipse can be found in an infinity of ways so that a segment of a straight line is their difference.

**1718**

Abraham de Moivre (1667-1754), French-English mathematician, publishes *Doctrine of Chances*, which treats the notion of probability in a systematic manner.

**1719**

Thomas Fantet De Lagny (1660-1734), French mathematician, calculates the value of $\pi$ to 112 correct places.

**1719**

The first arithmetic printed in the American colonies is James Hodder's *Arithmetick; or, that Necessary Art Made Easy,* published in Boston. It was originally published in London in 1661.

**1720**

Colin Maclaurin (1698-1746), Scottish geometer, publishes *Geometria Organica* in which he discusses the general properties of planar curves. Maclaurin is also the first to give a correct theory for distinguishing maxima from minima.

**1722**

English trigonometer Roger Cotes (1682-1716), has his *Harmonia Mensurarum* posthumously published. It is one of the earliest works to recognize the periodicity of the trigonometric functions and contains the first publication of the cycles of the

tangent and secant functions. It is said that Issac Newton remarked on Cotes' early death that "If Cotes had lived, we might have known something."

**1727**

Leonhard Euler (1707-1783), Swiss mathematician, contributes to mathematical notation by suggesting symbolizing the base of the natural logarithms by the letter "e."

**1728**

Daniel Bernoulli (1700-1782), Swiss mathematician, studies the mathematics of oscillations and is the first to suggest the usefulness of resolving a compound motion into motions of translation and rotation.

**1729**

The first arithmetic book written by an American-born individual is the *Arithmetick* by Isaac Greenwood (1701-1745). It is published in Boston.

**1730**

Abraham de Moivre (1667-1754), English mathematician, publishes his *Miscellanea Analytica* in which he invents what is now called De Moivre's Theorem. This discovery revolutionizes higher trigonometry and creates an imaginary trigonometry using recurring series.

**1730**

James Stirling (1692-1770), Scottish mathematician, publishes *Methodus Differentialis* in which he makes important contributions to infinitesimal calculus and the infinite series.

**1730**

The first algebra book printed in the American colonies is Pieter Venema's *Arithmetic; or, the Art of Ciphering*. It is published in New York.

**1731**

Alexis-Claude Clairaut (1713-1765), French geometer, publishes *Recherches sue les Courbes. . .Double Courbure*, which constitutes the first treatise on solid analytic geometry. This elegant work was written by the prodigious Clairaut at age 16, but he waits until he is 18 years old to publish it. Clairaut is elected to the Académie Royale des Sciences the same year.

**1733**

Girolamo Saccheri (1667-1733), Italian mathematician, publishes *Euclides ab Omni Naevo Vindicatus* in which he is the first to discuss the consequences of denying Euclid's fifth postulate (the parallel axiom). Saccheri's work is also the first to suggest the construction of a non-Euclidian geometry independent of that axiom, but he is unable and unwilling to go that far.

**1734**
George Berkeley (1685-1753), Irish philosopher and bishop, publishes *Analyst* in which he attempts to disprove the foundations of calculus and to show that they are no clearer than the principles of religious faith.

**1735**
Leonhard Euler solves the famous Königsberg bridge problem. He proves that it is impossible to cross the seven bridges, once each, in a single, continuous walk. In doing this, Euler proves that a network can be traversed without overlap only if the number of its odd vertices is either two or zero.

**1736**
John Colson edits and translates Isaac Newton's 1671 work on the calculus. Entitled *The Method of Fluxions and Infinite Series*, it is so technical that no printer will publish it until nine years after Newton's death.

**1737**
Leonhard Euler offers the first systematic discussion of continued fractions.

**1737**
Leonhard Euler adopts the 16th letter of the Greek alphabet ($\pi$) as the symbol for the ratio of the circumference to the diameter of a circle. Following his adoption and use, it is generally accepted. It was first suggested in 1706 by English mathematician William Jones.

**1737**
Thomas Simpson (1710-1761), English mathematician, publishes *A New Treatise on Fluxions* in which he expounds the fluxions method using only the procedures of classical geometry.

**1739**
Leonhard Euler invents the method of variation of parameters for solving differential equations.

**1742**
Colin Maclaurin publishes *Treatise of Fluxions* in which he advances the study of higher plane curves and applies classical geometry to physical problems. This is also the first systematic exposition on Newtonian methods, which Maclaurin defends.

**1742**
Christian Goldbach (1690-1764), German-Russian

mathematician, formulates what comes to be known as Goldbach's Theorem. Dealing with number theory, it states that every even number greater than two is the sum of two prime numbers. (See 1937)

**1744**
Leonhard Euler publishes *Methodus Inveniendi Lineas Curvas Maximi Minimive Proprietate Gaudentes* in which he systematically treats the calculus of variations and creates the new field of analytical mechanics. It becomes the first textbook of this new field.

**1748**
Leonhard Euler publishesthe two-volume *Introductio in Analysi Infinitorum*, a work that revolutionizes mathematical analysis and becomes its keystone work. From this time on, Euler's notion of "function" becomes fundamental in analysis.

**1748**
Maria Gaetana Agnesi (1718-1799), Italian mathematician, publishes *Instituzioni Analitiche*, a large, two-volume work that surveys elementary and advanced mathematics. Agnesi is best known for her consideration of the cubic curve or what comes to be translated as the "witch of Agnesi."

**1749**
Gottfried Achenwall (1719-1772), German statistician, introduces and first uses the word "statistik" in a work in which he employs statistical methods to describe the constituents and economic conditions of major countries. Achenwall is considered one of the founders of statistics.

**1750**
Gabriel Cramer (1704-1752), Swiss physicist, publishes *Introduction a l'Analyse des Ligne Courbes Algébriques* in which he amplifies the rule for solving simultaneous linear equations. He also discovers Cramer's Paradox involving curves of the nth order.

**1750**
Leonhard Euler introduces several new functional notations, including the letter "I" for imaginaries and the Greek Sigma ($\Sigma$) for the summation symbol.

**1755**
Nineteen-year-old Joseph-Louis Lagrange (1736-1813), Italian-French algebraist, number theorist, and astronomer, sends a paper to Leonhard Euler concerning his "calculus of variations." Euler is so impressed with the young man's work that he holds back his own writings on the subject, thus allowing Lagrange priority of publication.

**1755**

Leonhard Euler publishes *Institutiones Calculi Differentialis.* Thirteen years later, he publishes *Institutiones Calculi Integralis* (1768). In these two works Euler makes numerous improvements in differential and integral calculus. The system Euler offers is essentially the one still taught in college calculus courses today.

**1756**

The first French version of Issac Newton's *Principia* is published. It is translated some years earlier by Gabrielle-Émilie de Breteuil, the marquise du Châtelet (1706-1749), French mathematician, physicist, and linguist.

**1758**

Jean-Etienne Montucla (1725-1799), French mathematician and the first mathematics historian in France, publishes *Histoire des Mathématiques.* A second edition appears in 1799 and a third is completed by French mathematician Joseph-Jerome le François de Lalande (1732-1807) and published in 1802.

**1758**

John Landen (1719-1790), English mathematician, publishes *A Discourse Concerning the Residual Analysis* in which he argues for a purely algebraical method of solving problems in the calculus.

**1759**

Johann Heinrich Lambert (1728-1777), German geometer, publishes *Die Freye Perspektive* which, although intended for the artist who wants to draw with perspective, is a masterpiece in descriptive geometry.

**1763**

The *Essay Towards Solving a Problem in the Doctrine of Chances,* by English mathematician Thomas Bayes (1702-1761), is published posthumously. In this work, he proposes his theorem of inverse probability whereby probabilities of unknown causes are inferred from observed events. Bayes is also the first to use probability theory inductively.

**1765**

Johann Heinrich Lambert (1728-1777), German geometer, publishes *Beitrage zur Gebrauch der Mathematik und deren Anwendung* in which he gives the first systematic treatment of hyperbolic functions and the notation of these functions. He is also the first to prove rigorously that the number for π is irrational.

**1765**

Vincenzo Riccati (1707-1775), Italian mathematician, publishes *Institutiones Analyticae,* which is the first extensive treatment on the integral calculus. It also introduces the use of hyperbolic functions.

**1766**

Johann Heinrich Lambert questions the validity of Euclid's parallel postulate. Published posthumously in 1781 as *Die Theorie der Parallelinien,* it makes him a forerunner of the discovery of non-Euclidean geometry.

**1767**

Joseph-Louis Lagrange publishes *Trait de Résolution des Equations Numériques de Tous Degres* in which he gives a method of approximating the real roots of an equation by means of continued fractions.

**1770**

Leonhard Euler publishes *Vollständig Anleitung zur Algebra* in which he offers his theories of polynomial and indeterminate equations.

**1770**

Edward Waring (1734-1798), English algebraist, publishes *Meditationes Algebraicae* in which he states a number theory that becomes known as Waring's Problem. He also is the first to set forth a method of approximating the values of imaginary roots.

**1772**

Alexandre Théophile Vandermonde (1735-1796), French mathematician, obtains the formulae for solving general quadratic, cubic, and quartic equations.

**1777**

Georges-Louis Leclerc Buffon (1707-1788), French naturalist, devised his famous "needle problem" by which π may be approximated by probability methods.

**1779**

Étienne Bézout (1730-1783), French mathematician, publishes his *Théorie Générale des Equations Algébriques* in which he studies the theory of equations and offers a method of elimination by symmetric functions.

**1782**

Simon-Antoine-Jean Lhullier (1750-1840), Swiss mathematician, offers the first part of his theory on isoperimeters in which the use of differential methods makes it possible to rediscover classical results and extend them.

**1784**

Three volumes and a supplement on mathematics is begun this year as part of the large, topically-arranged *Encyclopédie.* This new encyclopedia is the brainchild of French encyclopedist Denis Diderot (1713-1784) and becomes the centerpiece of the French Enlightenment.

**1785**

Marie-Jean-Antoine-Nicolas Caritat de Condorcet (1743-1794), French mathematician and social thinker, publishes *Essai sur l'Application de l'Analyse a la Probabilit, des Decisions Rendues a la Pluralit, des voix* in which he contributes to a theory of probability and takes significant steps toward the use of mathematics in the social sciences.

**1785**

Adrien-Marie Legendre (1752-1833), French geometer and number theorist, begins publishing several papers that introduce the special functions now known as Legendre polynomials.

**1788**

Joseph-Louis Lagrange publishes his *Mécanique Analytique* whose mathematical significance is found in its general equations of motion of a dynamical system. These are known today as Lagrange's equations.

**1788**

The first algebra textbook written by an American-born individual is *A New and Complete System of Arithmetic* by Nicholas Pike (1743-1819). It is published in Newburyport, Massachusetts.

**1794**

Adrien-Marie Legendre publishes *Eléments de Géometrie* in which he rearranges and simplifies many of the propositions in Euclid's *Elements*. It is favorably received and is lated translated into several languages.

**1795**

Gaspar Monge (1746-1818), French geometer, makes public his discovery of descriptive geometry. Although as a young man he had invented an original method of using geometry to work out quickly constructional details that ordinarily required long and tedious arithmetical calculations, his methods were guarded until now by the French as a military secret. Monge's descriptive method for representing a solid in three-dimensional space on a two-dimensional plane eventually becomes a part of modern projective geometry. (See 1799)

**1795**

Johann Karl Friedrich Gauss (1777-1855), German mathematician, discovers and gives the first proof of the law of quadratic reciprocity.

**1796**

Johann Karl Friedrich Gauss enters into his diary his discovery that a regular polygon of 17 sides can be constructed with compasses and straightedge. This had defied discovery for more than 2,000 years.

**1797**

Joseph-Louis Lagrange publishes *Théorie des Fonctions Analytiques Contenant les Principes du Calcul Différentiel* in which he attempts to make the calculus more logically satisfying rather than more utilitarian. Although Lagrange does not achieve his goal of removing the concept of limit, his work eventually achieves a broader influence by initiating a new subject called the theory of functions of a real variable.

**1797**

Lorenzo Mascheroni (1750-1800), Italian mathematician, publishes *Geometria del Compasso* in which he reveals his surprising discovery that all Euclidean constructions possible with a rule and compass are possible with a compass alone.

**1797**

Caspar Wessel (1745-1818), Norwegian mathematician, produces the first graphical representation of complex numbers that is useful and consistent. His work goes virtually unnoticed until published in 1897 by the Danish Academy of Sciences.

**1797**

Johann Karl Friedrich Gauss writes his doctoral dissertation in which he gives the first wholly satisfactory proof of the fundamental theorem of algebra (that every algebraic equation must have at least one root, real or imaginary).

**1797**

Lazare-Nicolas-Marguerite Carnot (1753-1823), French geometer and military engineer, publishes *Réflexions sur la Métaphysique du Calcul Infinitesimal* in which he attempts to make the logic of calculus more rigorous.

**1797**

Sylvestre François Lacroix (1765-1843), French mathematician, publishes *Trait, du Calcul Différential et du Calcul Intégral* in which he synthesizes the work of his predecessors and also gives the first definition of a definite and indefinite integral.

**1798**

Adrien Marie Legendre (1752-1833), French number theorist, publishes his two-volume *Essai sur la Theorie des Nombres*, which is the first treatise devoted exclusively to number theory. It is also the first to publish the method of least squares as a way of calculating orbits.

**1799**

Pierre-Simon de Laplace publishes the first of his five-volume classic, *Mécanique Céleste*. From a mathematical standpoint, he develops the very useful concept of potential which is a function whose directional derivative at every point is equal to the component of the field of intensity in the given direction. Also of fundamental importance to mathematical physics is his equation called the Laplacian of a function.

**549**

**1799**

Gaspar Monge publishes *Géometrie Descriptive*, which is the first published exposition of his invention of descriptive geometry. This work revolutionizes engineering design.

**1800**

Louis-François-Antoine Arbogaste (1759-1803), French mathematician, publishes *Calcul des Dérivations* in which he develops what comes to be known as the Arbogast method by which the successive coefficients of a development are derived from one another when the expression is complicated. This work also for the first time separates the symbols of operation from those of quantity.

**1801**

Johann Karl Friedrich Gauss publishes *Disquisitiones Arithmeticae* which is a work of fundamental importance in the modern theory of numbers. In this classic of mathematical literature, he gives the first proof of the law of quadratic reciprocity.

**1803**

Lazare-Nicolas-Marguerite Carnot publishes *Géometrie de Position* in which sensed magnitudes are first systematically employed in synthetic geometry. By means of sensed magnitudes, several separate statements or relations can often be combined into a single inclusive statement or relation, and a single proof can frequently be formulated that would otherwise require the treatment of a number of different cases.

**1803**

Paolo Ruffini (1765-1822), Italian mathematician, makes the first noteworthy attempt to prove that no algebraic solution of the quintic equation (fifth degree or higher) is possible by means of radicals. (See 1824)

**1806**

Lazare-Nicolas-Marguerite Carnot publishes *Essai sur la Théorie des Transversals* in which he extends the theorem of Menelaus to the situation where the transversal in that theorem is replaced by an arbitrary algebraic curve.

**1806**

Jean-Robert Argand writes an essay in which he develops the Argand diagram. This is a method for graphically representing complex numbers and the operations upon them.

**1807**

Gaspar Monge publishes *Application de l'Analyse a la Géometrie* which is considered one of the most important early treatments of the differential geometry of surfaces. Monge's contributions make him the recognized founder of differential geometry.

**1807**

Jean-Baptiste-Joseph Fourier (1768-1830), French mathematician, announces what becomes known as Fourier's theorem. He states that any periodic oscillation, however complex, can be expressed as a mathematical series in which the terms are made up of trigonometric functions. His theorem proves valuable in the study of any wave phenomenon, and the mathematical treatment of such phenomena is called harmonic analysis.

**1810**

Joseph-Diaz Gergonne (1771-1859), French mathematician, founds the journal *Annales de Mathématiques*. For its first 15 years, it is the only journal in the world devoted exclusively to mathematics.

**1812**

Pierre-Simon de Laplace publishes *Théorie Analytique des Probabilités* in which he considers the theory of probability from all aspects and at all levels. This work gives it its modern form.

**1813**

The Analytical Society is founded at Trinity College, Cambridge, England, for the purpose of reforming the teaching and notation of calculus and toward the general revival of mathematical analysis in England. Its founders—George Peacock (1791-1858), John Herschel (1792-1871), and Charles Babbage (1792-1871)—are devoted to the introduction of Leibniz's notation in the calculus.

**1814**

Augustin–Louis Cauchy (1789-1857) French number theorist, publishes a memoir that becomes the basis of the theory of complex functions.

**1816**

Sophie Germain (1776-1831) French mathematician, is awarded a prize by the French Académie Royale des Sciences for her paper on the mathematical theory of elasticity. (See 1831)

**1817**

Bernhard Bolzano (1781-1848), Austrian mathematician, publishes his *Rein Analytischer Beweis* in which he pursues a purely arithmetic proof of the location theorem in algebra. Doing this requires a non-geometric approach to the continuity of a curve or function. This is also the first successful attempt to eliminate the use of infinitesimals from the differential calculus.

**1821**

Augustin Louis Cauchy publishes *Cours d'Analyse de l'École Polytechnique*, the first of three books that gives elementary calculus the rigorous character that it bears today.

**1822**
Jean-Victor Poncelet (1788-1867), French geometer, publishes his *Traité des Proprietés Projectives des Figures*, which gives great impetus to the study of projective geometry and inaugurates its so-called great period. Poncelet is also the first mathematician of stature to argue for thedevelopment of projective geometry as a separate branch of mathematics.

**1822**
Karl Wilhelm Feuerbach (1800-1834), German geometer, publishes a pamphlet in which he not only arrives at the nine-point circle but also proves that it is tangent to the inscribed and three inscribed circles of the given triangle. This last fact is known as Feuerbach's theorem, one of the more elegant theorems in the modern geometry of the triangle.

**1822**
Augustin- Louis Cauchy lays the foundations of the mathematical theory of elasticity.

**1824**
Niels Henrik Abel (1802-1829), Norwegian mathematician, publishes a paper in which he first proves the impossibility of solving the general quintic equation by means of radicals. This problem had puzzled mathematicians for two and a half centuries. Abel's proof is more rigorous than Ruffini's (1803), and his work also contains the first rigorous proof of the general binomial equation.

**1824**
Friedrich Wilhelm Bessel (1784-1846), German astronomer, publishes a paper in which he introduces a class of transcendental functions for applied mathematics that become known as Bessel's functions. He gives their principal properties and constructs tables for their evaluation.

**1825**
Adrien-Marie Legendre publishes *Traité des Fonctions Elliptiques et des Intégrales Euleriennes*, a three-volume work in which heintroduces the term "Eulerian integrals" for the beta and gamma functions and provides some basic tools of analysis that are very helpful to mathematical physicists.

**1825**
Joseph-Diaz Gergonne gives the principle of duality in geometry its first clear expression.

**1826**
August Leopold Crelle (1780-1855), German mathematician, founds the *Journal fur die Reine und Angewandte Mathematik*. It comes to be known as *Crelle's Journal*.

**1826**
Niels Henrik Abel publishes papers on elliptic functions and Abelian integrals in the newly-formed *Crelle's Journal*.

**1827**
Johann Karl Friedrich Gauss publishes *Disquisitones Generales circa Superfices Curvas* which becomes the basic work on the theory of surfaces. It is also the first systematic study of quadratic differential forms in two variables.

**1827**
August Ferdinand Möbius (1790-1868), German mathematician, publishes his *Der Barycentrische Calcul* which is a new treatment of analytical geometry by the introduction of homogeneous coordinates and the principle of duality.

**1828**
George Green (1793-1841), English mathematician, publishes his *Essay on the Application of Mathematical Analysis to the Theory of Electricity and Magnetism* in which he is the first to apply the concept of "potential function" to other than gravitational problems. Green introduces it into the mathematical theory of electricity and magnetism.

**1829**
Karl Gustave Jacob Jacobi (1804-1851), German mathematician, publishes his *Fundamenta Nova Theoriae Functionum Ellipticarum*in which he develops, independent of Abel, the theory of elliptic functions. He also is one of the early founders of determinant theory.

**1829**
Lambert-Adolphe-Jacques Quetelet (1796-1874), Belgian statistician and astronomer, gives the first statistical breakdown of a national census. He correlates death with age, sex, occupation, and economic status in the Belgian census.

**1829**
Jacques-Charles-François Sturm (1803-1855), Swiss algebraist, originates Sturm's theorem for determining the number and position of real roots of an algebraic equation between given limits.

**1829**
Nicolai Ivanovich Lobachevski (1793-1856), Russian geometer, publishes the first of a series of five papers in the *Kazan University Courier* in which he describes his discovery of non-Euclidean geometry. As early as 1826, Lobachevski had conceived this self-consistent mathematical system that includes the concept that an indefinite number of lines can be drawn in a plane parallel to a given line through a given point. This profoundly significant mathematical event led eventually to the mathematics of curved surfaces and Albert Einstein's

demonstration that the universe is non-Euclidean in structure. (See 1832)

**1830**
Évariste Galois (1811-1832), French algebraist and group theorist, publishes a short paper that essentially creates the study of groups. He is the first to use the word "group" in the technical sense and to apply groups of substitutions to the question of reducibility of algebraic equations.

**1830**
Augustus De Morgan (1806-1871), English logician, publishes *Elements of Arithmetic* in which he first introduces the term "mathematical induction," describing the logical process traditionally used by mathematicians but never rigorously considered or clarified.

**1831**
Sophie Germain introduces her notion of mean curvature (the arithmetic average of the two principal curvatures) into the differential geometry of surfaces.

**1832**
János Bolyai (1802-1860), Hungarian mathematician, appends a 26-page work titled *Appendix Scientiam Spatii Absolute Veram Exhibens* to his father Farkas's two-volume semi-philosophical work on elementary mathematics. In this short work, Bolyai announces his discovery of non-Euclidean geometry which he made at about the same time as N. I. Lobachevski. His discovery is totally independent of the Russian's, and when Bolyai finally sees Lobachevski's work, he thinks it has been plagarized from his own. (See 1829)

**1832**
Jakob Steiner (1796-1863), Swiss geometer, publishes *Systematische Entwickelungen*, which contains a complete discussion of reciprocation and the principle of duality and helps lay the foundation for synthetic geometry of the future.

**1833**
George Peacock (1791-1858), English algebraist, states that results demonstrated as valid for arithmetical algebra are also valid for symbolic algebra.

**1833**
William Rowan Hamilton (1805-1865), Irish algebraist, makes one of the first attempts at analyzing the basis of the irrational numbers. His theory views both rationals and irrationals as based on algebraic number couples.

**1833**
Gabriel Lam, (1795-1870), French mathematician, introduces to

applied mathematics his use of curvilinear coordinates.

**1835**
Julius Plücker (1801-1868), German geometer, publishes his *Systeme der Analytischen Geometrie* which lays the foundations of modern analytic geometry.

**1836**
Joseph Liouville (1809-1882), French number theorist, founds the mathematical journal *Journal de Mathématiques Pures et Appliquées*. It is known as *Liouville's Journal*.

**1837**
Peter Gustav Lejeune Dirichlet (1805-1859), German mathematician, proves that the sum of an absolutely convergent series is not altered by a reordering of the terms in the series.

**1837**
Pierre-Laurent Wantzel (1814-1848), French geometer, gives the first rigorous proofs of the impossibility of duplicating the cube or trisecting angles by the use of only a ruler and compasses.

**1837**
William Whewell (1794-1866), English mathematician and philosopher, publishes *The Mechanical Euclid* in which he expresses the need to axiomatize branches of mathematics other than geometry.

**1838**
Antoine-Augustin Cournot (1801-1877), French mathematician and economist, publishes his *Recherches sur Les Principes Mathématiques de la Théorie des Richesses* in which he is the first economist to apply mathematics to the study of economics.

**1839**
Peter Gustav Lejeune Dirichlet publishes *Vorlesungen Über Zahlentheorie*, which constitutes one of the most lucid introductions to the number theory of Johann Karl Friedrich Gauss.

**1839**
In England, the *Cambridge Mathematical Journal* is founded. In 1846 it becomes the *Cambridge and Dublin Mathematical Journal* and then changes its title again in 1855 to the *Quarterly Journal of Pure and Applied Mathematics*.

**1840**
Pierre-Charles-Alexandre Louis (1787-1872), French physician, pioneers medical statistics, being the first to compile systematically records of diseases and treatments.

**1841**

William Rutherford of England calculates the value of $\pi$ to 208 places. (See 1853)

**1841**

The first permanent journal devoted to the interests of teachers of mathematics rather than to mathematical research, the *Archiv der Mathematik und Physik,* is founded.

**1841**

Karl Gustav Jacob Jacobi is an early founder of the theory of determinants and invents the functional determinant or what is called the Jacobian.

**1842**

Jakob Steiner proves that for any given perimeter, the triangle with the greatest area is the equilateral triangle.

**1843**

William Rowan Hamilton discovers quaternions and invents a new branch of mathematics known as the algebra of quaternions. A quaternion is a type of complex number composed of one real number, and Hamilton uses geometric principles to manipulate complex numbers.

**1844**

Hermann Gunther Grassmann (1809-1877), German algebraist, publishes *Die Ausdehnungslehre* in which he pioneers modern vector analysis and creates a new algebra of n-dimensional space called the theory of extension. (See 1862)

**1844**

Arthur Cayley (1821-1895), English mathematician, creates and develops his theory of invariants which creates a new branch of analysis. (See 1878)

**1844**

Ferdinand Gotthold Eisenstein (1823-1852), German number theorist, helps develop the theory of complex numbers and also discovers the first covariant used in analysis.

**1844**

Joseph Liouville proves the existence of transcendental numbers.

**1847**

Karl Georg Christian von Staudt (1798-1867), German geometer, publishes his *Geometrie der Lage* which finally frees projective geometry of any metrical basis. He also offers the first complete theory of imaginary points, lines, and planes in projective geometry. (See 1856)

**1847**

George Boole (1815-1864), English logician, publishes *The Mathematical Analysis of Logic* in which he maintains that the essential character of mathematics lies in its form rather than in its content. This epoch-making work focuses attention on mathematics as symbol rather than only number and measurement. (See 1854)

**1847**

Ernst Eduard Kummer (1810-1893), German number theorist, originates the system of ideal numbers leading to the developing field of the theory of numbers.

**1847**

Augustus De Morgan publishes *Formal Logic; or, the Calculus of Formal Inference, Necessary and Probable* in which he develops a new logic of relations and offers a new system of nomenclature for logical expression.

**1847**

Johann Benedict Listing (1808-1882), German topologist, publishes *Vorstudien zur Topologie* which is the first book devoted to the systematic study of topology as a branch of geometry. He is also the first to introduce the term topology.

**1849**

Pafnuty Lvovich Chebyshev (1821-1894), Russian number theorist, publishes *Teoria Sravneny* which becomes a standard Russian textbook on the theory of numbers.

**1849**

William Whewell first offers the name "intrinsic equation" and points out its use in studying successive evolutes and involutes. This is a reaction to the use of arbitrary Cartesian and polar coordinates.

**1850**

Amede Mannheim (1831-1906), French mathematician, standardizes the modern slide rule.

**1850**

Bernhard Bolzano (1781-1848), Austrian mathematician, postulates means of distinguishing between finite and infinite classes in his posthumously published *Paradoxien des Unendlichen.*

**1852**

Michel Chasles (1793-1880), French geometer, publishes his *Traité de Géométrie Supérieure* in which he establishes the use of directed line segments in pure geometry.

**1853**
William Rutherford of England calculates the value of π to 400 correct decimal places.

**1853**
William Rowan Hamilton publishes *Lectures on Quaternions* in which he further develops his theory of quaternions and applies them to geometry and spherical trigonometry.

**1854**
George Boole publishes *Investigation of the Laws of Thought*, which establishes both formal logic and Boolean algebra. In his mathematization of logic, Boole elaborates on his work of 1847.

**1854**
Georg Friedrich Bernhard Riemann (1826-1866), German geometer, gives a probationary lecture at the University of Gottigen and presents a thesis offering a global view of geometry as a study of manifolds of any number of dimensions in any kind of space. He develops further the ideas of N. I. Lobachevski and János Bolyai and introduces a new, non-Euclidean system of geometry (now known as Riemann geometry) and a theory of space that provides a geometric foundation for modern physical theory.

**1854**
Arthur Cayley offers thefirst technical definition of a group and states rules that govern it.

**1856**
Karl Georg Christian von Staudt revises his earlier work and publishes *Beitrage zur Geometrie der Lage* in which he attempts unsuccessfully to show that analytic methods are superfluous to geometry. It also contains the first general theory of imaginary points, lines, and planes in projective geometry.

**1857**
Arthur Cayley invents the algebra of matrices, which will eventually be used by German physicist Werner Karl Heisenberg (1901-1976) in his quantum mechanics.

**1858**
Thomas Penyngton Kirkman (1806-1895), English group theorist, makes an early attempt at determining all permutation groups of given orders.

**1858**
Charles Hermite (1822-1901), French mathematician, publishes his *Sur la Résolution de l'Equation du Cinquième degré*, in which he offers the first solution of the general equation of the fifth

degree which he achieves by means of elliptic functions.

**1860**
Georg Friedrich Bernhard Riemann uses complex number theory and offers what becomes known as Riemann's Hypothesis. This will form the basis for most of the research in the theory of prime numbers for the next century.

**1860**
Karl Theodor Wilhelm Weierstrass (1815-1897), German mathematician, makes irrational numbers understandable in terms of sequences of rational numbers.

**1862**
Hermann Gunther Grassmann develops further his theory of extension and develops the algebra of tensor calculus, which becomes important to the later theory of general relativity.

**1862**
Luigi Cremona (1830-1903), Italian mathematician, publishes *Introduzione a Una Teoria Geometrica delle Curve Piane* in which he reconstructs the general theory of algebraic plane curves in a geometrical form. (See 1863)

**1863**
Luigi Cremona extends the theory of the transformation of curves and the correspondence of curves to three dimensions.

**1865**
The London Mathematical Society is founded and is the first of its kind. It immediately begins publishing its *Proceedings*.

**1865**
Ludwig Otto Hesse (1811-1874), German geometer, publishes his *Analytische Geometrie der Ebene* in which he applies the newly-developed theory of algebraic forms to the analytic geometry of curves and surfaces.

**1866**
Rudolf Friedrich Alfred Clebsch (1833-1872), German geometer, publishes *Theorie der Abelschen Funktionen* in which he applies elliptic theory and Abelian functions to geometry and the study of rational and elliptic curves.

**1867**
Hermann Hankel (1839-1873), German mathematician, publishes his *Theorie der Complexen Zahlensysteme* which includes the principle of permanency of the formal laws of calculus.

**1868**

Julius Plücker publishes *Neue Geometrie des Raumes* in which he elaborates the geometry of Cartesian three-dimensional space using straight lines and not points as the basic elements.

**1868**

Eugenio Beltrami (1835-1900), Italian geometer, publishes *Saggio di Interpretazione della Geometria Non Euclidea* in which he proposes the first Euclidean model of non-Euclidean (hyperbolic) geometry. This opens the way to the admission of the relative non-contradictoriness of non-Euclidean geometry.

**1869**

Francis Galton (1822-1911), English anthropologist, publishes *Hereditary Genius* in which he introduces the correlation coefficient. Galton is recognized as the founder of the statistical school of genetics. (See 1889)

**1869**

Elwin Bruno Christoffel (1829-1900), German mathematician, studies the problem of the equivalence of quadratic forms and also is the first to discuss the mapping of polygons.

**1869**

Hermann Amandus Schwarz (1843-1921), German geometer, conducts his significant studies on the theory of minimal surfaces.

**1870**

Benjamin Peirce (1809-1880), American algebraist, publishes *Linear Associative Algebra* in which he works out the multiplication tables for 162 algebras. Peirce also gives the famous definition, "Mathematics is the science which draws necessary conclusions."

**1870**

Karl Theodor Wilhelm Weierstrass makes substantial contributions to the arithmetization of analysis. His work results in the fact that today, all of analysis can be logically derived from a postulate set characterizing the real number system. For this, Weierstrass is called the "father of modern analysis."

**1870**

Camille Jordan (1838-1922), French number theorist, publishes *Trait des Substitutions et des Equations Algèbriques* in which he offers the fundamental concept of class of a substitution group and also proves the constancy of the factors of composition. He is also the first to systematically develop Galois' theory of groups.

**1871**

Carl Ohrtmann (1839-1885) founds the mathematical abstracting journal *Jahrbuch Über die Fortschritte*

*der Mathematik.*

**1871**

Enrico Betti (1823-1892), Italian mathematician, lays the basis for topology through his work on multidimensional spaces and his Betti numbers. These are the largest number of crosscuts that can be made without dividing a surface into more than one piece.

**1872**

La Société Mathématique de France is established in Paris, with its official journal known as its *Bulletin.*

**1872**

Felix Klein (1849-1925), German geometer, gives an address at the University of Erlangen which sets forth a new definition of geometry. This address comes to be known as the "Erlanger Program" and it serves to unify the many diverse forms of geometry.

**1872**

Julius Wilhelm Richard Dedekind (1831-1916), German mathematician, publishes his *Stetigkeit und Irrationale Zahlen* in which he offers the modern exposition of irrational numbers. Using his "Dedekind cuts" his methods make irrational as useful as rational numbers.

**1872**

Ludwig Sylow (1832-1918), Norwegian mathematician, originates Sylow subgroups and develops a fundamental theorem defining groups in an accepted technical sense.

**1872**

Heinrich Eduard Heine (1821-1881), German mathematician, develops what comes to be called the Heine-Borel Theorem on point set theory.

**1873**

Charles Hermite (1822-1901), French mathematician, publishes an article in which he proves that the number "e" is transcendental. Hermite shows that "e" cannot be the root of any polynomial equation with integral coefficients. This is the first familiar number that is shown to be transcendental.

**1873**

William Shanks (1812-1882), English mathematician, calculates the value of π to 707 decimal places. (See 1948)

**1873**

William Kingdon Clifford (1845-1879), English algebraist, develops the theory of biquaternions and links them with more general associative algebras. He uses biquaternions to study motion in non-Euclidean spaces.

**1874**
Georg Ferdinand Ludwig Philipp Cantor (1845-1918), German mathematician, begins his revolutionary work on set theory and the theory of the infinite and creates a whole new field of mathematical research. His set theory forms the basis of modern analysis and his new approach to the concept of infinity brings a new critical examination of the foundations of mathematics. (See 1883)

**1874**
Leon Walras (1834-1910), French economist, publishes his *Éléments d'Économie Pure* in which he uses mathematics in economic analysis and attempts to model his economics on Newtonian mechanics.

**1875**
Paul Gordan (1837-1912), German mathematician, publishes his *Über des Formensystem Binarer Formen* in which he proves the finiteness of any invariant system for any given binary form.

**1876**
Rudolf Lipschitz (1832-1903), German mathematician, further develops the existence theorem for solutions of first order differential equations of the French mathematician, Augustin-Louis Cauchy and formulates what comes to be known as the Cauchy-Lipschitz Theorems.

**1877**
George William Hill (1838-1914), American astronomer and mathematician, studies the trajectory of the moon's perigee and uses infinite determinants for the first time in analyzing its motion.

**1878**
James Joseph Sylvester (1814-1897), English-American algebraist, becomes the first editor of the *American Journal of Mathematics*. During his career, he founds with Arthur Cayley, the theory of algebraic invariants. This becomes essential to the theory of relativity.
(See 1844)

**1878**
Arthur Cayley states that the properties of groups are defined by their multiplication tables, thus asserting the essential equivalence of isomorphic groups.

**1878**
Ulisse Dini (1845-1918), Italian mathematician, publishes *Fondamenti per La Teoria delle Funzioni di Variabili Reali* in which he critically revises the foundations of analysis.

**1879**
Friedrich Ludwig Gottlob Frege (1848-1925), German logician and philosopher, publishes *Begriffsschrift,* a pioneering work in modern mathematical logic. (See 1884)

**1880**
Morris Benedikt Cantor (1829-1920), German mathematician, begins writing his four-volume history of mathematics entitled *Vorlesungen über Geschichte der Mathematik.*

**1880**
Jules-Henri Poincaré, (1854-1912), French mathematician and physicist, studies combinational topology using qualitative integrations of differential equations, and his work becomes the basis of 20th century combinatorial or algebraic topology. Poincaré is called the last of the universal mathematicians because of his first-class creative work in any branch of mathematics.

**1881**
Josiah Willard Gibbs (1839-1903), American mathematical physicist, publishes his *Vector Analysis* in which he begins the development of vector analysis and applies it to the problems of crystallography and to the computation of planetary and cometary orbits.

**1882**
Ferdinand Lindemann (1852-1939), German mathematician, first proves that $\pi$ is transcendental and that therefore it is impossible to square the circle (quadrature) using a ruler-and-compass construction. A complex number is said to be algebraic if it is a root of some polynomial having rational coefficients. Otherwise it is said to be transcendental.

**1882**
Leopold Kronecker (1823-1891), German algebraist, publishes his work on algebraic fields entitled *Theorie der Algebraischen Grossen.* He is best known for vehemently condemning the revolutionary work of Georg Cantor (and eventually being proved wrong). (See 1874)

**1882**
Moritz Pasch (1843-1930), German geometer, publishes *Vorlesungen über Neuere Geometrie,* which contains a system of axioms for descriptive geometry. He becomes the first geometer since Euclid to present elements of geometry as abstract postulates of their relations and helps lay the foundations of modern geometry.

**1882**
Giuseppe Veronesi (1854-1917), Italian geometer, publishes a paper in which he discusses the possibility of extending the

principle theories of elementary and analytical geometry to hyperspaces without the aid of coordinates. This marks the beginnings of this important line of research.

**1883**
The Edinburgh Mathematical Society is founded in Scotland and begins publishing its *Proceedings.*

**1883**
Georg Ferdinand Ludwig Philipp Cantor (1845-1918), German mathematician, publishes his *Grundlagen einer Allgemeinen Mannigfältigkeitslehre* in which he develops transfinite set theory and introduces transfinite ordinal numbers as a natural extension of the number concept.

**1884**
The Circolo Matematico di Palermo is organized in Italy and begins publishing its *Rendiconti* three years later.

**1884**
Friedrich Ludwig Gottlob Frege publishes *Die Grundlagen der Arithmetik* in which he studies the relation of logic to mathematics and develops a new set of mathematical definitions and symbols. Frege's work helps found modern formal logic. (See 1893)

**1886**
Luigi Bianchi (1856-1928), Italian geometer, publishes his *Lezioni de Geometria Differenziale* in which he reorganizes and systematizes differential geometry.

**1887**
Gustav Theodore Fechner (1801-1887), German experimental psychologist, publishes a paper in which he founds psychophysics. He attempts to make psychology an exact science by applying mathematical probability and statistics to experimental psychology.

**1887**
Vito Volterra (1860-1940), Italian mathematician, begins his study of functionals ( functions depending on other functions). His work expands the theory of functional analysis.

**1887**
Curbastro Gregorio Ricci (1853-1925), Italian mathematician, begins writing a series of papers in which he lays the foundations of absolute differential calculus or what is now called tensor analysis.

**1888**
The New York Mathematical Society is formed and begins issuing its *Bulletin.* (See 1894).

**1888**
Émile Lemoine (1840-1912), French mathematician, introduces his science of geometrography for quantitatively comparing one construction with another.

**1888**
Sonya Kovalevskaya (1850-1891), Russian mathematician also known as Sofia Kovalevskaia, receives the prestigious Prix Bordin from the French Académie Royale des Sciences for her theory of movement of a solid body around an immovable point.

**1888**
Marius Sophus Lie (1842-1899), Norwegian algebraist, publishes his *Theorie der Transformationsgruppen* in which he discovers contact transformations. These are analytic transformations that carry tangent surfaces into tangent surfaces (or change the straight lines of ordinary space into spheres).

**1888**
Julius Wilhelm Richard Dedekind publishes *Was Sind und Was Sollen die Zahlen* in which he claims that all of mathematics is a branch of logic.

**1889**
Francis Galton publishes *Natural Inheritance* in which he further develops his concept of correlation coefficients and offers the formula for the standard error of estimate.

**1889**
Giuseppe Peano (1858-1932), Italian logician, publishes *Arithmetices Principia, Novo Methodo Exposita* which is a treatment of his system of mathematical logic. (See 1894)

**1890**
The Deutsche Mathematiker-Vereinigung is organized in Germany and begins publishing the *Jahresbericht* in 1892.

**1890**
Giuseppe Peano builds a continuous plane curve that completely fills a square, thus constructing the first "space-filling" curve.

**1890**
Ernst Schröder (1841-1902), German logician, publishes the initial volume of his four-volume *Vorlesungen über die Algebra* which is a treatment of the symbolic logic of his time.

**1892**
Federigo Enriques (1871-1949), Italian algebraist, elaborates the general theory of the invariants of algebraic surfaces.

**1893**

Eliakim Hastings Moore (1862-1932), American mathematician, proves that every finite commutative field contains a finite number of distinct elements.

**1893**

Friedrich Ludwig Gottlob Frege publishes *Grundgesetze der Arithmetik* in which he develops further his concept of the axiomatization of logic. Just as he is about to publish his second volume of this work in 1902, Frege receives a letter from English mathematician and philosopher Bertrand Russell, telling him that his work contains a paradox or contradiction, thus invalidating nearly ten years of work.

**1894**

The New York Mathematical Society changes its name to the American Mathematical Society and begins issuing its *Transactions*.

**1894**

Giuseppe Peano collaborates in the publication of *Formulaire de Mathématiques*, which attempts to develop a formalized language that would contain not only mathematical logic but all the most important branches of mathematics. Peano is considered one of the founders of symbolic logic.

**1894**

John Venn (1834-1923), English logician, publishes his *Symbolic Logic* in which he graphically illustrates the algebra of classes by means of what come to be known as Venn diagrams.

**1895**

Jules-Henri Poincaré publishes *Analysis Situs* and founds combinatorial topology. He also introduces the notion of the Poincaré group also known as the first homotropy group.

**1895**

Grace Emily Chisholm (1868-1944), English mathematician, becomes the first woman to receive a German doctorate in mathematics through the regular examination process. She attends Göttingen University in Germany because, at this time, females were not admitted into British graduateschools. Chisholm later marries English mathematician William Henry Young. (See 1906)

**1896**

Jacques-Salomon Hadamard (1865-1963), French mathematician, and Belgian mathematician Charles-Jean-Gustave de la Vallée-Poussin, prove the prime number theorem independently.

**1896**

Hermann Minkowski (1864-1909), German mathematician,

publishes *Geometrie der Zahlen* in which he devises a geometrical method for number-theory study. This is the first major work on the geometrical theory of numbers.

**1896**

Ferdinand Georg Frobenius (1849-1917), German group theorist, works on theories of finite groups and creates theorems bearing his name.

**1897**

The first "official" international mathematical congress is held at Zurich, Switzerland. These will be held every four years.

**1897**

David Hilbert (1862-1943), German algebraist, publishes *Der Zahlbericht* in which he reorganizes algebraic number theory.

**1898**

Felix Klein (1849-1925), German mathematician, founds the great mathematical work *Enzyklopädie der Mathematischen Wissenschaften*. This large, cooperative work is intended to digest the now-burgeoning results of modern mathematical research.

**1899**

David Hilbert publishes *Grundlagen der Geometrie* which is the first book to contain a genuinely satisfactory set of axioms for geometry. It also founds the formalist school whose thesis is that mathematics is concerned with formal symbolic systems.

**1900**

Frank Morley (1860-1937), American geometer, proves a result concerned with entities in Euclidean geometry that cannot beconstructed with straightedge and compass alone.

**1900**

David Hilbert presents a list of 23 unsolved problems to the International Congress of Mathematicians in Paris which stimulates continued research for decades.

**1901**

Karl Pearson (1857-1936), English mathematician, founds the statistical journal *Biometrika*. He is one of the founders of modern statistics and works to apply statistics to the biological and social sciences.

**1902**

Johannes Tropfke (1866-1939), German mathematician, publishesthe first edition of his seven-volume *Geschichte der Elementarmathematik*. This is an especially significant work on the history of mathematics' elementary branches.

**1903**
Bertrand Russell (1872-1970), English mathematician and philosopher, publishes *The Principles of Mathematics* in which he outlines his plan to build a pure mathematics using a small number of logical concepts and principles. (See 1910)

**1904**
Henri-Léon Lebesgue (1875-1941), French mathematician, publishes *Leçons sur l'Intégration et la Recherche des Fonctions Primitifs* in which he offers his theory on the functions of real variables.

**1904**
Ernst Friedrich Ferdinand Zermelo (1871-1953), German mathematician, formulates an axiom of choice to prove that every set can be well-ordered. This is now known as Zermelo's axiom. (See 1908)

**1905**
Maurice-René Fréchet (1878-1973), French mathematician, publishes *Sur Quelques Points du Calcul Fonctionnel,* which systematically establishes thefundamental principles of functional calculus and introduces much of the terminology in use today.

**1905**
Karl Pearson publishes a letter in *Nature* in which he introduces the stochastic model called the "random walk." Pearson discusses the situation of a hypothetical walker whose direction and number of steps are unknown, and only the conditions of transit are considered. This model is later used in chemistry, epidemiology, and sociology.

**1906**
Maurice-René Fréchet inaugurates the study of abstract spaces and gives the first definition and theory of abstract sets. Geometry now becomes more generalized as the study of a set of points with some superimposed structure.

**1906**
Andrei Andreyevich Markov (1856-1922), Russian probabilist, begins work toward his development of the theory of stochastic processes, especially those called "Markov chains." In mathematics, this is a sequence of random events in which the probability of each event depends upon the outcome of previous trials.

**1906**
Ernest Julius Wilczynski (1876-1932), German-American geometer, publishes *Projective Differential Geometry of Curves and Ruled Surfaces.* His theory of projective differential geometry is a major development in that field.

**1906**

Thomas Muir (1844-1934), Scottish mathematician, publishes the first of his five-volume history of determinants entitled *Theory of Determinants in Their Historical Order of Development.*

**1906**
William Henry Young (1863-1942), English number theorist, and his wife, Grace Chisholm Young, publish *The Theory of Set Points* which is the first comprehensive textbook on set theory and its applications to function theory.

**1908**
Luitzen Egbertus Jan Brouwer (1881-1966), Dutch mathematician, founds the intuitionist school of mathematics. This revolutionary doctrine views the nature of mathematics as mental constructions governed by self-evident laws. (See 1912)

**1908**
Ernst Friedrich Ferdinand Zermelo provides a set of seven axioms which places set theory on an axiomatic basis.

**1909**
Edmund Landau (1877-1938), German number theorist, publishes *Handbuch der Lehre von der Verteilung der Primzahlen,* which is one of the first treatises on the analytical theory of numbers.

**1910**
Ernst Steinitz (1871-1928), German algebraist, generalizes Leopold Kronecker's theory of algebraic magnitudes and provides a method of introducing positive rational numbers that is free from intuition.

**1910**
Alfred North Whitehead and Bertrand Russell publish their classic three-volume *Principia Mathematica,* in which they propound that all mathematics is a subset of formal logic and in which they try unsuccessfully to make all of mathematics completely rigorous.

**1912**
Luitzen Egbertus Jan Brouwer attacks formalism and logicism in mathematics and offers his own theory of intuitionism. He suggests that humans are born with a innate sense of the natural numbers and that logic is a subset of mathematics and not the opposite.

**1914**
Felix Hausdorff (1868-1942), German topologist, publishes his *Grundzüge der Mengenlehre* in which he offers the classical formulation of set topology. In this systematic exposition, he develops topological spaces that become known as Hausdorff spaces.

559

**1915**

Arthur Lyon Bowley (1869-1957), English statistician and economist, publishes *Livelihood and Poverty*. This is a pioneering work in sample surveys and does much to advance the proper application of statistics to social studies.

**1918**

Hermann Weyl (1885-1955), German-American geometer, creates the first non-Riemannian geometry while attempting to unify James Clerk Maxwell's electromagnetic field theory and Albert Einstein's gravitational field theory.

**1921**

Emmy Noether (1882-1935), German-American algebraist, publishes her studies on abstract rings and ideal theory which become particularly important in the development of modern algebra.

**1921**

Émile Felix-Édouard-Borel (1871-1956), French number theorist, initiates the modern mathematical theory of games by closely analyzing several familiar games.

**1922**

Ludwig Josef Johann Wittgenstein (1889-1951), Austrian philosopher, suggests that all mathematics is a tautology.

**1922**

Oswald Veblen (1880-1960) and Luther Pfahler Eisenhart (1876-1965), American geometers, generalize the Riemannian theory of geodesics into a geometry of paths.

**1925**

Ronald Alymer Fisher (1890-1962), English statistician, publishes his *Statistical Methods for Research Workers* in which he develops statistical techniques for the analysis of variance as well as for the use and validation of small samples.

**1925**

David Hilbert (1862-1943) publishes his paper "On the Infinite" in which he elaborates on his formalist view of mathematics. Hilbert views it as a formal system based on abstract symbols that do not have a physical reality.

**1927**

Stanisław Leshniewski (1886-1939), Polish logician, begins publishing his theories. He develops one of the most complete, precise, and rigorous systems of logic and of the foundations of mathematics that have yet been devised. His work sets a new standard for mathematical rigor.

**1928**

John von Neumann (1903-1957), Hungarian-American number theorist, and Oskar Morgenstren (1902-1977), German-American economist, write *The Theory of Games and Economic Behavior* in which they develop game theory or a mathematical theory of games. This work is not published until 1944.

**1928**

Richard von Mises (1883-1953), Austrian-American probabilist, publishes *Probability, Statistics and Truth* in which he offers a philosophical approach to probability theory.

**1931**

Stefan Banach (1892-1945), Polish mathematician, publishes *Théorie des Opérations Linéaires* in which he founds modern functional analysis and helps develop the theory of topological vector spaces.

**1931**

Kurt Gödel (1906-1978), Austrian-American mathematician, publishes a paper whose incompleteness theorem startles the mathematical community. Called Gödel's proof, it states that within any rigidly logical mathematical system there are propositions that cannot be proved or disproved on the basis of the axioms within that system. Ironically, his proof that there is no certainty in mathematics proves to be a stimulus to research.

**1931**

George David Birkhoff (1884-1944), American probabilist, formulates the ergodic theorem which concerns limits of probability and weighted means.

**1931**

Richard von Mises (1883-1953), Austrian-American probabilist, publishes his *Wahrscheinlichkeits-rechnung* in which he introduces the concept of sample space into probability theory.

**1932**

The Fields Medal of the International Mathematical Congress is established on the behest of Canadian mathematician John Charles Fields upon his death. This prize becomes the equivalent of the Nobel Prize in mathematics.

**1932**

Abraham Adrian Albert, American algebraist, proves the existence of noncyclic division algebras of degree four.

**1934**

Aleksandr Osipovich Gelfond (1906-1968), Russian algebraist and number theorist, originates basic techniques in the study of transcendental numbers (numbers that cannot be expressed as the root or solution of an algebraic equation with rational coefficients), and profoundly advances transcendental number theory.

**1936**
Alonzo Church, American mathematician, offers Church's theorem, which states that there is no single method for determining whether a mathematical statement is provable or even true.

**1937**
Alan Mathison Turing (1912-1954), English mathematician, publishes his paper "On Computable Numbers" in which he describes the imaginary "Turing machine" that can solve all computable problems and with which he helped prove the existence of undecidable mathematical statements.

**1937**
Ivan Matveevich Vinogradov (1891-1983), Russian number theorist, publishes *New Method in the Analytical Theory of Numbers* in which he develops a formula for number representations that leads to his solution of the Goldbach problem. (See 1742)

**1939**
The Bourbaki group is formed in France and begins publishing its on-going *Eléments de Mathématiques*. Working under the collective pseudonym of Nicolas Bourbaki, a group of mostly French mathematicians begins writing what it considers to be a definitive synthesis and survey of all mathematics.

**1940**
Abraham Adrian Albert publishes *Introduction to Algebraic Theories* and begins his work on non-associative algebras.

**1942**
The number of pages published in mathematical journals falls to one-half its pre-war volume, and graduate work in pure mathematics in England and the United States approaches the vanishing point.

**1943**
The first volume of the *Quarterly of Applied Mathematics* appears. It is sponsored by Brown University in Providence, Rhode Island.

**1945**
The first Canadian Mathematical Congress is held in Montreal, Quebec.

**1948**
D. F. Ferguson of England and J. W. Wrench of the United States jointly publish a corrected version of William Shanks' 1873

calculations for π and offer its value up to 808 places.

**1948**
A completely elementary proof of the prime number theorem is given by Norwegian Atle Selberg (1917- ) and Hungarian Paul Erdös (1913-1996), and also independently by the Czech mathematician P. Kuhn. Until now, all known proofs depended in an essential way on the theory of functions of a complex variable.

**1949**
The ENIAC computer calculates the value of π to 2,037 decimal places.

**1950**
Laurent Schwartz (1915- ), French probabilist, develops a new branch of analysis he calls the theory of distributions. A distribution is a generalization of the classical concept of a function. The new theory has applications in potential theory, spectral theory, partial differential equations, and other aspects of pure and applied analysis.

**1951**
The National Research Council of the U. S. National Academy of Sciences creates a new, separate division of mathematics in order to better represent its growing needs.

**1951**
The International Mathematical Union is established by 15 nations. It is founded to promote cooperation among the world's mathematicians as well as wider dissemination of the results of mathematical research.

**1955**
Henri-Paul Cartan (1904- ), French algebraist, and Samuel Eilenberg (1913- ), Polish-American algebraist, publish their *Homological Algebra*, which begins a new field of study. It is a development of abstract algebra that is concerned with results valid for many different kinds of spaces. It becomes a powerful cross between abstract algebra and algebraic topology.

**1959**
An IBM 704 computer calculates the value of π to 16,167 decimal places.

**1961**
An IBM 7090 computer calculates the value of π to 100,265 decimal places.

**1964**
Paul Joseph Cohen (1934- ), American number theorist, proves

that Georg Cantor's continuum hypothesis is independent of other axioms of set theory and therefore demonstrates that it can be introduced to set theory as a separate axiom.

**1965**
Z. Janke discovers an isolated simple group of high order. This is the first nonclassical finite simple group to be found since E. Mathieu discovered five such groups in 1861.

**1966**
A STRETCH computer calculates the value of π to 250,000 decimal places.

**1966**
Lennart Axel Edvard Carleson, Swedish mathematician, solves a long-standing problem in the theory of Fourier series, demonstrating that the Fourier series of square integrable functions converge outside sets of measure zero.

**1967**
A CDC 6600 computer calculates the value of π to 500,000 decimal places.

**1969**
Robion C. Kirby, American topologist, makes fundamental advances in the theory of manifolds which are of major concern to modern topology.

**1969**
The U. S. National Academy of Sciences publishes its long-awaited, three-volume report on the state of American mathematics, entitled *The Mathematical Sciences*. It documents the assumption of mathematical leadership by the U. S. since World War II, reviews new applications of mathematics in the biological and social sciences, and forecasts future needs.

**1971**
American mathematicians John Billhart and Michael Morrison work with the Fermat number F7 and accomplish its factorization by using a computer.

**1972**
Michio Suzuki, Japanese-Canadian group theorist, discovers a family of mathematical groups for which divisibility by three does not occur. Until now, all known finite simple groups had the property that the number of elements in the group were divisible by three.

**1973**
Russian mathematician M. Lomonosov offers a simple proof

demonstrating that a continuous linear transformation admits a closed invariant subspace (that it can be broken down into simpler constituents).

**1974**
Pierre Deligne, Belgian mathematician, completely proves the French mathematician André Weil's Riemann hypothesis. His proof of this conjecture as well as two others by Weil all relate on a very abstract level certain properties of prime numbers to the solutions of equations in finite systems of arithmetic.

**1974**
A CDC 7600 computer calculates the value of π to 1,000,000 places.

**1975**
Stephen Cook of Canada and Richard Kays of the U.S. advance the study of combinatorics by classifying many combinatorial problems as essentially equivalent in the sense that an algorithm that would solve one of them can be adapted to solve them all.

**1976**
Kenneth I. Appel, American mathematician, and Wolfgang Haken, German-American mathematician, resolve the century-old four-color conjecture first posed in the early 1850s by English mathematician Francis Guthrie. Employing a combination of computer methods and theoretical reasoning, they prove for the first time that the four-color conjecture is true. This conjecture posed the question: can every map on a plane be colored with four colors so that adjacent regions receive differentcolors?

**1977**
Daniel Quillen, American mathematician, and A. A. Suslin, Russian mathematician, independently discover two very different proofs of the twenty-year old conjecture concerning the structure of generalized vector spaces. These proofs confirm that all abstract spaces of a certain common type are constructed in direct analogy with two- and three-dimensional Euclidean space.

**1979**
William P. Thurston, American topologist, receives the Alan T. Waterman Award of the U. S. National Science Foundation for his work that revolutionizes mathematicians' understanding of the way surfaces interleave and the way space curves, not just in two and three dimension but in higher dimensions as well.

**1980**
Close cooperation among several mathematicians in different institutions result in the complete classification of the finite simple groups, the basic building blocks of a major part of modern algebra. This is a major achievement in mathematics research.

**1981**

Hendrik W. Lenstra, Dutch mathematician, demonstrates that a certain class of integer programming problems do not exhibit a pattern of exponential growth. It is important to know if a particular problem has a solution time that grows exponentially with the size of the problem.

**1982**

Benoit B. Mandelbrot (1924- ), Polish-American mathematician, publishes *The Fractal Geometry of Nature* in which he founds a new branch of mathematics based on the study of the irregularities in nature. A fractal is a figure that is self-similar at varying scales.

**1982**

Michael Hartley Freedman, American mathematician, announces his verification of the four-dimensional case of the Poincaré conjecture, one of mathematics' most famous unsolved problems. However, his proof is later shown to be incorrect. Poincaré's conjecture deals with topology and the mathematics of spaces.

**1983**

Gerd Faltings, German-American mathematician, proves the Mordrell conjecture which suggests a relationship between the number of solutions that equations may have and the geometry of certain surfaces determined by these equations. This is an important step toward verifying Fermat's Last Theorem.

**1984**

Louis de Branges, French-American mathematician, proves the Bieberbach conjecture in the field of complex analysis.

**1985**

Hendrik W. Lenstra devises a method of factoring based on so-called elliptic curves, the behavior of which makes it possible to factor large numbers that had resisted all other methods. Prior to this, all theories of factoring were based on the properties of certain quadratic equations introduced by Johann Karl Friedrich Gauss in the 18th century.

**1988**

Yoichi Miyaoka, Japanese mathematician, announces but later retracts his claim that he has proved Fermat's Last Theorem. After announcing his proof early in the year, he realizes there is an essential flaw in his argument. (See 1993)

**1989**

The emerging theory of chaos grows in popularity and takes on elements of controversy. Mathematicians debate whether chaos is really a new mathematics or merely old mathematics in a modern package.

**1991**

Wu-Yi Hsiang, American mathematician, solves Johannes Kepler's "sphere-packing" problem which was posed in 1611 and which has remained unsolved until now. Hsiang's solution has implications for physics, for if it were possible to deduce from quantum mechanics that the atoms in a real crystal must pack together with maximum density, then Hisang's theorem would prove that the regular atomic structure of crystals is also a consequence of quantum mechanics.

**1992**

German mathematician Victor Bangert and John Franks, an American, achieve a major breakthrough in the theory of geodesics, achieving a better understanding of the shortest curves joining two points on surfaces.

**1993**

Andrew John Wiles (1953- ), English-born American mathematician, announces his proof of Fermat's Last Theorem. His 200-page paper is the result of a seven-year attack on the 325-year old problem that many mathematicians had declared unsolvable.

**1994**

English-born American mathematician Andrew John Wiles publishes a corrected, improved version of his proof of Fermat's Last Theorem.

**1994**

University of Southern California computer scientist Leonard M. Adleman uses simple DNA manipulations to solve a mathematical routing problem, the first instance of actual computation being carried out at the molecular level.

**1995**

Columbia University mathematicians David V. And Gregory V. Chudnovsky calculate the value of $\pi$ to more than four billion places; they also devise a method to ensure that all the decimals are correct.

**1996**

Mathematician Jeong-Hon Kim of AT&T Bell Labs devises a theoretical method to analyze large structures which normally require computer analysis. Based on asymptotic theories, which target the boundaries of large systems instead of analyzing them in their entirety, Kim's approach constitutes a significant contribution to Ramsey theory, a branch of mathematics devoted to unavoidable structures in large systems.

**1996**

Lov K. Grover, a mathematician at AT&T Bell Laboratories in Murray Hill, NJ, proposes an algorithm based on quantum logic that will speed up the time required to find information on an unsorted list. Unfortunately, Grover's method requires a yet-to-be-created computer that behaves according to the laws of quantum mechanics.

**1997**
University of Tokyo computer scientist Yasumasa Kanada and associates calculate $\pi$ to 51.5396 billion digits by dividing the task among 1,024 processors on a Hitachi SR2201 computer.

# Selected Mathematics Awards and Prizes

## Carl B. Allendoerfer Award

Established in 1976 in honor of Carl B. Allendoefer, an American mathematician who served as president of the Mathematical Association of America in 1959-60, this award is given for articles "of expository excellence" published in *Mathematics Magazine*. For additional information about this award, please contact: The Mathematical Association of America, 1529 18th Street, N.W., Washington, D.C., 20036.

## Recipients

**1977**
Joseph A. Gallian
B.L. van der Waerden

**1978**
Geoffrey C. Shepard and
Branko Grnbaum

**1979**
Doris W. Shattschneider
Bruce C. Berndt

**1980**
Victor L. Klee, Jr.
Ernst Snapper

**1981**
Stephen B. Maurer
Donald E. Sanderson

**1982**
J. Ian Richards
Marjorie Senechal

**1983**
Donald O. Koehler
Clifford H. Wagner

**1984**
Judith V. Grabiner

**1985**
Frederick S. Gass
Philip D. Straffin, Jr., and
Bernard Grofman

**1986**
Bart Braden
Saul Stahl

**1987**
Israel Kleiner
Paul Zorn

**1988**
Bart Braden
Steven Galovich

**1989**
W.B. Raymond Lickorish and
Kenneth C. Millett
Judith V. Grabiner

**1990**
Thomas Archibald
Fan Chung, Martin Gardner, and
Robert Graham

**1991**
Ranjan Roy

**1992**
Israel Kleiner
Gulbank D. Chakerian and
David E. Logothetti

**1993**
no award given

**1994**
Joan Hutchinson

**1995**
Lee Badger
Tristan Needham

**1996**
Daniel J. Velleman and Gregory
S. Call

**1997**
Colm Mulcahey
Lin Tan

## The George David Birkhoff Prize in Applied Mathematics

Established in 1967 and jointly administered by the American Mathematical Society and the Society for Industrial and Applied Mathematics, this award is given in recognition of "an outstanding contribution to applied mathematics in the highest and broadest sense." Named in honor of American mathematician George David Birkhoff, this prize is normally awarded every five years. For more information about this prize, please contact the American Mathematical Society, P.O. Box 6248, Providence, Rhode Island, 02940.

### Recipients

**1968**
Jürgen K. Moser

**1973**
Fritz John
James B. Serrin

**1978**
Garrett Birkhoff
Mark Kac
Clifford A. Truesdell

**1983**
Paul R. Garabedian

**1988**
Elliot H. Lieb

**1994**
Ivo Babuska
S.R.S. Varadhan

## Chauvenet Prize

Sponsored by the Mathematical Association of America, the Chauvenet Prize was established in 1925 in honor of William Chauvenet, a professor of mathematics at the United States Navel Academy. The award is given annually to an author of "an outstanding expository article on a mathematical topic by a member of the Association." To learn more about this prize, contact the Mathematical Association of America, 1529 18th Street, N.W., Washington, D.C. 20036.

### Recipients

**1925**
G.A. Bliss

**1929**
T.H. Hildebrandt

**1932**
G.H. Hardy

**1935**
Dunham Jackson

**1938**
G.T. Whyburn

**1941**
Saunders MacLane

**1944**
R.H. Cameron

**1947**
Paul R. Halmos

**1950**
Mark Kac

**1953**
E.J. McShane

**1956**
R.H. Bruck

**1960**
Cornelius Lanczos

**1963**
Philip J. Davis

**1964**
Leon Henkin

**1965**
Jack K. Hale and Joseph P. LaSalle

**1966**
no award given

**1967**
Guido Weiss

**1968**
Mark Kac

**1969**
no award given

**1970**
Shiing Shen Chern

**1971**
Norman Levinson

**1972**
Jean François TrSves

**1973**
C.D. Olds

**1974**
Peter D. Lax

**1975**
M.D. Davis and Reuben Hersh

**1976**
Lawrence Zalcman

**1977**
Gilbert Strang

**1978**
Shreeram Abhyankar

**1979**
Neil J. A. Sloane

**1980**
Heinz Bauer

**1981**
Kenneth I. Gross

**1984**
R. Arthur Knoebel

**1985**
Carl Pomerance

**1986**
George Miel

**1987**
James H.Wilkinson

**1988**
Steve Smale

**1989**
Jacob Korevaar

**1990**
David Allen Hoffman

**1991**
W.B. Raymond Lickorish
and Kenneth C. Millett

**1992**
Steven G. Krantz

**1993**
David H. Bailey, Jonathan
M. Borwein, and Peter B.
Borwein

**1994**
Barry Mazur

**1995**
Donal G. Saari

**1996**
Joan Birman

**1997**
Tom Hawkins

## The Frank Nelson Cole Prize in Algebra
## The Frank Nelson Cole Prize in Number Theory

These prizes were established in 1928 in honor of Professor Frank Nelson Cole on the occasion of his retirement as secretary of the American Mathematical Society and editor-in-chief of the AMS *Bulletin*. Awards in algebra and number theory are given separately every five years. For further information, please contact the American Mathematical Society, P.O. Box 6248, Providence, Rhode Island, 02940.

**Recipients**

**1928**
L.E. Dickson

**1931**
H.S. Vandiver

**1939**
A. Adrian Albert

**1941**
Claude Chevalley

**1944**
Oscar Zariski

**1946**
H.B. Mann

**1949**
Richard Brauer

**1951**
Paul Erdös

**1954**
Harish-Chandra

**1956**
John T. Tate

**1960**
Serge Lang
Maxell A. Rosenlicht

**1962**
Kenkichi Iwasawa
Bernard M. Dwork

**1965**
Walter Feit and
John G. Thompson

**1967**
James B. Ax and
Simon B. Kochen

**1970**
John R. Stallings
Richard G. Swan

**1972**
Wolfgang M.Schmidt

**1975**
Hyman Bass
Daniel G. Quillen

**1977**
Goro Shimura

**1980**
Michael Aschbacher
Melvin Hochster

**1982**
Robert P. Langlands

**1985**
George Lusztig

**1987**
Dorian M. Goldfeld
Benedict H. Gross and
Don B. Zagier

**1990**
Shigefumi Mori

**1992**
Karl Rubin

**1995**
Michel Raynaud and
David Harbater

**1997**
Andrew J. Wiles

## The Fields Medal

The first Fields Medal, named after Canadian algebraist John Charles Fields, was awarded in 1936. Often described as the "Nobel Prize in Mathematics," the Fields Medal is given to a mathematician under the age of forty for "both existing work and the promise of future achievement." Additional information about the Fields Medal may be obtained by writing the International Mathematical Union, Estrada Dona Castoria 110, Jardim Botanico, 22460 Rio de Janeiro, Brazil. (Fields Medals were not awarded during World War II.)

### Recipients

**1936**
Lars Valerian Ahlfors
Jesse Douglas

**1950**
Laurent Schwartz
Alte Selberg

**1954**
Kunihiko Kodaira
Jean-Pierre Serre

**1958**
Klaus Friedrich Roth
René Thom

**1962**
Lars V. Hörmander
John Williard Milnor

**1966**
Michael Francis Atiyah
Paul Joseph Cohen
Alexander Grothendieck
Stephen Smale

**1970**
Alan Baker
Heisuke Hironaka
Serge P. Novikov
John Griggs Thompson

**1974**
Enrico Bombieri
David Bryant Mumford

**1978**
Pierre René Deligne
Charles Louis Fefferman
Gregori Alexandrovitch
Margulis
Daniel G. Quillen

**1982**
Alain Connes
William P. Thurston
Shing-Tung Tau

**1986**
Simon Donaldson
Gerd Faltings
Michael Freedman

**1990**
Vladimir Drinfeld
Vaughan Jones
Shigefumi Mori
Edward Witten

**1994**
Pierre-Louis Lions
Jean-Christophe Yoccoz
Jean Bourgain
Efim Zelmanov

## Lester R. Ford Award

In 1964, the Mathematical Association of America established this Award "to recognize authors of articles of expository excellence" published in the *American Mathematical Monthly* or the *Mathematics Magazine*. Ford Awards given from 1976 on are for articles published in the *American Mathematical Monthly* only. A separate prize, the Allendoerfer Award, was created for articles appearing in *Mathematics Magazine* in 1976. The Ford Award is named for Lester R. Ford, Sr., an American mathematician who served as editor of the *American Mathematical Monthly* (1942-46) and as president of the Mathematical Association of America (1947-48). For additional information about this award, please contact the Mathematical Association of America, 1529 18th Street, N.W., Washington, D.C., 20036.

### Recipients

**1965**
R. H. Bing
Louis Brand
Robert G. Kuller
R. Duncan Luce
Hartley Rogers, Jr.
Elmer Tolsted

**1966**
Carl B. Allendoerfer
Peter D. Lax
Marvin Marcus and Henryk Minc

**1967**
Wai-Kai Chen
D.R. Fulkerson
Mark Kac
M. Zuhair Nashed
Paul B. Yale

**1968**
Frederick Cunningham, Jr.
W.F. Newns
Daniel Pedoe
Keith L. Phillips
F.V. Waught and Margaret W. Maxfield
Hans J. Zassenhaus

**1969**
Harley Flanders
George Forsythe
Marcel F. Neuts
Pierre Samuel
Hassler Whitney

Albert Wilansky

**1970**
Henry L. Alder
Ralph P. Boas
William A. Coppel
Norman Levinson
John Milnor
Ivan Niven

**1971**
Jean A. Dieudonné
George Forsythe
Paul R. Halmos
Peter V. O'Neil
Olga Taussky

**1972**
Gulbank D. Chakerian and Lester H. Lange
Paul M. Cohn
Frederick Cunningham, Jr.
W. J. Ellison
Leon Henkin
Victor Klee
Eric Langford

**1973**
Jean Dieudonné
Samuel Karlin
Peter D. Lax
Thomas L. Saaty
Lynn A. Steen
R. L Wilder

**1974**

Patrick Billingsley
Garrett Birkhoff
Martin D. Davis
I.J. Schoenberg
Lynn A. Steen
R. J. Wilson

**1975**
Raymond Ayoub
J. Callahan
Donald E. Knuth
Johannes C.C. Nitsche
Sherman K. Stein
Lawrence Zalcman

**1976**
Michel L. Balinski and H.P. Young
Edward A. Bender and J. R. Goldman
Branko Grunbaum
James E. Humphreys
Joseph B. Keller and David . McLaughlin
Justin J. Price

**1977**
Shreeram Abhyankar
Joseph B. Keller
Donald S. Passman
Douglas Wiens, Hideo Wada, Daihachiro Sato, and James P. Jones
William P. Ziemer, William H. Wheeler, S.H. Moolgavkar, Paul R. Halmos,

John H. Ewing, William H. Gustafson

**1978**
Ralph Boas
Louis H. Kauffman and Thomas F. Banchoff
Neil J.A.Sloane

**1979**
Bradley Efron
Ned Glick
Kenneth I. Gross
Joseph B. Kruskal and Lawrence A. Shepp

**1980**
Desmond Fearnley-Sander
David Gale
Karel Hrbacek
Cathleen S. Morawetz
Robert Osserman

**1981**
R. Creighton Buck
Bruce H. Pourciau
Alan H. Schoenfeld
Edward R. Swart
Lawrence A. Zalcman

**1982**
Philip Davis
R. Arthur Knoebel

**1983**
Robert F. Brown
Tony Rothman
Robert S. Strichartz

**1984**
Judith Grabiner
Roger Howe
John Milnor
Joel Spencer
William C. Waterhouse

**1985**
John D. Dixon
Donald G. Saari and John B. Urenko

**1986**
Jeffrey C. Lagarias
Michael E. Taylor

**1987**
Stuart S. Antman
Howard Hiller
Jacob Korevaar
Peter M. Neumann

**1988**
James Epperson
Stan Wagon

**1989**
Gert Almkvist and Bruce Berndt
Richard K. Guy

**1990**
Jacob Goodman, Janos Pach, Chee K. Yap
Doron Zeilberger

**1991**
Marcel Y. Berger
Ronald Graham and Frances Yao
Joyce Justicz, Edward R. Scheinerman, and Peter Winkler

**1992**
Clement W.H. Lam

**1993**
no award given

**1994**
Bruce C. Berndt and S. Bhargava
Reuben Hersh
Joseph H. Silverman
Dan Velleman and Istvan Szalkai

**1995**
Fernando Q. Gouvea
Jonathan L. King
I. Kleiner and N. Movshovitz-Hadar

William C. Waterhouse

**1996**
Martin Aigner
Sheldon Axler
John Oprea

**1997**
Robert G. Bartle
A.F. Beardon
John Brillhart and Patrick-Morton

**571**

### Krieger-Nelson Prize

This Prize is named after Canadian mathematicians Cecilia Krieger and Evelyn Nelson. Established in 1995 through the auspices of the Canadian Mathematical Society, the prize recognizes "outstanding research by a female mathematician." For additional information, please contact The Canadian Mathematical Society, 577 King Edward, Suite 109, P.O. Box 450, Station A, Ottawa, Ontario, Canada, K1N 6N5.

### Recipients

| **1995** | **1996** | **1997** |
|---|---|---|
| Nancy Reid | Olga Kharlampovich | Cathleen Morawetz |

### Rolf Nevanlinna Prize

Sponsored by the International Mathematical Union, the Rolf Nevanlinna Prize is awarded to mathematicians working in informational science. The first Nevanlinna Prize was given in 1982.
To learn more about this prize, please contact the International Mathematical Union, Estrada Dona Castoria 110, Jardim Botanico, 22460 Rio de Janeiro, Brazil.

### Recipients

| **1982** | **1986** | **1990** |
|---|---|---|
| Robert Trajan | Leslie Valiant | A.A. Razborov |

**1994**
Avi Widgerson

### The Ruth Lyttle Satter Prize in Mathematics

The Ruth Lyttle Satter Prize is awarded biennially "to recognize an outstanding contribution to mathematics research by a woman in the previous five years." This award was established in 1990 by American mathematician Joan S. Birman in memory of her sister, Ruth Lyttle Satter. For additional information about the Ruth Lyttle Satter Prize, please contact the American Mathematical Society, P.O. Box 6248, Providence, Rhode Island, 02940.

### Recipients

| **1991** | **1993** | **1995** |
|---|---|---|
| Margaret Dusa McDuff | Lai-Sang Young | Sun-Yung Alice Chang |

**1997**
Ingrid Daubechies

## The Leroy P. Steele Prizes

These prizes have been awarded since 1970 for the recognition of outstanding published mathematical research. Since 1977, up to three prizes have been awarded each year to authors for works in the following categories: the cumulative influence of the total mathematical work of the recipient; a book or substantial survey of an expository-research paper; and a paper, whether recent or not, that has proved to be of fundamental or lasting importance in its field, or a model of important research. The prizes were established in honor of George David Birkhoff, William Fogg Osgood, and William Caspar Graustein. Additional information about the prize may be obtained by writing The American Mathematical Society, P.O. Box 6248, Providence, Rhode Island, 02940.

## Recipients

**1970**
Solomon Lefschetz

**1971**
James B. Carrell
Jean A. Dieudonné
Phillip A. Griffiths

**1972**
Edward B. Curtis
William J. Ellison
Lawrence F. Payne
Dana S. Scott

**1975**
Lipman Bers
Martin D. Davis
Joseph L. Taylor
George W. Mackey
H. Blaine Lason

**1976**
no awards given

**1977**
no awards given

**1978**
no awards given

**1979**

Salomon Bochner
Hans Lewy
Antoni Zygmund
Robin Hartshorne
Joseph J. Kohn

**1980**
André Weil
Harold M. Edwards
Gerhard P. Hochschild

**1981**
Oscar Zariski
Eberhard Hopf

**1982**
Lars V. Ahlfors
Tsit-Yuen Lam
John W. Milnor
Fritz John

**1983**
Paul R. Halmos
Steven C. Kleene
Shiing-Shen Chern

**1984**
Elias M. Stein
Lennart Carleson
Joseph L. Doob

**1985**
Michael Spivak
Robert Steinberg
Hassler Whitney

**1986**
Donald E. Knuth
Rudolf E. Kalman
Saunders MacLane

**1987**
Martin Gardner
Herbert Federer and Wendell Fleming
Samuel Eilenberg

**1988**
Sigurdur Helgason
Gian-Carlo Rota
Deane Montgomery

**1989**
Daniel Gorenstein
Alberto P. Calderón
Irving Kaplansky

**1990**
R.D. Richtmyer
Bertram Kostant
Raoul Bott

**1993**
Jacques Diximier
James Glimm
Peter D. Lax
Walter Rudin
George Daniel Mostow
Eugene B. Dynkin

**1994**
Ingrid Daubechies
Louis de Branges
Louis Nirenberg

**1995**
Jean-Pierre Serre
Edward Nelson
John T. Tate

**1996**
Bruce C. Berndt
William Fulton
Daniel Stroock and
S.R.S. Varadhan
Goro Shimura

**1997**
Anthony W. Knapp
Mikhael Gromov
Ralph S. Phillips

## Sylvester Medal ( The Royal Society)

The Sylvester Medal is sponsored by the Royal Society, the oldest extant scientific organization in the United Kingdom. The Medal is awarded every three years for "the encouragement of Mathematical Research, irrespective of nationality."

### Recipients

**1901**
Henri Poincaré

**1904**
Georg Cantor

**1907**
Wilhelm Wirtinger

**1910**
Henry Frederick Baker

**1913**
James Whitbread Lee Glaisher

**1916**
Jean Gaston Darboux

**1919**
Percy Alexander MacMahon

**1922**
Tullio Levi-Civita

**1925**
Alfred North Whitehead

**1928**
William Henry Young

**1931**
Edmund Taylor Whittaker

**1934**
Russell Bertrand

**1937**
Edward Hough Augustus Love

**1940**
Godfrey Harold Hardy

**1943**
John Edensor Littlewood

**1952**
A. S. Besicovitch

**1955**
Edward Charles Titchmarsh

**1958**
Maxwell Herman
Alexander Newman

**1961**
Philip Hall

**1964**
Mary Lucy Cartwright

**1967**
Harold Davenport

**1970**
George Frederick James Temple

**1973**
John William Scott Cassels

**1976**
David George Kendall

**1979**
Graham Higman

**1985**
John Griggs Thompson

**1988**
Charles Terence Clegg Wall

**1991**
Klaus Friedrich Roth

**1994**
Peter Whittle

**1997**
Harold Scott Macdonald Coxeter

## Third World Academy of Sciences Award in Mathematics

The Third World Academy of Sciences (TWAS) was founded in Trieste, Italy, in 1983 by a group of scientists under the leadership of Nobel laureate Abdus Salam. One of its major directives "is to give recognition and support to high calibre scientific research performed by individual scientists from developing countries, and to foster it for the benefit of human welfare and the development of the Third World." This Award, one of five in Basic Sciences (biology, chemistry, physics, and medical sciences) was established in 1985. For additional information, please write: Third World Academy of Sciences (TWAS), c/o International Centre for Theoretical Physics (ICTP), P.O. Box 586, 34100 Trieste, Italy.

### Recipients

**1985**
Liao Shan Tao

**1986**
Mauricio M. Peixoto

**1987**
Mudambai S. Narasimhan

**1988**
Jacob Palis

**1989**

no award given

**1990**
Wu Wen-Tsun

**1991**
Madabusi S. Raghunathan

**1992**
Manfredo do Carmo

**1993**
Chang Kung-China

**1994**
Ricardo Mañé

**1995**
Carlos Segovia
Ibrahim A. Eltayeb

**1996**
César L. Camacho

**1997**
no award given

## Oswald Veblen Prize in Geometry

Established under the auspices of the American Mathematical Society in honor of Professor Oswald Veblen in 1961, this Prize recognizes outstanding research in geometry or topology. Beginning in 1966, it has been given every five years. To learn more about this prize, please contact the American Mathematical Society, P.O. Box 6248, Providence, Rhode Island, 02940.

### Recipients

**1964**
C. D. Papakyriakopoulos
Raoul Bott

**1966**
Steven Smale
Morton Brown
Barry Mazur

**1971**
Robion C. Kirby
Dennis P. Sullivan

**1976**
William P. Thurston
James Simons

**1981**
Mikhael Gromov
Shing-Tung Yau

**1986**
Michael H. Freedman

**1991**
Andrew J. Casson
Clifford H. Taubes

**1996**
Richard Hamilton
Gang Tian

## The Wolf Prize in Mathematics

The Wolf Prize is sponsored by the Wolf Foundation, an organization founded by Israeli chemist and philanthropist Ricardo Wolf in 1976. Its mission is "to promote science and art for the benefit of mankind." The Foundation also gives awards for achievements in the arts, agriculture, chemistry, and medicine. For more information about the Wolf Prize, contact The Wolf Foundation, 56 Kidushei Ha'shoa St., P.O. Box 398, 46103 Herzlia Bet, Israel.

### Recipients

**1978**
Izrail M. Gelfand
Carl L. Siegel

**1979**
Jean Leray
Andre Weil

**1980**
Henri Cartan
Andrei N. Kolmogorov

**1981**

Lars V. Ahlors
Oscar Zariski

**1982**
Hassler Whitney
Mark Grigor'evich Krein

**1983-84**
Shing S. Chern
Paul Erdös

**1984-85**
Kunihiko Kodaira

Hans Lewy

**1986**
Samuel Eilenberg
Atle Selberg

**1987**
Kiyoshi Ito
Peter D. Lax

**1988**
Friedrich Hirzebruch
Lars Hormander

**1989**
Alberto P. Calderón
John W. Milnor

**1990**
Ennio De Giogri
Ilya Piatetski-Shapiro

**1991**
No awards given

**1992**
Lennart A.E. Carleson
John G. Thompson

**1993**
Mikhael Gromov
Jacques Tits

**1994-95**
Jurgen K. Moser

**1995-96**
Robert Langlands
Andrew J. Wiles

**1996-97**
Joseph B. Keller
Yakov G. Sinai

# Selected Bibliography

## Biographies

Albers, Donald J., and G.L. Alexanderson. *Mathematical People: Profiles and Interviews.* New York: Birkhäuser, 1985.

Albers, Donald J., Gerald L. Alexanderson, and Constance Reid. *More Mathematical People.* New York: Harcourt, 1991.

Bell, Eric T. *Men of Mathematics.* New York: Simon and Schuster, 1986.

Gillispie, Charles Coulston, editor-in-chief. *Dictionary of Scientific Biography.* Sixteen volumes. New York: Charles Scribner's Sons, 1970-1980.

Grinstein, Louise, and Paul J. Campbell. *Women of Mathematics: A Biobibliographic Sourcebook.* Westport, CT: Greenwood Press, 1987.

Osen, Lynn M. *Women in Mathematics.* Cambridge, MA: The MIT Press, 1995.

Peri, Teri. *Math Equals: Biographies of Women Mathematicians.* Menlo Park, CA: Addison-Wesley, 1978.

Ritchie, David. *The Computer Pioneers.* New York: Simon and Schuster, 1986.

Terry, Leon. *The Mathmen.* New York: McGraw-Hill Book Company, 1964.

## Dictionaries and Encyclopedias

Abbott, David, editor. *The Biographical Dictionary of Scientists: Mathematicians.* New York: P. Bedrick Books, 1985.

American Council of Learned Societies. *Biographical Dictionary of Mathematicians.* New York: Charles Scribner's Sons, 1991.

*Dictionary of Mathematics Terms.* New York: Barron's Educational Series, Inc., 1987.

Dunham, W. *The Mathematical Universe: An Alphabetical Journey through the Great Proofs, Problems, and Personalities.* New York: John Wiley & Sons, Inc., 1994.

Itô, Kiyosi, editor. *Encyclopedia Dictionary of Mathematics.* Cambridge, MA: MIT Press, 1987.

## General

Hoffman, Paul. *Archimedes' Revenge: The Joys and Perils of Mathematics.* New York: Ballantine, 1989.

Immergut, Brita. *Arithmetic and Algebra—Again.* New York: McGraw-Hill, 1994.

Kershner, R. B. *The Anatomy of Mathematics.* New York: Ronald Press Co., 1950.

Kramer, Edna E. *The Nature and Growth of Modern Mathematics.* Princeton, NJ: Princeton University Press, 1981.

Moore, Gregory H. *Zermelo's Axiom of Choice, Its Origins, Development, and Influence.* New York: Springer-Verlag, 1982.

Slavin, Steve. *All the Math You'll Ever Need.* New York: John Wiley and Sons, 1989.

Temple, George. *100 Years of Mathematics.* New York: Springer-Verlag, 1981.

Wulforst, Harry. *Breakthrough to the Computer Age.* New York: Charles Scribner's Sons, 1982.

Zwillinger, Daniel, editor. *CRC Standard Mathematical Tables and Formulae.* 30th edition. Boca Raton: CRC Press, 1996.

## Historical Surveys

Alic, Margaret. *Hypatia's Heritage: A History of Women Scientists from Antiquity through the Nineteenth Century.* Boston: Beacon Press, 1986.

Allman, George Johnson. *Greek Geometry from Thales to Euclid.* New York: Longmans, Green, & Co., 1889.

Aspray, William and Philip Kitcher, editors. *History and Philosophy of Modern Mathematics.* Minneapolis: University of Minnesota Press, 1988.

Ball, W.W.R. *A Short Account of the History of Mathematics.* New York: Dover Publications, Inc., 1960.

Boyer, Carl B. *A History of Mathematics.* Second edition. Revised by Uta C. Merzbach. New York: John Wiley & Sons, Inc., 1991.

Bunt, Lucas N. H., et al. *The Historical Roots of Elementary Mathematics.* Englewood Cliffs, NJ: Prentice Hall, Inc., 1976.

Cajori, F. *A History of Mathematics.* New York: Chelsea, 1985.

Dieudonné, Jean. *A History of Algebraic and Differential Topology.* Boston: Birkhäuser, 1989.

Eaves, Howard. *An Introduction to the History of Mathematics.* Fourth edition. New York: Holt, Rinehart, & Winston, 1976.

Fang, J. *Mathematicians from Antiquity to Today.* Hauppauge, NY: Paideia Press, 1972.

Green, Judy, and Jeanne Laduke. *A Century of Mathematics in America.* Providence, RI: American Mathematical Society, 1989.

Heath, T. L. *A History of Greek Mathematics.* New York: Dover, 1981.

Katz, Victor J. *A History of Mathematics.* Reading, PA: Addison-Wesley, 1993.

Smith, D.E. *History of Mathematics.* New York: Dover, 1958.

Stigler, Stephen M. *The History of Statistics: The Measurement of Uncertainty Before 1990.* Cambridge, MA: Harvard University Press, 1986.

Stillwell, John. *Mathematics and Its History.* New York: Springer-Verlag, 1989.

Struik, Dirk J. *A Concise History of Mathematics.* Fourth revised edition. New York: Dover Publications, 1987.

Mikami, Yoshio. *The Development of Mathematics in China and Japan,* 1913. Reprint, Chelsea Publishing Company, 1961.

## Newsletters and Periodicals

*Acta Mathematica*
Mittag-Leffler Institute
Auravaegen 17
S-162
Djursholm, Sweden
**Frequency:** Quarterly
Interantional coverage of pure and applied mathematics.

*Advances in Mathematics*
Academic Press, Inc.
525 B Street, Suite 1900
San Diego, CA 92101-4411
**Frequency:** Quarterly
Reports significant advances in all areas of pure mathematics.

*American Journal of Mathematics*
The Johns Hopkins University Press
John Hopkins University
Department of Mathematics
Baltimore, MD 21218
**Frequency:** Bimonthly
Covers applied and pure mathematics.

*American Mathematical Monthly*
Mathematical Association of America
1529 18th Street N.W.
Washington, D.C. 20036-1358
**Founded :** 1894
**Frequency:** 10 issues per year
A journal of mathematical exposition.

*Annals of Mathematics*
Princeton University Press
309 Fine Hall
Princeton, NJ 08540-0001
**Frequency:** Bimonthly
A journal of research in pure mathematics.

*Association for Women in Mathematics Newsletter*
Association for Women in Mathematics
4114 Computer & Space Sciences Bldg.
University of Maryland
College Park, MD 20742-2461
**Frequency:** 6 issues per year
Concerned with the process of women in professional fields, particularly in mathematics and related careers.

*Bulletin (New Series) of the American Mathematical Society*
American Mathematical Society
201 Charles St.
P.O. Box 6248
Providence, RI 02904
**Frequency:** Quarterly
Features research-expository papers, research announcements, and book reviews.

*Canadian Journal of Mathematics/Journal Canadien de Mathématiques*
Department of Mathematics
University of British Columbia
Vancouver, BC, Canada V6T 1Y4

**Frequency:** Bimonthly
Covers research in mathematics (English and French).

*Canadian Mathematical Bulletin*
Department of Mathematics and Statistics
McMaster University
Hamilton, ON, Canada L8S 4K1
**Frequency:** Quarterly

*College Mathematics Journal*
Mathematical Association of America
1529 18th Street, N.W.
Washington, D.C. 20036-1358
**Frequency:** 5 issues per year
Journal providing articles for undergraduate college math students.

*Duke Mathematical Journal*
Duke University Press
P.O. Box 90660
Dunham, NC 27708-0660
**Frequency:** Monthly
A leading journal about mathematical research.

*Historia Mathematica: International Journal of the History of Mathematics*
Academic Press, Inc.
525 B Street, Suite 1900
San Diego, CA 92101-4411
**Frequency:** Quarterly
Chronicles the history of mathematical sciences worldwide.

*Mathematical Intelligencer*
Springer-Verlag
175 Fifth Avenue
New York, NY 10010
**Frequency:** Quarterly
General interest magazine for academic and non academic audiences.

*Mathematical Reviews*
American Mathematical Society
P.O. Box 6248
Providence, RI 02940-6248
**Frequency:** 13 issues per year

*Mathematics Magazine*
Mathematical Association of America
1529 18th Street, N.W.
Washington, D.C. 20036-1358
**Frequency:** Bimonthly
Features informal mathematical articles designed to appeal to faculty and students at the undergraduate level.

*Modern Logic: International Journal for the History of Mathematical Logic*
Modern Logic Publishing
2408-1/2 Lincoln Way
Ames, IA 50014-7217
**Frequency:** Quarterly
Covers historical studies of 19th- and 20th-century mathematical logic.

*SIAM Review*
Society for Industrial and Applied Mathematics
3600 University City Science Center
Philadelphia, PA 19104-2688
**Frequency:** Quarterly
Covers current papers on applied mathematics.

*Wonderful Ideas*
Nancy Segal James
P.O. Box 64691
Burlington, VT 05406
**Frequency:** 5 issues per year
Provides creative math activities, games, and lessons, with a focus on manipulatives and problem solving.

## Internet Sites

As the Internet is currently expanding, addresses of web sites listed below may be altered and/or changed.

**American Mathematical Society (AMS)**
http://e-math.ams.org/

**Ask Dr. Math!**
http://forum.swarthmore.edu/dr.math/dr-math.html

**Biographies of Women Mathematicians**
http://www.scottlan.edu/lriddle.women/women.htm

**Canadian Mathematical Society**
http://camel.cecm.sfu.ca/

**EINet Galaxy Mathematics Node**
http://galaxy.einet.net/galaxy/Science/Mathematics.html

**The Geometry Center**
http://www.geom.umn.edu/

**MAA Online (Mathematical Association of America)**
http://www.maa.org/

**MacTutor History of Mathematics Archive**
http://www-groups.dcs.st-and.ac.uk/~history/index.html

**The Math-Forum**
http://forum.swarthmore.edu/

**The Mathematics Archives**
http://archives.math.utk.edu/

**Mathematics Information Servers**
http://www.math.psu.edu/OtherMath.html

**Mathematical Programming Glossary**
http://www-math.cudenver.edu/~hgreenbe/glossary/glossary.html

**Past Notable Women of Computing and Mathematics**
http://www.cs.yale.edu/HTML/YALE/CS/HyPlans/tap/past-women.html

**Society for Industrial and Applied Mathematics (SIAM)**
http://www.siam.org

**Women In Math Project**
http://darkwing.uoregon.edu/~wmnmath/

**The World Wide Web Virtual Library: Mathematics**
http://euclid.math.fsu.edu/Science/math.html

# Field of Specialization Index

# Gender Index

# Nationality/Ethnicity Index

Scientists are listed by country of origin and/or citizenship as well as by ethnicity (see African American, Asian American, Hispanic American).

# Subject Index